Springer Proceedings in Mathematics & Statistics

Volume 344

This book series features volumes composed of selected contributions from workshops and conferences in all areas of current research in mathematics and statistics, including operation research and optimization. In addition to an overall evaluation of the interest, scientific quality, and timeliness of each proposal at the hands of the publisher, individual contributions are all refereed to the high quality standards of leading journals in the field. Thus, this series provides the research community with well-edited, authoritative reports on developments in the most exciting areas of mathematical and statistical research today.

More information about this series at http://www.springer.com/series/10533

R. N. Mohapatra · S. Yugesh · G. Kalpana ·
C. Kalaivani
Editors

Mathematical Analysis and Computing

ICMAC 2019, Kalavakkam, India, December 23–24

Springer

Editors
R. N. Mohapatra
Department of Mathematics
University of Central Florida
Orlando, FL, USA

S. Yugesh
Department of Mathematics
Sri Sivasubramaniya Nadar
College of Engineering
Kalavakkam, Tamil Nadu, India

G. Kalpana
Department of Mathematics
Sri Sivasubramaniya Nadar
College of Engineering
Kalavakkam, Tamil Nadu, India

C. Kalaivani
Department of Mathematics
Sri Sivasubramaniya Nadar
College of Engineering
Kalavakkam, Tamil Nadu, India

ISSN 2194-1009 ISSN 2194-1017 (electronic)
Springer Proceedings in Mathematics & Statistics
ISBN 978-981-33-4645-1 ISBN 978-981-33-4646-8 (eBook)
https://doi.org/10.1007/978-981-33-4646-8

Mathematics Subject Classification: 37C25, 37-XX

This Springer imprint is published by the registered company Springer Nature Singapore Pte Ltd.
The registered company address is: 152 Beach Road, #21-01/04 Gateway East, Singapore 189721, Singapore

Preface: International Conference on Mathematical Analysis and Computing

It is indeed a pleasure and privilege to introduce the proceedings of the selected contributions of the participants of the International Conference on Mathematical Analysis and Computing 2019 (ICMAC 2019) which took place at Sri Sivasubramaniya Nadar College of Engineering, Chennai, India, during December 23–24, 2019.

The main goal of the conference was to provide a platform for experts in mathematical analysis, geometric function theory, and soft computing to meet and interact their results and ideas with each other. Mathematical analysis is one of the most compelling areas of research because of its rich applications. Soft computing is a main pillar of most of the recent research, industrial, and commercial activities. The lectures in the conference were well received by the participants and discussions were fruitful.

The main objective of this "proceedings" is to disseminate recent advances in the studies of diverse areas of mathematical analysis, geometric function theory, and soft computing. These are crystallized in the form of original research articles and expository survey papers.

The papers included in this volume are based on a rigorous peer-review process by the committee of experts in various disciplines. Every submitted paper was first screened by the members of the editorial board, and once it clears the initial screening, it was peer reviewed by at least two potential reviewers in the related area of expertise from the pool of potential reviewers. The paper is accepted if at least two reviewers recommend it for acceptance.

We thank all the invited speakers, reviewers, and the authors who made their valuable contribution toward the success of the conference ICMAC 2019. We would like to thank DST SERB and SSN Trust for the financial support rendered to organize ICMAC 2019. Our sincere gratitude to Mr. Shamim Ahmad, Senior Editor, Springer Nature for his collaboration and timely cooperation.

We express our sincere thanks and gratitude to Dr. Shiv Nadar, Founder, SSN Institutions and Chairman, HCL Technologies; Ms. Roshni Nadar Malhotra, Executive Director and CEO, HCL Corporation, Vice Chairperson, HCL Technologies, Trustee, Shiv Nadar Foundation, Founder and Trustee, The Habitats

Trust; Mr. R. Srinivasan, Director and CEO Redington Ltd. Global Management and Business Leader, Chairman, SSN Institutions; Ms. Kala Vijayakumar, President, SSN Institutions; Members of the Board of Management; Dr. S. Salivahanan, Former Principal, SSN College of Engineering; Dr. V. E. Annamalai, Principal, SSN College of Engineering; Dr. S. Narasimman, COE and Professor, Department of Mathematics, SSN College of Engineering; and Dr. P. Venugopal, Head, Department of Mathematics, SSN College of Engineering for their continuous support and encouragement.

Orlando, USA	Dr. R. N. Mohapatra
Kalavakkam, India	S. Yugesh
Kalavakkam, India	G. Kalpana
Kalavakkam, India	C. Kalaivani

Contents

About the Editors

Prof. R. N. Mohapatra is a Professor of Mathematics at the University of Central Florida, Orlando, USA. He joined the department in August 1984 after spending a year at the York University, Downsview, Ontario, Canada. Prior to that he served as a Professor of Mathematics at the American University of Beirut, Lebanon. He has visited University of Alberta, Edmonton, Canada; Memorial University, Newfound-land, Canada; and the University of Arkansas, Fayetteville as a Visiting Professor to teach teachers during a summer institute supported by NSF. He has published 167 research papers in refereed journals and refereed conference proceedings. He has written two books, one entitled "Biomedical Statistics with Computing" with Dr. Surya Mohapatra and Prof. Mary H. Regier, and the other entitled "Fuzzy Differential Equations" in collaboration with late Prof. V. Lakshmikantham. The two books were published by Research Studies Press and Taylor Francis, respectively. He has also edited eight conference proceedings published by Springer. His areas of current research interest are Approximation Theory, Equilibrium Problems, and Variational Inequalities, Frames in Hilbert C* modules.

Dr. S. Yugesh is Assistant Professor at the Department of Mathematics, Sri Sivasubramaniya Nadar College of Engineering, Tamil Nadu, India. He obtained his Ph.D. degree from Anna University, Chennai, India in 2016. He has written 10 research papers published in peer-reviewed journals, 12 papers in international conference proceedings, and 1 edited volume book. His areas of research are sampling theory, frame theory, splines, and wavelets. He referees articles for professional journals.

Dr. G. Kalpana is Assistant Professor at the Department of Mathematics, Sri Sivasubramaniya Nadar College of Engineering, Tamil Nadu, India. She obtained her Ph.D. degree from Anna University, Chennai, India. She has written some research papers published in peer-reviewed journals and in the national and international conference proceedings. Her current research work includes fuzzy metric spaces and supremum metric on fuzzy sets.

Dr. C. Kalaivani is Assistant Professor at the Department of Mathematics, Sri Sivasubramaniya Nadar College of Engineering, Tamil Nadu, India. She obtained her Ph.D. degree from Alagappa University, Karaikudi, India. She has written some research papers published in peer-reviewed journals and in the national and international conference proceedings. Her current research work includes topology and fuzzy set theory.

A Primality Test and a Theorem on Twin Primes

B. N. Prasad Rao and M. Rangamma

Abstract Any number greater than one can be represented as a product of prime numbers in a unique way, apart from the order in which the prime factors occur. This is known as the Fundamental Theorem of Arithmetic. Thus, prime numbers are solely used to represent prime numbers or composite numbers. There is a difficulty to prove a number as a prime number by using the canonical representation of numbers. We have put a small step in overcoming this difficulty. In this paper, we made a basic and fundamental change in representing positive integers as the sum of a composite number and a prime number. This basic and fundamental change in the representation of numbers helps in testing the primality of numbers through representation of numbers as the sum of a composite number and prime number. By using this representation, we have also proved a theorem on twin primes. Goldbach's conjecture is about the representation of an even number greater than 4 as the sum of two prime numbers. In this paper, we have proved that an odd number under certain conditions get represented as the sum of two prime numbers.

Keywords Primality test · Twin primes · Representation of positive integers · Composite number

Mathematics Subject Classification (2010) 11A41 · 11A51

I thank the Government of India for supporting my research through the University Grants Commission B.S.R.F fellowship.

B. N. Prasad Rao (✉) · M. Rangamma
Department of Mathematics, Osmania University, Hyderabad, Telangana, India
e-mail: prasadrao4@gmail.com

M. Rangamma
e-mail: rangamma1999@gmail.com

R. N. Mohapatra et al. (eds.), *Mathematical Analysis and Computing*,
Springer Proceedings in Mathematics & Statistics 344,
https://doi.org/10.1007/978-981-33-4646-8_1

1 Introduction

A prime number is represented as a product of 1 and itself. A composite number is represented as a product of two or more prime numbers. Thus, prime numbers are used in representing integers. If a number is a prime number, then it will show certain characteristics. For example, we have Wilson's Theorem, which states as "If p is a prime number, then $(p-1)! \equiv -1(modp)$". The converse of this statement is also true. AKS primality test [4] and Miller test [6] uses nature of a prime number in testing the primality of a number. Lucas test [5] and Pepin primality test [7] are based on the form of the given number n. Pocklington primality test [8] requires the number n to satisfy certain conditions before it becomes a prime number. Goldwasser and Kilian primality test [3] is based on Elliptic curves and this idea was modified by Adleman and Huang [1]. Thus, to prove a number as a prime number, we have two choices. The first one is the choice of using the definition of a prime number and the second one is the choice of using the characteristics of a prime number. There are no tests based on the definition of a prime number. Mostly, we prove the number as a prime number by assuming the number as a composite number. Thus, the definition of a prime number as the representation of the product of 1 and itself is not helping in proving the number as a prime number. Thus, our paper aims at proving a number as a prime number by representing the number as the sum of a composite number and prime number.

The unsolved Goldbach's conjecture states that, "every even integer greater than 4 is the sum of two odd prime numbers". Thus, even numbers greater than 4 has a prime number + prime number representation. This unsolved problem suggests a need to study positive integers in terms of sum of two numbers, where either number in the sum is a composite number or prime number or both numbers are prime numbers or both numbers are composite numbers. This paper is the first step in this direction. In Sect. 2 of this paper, we have proved that every integer greater than 5 can be represented as the sum of a composite number and a prime number. Using this representation to natural numbers, we proved in Sect. 4 a criterion for prime number and a criterion for composite number. Clement [2] has given a criterion for twin primes. His criterion is based on Wilson's theorem. In this paper, we have proved a theorem for twin primes, in terms of the representation of a number as the sum of a composite number and a prime number. The theorem on twin prime shows that an odd number under certain conditions get represented as the sum of two prime numbers, which is contrary to Goldbach conjecture. In Sect. 3, we have introduced new concepts in the form of definitions.

2 Representation of Positive Integers

Theorem 1 *Let $n > 5$, $n \in \mathbb{N}$. Then n can be represented as a sum of a composite number and a prime number.*

Proof Let $n > 5, n \in \mathbb{N}$.

Case 1. Let n be any composite number and p be any prime factor of n. Then $n = pb$ where $b \in \mathbb{N}, b > 1$.
When $b = 2$ then $n = 2p$.

$$n = 2(p - 1 + 1)$$

$$n = 2(p - 1) + 2$$

when $b > 2$. Since $n = bp$, and hence

$$n = p(b - 1 + 1)$$

$$= p(b - 1) + p$$

Thus, any composite number n is represented as a sum of a composite number and a prime number.

Case 2. Let n be any prime number and p be any prime number less than n. Then $n - p = 2k$ for some $k \in \mathbb{N}$.
For $k = 1$, we have, $n = p + 2 = (p - 1) + 3 = 2m + 3$ where $m > 1, m \in \mathbb{N}$.
For $k > 1$ we have,

$$n - p = 2k$$

$$n = 2k + p$$

Thus, any prime number n is represented as a sum of a composite number and a prime number.

3 Definitions

Natural numbers are represented as the sum of a composite number and a prime number or sum of two composite numbers or sum of two prime numbers. In view of this, we have the following new concepts in the form of definitions that follow.

Definition 1 Let the representation of a positive integer n be $n = m + p$, where m is a composite number and p a prime number. We call this representation of n as *cp-representation* of n.

Definition 2 Let the representation of a positive integer n be $n = p + q$, where p, q are prime numbers. We call this representation of n as *pp-representation* of n.

Definition 3 Let the representation of a positive integer n be $n = m_1 + m_2$, where m_1, m_2 are composite numbers. We call this representation of n as *cc-representation* of n.

4 Theorem on Primality Test and Twin Primes

Let us consider the following example:

Example 1 Let us consider the natural number $n = 17$. We write various *cp-representations* of the number 17 as follows:

$17 = 15 + 2 = 14 + 3 = 12 + 5 = 10 + 7 = 6 + 11 = 4 + 13$. In each *cp-representation* of 17, we observe that, no prime number divides the corresponding composite number. This motivates us to the following theorem.

Theorem 2 *Let $q > 5$, $q \in \mathbb{N}$. Then q is a prime number if and only if for each cp-representation of q as $q = m + p$, we have the condition $p \nmid m$, where m is a composite number and p a prime number.*

Proof Let $q > 5, q \in \mathbb{N}$ be a prime number.
Let us suppose that, for all *cp-representations* of q as $q = m + p$,where m is a composite number and p a prime number, the condition $p \nmid m$ is false.
Let $q = m + p$ be a *cp-representation* of q such that $p \mid m$.
Then, $q = kp + p$ where $k > 1, k \in \mathbb{N}$.
Thus, $q = (k + 1)p$ and hence q is a composite number.
This contradicts our hypothesis and hence our supposition is false and the theorem must be true.
Conversely. Let $q = m + p$ be any *cp-representation* of q with the condition that $p \nmid m$. We prove that q is a prime number.
Also, let $p_1 < p_2 < \cdots < p_r < q$ be prime numbers less than q.
Suppose, if possible, that q is a composite number, and hence it has some prime factor $p_j \leqslant \sqrt{q}$, where $1 \leqslant j < r$.
$p_j \mid q$, and hence $q = lp_j$, where $l > 1, l \in \mathbb{N}$.
If $l = 2$ then $q = 2(p_j - 1) + 2$. ($p_j \neq 2$, because $q > 5$)
If $l > 2$ then $q = lp_j$.
$q = (l - 1)p_j + p_j$
Thus, q is represented as a sum of a composite and a prime number where the prime number divides the corresponding composite number in the sum. This contradicts our hypothesis.
This completes the proof.

Corollary 1 *Let $q > 8$, $q \in \mathbb{N}$ has a cp-representation $q = m + p$ where m is a composite number, p a prime number and $p \leq \sqrt{q}$. Then q is a prime number if and only if for each cp-representation of q as $q = m + p$, we have the condition $p \nmid m$.*

Corollary 2 *Let $q > 8$, $q \in \mathbb{N}$ has a cp-representation $q = m + p$ where m is a composite number, p a prime number. Then q is a composite number if and only if there exists atleast one cp-representation of q as $q = m + p$ such that $p \mid m$.*

Example 2 Let us consider the odd number 19. The various representations of the number 19 are as follows:

$$19 = 2 + 17 = 3 + 16 = 5 + 14 = 7 + 12 = 9 + 10 = 11 + 8 = 13 + 6 = 15 + 4$$

We note that the number 19 has a *pp-representation.* Also, 19 and 17 are twin prime numbers. When two numbers are not twin primes, then the biggest of the two odd numbers has only *cc-rpresentation* and *cp-representation.* We also note that Goldbach's conjecture states that every even number greater than 4 is the sum of two primes. The same is observed for some odd numbers under certain conditions. We have the following theorem in this regard:

Theorem 3 *Let p, q be prime numbers where $p > q, p > 5$. Then p, q are twin primes if and only if p has a pp-representation.*

Proof Let p, q be twin primes. Then p = 2 + q and hence, the prime number p has a *pp-representation.*
Conversely. Let the representation of a prime number p be *pp-representation.*
Let $p_1 < p_2 < \cdots < p_r < p$ be prime number less than p and $p > q$.
Since given that $q < p$, and hence q must be one of the primes $p_i \in \{p_1, p_2, \ldots, p_r\}$
Since $p > 5$ is an odd prime number, and hence it has a *cp-representation* given by $p = 2^k l + p_i$ for some prime number $p_i < p, k \geqslant 1, k, l \in \mathbb{N}$.
Since, given that the prime number p has a *pp-representation,* and hence this is possible only when $k = l = 1$ is substituted in above equation
Putting $k = l = 1$ in the equation $p = 2^k l + p_i$, we get $p = 2 + p_r = 2 + q$ where $q = p_r$.
This completes the proof.

Acknowledgements I thank Prof. P. V. Arunachalam, Dr. C. Goverdan, and Prof. V. Darmiah for their encouragement and valuable suggestions. I also thank reviewers for reviewing this article.

References

1. Adleman, L. M., Huang, M.D.: Primality testing and Abelian varieties over finite fields fields. Lecture Notes in Math, Springer, p. 1512 (1992)
2. Clement, P.A.: Congruences for sets of primes. Am. Math. Monthly **56**, 0023–0025 (1949)
3. Goldwasser, S., Kilian, J.: Almost all primes can be quickly certified. Proc. Ann. ACM Symp. Theory Comput. 316–329 (1986)
4. Manindra, A., Neeraj, K., Nitin, S.: Primes is in P. Ann. Math. **160**, 0781–0791 (2004)
5. Michal, K., Florian, L., Lawrence, S.: 17 Lectures on Fermat Numbers: From Number Theory to Geometry, CMS Books in Mathematics, Canadian Mathematical Society, vol. 9, p. 41. Springer (2001)

6. Miller Gary, L.: Riemann's hypothesis and test s for primality. J. Comput. Syst. Sci. **13**, 0300–0317 (1976)
7. Pépin, P.: Sur la formule $2^{2^n} + 1$. Comptes rendus de l'Académie des Sciences de Paris **85**, 0329–0333 (1877)
8. Pocklington H.C.: The determination of the prime or composite nature of large numbers by Fermat's theorem. In: Proceedings of the Cambridge Philosophical Society, vol. 18, pp. 0029–0030

Choosing the Best Copula Function in Mathematical Modeling

Pranesh Kumar and Cigdem Topcu Guloksuz

Abstract Multivariate distributions with any type of marginal distributions can be easily constructed using Sklar's theorem and copula functions. Precisely, a copula function is a parametric representation of a multivariate distribution function which is expressed in terms of marginally uniform random variables on the unit interval. Copula functions possess some appealing mathematical properties, such as they allow scale-free measures of linear/nonlinear stochastic dependence and are useful in simulating families of multivariate distributions. The concept of stochastic tail dependence refers to the clustering of extreme events. Extreme events, for example, in economic systems and in natural hazards contexts generally exhibit tail dependence, and thus it becomes very important to analyze the extreme behavior. Copula functions can measure nonparametric, distribution-free, or scale-invariant nature of dependence and extreme events. Over the past few decades, several examples of copula functions and copula families with one or more real parameters are studied and are applied in various disciplines like statistics, insurance, finance, economics, survival analysis, information theory, image processing, and engineering. In this paper, we aim to discuss copula functions, copula properties, copula families, and simulations using copula functions. Given a large number of copulas available, we will address an important question from the practitioner's points of view as how to choose the most appropriate copula from the family of Archimedean copulas, namely, Clayton, Gumbel, Frank, and Guloksuz–Kumar for fitting the prediction models.

Keywords Copula functions · Multivariate distributions · Extreme events · Dependence · Simulation

P. Kumar (✉)
University of Northern British Columbia, Prince George, BC V2N 4Z9, Canada
e-mail: pranesh.kumar@unbc.ca

C. Topcu Guloksuz
Ankara Science University, 06200 Ankara, Turkey

R. N. Mohapatra et al. (eds.), *Mathematical Analysis and Computing*,
Springer Proceedings in Mathematics & Statistics 344,
https://doi.org/10.1007/978-981-33-4646-8_2

1 Introduction: Copulas and Archimedean Copulas

Sklar [17] introduced copula function and copula in his article in 1959. Copulas describe dependence structure among random variables and have some advantages provided by copula modeling. One of the advantages is constructing the joint probability distribution function independent of its marginal. Further, copula functions can model both symmetric and asymmetric dependence unlike linear correlation measures. For more details about copulas, references are made to Frees and Valdez [4], Matteis [15], Nelsen [16], Genest and Favre [7], Belgorodski [1], Cherubini et al. [2]. In bivariate context, the dependence structure between two random variables is completely described by known bivariate distributions. However, when different types of dependence structures prevail, bivariate copula functions are used to obtain more efficient models.

Precisely, a copula function is a parametric representation of a multivariate probability distribution function which is expressed in terms of marginally uniform random variables on the unit interval. For any pair (*X*, *Y*) of continuous random variables, the bivariate probability distribution function (*H*) can be expressed in the form $H(x, y) = C(u, v), u, v \in (0, 1)$. Equivalently, *u* and *v* can be considered as the continuous marginal distributions *F*(*x*) and *G*(*y*), respectively. Here, *C* is the copula function with $C : [0, 1]^2 \rightarrow [0, 1]$. Thus, it can be rewritten that $H(x, y) = C(F(x), G(y))$ and that copula function *C* is a joint probability distribution function with marginal also being probability distribution functions. It should be noted that, if the marginal are continuous, there is a unique copula representation.

The joint probability density function *f* (*x*, *y*), in terms of copula, is expressed by

$$f(x, y) = \Pi_{i=1}^{k} f(x) \times f(y) \times c(F(x), G(y)), \quad (1)$$

where *f*(*x*) and *f*(*y*) are the marginal density function of *X* and *Y* and coupling is provided by the copula density

$$c(u, v) = c(F(x), G(y)) = \partial^2 C(u, v)/\partial u \partial v, \quad (2)$$

if it exists. In case of independent random variables, copula density *c*(*u*, *v*) is identically equal to one. It may be noted that the importance of the above equation *f*(*x*, *y*) is that the independent portion expressed as the product of the marginals can be separated from the function *c*(*u*, *v*) describing the dependence structure or shape. The dependence structure summarized by a copula is invariant under increasing and continuous transformations of the marginals.

A special class of copulas which is being studied extensively is referred to as the Archimedean copulas which can be generated from a function ϕ, called a generator function. For $0 < t < 1$, the generator function ϕ is such that $\phi(1) = 0$, $\lim_{t \to \infty} (\phi(t)) = \infty$, $\phi'(t) < 0$, $\phi''(t) > 0$. Thus, ϕ is a continuous, strictly decreasing, and convex function and always has an inverse, ϕ^{-1}. Additionally, every ϕ that satisfies these conditions can generate a bivariate copula *C*, known as the Archimedean copula,

$$C(F(x), G(y)) = \phi^{-1}\{\phi(F(x)) + \phi(G(y))\}. \tag{3}$$

It may be noted that the different generator functions result in different Archimedean copulas. The Archimedean copula is indexed by a parameter (θ), called copula parameter. Copula parameter can be real or multidimensional and is embedded in the generator function ϕ. For a given random sample data ($X_i, Y_i; i = 1, 2, \ldots, n$), the copula parameter θ should be estimated to specify the copula function that models the dependence structure. In literature, there exist various techniques to estimate copula parameters [Nelson (16)]. The one-parameter Archimedean copulas have another useful property about the relationship between the copula parameter and nonparametric association measures like the Kendall τ, Spearman ρ. Scale-invariant-dependent measures can be expressed as copula functions of random variables. Two standard nonparametric dependence measures expressed in copula form are as follows:

$$\text{Kendall's tau}: \ \tau = 4 \iint_{I^2} C(u, v) \mathrm{d}C(u, v) - 1 \tag{4}$$

$$\text{Spearman's Rho}: \ \rho = 12 \iint_{I^2} C(u, v) \mathrm{d}u\mathrm{d}v - 3. \tag{5}$$

Kendall's τ can be estimated from the data by

$$\tau = 2 \sum_{i<j} \mathrm{Sign}\big[(x_i - x_j)(y_i - y_j)\big]/n(n-1). \tag{6}$$

It may be noted that the linear correlation coefficient is not expressible in a copula form. For Archimedean copulas with generator function ϕ, there is an alternative expression for τ:

$$\tau = 4 \int_0^1 \big((\phi(t))/(\phi'(t))\big) dt + 1. \tag{7}$$

The asymptotic properties of this estimator are studied in Kojadinovic and Yan [14]. Apart from rank-based nonparametric strategies, the maximum likelihood methods to estimate the copula parameter are given by Genest and Rivest [6] and Joe [11].

Copula functions, as scale-free measures of linear/nonlinear stochastic dependence, are useful in simulating families of multivariate probability distributions. One of the advantages of working with Archimedean copulas is that an Archimedean copula can be uniquely determined by the function

$$K\phi(t) = t - \big((\phi(t))/(\phi'(t))\big), \text{ for } 0 < t < 1. \tag{8}$$

$K_\phi(t)$ is the distribution function of $C(u, v)$ and a bivariate Archimedean copula is determined by a univariate function $K_\phi(t)$ which is also called Kendall distribution function.

The empirical estimate of copula, $K_n(t)$ from a random sample of size n [16] is given by

$$K_n(t) = \frac{\#(T_i \leq t)}{n+1}, \tag{9}$$

where pseudo-observations $T_i's$ are defined as

$$T_i = H_n(X_i, Y_i) = \frac{\sum_{j=1}^{n} I[\{X_j \leq X_i\, Y_j \leq Y_i\}]}{\mathrm{n}+1}, i = 1, 2, \ldots n \tag{10}$$

To select the suitable Archimedean copula function $K_\phi(t)$ from the family of Archimedean copulas, Frees and Valdez [4] considered the degree of closeness between $K_n(t)$ and $K_\phi(t)$ and suggested minimum distance measure

$$MD = \int \left[K_n(t) - K_\phi(t)\right]^2 dK_n(t). \tag{11}$$

Other copula selection criteria are based on information measures such as Akaike information criteria (AIC) and Bayesian information criteria (BIC), both being dependent on likelihood functions and are

$$AIC = -2\sum_{i=1}^{n} \ln\left[c\left(u_{i,1}, u_{i,2}\right); \theta\right] + 2k, \tag{12}$$

$$BIC = -2\sum_{i=1}^{n} \ln\left[c\left(u_{i,1}, u_{i,2}\right); \theta\right] + \ln(nk). \tag{13}$$

The preferred copula is with the lowest AIC/BIC values. Also, some derivations of AIC and other information criteria for model selection are given in Grønneberg and Hjort [8].

What follows now, the bivariate Archimedean copulas studied in this paper are given in Table 1. The expressions of Kendall's τ in terms of copula parameters are provided in the last column. The copula parameter in each case measures the degree of dependence and controls the association between two variables. In Table 2, the expressions of Archimedean copulas, their generator functions, and distribution functions are given. As an example of the graphic representation of copula, the scatterplot of the Guloksuz–Kumar (NewCop) copula [9] and its distribution function are shown in Fig. 1.

Table 1 Archimedean copulas and Kendall's τ

Family	Bivariate copula $C(u,v), 0 < u, v < 1$	Copula parameter space	τ
Clayton	$\left(u^{-\theta} + v^{-\theta} - 1\right)^{-1/\theta}$	$\theta > 1$	$\frac{\theta}{\theta+2}$
Gumbel	$\exp\left[-\left[(-\ln u)^{\theta} + (-\ln v)^{\theta}\right]^{1/\theta}\right]$	$\theta \geq 1$	$\frac{\theta-1}{\theta}$
Frank	$\left(-\frac{1}{\theta}\right)\ln\left\{\frac{(1-e^{-\theta})-(1-e^{-\theta u})(1-e^{-\theta v})}{(1-e^{-\theta})}\right.$	$-\infty < \theta < \infty$	$1 - \frac{4}{\theta}\left[1 - D_1^*(\theta)\right]$
Guloksuz–Kumar	$1 - \frac{1}{\theta}\ln\left(e^{\theta(1-u)} + e^{\theta(1-v)} - 1\right)$	$\theta > 0$	$\frac{4\left(1-e^{-\theta}-\theta\right)}{\theta^2} + 1 +$ 1

$D_n^*(\theta) = n/\theta^n \int_0^{\theta} \frac{t^n}{(e^t-1)dt}$, $n > 0$ is a Debye function

Table 2 Archimedean copulas, generator functions, and Copula distribution function

Family	Generator $\phi(t), 0 < t < 1$	Distribution function $K_{\phi}(t) = t - \frac{\phi(t)}{\phi'(t)}, 0 < t < 1$
Clayton	$t^{-\theta} - 1$	$t - \frac{t^{\theta+1}-t}{\theta}$
Gumbel	$(-\ln t)^{\theta}$	$t - \frac{t\ln t}{\theta}$
Frank	$-\ln\frac{e^{-\theta t}-1}{e^{-\theta}-1}$	$t - \frac{\ln\frac{e^{-\theta t}-1}{e^{-\theta}-1}}{\theta}\left(e^{\theta t} - 1\right)$
Guloksuz–Kumar	$e^{\theta(1-t)} - 1$	$t - \frac{e^{\theta(1-t)}-1}{-\theta e^{-\theta(t-1)}}$

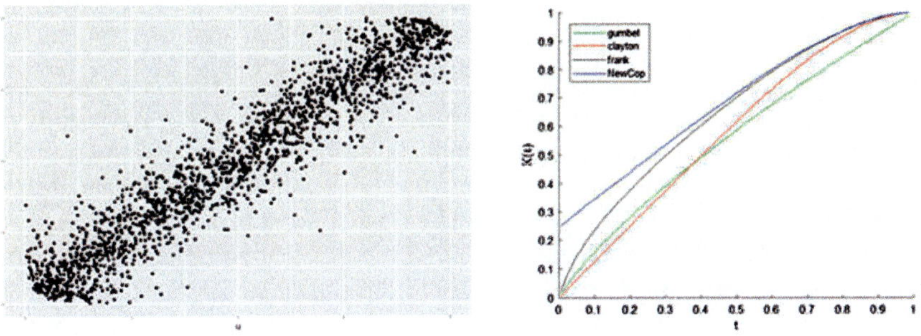

Fig. 1 Scatterplot and distribution function of the Guloksuz–Kumar (NewCop) copula

2 Copula Tail Dependence Measures

The concept of stochastic tail dependence refers to the clustering of extreme events. Extreme events, for example, in economic systems and in natural hazards contexts generally exhibit tail dependence, and thus it becomes very important to analyze the extreme behavior. The definition of tail dependence is the limiting probability that a random variable exceeds a certain threshold, given that another random variable already exceeds that threshold. Copula functions can measure nonparametric, distribution-free, or scale-invariant nature of dependence and extreme events. Joe [11] defines the upper tail dependence:

$$\lambda_U := \lim_{u \to 1} [\{1 - 2u + C(u, u)\}/(1 - u)] \tag{14}$$

$C(u, v)$ has upper tail dependence for $\lambda_U \in (0, 1]$ and no upper tail dependence for $\lambda_U = 0$.

Similarly, lower tail dependence in terms of copula is defined as follows:

$$\lambda_L := \lim_{u \to 0} [C(u, u)/u], \tag{15}$$

and copula $C(u, v)$ has lower tail dependence for $\lambda_L \in (0, 1]$ and no lower tail dependence for $\lambda_L = 0$. This measure is extensively used in extreme value theory. It is the probability that one variable is extreme given that other is extreme. Tail measures are copula based and copula is related to the full distribution via quantile transformations, i.e.,

$$C(u, v) = F\left(F_1^{-1}(u), F_2^{-1}(v)\right), \tag{16}$$

for all $u, v \in (0, 1]$ in the bivariate case.

3 Copula Simulation

The following steps are used to simulate data from a bivariate copula. The procedure mainly consists of first to estimate the marginal distributions and then to specify the copula function. Marginal distributions can be estimated by empirical or parametric ways. For simulation, let (U, V) be a random pair from a bivariate Archimedean copula, $\phi(t)$ the generator function, and $K(t)$ the distribution function of copula. Then, n-pairs of data $(X_i, Y_i; i = 1, 2, \ldots, n)$ from a bivariate Archimedean copula can be simulated from the following steps:

(1) Draw two independent random numbers, p and q, from Uniform (0, 1).
(2) $t = K^{-1}(q)$.
(3) $u = \phi^{-1}[p\phi(t)]$ and $v = \phi^{-1}[(1 - p)\phi(t)]$.
(4) $X = F^{-1}(u)$ and $Y = F^{-1}(v)$.
(5) Repeat the above steps n times to get n pairs of data $(X_i, Y_i; i = 1, 2, \ldots, n)$.

It should be also noted that for the step 2, when the inverse of $K(t)$ does not have the closed form, the following equation can be solved by numerical methods, like the Newton–Raphson method:

$$\left|t - ((\phi(t))/(\phi'(t)))\right| - q = 0. \tag{17}$$

4 Empirical Study

Several applications of copula functions and copula families with one or more real parameters are studied in various disciplines like statistics, insurance, finance, economics, survival analysis, information theory, image processing, and engineering. We now address an important concern from the practitioner's point of view as how to choose the most appropriate copula as a model from the large family of Archimedean copulas for a given sample dataset.

4.1 *Modeling and Simulation*

For this objective, we begin with fitting the linear model for making predictions and analyzing the prediction errors for choosing the best copula among the Archimedean copulas, namely, Clayton [3], Gumbel [10], Frank [5], and Guloksuz–Kumar copulas [9]. For more details on copula-based prediction, refer to Kumar and Shoukri [12, 13]. The calculation steps are as follows:

i. Randomly generate 1000 pairs of (X, Y) data from the bivariate standard normal distribution and for the Pearson correlation coefficient $= 0.1, 0.2, 0.3, 0.4, 0.5, 0.6, 0.7, 0.75, 0.8, 0.85, 0.9, 0.95, 0.99$.
ii. Calculate Kendall's τ and the copula parameter θ using the expressions in the last column of Table 1.
iii. For each copula, simulate 1000 pairs of (X, Y) and repeat simulations 100 times.
iv. For the simulated data in the previous step (iv), obtain least squares estimates a and b to fit the linear regression model to predict Y from X, i.e., $E(Y) = \alpha + \beta X + \varepsilon$.
v. The estimated intercept and slope coefficient values (a and b), of the parameters (α and β), for the copula-based models are, respectively, the averages of the 100 (a and b) values from the fitted models in step (iv).
vi. Calculate the 95% confidence intervals (CIs) of the slope parameter and the mean absolute prediction error (*MAPE*)

$$MAPE = \sum_{i=1}^{n} 100\left|\frac{y_i - \hat{y}_i}{ny_i}\right| \tag{18}$$

In Table 3, values of the Pearson correlation coefficient r, Kendall τ, and copula parameters θ of Clayton, Gumbel, Frank, and Guloksuz–Kumar copulas are provided.

It is noted that for the range of values of the correlation coefficient r between 0.1 and 0.99, the range of values of Kendall τ is between 0.0852 and 0.9132 and the range of copula parameters are Clayton (0.1864, 21.0397), Gumbel (1.0932, 11.5198), Frank (0.717, 44.3712), and Guloksuz–Kumar (2.9798, 45.0566). Since copula parameter is a measure of the dependence or association between two variables like Pearson correlation coefficient r and Kendall τ, the range of Frank copula parameter (43.5995) is the maximum, followed by the range of Guloksuz–Kumar copula parameter (42.0768) and is the minimum (10.4266) for the Gumbel copula parameter.

4.2 *Choosing Copula for the Prediction Modeling Using Mean Absolute Prediction Error (MAPE)*

We summarize fitted prediction models, 95% confidence intervals (CI) of regression coefficients, and mean absolute prediction error ($MAPE$) of the models in Table 4. It is clearly evident from the minimum distance measure ($MAPE$) values that the Guloksuz–Kumar-copula-based prediction model has the minimum prediction error for all the Pearson correlation coefficient (r) considered except $r = 0.1$, where the Clayton-copula-based prediction model has the minimum prediction error (Table 4).

Table 3 Kendall τ and copula parameters θ

Correlation r	Kendall τ	Clayton θ	Gumbel θ	Frank θ	Guloksuz–Kumar θ
0.1	0.0852	0.1864	1.0932	0.7717	2.9798
0.2	0.1199	0.2724	1.1362	1.0916	3.1721
0.3	0.1726	0.4172	1.2086	1.5920	3.4921
0.4	0.2319	0.6038	1.3019	2.1834	3.8991
0.5	0.3415	1.0373	1.5186	3.4046	4.8259
0.6	0.4190	1.4426	1.7213	4.4295	5.6764
0.7	0.4940	1.9524	1.9762	5.6278	6.7319
0.75	0.5252	2.2126	2.1063	6.2148	7.2666
0.8	0.5810	2.7732	2.3866	7.4430	8.4118
0.85	0.6587	3.8592	2.9296	9.7401	10.6144
0.9	0.7037	4.7493	3.3746	11.5815	12.4109
0.95	0.8060	8.3070	5.1535	18.8115	19.5602
0.99	0.9132	21.0397	11.5198	44.3712	45.0566

Table 4 Models for prediction and mean absolute prediction error (MAPE)

Correlation	Data	Intercept	Slope (std. error)	95% CI (slope)	MAPE
$r = 0.1$	BiNormal (0, 1)	−0.0444	0.1270 (0.0318)	0.0646–0.1895	1.3121
	Clayton	−0.0033	0.1380 (0.0312)	0.0768–0.1992	1.2090
	Gumbel	−0.0012	0.1399 (0.0311)	0.0789–0.2008	1.2109
	Frank	−0.0015	0.1218 (0.0314)	0.0602–0.1833	1.3096
	Guloksuz–Kumar	0.1246	−0.0952 (0.0315)	−0.1569–0.0336	1.5400
$r = 0.2$	BiNormal (0, 1)	0.0300	0.1824 (0.0316)	0.1203–0.2445	1.7407
	Clayton	0.001	0.1925 (0.0311)	0.1315–0.2534	1.7874
	Gumbel	0.0020	0.1961 (0.0312)	0.1349–0.2573	1.8037
	Frank	0.0035	0.1702 (0.0310)	0.1095–0.2310	1.6862
	Guloksuz–Kumar	0.1029	−0.0383 (0.0315)	−0.1000–0.0234	1.3838
$r = 0.3$	BiNormal (0, 1)	−0.0194	0.2677 (0.0305)	0.2079–0.3275	2.0681
	Clayton	0.0035	0.2778 (0.0305)	0.2180–0.3374	2.1367
	Gumbel	0.0035	0.2801 (0.0306)	0.2201–0.3399	2.1475
	Frank	0.0026	0.2443 (0.0307)	0.1840–0.3045	1.9797
	Guloksuz–Kumar	0.0813	0.0471 (0.0315)	−0.0146–0.1089	1.3641
$r = 0.4$	BiNormal (0, 1)	0.0286	0.3636 (0.0284)	0.3078–0.4193	1.8016
	Clayton	0.0010	0.3574 (0.0295)	0.2996–0.4153	1.7912
	Gumbel	0.0047	0.3655 (0.0295)	0.3077–0.4233	1.8125
	Frank	0.0073	0.3209 (0.0299)	0.2622–0.3797	1.6782
	Guloksuz–Kumar	0.0578	0.1405 (0.0313)	0.0791–0.2019	1.2514
$r = 0.5$	BiNormal (0, 1)	0.0324	0.4706 (0.0261)	0.4195–0.5217	2.8881
	Clayton	0.0074	0.5017 (0.0275)	0.4479–0.5556	3.0339
	Gumbel	0.0013	0.5037 (0.0272)	0.4504–0.5569	3.0441
	Frank	0.0037	0.4708 (0.0280)	0.4159–0.5257	2.8890
	Guloksuz–Kumar	0.0290	0.3395 (0.0298)	0.2811–0.3979	2.2896
$r = 0.6$	BiNormal (0, 1)	0.0548	0.5927 (0.0242)	0.5453–0.6401	2.1512
	Clayton	0.0023	0.6031 (0.0254)	0.5533–0.6529	2.1413
	Gumbel	0.0004	0.6064 (0.0251)	0.5572–0.6556	2.1488
	Frank	0.0016	0.5702 (0.0263)	0.5187–0.6216	2.0580
	Guloksuz–Kumar	0.0144	0.4665 (0.0280)	0.4117–0.5213	1.8102
$r = 0.7$	BiNormal (0, 1)	−0.0081	0.7132 (0.0227)	0.6687–0.7577	2.9027
	Clayton	−0.0015	0.6773 (0.0233)	0.6316–0.7229	2.7072
	Gumbel	−0.0006	0.6936 (0.0227)	0.6491–0.7380	2.7499
	Frank	−0.0002	0.6522 (0.0241)	0.6050–0.6994	2.6106
	Guloksuz–Kumar	0.0072	0.5814 (0.0256)	0.5312–0.6316	2.2993
$r = 0.75$	BiNormal (0, 1)	−0.0272	0.7293 (0.0210)	0.6880–0.7706	3.9980

(continued)

Table 4 (continued)

Correlation	Data	Intercept	Slope (std. error)	95% CI (slope)	MAPE
	Clayton	−0.0003	0.7065 (0.0223)	0.6629–0.7502	3.9776
	Gumbel	−0.0015	0.7192 (0.0218)	0.6764–0.7620	4.0356
	Frank	−0.0021	0.6870 (0.0231)	0.6417–0.7322	3.8765
$r = 0.8$	BiNormal (0, 1)	0.0140	0.7801 (0.0189)	0.7430–0.8173	2.6190
	Clayton	0.0012	0.7634 (0.0205)	0.7231–0.8036	2.5708
	Gumbel	0.0009	0.7830 (0.0196)	0.7445–0.8214	2.6253
	Frank	0.0034	0.7479 (0.0212)	0.7064–0.7895	2.5276
	Guloksuz–Kumar	0.0044	0.7033 (0.0224)	0.6593–0.7472	2.4066
$r = 0.85$	BiNormal (0, 1)	0.00001	0.8514 (0.0162)	0.8197–0.8831	2.6908
	Clayton	0.0009	0.8217 (0.0180)	0.7864–0.8570	2.6088
	Gumbel	0.0003	0.8498 (0.0166)	0.8173–0.8824	2.6861
	Frank	0.0003	0.8138 (0.0185)	0.7776–0.8499	2.5881
	Guloksuz–Kumar	0.0029	0.7942 (0.0195)	0.7559–0.8324	2.5334
$r = 0.9$	BiNormal (0, 1)	0.0146	0.9014 (0.0142)	0.8736–0.9292	1.4276
	Clayton	0.0016	0.8576 (0.0164)	0.8255–0.8896	1.3725
	Gumbel	0.0011	0.8833 (0.0146)	0.8546–0.9120	1.4035
	Frank	0.0010	0.8469 (0.0167)	0.8142–0.8797	1.3613
	Guloksuz–Kumar	0.0007	0.8339 (0.0175)	0.7997–0.8681	1.3468
$r = 0.95$	BiNormal (0, 1)	0.0056	0.9654 (0.0096)	0.9466–0.9841	2.0714
	Clayton	0.0016	0.9199 (0.0126)	0.8953–0.9445	1.9683
	Gumbel	0.0001	0.9491 (0.0100)	0.9294–0.9687	2.0163
	Frank	0.0001	0.9191 (0.0125)	0.8946–0.9436	1.9618
	Guloksuz–Kumar	0.0015	0.9172 (0.0128)	0.8922–0.9422	1.9603
$r = 0.99$	BiNormal (0, 1)	−0.0027	0.9899 (0.0043)	0.9815–0.9983	0.9086
	Clayton	−0.0008	0.9717 (0.0074)	0.9573–0.9861	0.8891
	Gumbel	−0.0008	0.9898 (0.0046)	0.9807–0.9989	0.9024
	Frank	0.8405	2.1743 (0.0460)	2.0842–2.2644	5.7300
	Guloksuz–Kumar	−0.0010	0.9726 (0.0072)	0.9585–0.9867	0.8929

5 Concluding Remarks

Copula functions and copula families with one or more real parameters have become very popular recently for the modeling and simulation in various disciplines like statistics, insurance, finance, economics, survival analysis, information theory, image processing, and engineering. Since there are a large number of copulas available, we investigated an important question as how to choose the most appropriate copula from the family of Archimedean copulas, namely, Clayton, Gumbel, Frank, and Guloksuz–Kumar for making the predictions from the linear models.

We have focused at the bivariate normal population, $N(0, 1)$, since most of the application data do follow the exact or approximate multivariate normal distributions and data can easily be standardized. Further, we have considered the bivariate normal populations with the Pearson correlation coefficient $r = 0.1, 0.2, 0.3, 0.4, 0.5, 0.6, 0.7, 0.75, 0.8, 0.85, 0.9, 0.95, 0.99$. The Guloksuz–Kumar-copula-based prediction models have attained the minimum prediction error for all the Pearson correlation coefficient (r) values considered, except $r = 0.1$, where the Clayton-copula-based prediction model resulted in the minimum prediction error. However, it may further be noted that when response and explanatory variable have a correlation of very small magnitude like $r = 0.1$, it implies that there is almost no association and the explanatory variable is of no use in making predictions about the response variable, and hence no need for fitting the model.

References

1. Belgorodski, N.: Selecting pair copula families for regular vines with application to the multivariate analysis of European stock market indices. Master Thesis, Technische Universitt Munchen (2010)
2. Cherubini, U., Luciano, E.E., Vecchiato, W.: Copula Methods in Finance. John Wiley, NY (2004)
3. Clayton, D.G.: A Model for association in bivariate life tables and its application in epidemiological studies of familial tendency in chronic disease incidence. Biometrika **65** (141–151 (1978)
4. Frees, E.W., Valdez, E.A.: Understanding relationships using copulas. North Am. Actuar. J. **2**, 1–25 (1998)
5. Frank, M.J.: On the simultaneous associativity of F(x, y) and x + y-F(x, y). Aequ. Math. **19**, 194–226 (1979)
6. Genest, C., Rivest, L.P.: Statistical inference procedures for bivariate Archimedean copulas. Am. Stat. **88**(423), 1034–1043 (1993)
7. Genest, C., Favre, A.: Everything you always wanted to know about copula modeling but were afraid to ask. J. Hydrol. Eng. **12**, 347–368 (2007)
8. Grønneberg, S., Hjort, N.L.: The copula information criteria. Scand. J. Stat. **41**, 436–459 (2014)
9. Guloksuz, C.T., Kumar, P.: On a new bivariate one parameter Archimedean copula function and its application to modeling dependence in climate and life sciences. Stat. Appl. **17**(2), 23–36 (2019)
10. Gumbel, E.J.: Bivariate exponential distributions. J. Am. Stat. Assoc. **55**, 698–707 (1960)
11. Joe, H.: Multivariate Models and Dependence Concepts, 1st edn. Chapman Hall Ltd, U.K (1997)
12. Kumar, P., Shoukri, M.M.: Aortic stenosis using the Archimedean copula methodology. J. Data Sci. **6**, 173–187 (2008)
13. Kumar, P., Shoukri, M.M.: Copula based prediction models: an application to aortic regurgitation study. BMC Med. Res. Methodol. **7**(21), 1–9 (2007)
14. Kojadinovic, I., Yan, J.: Comparison of three semiparametric methods for estimating dependence parameters in copula models. Insur. Math. Econ. **47**, 52–63 (2010)
15. Matteis De, R.: Fitting copulas to data. Diploma Thesis, Institute of Mathematics of University Zurich (2001)
16. Nelsen, R.: An Introduction to Copulas, 2nd edn. Springer, New York (2006)
17. Sklar, A.: Fonctions de repartition a n dimensions et leurs marges. Publications del Institut de Statistique del Universite de Paris **8**, 229–231 (1959)

On Generalized Kantorovich Sampling Type Series

A. Sathish Kumar and Prashant Kumar

Abstract In the present article, we analyse the behaviour of a new family of Kantorovich type sampling operators $(K_w^\varphi f)_{w>0}$. We obtain an asymptotic formula for $(K_w^\varphi f)_{w>0}$ involving the first and second-order derivatives of f and the moments of the kernel φ. We also give a quantitative version of asymptotic formula which shows that when f belongs to $C^2(\mathbb{R})$ (the set of all $f \in C(\mathbb{R})$ such that f is twice differentiable functions), the convergence is uniform. Finally, we give some examples for the kernel to which the theory can be applied.

Keywords Generalized sampling series · Average Kantorovich type sampling series · Pointwise convergence · Fourier transform · B-spline kernel · Graphical representation

Mathematics Subject Classification (2010) 41A35 · 47A58 · 94A20 · 41A25

1 Introduction

The theory of generalized sampling series was first initiated by P. L. Butzer and his school [6] and [7]. In recent years, it is an attractive topic in approximation theory due to its wide range of applications, especially in signal and image processing. For $w > 0$, a generalized sampling series of a function $f : \mathbb{R} \to \mathbb{R}$ is defined by

$$(T_w^\varphi f)(x) = \sum_{k=-\infty}^{\infty} \varphi(wx - k) f\left(\frac{k}{w}\right), \quad x \in \mathbb{R},$$

A. S. Kumar (✉) · P. Kumar
Department of Mathematics, Visvesvaraya National Institute of Technology,
Nagpur 440010, India
e-mail: mathsatish9@gmail.com

P. Kumar
e-mail: pranwd92@gmail.com

R. N. Mohapatra et al. (eds.), *Mathematical Analysis and Computing*,
Springer Proceedings in Mathematics & Statistics 344,
https://doi.org/10.1007/978-981-33-4646-8_3

where φ is a kernel function on $\mathbb{R}$. These type of operators have been studied by many authors (see, e.g. [2, 3, 12, 13, 16]). The Kantorovich type generalizations of approximation operators is an important subject in approximation theory and they are the method to approximate Lebesgue integrable functions. The Kantorovich type generalizations of the generalized sampling operators were introduced by P. L. Butzer and his school (see, e.g. [6, 8, 10, 18] etc.). In [4], the authors have introduced the sampling Kantorovich operators and studied their rate of convergence in the general settings of Orlicz spaces. Also, the nonlinear version of sampling Kantorovich operators has been studied in [9] and [19].

We consider the generalized Kantorovich type sampling series. In this paper, we analyse the approximation properties of the following type of generalized Kantorovich sampling series

$$(K_w^{\varphi} f)(x) = \sum_{k=-\infty}^{\infty} \varphi(wx-k) w \int_{\frac{k-\frac{1}{2}}{w}}^{\frac{k+\frac{1}{2}}{w}} f(u)du,$$

$f \in C(\mathbb{R})$ (The class of all uniformly continuous and bounded functions on $\mathbb{R}$). The above series represents the generalized sampling series in the L^1 setting, where we considered the samples values as an average of f on a small interval around k/w, that is, the mean value $w \int_{k-(1/2)/w}^{k+(1/2)/w} f(u)du$, instead of the sampling values $f(k/w)$. Usually, more information is known around a point than precisely at that point; this procedure simultaneously reduces jitter errors.

Let $\varphi \in C(\mathbb{R})$ be fixed. For every $\nu \in \mathbb{N}_0 = \mathbb{N} \cup \{0\}$, $u \in \mathbb{R}$ we define the algebraic moments as

$$m_\nu(\varphi, u) := \sum_{k=-\infty}^{\infty} \varphi(u-k)(k-u)^\nu \quad \text{and the absolute moments by}$$

$$M_\nu(\varphi) := \sup_{u \in \mathbb{R}} \sum_{k=-\infty}^{\infty} |\varphi(u-k)||(k-u)|^\nu.$$

Also, note that for $\mu, \nu \in \mathbb{N}_0$ with $\mu < \nu$, $M_\nu(\varphi) < +\infty$ implies $M_\mu(\varphi) < +\infty$. Indeed for $\mu < \nu$, we have

$$\begin{aligned}\sum_{k=-\infty}^{\infty} |\varphi(u-k)||(k-u)|^\nu &= \sum_{|u-k|<1} |\varphi(u-k)||(k-u)|^\nu + \sum_{|u-k|\geq 1} |\varphi(u-k)||(k-u)|^\nu \\ &\leq 2\|\varphi\|_\infty + \sum_{|u-k|\geq 1} |\varphi(u-k)| \frac{|(k-u)|^\nu}{|(k-u)|^{\nu-\mu}} \\ &\leq 2\|\varphi\|_\infty + M_\nu(\varphi).\end{aligned}$$

When φ is compactly supported, then we have $M_\nu(\varphi) < +\infty$ for every $\nu \in \mathbb{N}_0$.

We suppose that the following assumptions hold:

(i) for every $u \in \mathbb{R}$, we have

$$\sum_{k=-\infty}^{\infty} \varphi(u-k) = 1,$$

(ii) for every $r > 0$, there holds

$$\lim_{r\to\infty} \sum_{|u-k|>r} |\varphi(u-k)|(k-u)^2 = 0$$

uniformly with respect to $u \in \mathbb{R}$,

(iii) for every $u \in \mathbb{R}$, we have

$$m_1(\varphi, u) = m_1(\varphi) = \sum_{k=-\infty}^{\infty} \varphi(u-k)(k-u) = 0,$$

(iv) for every $u \in \mathbb{R}$, we have

$$M_2(\varphi) = \sup_{u\in\mathbb{R}} \sum_{k=-\infty}^{\infty} |\varphi(u-k)|(k-u)^2 < +\infty$$

$$M_3(\varphi) = \sup_{u\in\mathbb{R}} \sum_{k=-\infty}^{\infty} |\varphi(u-k)||(k-u)|^3 < +\infty.$$

We study here the behaviour of the point-wise convergence at a point in which f is sufficiently regular, giving an exact order of approximation. In particular, we obtain an asymptotic formula for $(K_w^{\varphi} f)_{w>0}$ involving the first- and second-order derivatives of f and the moments of the kernel φ. We also give a quantitative version of asymptotic formula which shows that when f belongs to $C^2(\mathbb{R})$, the convergence is uniform. An important tool in order to obtain this estimate is Peetre K-functional (see [14, 15]). We remark that quantitative Voronovskaja formula has important links with the theory of semi-groups of operators (see [1, 5]). Finally, we present some examples of kernel φ which satisfy the assumptions used in the general theory.

Theorem 1 *For $f \in L^{\infty}(\mathbb{R})$, $\lim_{w\to\infty} (K_w^{\varphi} f)(x) = f(x)$ at every point of x of continuity of f. Moreover, if the function is uniformly continuous and bounded on $\mathbb{R}$, then $\lim_{w\to\infty} \|K_w^{\varphi} f - f\|_{\infty} = 0$.*

Proof Let $\epsilon > 0$ be fixed. By the continuity of f at the point x, there exists $\delta > 0$ such that $|f(u) - f(x)| \leq \epsilon$, whenever $|u - x| \leq \delta$. We can write

$$|(K_w^{\varphi}f)(x) - f(x)| \le \sum_{|k-wx|<\delta w/2} |\varphi(wx-k)| \int_{\frac{k-\frac{1}{2}}{w}}^{\frac{k+\frac{1}{2}}{w}} |f(u) - f(x)|du$$
$$+ \sum_{|k-wx|\ge\delta w/2} |\varphi(wx-k)| \int_{\frac{k-\frac{1}{2}}{w}}^{\frac{k+\frac{1}{2}}{w}} |f(u) - f(x)|du = I_1 + I_2, \text{(say)}.$$

By the continuity of f at the point x, we get

$$I_1 \le \epsilon \sum_{|k-wx|<\delta w/2} |\varphi(wx-k)| \le \epsilon.M_0(\varphi),$$

for sufficiently large $w > 0$. Also, let w be such that $\delta w/2 > R$. Then

$$\sum_{|k-wx|>\delta w/2} |\varphi(wx-k)| < \epsilon, \text{ for every } x \in \mathbb{R}.$$

Hence, we get

$$I_2 \le 2\|f\|_\infty \sum_{|k-wx|\ge\delta w/2} |\varphi(wx-k)| \le \epsilon.2\|f\|_\infty,$$

for sufficiently large $w > 0$ and hence the first part of the theorem follows. The second part can be proved similarly.

Remark 1 Since the functions in $L^\infty(\mathbb{R})$ are locally Lebesgue integrable, the generalized Kantorovich sampling series used in the above theorem are well defined for $f \in L^\infty(\mathbb{R})$.

Theorem 2 *Let $f \in L^\infty(\mathbb{R})$. If $f''(x)$ exists at a point $x \in \mathbb{R}$, then*

$$\lim_{w\to\infty} w^2[(K_w^{\varphi}f)(x) - f(x)] = \frac{f''(x)}{24}.$$

Proof From Taylor's theorem, we have

$$f(u) = f(x) + f'(x)(u-x) + \frac{f''(x)}{2}(u-x)^2 + h(u-x)(u-x)^2,$$

for some bounded function h such that $\lim_{t\to 0} h(t) = 0$. Thus, we have
$(K_w^{\varphi}f)(x) - f(x)$

$$= f'(x) \sum_{k=-\infty}^{\infty} \varphi(wx-k)w \int_{\frac{k-\frac{1}{2}}{w}}^{\frac{k+\frac{1}{2}}{w}} (u-x)du + \frac{f''(x)}{2} \sum_{k=-\infty}^{\infty} \varphi(wx-k)w \int_{\frac{k-\frac{1}{2}}{w}}^{\frac{k+\frac{1}{2}}{w}} (u-x)^2du$$
$$+ \sum_{k=-\infty}^{\infty} \varphi(wx-k)w \int_{\frac{k-\frac{1}{2}}{w}}^{\frac{k+\frac{1}{2}}{w}} h(u-x)(u-x)^2du := I_1 + I_2 + I_3.$$

First, we evaluate I_1.

$$I_1 = \frac{wf'(x)}{2} \sum_{k=-\infty}^{\infty} \varphi(wx-k)\left[\left(\frac{k+1/2}{w}-x\right)^2 - \left(\frac{k-1/2}{w}-x\right)^2\right]$$
$$= \frac{f'(x)}{w} \sum_{k=-\infty}^{\infty} \varphi(wx-k)(k-wx) = \frac{f'(x)}{w} m_1(\varphi) = 0.$$

As to I_2, we have analogously

$$I_2 = \frac{wf''(x)}{6} \sum_{k=-\infty}^{\infty} \varphi(wx-k)\left[\left(\frac{k+1/2}{w}-x\right)^3 - \left(\frac{k-1/2}{w}-x\right)^3\right]$$
$$= \frac{f''(x)}{24w^2} \sum_{k=-\infty}^{\infty} \varphi(wx-k) = \frac{f''(x)}{24w^2}.$$

In order to obtain an estimate of I_3, let $\epsilon > 0$ be fixed. Then, there exists $\delta > 0$ such that $|h(t)| \leq \epsilon$ for $|t| \leq \delta$. Then, we have

$$|I_3| \leq \sum_{|k-wx|<\delta w/2} |\varphi(wx-k)| \int_{\frac{k-\frac{1}{2}}{w}}^{\frac{k+\frac{1}{2}}{w}} |h(u-x)|(u-x)^2 du$$
$$+ \sum_{|k-wx|\geq\delta w/2} |\varphi(wx-k)| \int_{\frac{k-\frac{1}{2}}{w}}^{\frac{k+\frac{1}{2}}{w}} |h(u-x)|(u-x)^2 du := J_1 + J_2.$$

We have

$$J_1 \leq w\epsilon \sum_{|k-wx|<\delta w/2} |\varphi(wx-k)| \int_{\frac{k-\frac{1}{2}}{w}}^{\frac{k+\frac{1}{2}}{w}} (u-x)^2 du \leq \frac{\epsilon}{12w^2} \sum_{k=-\infty}^{\infty} |\varphi(wx-k)| = \frac{\epsilon.M_0(\varphi)}{12w^2}.$$

Next, we obtain J_2. Let $R > 0$ be such that $\sum_{|u-k|>R} |\varphi(u-k)|(u-k)^2 < \epsilon$ uniformly with respect to $u \in \mathbb{R}$. Also, let w be such that $\delta w/2 > R$. Then

$$\sum_{|k-wx|>\delta w/2} |\varphi(wx-k)|(wx-k)^2 < \epsilon$$

for every $x \in \mathbb{R}$. The same inequality holds also for the series

$$\sum_{|k-wx|>\delta w/2} |\varphi(wx-k)||wx-k|^j < \epsilon, \quad \text{for} \quad j = 0, 1.$$

Hence, we get

$$J_2 \leq \frac{w\|h\|_\infty}{3} \sum_{|k-wx|\geq \delta w/2} |\varphi(wx-k)| \int_{\frac{k-\frac{1}{2}}{w}}^{\frac{k+\frac{1}{2}}{w}} (u-x)^2 du$$
$$\leq \frac{\|h\|_\infty}{12w^2} \sum_{|k-wx|\geq \delta w/2} |\varphi(wx-k)| \leq \frac{\epsilon \|h\|_\infty}{12w^2}.$$

Hence, $\lim_{w\to\infty} w^2 I_3 = 0$. This completes the proof.

Remark 2 The boundedness assumption on f can be relaxed by assuming that there are two positive constant a, b such that $|f(x)| \leq a + bx^2$, for every $x \in \mathbb{R}$. We have

$$|K_w^\varphi f)(x)| \leq \sum_{k=-\infty}^{\infty} |\varphi(wx-k)| w \int_{\frac{k-\frac{1}{2}}{w}}^{\frac{k+\frac{1}{2}}{w}} |f(u)| du$$
$$\leq \sum_{k=-\infty}^{\infty} |\varphi(wx-k)| w \int_{\frac{k-\frac{1}{2}}{w}}^{\frac{k+\frac{1}{2}}{w}} (a+bu^2) du$$
$$\leq M_0(\varphi)\left(a + \frac{b}{12w^2} + bx^2\right) + M_1(\varphi)\frac{2bx}{w} + M_2(\varphi)\frac{b}{w^2},$$

and hence the series $K_w^\varphi f$ is absolutely convergent for every $x \in \mathbb{R}$. Moreover, for a fixed $x_0 \in \mathbb{R}$,

$$P_2(x) = f(x_0) + f'(x_0)(x-x_0) + \frac{f''(x_0)}{2}(x-x_0)^2,$$

Taylor's polynomial of second order centred at the point x_0, by Taylor's formula we can write

$$\frac{f(x) - P_2(x)}{(x-x_0)^2} = h(x-x_0),$$

where h is a function such that $\lim_{t\to 0} h(t) = 0$. Then, h is bounded on $[x_0 - \delta, x_0 + \delta]$, for some $\delta > 0$. For $|x - x_0| > \delta$, we have

$$|h(x-x_0)| \leq \frac{a+bx^2}{(x-x_0)^2} + \frac{|P_2(x)|}{(x-x_0)^2},$$

and the terms on the right-hand side of the above inequality are all bounded for $|x - x_0| > \delta$. Hence, $h(. - x_0)$ is bounded on $\mathbb{R}$. Along the lines of the proof of Theorem 2, the same Voronovskaya formula can be obtained.

Let $C^{(m)}$ denote the set of all $f \in C(\mathbb{R})$ such that f is m times continuously differentiable and $\|f^{(m)}\|_\infty < \infty$. Let $\delta > 0$. For $f \in C(\mathbb{R})$, Peetre's K-functional is defined as

$$K(\delta, f, C^{(0)}, C^{(1)}) := \inf\{\|f - g\|_\infty + \delta\|g'\|_\infty : g \in C^{(1)}\}.$$

For a given $\delta > 0$, the usual modulus of continuity of a given uniformly continuous function $f : \mathbb{R} \to \mathbb{R}$ is defined as

$$\omega(f, \delta) := \sup_{|x-y|\leq\delta} |f(x) - f(y)|.$$

It is well known that, for any positive constant $\lambda > 0$, the modulus of continuity satisfies the following property

$$\omega(f, \lambda\delta) \leq (\lambda + 1)\omega(f, \delta).$$

For a function $f \in C^{(m)}$, $x_0, x \in \mathbb{R}$ and $m \geq 1$, Taylor's formula is given by

$$f(x) = \sum_{k=0}^{m} \frac{f^{(k)}(x_0)}{k!}(x - x_0)^k + R_m(f; x_0, x)$$

and the remainder term $R_m(f; x_0, x)$ is estimated by

$$|R_m(f; x_0, x)| \leq \frac{|x - x_0|^m}{m!}\omega(f^{(m)}, |x - x_0|).$$

The following estimate for the remainder $R_m(f; x_0, x)$ in terms of $\overline{\omega}$ was proved in [11].

Lemma 1 *Let $f \in C^{(m)}$, $m \in \mathbb{N}^0$ and $x_0, x \in \mathbb{R}$. Then, we have*

$$|R_m(f; x_0, x)| \leq \frac{|x - x_0|^m}{m!}\overline{\omega}\left(f^{(m)}, \frac{|x - x_0|}{m + 1}\right).$$

Theorem 3 *For $f \in C(\mathbb{R})$ and for $\delta > 0$, we have*

$$|(K_w^{\varphi} f)(x) - f(x)| \leq \omega(f, \delta)\left(\left(1 + \frac{1}{\delta w}M_0(\varphi)\right) + \frac{2}{\delta w}M_1(\varphi)\right).$$

Proof Since $\omega(f, \lambda\delta) \leq (\lambda + 1)\omega(f, \delta)$, for $\lambda > 0$ and $|u - x| \leq 2|x - \frac{k}{w}| + \frac{1}{w}$, for $u \in \left[\frac{k - 1/2}{w}, \frac{k + 1/2}{w}\right]$, we have

$$
\begin{aligned}
|(K_w^{\varphi}f)(x)-f(x)| &\leq \sum_{k=-\infty}^{\infty}|\varphi(wx-k)|w\int_{\frac{k-\frac{1}{2}}{w}}^{\frac{k+\frac{1}{2}}{w}}\omega(f,|u-x|)du\\
&\leq \sum_{k=-\infty}^{\infty}|\varphi(wx-k)|w\int_{\frac{k-\frac{1}{2}}{w}}^{\frac{k+\frac{1}{2}}{w}}\left(1+\frac{|u-x|}{\delta}\right)\omega(f,\delta)du\\
&\leq \omega(f,\delta)M_0(\varphi)+\sum_{k=-\infty}^{\infty}|\varphi(wx-k)|w\int_{\frac{k-\frac{1}{2}}{w}}^{\frac{k+\frac{1}{2}}{w}}\left(2\left|\frac{k}{w}-x\right|+\frac{1}{w}\right)du\\
&\leq \omega(f,\delta)M_0(\varphi)+\frac{2\omega(f,\delta)}{w\delta}M_1(\varphi)+\frac{\omega(f,\delta)}{w\delta}M_0(\varphi)
\end{aligned}
$$

and hence the result follows.

Remark 3 If we choose $\delta=\frac{1}{w}$ in the above Theorem 3, then, we obtain the estimate

$$\|(K_w^{\varphi}f)-f\|_{\infty}\leq C\omega(f,1/w),$$

where $C=2(M_0(\varphi)+M_1(\varphi))$.

Theorem 4 *Let $f\in C^{(2)}(\mathbb{R})$ be fixed. Then, for every $x\in\mathbb{R}$, we have*

$$\left|w^2((K_w^{\varphi}f)(x)-f(x))-\frac{f''(x)}{24}\right|\leq\frac{A}{24}\overline{\omega}\left(f'',\frac{4B}{Aw}\right),$$

where $A=M_0(\varphi)+12M_2(\varphi)$ and $B=M_0(\varphi)+6M_1(\varphi)+12M_2(\varphi)+8M_3(\varphi)$.

Proof Let $f\in C^{(2)}(\mathbb{R})$ be fixed. Then, we can write
$\left|w^2[(K_w^{\varphi}f)(x)-f(x)]-\frac{f''(x)}{24}\right|$

$$
\begin{aligned}
&=\Bigg|w^3f'(x)\sum_{k=-\infty}^{\infty}\varphi(wx-k)\int_{\frac{k-\frac{1}{2}}{w}}^{\frac{k+\frac{1}{2}}{w}}(u-x)du+\frac{w^3f''(x)}{2}\sum_{k=-\infty}^{\infty}\varphi(wx-k)\int_{\frac{k-\frac{1}{2}}{w}}^{\frac{k+\frac{1}{2}}{w}}(u-x)^2du\\
&\quad+w^3\sum_{k=-\infty}^{\infty}\varphi(wx-k)\int_{\frac{k-\frac{1}{2}}{w}}^{\frac{k+\frac{1}{2}}{w}}h(u-x)(u-x)^2du-\frac{f''(x)}{24}\Bigg|\\
&\leq w^3\sum_{k=-\infty}^{\infty}|\varphi(wx-k)|\int_{\frac{k-\frac{1}{2}}{w}}^{\frac{k+\frac{1}{2}}{w}}|h(u-x)|(u-x)^2du:=J.
\end{aligned}
$$

Putting $R_2(f;u,x)=h(u-x)(u-x)^2$, using Lemma 1, we obtain

$$J \le \frac{w^3}{2} \sum_{k=-\infty}^{\infty} |\varphi(wx-k)| \int_{\frac{k-\frac{1}{2}}{w}}^{\frac{k+\frac{1}{2}}{w}} (u-x)^2 \overline{\omega}\left(f'', \frac{|u-x|}{3}\right) du$$

$$\le w^3 \sum_{k=-\infty}^{\infty} |\varphi(wx-k)| \int_{\frac{k-\frac{1}{2}}{w}}^{\frac{k+\frac{1}{2}}{w}} (u-x)^2 K\left(\frac{|u-x|}{6}, f''\right) du.$$

For $g \in C^{(3)}$, we have

$$J \le w^3 \sum_{k=-\infty}^{\infty} |\varphi(wx-k)| \int_{\frac{k-\frac{1}{2}}{w}}^{\frac{k+\frac{1}{2}}{w}} (u-x)^2 \left(\|(f-g)''\|_\infty + \frac{|u-x|}{6} \|g'''\|_\infty \right) du$$

$$\le w^3 \|(f-g)''\|_\infty \sum_{k=-\infty}^{\infty} |\varphi(wx-k)| \int_{\frac{k-\frac{1}{2}}{w}}^{\frac{k+\frac{1}{2}}{w}} (u-x)^2 du$$

$$+ \frac{w^3 \|g'''\|_\infty}{6} \sum_{k=-\infty}^{\infty} |\varphi(wx-k)| \int_{\frac{k-\frac{1}{2}}{w}}^{\frac{k+\frac{1}{2}}{w}} |u-x|^3 du := I_1 + I_2.$$

It is easy to see that $I_1 \le \|(f-g)''\|_\infty \left(\frac{M_0(\varphi}{12} + M_2(\varphi) \right)$.

Since for $u \in \left[\frac{k-1/2}{w}, \frac{k+1/2}{w} \right]$, $|u-x| \le 2|x - \frac{k}{w}| + \frac{1}{w}$, we obtain

$$I_2 \le \frac{\|g'''\|_\infty}{6w} (M_0(\varphi) + 6M_1(\varphi) + 12M_2(\varphi) + 8M_3(\varphi)).$$

Thus, we have

$$J \le \frac{A}{12} \left(\|(f-g)''\|_\infty + \frac{2B}{Aw} \|g'''\|_\infty \right).$$

Taking the infimum over all $g \in C^{(3)}$, we get the desired result.

We discuss the average sampling Kantorovich operators based upon the combinations of spline functions. The $B-$spline of order $h \in \mathbb{N}$ is defined as ([7, 17])

$$B_h(x) := \chi_{[-\frac{1}{2},\frac{1}{2}]} \star \chi_{[-\frac{1}{2},\frac{1}{2}]} \star \chi_{[-\frac{1}{2},\frac{1}{2}]} \star \cdots \star \chi_{[-\frac{1}{2},\frac{1}{2}]}, \text{ (h times)}$$

where

$$\chi_{[-\frac{1}{2},\frac{1}{2}]} = \begin{cases} 1, & \text{if } -\frac{1}{2} \le x \le \frac{1}{2} \\ 0, & \text{otherwise} \end{cases}$$

and $*$ denotes the convolution.

Let $w \in \mathbb{R}$ and $h \in \mathbb{N}$. The Fourier transform of the functions $B_h(x)$ is given by $\widehat{B}_h(w) = \left(\chi_{[\frac{-1}{2},\frac{1}{2}]}(w)\right)^h = \left(\frac{\sin w/2}{w/2}\right)^h$. Given real numbers ψ_0, ψ_1 with $\psi_0 < \psi_1$ we will construct the linear combination of translates of B_h, with $h \geq 2$ of type

$$\varphi(x) = a_0 B_h(x - \psi_0) + a_1 B_h(x - \psi_1).$$

The Fourier transform of φ is $\widehat{\varphi}(w) = \left(a_0 e^{-i\psi_0 w} + a_1 e^{-i\psi_1 w}\right)\widehat{B}_h(w)$. Using the Poisson summation formula, we obtain

$$\sum_{k=-\infty}^{\infty} \varphi(u-k) = \sum_{k=-\infty}^{\infty} \widehat{\varphi}(2\pi k) e^{i2\pi k u}.$$

We have

$$\widehat{B_h}(2\pi k) = \left(\frac{\sin(\pi k)}{\pi k}\right)^h = \begin{cases} 1, \text{ if } k = 0 \\ 0, \text{ if } k \neq 0 \end{cases}$$

and hence

$$\widehat{\varphi}(2\pi k) = \begin{cases} a_0 + a_1, & \text{if } k = 0 \\ 0, & \text{if } k \neq 0 \end{cases}$$

Therefore, condition (i) is satisfied if $a_0 + a_1 = 1$. From Poisson summation formula, we obtain

$$(-i) \sum_{k=-\infty}^{\infty} \varphi(u-k)(u-k) = \sum_{k=-\infty}^{\infty} \widehat{\varphi}'(2\pi k) e^{i2\pi k u}.$$

We have

$$\widehat{\varphi}'(w) = (-i\psi_0 a_0 e^{-i\psi_0 w} - i\psi_1 a_1 e^{-i\psi_1 w})\widehat{B}_h(w) + (a_0 e^{-i\psi_0 w} + a_1 e^{-i\psi_1 w})\widehat{B}'_h(w).$$

Since $\widehat{B}'_h(2\pi k) = 0$, $\forall k$, we get $\widehat{\varphi}'(2\pi k) = 0$. Hence, we obtain

$$\widehat{\varphi}(0) = a_0 + a_1 = 1, \quad \widehat{\varphi}'(0) = \psi_0 a_0 + \psi_1 a_1 = 0.$$

Solving the linear system, we obtain the unique solution

$$a_0 = \frac{\psi_1}{\psi_1 - \psi_0}, \; a_1 = -\frac{\psi_0}{\psi_0 - \psi_1}.$$

It is easy to see that the support of φ is contained in the interval $[\psi_0 - \frac{h}{2}, \psi_1 - \frac{h}{2}]$. Since $\varphi(u-k) = 0$ if $|u-k| > r$ for r sufficiently large, we have

$$\lim_{r\to\infty} \sum_{|k-u|>r} \varphi(u-k)(k-u)^2 = 0.$$

Now, we have

$$\sum_{k\in\mathbb{Z}} |\varphi(u-k)||(k-u)|^2 = \sum_{|k-u|<R} |\varphi(u-k)||(k-u)|^2 + \sum_{|k-u|\geq R} |\varphi(u-k)||(k-u)|^2.$$

We can see that $\sup_u |\{k : |u-k| < R\}| \leq N_0$. Thus, we get

$$M_2(\varphi) = \sum_{k\in\mathbb{Z}} |\varphi(u-k)||(k-u)|^2 < \infty.$$

Similarly, we can show that $M_3(\varphi) = \sum_{k\in\mathbb{Z}} |\varphi(u-k)||(k-u)|^3 < \infty$. Now, we show the approximation of functions f_1, f_2 by the average Kantorovich type sampling series based on the B-spline kernel.

$$\text{Let} \quad f_1(u) = \begin{cases} \frac{1}{5}, \text{ if, } -1 \leq u < 0 \\ \frac{1}{6}, \text{ if, } 0 \leq u < 1 \end{cases}$$

and

$$f_2(u) = \begin{cases} \frac{9}{u^6}, \text{ if, } u < -3 \\ 1, \text{ if, } -3 \leq u < -2 \\ \frac{-1}{2}, \text{ if, } -2 \leq u < -1 \\ \frac{-3}{2}, \text{ if, } -1 \leq u < 0 \\ 1, \text{ if, } 0 \leq u < 1 \\ \frac{-2}{u}, \text{ if, } u > 1. \end{cases}$$

Now, we show and compare the approximation of average Kantorovich type sampling series $(K_w^{\varphi} f)_{w>0}$ for different values of w (Figs. 1 and 2).

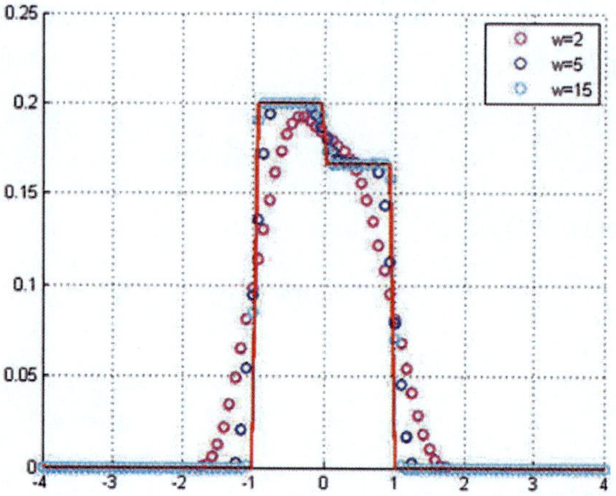

Fig. 1 This graph shows the approximation of the function f_1 on $[-1, 1]$ by average Kantorovich sampling series $K_w^{\varphi} f_1$ based upon B-spline type kernel for $w = 2, 5, 15$

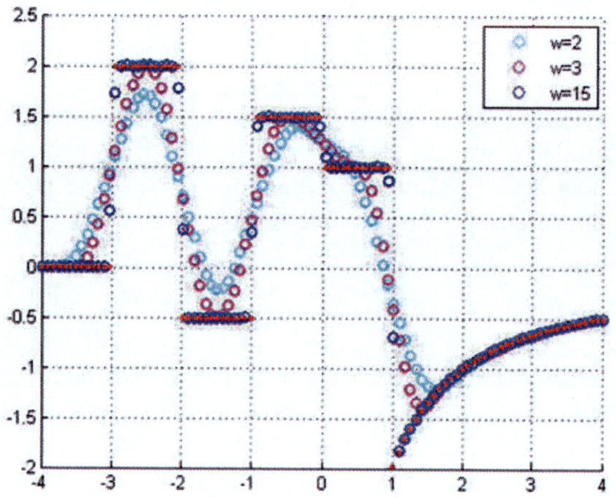

Fig. 2 This graph shows the approximation of the function f_2 on $[-3, 1]$ by average Kantorovich type sampling series $K_w^{\varphi} f_1$ based upon B-spline type kernel for $w = 2, 3, 15$

2 Final Remarks and Conclusion

In this article, we introduce a new family of average Kantorovich type sampling series. We have analysed the approximation properties of the operators, e.g. pointwise convergence theorem, Voronovskaja type theorem and its quantitative version for these operators. In the last section, we present some examples of kernel which satisfies the assumptions of our theory. It is also observed that the average Kantorovich type sampling series approximates $f \in C(\mathbb{R})$ in a better way than the classical generalized sampling series. Finally, we conclude that our operator $(K_w^{\varphi} f)_{w>0}$ gives the better convergence as compared to the classical generalized sampling series $(T_w^{\varphi} f)_{w>0}$.

References

1. Altomare, F.: Asymptotic formula for Bernstein-Schnabl operators and smoothness. Boll. U.M.I. Ser. II. **9**, 135–150 (2009)
2. Bardaro, C., Mantellini, I.: A Voronovskaya type theorem for a general class of discrete operators. Rocky Mountain J. Math. **39** 1411–1442(2009)
3. Bardaro, C., Mantellini, I.: A quantitative Voronovskaja formula for generalized sampling operators. East. J. Approx. **15**(4), 429–441 (2009)
4. Bardaro, C., Vinti, G., Butzer, P.L., Stens, R.L.: Kantorovich-type generalized sampling series in the setting of Orlicz spaces. Sampl. Theor. Signal Image Process. **6**, 29–52 (2007)
5. Butzer, P.L., Berens, H.: Semi-Groups of Operators and Approximation, Die Grundlehren der mathematischen Wissenschften 145. Springer, Berlin (1967)
6. Butzer, P.L., Stens, R.L.: Sampling theory for not necessarily bandlimited functions: an historical overview. SIAM Rev. **34**, 40–53 (1992)
7. Butzer, P.L., Stens, R.L.: Linear prediction by samples from the past. In: Marks II, R.J. (ed.) Advanced Topics in Shannon Sampling and Interpolation Theory. Springer Texts Electrical Engineering. Springer, New York (1993)
8. Butzer, P.L., Nessel, R.J.: Fourier Analysis and Approximation, vol. 1: One-Dimensional Theory. Pure and Applied Mathematics, vol. 40. Academic Press, New York and London (1971)
9. Costarelli, D., Vinti, G.: Approximation by nonlinear multivariate sampling Kantorovich type operators and applications to image processing. Num. Funct. Anal. Optim. **34**, 819–844 (2013)
10. Donnini, C., Vinti, G.: Approximation by means of Kantorovich generalized sampling operators in Musielak-Orlicz spaces. PanAmer. Math. J. **18**(2), 1–18 (2008)
11. Gonska, H.H., Pitul, P., Raşa, I.: On Peano's form of the Taylor remainder, Voronovskaja's theorem and the commutator of positive linear operators. In: Proceedings of the International Conference on Numerical Analysis and Approximation Theory, pp. 55–80. Cluj-Napoca (2006)
12. Kivinukk, A., Tamberg, G.: Interpolating generalized Shannon sampling operators, their norms and approximation properties. Sampl. Theor. Signal Image Process. **8**, 77–95 (2009)
13. Mantellini, I.: Generalized sampling operators in modular spaces. Comment. Math. **38**, 77–92 (1998)
14. Peetre, J.: A Theory of Interpolation of Normed Spaces, vol. 39. Lecture Notes, Brasilia (Notas de Mathematica) (1963)
15. Peetre, J.: Exact interpolation theorems for Lipshitz continuous functions. Ricerche Mat. **18**, 239–259 (1969)
16. Ries, S., Stens, R.L.: Approximation by generalized sampling series. In: Sendov, Bl., Petrushev, P., Maalev, R., Tashev, S. (eds.) Constructive Theory of Functions, pp. 746–756. Publ. House Bulgarian Academy of Sciences, Sofia (1984)
17. Schonberg, I.J.: Cardinal Spline Interpolation. Regional Conference series in Applied Mathematics, SIAM, Philadelphia (1973)
18. Vinti, G.: A general approximation result for nonlinear integral operators and applications to signal processing. Appl. Anal. **1–2**(79), 217–238 (2001)
19. Vinti, G., Zampogni, L.: Approximation by means of nonlinear Kantorovich sampling type operators in Orlicz spaces. J. Approx. Theory **161**, 511–528 (2009)

The Fekete-Szegö Problem for Bazilevič Function

Sarbeswar Barik

Abstract In this paper, a new subclass of bi-univalent functions is introduced using subordination and also the bounds for the Fekete-Szegö problem are obtained by refining the well known estimates for the initial coefficients of the Carathéodory functions. The method of proof is different, the results obtained in this paper are improved and μ varies from $-\infty$ to ∞.

Keywords Analytic function · Bi-univalent function · Coefficient bound · Fekete-Szegö

1 Preliminaries and Definitions

Let $\mathcal{A}$ denote the family of functions $f(z)$ represented by the following Taylor-Maclaurin's series:

$$f(z) = z + \sum_{n=2}^{\infty} a_n z^n \tag{1}$$

which are analytic in the open unit disc

$$\mathbb{U} = \{z \in \mathbb{C} : |z| < 1\}.$$

A function $f \in \mathcal{A}$ is said to be *univalent* in $\mathbb{U}$ if $f(z)$ is one to one in $\mathbb{U}$. As usual, we denote by $\mathcal{S}$ the subclass of functions in $\mathcal{A}$ which are univalent in $\mathbb{U}$. The function $f \in \mathcal{S}$ is said to be star-like in $\mathbb{U}$ [5] if $\Re\left(\frac{zf'(z)}{f(z)}\right) > 0$ $(z \in \mathbb{U})$. The class of such function is denoted by $\mathcal{S}^*$.

The function $f \in \mathcal{S}$ has an inverse f^{-1}, defined by

S. Barik (✉)
Department of Computer Science, Central University of Odisha,
Koraput 7643004, Odisha, India
e-mail: sarbeswar1204@gmail.com

R. N. Mohapatra et al. (eds.), *Mathematical Analysis and Computing*,
Springer Proceedings in Mathematics & Statistics 344,
https://doi.org/10.1007/978-981-33-4646-8_4

$$f^{-1}(f(z)) = z, \qquad (z \in \mathbb{U})$$

and

$$f(f^{-1}(w)) = w, \quad (w \in \text{range of } f).$$

It is well known that for every function $f \in \mathcal{S}$, the inverse function $f^{-1}(w)$ is analytic in some disc $|w| < r_0(f),\ r_0(f) \geq \frac{1}{4}$. Moreover, $f^{-1}(w)$ has the Taylor-Maclaurin series expansion of the form:

$$f^{-1}(w) = w + \sum_{n=2}^{\infty} b_n w^n \quad \left(|w| < r_0(f), r_0(f) \geq \frac{1}{4}\right).$$

For initial values of n, one can easily obtain the following:

$$b_2 = -a_2, \quad b_3 = 2a_2^2 - a_3, \quad b_4 = 5a_2a_3 - 5a_2^3 - a_4$$

and so on. We say that the function f is bi-univalent in $\mathbb{U}$ if $f^{-1}(w)$ has analytic continuation to $\mathbb{U}$. The study of bi-univalent analytic functions on the unit disc was initiated by Lewin [9]. Seminal work on bi-univalent functions can be found in [2, 3, 9, 12, 20].

If F and ϕ are analytic in $\mathbb{U}$, we say that F is subordinate to ϕ written as $F \prec \phi$ if and only if $F(z) = \phi(u(z))$ for some function $u(z)$ satisfying the conditions of Schwarz lemma ($u(z)$ is analytic in $\mathbb{U}$, $u(0) = 0$, $|u(z)| \leq 1$).

The determination of upper bounds for the nonlinear functionals $|\ a_3 - \mu a_2^2\ |$ for any given family $\mathcal{F}$ of a normalized analytic function is popularly known as the Fekete-Szegö problem for the family $\mathcal{F}$. Many authors [8, 11, 18] including [6] studied the Fekete-Szegö problem for different subclasses of $\mathcal{S}$. We shall need the following definitions for our investigation.

Definition 1 The function ϕ is said to be a member of the the family $\mathcal{R}$ if the following conditions are satisfied: (i) $\phi(0) = 1$, (ii) $\phi(z)$ is analytic and one to one in $\mathbb{U}$ and (iii) The range of $\phi(z)$, i.e. $\phi(\mathbb{U})$ is a region in the right half plane containing the point 1. We furthermore assume that functions $\phi \in \mathcal{R}$ are represented by the following series:

$$\phi(z) = 1 + B_1 z + B_2 z^2 + \cdots, \quad 0 \leq B_2 \leq B_1, \quad z \in \mathbb{U}. \tag{2}$$

The following are some examples of members of $\mathcal{R}$, which are extensively discussed in the literature.

Example 1

$$\phi_1(z) = \frac{1+z}{1-z} = 1 + 2z + 2z^2 + \cdots \quad (z \in \mathbb{U})$$

maps $\mathbb{U}$ onto the right half plane. Moreover, the class $\mathcal{P}$, consisting of functions p satisfying $p(z) \prec \phi_1(z)$, is the well studied class of Carathéodory functions.

Example 2

$$\phi_2(z) = \left(\frac{1+z}{1-z}\right)^{\alpha} = 1 + 2\alpha z + 2\alpha^2 z^2 + \cdots \quad (0 < \alpha \le 1,\ z \in \mathbb{U}).$$

The function $\phi_2(z)$ maps $\mathbb{U}$ one to one onto the sector in the right half plane with opening of angle $\alpha\pi$ at the origin and is symmetric with respect to the real axis. Moreover, for $\alpha = 1,\ \phi_2(z) = \phi_1(z)$.

Example 3

$$\phi_3(z) = \frac{1 + (1-2\beta)z}{1-z} = 1 + 2(1-\beta)z + 2(1-\beta)z^2 + \cdots \quad (0 \le \beta < 1,\ z \in \mathbb{U}).$$

The function $\phi_3(z)$ maps $\mathbb{U}$ one to one onto the region $\Re(w) > \beta$. In the particular case $\beta = 0,\ \phi_3(z) = \phi_1(z)$.

Example 4 [10, 17]

$$\phi_4(z) = 1 + \frac{2}{\pi^2}\left(\log\frac{1+\sqrt{z}}{1-\sqrt{z}}\right)^2 = 1 + \frac{8}{\pi^2}z + \frac{16}{3\pi^2}z^2 + \cdots \quad (z \in \mathbb{U}).$$

The function $\phi_4(z)$ is the Riemann map of $\mathbb{U}$ onto the interior of the parabola

$$\{w : w = x + iy \ and\ y^2 = 2x - 1\}$$

which has vertex at the point $\left(\frac{1}{2}, 0\right)$ in the right half plane.

Let $h \in \mathcal{S}^*$ a function $f \in \mathcal{A}$ is said to be Bazilevič of function type γ with respect to the function h [16, 21] if

$$\left(\frac{zf'(z)}{(f(z))^{1-\gamma}(h(z))^{\gamma}}\right) > 0 \quad \gamma \ge 0,\ (z \in \mathbb{U}). \tag{3}$$

It is well known that the functions that satisfy (3) are univalent. In the particular case, $h(z) = z$. Let $\mathcal{B}(\gamma)$ be the class of functions in $\mathcal{A}$ satisfying

$$\left(\frac{z}{f(z)}\right)^{1-\gamma} f'(z) > 0 \quad (z \in \mathbb{U}). \tag{4}$$

In the present paper, we introduce the following class of bi-univalent functions using subordination And also obtain the Fekete-Sezegö inequalities for three subclasses of bi-univalent functions. Thus, we have the following.

Definition 2 Let ϕ be a member of the class $\mathcal{R}$. The function $f \in \mathcal{A},\ \gamma \ge 0$, is said to be in the class $\sigma\mathcal{B}_{\gamma}(\phi)$ if the following conditions are satisfied:

$$\left(\frac{z}{f(z)}\right)^{1-\gamma} f'(z) \prec \phi(z) \quad (z \in \mathbb{U}), \tag{5}$$

and

$$\left(\frac{w}{g(w)}\right)^{1-\gamma} g'(w) \prec \phi(w) \quad (w \in \mathbb{U}), \tag{6}$$

where g is the analytic continuation of f^{-1} to $\mathbb{U}$.

Taking $\gamma = 1$, in Definition 2 we get the following class studied earlier in [19].

The problem of determination of bounds for the initial coefficients of functions in different subclasses of bi-univalent functions has been widely investigated in the literature. However, not much is known on the Fekete-Szegö problem for bi-univalent functions. Recently, Orhan et al. [14] extended the work of Zaprawa [22] and obtained bounds on $| a_3 - \mu a_2^2 |$ for some subclasses of bi-univalent functions. Jahangiri et al. [7] also obtained bounds for the Fekete-Szegö problem for the class $\mathcal{S}_\sigma^\star(\phi)$ and $\mathcal{CV}_\sigma^\star(\phi)$.

In this paper, we adapt a method of proof different from Zaprawa and Jahangiri et al. and also find estimates on $| a_3 - \mu a_2^2 |$ where μ is real for the function class $\mathcal{H}_\sigma(\phi)$ and $\mathcal{S}_\sigma^\star(\phi)$. Our bounds improve upon results of Zaprawa for the class $\mathcal{H}_\sigma(\phi)$. Our method of proof can be adopted for the variety of subclasses of bi-univalent functions studied in the literature. Recent works on the Fekete-Szegö problem for some subclasses of $\mathcal{S}$ can be found on [1, 4, 7, 13, 15, 18, 22]. We need the following result for our investigation.

Lemma 1 *(See [5].) Let $p(z) \in \mathcal{P}$ where*

$$p(z) = 1 + c_1 z + c_2 z^2 + c_3 z^3 + \cdots \quad (z \in \mathbb{U}).$$

Then

$$| c_n | \leq 2 \quad (n \in \mathbb{N}),$$

$$| c_2 - c_1^2 | \leq 2 \quad \textit{and} \quad | c_2 - \frac{1}{2} c_1^2 | \leq 2 - \frac{|c_1|^2}{2}.$$

2 Proof of the Theorem

The following calculations shall be used in the proof of our Theorem. Let u and v be analytic functions satisfying the conditions of the Schwarz lemma. Define the function p and q in $\mathcal{R}$ by the following.

$$p(z) = \frac{1 + u(z)}{1 - u(z)} = 1 + c_1 z + c_2 z^2 + \cdots \quad (z \in \mathbb{U}),$$

and

$$q(w) = \frac{1 + v(w)}{1 - v(w)} = 1 + l_1 w + l_2 w^2 + \cdots \quad (w \in \mathbb{U}).$$

This gives an expansion for $u(z)$ and $v(z)$ in terms of coefficients of the corresponding Carathéodory functions p and q. That is,

$$u(z) = \frac{c_1}{2} z + \frac{1}{2}\left(c_2 - \frac{c_1^2}{2}\right) z^2 + \cdots . \tag{7}$$

and

$$v(w) = \frac{l_1}{2} w + \frac{1}{2}\left(l_2 - \frac{l_1^2}{2}\right) w^2 + \cdots . \tag{8}$$

Taking the composition of the series (2) for ϕ and the series (7) for $u(z)$, we have

$$\phi(u(z)) = 1 + \frac{1}{2} B_1 c_1 z + \left\{\frac{1}{4} B_2 c_1^2 + \frac{1}{2} B_1 \left(c_2 - \frac{c_1^2}{2}\right)\right\} z^2 + \cdots . \tag{9}$$

Similarly, from the series (2) and (8) together, we get

$$\phi(v(w)) = 1 + \frac{1}{2} B_1 l_1 w + \left\{\frac{1}{4} B_2 l_1^2 + \frac{1}{2} B_1 \left(l_2 - \frac{l_1^2}{2}\right)\right\} w^2 + \cdots . \tag{10}$$

Theorem 1 *Let the function f given by* (1) *be in the class $\sigma\mathcal{B}_\gamma(\phi)$. Then for $-\infty < \mu < \infty$*
(i) if $\mu > \delta_1$ then

$$|a_3 - \mu a_2^2| \le \frac{2B_1}{(1+\gamma)(2+\gamma)} \begin{cases} 1 + \frac{\mu(2+\gamma)B_1^2 - 2(1+\gamma)(B_1+B_2)}{(2+\gamma)B_1^2 + 2(1+\gamma)(B_1-B_2)} & \frac{(2+\gamma)B_1^2}{2(1+\gamma)B_2} > 1, \gamma \le 1 \\ 1 + \frac{\mu(2+\gamma)B_1^2 - 2(1+\gamma)(B_1+B_2)}{2(1+\gamma)B_1} & \frac{(2+\gamma)B_1^2}{2(1+\gamma)B_2} \le 1, \gamma \le 1, \end{cases} \tag{11}$$

and

$$|a_3 - \mu a_2^2| \le \frac{B_1}{(2+\gamma)} \begin{cases} 1 + \frac{2\{\mu(2+\gamma)B_1^2 - 2(1+\gamma)(B_1+B_2)\}}{(1+\gamma)\{(2+\gamma)B_1^2 + 2(1+\gamma)(B_1-B_2)\}} & \frac{(2+\gamma)B_1^2}{2(1+\gamma)B_2} > 1, \gamma \ge 1 \\ 1 + \frac{\mu(2+\gamma)B_1^2 - 2(1+\gamma)(B_1+B_2)}{(1+\gamma)^2 B_1} & \frac{(2+\gamma)B_1^2}{2(1+\gamma)B_2} \le 1, \gamma \ge 1. \end{cases} \tag{12}$$

(ii) if $\delta_2 \le \mu \le \delta_1$

$$|a_3 - \mu a_2^2| \leq \frac{B_1}{2+\gamma} \quad (\delta_2 \leq \mu \leq \delta_1). \tag{13}$$

(iii) if $\mu < \delta_2$ then

$$|a_3 - \mu a_2^2| \leq \frac{2B_1}{2+\gamma} \begin{cases} 1 + \frac{2(1+\gamma)(B_2-B_1)-\mu(2+\gamma)B_1^2}{(2+\gamma)B_1^2+2(1+\gamma)(B_1-B_2)} & \frac{(2+\gamma)B_1^2}{2(1+\gamma)B_2} > 1, \gamma \leq 1 \\ 1 + \frac{2(1+\gamma)(B_2-B_1)-\mu(2+\gamma)B_1^2}{2(1+\gamma)B_1} & \frac{(2+\gamma)B_1^2}{2(1+\gamma)B_2} \leq 1, \gamma \leq 1 \end{cases} \tag{14}$$

and

$$|a_3 - \mu a_2^2| \leq \frac{B_1}{2+\gamma} \begin{cases} 1 + \frac{4(1+\gamma)(B_2-B_1)-2\mu(2+\gamma)B_1^2}{(1+\gamma)\{(2+\gamma)B_1^2+2(1+\gamma)(B_1-B_2)\}} & \frac{(2+\gamma)B_1^2}{2(1+\gamma)B_2} > 1, \gamma \geq 1 \\ 1 + \frac{2(1+\gamma)(B_2-B_1)-\mu(2+\gamma)B_1^2}{(1+\gamma)^2 B_1} & \frac{(2+\gamma)B_1^2}{2(1+\gamma)B_2} \leq 1, \gamma \geq 1 \end{cases} \tag{15}$$

where $\delta_1 = \frac{2(1+\gamma)(B_1+B_2)}{(2+\gamma)B_1^2}$ $\delta_2 = -\frac{2(1+\gamma)(B_1-B_2)}{(2+\gamma)B_1^2}$.

Proof Let the function f be a member of class $\sigma\mathcal{B}_\gamma(\phi)$. Then by Definition 2, we have the following:

$$\left(\frac{z}{f(z)}\right)^{1-\gamma} f'(z) = \phi(u(z)), \tag{16}$$

and

$$\left(\frac{w}{g(w)}\right)^{1-\gamma} g'(w) = \phi(v(w)), \tag{17}$$

where $u(z)$ and $v(w)$ are members of the class $\mathcal{A}$ and satisfy the conditions of the Schwarz lemma. Now equating the coefficients of $\left(\frac{z}{f(z)}\right)^{1-\gamma} f'(z)$ with the coefficient of $\phi(u(z))$ from (9), we get the following:

$$(1+\gamma)a_2 = \frac{1}{2}B_1 c_1, \tag{18}$$

$$(2+\gamma)a_3 + \frac{-2+\gamma+\gamma^2}{2}a_2^2 = \frac{1}{4}B_2 c_1^2 + \frac{1}{2}B_1\left(c_2 - \frac{c_1^2}{2}\right). \tag{19}$$

Similarly, a comparison of coefficients of both sides of (17) yields

$$-(1+\gamma)a_2 = \frac{1}{2}B_1 l_1, \tag{20}$$

$$-(2+\gamma)a_3 + \frac{6+5\gamma+\gamma^2}{2}a_2^2 = \frac{1}{4}B_2 l_1^2 + \frac{1}{2}B_1\left(l_2 - \frac{l_1^2}{2}\right). \tag{21}$$

From Eqs. (18) and (20), it is clear that

$$c_1^2 = l_1^2.$$

We add (19) with (21) and then use the relation $c_1^2 = l_1^2$. After simplification, we get

$$(2 + 3\gamma + \gamma^2)a_2^2 = \frac{B_1}{2}(c_2 + l_2) - \frac{B_1 - B_2}{2}c_1^2. \tag{22}$$

Again using (18), we get

$$a_2^2 = \frac{B_1^3}{2(1+\gamma)\{(2+\gamma)B_1^2 + 2(1+\gamma)(B_1 - B_2)\}}(c_2 + l_2). \tag{23}$$

We, next, express a_3 in terms of a_2^2 and the coefficients of p and q. For this, we subtract (21) from (19) and then use the relation $c_1^2 = l_1^2$. This gives the following after simplification:

$$a_3 = a_2^2 + \frac{B_1}{4(2+\gamma)}(c_2 - l_2). \tag{24}$$

Using (23) and (24), we get

$$\begin{aligned} a_3 - \mu a_2^2 &= \frac{B_1}{4(2+\gamma)}(c_2 - l_2) + (1-\mu)a_2^2 \\ &= \frac{B_1}{4(2+\gamma)}\left\{\left(\frac{2(1-\mu)(2+\gamma)B_1^2}{(1+\gamma)\{(2+\gamma)B_1^2 + 2(1+\gamma)(B_1 - B_2)\}} + 1\right)c_2\right. \\ &\quad \left. + \left(\frac{2(1-\mu)(2+\gamma)B_1^2}{(1+\gamma)\{(2+\gamma)B_1^2 + 2(1+\gamma)(B_1 - B_2)\}} - 1\right)l_2\right\} \\ &= \frac{B_1}{4(2+\gamma)}\left\{(h(\mu)+1)c_2 + (h(\mu)-1)l_2\right\}, \end{aligned}$$

where $h(\mu) = \frac{2(1-\mu)(2+\gamma)B_1^2}{(1+\gamma)\{(2+\gamma)B_1^2+2(1+\gamma)(B_1-B_2)\}}$.

The application of the triangle inequality and the estimate $|c_2| \leq 2$ and $|l_2| \leq 2$ give

$$|a_3 - \mu a_2^2| \leq \frac{B_1}{2(2+\gamma)}\left\{|h(\mu)+1| + |h(\mu)-1|\right\}.$$

Since $h(\mu)$ is real for all $\gamma \geq 0$, the above simplifies to the following:

$$|a_3 - \mu a_2^2| \leq \frac{B_1}{2+\gamma} \qquad (-1 \leq h(\mu) \leq 1)$$
$$= \frac{B_1}{2+\gamma} \qquad (\delta_2 \leq \mu \leq \delta_1)\,,$$

where $\delta_1 = \frac{2(1+\gamma)(B_1+B_2)}{(2+\gamma)B_1^2}$ $\delta_2 = -\frac{2(1+\gamma)(B_1-B_2)}{(2+\gamma)B_1^2}$.

This proves (13).

Now we shall obtain a refined estimate of $|c_1|$ for our use. Substituting $a_2^2 = \frac{B_1^2}{4(2-\gamma)^2}c_1^2$ from (18), in the relation (23) we get

$$c_1^2 = \frac{2(1+\gamma)B_1}{(2+\gamma)B_1^2 + 2(1+\gamma)(B_1 - B_2)}(c_2 + l_2).$$

Now, using the well known estimate $|c_2| \leq 2$ and $|l_2| \leq 2$, we obtain

$$|c_1| \leq \begin{cases} \frac{2\sqrt{2}\sqrt{(1+\gamma)B_1}}{\sqrt{(2+\gamma)B_1^2+2(1+\gamma)(B_1-B_2)}} & \frac{(2+\gamma)B_1^2}{2(1+\gamma)B_2} > 1 \\ 2 & \frac{(2+\gamma)B_1^2}{2(1+\gamma)B_2} \leq 1. \end{cases} \tag{25}$$

Using the relation (22) and (24), we get after simplification

$$2(2+\gamma)a_3 = \frac{(6+5\gamma+\gamma^2)B_1}{2(2+3\gamma+\gamma^2)}c_2 + \frac{(2-\gamma-\gamma^2)B_1}{2(2+3\gamma+\gamma^2)}l_2 - \frac{(2+\gamma)(B_1-B_2)}{2+3\gamma+\gamma^2}c_1^2. \tag{26}$$

Using (18) and (26), we get

$$\begin{aligned} 2(2+\gamma)(a_3 - \mu a_2^2) &= \frac{(3+\gamma)B_1}{2(1+\gamma)}c_2 + \frac{(1-\gamma)B_1}{2(1+\gamma)}l_2 \\ &\quad - \frac{(B_1 - B_2)}{1+\gamma}c_1^2 - \frac{\mu(2+\gamma)B_1^2}{2(1+\gamma)^2}c_1^2 \\ &= \frac{-B_1}{2(1+\gamma)}\left\{\frac{2(1+\gamma)(B_1-B_2)+\mu(2+\gamma)B_1^2}{(1+\gamma)B_1}c_1^2 \right. \\ &\quad \left. - (3+\gamma)c_2 - (1-\gamma)l_2\right\}. \end{aligned} \tag{27}$$

The application of the triangle inequality in the relation (27) gives

$$2(2+\gamma)|a_3-\mu a_2^2| \le \frac{B_1}{2(1+\gamma)}\left|\frac{2(1+\gamma)(B_1-B_2)+\mu(2+\gamma)B_1^2}{(1+\gamma)B_1}c_1^2 - (3+\gamma)c_2-(1-\gamma)l_2\right|$$

$$= \frac{B_1}{2(1+\gamma)}\left|\frac{2(1+\gamma)(B_1-B_2)+\mu(2+\gamma)B_1^2}{(1+\gamma)B_1}c_1^2 - 4c_1^2 - (3+\gamma)(c_2-c_1^2)-(1-\gamma)(l_2-l_1^2)\right|$$

$$= \frac{B_1}{2(1+\gamma)}\left|\frac{\mu(2+\gamma)B_1^2-2(1+\gamma)(B_1+B_2)}{(1+\gamma)B_1}\right||c_1^2| + (3+\gamma)\left|c_2-c_1^2\right| + |1-\gamma|\left|l_2-l_1^2\right|. \tag{28}$$

If $\mu > \delta_1$ then $\mu(2+\gamma)B_1^2 - 2(1+\gamma)(B_1+B_2) \ge 0$. Then applying Lemma 1 with refined bounds of $|c_1|$ from (25), we get

$$|a_3-\mu a_2^2| \le \frac{2B_1}{(1+\gamma)(2+\gamma)}\begin{cases} 1+\frac{\mu(2+\gamma)B_1^2-2(1+\gamma)(B_1+B_2)}{(2+\gamma)B_1^2+2(1+\gamma)(B_1-B_2)} & \frac{(2+\gamma)B_1^2}{2(1+\gamma)B_2} > 1, \gamma \le 1 \\ 1+\frac{\mu(2+\gamma)B_1^2-2(1+\gamma)(B_1+B_2)}{2(1+\gamma)B_1} & \frac{(2+\gamma)B_1^2}{2(1+\gamma)B_2} \le 1, \gamma \le 1. \end{cases}$$

This is precisely our estimate at (11).

For $\mu > \delta_1$ and $\gamma \ge 1$, we can write (28) as

$$2(2+\gamma)|a_3-\mu a_2^2| \le \frac{B_1}{2(1+\gamma)}\left|\frac{\mu(2+\gamma)B_1^2-2(1+\gamma)(B_1+B_2)}{(1+\gamma)B_1}\right||c_1^2| + (3+\gamma)\left|c_2-c_1^2\right| + |\gamma-1|\left|l_2-l_1^2\right|.$$

The application of Lemma 1 with refined bounds of $|c_1|$ from (25) gives

$$|a_3-\mu a_2^2| \le \frac{B_1}{2+\gamma}\begin{cases} 1+\frac{2\mu(2+\gamma)B_1^2-4(1+\gamma)(B_1+B_2)}{(1+\gamma)\{(2+\gamma)B_1^2+2(1+\gamma)(B_1-B_2)\}} & \frac{(2+\gamma)B_1^2}{2(1+\gamma)B_2} > 1, \gamma \ge 1 \\ 1+\frac{\mu(2+\gamma)B_1^2-2(1+\gamma)(B_1+B_2)}{(1+\gamma)^2B_1} & \frac{(2+\gamma)B_1^2}{2(1+\gamma)B_2} \le 1, \gamma \ge 1 \end{cases}$$

which is our assertion at (12).

Next, if $\mu < \delta_2$, then we write equation (27) as

$$2(2+\gamma)(a_3-\mu a_2^2) = \frac{B_1}{2(1+\gamma)}\left\{(3+\gamma)c_2+(1-\gamma)l_2+\frac{2(1+\gamma)(B_2-B_1)-\mu(2+\gamma)B_1^2}{(1+\gamma)B_1}c_1^2\right\}. \tag{29}$$

Using triangle inequality with Lemma 1 and refined bounds of $|c_1|$ from (25), we get

$$|a_3-\mu a_2^2| \leq \frac{2B_1}{2+\gamma}\begin{cases}1+\frac{2(1+\gamma)(B_2-B_1)-\mu(2+\gamma)B_1^2}{(2+\gamma)B_1^2+2(1+\gamma)(B_1-B_2)} & \frac{(2+\gamma)B_1^2}{2(1+\gamma)B_2}>1, \gamma\leq 1\\ 1+\frac{2(1+\gamma)(B_2-B_1)-\mu(2+\gamma)B_1^2}{2(1+\gamma)B_1} & \frac{(2+\gamma)B_1^2}{2(1+\gamma)B_2}\leq 1, \gamma\leq 1.\end{cases}$$

This is precisely the inequality in (14).

For $\mu<\delta_2$ and $\gamma\geq 1$, we can write (29) as

$$2(2+\gamma)(a_3-\mu a_2^2)=\frac{B_1}{2(1+\gamma)}\left\{(3+\gamma)c_2-(\gamma-1)l_2+\frac{2(1+\gamma)(B_2-B_1)-\mu(2+\gamma)B_1^2}{(1+\gamma)B_1}c_1^2\right\}.$$

Now, using triangle inequality with Lemma 1 and refined bounds of $|c_1|$ from (25), we get

$$|a_3-\mu a_2^2| \leq \frac{B_1}{2+\gamma}\begin{cases}1+\frac{4(1+\gamma)(B_2-B_1)-2\mu(2+\gamma)B_1^2}{(1+\gamma)\{(2+\gamma)B_1^2+2(1+\gamma)(B_1-B_2)\}} & \frac{(2+\gamma)B_1^2}{2(1+\gamma)B_2}>1, \gamma\geq 1\\ 1+\frac{2(1+\gamma)(B_2-B_1)-\mu(2+\gamma)B_1^2}{(1+\gamma)^2B_1} & \frac{(2+\gamma)B_1^2}{2(1+\gamma)B_2}\leq 1, \gamma\geq 1\end{cases}$$

which is our last assertion at (15). The proof of Theorem 1 is thus completed.

Corollary 1 *If f represented by the series (1) is a bi-starlike function in $\mathbb{U}$, then*

$$|a_3-\mu a_2^2| \leq \begin{cases}2\mu & \mu>1\\ 1 & 0\leq\mu\leq 1\\ 2(1-\mu) & \mu<0.\end{cases}$$

Proof By taking $\gamma=0$ and $\phi(z)=\phi_1(z)=1+2z+2z^2+\ldots$ and putting the value $B_1=2$, $B_2=2$ in Theorem 1, we get the result after simplification.

Corollary 2 *If f represented by the series (1) is a strongly bi-starlike function of order α $(0<\alpha\leq 1)$ in $\mathbb{U}$, then*

$$|a_3-\mu a_2^2| \leq 2\alpha\begin{cases}1+\frac{2\mu\alpha-(1+\alpha)}{\alpha+1} & \mu>\frac{1+\alpha}{2\alpha}\\ \frac{1}{2} & -\frac{1-\alpha}{2\alpha}\leq\mu\leq\frac{1+\alpha}{2\alpha}\\ 1+\frac{\alpha-1-2\mu\alpha}{\alpha+1} & \mu<-\frac{1-\alpha}{2\alpha}\end{cases}$$

Proof By taking $\gamma=0$ and $\phi(z)=\phi_2(z)=1+2\alpha z+2\alpha z^2+\ldots$ and putting the value $B_1=2\alpha$, $B_2=2\alpha^2$ in Theorem 1, we get the result after simplification.

Corollary 3 *If f represented by the series (1) is a bi-starlike function of order β $(0\leq\beta<1)$ in $\mathbb{U}$, then*

$$|a_3 - \mu a_2^2| \leq 2(1-\beta) \begin{cases} \mu & \mu > 1, \beta < \frac{1}{2} \\ 1 + 2(\mu - 1)(1-\beta) & \mu > 1, \beta \geq \frac{1}{2} \\ \frac{1}{2} & 0 \leq \mu \leq 1 \\ (1-\mu) & \mu < 0, \beta < \frac{1}{2} \\ 1 - 2\mu(1-\beta) & \mu < 0, \beta \geq \frac{1}{2} \end{cases}$$

Proof By taking $\gamma = 1$ and $\phi(z) = \phi_3(z) = 1 + 2(1-\beta)z + 2(1-\beta)z^2 + \ldots$ and putting the value $B_1 = 2(1-\beta)$, $B_2 = 2(1-\beta)$ in Theorem 1, we get the result after simplification.

Corollary 4 *If f represented by the series* (1) *is a bi-parabolic starlike function of order in* $\mathbb{U}$*, then*

$$|a_3 - \mu a_2^2| \leq \frac{8}{\pi^2} \begin{cases} 1 + \frac{24\mu - 5\pi^2}{3\pi^2} & \mu > \frac{5\pi^2}{24} \\ \frac{1}{2} & -\frac{\pi^2}{24} \leq \mu \leq \frac{5\pi^2}{24} \\ 1 - \frac{\pi^2 + 24\mu}{3\pi^2} & \mu < -\frac{\pi^2}{24} \end{cases}$$

Proof By taking $\gamma = 0$ and $\phi(z) = \phi_4(z) = 1 + \frac{8}{\pi^2}z + \frac{16}{3\pi^2}z^2 + \ldots$ and putting the value $B_1 = \frac{8}{\pi^2}$, $B_2 = \frac{16}{3\pi^2}$ in Theorem 1, we get the result after simplification.

Corollary 5 *If f represented by the series* (1) *is a bi-close-to-convex function in* $\mathbb{U}$*, then*

$$|a_3 - \mu a_2^2| \leq \frac{4}{\pi^2} \begin{cases} 1 + \frac{12\mu - 5\pi^2}{3\pi^2} & \mu > \frac{1+\alpha}{\alpha} \\ \frac{1}{3} & \frac{\alpha - 1}{\alpha} \leq \mu \leq \frac{1+\alpha}{\alpha} \\ 1 - \frac{\pi^2 + 12\mu}{3\pi^2} & \mu < \frac{\alpha - 1}{\alpha} \end{cases}$$

Proof By taking $\gamma = 1$ and $\phi(z) = \phi_4(z) = 1 + \frac{8}{\pi^2}z + \frac{16}{3\pi^2}z^2 + \ldots$ and putting the value $B_1 = \frac{8}{\pi^2}$, $B_2 = \frac{16}{3\pi^2}$ in Theorem 1, we get the result after simplification.

References

1. Al-Hawary, T., Frasin, B.A., Darus, M.: Fekete-Szegö problem for certain classes of analytic functions of complex order defined by the Dziok-Srivastava operator. Acta Math. Vietnam. **39**(2), 185–192 (2014)
2. Brannan, D.A., Cluine, J.G. (eds.): Aspects of contemporary complex analysis. In: Proceedings of the NATO Advanced Study Institute held at the University of Durham, Durham, 1–20 July 1979. New York and London. Academic Press (1980)
3. Brannan, D.A., Taha, T.S.: On some classes of bi-univalent functions. In: Mazhar, S.M., Hamoui, A., Faour, N.S. (eds.) Mathematical Analysis and its Applications, Kuwait, 18–21 February 1985, KFAS Proceedings Series, vol. 3. Pergamon Press, Elsevier Science Limited, Oxford (1998), pp. 53–60. See also Studia Univ. Babes-Bolyai Math. **31**(2), 70–77 (1986)

4. Bulut, S.: Fekete-Szegö problem for subclasses of analytic functions defined by Komatu integral operator. Arabian J. Math. **2**, 177–183 (2013)
5. Duren, P.L.: Univalent functions. In: Grundlehren der Mathematischen Wissenschaften. Band 259. Springer, New York, Berlin, Heidelberg and Tokyo (1983)
6. Fekete, M., Szegö, G.: Eine Bemerkung über ungerade schlichte Funktionen. J. Lond. Math. Soc. **8**, 85–89 (1933)
7. Jahangiri, J.M., Magesh, N., Yamini, J.: Fekete-Szegö inequality for classes of bi-starlike and bi-convex functions. Electron. J. Math. Anal. Appl. **3**(1), 133–140 (2015)
8. Kanasa, S., Darwish, H.E.: The Fekete-Szegö problem for starlike and convex functions of complex order. Appl. Math. Lett. **23**, 777–782 (2010)
9. Lewin, M.: On the coefficient problem for bi-univalent functions. Proc. Am. Math. Soc. **18**, 63–68 (1967)
10. Ma, W., Minda, D.: Uniformly convex functions II. Ann. Pol. Math. **58**, 275–285 (1993)
11. Mishra, A.K., Gochhayat, P.: The Fekete-Szegö problem for k-uniformly convex functions and for a class defined by the Owa-Srivastava operator. J. Math. Anal. Appl. **347**, 563–572 (2008)
12. Natanyahu, E.: The minimal distance of image boundary from the origin and the second coefficient of a univalent function in $|z|<1$. Arch. Rat. Mech. Anal. **32**, 100–112 (1969)
13. Orhan, H., Rŏducanu, D.: Fekete-Szegö problem for strongly starlike functions associated with generalized hypergeometric functions. Math. Comput. Model. **50**, 430–438 (2009)
14. Orhan, H., Magesh, N., Balaji, V.K.: Fekete-Szegö problem for certain classes of Ma-Minda bi-univalent functions. Afr. Mat. **27**, 889–897 (2016)
15. Sharma, P., Ankita, Raina, R.K.: Coefficient estimate and Fekete-Szegö inequality for a unified class of bi-univalent functions. Math. Sci. Res. J. **17**, 292–299 (2013)
16. Singh, R.: On Bazilevič functions. Proc. Am. Math. Soc. **38**, 261–271 (1973)
17. Srivastava, H.M., Mishra, A.K.: Applications of fractional calculus to parabolic starlike and uniformly convex functions. Comput. Math. Appl. **39**(3), 57–69 (2000)
18. Srivastava, H.M., Mishra, A.K., Das, M.K.: The Fekete-Szegö problem for a subclass of close-to-convex functions. Complex Var. **44**, 145–163 (2001)
19. Sivaprasad Kumar, S., Kumar, V., Ravichandran, V.: Estimates for the initial coefficients of bi-univalent functions. Tamsui Oxf. J. Inf. Math. Sci. **29**(4), 487–504 (2013)
20. Taha, T.S.: Topics in univalent function theory. Ph.D. Thesis, University of London (1981)
21. Thomas, D.K.: On Bazilevič functions. Trans. Am. Math. Soc. **132**, 353–361 (1968)
22. Zaprawa, P.: On the Fekete-Szegö problem for classes of bi-univalent functions. Bull. Belg. Math. Soc. Simon Stevin **21**, 169–178 (2014)

Study of Absolute Cesàro Summable Factor with Quasi-f-Power Increasing Sequences for an Infinite Series

Absolute Summable Factor for Infinite Series

Smita Sonker and Alka Munjal

Abstract In the present study, a wider class of sequence used for a least set of sufficient conditions for absolute Cesáro $\varphi - |C, \alpha; \delta; l|_k$ summable factor for an infinite series. Many corollaries have been determined by using suitable conditions in the main theorem. Validation of the theorem done by the previous findings of summability. In this way, the system's stability can be improved by finding the conditions for absolute summability.

Keywords Hölder's inequality · Abel's transform · Indexed summable factor · Minkowski's inequality

1 Introduction

Let partial sums's sequence of $\sum a_n$ is given by $s_n = \sum_{k=0}^{n} a_k$ and nth sequence to sequence transform of the sequence $\{s_n\}$ is determined by u_n, s.t.,

$$u_n = \sum_{k=0}^{\infty} u_{nk} s_k \tag{1}$$

S. Sonker (✉)
Department of Mathematics, National Institute of Technology Kurukshetra,
Kurukshetra 136119, India
e-mail: smita.sonker@gmail.com

A. Munjal
Department of Mathematics, Akal University,
Talwandi Sabo, Bathinda 151302, Punjab, India
e-mail: alkamunjal8@gmail.com

R. N. Mohapatra et al. (eds.), *Mathematical Analysis and Computing*,
Springer Proceedings in Mathematics & Statistics 344,
https://doi.org/10.1007/978-981-33-4646-8_5

An infinite series $\sum a_n$ is absolutely summable, if

$$\lim_{i\to\infty} u_i = s \tag{2}$$

and

$$\sum_{i=1}^{\infty} |u_i - u_{i-1}| < \infty. \tag{3}$$

Definition 1 [1] Let $\{na_n\}$ is a sequence. The nth Cesáro mean of this sequence is represented by t_n^α. This mean is of order α $(0 < \alpha \leq 1)$. The $\sum a_n$ is summable $|C, \alpha; \delta|_k$ for $\delta \geq 0$ and $k \geq 1$, if

$$\sum_{n=1}^{\infty} n^{\delta k-1} |t_n^\alpha|^k < \infty, \tag{4}$$

where t_n^α is

$$t_n^\alpha = \frac{1}{A_n^\alpha} \sum_{p=1}^{n} A_{n-p}^{\alpha-1} p a_p, \tag{5}$$

and

$$A_n^\alpha = \begin{cases} O(n^\alpha), & \text{for } n > 0, \\ 1, & \text{for } n = 0, \\ 0, & \text{for } n < 0. \end{cases} \tag{6}$$

Definition 2 If sequence of means $\{t_n^\alpha\}$ satisfies:

$$\sum_{n=1}^{\infty} \frac{\varphi_n^{k-1}}{n^{k-\delta k}} |t_n^\alpha|^k < \infty, \tag{7}$$

then $\sum a_n$ is $\varphi - |C, \alpha; \delta|_k$ summable. Where $\{\varphi_n\}$ is a positive real number sequence, $\delta \geq 0$ and $k \geq 1$.

Definition 3 If a sequence of means $\{t_n^\alpha\}$ satisfies

$$\sum_{n=1}^{\infty} \frac{\varphi_n^{l(k-1)}}{n^{l(k-\delta k)}} |t_n^\alpha|^k < \infty, \tag{8}$$

then $\sum a_n$ is $\varphi - |C, \alpha; \delta; l|_k$ for l is a real number, $k \geq 1$ and $\delta \geq 0$.

Bor [2–6] has determined various important results by using absolute summability factors for infinite series with the application of different classes of sequences. Özarslan derived the theorems on absolute matrix summable factors [7, 8] and

$(K, 1, \alpha)$ summable factor has been used by Parashar in [9]. Absolute product summability has been used by Chandra and Jain [10] for Fourier series. Various theorems on absolute Cesàro summability have been established by Sonker and Munjal in [11, 12] and they used triangle matrices for infinite series in [13].

2 Known Results

A positive sequence $B = \{B_n\}$ is quasi-f-power increasing sequence [14] with $K = K(B, f) \geq 1$ for all $1 \leq m \leq n$ s.t.

$$Kf_nB_n \geq f_mB_m \tag{9}$$

and

$$f = \left[f_n\left(\varsigma,\ \eta\right)\right] = \{n^{\varsigma}(\log n)^{\eta},\ 0 < \varsigma < 1, \eta \geq 0\}. \tag{10}$$

A wider class has been used in [15] and Bor [16] used absolute summable factor of order α for the result.

Theorem 2.1 *Let $\{B_n\}$ is a wider class (a quasi-f-power sequence), which is an increasing sequence for ς ($0 < \varsigma < 1$). Assume $\exists$ a sequence $\{D_n\}$ s.t. it is ξ-quasi-monotone with conditions:*

$$\sum n\xi_nB_n = O\left(1\right), \tag{11}$$

$$\Delta D_n \leq \xi_n, \tag{12}$$

$$\left|\Delta\lambda_n\right| \leq \left|D_n\right|, \tag{13}$$

$$\sum D_nB_n \textit{ is convergent for all } n. \tag{14}$$

If the following two conditions

$$\sum_{i=1}^{p} \frac{(w_i^{\alpha})^k}{i} = O\left(B_p\right) \quad as \quad p \to \infty, \tag{15}$$

$$\left|\lambda_i\right| B_i = O\left(1\right) \quad as \quad i \to \infty, \tag{16}$$

are satisfied, then $|C, \alpha|_k$ summable factor has been followed by infinite series $\sum a_n\lambda_n$ with $0 < \alpha \leq 1$ and $k \geq 1$.

3 Main Results

Increasing sequences are very useful for establish a number of results on absolute summable factor. In present study, quasi-f-power sequence is playing an important role for summable factor of a generalized series. Conditions are determined on absolute summable factor which are sufficient for an infinite series to make it absolute summable.

Theorem 3.1 *Let $\{B_n\}$ is a wider class of sequence (A quasi-f-power sequence), which is a increasing sequence for ς $(0 <\varsigma <1)$. Assume $\exists$ a sequence $\{D_n\}$ s.t. it is ξ-quasi-monotone with conditions:*

$$\sum n\xi_n B_n = O\left(1\right), \tag{17}$$

$$\Delta D_n \leq \xi_n, \tag{18}$$

$$\left|\Delta\lambda_n\right| \leq \left|D_n\right|, \tag{19}$$

$$\left|\lambda_n\right| B_n = O\left(1\right) \quad as \quad n \to \infty. \tag{20}$$

If the following two conditions

$$\sum D_n B_n < \infty \, for \; all \; n, \tag{21}$$

$$\sum_{i=1}^{p} \frac{\varphi_i^{l(k-1)} (w_i^{\alpha})^k}{i^{l(k-\delta k)}} = O\left(B_p\right) \quad as \quad p \to \infty, \tag{22}$$

$$\sum_{n=v}^{m} \frac{\varphi_n^{l(k-1)}}{n^{(\alpha-l\delta+l)k}} = O\left(\frac{\varphi_v^{l(k-1)}}{v^{(\alpha-l\delta+l)k-1}}\right), \tag{23}$$

are satisfied, then generalized summable factor $\varphi - |C, \alpha; \delta, l|_k$ has been followed by infinite series $\sum a_n\lambda_n$, where $k \geq 1, 0 < \alpha \leq 1, \delta \geq 0,$ l be a real number and w_n^{α} is

$$w_n^{\alpha} = \begin{cases} \max\limits_{1\leq v\leq n} \left|t_v^{\alpha}\right|, & 0 < \alpha < 1, \\ \left|t_n^{\alpha}\right|, & \alpha = 1. \end{cases} \tag{24}$$

4 Proof of the Main Theorem

The $\sum a_n\lambda_n$ will follow $\varphi - |C, \alpha; \delta, l|_k$ summable factor, if the nth mean T_n^α for α of $\{na_n\lambda_n\}$ satisfies the condition

$$\sum_{n=1}^{\infty} \frac{\varphi_n^{l(k-1)}}{n^{l(k-\delta k)}} |T_n^\alpha|^k < \infty. \tag{25}$$

The nth sequence to sequence transform T_n^α of $\{na_n\lambda_n\}$ is

$$\begin{aligned} T_n^\alpha &= \frac{1}{A_n^\alpha} \sum_{v=1}^{n} A_{n-v}^{\alpha-1} v a_v \lambda_v \\ &= \frac{1}{A_n^\alpha} \sum_{v=1}^{n-1} \Delta\lambda_v \sum_{p=1}^{v} A_{n-p}^{\alpha-1} p a_p + \frac{\lambda_n}{A_n^\alpha} \sum_{v=1}^{n} A_{n-v}^{\alpha-1} v a_v \end{aligned}$$

By taking modulus value of both side and using the concept of modulus,

$$\begin{aligned} |T_n^\alpha| &\le \frac{|\lambda_n|}{A_n^\alpha} \left| \sum_{v=1}^{n} v a_v A_{n-v}^{\alpha-1} \right| + \frac{1}{A_n^\alpha} \sum_{v=1}^{n-1} \left| \sum_{p=1}^{v} p a_p A_{n-p}^{\alpha-1} \right| |\Delta\lambda_v| \\ &\le |\lambda_n| w_n^\alpha + \frac{1}{A_n^\alpha} \sum_{v=1}^{n-1} |\Delta\lambda_v| A_v^\alpha w_v^\alpha \\ &= |T_{n,1}^\alpha| + |T_{n,2}^\alpha| \quad (say). \end{aligned}$$

With the use of Minkowski's inequality's concept,

$$|T_n^\alpha|^k \le 2^k \left(|T_{n,1}^\alpha|^k + |T_{n,2}^\alpha|^k \right). \tag{26}$$

It is enough to prove that

$$\sum_{n=1}^{\infty} \frac{\varphi_n^{l(k-1)}}{n^{l(k-\delta k)}} |T_{n,r}^\alpha|^k < \infty, \tag{27}$$

where $r = 1, 2$. By applying Abel's transformation and Hölder's inequality, we have

$$\sum_{n=2}^{m} \frac{\varphi_n^{l(k-1)}}{n^{l(k-\delta k)}} |T_{n,1}^{\alpha}|^k = O(1) \sum_{n=1}^{m} |\lambda_n| \, (w_n^{\alpha})^k \frac{\varphi_n^{l(k-1)}}{n^{l(k-\delta k)}}$$

$$= O(1) \, |\lambda_m| \sum_{n=1}^{m} \frac{\varphi_n^{l(k-1)}}{n^{l(k-\delta k)}} \left(w_n^{\alpha}\right)^k + O\,(1) \sum_{n=1}^{m-1} \Delta \, |\lambda_n| \sum_{v=1}^{n} \frac{\varphi_v^{l(k-1)}}{v^{l(k-\delta k)}} \left(w_v^{\alpha}\right)^k$$

$$= O(1) B_m \, |\lambda_m| + O\,(1) \sum_{n=1}^{m-1} B_n \Delta \, |\lambda_n|$$

$$= O(1) B_m \, |\lambda_m| + O\,(1) \sum_{n=1}^{m-1} B_n |D_n|$$

$$= O\,(1) \qquad as \qquad m \to \infty, \tag{28}$$

$$\sum_{n=2}^{m+1} \frac{\varphi_n^{l(k-1)}}{n^{l(k-\delta k)}} |T_{n,2}^{\alpha}|^k \leq \left(\sum_{v=1}^{n-1} |\Delta \lambda_v| \, A_v^{\alpha} w_v^{\alpha} \right)^k \sum_{n=2}^{m+1} \frac{\varphi_n^{l(k-1)}}{n^{l(k-\delta k)}} \frac{1}{(A_n^{\alpha})^k}$$

$$\leq \sum_{n=2}^{m+1} \frac{\varphi_n^{l(k-1)}}{n^{(\alpha-l\delta+l)k}} \left(\sum_{v=1}^{n-1} |D_v| \right)^{k-1} \sum_{v=1}^{n-1} v^{\alpha k} |D_v| \left(w_v^{\alpha}\right)^k$$

$$= O(1) \sum_{v=1}^{m} v^{\alpha k} |D_v| \left(w_v^{\alpha}\right)^k \sum_{n=v+1}^{m+1} \frac{\varphi_n^{l(k-1)}}{n^{(\alpha-l\delta+l)k}}$$

$$= O(1) \sum_{v=1}^{m} \frac{\varphi_v^{l(k-1)}}{v^{(\alpha-l\delta+l)k-1}} v^{\alpha k} |D_v| \left(w_v^{\alpha}\right)^k$$

$$= O\,(1) \sum_{v=1}^{m} \frac{\varphi_v^{l(k-1)}}{v^{l(k-\delta k)}} v \, |D_v| \left(w_v^{\alpha}\right)^k$$

$$= O(1) m \, |D_m| \sum_{v=1}^{m} \frac{\varphi_v^{k-1}}{v^{k-\delta k}} \left(w_v^{\alpha}\right) + O\,(1) \sum_{v=1}^{m-1} \Delta (v \, |D_v|) \sum_{r=1}^{v} \frac{\varphi_r^{k-1}}{r^{k-\delta k}} \left(w_r^{\alpha}\right)^k$$

$$= O(1) m B_m \, |D_m| + O\,(1) \sum_{v=1}^{m-1} |(v+1) \, \Delta \, |D_v| - |D_v|| B_v$$

$$= O(1) m \, |D_m| \, B_m + O(1) \sum_{v=1}^{m-1} |D_v| B_v + O\,(1) \sum_{v=1}^{m-1} v \, |\Delta D_v| \, B_v$$

$$= O(1) m \, |D_m| \, B_m + O(1) \sum_{v=1}^{m-1} |D_v| B_v + O\,(1) \sum_{v=1}^{m-1} v \xi_v B_v$$

$$= O\,(1) \qquad as \qquad m \to \infty, \tag{29}$$

Collecting (25)–(29), we have

$$\sum_{n=1}^{\infty} \frac{\varphi_n^{l(k-1)}}{n^{l(k-\delta k)}} \left|T_n^\alpha\right|^k < \infty. \tag{30}$$

Hence the proof of the theorem is completed.

5 Corollaries

Corollary 5.1 *Let $\{B_n\}$ is a wider class of sequence (quasi-f-power sequence), which is a increasing sequence for ς $(0 <\varsigma <1)$. Assume $\exists$ a sequence $\{D_n\}$ s.t. it is ξ-quasi-monotone with conditions (17–21) and*

$$\sum_{i=1}^{p} \frac{\varphi_i^{(k-1)} (w_i^\alpha)^k}{i^k} = O\left(B_p\right) \quad as \quad p \to \infty, \tag{31}$$

$$\sum_{i=v}^{p} \frac{\varphi_i^{(k-1)}}{i^{(\alpha+1)k}} = O\left(\frac{\varphi_v^{(k-1)}}{v^{(\alpha+1)k-1}}\right), \tag{32}$$

then $\varphi - |C, \alpha|_k$ summable factor followed by the series $\sum a_n \lambda_n$ with $k \geq 1, 0 < \alpha \leq 1$ and w_n^α is

$$w_n^\alpha = \begin{cases} \max\limits_{1\leq v\leq n} \left|t_v^\alpha\right|, & 0 < \alpha < 1, \\ \left|t_n^\alpha\right|, & \alpha = 1. \end{cases} \tag{33}$$

Proof Use $\delta = 0$ and $l = 1$ in the present result.

Corollary 5.2 *Let $\{B_n\}$ is a wider class of sequence (quasi-f-power sequence), which is a increasing sequence for ς $(0 <\varsigma <1)$. Assume $\exists$ a sequence $\{D_n\}$ s.t. it is ξ-quasi-monotone with conditions (17–21) and*

$$\sum_{i=1}^{p} \frac{(w_i^\alpha)^k}{i} = O\left(B_p\right) \quad as \quad p \to \infty, \tag{34}$$

then $|C, \alpha|_k$ summable factor followed by $\sum a_n \lambda_n$ series with $k \geq 1, 0 < \alpha \leq 1$ and w_n^α is

$$w_n^\alpha = \begin{cases} \max\limits_{1\leq v\leq n} \left|t_v^\alpha\right|, & 0 < \alpha < 1, \\ \left|t_n^\alpha\right|, & \alpha = 1. \end{cases} \tag{35}$$

Proof By using $l = 1,\ \varphi_n = n$ and $\delta = 0$ in the present resent.

6 Conclusion

Present work is on the absolute summability factor which makes the system stable. If an impulse response be absolutely summable, then the system be BIBO stable, i.e.,

$$\textit{BIBO stable} \iff \sum_{n=-\infty}^{\infty} |h(n)| < \infty. \tag{36}$$

With the help of a summable factor, the error can be minimized and the output can be made stable. Absolute summable factor can be used to predict the input data and the complete changes in the process.

Present work is very applicable in the rectification of signals in Filter. By finding the corollary, we can be concluded that the present result is very important and generalized research on absolute summability which can be used to find various previous results. Validation of present work is done by Corollary 5.1, which is established by Bor [16] for infinite series to be absolute summable.

Acknowledgements This work has been financially supported by the Science and Engineering Research Board (SERB) through Project No.: EEQ/2018/000393. The authors offer their true thanks to Science and Engineering Research Board for giving financial support.

References

1. Flett, T.M.: On an extension of absolute summability and some theorems of Littlewood and Paley. Proc. Lond. Math. Sci. **3**(1), 113–141 (1957)
2. Bor, H.: Factors for generalized absolute Cesàro summability. Math. Comput. Model. **53**(5), 1150–1153 (2011)
3. Bor, H.: An application of almost increasing sequences. Appl. Math. Lett. **24**(3), 298–301 (2011)
4. Bor, H.: Generalized absolute Cesàro summability factors. Bull. Math. Anal. Appl. **8**(1), 6–10 (2016)
5. Bor, H.: On absolute summability factors. Proc. Am. Math. Soc. **118**(1), 71–75 (1986)
6. Bor, H.: Almost increasing sequences and their new applications II. Filomat **28**(3), 435–439 (2014)
7. Özarslan, H.S., Ari, T.: Absolute matrix summability methods. Appl. Math. Lett. **24**(12), 2102–2106 (2011)
8. Özarslan, H.S., Yavuz, E.: A new note on absolute matrix summability. J. Inequal. Appl. **2013**(474), 1–7 (2013)
9. Parashar, V.K.: On (N, P_n) and $(K, 1, \alpha)$ summability methods. Publ. L'Institut Mathe'matique **29**(43), 145–158 (1981)
10. Chandra, P., Jain, H.C.: Absolute product summability of the Fourier series and its applied series. Commun. Fac. Sci. Uni. Ank. Ser. **A1**(37), 95–107 (1988)
11. Sonker, S., Munjal, A.: Absolute summability factor $\varphi - |C, 1; \delta|_k$ of infinite series. Int. J. Math. Anal. **10**(23), 1129–1136 (2016)
12. Sonker, S., Munjal, A.: Absolute $\varphi - |C, \alpha, \beta; \delta|_k$ summability of infinite series. J. Inequal. Appl. **2017**(168), 1–7 (2017)

13. Sonker, S., Munjal, A.: Sufficient conditions for triple matrices to be bounded. Nonlinear Stud. **23**(4), 533–542 (2016)
14. Sulaiman, W.T.: Extension on absolute summability factors of infinite series. J. Math. Anal. Appl. **322**(2), 1224–1230 (2006)
15. Leindler, L.: A new application of quasi power increasing sequences. Publ. Math. Debr. **58**(4), 791–796 (2001)
16. Bor, H.: Some new results on infinite series and Fourier series. Positivity **19**(3), 467–473 (2015)

Sensitivity and Stability Analysis in the Transmission of Japanese Encephalitis with Logistic Growing Mosquito Population

Vinod Baniya and Ram Keval

Abstract In this manuscript, mathematical modeling for the spread of vector-borne disease, Japanese encephalitis in humans with constant recruitment from a pig population through the mosquito population has been proposed and analyzed. In this process, we assumed that the disease spread to the susceptible class is only due to exposure to infected mosquitoes. It is also assumed that the population of mosquitoes follows the logistic differential equation with a carrying capacity of mosquitoes, where the pig population has a constant rate. In order to perform, equilibrium, stability, and sensitivity analysis, we find a threshold condition, R_0 called basic reproduction number. If R_0 is exceeded, there is currently an equilibrium with the disease that is locally asymmetrically stable under certain conditions. A sensitivity analysis of the model is performed to find out the relative importance of various parameters responsible for the disease transmission.

Keywords Japanese encephalitis · Logistic growth model · Threshold condition · Stability · Sensitivity analysis

1 Introduction

The major anxiety of the present time is the birth of a new infectious disease, one of them is Japanese encephalitis (JE). JE is a viral infection of mosquito-borne disease and caused by the Japanese encephalitis virus (JEV), World Health Organization (WHO). The virus exists in a transmission cycle between vectors (mosquitoes) and reservoirs (pigs) [1]. A human gets infected when bitten by an infected mosquito. The JE disease cannot spread from men to men. The incubation period of the JE

V. Baniya · R. Keval (✉)
Department of Mathematics and Scientific Computing, Madan Mohan Malaviya University of Technology, Gorakhpur, U.P., India
e-mail: ramkeval@gmail.com

V. Baniya
e-mail: vinod2638@gmail.com

R. N. Mohapatra et al. (eds.), *Mathematical Analysis and Computing*,
Springer Proceedings in Mathematics & Statistics 344,
https://doi.org/10.1007/978-981-33-4646-8_6

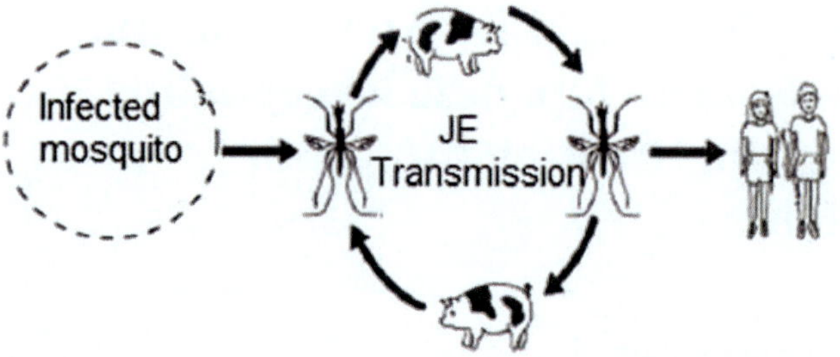

Fig. 1 Transmission cycle of Japanese encephalitis between mosquitoes and pigs, (WHO)

virus is 5–15 days, Centers for Disease Control, (CDC). The transmission cycle of Japanese encephalitis between mosquitoes and pigs is shown in Fig. 1.

Globally, 30,000–50,000 new cases of JE are reported every year and more than 3 billion people are at risk of developing the JE, National Vector Borne Disease Control Programme (NVBDCP). At present, there is no specific treatment available against JEV [1, 3, 7]. Therefore, the strategy for its control and prevention is necessary. There are some research articles on mathematical modeling of JE that have been published till now in which the first article was written by Mukhopadhyay et al. (1993). They have discussed JE model and its stability properties [2]. Panja et al. [2] have developed a mathematical model on JE and investigate their stability and bifurcation with/without the effect of some controllable parameters. Naresh et al. [3] presented a non-linear mathematical model for the spread of JE and analyzed their stability with environmental effects. Tapaswi et al. [5] used the basic model to analyze the transmission of JE in a 3-population to obtain an equilibrium, stability using threshold condition, R_0. De et al. [7] formulated a mathematical model for optimal control analysis and its dynamical behavior with humans, pigs, and mosquitoes. Throughout the world, JE is a growing and dangerous public health problem. Since the transmission dynamics of JE is still unclear, therefore due to this infectious disease, many people die every year [2]. Therefore, it is necessary to study the dynamics of JEV transmission and the strategy for its control. In this investigation, a JE disease model is proposed by considering three-compartment of humans, two-compartment of mosquitoes, and also two-compartment of pigs. Our objective of the present work is to investigate the dynamical behavior of the model in the transmission of JE with the logistic growing mosquito population. This is on the grounds that the size of the mosquito population changes quickly in its size. While the number of inhabitants in pigs that show no sickness side effects are commonly not dependent upon any savage conditions due to JE and in this manner, it is considered in a consistent size.

In this manuscript, the remaining work is organized as follows. In Sect. 2, we formulate a mathematical model which is based on some basic assumption. Section 3 contains the positivity and boundedness of the model. In Sect. 4, we have shown that the dynamic behavior of the system. Numerical experiments are performed to validate our analytical results in Sect. 5. The sensitivity analysis of the model has been discussed in Sect. 6. Finally, the conclusions of our present work are discussed in Sect. 7.

2 Model Analysis

In this work, we considered that the total human population (N_2) is constant, which is divided into three sub-populations, susceptible (H_1), vaccinated (H_2), and infected human population (H_3), such that $N_2 = H_1(t) + H_2(t) + H_3(t)$. In the human compartment, the recruited rate of susceptible population is $\mu_h(1-v)N_2$ where μ_h is the decay rate of humans and v in vaccinated humans. Susceptible humans acquire JEV at rate $\frac{\beta_3 M_2 H_1}{N_2}$ through contact with infected mosquitoes. γ_2 is the recovery rate from infected humans to the susceptible class of humans. The transfer rate δ from vaccinated class to susceptible, this is because vaccination does not give 100% protection. Non-treated infected humans die at a rate of ϵ. The mosquito population follows the logistic model $r(1-\frac{M}{K})M$, where r is intrinsic growth rate (difference of birth (α_1) and death rate (α_2) of mosquito population) and K is carrying capacity of mosquito population (M(t)). So that $M(t) = M_1(t) + M_2(t)$. Susceptible mosquitoes acquire JEV through contact with infected pigs at a rate of $\frac{\beta_1 M_1 P_2}{M}$. Susceptible pigs are recruited at a constant rate of $\mu_p N_1$, where μ_p is the death rate of pigs. The total pig population (N_1) is constant and is equal to sum of susceptibles ($P_1(t)$) and infected pigs ($P_2(t)$). Hence, the susceptible pigs are acquire JEV through contact with infected mosquitoes at rate $\frac{\beta_2 M_2 P_1}{N_1}$. Moreover, we accept that transmission of the infection is conceivable just to the class of defenseless people in the whole displaying process, for different classes JE spreads among those when they become just helpless classes. Along these lines, we don't consider direct JE transmission from the immunized human class to the tainted human class. From the above considerations, the dynamics of disease transmission obey the following differential equations as follows:

$$\frac{dH_1}{dt} = \mu_h(1-v)N_2 + \delta H_2 - \frac{\beta_3 M_2 H_1}{N_2} - \mu_h H_1 + \gamma_2 H_3 \tag{1}$$

$$\frac{dH_2}{dt} = v\mu_h N_2 - (\mu_h + \delta)H_2 \tag{2}$$

$$\frac{dH_3}{dt} = \frac{\beta_3 M_2 H_1}{N_2} - (\gamma_2 + \mu_h + \epsilon)H_3 \tag{3}$$

$$\frac{dP_1}{dt} = \mu_p N_1 - \frac{\beta_2 P_1 M_2}{N_1} - \mu_p P_1 + \gamma_1 P_2 \tag{4}$$

$$\frac{dP_2}{dt} = \frac{\beta_2 P_1 M_2}{N_1} - (\mu_p + \gamma_1)P_2 \tag{5}$$

$$\frac{dM_1}{dt} = \left(\alpha_1 - \frac{rM}{K}\right)M - \frac{\beta_1 P_2 M_1}{M} - \alpha_2 M_1 \tag{6}$$

$$\frac{dM_2}{dt} = \frac{\beta_1 P_2 M_1}{M} - \alpha_2 M_2 \tag{7}$$

$$\frac{dM}{dt} = r\left(1 - \frac{M}{K}\right) M \tag{8}$$

where,

$$H_1(t) + H_2(t) + H_3(t) = N_2(\text{constant})$$
$$P_1(t) + P_2(t) = N_1(\text{constant})$$
$$M_1(t) + M_2(t) = M(t)$$

3 Positive Invariance and Boundedness

In this section, the system of Eqs. 1–8 describes the population of human, mosquito, and pig and therefore it is necessary to prove that all the state variables are non-negative for all time t. This analysis guarantees that the system is well behaved, therefore realistic in representing populations with non-negative values. This can be proved by following theorem:

Theorem 1 *If* $H_1(0),\ H_2(0),\ H_3(0),\ P_1(0),\ P_2(0),\ M_1(0),\ M_2(0)$ *are non-negative, then* $H_1(t),\ H_2(t),\ H_3(t),\ P_1(t),\ P_2(t),\ M_1(t),\ M_2(t)$ *are non-negative for all time t>0. Moreover,*

$$\lim_{t\to\infty} sup \sum H_j(t) \leq N_2(= Constant),\quad \lim_{t\to\infty} sup \sum P_i(t) \leq N_1(=Constant)$$
$$\lim_{t\to\infty} sup \sum M_i(t) \leq M(t) \leq K.$$

Thus, the feasible region $\Omega = \{(H_1,\ H_2, H_3,\ P_1,\ P_2,\ M_1,\ M_2,\ M_3) | M_i(t) \geq 0,\ \ P_i(t) \geq 0,\ \ H_j \geq 0,\ \ i = 1, 2,\ \ j = 1, 2, 3 : \sum M_i(t) \leq K,\ \ \sum P_i(t) \leq N_1,\ \sum H_j(t) \leq N_2\}$

The proof is omitted. □

4 Dynamical Behavior of the System

4.1 Equilibrium and Basic Reproduction Number

In order to find the equilibrium points of the system of Eqs. 1–8, we equate to zero the R.H.S. of system of Eqs. 1–8, we get

(1) **Vector-free equilibrium point** $E_0 = (\bar{H}_1, \bar{H}_2, \bar{H}_3, \bar{P}_1, \bar{P}_2, \bar{M}_1, \bar{M}_2)$. where,

$$\bar{H}_1 = \frac{N_2((1-v)(\mu_h+\delta)+\delta v)}{\mu_h+\delta},\ \bar{H}_2 = \frac{N_2\mu_h v}{\mu_h+\delta},\ \bar{H}_3 = 0,\ \bar{P}_1 = N_1,\ \bar{P}_2 = 0$$
$$\bar{M}_1 = 0,\ \bar{M}_2 = 0.$$

The equilibrium E_0 exist if $M = 0$

(2) **Disease-free equilibrium point** $E_0' = (\bar{H}_1', \bar{H}_2', \bar{H}_3', \bar{P}_1', \bar{P}_2', \bar{M}_1', \bar{M}_2')$ where,

$$\bar{H}_1' = \frac{N_2((1-v)(\mu_h+\delta)+\delta v)}{\mu_h+\delta},\ \bar{H}_2' = \frac{N_2\mu_h v}{\mu_h+\delta},\ \bar{H}_3' = 0,\ \bar{P}_1' = N_1,\ \bar{P}_2' = 0$$
$$\bar{M}_1' = K,\ \bar{M}_2' = 0.$$

The equilibrium E_0' exist if $M = K$.

To measure disease transmission potential, "the basic reproduction number (R_0) can be established by using the next generation matrix approach". The potent characteristic root of matrix FV^{-1} is R_0, [2]. It is calculated as

$$R_0 = R_1 R_2 = \frac{\beta_1\beta_2}{\alpha_2(\mu_p+\gamma_1)}$$

where,

$$R_1 = \frac{\beta_1}{\mu_p+\gamma_1} \text{ and } R_2 = \frac{\beta_2}{\alpha_2}$$

Biologically, $R_1(R_2)$ is the normal number of mosquito (pig) populace reached by contaminated pigs (mosquitoes) during its hatching time of JEV [5]. Along these lines, R_0 is the normal number of auxiliary diseases produced by a solitary JEV tainted individual, which has been presented in a susceptible population, in which a few people have been immunized. If $R_0 < 1$, the disease cannot run at population and the infection will end over time. If $R_0 > 1$, then a disease is present and the disease can spread through the population.

(3) **Endemic equilibrium point** $E_1 = (H_1^*, H_2^*, H_3^*, P_1^*, P_2^*, M_1^*, M_2^*)$. where,

$$H_1^* = \frac{N_2(\mu_h+\gamma_2+\epsilon)(\delta H_2^* + N_2\mu_h(1-v))}{(\mu_h+\gamma_2+\epsilon)(N_2\mu_h+\beta_2 M_2^*) - \beta_3\gamma_2 M_2^*},\quad H_3^* = \frac{\beta_3 M_2^* H_1^*}{N_2(\mu_h+\gamma_2+\epsilon)}$$

$$P_1^* = \frac{N_1(\mu_p+\gamma_1)P_2^*}{\beta_2 M_2^*}, \quad P_2^* = \frac{KN_1\alpha_2(\mu_p+\gamma_1)(R_0-1)}{N_1(\mu_p+\gamma_1)+\beta_2 K}, \quad M_2^* = \frac{\beta_1 P_2^* M_1^*}{K\alpha_2}$$

$$M_1^* = \frac{K(\alpha_2 N_1(\gamma_1+\mu_p)+\beta_2\alpha_2 K)}{\beta_2(\alpha_2 K+\beta_1 N_1)}, \quad H_2^* = \frac{N_2\mu_h v}{\mu_h+\delta}$$

The system of Eqs. 1–8 has unique endemic equilibrium point E_1 if $R_0 > 1$.

4.2 Stability Analysis Around Equilibrium Points

In this subsection, we have discussed stability analysis of equilibrium points.

Theorem 2 *The vector-free equilibrium point (E_0) is saddle-node point.*

Proof The Jacobian matrix $J(E_0)$ of the system of Eqs. 1–8 is given by

$$J(E_0) = \begin{pmatrix} -\mu_h & \delta & \gamma_2 & 0 & 0 & 0 & \frac{-\beta_3 H_1}{N_2} \\ 0 & -(\mu_h+\delta) & 0 & 0 & 0 & 0 & 0 \\ 0 & 0 & -(\gamma_2+\mu_h+\epsilon) & 0 & 0 & 0 & \frac{\beta_3 H_1}{N_2} \\ 0 & 0 & 0 & -\mu_p & \gamma_1 & 0 & 0 \\ 0 & 0 & 0 & 0 & -(\mu_p+\gamma_1) & 0 & 0 \\ 0 & 0 & 0 & 0 & 0 & r & \alpha_1 \\ 0 & 0 & 0 & 0 & 0 & 0 & -\alpha_2 \end{pmatrix}$$

The characteristic roots of $J(E_0)$ are as

$$\phi_1 = -\mu_h, \phi_2 = -(\mu_h+\delta), \phi_3 = -(\mu_h+\gamma_2+\epsilon), \phi_4 = -\mu_p, \phi_5 = -(\mu_p+\gamma_1)$$
$$\phi_6 = r, \ \phi_7 = -\alpha_2$$

Since all the characteristic roots of $J(E_0)$ are real but have different signs, the equilibrium point E_0 is saddle-node point. □

Theorem 3 *The disease-free equilibrium point ($E_0^{'}$) is locally asymptotically stable if $R_0 < 1$, unstable if $R_0 > 1$.*

Proof The Jacobian matrix $J(E_0^{'})$ of the system of Eqs. 1–8 is given by

$$J(E_0^{'}) = \begin{pmatrix} J_{11} & J_{12} \\ O & J_{22} \end{pmatrix}$$

where

$$J_{11} = \begin{pmatrix} -\mu_h & \delta & \gamma_2 \\ 0 & -(\mu_h+\delta) & 0 \\ 0 & 0 & -(\mu_h+\gamma_2+\epsilon) \end{pmatrix}, \quad J_{12} = \begin{pmatrix} 0\,0\,0 & \frac{-\beta_3 H_1}{N_2} \\ 0\,0\,0 & 0 \\ 0\,0\,0 & \frac{\beta_3 H_1}{N_2} \end{pmatrix}$$

$$J_{22} = \begin{pmatrix} -\mu_p & \gamma_1 & 0 & \frac{-\beta_2 P_1}{N_1} \\ 0 & -(\mu_p + \gamma_1) & 0 & \frac{\beta_2 P_1}{N_1} \\ 0 & -\beta_1 & -r\ (\alpha_1 - 2) \\ 0 & \beta_1 & 0 & -\alpha_2 \end{pmatrix}$$

From above, we see that all the characteristic roots of block matrix J_{11} are negative where as the characteristic polynomial of the block matrix J_{22} is

$$(\mu_p + \lambda)(r + \lambda)[\lambda^2 + (\mu_p + \gamma_1 + \alpha_2)\lambda + \alpha_2(\mu_p + \gamma_1)(1 - R_0)] = 0 \qquad (9)$$

The roots of characteristic Eq. 9 of the Jacobian matrix $J(E_0^{'})$ are have negative real part if $R_0 < 1$. Thus, we have to prove that if $R_0 < 1$ then equilibrium point $(E_0^{'})$ is locally asymptotically stable (LAS) and unstable if $R_0 > 1$. □

Theorem 4 *If $R_1 \leq 1$ and $R_2 \leq 1$, the disease-free equilibrium point $(E_0^{'})$ is globally asymptotically stable in feasible region Ω.*

Proof To establish the global stability of $E_0^{'}$, consider the Lyapunov function,

$$L(t) = (H_1 - \bar{H}_1^{'} ln H_1) + H_2 + H_3 + (P_1 - \bar{P}_1^{'} ln P_1) + P_2 + (M_1 - \bar{M}_1^{'} ln M_1) + M_2$$

$$\begin{aligned} \dot{L(t)} &= \dot{H}_1\left(1 - \frac{\bar{H}_1^{'}}{H_1}\right) + \dot{H}_2 + \dot{H}_3 + \dot{P}_1\left(1 - \frac{\bar{P}_1^{'}}{P_1}\right) \\ &+ \dot{P}_2 + \dot{M}_1\left(1 - \frac{\bar{M}_1^{'}}{M_1}\right) + \dot{M}_2 \\ &= \left(\mu_h(1 - v)N_2 + \delta H_2 - \frac{\beta_3 M_2 H_1}{N_2} - \mu_h H_1 + \gamma_2 H_3\right)\left(1 - \frac{\bar{H}_1^{'}}{H_1}\right) \\ &+ v\mu_h N_2 - (\mu_h + \delta)H_2 + \frac{\beta_3 M_2 H_1}{N_2} - (\gamma_2 + \mu_h + \epsilon)H_3 \\ &+ \left(\mu_p N_1 - \frac{\beta_2 P_1 M_2}{N_1} - \mu_p P_1 + \gamma_1 P_2\right) \\ &\times \left(1 - \frac{\bar{P}_1^{'}}{P_1}\right) + \frac{\beta_2 P_1 M_2}{N_1} - (\mu_p + \gamma_1)P_2 \\ &+ \left((\alpha_1 - \frac{rM}{K})M - \frac{\beta_1 P_2 M_1}{M} - \alpha_2 M_1\right) \\ &\times \left(1 - \frac{\bar{M}_1^{'}}{M_1}\right) + \frac{\beta_1 P_2 M_1}{M} - \alpha_2 M_2 \end{aligned}$$

Using feasible region Ω, above equation can be written as

$$\leq -\mu_h(1-v)N_2\frac{\bar{H_1'}}{H_1} - \mu_h H_2 - H_3(\gamma_2+\mu_h+\epsilon) - \delta H_2\frac{\bar{H_1'}}{H_1} - M_2(1-R_2)$$
$$-\frac{P_2}{\mu_p+\gamma_1}(1-R_1) - \mu_h H_1\left(1-\frac{\bar{H_1'}}{H_1}\right) - \frac{\mu_p \bar{P_1'}\left(P_1-\bar{P_1'}\right)^2}{P_1\bar{P_1'}}$$

Thus, $\dot{L}(t)$ is negative if $R_1 \leq 1$ and $R_2 \leq 1$. Therefore, by LaSalle's invariance principle, [13] E_0' is globally asymptotically stable in feasible region Ω. □

Theorem 5 *The positive equilibrium point E_1 is LAS if $R_0 > 1$ and $2M_2^* < K$ in feasible region Ω (Fig. 6).*

Proof The Jacobian matrix $J(E_1)$ of the system of Eqs. 1–8 is given by

$$J(E_1) = \begin{pmatrix} J_{01} & J_{02} \\ O & J_{03} \end{pmatrix}$$

where,

$$J_{01} = \begin{pmatrix} -(\frac{\beta_3 M_2^*}{N_2}+\mu_h) & \delta & \gamma_2 \\ 0 & -(\mu_h+\delta) & 0 \\ \frac{\beta_3 M_2^*}{N_2} & 0 & -(\mu_h+\gamma_2+\epsilon) \end{pmatrix},\ J_{02} = \begin{pmatrix} 0\,0\,0 & \frac{-\beta_3 H_1^*}{N_2} \\ 0\,0\,0 & 0 \\ 0\,0\,0 & \frac{\beta_3 H_1^*}{N_2} \end{pmatrix}$$

$$J_{03} = \begin{pmatrix} -(\frac{\beta_2 M_2^*}{N_1}+\mu_p) & \gamma_1 & 0 & \frac{-\beta_2 P_1^*}{N_1} \\ \frac{\beta_2 M_2^*}{N_1} & -(\mu_p+\gamma_1) & 0 & \frac{\beta_2 P_1^*}{N_1} \\ 0 & -\frac{\beta_1 M_1^*}{M^*} & r-\frac{2rM_1^*+2M_2^*}{K} & \alpha_1-\frac{2rM_1^*+2M_2^*}{K} \\ 0 & \frac{\beta_1 M_1^*}{M^*} & \frac{(M_1^*+M_2^*)\beta_1 P_2^* - \beta_1 P_2^*}{(M_1^*+M_2^*)^2} & -\left(\frac{\beta_1 P_2^* M_2^*}{(M_1^*+M_2^*)^2}+\alpha_2\right) \end{pmatrix}$$

The characteristic equation of the block matrix J_{01} is

$$x^2 + a_1 x^2 + a_2 x + a_3 = 0 \tag{10}$$

where

$$a_1 = \frac{\beta_3 M_2^*}{N_2} + 3\mu_h + \delta + \epsilon$$
$$a_2 = (\mu_h+\delta)(\mu_h+\epsilon) + \frac{a_3}{\mu_h+\delta} + \left(\frac{\beta_3 M_2^*}{N_2}+\mu_h\right)(\mu_h+\delta)$$
$$a_3 = (\mu_h+\delta)[(\mu_h+\epsilon)\left(\frac{\beta_3 M_2^*}{N_2}+\mu_h\right) + \mu_h\gamma_2]$$

Table 1 Parametric values for human compartment in the model

Parameter	Value	Source
μ_h	0.0153	WHO
N_2	350	Assumed
β_3	0.6	[7]
v	0.5	[7]
δ	0.001	[3]
ϵ	0.2	[2]
γ_2	0.3	[7]

The Eq. 10 has negative real roots if and only if $a_1 > 0$ and $a_1 a_2 - a_3 > 0$.

After applying primary row and column operations, we obtain the following characteristic equation of the block matrix J_{03}:

$$(x + b_1)(x + b_2)(x + b_3)(x + b_4) = 0 \tag{11}$$

where

$$b_1 = \frac{\beta_2 M_2^*}{N_1} + \mu_p + \gamma_1$$
$$b_2 = \mu_p$$
$$b_3 = \frac{(2rM_2^* - kr)(\beta_1 P_2^* M_2^* + K^2) + \alpha_2 K(K-1)\beta_1 P_2^*}{\beta_2 K P_2^* M_2^*}$$
$$b_3 = \frac{\beta_1 P_2^* M_2^*}{K^2}$$

Thus, from Eqs. 10 and 11, the endemic equilibrium point E_1 if LAS if $R_0 > 1$ and $2M_2^* < K$ (Fig. 6). □

5 Numerical Illustrations

In this section, we illustrate the feasibility of our system of Eqs. 1–8, consider the following set of hypothetical parametric values [2, 3, 7].

The equilibrium points and the basic reproduction number can be determined numerically by utilizing the parametric qualities in Tables 1, 2, 3. The basic reproduction number $(R_0) = 0.7$ and the equilibrium points are shown in Tables 4, 5, 6 and 7.

Table 2 Parametric values for mosquito compartment in the model

Parameter	Value	Source
α_1	2.3	[3]
α_2	1.5	[3]
r	0.8	[2]
β_1	0.9	[7]
K	1000	[3]

Table 3 Parametric values for pig compartment in the model

Parameter	Value	Source
μ_p	0.1	WHO
β_2	0.7	[7]
γ_1	0.5	[7]
N_1	65	Assumed

Table 4 The vector-free equilibrium points calculated with respect to the parametric values are given in Tables 1, 2, 3

$\bar{H_1}$	$\bar{H_2}$	$\bar{H_3}$	$\bar{P_1}$	$\bar{P_2}$	$\bar{M_1}$	$\bar{M_2}$
175.031	164.316	0	65	0	0	0

Table 5 The disease-free equilibrium points calculated with respect to the parametric values are given in Tables 1, 2, 3

$\bar{H_1'}$	$\bar{H_2'}$	$\bar{H_3'}$	$\bar{P_1'}$	$\bar{P_2'}$	$\bar{M_1'}$	$\bar{M_2'}$
175.031	164.316	0	65	0	1000	0

6 Sensitivity Analysis

Since JE is an infectious disease, therefore it can be controlled by decreasing the disease transmission rates β_1 and β_2. This can be obtained by controlling the population of pigs and mosquitoes. Ultimately, our goal is to reduce R_0, so that $R_0 < 1$ which will result in the eradication of the disease. To know the mortality and morbidity of the population, it is necessary to study the relative importance of various parameters

Table 6 The endemic equilibrium points exists if $\alpha_2 = 0.8$ and $R_0 = 1.3125$, calculated with respect to the parametric values are given in Tables 1, 2, 3

H_1^*	H_2^*	H_3^*	P_1^*	P_2^*	M_1^*	M_2^*
95.1529	164.316	4.62154	50.3405	13.1935	983.776	14.6019

Table 7 Sensitivity indices of R_0 calculated with respect to the parametric values are given in Tables 1, 2, 3

$N_{R_0}^{\beta_1}$	$N_{R_0}^{\beta_2}$	$N_{R_0}^{\alpha_2}$	$N_{R_0}^{\mu_p}$	$N_{R_0}^{\gamma_1}$
1	1	−1	−0.167	−0.834

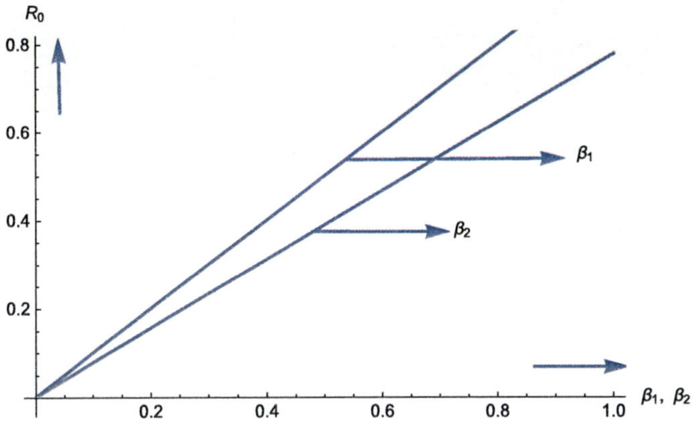

Fig. 2 The diagram shows the variation of R_0 with respect to β_1 and β_2

for JE transmission. In this section, the sensitivity index of R_0 calculated for various parameters related to disease transmission. The normalized forward sensitivity index defined by [8]

$$N_x^{\alpha} = \frac{\partial x}{\partial \alpha} . \frac{\alpha}{x}$$

where x is a variable which depends on a parameter α. Sensitivity index N_x^{α} measures the relative change in x when a parameter α changes. Now the sensitivity index of R_0 with respect to β_1, β_2, α_2, μ_p, and δ are as

The fact $N_{R_0}^{\beta_1} = 1$ and $N_{R_0}^{\beta_2} = 1$, means that increasing β_1 and β_2 by 10%, R_0 always increases by 10% (Fig. 2). Similarly, $N_{R_0}^{\alpha_2} = -1$, $N_{R_0}^{\mu_p} = -0.167$, $N_{R_0}^{\gamma_1} = -0.834$ means that increasing the parameters by 10%, R_0 always decreases by 10%, 1.67%, 8.34% respectively (Fig. 3). The parameters β_1 and β_2 have positive and high sensitivity index. Hence, the parameters β_1 and β_2 are more sensitive in the transmission of disease. Where as the parameters α_2, μ_p, and δ have negative sensitive index. The most negative sensitive parameter is α_2, with $N_{R_0}^{\alpha_2} = -1$. Thus, the parameter α_2 play very crucial role for the eradication of disease from the population. The effects on population are shown in Figs. 4, 5, 6, 7.

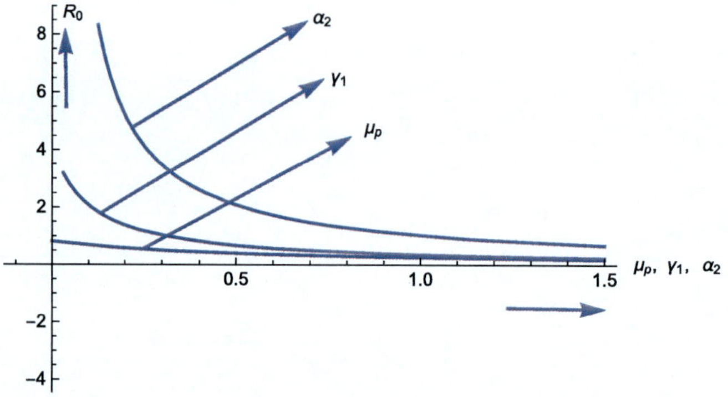

Fig. 3 The diagram shows the variation of R_0 with respect to μ_p, γ_1 and α_2

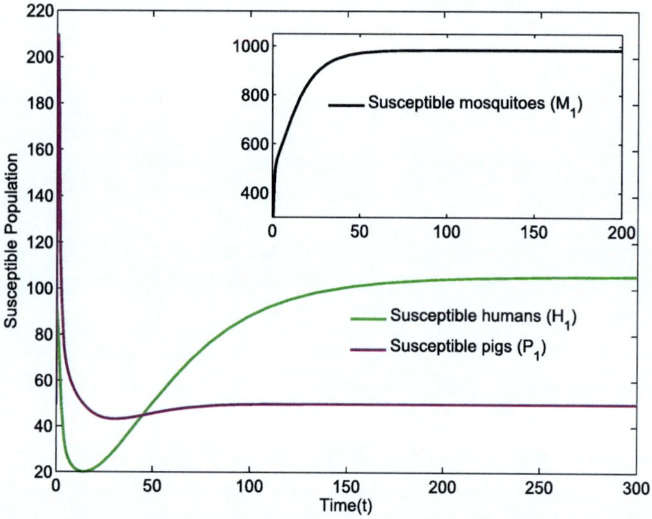

Fig. 4 The diagram shows the variation of susceptible population with respect to time, at endemic equilibrium point, parametric values are given in Tables 1, 2, 3

7 Conclusions

In this work, we used the basic model in the transmission of JE with a logistically growing mosquito population and analyzed its dynamics. In order to study the model's dynamic like equilibrium, stability, and sensitivity, we find a threshold parameter R_0, called basic reproduction number. If $R_0 \leq 1$, then there exist two equilibrium state. The mosquito-free equilibria state (E_0) which is always a saddle-note point. Naturally, this implies, this mosquito-free state with no mosquito populace

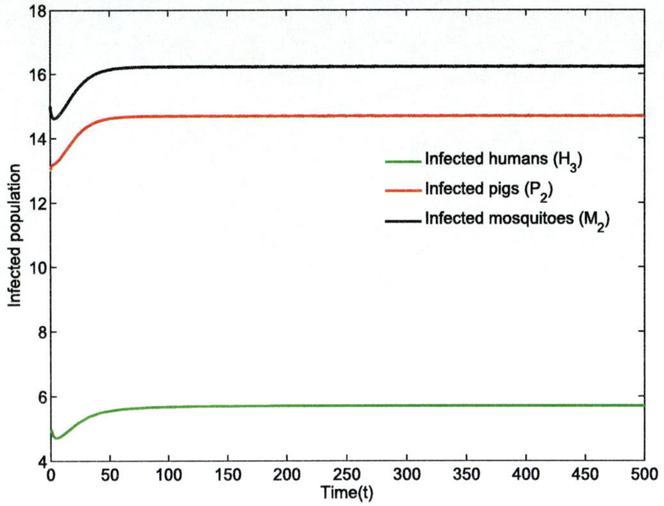

Fig. 5 The diagram shows the variation of infected population with respect to time, at endemic equilibrium point, parametric values are given in Tables 1, 2, 3

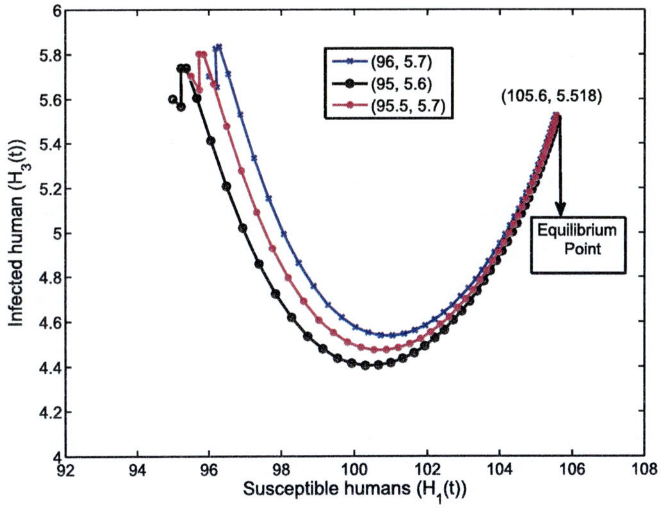

Fig. 6 The diagram shows the non-linear local stability of (H_1^*, H_3^*) in H_1^*-H_3^* plane

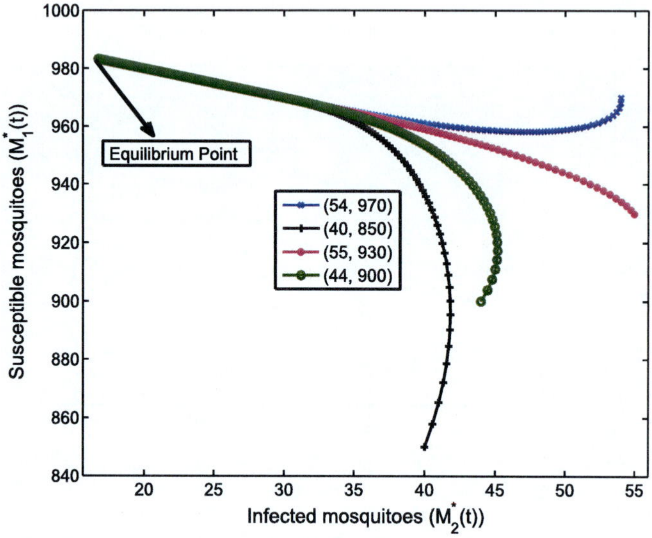

Fig. 7 The diagram shows the non-linear global stability of (M_2^*, M_1^*) in M_1^*-M_2^* plane

can never be achieved. Another harmony state $E_0^{'}$, called disease-free balanced state which is LAS if $R_0 < 1$ and unstable if $R_0 > 1$. When $R_0 > 1$, then there is an equilibrium state in which disease present, called endemic (or positive) equilibrium which is LAS if $2M_2^* < K$. To study the relative importance of various parameters of the model in the transmission of JE disease, sensitivity indexes of basic reproduction numbers were discussed. We believe that the work presented here for citizens affected by Japanese encephalitis disease may have an impact on both epidemic prevention and control.

References

1. World Health Organization: Japanese encephalitis fact sheets. Available from: https://www.who.int/news-room/fact-sheets/details/Japanese-encephalitis
2. Panja, P., Mondal, S.K., Chattopadhyay, J.: Stability and bifurcation analysis of Japanese encephalitis model with/without effect of some control parameters. Comp. Appl. Math. **32**(2), 155–175 (2016)
3. Naresh, R., Panday, S.: Modelling and analysis of the spread of Japanese encephalitis with environmental effects. Appl. Appl. Math. **4**(1), 155–175 (2009)
4. Agrawal, M., Verma, V.: The impact of media on the spreading and control of Japanese encephalitis. Int. J Math. Sci. Comput. **2**(2), 23–31 (2012)
5. Tapaswi, P.K., Ghosh, A.K., Mukhopadhyay, B.B.: Transmission of Japanese encephalitis in a 3-population model. Ecol. Model. **83**, 295–309 (1995)
6. Mukhopadhyay, B.B., Tapaswi, P.K.: An SIRS epidemic model of Japanese encephalitis. Math. Sci. **17**(2), 347–356 (1994)

7. De, A., Maity, K., Jana, S., Maiti, M.: Application of various control strategies to Japanese encephalitis: a mathematical study with human, pig and mosquito. Math. Biosci. **282**, 46–60 (2016)
8. Okosun, K.O., Mukamuri, M., Makinde, D.O.: Global stability analysis and control of leptospirosis. Open Math. **14**, 567–585 (2016)
9. Marsden, J., Sirovich, L., Golubitsky, M.: Mathematical Modeling in Population Biology and Epidemiology. Springer, New York (2000)
10. Ghosh, M., Chandra, P., Sinha, P., Shukla, J.B.: Modelling the spread of carrier dependent infectious diseases with environmental effect. Appl. Math. Comp. **152**, 385–402 (2004)
11. Singh, S., Shukla, J., Chandra, P.: Modelling and analysis of the spread of Malaria: environmental and ecological effects. Biol. Syst. **13**, 1–11 (2013)
12. Buonomo, B., Onofrio, A., Lacitignola, D.: Global stability of an SIR epidemic model with information dependent vaccination. Math. Biosci. **216**, 9–16 (2008)
13. Tewa, J.J., Dimi, L.J., Bowong, S.: Lyapunov function for a dengue disease transmission model. Chaos Solution and Fractols **39**, 936–941 (2007)
14. Naresh, R., Pandey, S., Misra,, A.K.: Analysis of a vaccination model for carrier dependent infectious disease with environmental effects. Nonlinear Model. Anal. **13**(3), 331–350 (2008)

A New Class of Incomplete Generalized Tri Lucas Polynomials

Suchita Arolkar and Y. S. Valaulikar

Abstract A new class of polynomials, namely, incomplete $h(x)$-B tri Lucas polynomials, is defined. Some identities involving these new polynomials are proved. A relation between incomplete $h(x)$-B tribonacci polynomials and incomplete $h(x)$-B tri Lucas polynomials is obtained. Some identities involving derivatives of these polynomials are also obtained.

Keywords $h(x)$-B tribonacci polynomials · $h(x)$-B tri Lucas polynomials · Incomplete $h(x)$-B tribonacci polynomials · Incomplete $h(x)$-B tri Lucas polynomials

Mathematics Subject Classification (2010) 11B39 · 11B37

1 Introduction

Fibonacci sequence is generated by a recurrence relation with seed values 0 and 1 and subsequent terms are obtained by adding preceding two terms [8, 13]. This sequence is extended and generalized in many ways, either by changing the recurrence equation or by changing seed values. Two such extensions to third order and qth order recurrence relations are introduced in [2] and [5], respectively. Fibonacci and Lucas polynomials [15] play a very important role in the theory of Fibonacci numbers. These polynomials are extended to B-tribonacci and B-tri Lucas polynomials in [1]. Identities of incomplete k-Fibonacci and k-Lucas numbers are obtained in [10]. The same author, in [12], has introduced the incomplete generalized Fibonacci and

S. Arolkar (✉)
Department of Mathematics and Statistics, Dnyanprassarak Mandal's College and Research Centre, Assagao, Goa, India
e-mail: suchita.golatkar@yahoo.com; suchita@dmscollege.ac.in

Y. S. Valaulikar
Department of Mathematics, Goa University, Taleigaon Plateau, Goa, India
e-mail: ysv@unigoa.ac.in; ysvgoa@gmail.com

R. N. Mohapatra et al. (eds.), *Mathematical Analysis and Computing*, Springer Proceedings in Mathematics & Statistics 344,
https://doi.org/10.1007/978-981-33-4646-8_7

Lucas polynomials and studied identities related to them. Another type of extension of Fibonacci numbers and polynomials called incomplete tribonacci-Lucas numbers and polynomials are studied in [14]. Incomplete Fibonacci and Lucas numbers, and incomplete tribonacci numbers are studied in [7] and [11], respectively. In [9], $h(x)$-Fibonacci and $h(x)$-Lucas polynomials are introduced. These are further extended to $h(x)$-B tribonacci polynomials and $h(x)$-B tri Lucas polynomials in [3]. We reproduce these definitions here.

$$({}^tB)_{h,n+2}(x) = h^2(x)\ ({}^tB)_{h,n+1}(x) + 2h(x)\ ({}^tB)_{h,n}(x) + \ ({}^tB)_{h,n-1}(x), \forall n \geq 1, \tag{1}$$

$$\text{with } ({}^tB)_{h,0}(x) = 0,\ ({}^tB)_{h,1}(x) = 0 \text{ and } ({}^tB)_{h,2}(x) = 1,$$

where $({}^tB)_{h,n}(x)$ is the nth $h(x)$-B tribonacci polynomial, and

$$({}^tL)_{h,n+2}(x) = h^2(x)\ ({}^tL)_{h,n+1}(x) + 2h(x)\ ({}^tL)_{h,n} + \ ({}^tL)_{h,n-1}(x), \forall n \geq 1, \tag{2}$$

$$\text{with } ({}^tL)_{h,0}(x) = 0,\ ({}^tL)_{h,1}(x) = 2, \text{ and } ({}^tL)_{h,2}(x) = h^2(x),$$

where $({}^tL)_{h,n}(x)$ is the nth $h(x)$-B tri Lucas polynomial. Observe that in (1) and in (2) the coefficients on the right-hand side are the terms of the binomial expansion of $(h(x)+1)^2$.

We list below some identities from [3]

(i) The nth term $({}^tB)_{h,n}(x)$ of (1) is given by

$$({}^tB)_{h,n}(x) = \sum_{r=0}^{\lfloor \frac{2n-4}{3} \rfloor} \frac{(2n-4-2r)^{\underline{r}}}{r!}\ h^{2n-4-3r}(x), \forall n \geq 2. \tag{3}$$

(ii) The nth term $({}^tL)_{h,n}(x)$ of (2) is given by

$$({}^tL)_{h,n}(x) = \sum_{r=0}^{\lfloor \frac{2n-2}{3} \rfloor} \Big(\frac{(2n-2)}{(2n-2-2r)} \frac{(2n-2-2r)^{\underline{r}}}{r!} \tag{4}$$

$$-r(r-1)\frac{(2n-4-2r)^{\underline{r-2}}}{r!} \Big)\ h^{2n-2-3r}(x), \forall n \geq 2.$$

(iii) The derivative of $({}^tB)_{h,n}(x)$ with respect to x is given by

$$\frac{d}{dx}[({}^tB)_{h,n}(x)] = 2\frac{d}{dx}(h(x)) \sum_{i=0}^{n} \Big(h(x)\ ({}^tB)_{h,n+1-i}(x) + ({}^tB)_{h,n-i}(x)\Big) ({}^tB)_{h,i}(x). \tag{5}$$

(iv) The derivative of $({}^tL)_{h,n}(x)$ with respect to x is given by

$$\frac{d}{dx}\left[({}^tL)_{h,n}(x)\right] = 2\frac{d}{dx}[h(x)] - \Big[h(x)\ ({}^tB)_{h,n}(x) + \\ + \sum_{i=0}^{n}\Big(h(x)\ ({}^tL)_{h,n+1-i}(x) + \ ({}^tL)_{h,n-i}(x)\Big)({}^tB)_{h,i}(x)\Big]. \tag{6}$$

Incomplete $h(x)$-B tribonacci polynomials [4] are the extension of incomplete $h(x)$-Fibonacci polynomials [12]. The $h(x)$-B-tribonacci polynomials [3] and incomplete $h(x)$-B-tribonacci polynomials in [4] are extended to the qth order relations in [6]. We reproduce the definition of incomplete $h(x)$-B tribonacci polynomials from [4] and list some identities related to these polynomials which are required to prove results of this paper.

Definition 1 The incomplete $h(x)$-B tribonacci polynomials are defined by

$$({}^tB)^l_{h,n}(x) = \sum_{r=0}^{l}\frac{(2n-4-2r)^{\underline{r}}}{r!}\ h^{2n-4-3r}(x),\ \forall\, 0 \le l \le \left\lfloor\frac{2n-4}{3}\right\rfloor \text{ and } n \ge 2. \tag{7}$$

Note that $({}^tB)^{\lfloor\frac{2n-4}{3}\rfloor}_{h,n}(x) = ({}^tB)_{h,n}(x)$.

For simplicity, we use $({}^tB)^l_{h,n}(x) = ({}^tB)^l_{h,n}$, $({}^tB)_{h,n}(x) = ({}^tB)_{h,n}$, $h(x) = h$ and list below the identities related to (7).

(1) For $n \ge 3$,

$$({}^tB)^{l+2}_{h,n+3} = h^2({}^tB)^{l+2}_{h,n+2} + 2h({}^tB)^{l+1}_{h,n+1} + ({}^tB)^{l}_{h,n}\,, \tag{8}$$

$0 \le l \le \lfloor\frac{2n-6}{3}\rfloor$.

(2) For $s \ge 1$,

$$\sum_{i=0}^{2s}\frac{(2s)^{\underline{i}}}{i!}\ ({}^tB)^{l+i}_{h,n+i}\ h^i = ({}^tB)^{l+2s}_{h,n+3s},\ 0 \le l \le \left\lfloor\frac{2n-2s-4}{3}\right\rfloor. \tag{9}$$

(3) For $s \ge 1$,

$$\sum_{i=0}^{s-1}\Big(2\,h^{2s-1-2i}({}^tB)^{l+1}_{h,n+1+i} + h^{2s-2-2i}({}^tB)^{l}_{h,n+i}\Big) \\ = ({}^tB)^{l+2}_{h,n+2+s} - h^{2s}({}^tB)^{l+2}_{h,n+2},\ \forall\, n \ge \left\lfloor\frac{3l+6}{2}\right\rfloor. \tag{10}$$

In this paper, we introduce the incomplete $h(x)$-B tri Lucas polynomials and study identities related to them. We also study their relation with the incomplete $h(x)$-B tribonacci polynomials.

2 Incomplete $h(x)$-B Tri Lucas Polynomials

In this section, we define the incomplete $h(x)$-B tri Lucas polynomials and obtain some recurrence relations involving these polynomials. We also obtain some identities for derivatives of these polynomials.

Definition 2 The incomplete $h(x)$-B tri Lucas polynomials are defined by

$$({}^tL)^l_{h,n}(x)$$

$$= \sum_{r=0}^{l} \left(\frac{(2n-2)}{(2n-2-2r)} \frac{(2n-2-2r)^{\underline{r}}}{r!} - \frac{(2n-4-2r)^{\underline{r-2}}}{(r-2)!} \right) h^{2n-2-3r}(x), \quad (11)$$

$\forall\, 0 \le l \le \lfloor \frac{2n-2}{3} \rfloor$ *and* $n \ge 2$.

Note that $({}^tL)^{\lfloor \frac{2n-2}{3} \rfloor}_{h,n}(x) = ({}^tL)_{h,n}(x)$.

For simplicity, we use $({}^tL)^l_{h,n}(x) = ({}^tL)^l_{h,n}$, $({}^tL)_{h,n}(x) = ({}^tL)_{h,n}$, and $h(x) = h$.

Following result gives a relation between incomplete $h(x)$-B tribonacci and incomplete $h(x)$-B tri Lucas polynomials.

Theorem 1 *For* $n \ge 4$,

$$({}^tL)^l_{h,n} = ({}^tB)^l_{h,n+1} + 2h\,({}^tB)^{l-1}_{h,n-1} + ({}^tB)^{l-2}_{h,n-2}, \quad (12)$$

$2 \le l \le \lfloor \frac{2n-2}{3} \rfloor$.

Proof From (7), we have

$$({}^tB)^l_{h,n+1} + 2h({}^tB)^{l-1}_{h,n-1} + ({}^tB)^{l-2}_{h,n-2}$$

$$= \sum_{r=0}^{l} \frac{(2n-2-2r)^{\underline{r}}}{r!} h^{2n-2-3r} + 2h \sum_{r=0}^{l-1} \frac{(2n-6-2r)^{\underline{r}}}{r!} h^{2n-6-3r}$$

$$+ \sum_{r=0}^{l-2} \frac{(2n-8-2r)^{\underline{r}}}{r!} h^{2n-8-3r}$$

$$= \sum_{r=0}^{l} \left[\frac{(2n-2-2r)^{\underline{r}}}{r!} + 2\left(\frac{(2n-4-2r)^{\underline{r-1}}}{(r-1)!} + \frac{(2n-4-2r)^{\underline{r-2}}}{(r-2)!} \right) - \frac{(2n-4-2r)^{\underline{r-2}}}{(r-2)!} \right] h^{2n-2-3r}$$

$$= \sum_{r=0}^{l} \left[\frac{(2n-2-2r)^{\underline{r}}}{r!} + 2\left(\frac{(2n-3-2r)^{\underline{r-1}}}{(r-1)!}\right) - \frac{(2n-4-2r)^{\underline{r-2}}}{(r-2)!} \right] h^{2n-2-3r}$$

$$= \sum_{r=0}^{l} \left[\frac{2n-2}{2n-2-2r}\left(\frac{(2n-2-2r)^{\underline{r}}}{r!}\right) - \frac{(2n-4-2r)^{\underline{r-2}}}{(r-2)!} \right] h^{2n-2-3r}$$

$= ({}^{t}L)^{l}_{h,n}$, from (11). □

Using (8) and (12), following corollary can be proved.

Corollary 1 *For $n \geq 1$,*

$$({}^{t}L)^{l}_{h,n} = 2\ ({}^{t}B)^{l}_{h,n+1} - h^{2}\ ({}^{t}B)^{l}_{h,n},\ 0 \leq l \leq \left\lfloor \frac{2n-2}{3} \right\rfloor. \tag{13}$$

Following theorem can be proved by using (8) and (12).

Theorem 2 *The recurrence relation of the incomplete $h(x)$-B tri Lucas sequence $({}^{t}L)^{l}_{h,n}$ is given by*

$$({}^{t}L)^{l+2}_{h,n+3} = h^{2}({}^{t}L)^{l+2}_{h,n+2} + 2h({}^{t}L)^{l+1}_{h,n+1} + ({}^{t}L)^{l}_{h,n}, \tag{14}$$

$0 \leq l \leq \lfloor \frac{2n-2}{3} \rfloor$ *and* $n \geq 1$.

Next, we have the following.

Theorem 3 *For $s \geq 0$ and $n \geq 2$,*

$$({}^{t}L)^{l+2s}_{h,n+3s} = \sum_{i=0}^{2s} \frac{(2s)^{\underline{i}}}{i!}\ ({}^{t}L)^{l+i}_{h,n+i}\ h^{i},\ 0 \leq l \leq \left\lfloor \frac{2n-2-2s}{3} \right\rfloor. \tag{15}$$

Theorem 4 *For $s \geq 1$ and $n \geq 3$,*

$$({}^tL)^{l+2}_{h,n+2+s} - h^{2s}\,({}^tL)^{l+2}_{h,n+2} = \sum_{i=0}^{s-1}\left(2\,h^{2s-1-2i}({}^tL)^{l+1}_{h,n+1+i} + h^{2s-2-2i}({}^tL)^{l}_{h,n+i}\right), \tag{16}$$

$0 \le l \le \lfloor \frac{2n-6}{3} \rfloor$.

We need the following Lemma.

Lemma 1 *For all $n \geq 2$,*

$$\sum_{r=0}^{\lfloor \frac{2n-2}{3} \rfloor} r\left(\frac{(2n-2)}{(2n-2-2r)}\frac{(2n-2-2r)^{\underline{r}}}{r!} - \frac{(2n-4-2r)^{\underline{r-2}}}{(r-2)!}\right) h^{2n-2-3r}$$

$$= \frac{(2n-2)}{3}\,({}^tL)_{h,n} - \frac{2}{3}\,h\left[\sum_{i=0}^{n}\left(h\,({}^tL)_{h,n+1-i} + ({}^tL)_{h,n-i}\right)({}^tB)_{h,i} - h\,({}^tB)_{h,n}\right]. \tag{17}$$

Proof Differentiating both sides (4) with respect to x and using (6), we get

$$2h\sum_{i=0}^{n}\left(h\,({}^tL)_{h,n+1-i} + ({}^tL)_{h,n-i}\right)({}^tB)_{h,i} - h\,({}^tB)_{h,n}$$

$$= (2n-2)({}^tL)_{h,n}$$

$$-3\sum_{r=0}^{\lfloor \frac{2n-2}{3} \rfloor} r\left(\frac{(2n-2)}{(2n-2-2r)}\frac{(2n-2-2r)^{\underline{r}}}{r!} - \frac{(2n-4-2r)^{\underline{r-2}}}{(r-2)!}\right) h^{2n-2-3r}$$

Therefore, by rearranging the terms, we get the required result. □

We use Lemma 1 to prove the next theorem.

Theorem 5 *For all $n \geq 2$,*

$$\sum_{l=0}^{\lfloor \frac{2n-2}{3} \rfloor} ({}^tL)^{l}_{h,n} = \left(\lfloor \frac{2n-2}{3} \rfloor - \frac{2n-5}{3}\right)({}^tL)_{h,n} \tag{18}$$

$$+\frac{2}{3}\,h\left[\sum_{i=0}^{n}\left(h\,({}^tL)_{h,n+1-i} + ({}^tL)_{h,n-i}\right)({}^tB)_{h,i} - h\,({}^tB)_{h,n}\right].$$

Proof From (11), we have

$$\sum_{l=0}^{\lfloor\frac{2n-2}{3}\rfloor} ({}^tL)^l_{h,n} = \left(\lfloor\tfrac{2n-2}{3}\rfloor + 1\right)\left(\tfrac{(2n-2)}{(2n-2)}\tfrac{(2n-2)^{\underline{0}}}{0!}\right) h^{2n-2}$$
$$+(\lfloor\tfrac{2n-2}{3}\rfloor)\left(\tfrac{(2n-2)}{(2n-4)}\tfrac{(2n-4)^{\underline{1}}}{1!}\right) h^{2n-5} + \cdots$$
$$+\left(\lfloor\tfrac{2n-2}{3}\rfloor + 1 - r\right)\left(\tfrac{(2n-2)}{(2n-2-2r)}\tfrac{(2n-2-2r)^{\underline{r}}}{r!} - \tfrac{(2n-4-2r)^{\underline{r-2}}}{(r-2)!}\right) h^{2n-2-3r} + \cdots$$
$$+\left(\tfrac{(2n-2)}{(2n-2-2\lfloor\frac{2n-2}{3}\rfloor)}\tfrac{(2n-2-2r)^{\underline{\lfloor\frac{2n-2}{3}\rfloor}}}{(\lfloor\frac{2n-2}{3}\rfloor)!} - \tfrac{(2n-4-2\lfloor\frac{2n-2}{3}\rfloor)^{\underline{\lfloor\frac{2n-2}{3}\rfloor-2}}}{(\lfloor\frac{2n-2}{3}\rfloor-2)!}\right) h^{2n-2-3\lfloor\frac{2n-2}{3}\rfloor}$$

$$= \sum_{r=0}^{\lfloor\frac{2n-2}{3}\rfloor} \left(\lfloor\tfrac{2n-2}{3}\rfloor + 1 - r\right)\left(\tfrac{(2n-2)}{(2n-2-2r)}\tfrac{(2n-2-2r)^{\underline{r}}}{r!} - \tfrac{(2n-4-2r)^{\underline{r-2}}}{(r-2)!}\right) h^{2n-2-3r}$$
$$= \left(\lfloor\tfrac{2n-2}{3}\rfloor + 1\right)\sum_{r=0}^{\lfloor\frac{2n-2}{3}\rfloor} \left(\tfrac{(2n-2)}{(2n-2-2r)}\tfrac{(2n-2-2r)^{\underline{r}}}{r!} - \tfrac{(2n-4-2r)^{\underline{r-2}}}{(r-2)!}\right) h^{2n-2-3r}$$
$$-\sum_{r=0}^{\lfloor\frac{2n-2}{3}\rfloor} r\left(\tfrac{(2n-2)}{(2n-2-2r)}\tfrac{(2n-2-2r)^{\underline{r}}}{r!} - \tfrac{(2n-4-2r)^{\underline{r-2}}}{(r-2)!}\right) h^{2n-2-3r}$$

$$= \left(\lfloor\tfrac{2n-2}{3}\rfloor + 1 - \tfrac{(2n-2)}{3}\right)({}^tL)_{h,n}$$
$$+\tfrac{2}{3}h\left[\sum_{i=0}^{n}\left(h\,({}^tL)_{h,n+1-i} + ({}^tL)_{h,n-i}\right)({}^tB)_{h,i} - h\,({}^tB)_{h,n}\right]$$

$$= \left(\lfloor\tfrac{2n-2}{3}\rfloor - \tfrac{2n-5}{3}\right)({}^tL)_{h,n}$$
$$+\tfrac{2}{3}h\left[\sum_{i=0}^{n}\left(h\,({}^tL)_{h,n+1-i} + ({}^tL)_{h,n-i}\right)({}^tB)_{h,i} - h\,({}^tB)_{h,n}\right].$$

□

Following identities involving derivatives of $({}^tB)^l_{h,n}$ and $({}^tL)^l_{h,n}$ with respect to x can be proved using (8), (12), and (14).

Theorem 6 *For* $n \geq 3$,

(1)

$$\frac{d^k}{dx^k}\left(({}^tB)^l_{h,n}\right)$$
$$= \sum_{s=0}^{k} \frac{k^{\underline{s}}}{s!}\left(\frac{d^s}{dx^s}(h^2)\frac{d^{k-s}}{dx^{k-s}}\left(({}^tB)^l_{h,n-1}\right) + 2\,\frac{d^s}{dx^s}(h)\frac{d^{k-s}}{dx^{k-s}}\left(({}^tB)^{l-1}_{h,n-2}\right)\right)$$
$$+\frac{d^k}{dx^k}\left(({}^tB)^{l-2}_{h,n-3}\right).$$

(2)

$$
\begin{aligned}
&\tfrac{d^k}{dx^k}\left(({}^tL)^l_{h,n}\right)\\
&\quad = \tfrac{d^k}{dx^k}\left(({}^tB)^l_{h,n+1}\right) + 2\sum_{s=0}^{k} \tfrac{k^{\underline{s}}}{s!}\left(\tfrac{d^s}{dx^s}(h)\tfrac{d^{k-s}}{dx^{k-s}}\left(({}^tB)^{l-1}_{h,n-1}\right)\right)\\
&\qquad + \tfrac{d^k}{dx^k}\left(({}^tB)^{l-2}_{h,n-2}\right).
\end{aligned}
$$

(3)

$$
\begin{aligned}
&\tfrac{d^k}{dx^k}\left(({}^tL)^l_{h,n}\right)\\
&\quad = \sum_{s=0}^{k} \tfrac{k^{\underline{s}}}{s!}\left(\tfrac{d^s}{dx^s}(h^2)\tfrac{d^{k-s}}{dx^{k-s}}\left(({}^tL)^l_{h,n-1}\right) + 2\,\tfrac{d^s}{dx^s}(h)\tfrac{d^{k-s}}{dx^{k-s}}\left(({}^tL)^{l-1}_{h,n-2}\right)\right)\\
&\qquad + \tfrac{d^k}{dx^k}\left(({}^tL)^{l-2}_{h,n-3}\right).
\end{aligned}
$$

References

1. Arolkar, S., Valaulikar, Y.S.: Generalized bivariate B-tribonacci and B-tri-Lucas polynomials. In: Proceedings of CPMSED-2015. Krishi Sanskriti Publications, pp. 10–13 (2015)
2. Arolkar, S., Valaulikar, Y.S.: On an extension of Fibonacci sequence. Bull. Marathwada Math. Soc. **17**(1), 1–8 (2016)
3. Arolkar, S., Valaulikar, Y.S.: $h(x)$-B tribonacci and $h(x)$-B tri Lucas polynomials. Kyungpook Math. J. **56**, 1125–1133 (2016)
4. Arolkar, S., Valaulikar, Y.S.: Incomplete $h(x)$-B tribonacci polynomials. Turkish J. Anal. Number Theory **4**(6), 155–158 (2016)
5. Arolkar, S., Valaulikar, Y.S.: On a B-q bonacci sequence. Int. J. Adv. Math. (1), 1–8 (2017)
6. Arolkar, S., Valaulikar, Y.S.: On two new classes of B-q bonacci polynomials. J. Int. Seq. **21** (2018). Article 18.4.1
7. Filipponi, P.: Incomplete Fibonacci and Lucas numbers. Rend. Circ. Mat. Palermo **45**(2), 37–56 (1996)
8. Koshy, T.: Fibonacci and Lucas Numbers with Applications. Wiley-Interscience (2001)
9. Nalli, A., Haukkane, P.: On generalized Fibonacci and Lucas polynomials. Chaos, Solitons Fractals **42**(5), 3179–3186 (2009)
10. Ramírez, J.: Incomplete k-Fibonacci and k-Lucas numbers. Chinese J. Math. (2013). Article ID 107145
11. Ramírez, J., Sirvent, V.F.: Incomplete tribonacci numbers and polynomials. J. Int. Seq. **17** (2014). Article 14.4.2
12. Ramírez, J.: Incomplete generalized Fibonacci and Lucas polynomials. Hacet. J. Math. Stat. **42**, 363–373 (2015)
13. Vajda, S.: Fibonacci and Lucas Numbers, and the Golden Section: Theory and Applications. Dover (2008)
14. Yilmaz, N., Taskara, N.: Incomplete Tribonacci-Lucas numbers and polynomials. Adv. Appl. Clifford Alge. **25**, 741–753 (2015)
15. Yu, H., Liang, C.: Identities involving partial derivatives of bivariate Fibonacci and Lucas polynomials. Fibonacci Quart. **33**(2), 19–23 (1997)

Hyers Stability of Additive Functional Equation in Banach Space

P. Agilan, John M. Rassias, and V. Vallinayagam

Abstract The purpose of this paper is to determine the general solution and generalized Ulam–Hyers stability concerning the additive functional equation in Banach space by using direct and fixed point method.

Keywords Additive functional equations · Generalized Ulam–Hyers stability · Fixed point

1 Introduction

The literature on the stability of functional equation is started in the year 1940 by posting the question of Ulam [16] in front of a Mathematical Colloquium at the University of Wisconsin. D. H. Hyers [10] who is the first author to derive the answer to the question of Ulam in Banach spaces. This stability problem was investigated in various normed spaces by various authors [2, 8, 9, 12, 13]. We refer also other researcher works [1, 5–7, 14, 15].

The following additive functional equations derive the solution and its Ulam stability in Banach Spaces (see [3, 4]).

P. Agilan (✉) · V. Vallinayagam
Department of Mathematics, St. Joseph's College of Engineering,
Chennai 600119, Tamilnadu, India
e-mail: agilram@gmail.com

V. Vallinayagam
e-mail: vngam19@gmail.com

J. M. Rassias
Pedagogical Department E.E., Section of Mathematics and Informatics, National and
Capodistrian University of Athens, Agamemnonos Str., 15342 Aghia Paraskevi, Athens, Greece
e-mail: jrass@otenet.gr

R. N. Mohapatra et al. (eds.), *Mathematical Analysis and Computing*,
Springer Proceedings in Mathematics & Statistics 344,
https://doi.org/10.1007/978-981-33-4646-8_8

$$\left(\lambda^n+1\right) f\left(e^{\frac{in\pi}{3}}\vartheta+e^{\frac{-in\pi}{3}}\omega\right)+\left(\lambda^n-1\right) f\left(e^{\frac{-in\pi}{3}}\vartheta+e^{\frac{in\pi}{3}}\omega\right)$$
$$+e^{\frac{-in\pi}{3}} f(\vartheta-\omega)+e^{\frac{in\pi}{3}} f(\omega-\vartheta)=2\lambda^n\cos\left(\frac{n\pi}{3}\right)[f(\vartheta)+f(\omega)] \quad (1)$$

$$\left[\cos h\left(\frac{n\pi}{2}\right)+\sin h\left(\frac{n\pi}{2}\right)\right] f\left(e^{\frac{n\pi}{2}}\alpha+e^{\frac{-n\pi}{2}}\beta\right)$$
$$+\left[\cos h\left(\frac{n\pi}{2}\right)-\sin h\left(\frac{n\pi}{2}\right)\right] f\left(e^{-\frac{n\pi}{2}}\alpha+e^{\frac{n\pi}{2}}\beta\right)$$
$$+e^{-\frac{n\pi}{2}}\sin h\left(\frac{n\pi}{2}\right) f(\alpha-\beta)+e^{\frac{n\pi}{2}}\sin h\left(\frac{n\pi}{2}\right) f(\beta-\alpha)$$
$$=2\cos^2 h\left(\frac{n\pi}{2}\right)[f(\alpha)+f(\beta)] \quad (2)$$

The main objective of this article is to derive the general solution and Hyers stability of the additive functional equation

$$\left(e^{2\left(\frac{n\pi}{3}\right)}+6e^{\left(\frac{n\pi}{3}\right)}\right) f\left(x\sin\left(\frac{n\pi}{3}\right)+y\cos\left(\frac{n\pi}{3}\right)\right)$$
$$+\left(9-6e^{\left(\frac{n\pi}{3}\right)}\right) f\left(y\sin\left(\frac{n\pi}{3}\right)+z\cos\left(\frac{n\pi}{3}\right)\right)$$
$$+6e^{\left(\frac{n\pi}{3}\right)} f\left(z\sin\left(\frac{n\pi}{3}\right)+x\cos\left(\frac{n\pi}{3}\right)\right)$$
$$-6e^{\left(\frac{n\pi}{3}\right)}\sin\left(\frac{n\pi}{3}\right) f(x-y)-6e^{\left(\frac{n\pi}{3}\right)}\cos\left(\frac{n\pi}{3}\right) f(y-z)$$
$$=\left(e^{2\left(\frac{n\pi}{3}\right)}\sin\left(\frac{n\pi}{3}\right)+6e^{\left(\frac{n\pi}{3}\right)}\cos\left(\frac{n\pi}{3}\right)\right) f(x)$$
$$+\left(9\sin\left(\frac{n\pi}{3}\right)+e^{2\left(\frac{n\pi}{3}\right)}\cos\left(\frac{n\pi}{3}\right)\right) f(y)$$
$$+\left(6e^{\left(\frac{n\pi}{3}\right)}\sin\left(\frac{n\pi}{3}\right)+9\cos\left(\frac{n\pi}{3}\right)\right) f(z) \quad (3)$$

where $n \in Z$ in Banach Spaces using direct and fixed point methods.

2 General Solution

We investigate the general solution of the additive FE (3) in this section. Let us consider (X, Y) be vector spaces and Functional Equation as FE.

Lemma 1 *Let $f : X \to Y$ be a odd mapping fulfills the additive functional equation*

$$f(x+y)=f(x)+f(y) \quad (4)$$

for all $x, y \in X$ if $f : X \to Y$ fulfills the additive mapping (3) for all $x, y, z \in X$.

Proof An odd function $f : X \to Y$ satisfies the FE (4). Considering (x, y) by $(0, 0)$ in (4), we obtain $f(0) = 0$. Let x by $-y$ in (4), we reach $f(-y) = -f(y)$ for all

$y \in X$. By interchanging y by x and y by $2x$ in (4), we arrive

$$f(2x) = 2f(x) \quad \text{and} \quad f(3x) = 3f(x) \tag{5}$$

for all $x \in X$. By induction of a, we have

$$f(ax) = af(x) \tag{6}$$

Replacing (x, y) by $\left(x \sin\left(\frac{n\pi}{3}\right), y \cos\left(\frac{n\pi}{3}\right)\right)$ in (4) and using (6), we get

$$f\left(x \sin\left(\frac{n\pi}{3}\right) + y \cos\left(\frac{n\pi}{3}\right)\right) = \sin\left(\frac{n\pi}{3}\right) f(x) + \cos\left(\frac{n\pi}{3}\right) f(y) \tag{7}$$

for all $x, y \in X$. Both sides multiply by $\left(e^{2\left(\frac{n\pi}{3}\right)} + 6e^{\left(\frac{n\pi}{3}\right)}\right)$ in (7), we arrive

$$\begin{aligned}&\left(e^{2\left(\frac{n\pi}{3}\right)} + 6e^{\left(\frac{n\pi}{3}\right)}\right) f\left(x \sin\left(\frac{n\pi}{3}\right) + y \cos\left(\frac{n\pi}{3}\right)\right)\\ &\quad = \left(e^{2\left(\frac{n\pi}{3}\right)} + 6e^{\left(\frac{n\pi}{3}\right)}\right) \sin\left(\frac{n\pi}{3}\right) f(x) + \left(e^{2\left(\frac{n\pi}{3}\right)} + 6e^{\left(\frac{n\pi}{3}\right)}\right) \cos\left(\frac{n\pi}{3}\right) f(y)\end{aligned} \tag{8}$$

Replacing (x, y) by $\left(y \sin\left(\frac{n\pi}{3}\right), z \cos\left(\frac{n\pi}{3}\right)\right)$ in (4) and using (6), we get

$$f\left(y \sin\left(\frac{n\pi}{3}\right) + z \cos\left(\frac{n\pi}{3}\right)\right) = \sin\left(\frac{n\pi}{3}\right) f(y) + \cos\left(\frac{n\pi}{3}\right) f(z) \tag{9}$$

for all $y, z \in X$. Both sides multiply by $\left(9 - 6e^{\left(\frac{n\pi}{3}\right)}\right)$ in (9), we arrive

$$\begin{aligned}&\left(9 - 6e^{\left(\frac{n\pi}{3}\right)}\right) f\left(y \sin\left(\frac{n\pi}{3}\right) + z \cos\left(\frac{n\pi}{3}\right)\right)\\ &\quad = \left(9 - 6e^{\left(\frac{n\pi}{3}\right)}\right) \sin\left(\frac{n\pi}{3}\right) f(y) + \left(9 - 6e^{\left(\frac{n\pi}{3}\right)}\right) \cos\left(\frac{n\pi}{3}\right) f(z)\end{aligned} \tag{10}$$

Replacing (x, y) by $\left(z \sin\left(\frac{n\pi}{3}\right), x \cos\left(\frac{n\pi}{3}\right)\right)$ in (4) and using (6), we get

$$f\left(z \sin\left(\frac{n\pi}{3}\right) + x \cos\left(\frac{n\pi}{3}\right)\right) = \sin\left(\frac{n\pi}{3}\right) f(z) + \cos\left(\frac{n\pi}{3}\right) f(x) \tag{11}$$

for all $x, z \in X$. Both sides multiply by $6e^{\left(\frac{n\pi}{3}\right)}$ in (11), we arrive

$$\begin{aligned}&6e^{\left(\frac{n\pi}{3}\right)} f\left(z \sin\left(\frac{n\pi}{3}\right) + x \cos\left(\frac{n\pi}{3}\right)\right)\\ &\quad = 6e^{\left(\frac{n\pi}{3}\right)} \sin\left(\frac{n\pi}{3}\right) f(z) + 6e^{\left(\frac{n\pi}{3}\right)} \cos\left(\frac{n\pi}{3}\right) f(x)\end{aligned} \tag{12}$$

for all $x, z \in X$. Setting y by $-y$ in (4), and using oddness of f we have

$$f(x-y)=f(x)-f(y) \tag{13}$$

for all $x, y \in X$. Both sides multiply by $-6e^{\left(\frac{n\pi}{3}\right)}\sin\left(\frac{n\pi}{3}\right)$ in (13), we arrive

$$-6e^{\left(\frac{n\pi}{3}\right)}\sin\left(\frac{n\pi}{3}\right)f(x-y)=-6e^{\left(\frac{n\pi}{3}\right)}\sin\left(\frac{n\pi}{3}\right)f(x)+6e^{\left(\frac{n\pi}{3}\right)}\sin\left(\frac{n\pi}{3}\right)f(y) \tag{14}$$

for all $x, z \in X$. Setting (x, y) by $(y, -z)$ in (4), and using oddness of f we have

$$f(y-z)=f(y)-f(z) \tag{15}$$

for all $y, z \in X$. Both sides multiply by $-6e^{\left(\frac{n\pi}{3}\right)}cos\left(\frac{n\pi}{3}\right)$ in (15), we arrive

$$-6e^{\left(\frac{n\pi}{3}\right)}cos\left(\frac{n\pi}{3}\right)f(y-z)=-6e^{\left(\frac{n\pi}{3}\right)}cos\left(\frac{n\pi}{3}\right)f(y)+6e^{\left(\frac{n\pi}{3}\right)}cos\left(\frac{n\pi}{3}\right)f(z) \tag{16}$$

for all $y, z \in X$. Adding (8), (10), (12), (14), and (16), we obtain (3).

3 Stability Results: A Classical Method

Let us assume X be a normed space and Y be a Banach space, respectively. The mapping is defined as $D\,F : X \to Y$ by

$$\begin{aligned}
D\,F(x,y,z) &= \left(e^{2\left(\frac{n\pi}{3}\right)}+6e^{\left(\frac{n\pi}{3}\right)}\right)f\left(x\sin\left(\frac{n\pi}{3}\right)+y\cos\left(\frac{n\pi}{3}\right)\right)\\
&+\left(9-6e^{\left(\frac{n\pi}{3}\right)}\right)f\left(y\sin\left(\frac{n\pi}{3}\right)+z\cos\left(\frac{n\pi}{3}\right)\right)\\
&+6e^{\left(\frac{n\pi}{3}\right)}f\left(z\sin\left(\frac{n\pi}{3}\right)+x\cos\left(\frac{n\pi}{3}\right)\right)\\
&-6e^{\left(\frac{n\pi}{3}\right)}\sin\left(\frac{n\pi}{3}\right)f(x-y)-6e^{\left(\frac{n\pi}{3}\right)}\cos\left(\frac{n\pi}{3}\right)f(y-z)\\
&-\left(e^{2\left(\frac{n\pi}{3}\right)}\sin\left(\frac{n\pi}{3}\right)-6e^{\left(\frac{n\pi}{3}\right)}\cos\left(\frac{n\pi}{3}\right)\right)f(x)\\
&-\left(9\sin\left(\frac{n\pi}{3}\right)+e^{2\left(\frac{n\pi}{3}\right)}\cos\left(\frac{n\pi}{3}\right)\right)f(y)\\
&-\left(6e^{\left(\frac{n\pi}{3}\right)}\sin\left(\frac{n\pi}{3}\right)+9\cos\left(\frac{n\pi}{3}\right)\right)f(z)
\end{aligned}$$

where $n \in Z$ for all $x, y, z \in X$.

Theorem 1 *Let $f : X \to Y$ be a function fulfilling the inequality*

$$\|D\,F\,(x,y,z)\| \leqslant \Delta\,(x,y,z) \tag{17}$$

for all $x, y, z \in X$, *where* $\Delta : X^3 \to [0, \infty)$ *be a function such that*

$$\lim_{n\to\infty} \frac{\Delta\left(T^{nj}x, T^{nj}y, T^{nj}z\right)}{T^{nj}} = 0 \tag{18}$$

for all $x, y, z \in X$. *There* $\exists$ *only one additive mapping* $A : X \to Y$ *and fulfilling the FE (3) such that*

$$\|f(x) - A(x)\| \leqslant \frac{1}{T\left(e^{\left(\frac{n\pi}{3}\right)} + 3\right)^2} \sum_{k=\frac{1-j}{2}}^{\infty} \frac{\Delta(T^{kj}x, T^{kj}x, T^{kj}x)}{T^{kj}} \tag{19}$$

for all $x \in X$ *with* $j \in \{-1, 1\}$. *The function* $A(x)$ *is defined as*

$$A(x) = \lim_{n\to\infty} \frac{f(T^{nj}x)}{T^{nj}} \tag{20}$$

for all $x \in X$.

Proof First, we prove when $j = 1$. Substituting (x, y, z) by (x, x, x) in (17), we have

$$\begin{aligned} &\left\| \left(e^{\left(\frac{n\pi}{3}\right)} + 3\right)^2 f\left(\left(\sin\left(\frac{n\pi}{3}\right) + \cos\left(\frac{n\pi}{3}\right)\right)x\right) \right. \\ &\quad \left. - \left(e^{\left(\frac{n\pi}{3}\right)} + 3\right)^2 \left(\sin\left(\frac{n\pi}{3}\right) + \cos\left(\frac{n\pi}{3}\right)\right) f(x) \right\| \leqslant \Delta\left(x, x, x\right) \end{aligned} \tag{21}$$

for all $x \in X$. Setting $T = \left(\sin\left(\frac{n\pi}{3}\right) + \cos\left(\frac{n\pi}{3}\right)\right)$ in (21), we arrive

$$\left\| \left(e^{\left(\frac{n\pi}{3}\right)} + 3\right)^2 Tf(x) - \left(e^{\left(\frac{n\pi}{3}\right)} + 3\right)^2 f\left(Tx\right) \right\| \leqslant \Delta\left(x, x, x\right) \tag{22}$$

for all $x \in X$. Both sides divide by $\left(e^{\left(\frac{n\pi}{3}\right)} + 3\right)^2 T$ in (22), we have

$$\left\| f(x) - \frac{f(Tx)}{T} \right\| \leqslant \frac{\Delta\left(x, x, x\right)}{T\left(e^{\left(\frac{n\pi}{3}\right)} + 3\right)^2} \tag{23}$$

for all $x \in X$. By considering x by Tx and dividing by T in (23), we get

$$\left\| \frac{f(Tx)}{T} - \frac{f(T^2x)}{T^2} \right\| \leqslant \frac{\Delta\left(Tx, Tx, Tx\right)}{T^2\left(e^{\left(\frac{n\pi}{3}\right)} + 3\right)^2} \tag{24}$$

for all $x \in X$. In (23) and (24), we reach

$$\left\| f(x) - \frac{f(T^2x)}{T^2} \right\| \leq \left\| f(x) - \frac{f(Tx)}{T} \right\| + \left\| \frac{f(Tx)}{T} - \frac{f(T^2x)}{T^2} \right\|$$
$$\leq \frac{1}{T\left(e^{\left(\frac{n\pi}{3}\right)} + 3\right)^2} \left[\Delta(x, x, x) + \frac{\Delta(Tx, Tx, Tx)}{T} \right] \quad (25)$$

for all $x \in X$. Generalizing for a positive integer n, we reach

$$\left\| f(x) - \frac{f(T^n x)}{T^n} \right\| \leqslant \frac{1}{T\left(e^{\left(\frac{n\pi}{3}\right)} + 3\right)^2} \sum_{k=0}^{n-1} \frac{\Delta(T^k x, T^k x, T^k x)}{T^k} \quad (26)$$
$$\leqslant \frac{1}{T\left(e^{\left(\frac{n\pi}{3}\right)} + 3\right)^2} \sum_{k=0}^{\infty} \frac{\Delta(T^k x, T^k x, T^k x)}{T^k}$$

for all $x \in X$. Hence

$$\left\{ \frac{f(T^n x)}{T^n} \right\},$$

is a Cauchy sequence and it converges to a point $A(x) \in X$. Indeed, consider x by $T^m x$ and dividing by T^m in (26),for any $m, n > 0$, we obtain

$$\left\| \frac{f(T^m x)}{T^m} - \frac{f(T^{n+m} x)}{T^{(n+m)}} \right\| = \frac{1}{T^m} \left\| f(T^m x) - \frac{f(T^n \cdot T^m x)}{T^n} \right\|$$
$$\leqslant \frac{1}{T\left(e^{\left(\frac{n\pi}{3}\right)} + 3\right)^2} \sum_{k=0}^{n-1} \frac{\Delta(T^{k+m} x, T^{k+m} x, T^{k+m} x)}{T^{k+m}}$$
$$\leqslant \frac{1}{T\left(e^{\left(\frac{n\pi}{3}\right)} + 3\right)^2} \sum_{k=0}^{\infty} \frac{\Delta(T^{k+m} x, T^{k+m} x, T^{k+m} x)}{T^{k+m}}$$
$$\to 0 \quad as \quad m \to \infty$$

for all $x \in X$. Thus, we define mapping $A : X \to Y$

$$A(x) = \lim_{n\to\infty} \frac{f(T^n x)}{T^n} \quad \forall \ x \in X.$$

Taking $n \to \infty$ in (26), we get (19) that holds for all $x \in X$. To show that A satisfies (3), consider (x, y, z) by $(T^n x, T^n y, T^n z)$ and dividing by T^n in (17), we have

$$\frac{1}{T^n} \left\| Df\left(T^n x, T^n y, T^n z\right)\right\| \leqslant \frac{1}{T^n} \Delta\left(T^n x, T^n y, T^n z\right)$$

for all $x, y, z \in X$. Taking $n \to \infty$ in the above inequality on both sides and by applying the definition of $A(x)$, we obtain

$$\begin{aligned}
&\left(e^{2\left(\frac{n\pi}{3}\right)} + 6e^{\left(\frac{n\pi}{3}\right)}\right) A\left(x \sin\left(\frac{n\pi}{3}\right) + y \cos\left(\frac{n\pi}{3}\right)\right) \\
&+ \left(9 - 6e^{\left(\frac{n\pi}{3}\right)}\right) A\left(y \sin\left(\frac{n\pi}{3}\right) + z \cos\left(\frac{n\pi}{3}\right)\right) \\
&+ 6e^{\left(\frac{n\pi}{3}\right)} A\left(z \sin\left(\frac{n\pi}{3}\right) + x \cos\left(\frac{n\pi}{3}\right)\right) \\
&- 6e^{\left(\frac{n\pi}{3}\right)} \sin\left(\frac{n\pi}{3}\right) A(x-y) - 6e^{\left(\frac{n\pi}{3}\right)} \cos\left(\frac{n\pi}{3}\right) A(y-z) \\
&= \left(e^{2\left(\frac{n\pi}{3}\right)} \sin\left(\frac{n\pi}{3}\right) + 6e^{\left(\frac{n\pi}{3}\right)} \cos\left(\frac{n\pi}{3}\right)\right) A(x) \\
&+ \left(9 \sin\left(\frac{n\pi}{3}\right) + e^{2\left(\frac{n\pi}{3}\right)} \cos\left(\frac{n\pi}{3}\right)\right) A(y) \\
&+ \left(6e^{\left(\frac{n\pi}{3}\right)} \sin\left(\frac{n\pi}{3}\right) + 9 \cos\left(\frac{n\pi}{3}\right)\right) A(z)
\end{aligned}$$

Hence A satisfies (3) for all $x, y, z \in X$. In order to prove the existence of $A(x)$ is unique, assume that $B(x)$ be another additive mapping satisfying (3) and (19). Now,

$$\begin{aligned}
\|A(x) - B(x)\| &= \frac{1}{T^n} \left\|A(T^n x) - B(T^n x)\right\| \\
&\leq \frac{1}{T^n} \left\{\left\|A(T^n x) - f(T^n x)\right\| + \left\|f(T^n x) - B(T^n x)\right\|\right\} \\
&\leqslant \frac{2}{T\left(e^{\left(\frac{n\pi}{3}\right)} + 3\right)^2} \sum_{k=0}^{\infty} \frac{\Delta(T^{k+n}x, T^{k+n}x, T^{k+n}x)}{T^{(k+n)}} \\
&\to 0 \quad as \ n \to \infty
\end{aligned}$$

for all $x \in X$. This proves that $A(x) = B(x)$. Therefore A is unique.

By applying the above procedure with $j = -1$, we reach similar type stability result.

Corollary 1 *Let us assume that $\mathcal{H}$ and s be nonnegative real numbers. Then the mapping $f : X \to Y$ fulfilling the inequality*

$$\|D\ F(x, y, z)\| \leqslant \begin{cases} \mathcal{H}, & \\ \mathcal{H}\{||x||^s + ||y||^s + ||z||^s\}, & s \neq 1; \\ \mathcal{H}||x||^s||y||^s||z||^s, & 3s \neq 1; \\ \mathcal{H}\left\{||x||^s||y||^s||z||^s + \left\{||x||^{3s} + ||y||^{3s} + ||z||^{3s}\right\}\right\}, & 3s \neq 1; \end{cases} \tag{27}$$

for all $x, y, z \in X$. There $\exists$ only one additive mapping $A : X \to Y$ such that

$$\|f(x)-A(x)\| \leqslant \begin{cases} \dfrac{\mathcal{H}}{\left(e^{\left(\frac{n\pi}{3}\right)}+3\right)^2 T-1|}, \\ \dfrac{3\,\mathcal{H}||x||^s}{\left(e^{\left(\frac{n\pi}{3}\right)}+3\right)^2 T-T^s|}, \\ \dfrac{\mathcal{H}||x||^{3s}}{\left(e^{\left(\frac{n\pi}{3}\right)}+3\right)^2 T-T^{3s}|}, \\ \dfrac{4\,\mathcal{H}||x||^{3s}}{\left(e^{\left(\frac{n\pi}{3}\right)}+3\right)^2 T-T^{3s}|} \end{cases} \tag{28}$$

4 Stability Results: Fixed-Point Method

In order to prove the fixed point stability results, we have to use the alternative of fixed point theorem [11].

Theorem 2 *Let $f : X \to Y$ be a odd function satisfying the functional inequality*

$$\|D\ \ F(x, y, z)\| \leq \Delta(x, y, z) \tag{29}$$

for all $x, y, z \in X$ and a mapping for which $\exists$ functions $\Delta, \beta, \lambda : X^3 \to [0, \infty)$ with the condition

$$\lim_{h\to\infty} \frac{\Delta(\mu_i^h x,\, \mu_i^h y,\, \mu_i^h z)}{\mu_i^h} = 0, \tag{30}$$

where

$$\mu_i = \begin{cases} T, & i = 0, \\ \frac{1}{T}, & i = 1 \end{cases}$$

Assume that there exists an $L = L(i) < 1$ such that the function

$$x \to \lambda(x) = \frac{1}{\left(e^{\left(\frac{n\pi}{3}\right)}+3\right)^2} \Delta\left(\frac{x}{T}, \frac{x}{T}, \frac{x}{T}\right),$$

we have the following the property

$$\lambda(x) = L\ \mu_i\ \lambda\left(\frac{x}{\mu_i}\right) \tag{31}$$

for all $x \in X$. There $\exists$ only one additive function $A : X \to Y$ satisfying the FE (3) and

$$\|f(x) - A(x)\| \leq \frac{L^{1-i}}{1-L}\lambda(x) \tag{32}$$

all $x \in X$.

Proof Let us take the set $X = \{N/N : X \to Y,\ N(0) = 0\}$ and introduce the generalized metric on $X, d(N, M) = \inf\{H \in (0, \infty) : \| N(x) - M(x) \| \leq H\lambda(x), x \in X\}$. It is easy to show that (X, d) is complete with respect to the defined metric. Let us define the mapping $\mathcal{D} : X \to X$ by $\mathcal{D}N(x) = \frac{1}{\mu_i}L(\mu_i x), \forall x \in X$. Now $N, M \in X$,

$$\begin{aligned}
d(N, M) \leq K &\Rightarrow \| N(x) - M(x) \| \leq H\lambda(x), x \in X.\\
&\Rightarrow \left\| \frac{1}{\mu_i}N(\mu_i x) - \frac{1}{\mu_i}M(\mu_i x) \right\| \leq \frac{1}{\mu_i}H\lambda(\mu_i x), x \in X,\\
&\Rightarrow \left\| \frac{1}{\mu_i}N(\mu_i x) - \frac{1}{\mu_i}M(\mu_i x) \right\| \leq LH\lambda(x), x \in X,\\
&\Rightarrow \| \mathcal{D}N(x) - \mathcal{D}M(x) \| \leq LH\lambda(x), x \in X,\\
&\Rightarrow d(\mathcal{D}N, \mathcal{D}M) \leq LH. \Rightarrow d(\mathcal{D}N, \mathcal{D}M) \leq Ld(N, M).
\end{aligned}$$

for all $N, M \in X$. That is, $\mathcal{D}$ with Lipschitz constant L and a strictly contractive mapping on X.

In (23), we obtain

$$\left\| f(x) - \frac{f(Tx)}{T} \right\| \leqslant \frac{\Delta(x, x, x)}{T\left(e^{\left(\frac{n\pi}{3}\right)} + 3\right)^2} \tag{33}$$

where

$$\beta(x) = \frac{\Delta(x, x, x)}{T\left(e^{\left(\frac{n\pi}{3}\right)} + 3\right)^2}$$

for all $x \in X$. Applying (31) for the case $i = 0$, which reduces to

$$\left\| \frac{1}{T}f(Tx) - f(x) \right\| \leq \frac{1}{T}\lambda(x)$$

for all $x \in X$.

$$\text{i.e.,}\quad d(\mathcal{D}f, f) \leq \frac{1}{T} = L = L^{1-0} = L^{1-i} < \infty.$$

Again considering $x = \frac{x}{T}$ in (33), we arrive

$$\left\| f(x) - Tf\left(\frac{x}{T}\right) \right\| \leq \frac{1}{\left(e^{\left(\frac{n\pi}{3}\right)} + 3\right)^2} \Delta\left(\frac{x}{T}\right).$$

for all $x \in X$. Applying (31) for the case $i = 1$, it reduces to

$$\left\| f(x) - Tf\left(\frac{x}{T}\right) \right\| \leq \lambda(x)$$

for all $x \in X$.

$$\text{i.e.,} \quad d(f, \mathcal{D}f) \leq 1 = L^0 = L^{1-1} = L^{1-i} < \infty.$$

By above two cases, we reach

$$d(f, \mathcal{D}f) \leq L^{1-i}.$$

Therefore (F_i) holds.

By (F_{ii}), there $\exists$ a fixed point A of $\mathcal{D}$ in X such that

$$A(x) = \lim_{h \to \infty} \frac{f(\mu_i^h x)}{\mu_i^h}, \qquad \forall\ x \in X. \tag{34}$$

In order to show $A(x)$ is additive FE. Considering (x, y, z) by $\left(\mu_i^h x, \mu_i^h y, \mu_i^h z\right)$ in (29) and dividing by μ_i^h, it follows from (30) and (34), A fulfilling the FE (3) for all $x, y, z \in X$.

By (F_{iii}), A is the only one fixed point of $\mathcal{D}$ in the set $Y = \{f \in X : d(\mathcal{D}f, A) < \infty\}$, applying the alternative fixed point result A is the only one mapping such that

$$\|f(x) - A(x)\| \leq H\lambda(x)$$

for all $x \in X$ and $H > 0$. At last (F_{iv}), we arrive

$$d(f, A) \leq \frac{1}{1 - L} d(f, \mathcal{D}f)$$

which implies

$$d(f, A) \leq \frac{L^{1-i}}{1 - L}.$$

Therefore we conclude that

$$\| f(x) - A(x) \| \leq \frac{L^{1-i}}{1 - L} \lambda(x).$$

for all $x \in X$.

Corollary 2 *Let $f : X \to Y$ be a mapping and there exist real numbers $\mathcal{H}$ and s such that the inequality (27) for all $x, y, z \in X$, then $\exists$ a only one additive function $A : X \to Y$ such that*

$$\|f(x) - A(x)\| \leqslant \begin{cases} \dfrac{\mathcal{H}}{\left(e^{\left(\frac{n\pi}{3}\right)} + 3\right)^2 T - 1|}, \\ \dfrac{3\,\mathcal{H}||x||^s}{\left(e^{\left(\frac{n\pi}{3}\right)} + 3\right)^2 T - T^s|}, \\ \dfrac{\mathcal{H}||x||^{3s}}{\left(e^{\left(\frac{n\pi}{3}\right)} + 3\right)^2 T - T^{3s}|}, \\ \dfrac{4\,\mathcal{H}||x||^{3s}}{\left(e^{\left(\frac{n\pi}{3}\right)} + 3\right)^2 T - T^{3s}|} \end{cases} \tag{35}$$

Proof Let us set

$$\Delta(x, y, z) = \begin{cases} \mathcal{H}, \\ \mathcal{H}\left\{||x||^s + ||y||^s + ||z||^s\right\}, \\ \mathcal{H}\ ||x||^s\ ||y||^s\ ||z||^s, \\ \mathcal{H}\left\{||x||^s||y||^s||z||^s + (||x||^{3s} + ||y||^{3s} + ||z||^{3s})\right\} \end{cases}$$

for all $x, y, z \in X$. Now

$$\begin{aligned} &\frac{\Delta(\mu_i^h x, \mu_i^h y, \mu_i^h z)}{\mu_i^h} \\ &= \begin{cases} \dfrac{\mathcal{H}}{\mu_i^h}, \\ \dfrac{\mathcal{H}}{\mu_i^h}\left\{||\mu_i^h x||^s + ||\mu_i^h y||^s + ||\mu_i^h z||^s\right\}, \\ \dfrac{\mathcal{H}}{\mu_i^h}||\mu_i^h x||^s\ ||\mu_i^h y||^s\ ||\mu_i^h z||^s, \\ \dfrac{\mathcal{H}}{\mu_i^h}\left\{||\mu_i^h x||^s\ ||\mu_i^h y||^s\ ||\mu_i^h z||^s\ \left\{||\mu_i^h x||^{3s} + ||\mu_i^h y||^{3s} + ||\mu_i^h z||^{3s}\right\}\right\} \end{cases} \\ &= \begin{cases} \to 0 \text{ as } h \to \infty, \\ \to 0 \text{ as } h \to \infty, \\ \to 0 \text{ as } h \to \infty, \\ \to 0 \text{ as } h \to \infty. \end{cases} \end{aligned}$$

Therefore, the inequality (30) holds. Then we get

$$\lambda(x) = \frac{1}{\left(e^{\left(\frac{n\pi}{3}\right)} + 3\right)^2}\left[\Delta\left(\frac{x}{T}, \frac{x}{T}, \frac{x}{T}\right)\right].$$

Hence

$$\lambda(x) = \frac{1}{\left(e^{\left(\frac{n\pi}{3}\right)} + 3\right)^2}\left[\Delta\left(\frac{x}{T}, \frac{x}{T}, \frac{x}{T}\right)\right] = \begin{cases} \dfrac{\mathcal{H}}{\left(e^{\left(\frac{n\pi}{3}\right)} + 3\right)^2 T}, \\ \dfrac{3\mathcal{H}}{\left(e^{\left(\frac{n\pi}{3}\right)} + 3\right)^2 T^s}||x||^s, \\ \dfrac{\mathcal{H}}{\left(e^{\left(\frac{n\pi}{3}\right)} + 3\right)^2 T^{3s}}||x||^{3s}, \\ \dfrac{4\mathcal{H}}{\left(e^{\left(\frac{n\pi}{3}\right)} + 3\right)^2 T^{3s}}||x||^{3s}. \end{cases}$$

Also,

$$\frac{1}{\mu_i}\lambda(\mu_i x) = \begin{cases} \dfrac{\mathcal{H}}{\mu_i \cdot \left(e^{\left(\frac{n\pi}{3}\right)} + 3\right)^2 T}, \\ \dfrac{3\mathcal{H}}{\mu_i \cdot \left(e^{\left(\frac{n\pi}{3}\right)} + 3\right)^2 T^s}||\mu_i x||^s, \\ \dfrac{\mathcal{H}}{\mu_i \cdot \left(e^{\left(\frac{n\pi}{3}\right)} + 3\right)^2 T^{3s}}||\mu_i x||^{3s}, \\ \dfrac{\mathcal{H}}{\mu_i \cdot \left(e^{\left(\frac{n\pi}{3}\right)} + 3\right)^2 T^{3s}}||\mu_i x||^{3s}. \end{cases}$$

$$= \begin{cases} \mu_i^{-1}\dfrac{\mathcal{H}}{\left(e^{\left(\frac{n\pi}{3}\right)} + 3\right)^2}, \\ \mu_i^{s-1}\dfrac{3\mathcal{H}}{\left(e^{\left(\frac{n\pi}{3}\right)} + 3\right)^2 T^s}||x||^s, \\ \mu_i^{3s-1}\dfrac{\mathcal{H}}{\left(e^{\left(\frac{n\pi}{3}\right)} + 3\right)^2 T^{3s}}||x||^{3s}, \\ \mu_i^{3s-1}\dfrac{4\mathcal{H}}{\left(e^{\left(\frac{n\pi}{3}\right)} + 3\right)^2 T^{3s}}||x||^{3s}. \end{cases}$$

$$= \begin{cases} \mu_i^{-1}\lambda(x), \\ \mu_i^{s-1}\lambda(x), \\ \mu_i^{3s-1}\lambda(x), \\ \mu_i^{3s-1}\lambda(x). \end{cases}$$

Now, from (32), we prove the following cases for conditions:
Case: 1 $L = T^{-1}$ if $i = 0$

$$\|f(x) - A(x)\| \le \frac{L^{1-i}}{1-L}\lambda(x) = \frac{\left(T^{-1}\right)^{1-0}}{1-(T)^{-1}} \cdot \frac{\mathcal{H}}{\left(e^{\left(\frac{n\pi}{3}\right)}+3\right)^2 T}$$
$$= \frac{\mathcal{H}}{\left(e^{\left(\frac{n\pi}{3}\right)}+3\right)^2 (T-1)}.$$

Case: 2 $L = T$ if $i = 1$

$$\|f(x) - A(x)\| \le \frac{L^{1-i}}{1-L}\lambda(x) = \frac{(T)^{1-1}}{1-T} \cdot \frac{\mathcal{H}}{\left(e^{\left(\frac{n\pi}{3}\right)}+3\right)^2 T}$$
$$= \frac{\mathcal{H}}{\left(e^{\left(\frac{n\pi}{3}\right)}+3\right)^2 (1-T)}.$$

Case: 1 $L = T^{s-1}$ if $i = 0$

$$\|f(x) - A(x)\| \le \frac{L^{1-i}}{1-L}\lambda(x) = \frac{\left(T^{s-1}\right)^{1-0}}{1-T^{s-1}} \frac{3\mathcal{H}}{\left(e^{\left(\frac{n\pi}{3}\right)}+3\right)^2 T^s}||x||^s$$
$$= \frac{T^s}{T-T^s} \frac{3\mathcal{H}}{\left(e^{\left(\frac{n\pi}{3}\right)}+3\right)^2 T^s}||x||^s$$
$$= \frac{3\mathcal{H}||x||^s}{\left(e^{\left(\frac{n\pi}{3}\right)}+3\right)^2 (T-T^s)}.$$

Case: 2 $L = \frac{1}{T^{s-1}}$ if $i = 1$

$$\|f(x)-A(x)\| \le \frac{L^{1-i}}{1-L}\lambda(x) = \frac{\left(\frac{1}{T^{s-1}}\right)^{1-1}}{1-\frac{1}{T^{s-1}}}\frac{3\mathcal{H}}{\left(e^{\left(\frac{n\pi}{3}\right)}+3\right)^2 T^s}||x||^s$$
$$= \frac{T^s}{T^s-T}\frac{3\mathcal{H}}{\left(e^{\left(\frac{n\pi}{3}\right)}+3\right)^2 T^s}||x||^s$$
$$= \frac{3\mathcal{H}||x||^s}{\left(e^{\left(\frac{n\pi}{3}\right)}+3\right)^2 (T^s-T)}.$$

Case: 1 $L = T^{3s-1}$ if $i = 0$

$$\|f(x)-A(x)\| \le \frac{L^{1-i}}{1-L}\lambda(x) = \frac{\left(T^{3s-1}\right)^{1-0}}{1-T^{3s-1}}\frac{\mathcal{H}}{\left(e^{\left(\frac{n\pi}{3}\right)}+3\right)^2 T^{3s}}||x||^{3s}$$
$$= \frac{T^{3s}}{T-T^{3s}}\frac{\mathcal{H}}{\left(e^{\left(\frac{n\pi}{3}\right)}+3\right)^2 T^{3s}}||x||^{3s}$$
$$= \frac{\mathcal{H}||x||^{3s}}{\left(e^{\left(\frac{n\pi}{3}\right)}+3\right)^2 \left(T-T^{3s}\right)}.$$

Case: 2 $L = \frac{1}{T^{3s-1}}$ if $i = 1$

$$\|f(x)-A(x)\| \le \frac{L^{1-i}}{1-L}\lambda(x) = \frac{\left(\frac{1}{T^{3s-1}}\right)^{1-1}}{1-\frac{1}{T^{3s-1}}}\frac{\mathcal{H}}{\left(e^{\left(\frac{n\pi}{3}\right)}+3\right)^2 T^{3s}}||x||^{3s}$$
$$= \frac{T^{3s}}{T^{3s}-T}\frac{\mathcal{H}}{\left(e^{\left(\frac{n\pi}{3}\right)}+3\right)^2 T^{3s}}||x||^{3s}$$
$$= \frac{\mathcal{H}||x||^{3s}}{\left(e^{\left(\frac{n\pi}{3}\right)}+3\right)^2 (T^{3s}-T)}.$$

Case: 1 $L = T^{3s-1}$ if $i = 0$

$$\|f(x) - A(x)\| \leq \frac{L^{1-i}}{1-L}\lambda(x) = \frac{\left(T^{3s-1}\right)^{1-0}}{1 - T^{3s-1}} \frac{4\mathcal{H}}{\left(e^{\left(\frac{n\pi}{3}\right)} + 3\right)^2 T^{3s}} ||x||^{3s}$$

$$= \frac{T^{3s}}{T - T^{3s}} \frac{4\mathcal{H}}{\left(e^{\left(\frac{n\pi}{3}\right)} + 3\right)^2 T^{3s}} ||x||^{3s}$$

$$= \frac{4\mathcal{H}||x||^{3s}}{\left(e^{\left(\frac{n\pi}{3}\right)} + 3\right)^2 \left(T - T^{3s}\right)}.$$

Case: 2 $L = \frac{1}{T^{3s-1}}$ if $i = 1$

$$\|f(x) - A(x)\| \leq \frac{L^{1-i}}{1-L}\lambda(x) = \frac{\left(\frac{1}{T^{3s-1}}\right)^{1-1}}{1 - \frac{1}{T^{3s-1}}} \frac{4\mathcal{H}}{\left(e^{\left(\frac{n\pi}{3}\right)} + 3\right)^2 T^{3s}} ||x||^{3s}$$

$$= \frac{T^{3s}}{T^{3s} - T} \frac{4\mathcal{H}}{\left(e^{\left(\frac{n\pi}{3}\right)} + 3\right)^2 T^{3s}} ||x||^{3s}$$

$$= \frac{4\mathcal{H}||x||^{3s}}{\left(e^{\left(\frac{n\pi}{3}\right)} + 3\right)^2 \left(T^{3s} - T\right)}.$$

References

1. Aczel, J., Dhombres, J.: Functional Equations in Several Variables. Cambridge Univ, Press (1989)
2. Aoki, T.: On the stability of the linear transformation in Banach spaces. J. Math. Soc. Japan, **2**, 64–66 (1950)
3. Agilan, P., Vallinayagam, V.: Generalized Ulam-Hyers stability of complex additive functional equation. J. Phys. Conf. Ser. **1377**, 012011 (2019)
4. Agilan, P., John, M., Rassias, Vallinayagam, V.: Stability of hyperbolic type functional equation in modular spaces. In: Proceedings of International Conference on Mathematical Computer Engineering. ICME (2020)
5. Arunkumar, M., Agilan, P.: Solution and Ulam-Hyers stability of a $n-$ dimensional additive functional equation in Banach space and Banach algebra: the direct and fixed point methods. Int. J. Pure Appl. Math. **120**(9), 93–104 (2018)
6. Czerwik, S.: Functional Equations and Inequalities in Several Variables. World Scientific, River Edge, NJ (2002)
7. Lee. D.O.: Hyers-Ulam stability of an addtiive type functional equation. J. Appl. Math. Comput. **13**(1-2), 471–477 (2003)
8. Găvrută, P.: A generalization of the Hyers-Ulam-Rassias stability of approximately additive mappings. J. Math. Anal. Appl. **184**, 431–436 (1994)
9. Găvrută, P.: An answer to a question of J.M.Rassias concerning the stability of cauchy functional equation. Adv. Equ. Inequal. Hadronic Math. Ser. 67–71 (1999)

10. Hyers, D.H.: On the stability of the linear functional equation. Proc. Nat. Acad. Sci. U.S.A. **27**, 222–224 (1941)
11. Margolis, B., Diaz, J.B.: A fixed point theorem of the alternative for contractions on a generalized complete metric space. Bull. Am. Math. Soc. **126**(74), 305–309 (1968)
12. Rassias, T.M.: On the stability of the linear mapping in Banach spaces. Proc. Am. Math. Soc. **72**, 297–300 (1978)
13. Rassias, J.M.: On approximately of approximately linear mappings by linear mappings. J. Funct. Anal. USA **46**, 126–130 (1982)
14. Rassias, J.M., Kim, H.M.: Generalized Hyers–Ulam stability for general additive functional equations in quasi-β-normed spaces. J. Math. Anal. Appl. **356**, 302–309 (2009)
15. Rassias, J.M., Jun, K.W., Kim, H.M.: Approximate $(m, n)-$ Cauchy-Jensen additive mappings in C*-algebras. Acta Mathematica Sinica, English Ser. **27**(10), 1907–1922 (2011)
16. Ulam, S.M.: Problems in Modern Mathematics. Science Editions, Wiley, NewYork (1964)

All Finite Topological Spaces are Weakly Reconstructible

A. Anat Jaslin Jini and S. Monikandan

Abstract The deck of a topological space X is the set $\mathscr{D}(X) = \{[X - \{x\}] : x \in X\}$, where $[Z]$ denotes the homeomorphism class of Z. A space X is topologically reconstructible if whenever $\mathscr{D}(X) = \mathscr{D}(Y)$ then X is homeomorphic to Y. A topological space X is said to be weakly reconstructible if it is reconstructible from its multi-deck. It is shown that all finite topological spaces are weakly reconstructible.

Keywords Reconstruction · Finite topological space · Homeomorphism

AMS Subject Classification (2010) Primary 05C60; Secondary 54A05

1 Introduction

A vertex-deleted subgraph or *card* $G - v$ of a graph G is obtained by deleting the vertex v and all edges incident with v. The collection of all cards of G is called the *deck* of G. A graph H is a *reconstruction* of G if H has the same deck as G. A graph is said to be *reconstructible* if it is isomorphic to all its reconstructions. A parameter p defined on graphs is reconstructible if, for any graph G, it takes the same value on every reconstruction of G. The graph reconstruction conjecture, posed by Kelly and Ulam [1] in 1941, asserts that every graph G on n ($\geq$3) vertices is reconstructible. More precisely, if G and H are finite graphs with at least three vertices such that $\mathscr{D}(H) = \mathscr{D}(G)$, then G and H are isomorphic.

In 2016, Pitz and Suabedissen [2] have introduced the concept of reconstruction in topological spaces as follows. For a topological space X, the subspace X_x is called a card of X. The set $\mathscr{D}(X) = \{[X_x] : x \in X\}$ of subspaces of X is called the deck of

A. Anat Jaslin Jini · S. Monikandan (✉)
Department of Mathematics, Manonmaniam Sundaranar University,
Tirunelveli 627012, Tamil Nadu, India
e-mail: monikandans@gmail.com

A. Anat Jaslin Jini
e-mail: jaslina.142@gmail.com

R. N. Mohapatra et al. (eds.), *Mathematical Analysis and Computing*,
Springer Proceedings in Mathematics & Statistics 344,
https://doi.org/10.1007/978-981-33-4646-8_9

X, where $[X_x]$ denotes the homeomorphism class of the card X_x. Given topological spaces X and Z, we say that Z is a reconstruction of X if their decks agree. A topological space X is said to be reconstructible if the only reconstructions of it are the spaces homeomorphic to X. Formally, a space X is reconstructible if $\mathscr{D}(X) = \mathscr{D}(Z)$ implies $X \cong Z$ and a property $\mathscr{P}$ of topological spaces is reconstructible if $\mathscr{D}(X) = \mathscr{D}(Z)$ implies "X has $\mathscr{P}$ if and only if Z has $\mathscr{P}$". The multi-deck of a topological space X is the multi-set $\mathscr{D}'(X) = \{X_x \ : \ x \in X\}$. In other words, the multi-deck not only knows which card occurs, but also how often they occur. If a space is reconstructible from its multi-deck, we will say that it is *weakly reconstructible*.

By *order of a set*, we mean the number of elements in the set. By *size of a space*, we mean the number of open sets in the space. Terms not defined here are taken as in [3]. Gartside et al. [2, 4, 5] have proved that the space of real numbers, the space of rational numbers, the space of irrational numbers, every compact Hausdorff space that has a card with a maximal finite compactification, and every Hausdorff continuum X with weight $\omega(X) < |X|$ are reconstructible. In their above paper, they also proved certain properties of a space, namely, all hereditary separation axioms and all cardinal invariants are reconstructible. All finite sequences are reconstructed by Manvel et al. [6]. In this paper, we show that all finite topological spaces of order $n(\geq 4)$ are weakly reconstructible. The condition $n \geq 4$ is needed because there are non-reconstructible topological spaces of size 2 or 3. For $n = 2$, the set $X = \{a, b\}$ endowed with any of the three topologies $\tau_1 = \{\emptyset, \{a\}, \{b\}, X\}$, $\tau_2 = \{\emptyset, \{a\}, X\}$ or $\tau_3 = \{\emptyset, X\}$ is not weakly reconstructible because all these topological spaces have the same multi-deck (subspaces) $\{(X_a, \tau_{X_a}), (X_b, \tau_{X_b})\}$, where $\tau_{X_a} = \{\emptyset, \{b\}\}$ and $\tau_{X_b} = \{\emptyset, \{a\}\}$. For $n = 3$, the set $X = \{a, b, c\}$ endowed with any of the two topologies $\tau_1 = \{\phi, \{c\}, X\}$, $\tau_2 = \{\phi, \{a, b\}, X\}$ is not reconstructible because all these topological spaces have the same multi-deck $\{(X_a, \tau_{X_a}), (X_b, \tau_{X_b}), (X_c, \tau_{X_c})\}$ where $\tau_{X_a} = \{\phi, \{c\}, \{b, c\}\}$, $\tau_{X_b} = \{\emptyset, \{c\}, \{a, c\}\}$ and $\tau_{X_c} = \{\phi, \{a, b\}\}$. Also, the set $X = \{a, b, c\}$ endowed with any of the two topologies $\tau_1 = \{\phi, \{c\}, \{a, c\}, \{b, c\}, X\}$, $\tau_2 = \{\phi, \{a\}, \{b\}, \{a, b\}, X\}$ is not reconstructible because all these topological spaces have the same multi-deck $\{(X_a, \tau_{X_a}), (X_b, \tau_{X_b}), (X_c, \tau_{X_c})\}$ where $\tau_{X_a} = \{\phi, \{c\}, \{b, c\}\}$, $\tau_{X_b} = \{\emptyset, \{c\}, \{a, c\}\}$ and $\tau_{X_c} = \{\phi, \{a\}, \{b\}, \{a, b\}\}$.

2 Finite Topological Spaces

Throughout the paper, we assume that X, where $X = \{x_1, x_2, \ldots, x_n\}$, is a finite topological space of size n, where $n \geq 4$. It is clear that any set X endowed with in-discrete topology is weakly reconstructible. The next theorem gives that every discrete topological space is weakly reconstructible. So, we can assume that the topology we consider after Theorem 1 is neither discrete nor in-discrete.

Theorem 1 *Every discrete topological space is weakly reconstructible.*

Proof Clearly, the size of the subspace topology endowed with each card is 2^{n-1}. Consider a subspace $(X_{x_i}, \tau_{X_{x_i}})$ of X for some i, $1 \leq i \leq n$. Now, the collection $\{U : U \in \tau_{X_{x_i}}\} \cup \{U \cup \{x\} : U \in \tau_{X_{x_i}}\}$ is clearly the discrete topology on X.

Lemma 1 *The property that whether a space X has an isolated point or not can be determined from the multi-deck of X.*

Proof If all the cards have no isolated point, then X has no isolated point since any isolated point x_r of X must occur precisely $n-1$ cards except X_{x_r} of X. So, we can assume that at least one of the cards contains an isolated point. We claim that if an isolated point x_r of a card, say X_{x_s} occurs as an isolated point in any other cards, then x_r is an isolated point of X. Suppose that the claim is not true and so x_r is not an isolated point of X. Then $\{x_r, x_s\}$ is open in X and hence x_r is an isolated point of exactly one card, that is, $\{x_r\}$ could occur in only one card, namely, X_{x_s}, giving a contradiction and proving the claim. Therefore, if an isolated point of any card satisfies the above claim, then X has an isolated point. If none of the isolated points of any cards satisfy the claim, then X has no isolated point.

If only one among all the isolated points of all cards occurs as an isolated point in any one of the remaining cards, then X has only one isolated point; if at least two among all the isolated points of all cards occur as isolated points in any of the remaining cards, then X has at least two isolated points. This proves the next lemma.

Lemma 2 *The property that whether a space X has one isolated point or more than one isolated points can be determined from the multi-deck of X.*

3 Reconstructing a Topological Space with an Isolated Point

A topological space X is said to have an *ascending chain* if all the open sets of X together form an ascending chain. By an m-open set, we mean an open set of order m. This section deals with a space with a unique isolated point, say x_1.

Lemma 3 *Let X be a finite topological space. Then X has an ascending chain if and only if every card has an ascending chain.*

Proof Necessity is obvious. For sufficiency, in each card, assume that all the open sets form an ascending chain. Suppose, to the contrary, that there exist two open sets A and B in X such that none of them is contained in the other. Then none of A and B is equal to X or ϕ. Also there exist two elements x_r and x_s such that $x_r \in A - B$ and $x_s \in B - A$. If $A \cup B \neq X$, then A and B are open sets in the card X_{x_t}, where $x_t \in X - (A \cup B)$, and hence all the open sets in the card X_{x_t} would not form an ascending chain, giving a contradiction. So, assume that $A \cup B = X$. Now, one of A and B, say A has at least two elements; let x_t be an element in A other than x_r.

Then, in the card X_{x_t}, the sets $A - \{x_t\}$ and $B - \{x_t\}$ are open and none of them contained in the other. Hence open sets in the card X_{x_t} would not form an ascending chain, again a contradiction to our assumption.

Lemma 4 *Let X be a finite topological space X of size n with a unique isolated point. Then, for $i = 1, 2, ..., n$, space X has only one open set of order i and X has an ascending chain if and only if each card has an ascending chain and any two cards are homeomorphic.*

Proof *Necessity*: Assume that X has only one open set of order i, $i = 1, 2, ..., n$ and it has an ascending chain. Then the open sets in X must be of the form $U_1 \subseteq U_2 \subseteq ... \subseteq U_n$, where $U_i = \{x_1, x_2, \ldots, x_i\}$. Now each card has only one open set of order j, where $j = 1, 2, ..., n - 1$. Now consider the cards X_{x_r} and X_{x_s}, where $1 \le r < s \le n$. Then $\tau_{X_{x_r}} = \{\phi, U_1, U_2, \ldots, U_{r-1} = U_r - \{x_r\}, \ldots, U_n - \{x_r\}\}$ and $\tau_{X_{x_s}} = \{\phi, U_1, U_2, \ldots, U_{s-1} = U_s - \{x_s\}, \ldots, U_n - \{x_s\}\}$. The mapping $h : X_{x_r} \to X_{x_s}$ defined by

$h(x_t) = \begin{cases} x_t & \text{if } t < r \text{ and } t > s \\ x_{t-1} & \text{if } r < t \le s \end{cases}$ is a desired homeomorphism from X_{x_r} onto X_{x_s}.

Also, by Lemma 3, each card has an ascending chain.

Sufficiency: Assume that any two cards are homeomorphic and each card has an ascending chain. By Lemma 3, X has an ascending chain. Clearly, X has an open set of order 1 and n. We shall now prove that X has an open set of order i for $2 \le i \le n - 1$. Suppose, to the contrary, that X has no open set of order i for some i, $2 \le i \le n - 1$ and i is the smallest among such integers. Let j be the smallest integer with $j > i$ such that X has an open set of order j and it has no open set of order $i, i + 1, ..., j - 1$. Now, we consider the two cards X_{x_r} and X_{x_1} where $x_r \in V_j - V_{i-1}$ and V_i represents the i-open set in X and x_1 is the isolated point of X. Then V_{i-1} is an open set of order $i - 1$ in the card X_{x_r}. On the other side, since X has an ascending chain, it follows that the element x_1 belongs to every open set in X. This together with our assumption that the space X has no open set of order i, where $i < j$ implies that the card X_{x_1} has no open set of order $i - 1$. Hence X_{x_1} is not homeomorphic with X_{x_r}, which is a contradiction to our assumption.

Lemma 5 *Let X be a finite topological space of size n with a unique isolated point. Then X has an ascending chain and X has no open set of order i for some i, $2 \le i \le n - 1$ if and only if each card has an ascending chain and at least two cards are non-homeomorphic.*

Proof If all the open sets in X form an ascending chain, then it is clear that X does not have more than one open set of the same order and hence the lemma follows by Lemma 4.

Lemma 6 *Let X be a finite topological space of size n with a unique isolated point. Then the isolated point belongs to all the open sets in X and X does not have an ascending chain if and only if at least one card does not have an ascending chain and in exactly $n - 1$ cards, the isolated point belongs to all the open sets.*

Proof *Necessity*: Assume that the isolated point belongs to all the open sets in X and the open sets in X does not have an ascending chain. Then the card X_{x_1} does not have an ascending chain. In every card X_{x_i}, where $i \neq 1$, the isolated point x_1 belongs to all the open sets.
Sufficiency: Assume that at least one card does not have an ascending chain and in exactly $n - 1$ cards, the isolated point belongs to all the open sets. Suppose, to the contrary, that the isolated point of X does not belong to an open set, say A in X. Since $|A| \geq 2$, the isolated point does not belong to all the open sets in the cards X_{x_r}, where $x_r \in A$. Then at least two cards in X, the isolated point does not belong to all the open sets, a contradiction. Also, suppose, to the contrary, that X has an ascending chain. Then, by Lemma 3, each card has an ascending chain, a contradiction.

Lemma 7 *Let X be a finite topological space of size n with a unique isolated point. Then the isolated point of X does not belong to an open set in X if and only if the isolated point does not belong to at least one open set in exactly* $n - 1$ *cards.*

Proof *Necessity*: Assume that the isolated point x_1 of X does not belong to an open set, say U in X. Since x_1 is the unique isolated point, we have $|U| \geq 2$. Then the cards X_{x_i}, where $i \neq 1$ have two disjoint open sets, namely, $\{x_1\}$ and $U \cap X_{x_i}$.
Sufficiency: Assume that the isolated point does not belong to at least one open set in exactly $n - 1$ cards. Suppose, to the contrary, that the isolated point belongs to all the open sets in X. Then, by above Lemma, the isolated point belongs to all the open sets in every card X_{x_i}, where $i \neq 1$, a contradiction.

Theorem 2 *A finite topological space with a unique isolated point is weakly reconstructible.*

Proof Let X be a topological space of size n with unique isolated point x_1. We proceed by two cases depending upon the open sets in the ascending chain form or not.
Case 1. Each card has an ascending chain.
Case 1.1. Any two cards are homeomorphic.
By Lemma 4, X has an ascending chain and for $i = 1, 2, \ldots, n$, the space X has only open set of order i and hence the collection $\{U \cup \{x_r\} \ : \ U \in \tau_{X_{x_r}}\}$ obtained from any card X_{x_r} is the desired topology on X.
Case 1.2. At least two cards are non-homeomorphic.
By Lemma 5, X has an ascending chain and X does not have an i-open set for some i, $2 \leq i \leq n - 1$. Choose a card X_{x_1}, that is, the card which does not have the isolated point x_1. Since X has an ascending chain, the collection $\{U \cup \{x_1\} \mid U \in \tau_{X_{x_1}}\}$ is the required topology on X.
Case 2. At least one card does not have an ascending chain.
Case 2.1. The isolated point belongs to all the open sets in exactly $n - 1$ cards.
By Lemma 6, the isolated point belongs to all the open sets in X and they does not form an ascending chain. Choose a card X_{x_1}, that is, the card which does not have the isolated point x_1. Since the isolated point belongs to all the open sets in X, the collection $\{U \cup \{x_1\} \mid U \in \tau_{X_{x_1}}\}$ is the required topology on X.

Case 2.2. The isolated point does not belong to at least one open set in exactly $n-1$ cards.
Now, by Lemma 7, the isolated point of X does not belong to at least one open set in X and so the card, obtained by deleting a point in the open set containing no isolated point, has the same size as X. Choose a card X_{x_r} with maximum size with the following condition; for an open set V in the card X_{x_r} such that $V \subset W$, where W is an open set in the card X_{x_1}, which does not have the isolated point x_1 and $W - V = \{x_r\}$. Therefore, for a non-empty open set U in X_{x_r}, U is open in X when $U \cap V = \phi$; $U \cup \{x_r\}$ is open in X otherwise (that is $U \cap V = V$). Thus, the collection $\{U \cup \{x_r\} \ : \ U \cap V = V \text{ and } \phi \neq U \in \tau_{X_{x_r}}, \} \cup \{U \ : \ U \cap V = \phi$ and $\phi \neq U \in \tau_{X_{x_r}}\}$ is the desired topology on X.

Theorem 3 *A finite topological space with more than one isolated point is weakly reconstructible.*

Proof We can assume that the topology τ_X is not discrete. Let X_{x_i} be a card with maximum number of open sets. Since X has at least two isolated points, every card has at least one isolated point; let x_1 be an isolated point of X_{x_i}. If $\{x_1\}$ is open in precisely $n-1$ cards, then it must be open in X. Otherwise, $\{x_1\}$ is not open in X and so the set consists of x_1 and the deleted point x_i is open in X. Repeat these steps for the rest of the 1-open sets in X_{x_i} to identify all the 1-open sets in X. Next, consider a 2-open set $\{x_1, x_2\}$ in X_{x_i}. If $\{x_1, x_2\}$ is open in precisely $n-2$ cards, then it must be open in X. Otherwise, $\{x_1, x_2\}$ is not open in X and so the set $\{x_1, x_2\}$ along with the deleted point x_i is open in X. Proceeding so on, in general, an m-open set $(3 \leq m \leq n-2)$ in X_{x_i} is open in precisely $n-m$ cards of X, then it must be open in X. Otherwise, the m-open set along with the deleted point x_i is open in X.

If no cards other than X_{x_i} contains open sets of order at most $n-2$ other than the open sets in the card X_{x_i}, then all the open sets so formed from the card X_{x_i} consists of all open sets in X of order at most $n-2$ and possibly all open sets in X of order $n-1$. Now we shall find the remaining open sets, if any, in X of order $n-1$. In order to find the missing open sets (if any) of order $n-1$ in X, we consider a card, say X_{x_i} such that the unique $(n-1)$-open set $X - x_i$ in the card is not in the collection so formed. Note that each $(n-1)$-open set in a card contains at least one isolated point of X.

Now, let $\mathscr{C}(X_{x_i}) = \{X - \{x_i, x_r\} : x_r \text{ is an isolated point of } X\}$. If an element of $\mathscr{C}(X_{x_i})$ does not belong to any card, then $X - \{x_i\}$ is not open in X, since the element itself is not open in the space X. So, assume that each element of $\mathscr{C}(X_{x_i})$ must be an open set of at least one card of X. If some of the elements of the set $\mathscr{C}(X_{x_i})$ are open in X, then the set $X - x_i$ is open in X. So, assume that no element of the set $\mathscr{C}(X_{x_i})$ is not open in X. Then each element of $\mathscr{C}(X_{x_i})$ along with the deleted point of the card, in which the element is open, is open in X and hence $X - \{x_i\}$ is not open in X. Now, all the open sets so formed is the desired topology on X.

Suppose that at least one card, say X_{x_k} contains open sets other than that in X_{x_i}. If an m-open set $(1 \leq m \leq n-2)$ in X_{x_k} is open in precisely $n-m$ cards of X, then it must be open in X. Otherwise, the m-open set along with the deleted point x_k is open in X. We are continuing like this until all the distinct open sets in all the cards are considered. Now, all the open sets so formed is the desired topology on X.

4 Reconstructing the Topological Space with No Isolated Point

By X, we mean here a finite topological space with no isolated point. Now, the order of the smallest open set of X, say l is known from the multi-deck of X as l equals to one more than the order of the smallest open set among all cards of X.

Lemma 8 *For $l \geq 2$, the property that whether a topological space X has one open set of order l or at least two open sets of order l, where l is the smallest integer can be determined from the multi-deck of X.*

Proof If precisely l cards have an open set of order $l-1$, then the space has only one open set of order l. For all other possibilities, the space has more than one open set of order l.

Lemma 9 *Let X be a finite topological space of size n with no isolated point; let l be the smallest integer such that X has a unique open set of order l. Then each card has an ascending chain and the difference between any two consecutive non-empty open sets is just an element if and only if X has only one open set of order i, for $i = l, l+1, ..., n$ and X has an ascending chain.*

Proof *Necessity*: Assume that the difference between any two consecutive non-empty open sets is just an element and each card has an ascending chain. By Lemma 3, X has an ascending chain and so there cannot be more than one open set of same order. Clearly, X has open sets of order l and n. We shall now prove that X has an open set of order i for $l+1 \leq i \leq n-1$. Suppose, to the contrary, that X has no open set of order i for some $i, l+1 \leq i \leq n-1$ and i is the smallest among such integers. Let j be the smallest integer with $j > i$ such that X has an open set of order j and it has no open set of order $i, i+1, ..., j-1$. Now, we consider the card X_{x_r} where $x_r \in V_l$ and V_l represents the l-open set in X. Then $\tau_{X_{x_r}} = \{\phi, V_l - \{x_r\}, ..., V_{i-1} - \{x_r\}, V_j - \{x_r\}, ..., V_n - \{x_r\}\}$. Note that the difference between the two consecutive open sets $V_{i-1} - \{x_r\}$ and $V_j - \{x_r\}$ is $(j-i)+1 > 1$, which is a contradiction to our assumption.

Sufficiency: Assume that X has only one open set of order i, for $i = l, l+1, ..., n$ and all these open sets form an ascending chain. Then the open sets in X must be of the form $U_l \subseteq U_{l+1} \subseteq ... \subseteq U_n$, where $U_j = \{x_1, x_2, ..., x_j\}$ for $j = l, l+1, \ldots, n$. Let the l-open set in X be V. Then each card X_{x_r}, where $x_r \notin V$, has only one open set of order j, where $j = l, l+1, ..., n-1$ and each card X_{x_s}, where $x_s \in V$, has only one open set of order j, where $j = l-1, l, ..., n-1$. Also, by Lemma 3, each card has an ascending chain.

Lemma 10 *Let X be a finite topological space of size n with no isolated point; let l be the smallest integer such that X has a unique open set of order l. Then each card has an ascending chain and in some card, the difference between two consecutive non-empty open sets is at least two if and only if X has an ascending chain and X has no open set of order i for some i, $l+1 \leq i \leq n-1$.*

Proof If all the open sets in X form an ascending chain, then it is clear that X does not have more than one open set of the same order and hence the lemma follows by Lemma 9.

Lemma 11 *Let X be a finite topological space of size n with no isolated point; Let l be the smallest integer such that X has a unique open set of order l. Then at least one card does not have an ascending chain and in each card, all the non-empty open sets have an open set in common if and only if the l-open set intersect all the non-empty open sets in X and X does not have an ascending chain.*

Proof *Necessity*: Assume that at least one card does not have an ascending chain and in each card, all the non-empty open sets have an open set in common. Suppose, to the contrary, that the l-open set, say U of X does not intersect at least one open set, say V in X. Since U is the unique open set of order l, we have $|V| \geq l+1$. Now, the cards obtained by deleting a point x_r, where $x_r \in V$ has at least two disjoint open sets, namely, U and $V - \{x_r\}$, which is a contradiction. Also, suppose, to the contrary, that X has an ascending chain. Then by Lemma 3, each card has an ascending chain, a contradiction.
Sufficiency: Assume that the l-open set, say U of X intersect all the non-empty open sets in X and X does not have an ascending chain. Then the open sets in the card X_{x_r}, where $x_r \in U$, does not form an ascending chain. And the cards obtained by deleting a point x_r, where $x_r \in U$ has the open set $U - \{x_r\}$ as a common open set and the cards obtained by deleting a point x_s, where $x_s \notin U$ has the open set U as a common open set.

Lemma 12 *Let X be a finite topological space of size n with no isolated point; Let l be the smallest integer such that X has a unique open set of order l. Then the l-open set of X does not intersect with at least one open set in X if and only if each card has at least two disjoint open sets.*

Proof *Necessity*: Assume that the l-open set, say U of X does not intersect with at least one open set, say V in X. Since U is the unique open set of order l, we have $|V| \geq l+1 \geq 3$. Now the cards obtained by deleting a point x_r, where $x_r \notin U \cup V$, has two disjoint open sets, namely, U and V and the cards obtained by deleting the point x_s, where $x_s \in U \cup V$ has two disjoint open sets, namely, U, $V - \{x_s\}$ or $U - \{x_s\}$, V.
Sufficiency: Assume that each card of X has at least two disjoint open sets. Suppose, to the contrary, that the l-open set intersects all the non-empty open sets in X. Then, by Lemma 11, the open sets in each card have an open set in common, a contradiction.

A topology τ on X such that the union of all elements of order l in τ equals X, then we say τ is an l-modulo topology on X.

Lemma 13 *Let X be a finite topological space with more than one l-open set for some $l \geq 2$. Then X is not an l-modulo topological space if and only if at least two cards are non-homeomorphic.*

Proof *Necessity:* Assume that X is not an l-modulo topological space. Then there exists a point, say $x_k \in X$ such that it belongs to an $(l+i)$-open set, where $i \geq 1$. Let A be an l-open set and let B be an $(l+i)$-open set. Then either $A \subseteq B$ or $A \cap B = \phi$. If $A \subseteq B$, then choose two points x_r and x_s, where $x_r \in A$ and $x_s \in B - A$. Now consider the two cards X_{x_r} and X_{x_s}. Since the point x_r belongs to an l-open set, the card X_{x_r} has an $(l-1)$-open set. Since the point $x_s \in B - A$ belongs to an $(l+i)$-open set, the card X_{x_s} has no $(l-1)$-open set. Hence X_{x_r} is not homeomorphic with X_{x_s}. If $A \cap B = \phi$, then choose two points x_r and x_s such that $x_r \in A$ and $x_s \in B$. Then, in the corresponding two cards X_{x_r}, X_{x_s}, the card X_{x_r} has an $(l-1)$-open set and the card X_{x_s} has no $(l-1)$-open set. Hence X_{x_r} is not homeomorphic with X_{x_s}.
Sufficiency: Assume that at least two cards are non-homeomorphic. Suppose, to the contrary, that X is an l-modulo topological space. Then the order of each open set in X is ml, where $m \geq 1$. Consider any two cards X_{x_r} and X_{x_s}. Then, clearly these two cards have exactly one $(l-1)$-open set. Define a mapping from X_{x_r} to X_{x_s} by the elements in the $(l-1)$-open set in the card X_{x_r} to the elements in the $(l-1)$-open set in the card X_{x_s} and the elements in an l-open set in the card X_{x_r} to the elements in an l-open set in the card X_{x_s}. It is clearly a homeomorphism from X_{x_r} onto X_{x_s}, which is a contradiction.

Corollary 1 *Let X be a finite topological space having more than one l-open set for some $l \geq 2$. If X is an l-modulo topological space if and only if any two cards are non-homeomorphic.*

Theorem 4 *A finite topological space with no isolated point is weakly reconstructible.*

Proof Let X be a topological space of size n with no isolated point. We proceed by two cases as below.
Case 1. The space X has only one l-open set.

Three subcases arise. First we consider the case that each card has an ascending chain. Then, by Lemmas 9 and 10, X has only one open set of order i, for $i = l, l+1, ..., n$ or X has no open set of order i for some i, $l+1 \leq i \leq n-1$ and all these open sets form an ascending chain. Consider a card X_{x_r} with unique $(l-1)$-open set. Now, the collection $\{U \cup \{x_r\} \mid U \in \tau_{X_{x_r}}, U \neq \phi\}$, is the desired topology on X.

Next, we consider the case that at least one card does not have an ascending chain and in each card, all the non-empty open sets have an open set in common. Now, by Lemma 11, the l-open set intersects all the non-empty open sets in X and X does not have an ascending chain. Consider a card X_{x_r} with unique $(l-1)-$ point open set. Now, the collection $\{U \cup \{x_r\} \mid U \in \tau_{X_{x_r}}, U \neq \phi\}$, is the desired topology on X.

Finally, we consider the case that each card has at least two disjoint open sets. Then, by Lemma 12, the l-open set of X does not intersect with at least one open set in

X. Consider a card X_{x_r} with unique $(l-1)$-open set. Then, the $(l-1)$-open set does not intersect with at least one open set in X_{x_r} as the l-open set of X does not intersect with at least one open set in X. Now, for all non-empty open sets in X_{x_r}, if $U \cap V = \phi$, then U is open in X; if $U \cap V = V$, then $U \cup \{x_r\}$ is open in X, where V denotes the $(l-1)$-open set in the card X_{x_r}. Thus, the collection $\{U \cup \{x_r\} : U \cap V = V$ and $\phi \neq U \in \tau_{X_{x_r}},\} \cup \{U : U \cap V = \phi$ and $\phi \neq U \in \tau_{X_{x_r}}\}$ is the desired topology on X.

Case 2. The space X has more than one l-open set.

We proceed by two subcases as follows. We first consider the case that any two cards are homeomorphic. Then, by Corollary 1, X is an l-modulo topological space. Consider any one of the card, say X_{x_r}. Now, for all non-empty open sets in X_{x_r}, if $U \cap V = \phi$, then U is open in X; if $U \cap V = V$, then $U \cup \{x_r\}$ is open in X, where V denotes the $(l-1)$-open set in the card X_{x_r}. Thus, the collection $\{U \cup \{x_r\} : U \cap V = V$ and $\phi \neq U \in \tau_{X_{x_r}},\} \cup \{U : U \cap V = \phi$ and $\phi \neq U \in \tau_{X_{x_r}}\}$ is the desired topology on X.

Now, we assume that at least two cards are non-homeomorphic. By Lemma 13, X is not an l-modulo topological space. Consider a card X_{x_r} with unique $(l-1)$-open set. Then, the $(l-1)$-open set does not intersect with at least one open set in X_{x_r} as the l-open set of X does not intersect with at least one open set in X. Now, for all non-empty open sets in X_{x_r}, if $U \cap V = \phi$, then U is open in X; if $U \cap V = V$, then $U \cup \{x_r\}$ is open in X where V denotes the $(l-1)$-open set in the card X_{x_r}. Thus, the collection $\{U \cup \{x_r\} : U \cap V = V$ and $\phi \neq U \in \tau_{X_{x_r}},\} \cup \{U : U \cap V = \phi$ and $\phi \neq U \in \tau_{X_{x_r}}\}$ is the desired topology on X.

References

1. Ulam, S.M.: A Collection of Mathematical Problems. Interscience Publishers (1960)
2. Pitz, M.F., Suabedissen, R.: A topological variation of the reconstruction conjecture. Glasg. Math. J. **59**(1), 1–15 (2016)
3. Engelking, R.: General Topology. Heldermann Verlag, Berlin (1989)
4. Gartside, P., Pitz, M.F., Suabedissen, R.: Reconstructing topological graphs and continua. Colloq. Math. **148**, 107–122 (2017)
5. Gartside, P., Pitz, M.F., Suabedissen, R.: Reconstructing compact metrizable spaces. Proc. Am. Math. Soc. **145**, 429–443 (2017)
6. Manvel, B., Meyerowitz, A., Schwenk, A., Smith, K., Stockmeyer, P.: Reconstruction of sequences. Discrete Math. **94**, 209–219 (1991)

Certain Properties of a Subclass of Uniformly Spirallike Functions

Geetha Balachandar

Abstract The concepts of neighbourhoods of analytic functions are used to prove several inclusion relations associated with the δ-neighbourhood of a subclass of uniformly spirallike functions.

Keywords Analytic functions · Univalent functions · Uniformly convex functions · Uniformly spirallike functions · δ-neighbourhood · Integral operator.

Mathematics Subject Classification (2010) 30C45

1 Introduction and Definitions

Let S denote the class of functions of the form $f(z) = z + \sum_{n=2}^{\infty} a_n z^n$, which are analytic and univalent in the open unit disc $U = \{z \in \mathbb{C} : |z| < 1\}$. Also let S^* and C denote the subclasses of S that are, respectively, starlike and convex. Motivated by certain geometric conditions, Goodman [3, 4] introduced an interesting subclass of starlike functions called uniformly starlike functions denoted by UST and an analogous subclass of convex functions called uniformly convex functions, denoted by UCV. From [8, 10], we have

$$f \in UCV \Leftrightarrow Re\left\{1 + \frac{zf''(z)}{f'(z)}\right\} \geq \left|\frac{zf''(z)}{f'(z)}\right|, z \in U.$$

In [10], Ronning introduced a new class S_p of starlike functions which has more manageable properties. The classes UCV and S_p were further extended by Kanas and Wisniowska in [5, 6] as $k - UCV(\alpha)$ and $k - ST(\alpha)$. The classes of uniformly spirallike and uniformly convex spirallike were introduced by Ravichandran et al.

G. Balachandar (✉)
Department of Mathematics, R.M.K College of Engineering and Technology,
Puduvoyal 601206, Tamil Nadu, India
e-mail: gbalachandar1989@gmail.com

R. N. Mohapatra et al. (eds.), *Mathematical Analysis and Computing*,
Springer Proceedings in Mathematics & Statistics 344,
https://doi.org/10.1007/978-981-33-4646-8_10

[9]. This was further generalized in [13] as $UCSP(\alpha, \beta)$. In [14], Herb Silverman introduced the subclass T of functions of the form

$$f(z) = z - \sum_{n=2}^{\infty} a_n z^n, \tag{1}$$

which are analytic and univalent in the unit disc U. Motivated by [15], new subclasses with negative coefficients $UCSPT(\alpha, \beta)$ and $SP_pT(\alpha, \beta)$ were introduced and studied in [12]. A function $f(z)$ defined by (1) is in $UCSPT(\alpha, \beta)$ if

$$Re\left\{e^{-i\alpha}\left(1 + \frac{zf''(z)}{f'(z)}\right)\right\} \geq \left|\frac{zf''(z)}{f'(z)}\right| + \beta, \tag{2}$$

$|\alpha| < \frac{\pi}{2}, 0 \leq \beta < 1$. For the class $UCSPT(\alpha, \beta)$, [12] proved the following lemma.

Lemma 1 *A function* $f(z) = z - \sum_{n=2}^{\infty} a_n z^n$ *is in* $UCSPT(\alpha, \beta)$ *if and only if*

$$\sum_{n=2}^{\infty} (2n - \cos\alpha - \beta) n a_n \leq \cos\alpha - \beta. \tag{3}$$

Using (1), the functions $f(z) \in UCSPT(\alpha, \beta)$ will satisfy

$$a_2 \leq \frac{(\cos\alpha - \beta)}{2(4 - \cos\alpha - \beta)}. \tag{4}$$

The subclass $UCSPT_c(\alpha, \beta)$ is the class of functions in $UCSPT(\alpha, \beta)$ of the form

$$f(z) = z - \frac{c(\cos\alpha - \beta)z^2}{2(4 - \cos\alpha - \beta)} - \sum_{n=3}^{\infty} a_n z^n, \tag{5}$$

$(a_n \geq 0)$, where $0 \leq c \leq 1$ was studied in [1]. Let $SP_pT_c(\alpha, \beta)$ be the subclass of functions in $SP_pT(\alpha, \beta)$ of the form

$$f(z) = z - \frac{c(\cos\alpha - \beta)z^2}{(4 - \cos\alpha - \beta)} - \sum_{n=3}^{\infty} a_n z^n,$$

$(a_n \geq 0)$, where $0 \leq c \leq 1$ was studied in [2]. When c = 1, we get

$$UCSPT_1(\alpha, \beta) = UCSPT(\alpha, \beta) \text{ and } SP_pT_1(\alpha, \beta) = SP_pT(\alpha, \beta).$$

Lemma 2 *The function* $f(z)$ *defined by (5) belongs to* $UCSPT_c(\alpha, \beta)$ *if and only if*

$$\sum_{n=3}^{\infty}(2n - \cos\alpha - \beta)na_n \leq (1-c)(\cos\alpha - \beta). \tag{6}$$

The result is sharp.

Lemma 3 *The function $f(z)$ defined by (5) belongs to $SP_pT_c(\alpha, \beta)$ if and only if*

$$\sum_{n=3}^{\infty}(2n - \cos\alpha - \beta)a_n \leq (1-c)(\cos\alpha - \beta).$$

The result is sharp.

Definition 1 If $f(z) = \sum_{n=0}^{\infty} a_n z^n$ and $g(z) = \sum_{n=0}^{\infty} b_n z^n$ are analytic in U, then the Hadamard product of f and g is $f * g = \sum_{n=0}^{\infty} a_n b_n z^n$. Following the works of Goodman and Ruscheweyh, we define the δ-neighbourhood of f as

$$N_\delta(f) = \left\{ g(z) : g(z) = z + \sum_{n=2}^{\infty} b_n z^n and \sum_{k=2}^{\infty} k\,|a_k - b_k| \leq \delta \right\}. \tag{7}$$

For the identity function $e(z) = z$, we have

$$N_\delta(e) = \left\{ g(z) : g(z) = z + \sum_{n=2}^{\infty} b_n z^n and \sum_{k=2}^{\infty} k\,|b_k| \leq \delta \right\}. \tag{8}$$

2 Inclusion Relations Involving the δ-Neighbourhood $N_\delta(e)$

Theorem 1 *$UCSPT_c(\alpha, \beta) \subset N_\delta(e)$, where*

$$\delta = \left\{ \frac{1-c}{6 - \cos\alpha - \beta} + \frac{c}{4 - \cos\alpha - \beta} \right\} (\cos\alpha - \beta).$$

Proof Using Lemma 2 for $f \in UCSPT_c(\alpha, \beta)$,

$$\sum_{n=3}^{\infty}(2n - \cos\alpha - \beta)na_n \leq (1-c)(\cos\alpha - \beta) \tag{9}$$

which implies

$$\sum_{n=3}^{\infty} na_n \leq \frac{(1-c)(\cos\alpha - \beta)}{6 - \cos\alpha - \beta} \tag{10}$$

or

$$\sum_{n=3}^{\infty} n\,|a_n| \leq \frac{(1-c)(\cos\alpha-\beta)}{6-\cos\alpha-\beta}. \tag{11}$$

Hence

$$2\,|a_2| + \sum_{n=3}^{\infty} n\,|a_n| \leq \frac{(1-c)(\cos\alpha-\beta)}{6-\cos\alpha-\beta} + \frac{c(\cos\alpha-\beta)}{4-\cos\alpha-\beta} \tag{12}$$

or

$$\sum_{n=2}^{\infty} n\,|a_n| \leq \frac{(1-c)(\cos\alpha-\beta)}{6-\cos\alpha-\beta} + \frac{c(\cos\alpha-\beta)}{4-\cos\alpha-\beta} = \delta. \tag{13}$$

$\Longrightarrow f \in N_\delta(e)$ or $UCSPT_c(\alpha, \beta) \subset N_\delta(e)$.

Theorem 2 *$f \in UCSPT_c(\alpha_1, \beta_1)$ and $g \in UCSPT_c(\alpha_2, \beta_2)$ then $f * g \in N_{\delta_1,\delta_2}(e)$, where*

$$\delta_1 = \left\{\frac{1-c}{6-\cos\alpha_1-\beta_1} + \frac{c}{4-\cos\alpha_1-\beta_1}\right\}(\cos\alpha_1-\beta_1)$$

and

$$\delta_2 = \left\{\frac{1-c}{6-\cos\alpha_2-\beta_2} + \frac{c}{4-\cos\alpha_2-\beta_2}\right\}(\cos\alpha_2-\beta_2),$$

$0 \leq \beta_1, \beta_2 < 1$.

Proof By Theorem 1, since $f \in UCSPT_c(\alpha_1, \beta_1)$ and $g \in UCSPT_c(\alpha_2, \beta_2)$

$$\sum_{n=2}^{\infty} n\,|a_n| \leq \frac{(1-c)(\cos\alpha_1-\beta_1)}{6-\cos\alpha_1-\beta_1} + \frac{c(\cos\alpha_1-\beta_1)}{4-\cos\alpha_1-\beta_1} = \delta_1 \tag{14}$$

and

$$\sum_{n=2}^{\infty} n\,|b_n| \leq \frac{(1-c)(\cos\alpha_2-\beta_2)}{6-\cos\alpha_2-\beta_2} + \frac{c(\cos\alpha_2-\beta_2)}{4-\cos\alpha_2-\beta_2} = \delta_2. \tag{15}$$

Hence

$$\sum_{n=2}^{\infty} n\,|a_n b_n| < \sum_{n=2}^{\infty} n\,|a_n| \sum_{n=2}^{\infty} n\,|b_n| < \delta_1\delta_2. \tag{16}$$

$\Rightarrow f * g \in N_{\delta_1,\delta_2}(e)$.

3 Some Properties of $UCSPT_c(\alpha, \beta)$ and $SP_pT_c(\alpha, \beta)$

We discuss some properties of $UCSPT_c(\alpha, \beta)$ and $SP_pT_c(\alpha, \beta)$ with the following integral operators:

$$H(z) = \int_0^z \frac{f(t)}{t} dt \text{ (Alexander operator)} \tag{17}$$

and

$$H_\alpha(z) = \int_0^z (\frac{f(t)}{t})^\alpha dt, f \in S. \tag{18}$$

Theorem 3 *Suppose* $f \in SP_pT_c(\alpha, \beta)$ *then* $H(z) \in UCSPT_c(\alpha, \beta)$.

Proof Given $f \in SP_pT_c(\alpha, \beta)$. Therefore $Re\left\{e^{-i\alpha}\left(\frac{zf'(z)}{f(z)}\right)\right\} \geq \left|\frac{zf'(z)}{f(z)} - 1\right| + \beta.$

$$H(z) = \int_0^z \frac{f(t)}{t} dt$$

gives

$$H'(z) = \frac{f(z)}{z}, H''(z) = \frac{zf'(z) - f(z)}{z^2}$$

and

$$\frac{zH''(z)}{H'(z)} = \frac{zf'(z)}{f(z)} - 1.$$

Hence

$$\begin{aligned} Re\left\{e^{-i\alpha}\left(1 + \frac{zH''(z)}{H'(z)}\right)\right\} &= Re\left\{e^{-i\alpha}\frac{zf'(z)}{f(z)}\right\} \\ &\geq \left|\frac{zf'(z)}{f(z)} - 1\right| + \beta. \\ \Rightarrow Re\left\{e^{-i\alpha}\left(1 + \frac{zH''(z)}{H'(z)}\right)\right\} &\geq \left|\frac{zH''(z)}{H(z)}\right| + \beta. \\ &\Rightarrow H(z) \in UCSPT_c(\alpha, \beta). \end{aligned}$$

Theorem 4 *Suppose* $f \in SP_pT_c(\alpha, \beta)$ *then* $H_\alpha(z) \in UCSPT_c(\alpha, \beta)$.

The method of proving Theorem 4 is similar to that of Theorem 3.

4 Integral Means Inequalities

In [14], Silverman found that the function $f_2(z) = z - \frac{z^2}{2}$ is often extremal over the family T. He applied this function to resolve his integral means inequality conjectured in [16] and settled in [17], that

$$\int_0^{2\pi} |f(re^{i\theta})|^\eta \, d\theta \leq \int_0^{2\pi} |f_2(re^{i\theta})|^\eta \, d\theta,$$

for all $f \in T$, $\eta > 0$ and $0 < r < 1$.

In [17], he also proved his conjecture for some subclasses of T. Now, we prove Silverman's conjecture for the class of functions $UCSPT_c(\alpha, \beta)$. An analogous result is also obtained for the family of functions $SP_pT_c(\alpha, \beta)$. We need the concept of subordination between analytic functions and a subordination theorem of Littlewood [7]. Two given functions f and g, which are analytic in U, the function f is said to be subordinate to g in U if there exists a function w analytic in U with $w(0) = 0$, $|w(z)| < 1 (z \in U)$, such that $f(z) = g(w(z)) (z \in U)$. We denote this subordination by $f(z) \prec g(z)$.

Lemma 4 *If the functions f and g are analytic in U with $f(z) \prec g(z)$ then for $\eta > 0$ and $z = re^{i\theta} (0 < r < 1)$*

$$\int_0^{2\pi} |g(re^{i\theta})|^\eta \, d\theta \leq \int_0^{2\pi} |f(re^{i\theta})|^\eta \, d\theta. \tag{19}$$

We now discuss the integral means inequalities for $UCSPT_c(\alpha, \beta)$ and $SP_pT_c(\alpha, \beta)$.

Theorem 5 *Let $f \in UCSPT_c(\alpha, \beta)$, $|\alpha| < \pi/2$, $0 \leq \beta < 1$, $0 \leq c \leq 1$ and $f_2(z)$ be defined by*

$$f_2(z) = z - \frac{c(\cos\alpha - \beta)z^2}{2(4 - \cos\alpha - \beta)}.$$

Then for $z = re^{i\theta}$, $0 < r < 1$, we have

$$\int_0^{2\pi} |f(z)|^\eta \, d\theta \leq \int_0^{2\pi} |f_2(z)|^\eta \, d\theta. \tag{20}$$

Proof When $f(z) = z - \sum_{n=2}^{\infty} a_n z^n$, (20) is equivalent to

$$\int_0^{2\pi} \left|1 - \sum_{n=2}^{\infty} a_n z^{n-1}\right|^\eta d\theta \leq \int_0^{2\pi} \left|1 - \frac{c(\cos\alpha - \beta)z}{2(4 - \cos\alpha - \beta)}\right|^\eta d\theta.$$

By Lemma 4, it is enough to prove

$$1-\sum_{n=2}^{\infty} a_n z^{n-1} \prec 1-\frac{c(\cos\alpha-\beta)z}{2(4-\cos\alpha-\beta)}.$$

Assume $1-\sum_{n=2}^{\infty} a_n z^{n-1} = 1-\frac{c(\cos\alpha-\beta)w(z)}{2(4-\cos\alpha-\beta)}$.

By using $\sum_{n=2}^{\infty}(2n-\cos\alpha-\beta)na_n \le \cos\alpha-\beta$ with $a_2 = \frac{c(\cos\alpha-\beta)}{2(4-\cos\alpha-\beta)}$ we get

$$\begin{aligned}|w(z)| &= \left|\sum_{n=2}^{\infty}\frac{2(4-\cos\alpha-\beta)a_n z^{n-1}}{(\cos\alpha-\beta)}\right| \\ &\le |z|\sum_{n=2}^{\infty}\frac{2(4-\cos\alpha-\beta)a_n}{(\cos\alpha-\beta)} \\ &\le |z|.\end{aligned}$$

This completes the proof by Lemma 4.

Theorem 6 *Let* $f \in SP_pT_c(\alpha,\beta)$, $|\alpha| < \pi\ 2, 0 \le \beta < 1, 0 \le c \le 1$ *and* $f_2(z)$ *be defined by*

$$f_2(z) = z - \frac{c(\cos\alpha-\beta)z^2}{(4-\cos\alpha-\beta)}.$$

Then for $z = re^{i\theta}, 0 < r < 1$, *we have*

$$\int_0^{2\pi} |f(z)|^{\eta}\, d\theta \le \int_0^{2\pi} |f_2(z)|^{\eta}\, d\theta.$$

Proof Proceeding as in the proof of Theorem 5, we get the required result.

Acknowledgements This research was supported by the National Board for Higher Mathematics.

References

1. Balachandar, G.: Fixed coefficients for a new subclass of uniformly spirallike functions. Acta Univ. Apulensis **40**, 233–243 (2014)
2. Balachandar, G., Hema, P.: Fixed coefficients for a subclass of spirallike functions. IJMTT, **58**(4), 298–305 (2018)
3. Goodman, A.W.: On uniformly convex functions. Ann. Polon. Math. **56**(1), 87–92 (1991)
4. Goodman, A.W.: On uniformly starlike functions. J. Math. Anal. Appl. **155**(2), 364–370 (1991)
5. Kanas, S., Wisniowska, A.: Conic regions and k-uniform convexity. J. Comput. Appl. Math. **105**(1–2), 327–336 (1999)
6. Kanas, S., Wisniowska, A.: Conic domains and starlike functions. Rev. Roumaine Math. Pures Appl. **45**(4), 647–657 (2001)

7. Littlewood, J.E.: On inequalities in the theory of functions. Proc. London Math. Soc. **23**(1), 481–519 (1925)
8. Ma, W.C., Minda, D.: Uniformly convex functions. Ann. Polon. Math. **57**(2), 165–175 (1992)
9. Ravichandran, V., Selvaraj, C., Rajagopal, R.: On uniformly convex spiral functions and uniformly spirallike functions. Soochow J. Math. **29**(4), 393–405 (2003)
10. Ronning, F.: Uniformly convex functions and a corresponding class of starlike functions. Proc. Amer. Math. Soc. **118**(1), 189–196 (1993)
11. Schild, A., Silverman, H.: Convolutions of univalent functions with negative coefficients. Ann. Univ. Mariae Curie-Skłodowska Sect. A **29**, 99–107 (1977)
12. Selvaraj, C., Geetha, R.: On subclasses of uniformly convex spirallike functions and corresponding class of spirallike functions. Int. J. Contemp. Math. Sci. **5**(37–40), 1845–1854 (2010)
13. Selvaraj, C., Geetha, R.: On uniformly spirallike functions and a corresponding subclass of spirallike functions. Glo. J. Sci. Front. Res. **10**, 36–41 (2010)
14. Silverman, H.: Univalent functions with negative coefficients. Proc. Amer. Math. Soc. **51**, 109–116 (1975)
15. Silverman, H., Silvia, E.M.: Fixed coefficients for subclasses of starlike functions. Houston J. Math. **7**(1), 129–136 (1981)
16. Silverman, H.: A survey with open problems on univalent functions whose coefficients are negative. Rocky Mt. J. Math. **21**(3), 1099–1125 (1991)
17. Silverman, H.: Integral means for univalent functions with negative coefficients. Houston J. Math. **23**(1), 169–174 (1997)

Analysis of a Dengue Model with Climate Factors

Ananya Dwivedi and Ram Keval

Abstract Climate fluctuations in the environment displays a positive effect on vector-borne diseases and prove favourable for the dispersion of dengue. In this paper, we introduced a climate factor in the model of dengue and analyzed their stability. Therefore, our discussion focuses on a dengue transmission model which includes mosquito population and different classes of human population along with the addition of climate factors. The human population is divided into two parts: (1) Susceptible human at high risk, (2) Susceptible human at low risk. Biologically, feasible equilibria and their stability properties have been discussed. On the basis of these results, we concluded that change in climate is one of the greatest reasons for dengue transmission.

Keywords Dengue model · Basic reproduction number (R_0) · Local stability · Global stability · Lyapunov function

1 Introduction

During the twentieth century, research and development in medical sciences helped in the eradication of infectious diseases. However, at the beginning of the twenty-first century, the spread of infectious diseases due to unpredictable factors became painful for developing countries. Among these diseases, Dengue is one of the major concerns for the present era in tropical regions. It is caused by infectious bites of Aedes aegypti and Aedes albopictus female mosquitoes from Flaviviridae family. DEN-1, DEN-2, DEN-3 and DEN-4 are the reason behind the prevalence of dengue. Around 100 countries have become endemic, projecting half of the world's population at risk with

A. Dwivedi · R. Keval (✉)
Department of Mathematics and Scientific Computing, Madan Mohan Malaviya University of Technology Gorakhpur, Gorakhpur, India
e-mail: ramkeval@gmail.com

A. Dwivedi
e-mail: ananyadwivedi24@gmail.com

R. N. Mohapatra et al. (eds.), *Mathematical Analysis and Computing*,
Springer Proceedings in Mathematics & Statistics 344,
https://doi.org/10.1007/978-981-33-4646-8_11

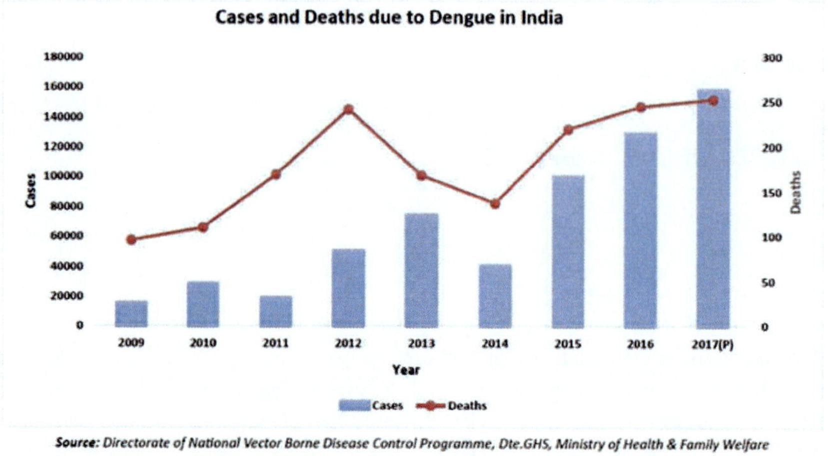

Fig. 1 Number of cases and deaths due to dengue appeared on National Health Profile

50–100 million infections occurring annually. Last few years, The National Vector Borne Disease Control Program-me (NVBDCP) and National Heath Profile 2018 [1], the increment in dengue was at extreme level. The dengue cases enhancement from 60,000 in 2009 to 75,808 in 2013, and subsequently to 188,401 in 2017, which shows the spike of dengue by 250% and 300%, respectively, as shown below in Fig. 1 [1].

Some authors have discussed the dengue transmission model including climate factors like rainfall, temperature humidity, etc. Nur et al. [2] discussed a SIR model with climate factors for transmission of dengue fever. Zhu et al. [3] studied an SIS-SI dengue model associated with a periodic environment to discuss temporal periodicity and spatial heterogeneity. Gao et al. [4] discussed a mixed vaccination strategy of a transmission model along with seasonal variations. Szu et al. [5] observed features of an epidemic and recognized that a major risk is involved when transmission of dengue occurs at 28 °C. Chavez et al. [6] discussed a control strategy with the help of the use of pesticides and also described variations in weather and their effect in every season. Feng et al. [7] analyzed the dengue epidemics model by varying the size of the vector and observed the strong influence of rainfall and other environmental factors.

From the above discussion, we conclude that seasonal effects are the greatest risk for the increment of dengue. Thus, we discussed the seasonal effect on the SIR model in climate change environments. In this paper, a dengue transmission model with various climatic factors is presented and the local and global stability analysis of the model with the help of Routh-Hurwitz criterion and Lyapunov function, respectively, has been discussed.

2 Mathematical Formulation and Its Description

Here, the total population of mosquito $N_m(t)$ is divided into two parts and human population $N_H(t)$ into four parts which are equal to

$$N_m(t) = S_m(t) + I_m(t)$$
$$N_H(t) = S_{H_1}(t) + S_{H_2}(t) + I_H(t) + R_H(t)$$

Based on transmission process of dengue, the following mathematical model has been proposed. Already, Tewa et al. [8] discussed that the infection rate per individual is given as $(\beta\alpha_H I_m)/(N_H + m)$ and infection rate per mosquito is given as $(\beta\alpha_m I_H)/(N_H + m)$. Where m is the number of alternative hosts available as blood sources and β is the biting rate of mosquito. The given model has been extended in the present work by adding climate factor γ in the model with the variable human population. Susceptible human is divided into two parts: (1) Susceptible human at high risk (S_{H_1}) means in this case humans are highly in contact with the disease (2) Susceptible human at low risk (S_{H_2}), i.e. in this area, humans are aware of the disease and can protect them by the use of control strategy. Recovered class does not exist for mosquitoes because they end their life-cycle with death. Therefore, by above assumptions, mathematical formulations for dengue model are

$$\frac{dS_m}{dt} = k - S_m I_H \left(\frac{\beta\,\alpha_m + \gamma}{N_H + m}\right) - \delta S_m \tag{1}$$

$$\frac{dI_m}{dt} = S_m I_H \left(\frac{\beta\,\alpha_m + \gamma}{N_H + m}\right) - I_m \tag{2}$$

$$\frac{dS_{H_1}}{dt} = r\,\pi_H - S_{H_1} I_m \left(\frac{\beta\alpha_H + \gamma}{N_H + m}\right) - \mu_H S_{H_1} \tag{3}$$

$$\frac{dS_{H_2}}{dt} = (1-r)\pi_H - \theta S_{H_2} I_m \left(\frac{\beta\alpha_H + \gamma}{N_H + m}\right) - \mu_H S_{H_2} \tag{4}$$

$$\frac{dI_H}{dt} = (S_{H_1} + \theta\, S_{H_2}) I_m \left(\frac{\beta\alpha_H + \gamma}{N_H + m}\right) - (q + \mu_H) I_H \tag{5}$$

$$\frac{dR_H}{dt} = q\, I_H - \mu_H R_H \tag{6}$$

The parameters and variables are non-negative positively invariant in the domain R^6_+.

Where

$S_m(0) > 0,\ I_m(0) > 0,\ S_{H_1}(0) > 0,\ S_{H_2}(0) > 0,\ I_H(0) > 0,\ R_H(0) > 0$

Table 1 Description of variables with respect to time (t)

Description	Variables
Total human population	$N_H(t)$
Susceptible human population at high risk	$S_{H_1}(t)$
Susceptible human population at low risk	$S_{H_2}(t)$
Infected human	$I_H(t)$
Recoverd human	$R_H(t)$
Total vector population	$N_m(t)$
Susceptible vector (mosquito)	$S_m(t)$
Infected vector (mosquito)	$I_m(t)$

The system (1–6) can also be written as

$$\frac{dS_m}{dt} = k - p_{11} S_m I_H - \delta S_m \tag{7}$$

$$\frac{dI_m}{dt} = p_{11} S_m I_H - \delta I_m \tag{8}$$

$$\frac{dS_{H_1}}{dt} = A - p_{12} S_{H_1} I_m - \mu_H S_{H_1} \tag{9}$$

$$\frac{dS_{H_2}}{dt} = B - p_{12} \theta S_{H_2} I_m - \mu_H S_{H_2} \tag{10}$$

$$\frac{dI_H}{dt} = p_{12}(S_{H_1} + \theta\, S_{H_2}) I_m - (q + \mu_H) I_H \tag{11}$$

$$\frac{dR_H}{dt} = q\, I_H - \mu_H R_H \tag{12}$$

where
$p_{11} = \left(\frac{\beta\alpha_m + \gamma}{N_H + m}\right)$, $p_{12} = \left(\frac{\beta\alpha_H + \gamma}{N_H + m}\right)$, $A = r\pi_H$, $B = (1 - r)\pi_H$.

3 Positivity and Boundedness of the System

In various steps, we have discussed the positivity and boundedness of the system here

Theorem *The existence region for all the solutions introducing oneself in the positive region is recommended by set Ω.*

$$\Omega = \{(S_{H_1}, S_{H_2}, N_m, N_H) : 0 \le N_m \le X_1,\ 0 \le S_{H_1} \le X_2,\ 0 \le S_{H_2} \le X_3,\ 0 \le N_H \le X_4\}$$

where

Table 2 Biological significance of parameters

Significance	Parameters	Dimension
Birth rate of vector population	k	Day^{-1}
Biting rate of vector per individual	β	Day^{-1}
Natural death rate of humans	μ_H	Day^{-1}
Death rate of vector population	δ	Day^{-1}
Mobilization rate from infected human to susceptible mosquito	α_H	Day^{-1}
Mobilization rate of suspectible human population	π_H	Day^{-1}
Intrinsic death rate of human	q	Day^{-1}
Section of first time admitted person joining the susceptible class at high risk	r	Day^{-1}
Transmission rate from infected mosquito (vector) to susceptible human	α_m	Day^{-1}
Low risk susceptible with relative chance of infection at high risk susceptible	θ	Day^{-1}

$$X_1 = max\left\{\frac{k}{\delta}, N_m(0)\right\}, X_2 = max\left\{\frac{r\pi_H}{\mu_H}, S_{H_1}(0)\right\},$$
$$X_3 = max\left\{\frac{(1-r)\pi_H}{\mu_H}, S_{H_2}(0)\right\}, X_4 = max\left\{\frac{\pi_H}{\mu_H}, N_H(0)\right\}$$

which is invariant and compact with respect to the system (1–6) [9].

Proof System (1–6) can be written as

$$\frac{dX}{dt} = CX + D \tag{13}$$

$$X = [S_m, I_m, S_{H_1}, S_{H_2}, I_H, R_H]^T \quad and$$

$$C = \begin{pmatrix} p_1 & 0 & 0 & 0 & 0 & 0 \\ p_2 & -\delta & 0 & 0 & 0 & 0 \\ 0 & 0 & p_3 & 0 & 0 & 0 \\ 0 & 0 & 0 & p_4 & 0 & 0 \\ 0 & p_5 & 0 & 0 & -(\mu_H + q) & 0 \\ 0 & 0 & 0 & 0 & q & -\mu_H \end{pmatrix}$$

where

$$p_1 = -\left[\delta + \left(\frac{(\beta\alpha_m + \gamma)I_H}{N_H + m}\right)\right], \quad p_2 = \left(\frac{(\beta\alpha_m + \gamma)I_H}{N_H + m}\right),$$
$$p_3 = -\left[\mu_H + \left(\frac{(\beta\alpha_H + \gamma)I_m}{N_H + m}\right)\right], p_4 = -\left[\mu_H + \left(\frac{(\beta\alpha_H + \gamma)\theta I_m}{N_H + m}\right)\right],$$
$$p_5 = (\beta\alpha_H + \gamma)\left(\frac{S_{H_1} + \theta S_{H_2}}{N_H + m}\right)$$

The vector $D = [k,\ 0,\ r\pi_H,\ (1-r)\pi_H,\ 0,\ 0]^T$ have positive nature. All the off-diagonal entries of $D(X)$ are non-negative. Therefore, the Metzler Matrix $D(X)$ is obtained $X \in R_+^6$ [10] which bounds in forever.

On adding the first two equations of the system (1–6) get

$$\frac{dN_m}{dt} = k - \delta N_m$$

Using a standard comparison theorem [11], we have,
$0 \le N_m(t) \le \frac{k}{\delta} + \left(N_m(0) - \frac{k}{\delta}\right)e^{-\frac{k}{\delta}}$. Thus, as $t \to \infty$, $0 \le N_m(t) \le \frac{k}{\delta}$, we have for any $t > 0$, $0 \le N_m \le X_1$, where $X_1 = max\left\{\frac{k}{\delta}, N_m(0)\right\}$.

Assume that $X_2 = max\left\{\frac{r\pi_H}{\mu_H}, S_{H_1}(0)\right\}$. Then $0 \le S_{H_1} \le X_2$. Similarly, let $X_3 = max\left\{\frac{(1-r)\pi_H}{\mu_H}, S_{H_2}(0)\right\}$. Then $0 \le S_{H_2} \le X_3$. By adding last four equations of the system (1–6), we get

$$\frac{dN_H}{dt} = \pi_H - \mu_H N_H$$

Assume that $X_4 = max\left\{\frac{\pi_H}{\mu_H}, N_H(0)\right\}$. Then $0 \le N_H \le X_4$.

Therefore all the feasible solution of the system (1–6) enter in the region Ω means this is attracting set in the region Ω.

4 Disease-Free Equilibrium and Its Stability

For system (7–12) the Disease-free equilibrium point is $W_0 = (\bar{S_m}, \bar{I_m}, \bar{S_{H_1}}, \bar{S_{H_2}}, \bar{I_H}, \bar{R_H})$, this equilibrium exist without any conditions [12].
Where

$$\bar{S_m} = \frac{k}{\delta}, \quad \bar{I_m} = 0, \quad \bar{S_{H_1}} = \frac{A}{\mu_H}, \quad \bar{S_{H_2}} = \frac{B}{\mu_H}, \quad \bar{I_H} = 0, \quad \bar{R_H} = 0.$$

4.1 Basic Reproduction Number

In this section using the next-generation matrix method [13, 14] found the basic reproduction number of the matrix. The process to find the basic reproduction number through Jacobian matrix F and V are given by [15]

$$F = \begin{pmatrix} \frac{\partial F_1}{\partial I_m} & \frac{\partial F_1}{\partial I_H} \\ \frac{\partial F_2}{\partial I_m} & \frac{\partial F_2}{\partial I_H} \end{pmatrix} \quad \text{and} \quad V = \begin{pmatrix} \frac{\partial V_1}{\partial I_m} & \frac{\partial V_1}{\partial I_H} \\ \frac{\partial V_2}{\partial I_m} & \frac{\partial V_2}{\partial I_H} \end{pmatrix}$$

$$F = \begin{pmatrix} 0 & \frac{\beta\alpha_m k}{\delta(N_H+m)} \\ \frac{\beta\alpha_H(A+\theta B)}{\mu_H(N_H+m)} & 0 \end{pmatrix} \quad \text{and} \quad V = \begin{pmatrix} \delta & 0 \\ 0 & \mu_H + q \end{pmatrix}$$

then $(F.V^{-1})$ are defined as

$$F.V^{-1} = \begin{pmatrix} 0 & \frac{\beta\alpha_m k}{\delta(N_H+m)(\mu_H+q)} \\ \frac{\beta\alpha_H(A+\theta B)}{\delta\mu_H(N_H+m)} & 0 \end{pmatrix}$$

where, $A = r\pi_H$, $B = (1-r)\pi_H$.

The basic reproduction number R_0 of the system (1–6) or (7–12) is defined by the maximum eigenvalues of the matrix $(F.V^{-1})$ [16] and it is given by

$$R_0 = \frac{\beta^2\alpha_m\alpha_H k(A+\theta B)}{\delta^2(N_H+m)^2\mu_H(\mu_H+q)} \quad OR \quad R_0 = \frac{p_{11}p_{12}k(A+\theta B)}{\delta^2\mu_H(\mu_H+q)}$$

4.1.1 Theorem

If $R_0 < 1$ the disease-free equilibrium W_0 is locally asymptotically stable and unstable if $R_0 > 1$.

Proof The Jacobian matrix for the model system (7–12) corresponding to equilibrium W_0 is given by

$$V_0 = \begin{pmatrix} -\delta & 0 & 0 & 0 & -\frac{p_{11}k}{\delta} & 0 \\ 0 & -\delta & 0 & 0 & \frac{p_{11}k}{\delta} & 0 \\ 0 & -\frac{p_{12}A}{\mu_H} & -\mu_H & 0 & 0 & 0 \\ 0 & -\frac{p_{12}\theta B}{\mu_H} & 0 & -\mu_H & 0 & 0 \\ 0 & \frac{p_{12}(A+\theta B)}{\mu_H} & 0 & 0 & -(\mu_H+q) & 0 \\ 0 & 0 & 0 & 0 & q & -\mu_H \end{pmatrix}$$

On applying the elementary row operation we get:

$$V_0 = \begin{pmatrix} -\delta & 0 & 0 & 0 & -\frac{p_{11}k}{\delta} & 0 \\ 0 & -\delta & 0 & 0 & \frac{p_{11}k}{\delta} & 0 \\ 0 & 0 & -\mu_H & 0 & -\frac{p_{11}p_{12}Ak}{\mu_H\delta^2} & 0 \\ 0 & 0 & 0 & -\mu_H & -\frac{p_{11}p_{12}\theta Bk}{\mu_H\delta^2} & 0 \\ 0 & 0 & 0 & 0 & -(\mu_H+q)+\frac{p_{11}p_{12}(A+\theta B)k}{\mu_H\delta^2} & 0 \\ 0 & 0 & 0 & 0 & 0 & -\mu_H \end{pmatrix}$$

which is upper triangular matrix, so the eigenvalues of Jacobian matrix V_0 are as:

$$\phi_1 = -\delta \text{ (multiplicity 2)}, \quad \phi_2 = -\mu_H \text{ (multiplicity 3)},$$

$$\phi_3 = -(\mu_H + q) + \frac{p_{11}p_{12}(A+\theta B)k}{\mu_H\delta^2}.$$

which is negative for all $\phi_i (i = 1, 2, 3) < 0$.

Then by, Routh-Hurwitz criterion [17] Jacobian matrix V_0 have a negative real part of all the eigenvalues are negative if $\phi_3 < 0$. Thus, the equilibrium V_0 is locally asymptotically stable if $\frac{p_{11}p_{12}(A+\theta B)k}{\mu_H\delta^2} < (\mu_H + q)$ and unstable if $\frac{p_{11}p_{12}(A+\theta B)k}{\mu_H\delta^2} > (\mu_H + q)$. This condition can also be written in the form of R_0.

4.1.2 Theorem

If $R_0 < 1$, the disease-free equilibrium $W_0 = (\bar{S}_m, \bar{I}_m, \bar{S}_{H_1}, \bar{S}_{H_2}, \bar{I}_H, \bar{R}_H) = (\frac{k}{\delta}, 0, \frac{A}{\mu_H}, \frac{B}{\mu_H}, 0, 0) \in \Omega$ is globally asymptotically stable in Ω with assumption.

$$p_{12}(\bar{S}_{H_1} + \theta\bar{S}_{H_2}) = \delta$$
$$p_{11}\bar{S}_m = \mu_H$$

Proof We consider the Lyapunov Function below

$$H(t) = (S_m - \bar{S}_m lnS_m) + I_m + (S_{H_1} - \bar{S}_{H_1} lnS_{H_1}) + (S_{H_2} - \bar{S}_{H_2} lnS_{H_2}) + I_H + R_H$$

Differentiating with respect to time t, we get

$$H'(t) = S'_m\left(1 - \frac{\bar{S}_m}{S_m}\right) + I'_m + S'_{H_1}(1 - \frac{\bar{S}_{H_1}}{S_{H_1}}) + S'_{H_2}(1 - \frac{\bar{S}_{H_2}}{S_{H_2}}) + I'_H + R'_H$$

$$H'(t) = (k - p_{11}S_m I_H - S_m)\left(1 - \frac{\bar{S}_m}{S_m}\right) + p_{11}S_m I_H - I_m$$

$$+(A - p_{12}S_{H_1}I_m - \mu_H S_{H_1})\left(1 - \frac{\bar{S}_{H_1}}{S_{H_1}}\right) + (B - p_{12}\theta S_{H_2}I_m - \mu_H S_{H_2})\left(1 - \frac{\bar{S}_{H_2}}{S_{H_2}}\right)$$
$$+p_{12}(S_{H_1} + \theta\, S_{H_2})I_m - (q + \mu_H)I_H + qI_H - \mu_H R_H$$

On solving further, we get

$$= k\left(1 - \frac{\bar{S}_m}{S_m}\right) + \bar{S}_m\left(1 - \frac{S_m}{\bar{S}_m}\right) + I_m(p_{12}(\bar{S}_{H_1} + \theta\bar{S}_{H_2}) -) + A\left(1 - \frac{\bar{S}_{H_1}}{S_{H_1}}\right)$$
$$+B\left(1 - \frac{\bar{S}_{H_2}}{S_{H_2}}\right) + \mu_H\bar{S}_{H_1}\left(1 - \frac{S_{H_1}}{\bar{S}_{H_1}}\right) + \mu_H\bar{S}_{H_2}\left(1 - \frac{S_{H_2}}{\bar{S}_{H_2}}\right) + I_H(p_{11}\bar{S}_m - \mu_H) - \mu_H R_H$$

Based on disease-free equillibrium, we have

$$W_0 = (\bar{S}_m, \bar{I}_m, \bar{S}_{H_1}, \bar{S}_{H_2}, \bar{I}_H, \bar{R}_H) = \left(\frac{k}{\delta}, 0, \frac{A}{\mu_H}, \frac{B}{\mu_H}, 0, 0\right)$$

and by assumption the above equation becomes

$$H'(t) = k\left(2 - \frac{\bar{S}_m}{S_m} - \frac{S_m}{\bar{S}_m}\right) + A\left(2 - \frac{\bar{S}_{H_1}}{S_{H_1}} - \frac{S_{H_1}}{\bar{S}_{H_1}}\right) + B\left(2 - \frac{\bar{S}_{H_2}}{S_{H_2}} - \frac{S_{H_2}}{\bar{S}_{H_2}}\right) - \mu_H R_H$$
$$= -k\frac{(S_m - \bar{S}_m)^2}{S_m\bar{S}_m} - A\frac{(S_{H_1} - \bar{S}_{H_1})^2}{S_{H_1}\bar{S}_{H_1}} - B\frac{(S_{H_2} - \bar{S}_{H_2})^2}{S_{H_2}\bar{S}_{H_2}} - \mu_H R_H$$

The result of $H'(t) \leq 0$ By using Lasalle's extension to Lyapunov's method [8] the limit set of each solution is obtained in the largest invariant set which $S_{H_1} = \bar{S}_{H_1}, S_{H_2} = \bar{S}_{H_2}, R_H = 0$ and $S_m = \bar{S}_m$ which is the singlten set W_0 and proves that disease-free equillibrium is Globally asymptotically stable.

5 Endemic Equilibrium and Its Stability

For the Eqs. (7–12), the Endemic-Equillibrium Point is $W_1 = (S_m^*, I_m^*, S_{H_1}^*, S_{H_2}^*, I_H^*, R_H^*)$, whose constituents are positive solutions of the equillibrium equation of the system (7–12).

$$S_m^* = \frac{(q + \mu_H)(k\, p_{12} + \mu_H)}{p_{12}\,(A\, p_{11} + q + B\, p_{11}\theta + \mu_H)} = C_1 \quad (14)$$

$$I_m^* = \frac{A\, k\, p_{11}\, p_{12} + B\, k\, p_{11}\, p_{12}\,\theta - \delta^2\, q\, \mu_H - \delta^2\, \mu_H^2}{\delta\, p_{12}(A\, p_{11} + \delta q + B\, p_{11}\,\theta + \delta\mu_H)} = C_2 \quad (15)$$

$$S_{H_1}^* = \frac{A}{p_{12}I_m^* + \mu_H} = C_3 \quad S_{H_2}^* = \frac{B}{p_{12}\theta I_m^* + \mu_H} = C_4 \quad (16)$$

$$I_H^* = \frac{p_{12}(S_{H_1}^* + \theta S_{H_2}^*)I_m^*}{\mu_H + q} = C_5, \quad R_H^* = \frac{qI_H^*}{\mu_H} = C_6 \quad (17)$$

the Total human population can be given by

$$N_H^* = \frac{A+B}{\mu_H} \tag{18}$$

5.1 Theorem

For $B_{22} < (\mu_H + q)$ the endemic equilibrium W_1 is locally asymptotically stable and unstable if $B_{22} > (\mu_H + q)$.
Where

$$B_{22} = \frac{p_{11}p_{12}S_m^*}{\delta}\left[(S_{H_1}^* + \theta S_{H_2}^*)B_{33} + \frac{p_{12}S_{H_1}^* I_m^*}{p_{12}I_m^* + \mu_H}B_{33} + \frac{p_{12}S_{H_2}^*\theta}{p_{12}\theta I_m^* + \mu_H}B_{33}\right]$$

$$B_{33} = \left(1 - \frac{p_{11}I_H^*}{p_{11}I_H^* + \delta}\right)$$

Proof The Jacobian matrix for the model system (7–12) corresponding to equilibrium W_1 is given by

$$V_1 = \begin{pmatrix} A_{11} & 0 & 0 & 0 & -p_{11}S_m^* & 0 \\ p_{11}I_H^* & -\delta & 0 & 0 & p_{11}S_m^* & 0 \\ 0 & -p_{12}S_{H_1} & -[p_{12}I_m + \mu_H] & 0 & 0 & 0 \\ 0 & -p_{12}\theta S_{H_2}^* & 0 & -[p_{12}\theta I_m^* + \mu_H] & 0 & 0 \\ 0 & A_{22} & p_{12}I_m^* & p_{12}I_m^*\theta & -(\mu_H + q) + B_{22} & 0 \\ 0 & 0 & 0 & 0 & q & -\mu_H \end{pmatrix}$$

where, $A_{11} = -\, p_{11}I_H^* - \delta$, $A_{22} = p_{12}(S_{H_1}^* + \theta S_{H_2}^*)$.

By applying elementry row operation the above system are: $B_{22} < (\mu_H + q)$ and $B_{22} > (\mu_H + q)$ then putting the values of S_m^*, I_m^*, $S_{H_1}^*$, $S_{H_2}^*$, I_H^* we get upper triangular matrix, and the eigenvalues of Jacobian matrix V_1 are negative. By, Routh-Hurwitz criterion [12] Jacobian matrix V_1 have negative real part or all the eigenvalues are negative. Thus, the equilibrium W_1 is locally asymptotically stable if $B_{22} < (\mu_H + q)$ and unstable if $B_{22} > (\mu_H + q)$.

5.2 Theorem

The endemic equillibrium is globally asymptotically stable in Ω provided the following inequalities hold:

$$Max\left\{\frac{(p_{11}X_4)^2}{3\delta}, \frac{p_{12}C_3^2}{2}, \frac{(p_{12}C_2)^2}{2\mu_H}, \frac{p_{12}\theta C_4^2}{2X_4}\right\} < \frac{\delta}{3} \tag{19}$$

$$Max\left\{p_{11}C_1, (p_{12}(C_3 + \theta C_4))^2\right\} < \frac{2}{3}\delta(\mu_H + q) \tag{20}$$

$$Max\left\{(p_{12}X_1)^2, q^2\right\} < \frac{2}{3}(\mu_H + q)\mu_H \tag{21}$$

$$Max\left\{p_{11}\right\} < \frac{2}{3}X_4(\mu_H + q) \tag{22}$$

Proof Let us Consider the following positive definite function of W_1 below

$$\begin{aligned} G(t) &= \frac{1}{2}(S_{H_1} - S_{H_2}^*)^2 + \frac{1}{2}(S_{H_2} - S_{H_2}^*)^2 + \frac{1}{2}(I_H - I_H^*)^2 + \frac{1}{2}(R_H - R_H^*)^2 \\ &+\frac{1}{2}(S_m - S_m^*)^2 + \frac{1}{2}(I_m - I_m^*)^2 \end{aligned}$$

Differentiating with respect to time $'t'$ along the solution

$$\begin{aligned} G'(t) &= S'_{H_1}(S_{H_1} - S_{H_1}^*) + S'_{H_2}(S_{H_2} - S_{H_2}^*) + I'_H(I_H - I_H^*) + R'_H(R_H - R_H^*) \\ &+S'_m(S_m - S_m^*) + I'_m(I_m - I_m^*) \end{aligned}$$

On rearranging the terms in the form of $-aX^2 + bXY - cY^2$. We get:

$$\begin{aligned} G'(t) &= -p_{11}X_4(S_m - S_m^*)^2 + p_{11}(I_H^* - I_H)(S_m - S_m^*) - \frac{\mu_H + q}{6}(I_H^* - I_H)^2 \\ &-\frac{\delta}{6}(S_m - S_m^*)^2 + p_{11}X_4(S_m - S_m^*)(I_m - I_m^*) - \frac{\delta}{6}(I_m - I_m^*)^2 \\ &-\frac{\delta}{6}(I_m - I_m^*)^2 + p_{11}C_1(I_H - I_H^*)(I_m - I_m^*) - \frac{(\mu_H + q)}{6}(I_H - I_H^*)^2 \\ &-p_{12}X_1(S_{H_1} - S_{H_1}^*)^2 + p_{12}C_3(S_{H_1} - S_{H_1}^*)(I_m - I_m^*) - \frac{\delta}{6}(I_m^* - I_m)^2 \\ &-\mu_H(S_{H_1} - S_{H_1}^*)^2 + p_{12}C_2(S_{H_1} - S_{H_1}^*)(I_H^* - I_H) - \frac{\mu_H + q}{6}(I_H^* - I_H)^2 \\ &-p_{12}\theta X_1(S_{H_2} - S_{H_2}^*)^2 + p_{12}\theta C_4(I_m^* - I_m)(S_{H_2} - S_{H_2}^*) - \frac{\delta}{6}(I_m^* - I_m)^2 \\ &-\mu_H(S_{H_2} - S_{H_2}^*)^2 + p_{12}\theta X_1(S_{H_2} - S_{H_2}^*)(I_H^* - I_H) - \frac{\mu_H + q}{6}(I_H^* - I_H)^2 \\ &-frac\mu_H + q6(I_H^* - I_H)^2 + p_{12}(C_3 + \theta C_4)(I_m - I_m^*)(I_H^* - I_H) - \frac{\delta}{6}(I_m - I_m^*)^2 \\ &-\frac{\mu_H + q}{6}(I_H - I_H^*)^2 + q(I_H - I_H^*)(R_H - R_H^*) - \mu_H(R_H - R_H^*)^2 \end{aligned}$$

Using region of attraction Ω, V(t) is Negative definite if the equation [9] $-aX^2 + bXY - cY^2$ provided $b^2 < 4ac$.

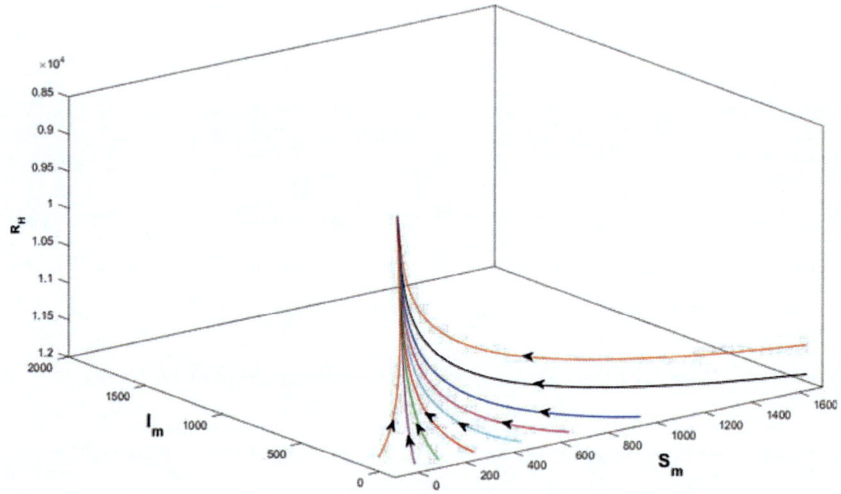

Fig. 2 Inside the region of attraction Ω global stability of endemic equilibrium in between $S_m - I_m - R_H$

Hence, from the above calculation, we made assertion that V(t) is Lyapunov function for the model system (7–12), provided the conditions (19–22) hold.

To verify the above theorem numerically, we choose the following set of hypothetical parameter values in the system (7–12) and the components of equillibrium W_1 are found to be (Table 1 and 2).

$k = 3.1, = 0.006$, $p_{11} = 0.101626$, $A = 0.0825$, $B = 0.0675$, $\mu_H = 0.00154$, $\theta = 0.85$, $p_{12} = 0.103659$, $q = 0.355$, $S_m = 67.5855$, $I_m = 449.081$, $S_{H_1} = 0.0017722$, $S_{H_2} = 0.00170585$, $I_H = 0.4207$, $R_H = 96.9795$. In Fig. 2 verifies the global stability of endemic equilibrium state.

6 Conclusion

In this article, we discussed a dengue transmission SIR model with the inclusion of climate as a factor along with a different class of human. For the discussion of local and global stability, we use the Routh-Hurwitz criterion and Lyapunov function which is a powerful technique for more than one-dimensional system. In general discussion, we concluded that if the condition is $R_0 \leq 1$, then the disease-free equilibrium is asymptotically stable, the infected population is at the recovered stage and if the condition is $R_0 > 1$, then there is a single endemic equilibrium which is globally asymptotically stable among all regions for which infection is present in human population and it will persist.

References

1. National Vector Borne Disease Control Programme: Down to earth. https://www.downtoearth.org.in/news/health
2. Nur, W., Rachman, H., Abdal, M.N.: SIR model analysis for transmission of dengue fever disease with climate Lyapunov function. J. Phys. **1028** (2018). https://doi.org/10.1088/1742-6596/1028/1/012117
3. Zhu, M., Xu, Y.: A time-periodic dengue fever model in a heterogeneous environment. Math. Comput. Simul. **155**, 115–129 (2019)
4. Gao, S., Liu, Y., Andrade, H., Nieto, J.J.: Seasonality and mixed vaccination strategy in epidemic model with vertical transmission. Math. Comput. Simul. **81**, 1855–1868 (2011)
5. Chen, S.-C., Hsieh Meng, H.: Modelling the transmission dynamics of dengue fever: implications of temperature effects. Sci. Total Environ. **431**, 385–391 (2012)
6. Chavez, J.P., Gotz, T., Siegmnud, S., Wijaya, K.P.: A SIR dengue transmission model with seasonal effects and impulsive control. Math. Biosci. **289**, 29–39 (2017)
7. Feng, Z., Vealsco-Hernandez, V.: Competitive exclusion in a vector-host model for dengue fever. J. Math. Biol. **35**, 523–544 (1997)
8. Tewa, J.J., Dimi, L.J., Bowong, S.: Lyapunov function for a dengue disease transmission model. Chaos Solitons Fractals **39**, 936–941 (2007)
9. Ghosh, I., Tiwari, P.K., Chattopadhyay, J.: Effect of active case finding on dengue control: implication from a mathematical model. J. Theor. Biol. **464**, 50–62 (2019)
10. Abate, A., Tiwari, A., Sastry, S.: Box invariance in biologically-inspired dynamical systems. Automatica **45**, 1601–1610 (2009)
11. Lakshmikantham, V., Leela, S., Martynyuk, A.A.: Stability Analysis of Nonlinear Systems. Springer (1989)
12. Esteva, L., Vargas, C.: Analysis of a dengue disease transnmission model. Math. Biol. **150**, 131–151 (1998)
13. Van den Driessch, P., Watmough, J.: Reproduction number and subthreshold endemic equilibria for compartmental models of disease transmission. Math. Biosci. **180**, 29–48 (2002)
14. Driessche, P.V.D.: Reproduction number of infection disease model. Inf. Dis. Model. **2**, 288–303 (2017)
15. Esteva, L., Vargas, C.: Influence of vertical and mechanical transmission on the dynamics of dengue disease. Math. Biol. **167**, 51–64 (2009)
16. Taghikhani, R., Gurmel, A.B.: Mathematics of dengue transmission dynamics: role of vector vertical transmission and temperature fluctuations. Commun. Nonlinear Sci. Num. Simul. **22**, 266–292 (2015)
17. Esteva, L., Vargas, C.: A model for dengue disease with variable human population. Math. Biol. **38**, 220–240 (1999)

A Theorem on Coupled Fixed Point and It's Application

A. Santhi and S. Muralisankar

Abstract In this paper, we define a new class of mappings and prove a common coupled fixed point theorem in the context of complex valued metric spaces; we give an application to show the significance of theory in solving the system of fractional differential equation with nonlocal multi-point integral boundary condition.

Keywords Complex valued metric space · Coupled fixed point · Fractional differential equation · Integral boundary condition

Mathematics Subject Classification (2010) 47H10 · 54H25

1 Introduction

Huang and Zhang [7] introduced the concept of cone metric spaces, by using a Banach space over $\mathbb{R}$, as the distance measure, instead of $\mathbb{R}$. Following him, Azam and Fisher [2] introduced the notion of complex valued metric spaces, which is a particular version of cone metric spaces. Sintunavarat et al. [12] and Rouzkard et al. [10] are some others who extended the results of Azam and Fisher, using various classes of control functions.

Bhaskar and Lakshmikantham [5] introduced the concept of coupled points and proved a fixed point theorem in the context of partially ordered metric spaces, using a mixed monotone property. As a sequel, Ciric and Lakshmikantham [3] posted some common coupled fixed points theorems in partially ordered metric spaces. Kutbi et al. [9] extended the theory in the context of complex valued metric spaces.

A. Santhi (✉) · S. Muralisankar
School of Mathematics, Madurai Kamaraj University, Madurai 625021, Tamilnadu, India
e-mail: santhi25.antony@gmail.com

S. Muralisankar
e-mail: muralimku@gmail.com

R. N. Mohapatra et al. (eds.), *Mathematical Analysis and Computing*,
Springer Proceedings in Mathematics & Statistics 344,
https://doi.org/10.1007/978-981-33-4646-8_12

The system of fractional differential equations plays a vital role in building biological models like SIR epidemic models. Withal, a system of fractional differential equations has its indispensable part, in constructing dynamic models, which are used in the treatment of cancer. Many research papers are posted in the turf of fractional differential equations involving fixed point theory (see [1, 4, 6, 8]).

In Sect. 2, give all the definitions and the results which we need in the sequel. In Sect. 3, we define a new class of mappings and prove a common coupled fixed point theorem in the context of complex valued metric spaces. In Sect. 4, we give an application to emphasize the significance of theory in solving the system of fractional differential equation with nonlocal multi-point integral boundary condition.

2 Preliminaries

Let us fix some notations here. $\mathbb{N}$, $\mathbb{R}$ and $\mathbb{C}$ denote the set of all natural numbers, real numbers and complex numbers respectively. Let $z_1, z_2 \in \mathbb{C}$ and let '$\precsim$' be a partial order on $\mathbb{C}$ defined as follows:

(i) $z_1 \precsim z_2$ if and only if $Re(z_1) \leq Re(z_2)$, $Im(z_1) \leq Im(z_2)$;
(ii) $z_1 \prec z_2$ if and only if $Re(z_1) < Re(z_2)$, $Im(z_1) < Im(z_2)$.

Definition 1 [2] Let X be a nonempty set. The function $d : X^2 \to \mathbb{C}$ is said to be a complex valued metric if for all $x, y, z \in X$,

(CM1) $0 \precsim d(x, y)$ and $d(x, y) = 0 \Leftrightarrow x = y$;
(CM2) $d(x, y) = d(y, x)$;
(CM3) $d(x, y) \precsim d(x, z) + d(z, y)$.

If d is a complex valued metric, then the pair (X, d) is called a complex valued metric space.

Definition 2 [2] Let (X, d) be a complex valued metric space. Then a sequence $\{x_n\}$ in X is said to be

(i) convergent to $x \in X$, if for every $0 \prec r \in \mathbb{C}$ there exists $N \in \mathbb{N}$ such that $d(x_n, x) \prec r$ for all $n \geq N$.
(ii) Cauchy, if for every $0 \prec r \in \mathbb{C}$ there exists $N \in \mathbb{N}$ such that $d(x_n, x_{n+m}) \prec r$ for all $n \geq N$ and $m \in \mathbb{N}$.

Lemma 1 *[2] Let (X, d) be a complex valued metric space and let $\{x_n\}$ be a sequence in X. Then*

(i) $\{x_n\}$ converges to x if and only if $|d(x_n, x)| \to 0$ as $n \to \infty$.
(ii) $\{x_n\}$ is Cauchy if and only if $|d(x_n, x_{n+m})| \to 0$ as $n \to \infty$.

Definition 3 [10] Let $\mathtt{F}, \mathtt{G}$ be mappings from X^2 to X and $\mathtt{f}$ be a self map on X. A point (x, y) is said to be a

(i) coupled coincidence point of $\mathtt{F}$, $\mathtt{G}$ and $\mathtt{f}$ if

$$\mathtt{F}(x, y) = \mathtt{G}(x, y) = \mathtt{f}(x) \text{ and } \mathtt{F}(y, x) = \mathtt{G}(y, x) = \mathtt{f}(y);$$

(ii) common coupled fixed point of $\mathtt{F}$, $\mathtt{G}$ and $\mathtt{f}$ if

$$\mathtt{F}(x, y) = \mathtt{G}(x, y) = \mathtt{f}(x) = x \text{ and } \mathtt{F}(y, x) = \mathtt{G}(y, x) = \mathtt{f}(y) = y.$$

3 Coupled Fixed Theorem

We start this section by defining a class of control functions as follows.

Definition 4 Let $\mathtt{F}, \mathtt{G} : X^2 \to X$ be 2-variable mappings and $\mathtt{f} : X \to X$ be a self map. Let Λ be a class of mappings $\lambda : X^4 \to [0, 1)$ satisfying the conditions:

(CO1) $\lambda(\mathtt{F}(x, y), \mathtt{F}(y, x), u, v) \leq \lambda(\mathtt{f}x, \mathtt{f}y, u, v)$;

(CO2) $\lambda(\mathtt{G}(x, y), \mathtt{G}(y, x), u, v) \leq \lambda(\mathtt{f}x, \mathtt{f}y, u, v)$;

(CO3) $\lambda(x, y, u, v) = \lambda(u, v, x, y) = \lambda(y, x, v, u)$.

Example 1 Let $X = [0, 1)$. Let $\mathtt{F}, \mathtt{G} : X^2 \to X$ be the mappings defined by

$$\mathtt{F}(x, y) = \frac{x + y}{2} \text{ for all } (x, y) \in X^2$$

and

$$\mathtt{G}(x, y) = \frac{\sin(x + y)}{2} \text{ for all } (x, y) \in X^2.$$

Let $\mathtt{f} : X \to X$ and $\lambda : X^4 \to [0, 1)$ be the mappings defined by

$$\mathtt{f}(x) = x \text{ and } \lambda(x, y, u, v) = \frac{x + y + u + v}{4}.$$

Then, we have

$$\begin{aligned}\lambda(\mathtt{F}(x, y), \mathtt{F}(y, x), u, v) &= \lambda\left(\frac{x + y}{2}, \frac{x + y}{2}, u, v\right)\\ &= \frac{x + y + u + v}{4}\\ &= \lambda(x, y, u, v)\\ &= \lambda(\mathtt{f}x, \mathtt{f}y, u, v)\end{aligned}$$

and

$$\begin{aligned}\lambda(\mathrm{G}(x,y),\mathrm{G}(y,x),u,v) &= \lambda\left(\frac{\sin(x+y)}{2},\frac{\sin(x+y)}{2},u,v\right)\\ &= \frac{\sin(x+y)+u+v}{4}\\ &\leq \frac{x+y+u+v}{4}\\ &= \lambda(x,y,u,v)\\ &= \lambda(\mathtt{f}x,\mathtt{f}y,u,v).\end{aligned}$$

Also by the construction of λ, it follows that

$$\lambda(x,y,u,v)=\lambda(u,v,x,y)=\lambda(y,x,v,u).$$

Thus, λ satisfies $(CO1)$, $(CO2)$ and $(CO3)$.

Theorem 1 *Let* $\mathrm{F},\mathrm{G}:X^2\to X$ *be two mappings. If there exist mappings* $\lambda_i\in\Lambda$, $i=1,2,\ldots,5$ *such that* $\sum\limits_{i=1}^{5}\lambda_i(x,y,u,v)<1$ *and*

$$\begin{aligned}d(\mathrm{F}(x,y),\mathrm{G}(u,v)) \precsim{}& \lambda_1(x,y,u,v)\left(\tfrac{d(x,u)+d(y,v)}{2}\right)\\ &+\lambda_2(x,y,u,v)\left(\tfrac{d(x,\mathrm{F}(x,y))d(u,\mathrm{G}(u,v))}{1+d(x,\mathrm{F}(x,y))+d(u,\mathrm{G}(u,v))}\right)\\ &+\lambda_3(x,y,u,v)\left(\tfrac{d(x,\mathrm{G}(u,v))d(u,\mathrm{F}(x,y))}{1+d(x,\mathrm{G}(u,v))+d(u,\mathrm{F}(x,y))}\right)\\ &+\lambda_4(x,y,u,v)\left(\tfrac{d(y,\mathrm{F}(y,x))d(v,\mathrm{G}(v,u))}{1+d(y,\mathrm{F}(y,x))+d(v,\mathrm{G}(v,u))}\right)\\ &+\lambda_5(x,y,u,v)\left(\tfrac{d(y,\mathrm{G}(v,u))d(v,\mathrm{F}(y,x))}{1+d(y,\mathrm{G}(v,u))+d(v,\mathrm{F}(y,x))}\right)\end{aligned}\qquad(1)$$

for all $x,y,u,v\in X$, *then* F *and* G *have a unique common coupled fixed point.*

We will deduce Theorem 1 from a more general version which involves three functions F, G, and f in a symmetric fashion.

Theorem 2 *Let* $\mathrm{F},\mathrm{G}:X^2\to X$ *be two mappings and let* $\mathtt{f}:X\to X$ *be an injective map such that* $\mathrm{F}(X^2)\subseteq \mathtt{f}(X)$, $\mathrm{G}(X^2)\subseteq \mathtt{f}(X)$, *and* $\overline{\mathtt{f}(X)}=\mathtt{f}(X)$. *If there exist mappings* $\lambda_i\in\Lambda$, $i=1,2,\ldots,5$ *such that* $\sum_{i=1}^{5}\lambda_i(x,y,u,v)<1$ *and*

$$d(\mathrm{F}(x,y),\mathrm{G}(u,v)) \precsim \lambda_1(fx,fy,fu,fv)\left(\frac{d(fx,fu)+d(fy,fv)}{2}\right)$$
$$+\lambda_2(fx,fy,fu,fv)\left(\frac{d(fx,\mathrm{F}(x,y))d(fu,\mathrm{G}(u,v))}{1+d(fx,\mathrm{F}(x,y))+d(fu,\mathrm{G}(u,v))}\right)$$
$$+\lambda_3(fx,fy,fu,fv)\left(\frac{d(fx,\mathrm{G}(u,v))d(fu,\mathrm{F}(x,y))}{1+d(fx,\mathrm{G}(u,v))+d(fu,\mathrm{F}(x,y))}\right)$$
$$+\lambda_4(fx,fy,fu,fv)\left(\frac{d(fy,\mathrm{F}(y,x))d(fv,\mathrm{G}(v,u))}{1+d(fy,\mathrm{F}(y,x))+d(fv,\mathrm{G}(v,u))}\right)$$
$$+\lambda_5(fx,fy,fu,fv)\left(\frac{d(fy,\mathrm{G}(v,u))d(fv,\mathrm{F}(y,x))}{1+d(fy,\mathrm{G}(v,u))+d(fv,\mathrm{F}(y,x))}\right)$$

for all $x, y, u, v \in X$, then F, G, *and* f *have a unique coupled coincidence point.*

Proof Let $(x_0, y_0) \in X^2$. Since the range of F and G are contained in the range of f, it is possible to construct two sequences $\{x_n\}$ and $\{y_n\}$ so that

$$fx_{2m+1} = \mathrm{F}(x_{2m}, y_{2m}) \text{ and } fy_{2m+1} = \mathrm{F}(y_{2m}, x_{2m})$$
$$fx_{2m+2} = \mathrm{G}(x_{2m+1}, y_{2m+1}) \text{ and } fy_{2m+2} = \mathrm{G}(y_{2m+1}, x_{2m+1}).$$

We wish to show that both $\{fx_n\}$ and $\{fy_n\}$ are Cauchy sequence. By the given contractive condition, we have

$$d(fx_{2m+1}, fx_{2m}) \precsim \lambda_1(fx_{2m}, fy_{2m}, fx_{2m-1}, fy_{2m-1})\left(\frac{d(fx_{2m}, fx_{2m-1})+d(fy_{2m}, fy_{2m-1})}{2}\right)$$
$$+\lambda_2(fx_{2m}, fy_{2m}, fx_{2m-1}, fy_{2m-1})\left(\frac{d(fx_{2m+1}, fx_{2m})d(fx_{2m-1}, fx_{2m})}{1+d(fx_{2m+1}, fx_{2m})+d(fx_{2m-1}, fx_{2m})}\right)$$
$$+\lambda_4(fx_{2m}, fy_{2m}, fx_{2m-1}, fy_{2m-1})\left(\frac{d(fy_{2m+1}, fy_{2m})d(fy_{2m-1}, fy_{2m})}{1+d(fy_{2m+1}, fy_{2m})+d(fy_{2m-1}, fy_{2m})}\right).$$

By taking absolute value on both sides of the above inequality, we get that

$$|d(fx_{2m+1}, fx_{2m})| \leq \lambda_1(fx_{2m}, fy_{2m}, fx_{2m-1}, fy_{2m-1})\left(\frac{|d(fx_{2m}, fx_{2m-1})| + |d(fy_{2m}, fy_{2m-1})|}{2}\right)$$
$$+\lambda_2(fx_{2m}, fy_{2m}, fx_{2m-1}, fy_{2m-1})|d(fx_{2m-1}, fx_{2m})|$$
$$+\lambda_4(fx_{2m}, fy_{2m}, fx_{2m-1}, fy_{2m-1})|d(fy_{2m-1}, fy_{2m})|.$$

Subsequently, by using $(CO2)$, we have

$$|d(\mathtt{f}x_{2m+1}, \mathtt{f}x_{2m})|$$
$$\leq \lambda_1(\mathtt{f}x_{2m-1}, \mathtt{f}y_{2m-1}, \mathtt{f}x_{2m-1}, \mathtt{f}y_{2m-1}) \left(\frac{|d(\mathtt{f}x_{2m}, \mathtt{f}x_{2m-1})| + |d(\mathtt{f}y_{2m}, \mathtt{f}y_{2m-1})|}{2}\right)$$
$$+\lambda_2(\mathtt{f}x_{2m-1}, \mathtt{f}y_{2m-1}, \mathtt{f}x_{2m-1}, \mathtt{f}y_{2m-1})|d(\mathtt{f}x_{2m-1}, \mathtt{f}x_{2m})|$$
$$+\lambda_4(\mathtt{f}x_{2m-1}, \mathtt{f}y_{2m-1}, \mathtt{f}x_{2m-1}, \mathtt{f}y_{2m-1})|d(\mathtt{f}y_{2m-1}, \mathtt{f}y_{2m})|.$$

By using $(CO1)$, $(CO2)$ and $(CO3)$ repeatedly as before, we get

$$|d(\mathtt{f}x_{2m+1}, \mathtt{f}x_{2m})| \leq \lambda_1(\mathtt{f}x_1, \mathtt{f}y_1, \mathtt{f}x_1, \mathtt{f}y_1) \left(\frac{|d(\mathtt{f}x_{2m}, \mathtt{f}x_{2m-1})| + |d(\mathtt{f}y_{2m}, \mathtt{f}y_{2m-1})|}{2}\right)$$
$$+\lambda_2(\mathtt{f}x_1, \mathtt{f}y_1, \mathtt{f}x_1, \mathtt{f}y_1)|d(\mathtt{f}x_{2m-1}, \mathtt{f}x_{2m})|$$
$$+\lambda_4(\mathtt{f}x_1, \mathtt{f}y_1, \mathtt{f}x_1, \mathtt{f}y_1)|d(\mathtt{f}y_{2m-1}, \mathtt{f}y_{2m})|.$$

Proceeding analogous to the above discussion, we get

$$|d(\mathtt{f}x_{2m}, \mathtt{f}x_{2m-1})| \leq \lambda_1(\mathtt{f}x_1, \mathtt{f}y_1, \mathtt{f}x_1, \mathtt{f}y_1) \left(\frac{|d(\mathtt{f}x_{2m-1}, \mathtt{f}x_{2m-2})| + |d(\mathtt{f}y_{2m-1}, \mathtt{f}y_{2m-2})|}{2}\right)$$
$$+\lambda_2(\mathtt{f}x_1, \mathtt{f}y_1, \mathtt{f}x_1, \mathtt{f}y_1)|d(\mathtt{f}x_{2m-1}, \mathtt{f}x_{2m-2})|$$
$$+\lambda_4(\mathtt{f}x_1, \mathtt{f}y_1, \mathtt{f}x_1, \mathtt{f}y_1)|d(\mathtt{f}y_{2m-1}, \mathtt{f}y_{2m-2})|.$$

Using the above two inequalities, we get that

$$|d(\mathtt{f}x_{n+1}, \mathtt{f}x_n)| \leq \lambda_1(\mathtt{f}x_1, \mathtt{f}y_1, \mathtt{f}x_1, \mathtt{f}y_1) \left(\frac{|d(\mathtt{f}x_n, \mathtt{f}x_{n-1})| + |d(\mathtt{f}y_n, \mathtt{f}y_{n-1})|}{2}\right)$$
$$+\lambda_2(\mathtt{f}x_1, \mathtt{f}y_1, \mathtt{f}x_1, \mathtt{f}y_1)|d(\mathtt{f}x_n, \mathtt{f}x_{n-1})|$$
$$+\lambda_4(\mathtt{f}x_1, \mathtt{f}y_1, \mathtt{f}x_2, \mathtt{f}y_1)|d(\mathtt{f}y_n, \mathtt{f}y_{n-1})|.$$

Similarly, we can prove that

$$|d(\mathtt{f}y_{n+1}, \mathtt{f}y_n)| \leq \lambda_1(\mathtt{f}x_1, \mathtt{f}y_1, \mathtt{f}x_1, \mathtt{f}y_1) \left(\frac{|d(\mathtt{f}x_n, \mathtt{f}x_{n-1})| + |d(\mathtt{f}y_n, \mathtt{f}y_{n-1})|}{2}\right)$$
$$+\lambda_2(\mathtt{f}x_1, \mathtt{f}y_1, \mathtt{f}x_1, \mathtt{f}y_1)|d(\mathtt{f}y_n, \mathtt{f}y_{n-1})|$$
$$+\lambda_4(\mathtt{f}x_1, \mathtt{f}y_1, \mathtt{f}x_1, \mathtt{f}y_1)|d(\mathtt{f}x_n, \mathtt{f}x_{n-1})|.$$

Now, by adding the above two inequalities, we get

$$|d(\mathtt{f}x_{n+1}, \mathtt{f}x_n)| + |d(\mathtt{f}y_{n+1}, \mathtt{f}y_n)| \le (\lambda_1 + \lambda_2 + \lambda_4)(\mathtt{f}x_1, \mathtt{f}y_1, \mathtt{f}x_1, \mathtt{f}y_1) \\ (|d(\mathtt{f}x_n, \mathtt{f}x_{n-1})| + |d(\mathtt{f}y_n, \mathtt{f}y_{n-1})|).$$

If we let $h = (\lambda_1 + \lambda_2 + \lambda_4)(\mathtt{f}x_1, \mathtt{f}y_1, \mathtt{f}x_1, \mathtt{f}y_1)$, then it can be seen clearly that $h < 1$. In addition, we have

$$\begin{aligned} &|d(\mathtt{f}x_{n+1}, \mathtt{f}x_n)| + |d(\mathtt{f}y_{n+1}, \mathtt{f}y_n)| \\ &\le h\,(|d(\mathtt{f}x_n, \mathtt{f}x_{n-1})| + |d(\mathtt{f}y_n, \mathtt{f}y_{n-1})|) \\ &\le h^2\,(|d(\mathtt{f}x_{n-1}, \mathtt{f}x_{n-2})| + |d(\mathtt{f}y_{n-1}, \mathtt{f}y_{n-2})|) \\ &\vdots \\ &\le h^{n-1}\,(|d(\mathtt{f}x_2, \mathtt{f}x_1)| + |d(\mathtt{f}y_2, \mathtt{f}y_1)|)\,. \end{aligned}$$

Let $n < k$, then

$$\begin{aligned} &|d(\mathtt{f}x_n, \mathtt{f}x_k)| + |d(\mathtt{f}y_n, \mathtt{f}y_k)| \\ &\le (|d(\mathtt{f}x_n, \mathtt{f}x_{n-1})| + |d(\mathtt{f}y_n, \mathtt{f}y_{n-1})|) \\ &\quad + (|d(\mathtt{f}x_{n-1}, \mathtt{f}x_{n-2})| + |d(\mathtt{f}y_{n-1}, \mathtt{f}y_{n-2})|) \\ &\quad + \cdots + (|d(\mathtt{f}x_{k-1}, \mathtt{f}x_k)| + |d(\mathtt{f}y_{k-1}, \mathtt{f}y_k)|) \\ &\le (h^{n-2} + h^{n-3} + \cdots + h^{k-3})\,(|d(\mathtt{f}x_2, \mathtt{f}x_1)| + |d(\mathtt{f}y_2, \mathtt{f}y_1)|) \\ &\le \frac{h^{n-2}}{1-h}\,(|d(\mathtt{f}x_2, \mathtt{f}x_1)| + |d(\mathtt{f}y_2, \mathtt{f}y_1)|)\,. \end{aligned}$$

By letting limit $n \to \infty$ in the above inequality, we have

$$|d(\mathtt{f}x_n, \mathtt{f}x_k)| + |d(\mathtt{f}y_n, \mathtt{f}y_k)| \to 0.$$

Since $h < 1$, it follows that

$$\lim_{n\to\infty} |d(\mathtt{f}x_n, \mathtt{f}x_k)| = 0 \text{ and } \lim_{n\to\infty} |d(\mathtt{f}y_n, \mathtt{f}y_k)| = 0,$$

which in turn implies that both $\{\mathtt{f}x_n\}$ and $\{\mathtt{f}y_n\}$ are Cauchy sequences. Thus, there exist some $r, s \in X$ such that

$$\lim_{n\to\infty} \mathtt{f}x_n = r \text{ and } \lim_{n\to\infty} \mathtt{f}y_n = s.$$

But since $\mathtt{f}(X)$ is closed, there exist some $x, y \in X$ such that

$$\mathtt{f}(x) = r \text{ and } \mathtt{f}(y) = s.$$

Thus, it follows that

$$\lim_{m\to\infty} \mathtt{f}x_{2m} = \lim_{m\to\infty} \mathtt{G}(x_{2m-1}, y_{2m-1}) = \mathtt{f}(x)$$

$$\lim_{m\to\infty} \mathtt{f}y_{2m} = \lim_{m\to\infty} \mathtt{G}(y_{2m-1}, x_{2m-1}) = \mathtt{f}(y)$$

and

$$\lim_{m\to\infty} \mathtt{f}x_{2m-1} = \lim_{m\to\infty} \mathtt{F}(x_{2m-2}, x_{2m-2}) = \mathtt{f}(x)$$

$$\lim_{m\to\infty} \mathtt{f}y_{2m-1} = \lim_{m\to\infty} \mathtt{F}(y_{2m-2}, x_{2m-2}) = \mathtt{f}(y).$$

We wish to show that (x, y) is a coupled coincidence point of $\mathtt{F}$, $\mathtt{G}$, and $\mathtt{f}$. For consider,

$d(\mathtt{F}(x, y), \mathtt{G}(x_{2m-1}, y_{2m-1}))$

$$\begin{aligned}
&\precsim d(\mathtt{F}(x, y), \mathtt{f}x_{2m})\\
&\precsim \frac{\lambda_1(z)(d(\mathtt{f}x, \mathtt{f}x_{2m-1}) + d(\mathtt{f}y, \mathtt{f}y_{2m-1}))}{2}\\
&\quad + \lambda_2(z)\left(\frac{d(\mathtt{f}x, \mathtt{F}(x, y))d(\mathtt{f}x_{2m-1}, \mathtt{f}x_{2m})}{1 + d(\mathtt{f}x, \mathtt{F}(x, y)) + d(\mathtt{f}x_{2m-1}, \mathtt{f}x_{2m})}\right)\\
&\quad + \lambda_3(z)\left(\frac{d(\mathtt{f}x, \mathtt{f}x_{2m})d(\mathtt{f}x_{2m-1}, \mathtt{F}(x, y))}{1 + d(\mathtt{f}x, \mathtt{f}x_{2m}) + d(\mathtt{f}x_{2m-1}, \mathtt{F}(x, y))}\right)\\
&\quad + \lambda_4(z)\left(\frac{d(\mathtt{f}y, \mathtt{F}(y, x))d(\mathtt{f}y_{2m-1}, \mathtt{f}y_{2m})}{1 + d(\mathtt{f}y, \mathtt{F}(y, x)) + d(\mathtt{f}y_{2m-1}, \mathtt{f}y_{2m})}\right)\\
&\quad + \lambda_5(z)\left(\frac{d(\mathtt{f}y, \mathtt{f}y_{2m})d(\mathtt{f}y_{2m-1}, \mathtt{F}(y, x))}{1 + d(\mathtt{f}y, \mathtt{f}y_{2m}) + d(\mathtt{f}y_{2m-1}, \mathtt{F}(y, x))}\right),
\end{aligned}$$

where $z = (\mathtt{f}x, \mathtt{f}y, \mathtt{f}x_{2m-1}, \mathtt{f}y_{2m-1})$ and therefore

$$\lim_{m\to\infty} d(\mathtt{F}(x, y), \mathtt{f}x) = 0,$$

which implies that, $\mathtt{F}(x, y) = \mathtt{f}x$. Similarly, we can prove that

$$\mathtt{F}(y, x) = \mathtt{f}y,\ \mathtt{G}(x, y) = \mathtt{f}x \text{ and } \mathtt{G}(y, x) = \mathtt{f}y.$$

Thus, (x, y) is the required coupled coincidence point of $\mathtt{F}$, $\mathtt{G}$, and $\mathtt{f}$. To prove the uniqueness, suppose (u, v) is an another coupled coincidence point of $\mathtt{F}$, $\mathtt{G}$, and $\mathtt{f}$. That is, if

$$\mathtt{F}(u, v) = \mathtt{G}(u, v) = \mathtt{f}u$$

and

$$\mathtt{F}(v, u) = \mathtt{G}(v, u) = \mathtt{f}v.$$

Then, by the given contractive condition, we have

$$\begin{aligned} d(\mathtt{f}x, \mathtt{f}u) &\precsim \lambda_1(\mathtt{f}x, \mathtt{f}y, \mathtt{f}u, \mathtt{f}v)\left(\tfrac{d(\mathtt{f}x,\mathtt{f}u)+d(\mathtt{f}y,\mathtt{f}v)}{2}\right) \\ &+\lambda_3(\mathtt{f}x, \mathtt{f}y, \mathtt{f}u, \mathtt{f}v)\left(\tfrac{d(\mathtt{f}x,\mathtt{f}u)d(\mathtt{f}u,\mathtt{f}x)}{1+d(\mathtt{f}x,\mathtt{f}u)+d(\mathtt{f}u,\mathtt{f}x)}\right) \\ &+\lambda_5(\mathtt{f}x, \mathtt{f}y, \mathtt{f}u, \mathtt{f}v)\left(\tfrac{d(\mathtt{f}y,\mathtt{f}v)d(\mathtt{f}v,\mathtt{f}y)}{1+d(\mathtt{f}y,\mathtt{f}v)+d(\mathtt{f}v,\mathtt{f}y)}\right). \end{aligned}$$

Taking modulus on both on sides,

$$\begin{aligned} |d(\mathtt{f}x, \mathtt{f}u)| \leq \lambda_1(\mathtt{f}x, \mathtt{f}y, \mathtt{f}u, \mathtt{f}v) &\left(\tfrac{|d(\mathtt{f}x,\mathtt{f}u)|+|d(\mathtt{f}y,\mathtt{f}v)|}{2}\right) \\ &+\lambda_3(\mathtt{f}x, \mathtt{f}y, \mathtt{f}u, \mathtt{f}v)|d(\mathtt{f}x, \mathtt{f}u)| \\ &+\lambda_5(\mathtt{f}x, \mathtt{f}y, \mathtt{f}u, \mathtt{f}v)|d(\mathtt{f}y, \mathtt{f}v)|. \end{aligned}$$

Similarly, we can prove that

$$\begin{aligned} |d(\mathtt{f}y, \mathtt{f}v)| \leq \lambda_1(\mathtt{f}x, \mathtt{f}y, \mathtt{f}u, \mathtt{f}v) &\left(\frac{|d(\mathtt{f}x, \mathtt{f}u)| + |d(\mathtt{f}y, \mathtt{f}v)|}{2}\right) \\ &+\lambda_3(\mathtt{f}x, \mathtt{f}y, \mathtt{f}u, \mathtt{f}v)|d(\mathtt{f}y, \mathtt{f}v)| \\ &+\lambda_5(\mathtt{f}x, \mathtt{f}y, \mathtt{f}u, \mathtt{f}v)|d(\mathtt{f}x, \mathtt{f}u)|. \end{aligned}$$

Adding the above two inequalities, we get

$$\begin{aligned} |d(\mathtt{f}x, \mathtt{f}u)| + |d(\mathtt{f}y, \mathtt{f}v)| = (\lambda_1 + \lambda_3 + \lambda_5)&(\mathtt{f}x, \mathtt{f}y, \mathtt{f}u, \mathtt{f}v) \\ &(|d(\mathtt{f}x, \mathtt{f}u)| + |d(\mathtt{f}y, \mathtt{f}v)|). \end{aligned}$$

But we know that $(\lambda_1 + \lambda_3 + \lambda_5)(\mathtt{f}x, \mathtt{f}y, \mathtt{f}u, \mathtt{f}v) < 1$, which implies

$$|d(\mathtt{f}x, \mathtt{f}u)| + |d(\mathtt{f}y, \mathtt{f}v)| = 0.$$

Thus, it follows that $|d(\mathtt{f}x, \mathtt{f}u)| = 0 = |d(\mathtt{f}x, \mathtt{f}u)|$. Hence, $\mathtt{f}x = \mathtt{f}u$ and $\mathtt{f}y = \mathtt{f}v$. Finally, since $\mathtt{f}$ is injective, we get $x = u$ and $y = v$, which proves the assertion.

Corollary 1 *Let (X, d) be complete metric space and let $\mathtt{F}, \mathtt{G} : X^2 \to X$ be two mappings. If there exists $\gamma \in [0, 1)$ such that*

$$d(\mathtt{F}(x, y), \mathtt{G}(u, v)) \precsim \gamma\left(\frac{d(x, u) + d(y, v)}{2}\right) \tag{2}$$

for all $x, y, u, v \in X$, then $\mathtt{F}$ and $\mathtt{G}$ have a unique common coupled fixed point.

Proof The proof follows instantly, if we let λ_i, $i = 1, 2, \ldots 5$, to be the constant functions, $\lambda_1 = \gamma$ and $\lambda_2 = \lambda_3 = \lambda_4 = \lambda_5 = 0$ in Theorem 1.

Example 2 Let $X = \{a + ia : 0 \leq a \leq 0.5\}$ be a complete complex valued metric space with the metric $d(x, y) = |Re(x) - Re(y)| + i|Im(x) - Im(y)|$. Define $\mathrm{F} : X^2 \to X$ by $\mathrm{F}(x, y) = (1 + i)\left(\frac{a}{4} + \frac{b^2}{2}\right)$ for all $(x, y) \in X^2$, where $x = a + ia$ and $y = b + ib$ for some $a, b \in [0, 0.5]$. Let $x = a + ia$, $y = b + ib$, $u = c + ic$, and $v = d + id$ for some $a, b, c, d \in [0, 0.5]$, then

$$\begin{aligned}
d(\mathrm{F}(x, y), \mathrm{F}(u, v)) &= (1 + i)\left|\frac{a}{4} + \frac{b^2}{2} - \frac{c}{4} - \frac{d^2}{2}\right| \\
d(x, u) &= (1 + i)|a - c| \\
d(y, v) &= (1 + i)|b - d| \\
d(x, \mathrm{F}(x, y)) &= (1 + i)\left|a - \frac{a}{4} - \frac{b^2}{2}\right| \\
d(u, \mathrm{F}(u, v)) &= (1 + i)\left|c - \frac{c}{4} - \frac{d^2}{2}\right| \\
d(x, \mathrm{F}(u, v)) &= (1 + i)\left|a - \frac{c}{4} - \frac{d^2}{2}\right| \\
d(u, \mathrm{F}(x, y)) &= (1 + i)\left|c - \frac{a}{4} - \frac{b^2}{2}\right| \\
d(y, \mathrm{F}(y, x)) &= (1 + i)\left|b - \frac{b}{4} - \frac{a^2}{2}\right| \\
d(v, \mathrm{F}(v, u)) &= (1 + i)\left|d - \frac{d}{4} - \frac{c^2}{2}\right| \\
d(y, \mathrm{F}(v, u)) &= (1 + i)\left|b - \frac{d}{4} - \frac{c^2}{2}\right|
\end{aligned}$$

and

$$d(v, \mathrm{F}(y, x)) = (1 + i)\left|d - \frac{b}{4} - \frac{a^2}{2}\right|.$$

Let $\lambda_i, i = 1, 2, \ldots 5$ be the constant functions defined by $\lambda_1 = 0.7$, $\lambda_2 = 0.1$, $\lambda_3 = 0.1$, $\lambda_4 = 0.05$, and $\lambda_5 = 0.04$. Then clearly F satisfies the contractive condition (1). Thus, by Theorem 1, F has a unique coupled fixed point.

4 Application

In this section, we apply Theorem 1 to study the existence of a solution of nonlinear fractional differential equations.

Definition 5 [6] The Riemann-Liouville fractional integral of order q for a continuous function f is defined as

$$I^q f(t) = \frac{1}{\Gamma(q)} \int_0^t \frac{f(s)}{(t-s)^{1-q}} ds, \ q > 0,$$

provided the integral exists, where $\Gamma(.)$ is the Gamma function, which is defined by $\Gamma(x) = \int_0^\infty t^{x-1} e^{-t} dt$.

Definition 6 [6] For atleast $n-$times continuously differentiable function $f : [0, \infty) \to \mathbb{R}$, the Caputo derivative of fractional order q is defined by

$$^cD^q f(t) = \frac{1}{\Gamma(n-q)} \int_0^t \frac{f^n(s)}{(t-s)^{q+1-n}} ds, \ n-1 < q < n, \ n = [q] + 1,$$

where $[q]$ denotes the integer part of the real number q.

Lemma 2 *[6] For $q > 0$, the general solution of the fractional differential equation $^cD^q x(t) = 0$ is given by $x(t) = c_0 + c_1 t + \cdots + c_{n-1} t^{n-1}$, where $c_i \in \mathbb{R}$, $i = 0, 1, \ldots, n-1$ $(n = [q] + 1)$.*

Lemma 3 *[6] If $\beta > \alpha > 0$ and $x \in L_1[0, 1]$, then*

i. $^cD^\alpha I^\beta x(t) = I^{\beta-\alpha} x(t)$ *holds almost everywhere on* $[0, 1]$, *and it is valid at any point* $t \in [0, 1]$ *if* $x \in C[0, 1]$; $^cD^\alpha I^\alpha x(t) = x(t)$, *for all* $t \in [0, 1]$.
ii. $^cD^\alpha t^{\lambda-1} = \frac{\Gamma(\lambda)}{\Gamma(\lambda-\alpha) t^{\lambda-\alpha-1}}$ *and* $^cD^\alpha t^{\lambda-1} = 0, \ \lambda \leq [\alpha]$

Let

$$\begin{aligned} ^cD^q x(t) &= f(t, x(t), y(t)) \\ ^cD^q y(t) &= g(t, x(t), y(t)), \ \ 1 < q \leq 2, \ \ t \in [0, 1] \end{aligned} \tag{3}$$

be the system of fractional differential equations with nonlocal multi-point Caputo derivative and integral boundary conditions

$$^cD^\sigma (x(\xi) + y(\xi)) = \sum_{i=1}^n \alpha_i \, ^cD^\nu (x(\eta_i) + y(\eta_i)) \tag{4}$$

$$x(1) + y(1) = \sum_{i=1}^n \beta_i \int_0^{\eta_i} (x(s) + y(s)) ds + \gamma_i (x(\eta_i) + y(\eta_i)) \tag{5}$$

$$x'(1) - y'(1) = 0 \tag{6}$$

$$\int_0^1 (x(t) - y(t))dt = 0 \tag{7}$$

where ${}^cD^{\mu}$ is the Caputo fractional derivative of order μ such that $1 < q \le 2$, $0 < \sigma, \nu \le 1$, and $f, g : [0, 1] \times \mathbb{R} \times \mathbb{R} \to \mathbb{R}$ is a given continuous function, $0 < \xi < \eta_1 < \eta_2 < \cdots < \eta_n < 1$ and $\alpha_i, \beta_i, \gamma_i, i = 1, 2, \ldots, n$ are appropriate real constants.

To avoid repeated use of symbols and terminologies, we fix some notations here. Throughout this section, these notations are used in the context defined below only. Let

$$\begin{aligned}
\Delta_1 &= \xi^{1-\sigma}\Gamma(2-\nu) - \Gamma(2-\sigma)\sum_{i=1}^{n}\alpha_i\eta_i^{1-\nu} \\
\Delta_2 &= 1 - \sum_{i=1}^{n}(\beta_i\eta_i + \gamma_i) \\
\Delta_3 &= -2 + \sum_{i=1}^{n}\eta_i(\beta_i\eta_i + 2\gamma_i) \\
G &= \Gamma(2-\sigma)\Gamma(2-\nu).
\end{aligned}$$

Here, note that none of the constants Δ_1, Δ_2, and Δ_3 are zero. Let

$$\begin{aligned}
I_1^{\theta} &= \int_0^t \frac{(t-s)^{q-1}}{\Gamma(q)}\theta(s)ds \\
I_2^{\theta} &= \int_0^{\eta_i} \frac{(\eta_i-s)^{q-\nu-1}}{\Gamma(q-\nu)}\theta(s)ds \\
I_3^{\theta} &= \int_0^{\xi} \frac{(\xi-s)^{q-\sigma-1}}{\Gamma(q-\sigma)}\theta(s)ds \\
I_4^{\theta} &= \int_0^1 \frac{(1-s)^{q-1}}{\Gamma(q)}\theta(s)ds \\
I_5^{\theta} &= \int_0^{\eta_i} \frac{(\eta_i-s)^{q}}{\Gamma(q+1)}\theta(s)ds \\
I_6^{\theta} &= \int_0^{\eta_i} \frac{(\eta_i-s)^{q-1}}{\Gamma(q)}\theta(s)ds
\end{aligned}$$

$$I_7^\theta = \int_0^1 \frac{(1-s)^q}{\Gamma(q+1)}\theta(s)ds$$

$$I_8^\theta = \int_0^1 \frac{(1-s)^{q-2}}{\Gamma(q-1)}\theta(s)ds.$$

Next, we present a lemma which states that, the solution of the system of linear fractional differential equations (3), supplemented with certain boundary conditions, is equivalent to the solution a system of the integral equations (9).

Lemma 4 *Let $h, k : [0,1] \to \mathbb{R}$ be continuous functions. Then, the solution of the system of linear fractional differential equations*

$$\begin{aligned} {}^cD^q x(t) &= h(t) \\ {}^cD^q y(t) &= k(t), \quad 1 < q \le 2, \quad t \in [0,1] \end{aligned} \tag{8}$$

supplemented with the boundary conditions (4–7) is equivalent to the solution of the integral equations

$$x(t) = \begin{cases} I_1^h + \frac{\Delta_3 G}{4\Delta_1\Delta_2}\left(\sum\limits_{i=1}^n \alpha_i I_2^{h+k} - I_3^{h+k}\right) \\ +\frac{1}{2\Delta_2}\left(-I_4^{h+k} + \sum\limits_{i=1}^n \beta_i I_5^{h+k} + \sum\limits_{i=1}^n \gamma_i I_6^{h+k}\right) \\ +\frac{1}{2}I_7^{k-h} - \frac{1}{4}I_8^{k-h} + \frac{t}{2}I_8^{k-h} + \frac{tG}{2\Delta_1}\left(\sum\limits_{i=1}^n \alpha_i I_2^{h+k} - I_3^{h+k}\right) \end{cases}$$

$$y(t) = \begin{cases} I_1^k + \frac{\Delta_3 G}{4\Delta_1\Delta_2}\left(\sum\limits_{i=1}^n \alpha_i I_2^{h+k} - I_3^{h+k}\right) \\ +\frac{1}{2\Delta_2}\left(-I_4^{h+k} + \sum\limits_{i=1}^n \beta_i I_5^{h+k} + \sum\limits_{i=1}^n \gamma_i I_6^{h+k}\right) \\ -\frac{1}{2}I_7^{k-h} + \frac{1}{4}I_8^{k-h} - \frac{t}{2}I_8^{k-h} + \frac{tG}{2\Delta_1}\left(\sum\limits_{i=1}^n \alpha_i I_2^{h+k} - I_3^{h+k}\right). \end{cases} \tag{9}$$

Proof From Lemma 2, we can reduce (8) to the system of integral equations

$$\begin{aligned} x(t) &= I^q h(t) + c_0 + c_1 t \\ y(t) &= I^q k(t) + c_2 + c_3 t \end{aligned}$$

where $c_0, c_1, c_2, c_3 \in \mathbb{R}$ are arbitrary constants. Thus, the general solution of the system of fractional differential equations (8) is given by

$$x(t) = \int_0^t \frac{(t-s)^{q-1}}{\Gamma(q)} h(s)ds + c_0 + c_1 t \tag{10}$$

$$y(t) = \int_0^t \frac{(t-s)^{q-1}}{\Gamma(q)} k(s)ds + c_2 + c_3 t. \tag{11}$$

We wish to show that (10) and (11) are the required system of integral equations, and hence all that remains is to find the constants c_0, c_1, c_2 and c_3. If now, we compute the integral of the sum of the functions $x(t)$ and $y(t)$, from 0 to η_i, we find that

$$\int_0^{\eta_i} (x(s) + y(s))ds = I_5^{h+k} + (c_0 + c_2)\eta_i + (c_1 + c_3)\frac{\eta_i^2}{2}.$$

In follow, we compute the Caputo fractional derivatives of order ν and σ, of the sum of the functions $x(t)$ and $y(t)$, as follows:

$$\begin{aligned} {}^cD^{\nu}(x(\eta_i) + y(\eta_i)) &= I_2^{h+k} + \frac{\eta_i^{1-\nu}}{\Gamma(2-\nu)}(c_1 + c_3) \\ {}^cD^{\sigma}(x(\xi) + y(\xi)) &= I_3^{h+k} + \frac{\xi^{1-\sigma}}{\Gamma(2-\sigma)}(c_1 + c_3). \end{aligned}$$

Next by using the boundary conditions (4) and (5), it is easy to derive the following expressions:

$$c_1 + c_3 = \frac{G}{\Delta_1}\left(\sum_{i=1}^{n} \alpha_i I_2^{h+k} - I_3^{h+k}\right) \tag{12}$$

and

$$\begin{aligned} I_4^{h+k} + c_0 + c_1 + c_2 + c_3 = \sum_{i=1}^{n} \beta_i I_5^{h+k} + (c_0 + c_2)\sum_{i=1}^{n} \beta_i \eta_i \\ +(c_1 + c_3)\sum_{i=1}^{n} \beta_i \frac{\eta_i^2}{2} + \sum_{i=1}^{n} \gamma_i I_6^{h+k} \\ +(c_0 + c_2)\sum_{i=1}^{n} \gamma_i + (c_1 + c_3)\sum_{i=1}^{n} \gamma_i \eta_i. \end{aligned}$$

Then, using (12), we have

$$c_0 + c_2 = \frac{1}{\Delta_1}\left(-I_4^{h+k} + \sum_{i=1}^{n} \beta_i I_5^{h+k} + \sum_{i=1}^{n} \gamma_i I_6^{h+k}\right) + \frac{\Delta_3 G}{2\Delta_1 \Delta_2}\left(\sum_{i=1}^{n} \alpha_i I_2^{h+k} - I_3^{h+k}\right). \tag{13}$$

From the boundary conditions (6) and (7), we get

$$c_1 - c_3 = I_8^{k-h} \tag{14}$$

and

$$\int_0^1 I_1^{h-k} dt + (c_0 - c_2) + \frac{(c_1 - c_3)}{2} = 0. \tag{15}$$

Therefore, we have

$$c_0 - c_2 = I_7^{k-h} - \frac{1}{2} I_8^{k-h}. \tag{16}$$

Consequently, by adding (13) and (16), we get

$$c_0 = \frac{1}{2\Delta_1}\left(-I_4^{h+k} + \sum_{i=1}^{n} \beta_i I_5^{h+k} + \sum_{i=1}^{n} \gamma_i I_6^{h+k}\right) + \frac{\Delta_3 G}{4\Delta_1 \Delta_2}\left(\sum_{i=1}^{n} \alpha_i I_2^{h+k} - I_3^{h+k}\right) + \frac{1}{2} I_7^{k-h} - \frac{1}{4} I_8^{k-h}.$$

Next, by adding (12) and (14), we get

$$c_1 = \frac{G}{2\Delta_1}\left(\sum_{i=1}^{n} \alpha_i I_2^{h+k} - I_3^{h+k}\right) + \frac{1}{2} I_8^{k-h}.$$

After all, by using (16) and (14), we get

$$c_2 = \frac{1}{2\Delta_1}\left(-I_4^{h+k} + \sum_{i=1}^{n} \beta_i I_5^{h+k} + \sum_{i=1}^{n} \gamma_i I_6^{h+k}\right) + \frac{\Delta_3 G}{4\Delta_1 \Delta_2}\left(\sum_{i=1}^{n} \alpha_i I_2^{h+k} - I_3^{h+k}\right) - \frac{1}{2} I_7^{k-h} + \frac{1}{4} I_8^{k-h}$$

and

$$c_3 = \frac{G}{2\Delta_1}\left(\sum_{i=1}^{n} \alpha_i I_2^{h+k} - I_3^{h+k}\right) - \frac{1}{2} I_8^{k-h}$$

as desired.

Theorem 3 *Let $f, g : [0, 1] \times \mathbb{R} \times \mathbb{R} \to \mathbb{R}$ be continuous functions satisfying the Lipschitz condition: there exists a constant $L > 0$ such that*

$$|f(t, x(t), y(t)) - g(t, u(t), v(t))| \leq L\{|x(t) - u(t)| + |y(t) - v(t)|\},$$

for each $t \in [0, 1]$ and $x, y \in \mathbb{R}$. Then, the system of fractional boundary value problem (3) has a unique solution on $[0, 1]$ if $LR < 1$, where

$$R = \frac{2}{\Gamma(q+1)} + \frac{2}{\Gamma(q+2)} + \frac{3}{\Gamma(q)} + \frac{2a_1}{|\Delta_2|} + \frac{Ga_2}{|\Delta_1|}\left(\frac{|\Delta_3|}{4|\Delta_2|} + 2\right),$$

with

$$a_1 = \sum_{i=1}^{n} |\beta_i| \frac{\eta_i^{q+1}}{\Gamma(q+2)} + \frac{1 + \sum_{i=1}^{n} |\gamma_i| \eta_i^{q}}{\Gamma(q+1)}$$

and

$$a_2 = \sum_{i=1}^{n} |\alpha_i| \frac{\eta_i^{q-\nu}}{\Gamma(q-\nu+1)} + \frac{\xi^{q-\sigma}}{\Gamma(q-\sigma+1)}. \quad (17)$$

Proof Let C be the complete metric space of all continuous functions from $[0, 1]$ to $\mathbb{R}$ with the metric $d(x(t), y(t)) = \sup_{t\in[0,1]} |x(t) - y(t)|$. Let

$$\psi_1 = |f(s, x(s), y(s)) - g(s, u(s), v(s))|$$
$$\psi_2 = |f(s, x(s), y(s)) + g(s, x(s), y(s)) - f(s, u(s), v(s)) - g(s, u(s), v(s))|.$$

Let $\mathrm{F} : C \times C \to C$ be the function defined by

$$\mathrm{F}(x(t), y(t)) = \begin{cases} I_1^f + \frac{\Delta_3 G}{4\Delta_1\Delta_2}\left(\sum_{i=1}^{n} \alpha_i I_2^{f+g} - I_3^{f+g}\right) \\ \quad + \frac{1}{2\Delta_2}\left(-I_4^{f+g} + \sum_{i=1}^{n} \beta_i I_5^{f+g} + \sum_{i=1}^{n} \gamma_i I_6^{f+g}\right) \\ \quad + \frac{1}{2} I_7^{g-f} - \frac{1}{4} I_8^{g-f} + \frac{t}{2} I_8^{g-f} + \frac{tG}{2\Delta_1}\left(\sum_{i=1}^{n} \alpha_i I_2^{f+g} - I_3^{f+g}\right) \end{cases}$$

and let $G : \mathcal{C} \times \mathcal{C} \to \mathcal{C}$ be a function defined by

$$G(x(t), y(t)) = \begin{cases} I_1^{g} + \frac{\Delta_3 G}{4\Delta_1\Delta_2}\left(\sum_{i=1}^{n} \alpha_i I_2^{f+g} - I_3^{f+g}\right) \\ \quad +\frac{1}{2\Delta_2}\left(-I_4^{f+g} + \sum_{i=1}^{n} \beta_i I_5^{f+g} + \sum_{i=1}^{n} \gamma_i I_6^{f+g}\right) \\ \quad -\frac{1}{2} I_7^{g-f} + \frac{1}{4} I_8^{g-f} - \frac{t}{2} I_8^{g-f} + \frac{tG}{2\Delta_1}\left(\sum_{i=1}^{n} \alpha_i I_2^{f+g} - I_3^{f+g}\right). \end{cases}$$

Then, we have

$$\begin{aligned} &|\mathrm{F}(x(t), y(t)) - \mathrm{G}(u(t), v(t))| \\ &\leq \begin{cases} I_1^{\psi_1} + \frac{|\Delta_3| G}{4|\Delta_1||\Delta_2|}\left(\sum_{i=1}^{n} |\alpha_i| I_2^{\psi_2} + I_3^{\psi_2}\right) \\ \quad +\frac{1}{2|\Delta_2|}\left(I_4^{\psi_2} + \sum_{i=1}^{n} |\beta_i| I_5^{\psi_2} + \sum_{i=1}^{n} |\gamma_i| I_6^{\psi_2}\right) \\ \quad +\frac{1}{2} I_7^{\psi_2} + \frac{1}{4} I_8^{\psi_2} + \frac{t}{2} I_8^{\psi_2} + \frac{tG}{2|\Delta_1|}\left(\sum_{i=1}^{n} |\alpha_i| I_2^{\psi_2} + I_3^{\psi_2}\right). \end{cases} \\ &\leq \begin{cases} L\left(|x(t) - u(t)| + |y(t) - v(t)|\right) \\ \quad \Bigg[I_1 + \frac{|\Delta_3| G}{4|\Delta_1||\Delta_2|}\left(\sum_{i=1}^{n} |\alpha_i| I_2 + I_3\right) \\ \quad +\frac{1}{2|\Delta_2|}\left(I_4 + \sum_{i=1}^{n} |\beta_i| I_5 + \sum_{i=1}^{n} |\gamma_i| I_6\right) \\ \quad +\frac{1}{2} I_7 + \frac{1}{4} I_8 + \frac{t}{2} I_8 + \frac{tG}{2|\Delta_1|}\left(\sum_{i=1}^{n} |\alpha_i| I_2 + I_3\right)\Bigg]. \end{cases} \\ &\leq \begin{cases} \frac{L}{2}\left(\|x(t) - u(t)\| + \|y(t) - v(t)\|\right)\Big[\frac{2}{\Gamma(q+1)} + \frac{2a_1}{|\Delta_2|} + \frac{|\Delta_3| G a_2}{4|\Delta_1||\Delta_2|} \\ \quad + \frac{2}{\Gamma(q+2)} + \frac{1}{\Gamma(q)} + \frac{2}{\Gamma(q)} + \frac{2Ga_2}{|\Delta_1|}\Big]. \end{cases} \end{aligned}$$

Therefore,

$$d(\mathrm{F}(x(t), y(t)), G(u(t), v(t))) \leq LR\left(\frac{d(x(t), y(t)) + d(u(t), v(t))}{2}\right).$$

Now since $LR < 1$, we can conclude that F and G have a unique common coupled fixed point, using Theorem 1. Thus, the system of fractional differential equation (3) has a unique solution.

Conclusively, we give an example in order to enhance the understanding of the above theorem.

Example 3 Consider the system of fractional boundary value problem given by

$$
\begin{aligned}
{}^cD^{\frac{3}{2}}x(t) &= t + t^3x + \sin y \\
{}^cD^{\frac{3}{2}}y(t) &= t + t^2x + \sin y \\
{}^cD^{0.5}(x(0.15) + y(0.15)) &= 0.8\,{}^cD^{0.75}(x(0.2) + y(0.2)) \\
&\quad +1.4\,{}^cD^{0.75}(x(0.7) + y(0.7)) \\
x(1) + y(1) &= 0.7\int_0^{0.2}(x(s) + y(s))ds + 1.2\int_0^{0.7}(x(s) + y(s))ds \\
&\quad +0.4(x(0.2) + y(0.2)) + 0.5(x(0.7) + y(0.7)) \\
x'(1) - y'(1) &= 0 \text{ and}
\end{aligned}
$$

$$\int_0^1 (x(t) + y(t))dt = 0,$$

where $q = 1.5$, $n = 2$, $\alpha_1 = 0.8$, $\alpha_2 = 1.4$, $\beta_1 = 0.7$, $\beta_2 = 1.2$, $\gamma_1 = 0.4$, $\gamma_2 = 0.5$, $\eta_1 = 0.2$, $\eta_2 = 0.7$, $\xi = 0.15$, $\nu = 0.75$, $\sigma = 0.5$, $f(t, x, y) = t + t^3x + \sin y$, and $g(t, x, y) = t + t^2x + \sin y$. Then, we have

$$
\begin{aligned}
|f(t, x, y) - g(t, x, y)| &= |t + t^3x + \sin y - t - t^2u - \sin v| \\
&\leq |t^2||tx - u| + |\sin y - \sin v| \\
&\leq |t^2||x - u| + |y - v| \\
&\leq |x - u| + |y - v|.
\end{aligned}
$$

Also it is easy to see that $\Delta_1 = -1.2579$, $\Delta_2 = -0.88$, and $\Delta_3 = -0.524$, by some simple calculation. Now by letting $L = 1$, we get $LR = 0.7123 < 1$. Thus, the system satisfies all the conditions of Theorem 3 and hence has a unique solution.

Conflict of interest The authors declare that they have no conflict of interest.

References

1. Ahmad, B., Ntouyas, S.K.: Existence results for a coupled system of Caputo type sequential fractional differential equations with nonlocal integral boundary conditions. Appl. Math. Comput. **266**, 615–622 (2015)
2. Azam, A., Fisher, B.: Common fixed point theorems in complex valued metric spaces. Numer. Funct. Anal. Optim. **32**(3), 243–253 (2011)
3. Ciric, L., Lakshmikantham, V.: Coupled fixed point theorems for nonlinear contractions in partially ordered metric spaces. Nonlinear Anal. TMA **70**(12), 4341–4349 (2009)
4. Fazli, H., Nieto, J.J., Bahrami, F.: On the existence and uniqueness results for nonlinear sequential fractional differential equations. Appl. Comput. Math. **17**, 36–47 (2018)
5. Bhaskar, T.G., Lakshmikantham, V.: Fixed point theorems in partially ordered metric spaces and applications. Nonlinear Anal. **65**, 1379–1393 (2006)

6. Henderson, J., Luca, R., Tudorache, A.: On a system of fractional differential equations with coupled integral boundary conditions. Fract. Calc. Appl. Anal. **18**, 361–386 (2015)
7. Huang, L.G., Zhang, X.: Cone metric spaces and fixed point theorems of contractive mappings. J. Math. Anal. Appl. **332**(2), 1468–1476 (2007)
8. Kirane, M., Ahmad, B., Alsaedi, A., Al-Yami, M.: Non-existence of global solutions to a system of fractional diffusion equations. Acta Appl. Math. **133**, 235–248 (2014)
9. Kutbi, M.A., Azam, A., Ahmad, J., Bari, C.D.: Some common coupled fixed point results for generalized contraction in complex valued metric spaces. J. Appl. Math. **1–10** (2013)
10. Rouzkard, F., Imdad, M.: Some common fixed point theorems on complex valued metric spaces. Comput. Math. Appl. **64**, 1866–1874 (2012)
11. Sintunavarat, W., Kumam, P., Cho, Y.J.: Coupled fixed point theorems for nonlinear contractions without mixed monotone property. Fixed Point Theory Appl. **170**, 1–16 (2012)
12. Sintunavarat, W., Kumam, P.: Generalized common fixed point theorems in complex valued metric spaces and applications. J. Inequalities Appl. **84** (2012)

Neutrosophic G^*-Closed Sets and Its Applications

A. Atkinswestley and S. Chandrasekar

Abstract Salama and Alblowi [1] developed Neutrosophic topological spaces by using Smarandache's Neutrosophic sets. Arokiarani et al. [2] introduced NS(α) closed sets. Ishwarya et al. [3] introduced and studied NS semi-open sets. Dhavaseelan and Jafari [4] introduced Generalized Neutrosophic Closed sets. The aim of this present paper is we introduce and study the concepts of Neutrosophic (G^*)-Closed sets and Neutrosophic (G^*)-open sets in Neutrosophic topological spaces. Also, we study the application of Neutrosophic (G^*)-Closed sets.

Keywords $NS(G)CS \cdot NS(G^*)CS \cdot NS(G^*)OS$

1 Preliminaries

In this part, we review the required essential definition and results of Neutrosophic

Definition 1 [5] Let $\mathbb{N}^*_{\mathcal{X}}$ be a non-empty fixed set. A Neutrosophic set $\mathcal{W}^*_1$ is an object having the form

$\mathcal{W}^*_1 = \{\langle w, \mu_{\mathcal{W}^*_1}(w), \sigma_{\mathcal{W}^*_1}(w), \gamma_{\mathcal{W}^*_1}(w)\rangle w \in \mathbb{N}^*_{\mathcal{X}}\}$,

$\mu_{\mathcal{W}^*_1}(w)$—mean membership function

$\sigma_{\mathcal{W}^*_1}(w)$—mean indeterminacy

$\gamma_{\mathcal{W}^*_1}(w)$—mean non-membership function.

Definition 2 [5] Neutrosophic set

$\mathcal{W}^*_1 = \{\langle w, \mu_{\mathcal{W}^*_1}(w), \sigma_{\mathcal{W}^*_1}(w), \gamma_{\mathcal{W}^*_1}(w)\rangle : w \in \mathbb{N}^*_{\mathcal{X}}\}$, on $\mathbb{N}^*_{\mathcal{X}}$ and $\forall w \in \mathbb{N}^*_{\mathcal{X}}$ then complement of $\mathcal{W}^*_1$ is

A. Atkinswestley
Department of Mathematics, Roever college of Engineering and Technology, Elambalur, Perambalur (Dt), Tamil Nadu, India
e-mail: ats.wesly@gmail.com

S. Chandrasekar (✉)
Department of Mathematics, Arignar Anna Government Arts College, Namakkal (Dt), Tamil Nadu, India
e-mail: chandrumat@gmail.com

R. N. Mohapatra et al. (eds.), *Mathematical Analysis and Computing*, Springer Proceedings in Mathematics & Statistics 344,
https://doi.org/10.1007/978-981-33-4646-8_13

$$\mathcal{W}_1^{*C} = \{\langle\ w, \gamma_{w_1^*}((w)), 1-\sigma_{w_1^*}(w), \mu_{w_1^*}(w)\rangle : w \in \mathbb{N}_{\mathcal{X}}^*\}$$

Definition 3 [5] Let $\mathcal{W}_1^*$ and $\mathcal{W}_2^*$ are Neutrosophic sets, $\forall w \in \mathbb{N}_{\mathcal{X}}^*$

$$\mathcal{W}_1^* = \{\langle\ w, \mu_{\mathcal{W}_1^*}(w), \sigma_{\mathcal{W}_1^*}(w), \gamma_{\mathcal{W}_1^*}(w)\rangle : w \in \mathbb{N}_{\mathcal{X}}^*\}$$

$$\mathcal{W}_2^* = \{\langle\ w, \mu_{\mathcal{W}_2^*}(w), \sigma_{\mathcal{W}_2^*}(w), \gamma_{\mathcal{W}_2^*}(w)\rangle : w \in \mathbb{N}_{\mathcal{X}}^*\}$$

$$\text{Then } \mathcal{W}_1^* \subseteq \mathcal{W}_2^* \Leftrightarrow \mu_{\mathcal{W}_1^*}(w) \le \mu_{\mathcal{W}_2^*}(w), \sigma_{\mathcal{W}_1^*}(w) \le \sigma_{\mathcal{W}_2^*}(w)$$

$$\&\gamma_{\mathcal{W}_1^*}(w) \ge \gamma_{\mathcal{W}_2^*}(w)$$

Definition 4 [5] Let $\mathcal{W}_1^*$and $\mathcal{W}_2^*$ be two Neutrosophic sets that are

$$\mathcal{W}_1^* = \{ < w, \mu_{w_1^*}(w), \sigma_{w_1^*}(w), \gamma_{w_1^*}(w) > : w \in \mathbb{N}_{\mathcal{X}}^*\},$$
$$\mathcal{W}_2^* = \{ < w, \mu_{w_2^*}(w), \sigma_{w_2^*}(w), \gamma_{w_2^*}(w) > : w \in \mathbb{N}_{\mathcal{X}}^*\}$$

Then

$$\mathcal{W}_1^* \cap \mathcal{W}_2^* = \{ < w, \mu_{w_1^*}(w) \cap \mu_{w_2^*}(w), \sigma_{w_1^*}(w) \cap \sigma_{w_2^*}(w), \gamma_{w_1^*}(w) \cup \gamma_{w_2^*}(w) > : w \in \mathbb{N}_{\mathcal{X}}^*\}$$
$$\mathcal{W}_1^* \cup \mathcal{W}_2^* = \{ < w, \mu_{w_1^*}(w) \cup \mu_{w_2^*}(w), \sigma_{w_1^*}(w) \cup \sigma_{w_2^*}(w), \gamma_{w_1^*}(w) \cap \gamma_{w_2^*}(w) > : w \in \mathbb{N}_{\mathcal{X}}^*\}$$

Definition 5 [1] Let $\mathbb{N}_{\mathcal{X}}^*$ be non-empty set and $\mathbb{N}_{\tau}^*$ be the collection of Neutrosophic subsets of $\mathbb{N}_{\mathcal{X}}^*$satisfying the accompanying properties:

1. $0_{NS}, 1_{NS} \in \mathbb{N}_{\tau}^*$
2. $NS_{T1} \cap NS_{T2} \in \mathbb{N}_{\tau}^*$ for any $NS_{T1}, NS_{T2} \in \mathbb{N}_{\tau}^*$
3. $\cup NS_{Ti} \in \mathbb{N}_{\tau}^*$ for every $\{NS_{Ti}: i \in j\} \subseteq \mathbb{N}_{\tau}^*$

Then space $(\mathbb{N}_{\mathcal{X}}^*, \mathbb{N}_{\tau}^*)$ is called a Neutrosophic topological space (NS-T-S). The component of $\mathbb{N}_{\tau}^*$ are called NS-OS (Neutrosophic open set) and its complement is NS-CS (Neutrosophic closed set)

Example 1 Let $\mathbb{N}_{\tau}^* = \{w\}$ and $\forall w \in \mathbb{N}_{\mathcal{X}}^*$,

$$\mathcal{W}_1^* = \left\langle w, \frac{6}{10}, \frac{6}{10}, \frac{5}{10} \right\rangle, \quad \mathcal{W}_2^* = \left\langle w, \frac{5}{10}, \frac{7}{10}, \frac{9}{10} \right\rangle$$
$$\mathcal{W}_3^* = \left\langle w, \frac{6}{10}, \frac{7}{10}, \frac{5}{10} \right\rangle, \quad \mathcal{W}_4^* = \left\langle w, \frac{5}{10}, \frac{6}{10}, \frac{9}{10} \right\rangle$$

Then the collection $\mathbb{N}_{\tau}^* = \{0_{NS}, \mathcal{W}_1^*, \mathcal{W}_2^*, \mathcal{W}_3^*, \mathcal{W}_4^* 1_{NS}\}$ is called a NS-T-S on $\mathbb{N}_{\mathcal{X}}^*$.

Definition 6 $(\mathbb{N}_{\mathcal{X}}^*, \mathbb{N}_{\tau}^*)$ be a NS-T-S and

$$\mathcal{W}_1^* = \{ < w, \mu_{w_1^*}(w), \sigma_{w_1^*}(w), \gamma_{w_1^*}(w) > : w \in \mathbb{N}_{\mathcal{X}}^*\}.$$

Then Neutrosophic closure of $\mathcal{W}_1^*$ is

$$NSCl(\mathcal{W}_1^*) = \cap\{\mathcal{L} : \mathcal{L} \text{ is a NS-CS in } \mathbb{N}_{\mathcal{X}}^* \text{ and } \mathcal{W}_1^* \subseteq \mathcal{L}\}$$

Neutrosophic interior of $\mathcal{W}_1^*$ is

$$NSInt(\mathcal{W}_1^*) = \cup\{\mathcal{J} : \mathcal{J} \text{ is a NS-OS in } \mathbb{N}_{\mathcal{X}}^* \text{ and } \mathcal{J} \subseteq \mathcal{W}_1^*\}.$$

Definition 7 Let $(\mathbb{N}_{\mathcal{X}}^*, \mathbb{N}_{\tau}^*)$ be a NS-T-S and
$\mathcal{W}_1^* = \{ < w, \mu_{w_1^*}(w), \sigma_{w_1^*}(w), \gamma_{w_1^*}(w) > : w \in \mathbb{N}_{\mathcal{X}}^*\}$
NS(S)int($\mathcal{W}_1^*$) = $\cup\{ \mathcal{J} / \mathcal{J}$ is a NS(S)OS in$\mathbb{N}_{\mathcal{X}}^*$ and $\mathcal{J} \subseteq \mathcal{W}_1^*\}$,
NS(S)cl($\mathcal{W}_1^*$) = $\cap\{ \mathcal{L} / \mathcal{L}$ is a NS(S)CS in $\mathbb{N}_{\mathcal{X}}^*$and $\mathcal{W}_1^* \subseteq \mathcal{L} \}$.
NS(α)int($\mathcal{W}_1^*$) = $\cup\{ \mathcal{J} / \mathcal{J}$ is a NS(α)OS in $\mathbb{N}_{\mathcal{X}}^*$and $\mathcal{J} \subseteq \mathcal{W}_1^*\}$,
NS(α)cl($\mathcal{W}_1^*$) = $\cap\{ \mathcal{L} / \mathcal{L}$ is a NS(α)CS in $\mathbb{N}_{\mathcal{X}}^*$ and $\mathcal{W}_1^* \subseteq \mathcal{L}\}$.

Definition 8 Let $(\mathbb{N}_{\mathcal{X}}^*, \mathbb{N}_{\tau}^*)$ be a NSTS and $\mathcal{W}_1^* =< w, \mu_{w_1^*}(w), \sigma_{w_1^*}(w), \gamma_{w_1^*}(w) >$ be a Neutrosophic set in $\mathbb{N}_{\mathcal{X}}^*$. Then $\mathcal{W}_1^*$ is said to be

(i) Neutrosophic b closed set [2] ($NS(b)CS$) if NScl(NSint($\mathcal{W}_1^*$)) $\cap$ NS int(NScl ($\mathcal{W}_1^*$)) $\subseteq \mathcal{W}_1^*$,
(ii) Neutrosophic α-closed set [2] (NS(α)CS) if NScl(NSint(NScl($\mathcal{W}_1^*$))) $\subseteq \mathcal{W}_1^*$,
(iii) Neutrosophic pre-closed set [7] ($N(P)CS$) if NScl(NSint($\mathcal{W}_1^*$)) $\subseteq \mathcal{W}_1^*$,
(iv) Neutrosophic regular closed set [5] (NS(R)CS) if NScl(NSint($\mathcal{W}_1^*$)) = $\mathcal{W}_1^*$,
(v) Neutrosophic semi-closed set [3] ($NS(S)CS$) if NSint(NScl($\mathcal{W}_1^*$)) $\subseteq \mathcal{W}_1^*$,
(vi) Neutrosophic generalized closed set [4] ($NS(G)CS$) if NScl($\mathcal{W}_1^*$) $\subseteq \mathcal{J}$ whenever $\mathcal{W}_1^* \subseteq \mathcal{J}$ and $\mathcal{J}$ is NSOS.

2 Neutrosophic G^* Closed Sets

In this section, we introduce Neutrosophic G^*-Closed sets and studied some of its basic properties.

Definition 9 A NS $\mathcal{W}_1^*$ in $(\mathbb{N}_{\mathcal{X}}^*, \mathbb{N}_{\tau}^*)$ is said to be a Neutrosophic G^* Closed set ($NS(G^*)CS$ in short) if NScl($\mathcal{W}_1^*$) $\subseteq \mathcal{K}$ whenever $\mathcal{W}_1^* \subseteq \mathcal{K}$ and $\mathcal{K}$ is $NS(G)OS$ in $(\mathbb{N}_{\mathcal{X}}^*, \mathbb{N}_{\tau}^*)$.

The family of all $NS(G^*)CS$ of a NTS $(\mathbb{N}_{\mathcal{X}}^*, \mathbb{N}_{\tau}^*)$ is denoted by NSG*C($\mathbb{N}_{\mathcal{X}}^*$).

Example 2 Let $NS_X^* = \{w_1, w_2\}$ and let $NS_t = \{0_{NS}, \mathcal{K}, 1_{NS}\}$is NT on $\mathbb{N}_{\mathcal{X}}^*$, where $\mathcal{K} = \langle w, (\frac{3}{10}, \frac{5}{10}, \frac{7}{10}), (\frac{4}{10}, \frac{5}{10}, \frac{6}{10})\rangle$. Then the NS $\mathcal{W}_1^* = \langle w, (\frac{7}{10}, \frac{5}{10}, \frac{1}{10}), (\frac{6}{10}, \frac{5}{10}, \frac{0}{10})\rangle$ is $NS(G^*)CS$ in $(\mathbb{N}_{\mathcal{X}}^*, \mathbb{N}_{\tau}^*)$.

Theorem 1 *Every NSCS is $NS(G^*)CS$.*

Proof Let $\mathcal{W}_1^*$ be a NSCS in $(\mathbb{N}_{\mathcal{X}}^*, \mathbb{N}_{\tau}^*)$. Then NScl($\mathcal{W}_1^*$) = $\mathcal{W}_1^*$. Let $\mathcal{W}_1^* \subseteq \mathcal{K}$ and $\mathcal{K}$ is $NS(G)OS$ in $(\mathbb{N}_{\mathcal{X}}^*, \mathbb{N}_{\tau}^*)$. Therefore NScl($\mathcal{W}_1^*$) = $\mathcal{W}_1^* \subseteq \mathcal{K}$. Thus $\mathcal{W}_1^*$ is $NS(G^*)CS$ in $\mathbb{N}_{\mathcal{X}}^*$.

Example 3 Let $\mathbb{N}^*_{\mathcal{X}} = \{w_1, w_2\}$ and let $NS_t = \{0_{NS}, \mathcal{K}, 1_{NS}\}$is NT on $\mathbb{N}^*_{\mathcal{X}}$, where

$$\mathcal{K} = \left\langle w, \left(\frac{4}{10}, \frac{5}{10}, \frac{6}{10}\right), \left(\frac{2}{10}, \frac{5}{10}, \frac{7}{10}\right)\right\rangle$$

Then the NS $\mathcal{W}^*_1 = \langle w, \left(\frac{6}{10}, \frac{5}{10}, \frac{1}{10}\right), \left(\frac{7}{10}, \frac{5}{10}, \frac{1}{10}\right)\rangle$ is $NS(G^*)CS$ but not NSCS in $\mathbb{N}^*_{\mathcal{X}}$.

Theorem 2 *Every $NS(G^*)CS$ is $NS(G)CS$.*

Proof Let $\mathcal{W}^*_1$ be a $NS(G^*)CS$ in ($\mathbb{N}^*_{\mathcal{X}}$, $\mathbb{N}^*_{\tau}$). Let $\mathcal{W}^*_1 \subseteq \mathcal{K}$ and $\mathcal{K}$ is NSOS in ($\mathbb{N}^*_{\mathcal{X}}$, $\mathbb{N}^*_{\tau}$). Since every NSOS is $NS(G)OS$ and since $\mathcal{W}^*_1$ is $NS(G^*)CS$ in $\mathbb{N}^*_{\mathcal{X}}$. Therefore, NScl($\mathcal{W}^*_1$) $\subseteq \mathcal{K}$ whenever $\mathcal{W}^*_1 \subseteq \mathcal{K}$, $\mathcal{K}$ is NSOS in $\mathbb{N}^*_{\mathcal{X}}$. Thus $\mathcal{W}^*_1$ is $NS(G)CS$ in $\mathbb{N}^*_{\mathcal{X}}$.

Example 4 Let $NS^*_X = \{w_1, w_2\}$ and let $\mathbb{N}^*_{\tau} = \{0_{NS}, \mathcal{K}, 1_{NS}\}$ is NT on $\mathbb{N}^*_{\mathcal{X}}$, where

$$\mathcal{K} = \left\langle w, \left(\frac{3}{10}, \frac{5}{10}, \frac{6}{10}\right), \left(\frac{7}{10}, \frac{5}{10}, \frac{3}{10}\right)\right\rangle.$$

Then the NS $\mathcal{W}^*_1$ =$\langle w, \left(\frac{3}{10}, \frac{5}{10}, \frac{7}{10}\right), \left(\frac{8}{10}, \frac{5}{10}, \frac{2}{10}\right)\rangle$ is $NS(G)CS$ but not $NS(G^*)CS$ in $\mathbb{N}^*_{\mathcal{X}}$.

Theorem 3 *Every $NS(G^*)CS$ is $NS(\alpha G)CS$.*

Proof Let $\mathcal{W}^*_1$ be a $NS(G^*)CS$ in ($\mathbb{N}^*_{\mathcal{X}}$, $\mathbb{N}^*_{\tau}$). $\mathcal{W}^*_1$ is $NS(G)CS$ in $\mathbb{N}^*_{\mathcal{X}}$. Since NS α cl($\mathcal{W}^*_1$) $\subseteq$ NScl($\mathcal{W}^*_1$) and $\mathcal{W}^*_1$ is a $NS(G)CS$ in $\mathbb{N}^*_{\mathcal{X}}$. Therefore NS ($\alpha$)cl($\mathcal{W}^*_1$) $\subseteq$ NScl($\mathcal{W}^*_1$) $\subseteq \mathcal{K}$ whenever $\mathcal{W}^*_1 \subseteq \mathcal{K}$, $\mathcal{K}$ is NSOS in $\mathbb{N}^*_{\mathcal{X}}$. Thus $\mathcal{W}^*_1$ is $NS(\alpha G)CS$ in $\mathbb{N}^*_{\mathcal{X}}$.

Example 5 Let $\mathbb{N}^*_{\mathcal{X}}$ = $\{w_1, w_2\}$ and let $\mathbb{N}^*_{\tau} = \{0_{NS}, \mathcal{K}, 1_{NS}\}$is NT on $\mathbb{N}^*_{\mathcal{X}}$, where $\mathcal{K}$ =$\langle w, \left(\frac{3}{10}, \frac{5}{10}, \frac{2}{10}\right), \left(\frac{4}{10}, \frac{5}{10}, \frac{5}{10}\right)\rangle$. Then the NS $\mathcal{W}^*_1$ =$\langle w, \left(\frac{1}{10}, \frac{5}{10}, \frac{6}{10}\right), \left(\frac{4}{10}, \frac{5}{10}, \frac{5}{10}\right)\rangle$ is NSaGCS but not $NS(G^*)CS$ in $\mathbb{N}^*_{\mathcal{X}}$.

Theorem 4 *Every NS(R)CS is $NS(G^*)CS$.*

Proof Let $\mathcal{W}^*_1$ be a *NS(R)CS* in ($\mathbb{N}^*_{\mathcal{X}}$, $\mathbb{N}^*_{\tau}$). Then $\mathcal{W}^*_1$ = NS cl(NS int($\mathcal{W}^*_1$)). Let $\mathcal{W}^*_1$ $\subseteq \mathcal{K}$ and $\mathcal{K}$ is $NS(G)OS$ in(($\mathbb{N}^*_{\mathcal{X}}$, $\mathbb{N}^*_{\tau}$). Therefore NS cl($\mathcal{W}^*_1$) $\subseteq$ NS cl(NS int($\mathcal{W}^*_1$)). This implies NS cl($\mathcal{W}^*_1$) $\subseteq \mathcal{W}^*_1 \subseteq \mathcal{K}$. Thus $\mathcal{W}^*_1$ is $NS(G^*)CS$ in $\mathbb{N}^*_{\mathcal{X}}$.

Example 6 Let $NS^*_X = \{w_1, w_2\}$ and let $\mathbb{N}^*_{\tau}$ = $\{0_{NS}, \mathcal{K}, 1_{NS}\}$is NT on $\mathbb{N}^*_{\mathcal{X}}$, where $\mathcal{K} = \langle w, \left(\frac{3}{10}, \frac{5}{10}, \frac{6}{10}\right), \left(\frac{7}{10}, \frac{5}{10}, \frac{3}{10}\right)\rangle$. Then NS $\mathcal{W}^*_1 = \langle w, \left(\frac{7}{10}, \frac{5}{10}, \frac{3}{10}\right), \left(\frac{6}{10}, \frac{5}{10}, \frac{4}{10}\right)\rangle$ is $NS(G^*)CS$ but not NS(R)CS in $\mathbb{N}^*_{\mathcal{X}}$

Remark 1 $NS(G^*)CS$ is independent from NS(α)CS, $NS(S)CS$, $N(SP)CS$, *NS(SG)CS*, and $NS(b)CS$ as observed from the subsequent example.

Example 7 Let $\mathbb{N}^*_{\mathcal{X}} = \{w_1, w_2\}$ and let $\mathbb{N}^*_{\tau} = \{0_{NS}, \mathcal{K}, 1_{NS}\}$is NT on$\mathbb{N}^*_{\mathcal{X}}$ where $\mathcal{K} = \langle w, (\frac{3}{10}, \frac{5}{10}, \frac{2}{10}), (\frac{4}{10}, \frac{5}{10}, \frac{5}{10})\rangle$. Then NS $\mathcal{W}^*_1 = \langle w, (\frac{1}{10}, \frac{5}{10}, \frac{6}{10}), (\frac{4}{10}, \frac{5}{10}, \frac{5}{10})\rangle$ is *NS(α)CS, NS(S)CS*, $N(SP)CS$, $NS(G)CS$ and $NS(b)CS$ but not $NS(G^*)CS$ in $\mathbb{N}^*_{\mathcal{X}}$

Example 8 Let $\mathbb{N}^*_{\mathcal{X}} = \{w_1, w_2\}$ and let $\mathbb{N}^*_{\tau}$ =$\{0_{NS}, \mathcal{K}, 1_{NS}\}$is NT on $\mathbb{N}^*_{\mathcal{X}}$, where

$$\mathcal{K} = \left\langle w, \left(\frac{4}{10}, \frac{5}{10}, \frac{6}{10}\right), \left(\frac{3}{10}, \frac{5}{10}, \frac{7}{10}\right)\right\rangle$$

Then NS $\mathcal{W}^*_1 = \langle w, (\frac{6}{10}, \frac{5}{10}, \frac{1}{10}), (\frac{7}{10}, \frac{5}{10}, \frac{1}{10})\rangle$ is $NS(G^*)CS$ but neither NSaCS nor $NS(S)CS$ in $\mathbb{N}^*_{\mathcal{X}}$.

Example 9 Let $\mathbb{N}^*_{\mathcal{X}} = \{w_1, w_2\}$ and let $\mathbb{N}^*_{\tau} = \{0_{NS}, \mathcal{K}_1, \mathcal{K}_2, 1_{NS}\}$ is NT on $\mathbb{N}^*_{\mathcal{X}}$, where

$$\mathcal{K}_1 = \left\langle w, \left(\frac{5}{10}, \frac{5}{10}, \frac{5}{10}\right), \left(\frac{2}{10}, \frac{5}{10}, \frac{7}{10}\right)\right\rangle,$$

$$\mathcal{K}_2 = \left\langle w, \left(\frac{8}{10}, \frac{5}{10}, \frac{2}{10}\right), \left(\frac{8}{10}, \frac{5}{10}, \frac{3}{10}\right)\right\rangle.$$

Then,

$$NS\mathcal{W}^*_1 = \left\langle w, \left(\frac{5}{10}, \frac{5}{10}, \frac{5}{10}\right), \left(\frac{7}{10}, \frac{5}{10}, \frac{3}{10}\right)\right\rangle$$

is $NS(G^*)CS$ but neither $N(SP)CS$ nor $NS(b)CS$ in $\mathbb{N}^*_{\mathcal{X}}$.

Example 10 Let $\mathbb{N}^*_{\mathcal{X}}= \{w_1, w_2\}$ and let $\mathbb{N}^*_{\tau} = \{0_{NS}, \mathcal{K}, 1_{NS}\}$ is NT on $\mathbb{N}^*_{\mathcal{X}}$, where

$$\mathcal{K} = \left\langle w, \left(\frac{0}{10}, \frac{5}{10}, \frac{5}{10}\right), \left(\frac{9}{10}, \frac{5}{10}, \frac{1}{10}\right)\right\rangle$$

Then

$$NS\mathcal{W}^*_1 = \left\langle w, \left(\frac{0}{10}, \frac{5}{10}, \frac{7}{10}\right), \left(\frac{3}{10}, \frac{5}{10}, \frac{7}{10}\right)\right\rangle$$

is $N(SP)CS$ and NS bCS but not $NS(G^*)CS$ in $\mathbb{N}^*_{\mathcal{X}}$.

Example 11 Let $\mathbb{N}^*_{\mathcal{X}} = \{w_1, w_2\}$ and let $\mathbb{N}^*_{\tau}$=$\{0_{NS}, \mathcal{K}_1, \mathcal{K}_2, 1_{NS}\}$ is NT on $\mathbb{N}^*_{\mathcal{X}}$ where

$$\mathcal{K}_1 = \left\langle w, \left(\frac{5}{10}, \frac{5}{10}, \frac{5}{10}\right), \left(\frac{2}{10}, \frac{5}{10}, \frac{7}{10}\right)\right\rangle$$

$$\mathcal{K}_2 = \left\langle w, \left(\frac{8}{10}, \frac{5}{10}, \frac{2}{10}\right), \left(\frac{8}{10}, \frac{5}{10}, \frac{2}{10}\right)\right\rangle$$

Then

$$NS\mathcal{W}_1^* = \left\langle w, \left(\frac{7}{10}, \frac{5}{10}, \frac{3}{10}\right), \left(\frac{7}{10}, \frac{5}{10}, \frac{3}{10}\right)\right\rangle$$

is $NS(G^*)CS$ but not $NS(G)CS$ in $\mathbb{N}_{\mathcal{X}}^*$.

Theorem 5 *The union of two $NS(G^*)CS$'s is NSG^*CS.*

Proof Let $\mathcal{W}_1^*$ and $\mathcal{W}_2^*$ be the two $NS(G^*)CS$'s in $\mathbb{N}_{\mathcal{X}}^*$ and let $\mathcal{W}_1^* \cup \mathcal{W}_2^* \subseteq \mathcal{K}$, where $\mathcal{K}$ is a $NS(G)OS$ in $\mathbb{N}_{\mathcal{X}}^*$hence $\mathcal{W}_1^* \subseteq \mathcal{K}$ or $\mathcal{W}_2^* \subseteq \mathcal{K}$ or both $\subseteq \mathcal{K}$. Since $\mathcal{W}_1^*$ and $\mathcal{W}_2^*$ are $NS(G^*)CS$. NScl($\mathcal{W}_1^*$) $\subseteq \mathcal{K}$ and NScl($\mathcal{W}_2^*$) $\subseteq \mathcal{K}$. Therefore, NScl($\mathcal{W}_1^*$ $\cup$ $\mathcal{W}_2^*$) $\subseteq \mathcal{K}$. Thus $\mathcal{W}_1^* \cup \mathcal{W}_2^*$ is$NS(G^*)CS$.

Remark 2 The intersection of any two $NS(G^*)CS$ s is not a $NS(G^*)CS$ in common use the next example.

Example 12 Let $\mathbb{N}_{\mathcal{X}}^* = \{w_1, w_2\}$ and let $\mathbb{N}_{\tau}^*$ =$\{0_{NS}, \mathcal{K}, 1_{NS}\}$is NT on $\mathbb{N}_{\mathcal{X}}^*$, where

$$\mathcal{K} = \left\langle w, \left(\frac{5}{10}, \frac{5}{10}, \frac{1}{10}\right), \left(\frac{0}{10}, \frac{5}{10}, \frac{1}{10}\right)\right\rangle.$$

Then

$$NS's\,\mathcal{W}_1^* = \left\langle w, \left(\frac{2}{10}, \frac{5}{10}, \frac{7}{10}\right), \left(\frac{1}{10}, \frac{5}{10}, \frac{0}{10}\right)\right\rangle$$

$$\mathcal{W}_2^* = \left\langle w, \left(\frac{6}{10}, \frac{5}{10}, \frac{3}{10}\right), \left(\frac{0}{10}, \frac{5}{10}, \frac{1}{10}\right)\right\rangle$$

are $NS(G^*)CS$'s in $\mathbb{N}_{\mathcal{X}}^*$ but $\mathcal{W}_1^*$ n $\mathcal{W}_2^*$ is not a $NS(G^*)CS$ in $\mathbb{N}_{\mathcal{X}}^*$.

Theorem 6 *If $\mathcal{W}_1^*$ is $NS(G^*)CS$ in ($\mathbb{N}_{\mathcal{X}}^*$, $\mathbb{N}_{\tau}^*$), such that $\mathcal{W}_1^* \subseteq \mathcal{W}_2^* \subseteq$ NScl($\mathcal{W}_1^*$). Then $\mathcal{W}_2^*$ is also a $NS(G^*)CS$ of ($\mathbb{N}_{\mathcal{X}}^*$, $\mathbb{N}_{\tau}^*$).*

Proof Let $\mathcal{K}$ be a $NS(G)OS$ in ($\mathbb{N}_{\mathcal{X}}^*$, $\mathbb{N}_{\tau}^*$) such that $\mathcal{W}_2^* \subseteq \mathcal{K}$, Since $\mathcal{W}_1^* \subseteq \mathcal{W}_2^*$, $\mathcal{W}_1^* \subseteq \mathcal{K}$ and $\mathcal{K}$ be a $NS(G)OS$. Also since $\mathcal{W}_1^*$ is NSG*CS, NScl($\mathcal{W}_1^*$) $\subseteq \mathcal{K}$. By hypothesis $\mathcal{W}_2^* \subseteq$ NScl($\mathcal{W}_1^*$). This implies NScl($\mathcal{W}_2^*$) $\subseteq$ (NS)cl(NS)cl($\mathcal{W}_1^*$)) $\subseteq \mathcal{K}$. Therefore NScl($\mathcal{W}_2^*$) $\subseteq \mathcal{K}$. Hence $\mathcal{W}_2^*$ is $NS(G^*)CS$ of $\mathbb{N}_{\mathcal{X}}^*$.

Theorem 7 *If $\mathcal{W}_1^*$ is both $NS(G)OS$ and $NS(G^*)CS$ of ($\mathbb{N}_{\mathcal{X}}^*$, $\mathbb{N}_{\tau}^*$), then $\mathcal{W}_1^*$ is NSCS in $\mathbb{N}_{\mathcal{X}}^*$.*

Proof Let $\mathcal{W}_1^*$ is $NS(G)OS$ in $\mathbb{N}_{\mathcal{X}}^*$. Since $\mathcal{W}_1^* \subseteq \mathcal{W}_1^*$, by hypothesis NScl($\mathcal{W}_1^*$) $\subseteq$ $\mathcal{W}_1^*$. But using the definition, $\mathcal{W}_1^* \subseteq$ NScl($\mathcal{W}_1^*$). Therefore NScl($\mathcal{W}_1^*$) $= \mathcal{W}_1^*$. Hence $\mathcal{W}_1^*$ is NSCS of $\mathbb{N}_{\mathcal{X}}^*$.

Theorem 8 *Let ($\mathbb{N}_{\mathcal{X}}^*$, $\mathbb{N}_{\tau}^*$) be a NTS. Then $NS(G)O$ ($\mathbb{N}_{\mathcal{X}}^*$) =$NS(G)C$ ($\mathbb{N}_{\mathcal{X}}^*$) if and only if every NS in ($\mathbb{N}_{\mathcal{X}}^*$, $\mathbb{N}_{\tau}^*$) is $NS(G^*)CS$ in $\mathbb{N}_{\mathcal{X}}^*$.*

Proof **Necessity**:
Suppose that $NS(G)O$ ($\mathbb{N}^*_{\mathcal{X}}$) =NSGC($\mathbb{N}^*_{\mathcal{X}}$). Let $\mathcal{W}^*_1 \subseteq \mathcal{K}$ and $\mathcal{K}$ is $NS(G)OS$ in $\mathbb{N}^*_{\mathcal{X}}$. This implies NScl($\mathcal{W}^*_1$) $\subseteq$ NScl($\mathcal{K}$). Since $\mathcal{K}$ is $NS(G)OS$ in $\mathbb{N}^*_{\mathcal{X}}$. Since by hypothesis $\mathcal{K}$ is $NS(G)CS$ in $\mathbb{N}^*_{\mathcal{X}}$, NScl($\mathcal{K}$) $\subseteq \mathcal{K}$. This implies NScl($\mathcal{W}^*_1$) $\subseteq \mathcal{K}$. Therefore $\mathcal{W}^*_1$ is $NS(G^*)CS$ in $\mathbb{N}^*_{\mathcal{X}}$
Sufficiency: Suppose that each NS in ($\mathbb{N}^*_{\mathcal{X}}$, $\mathbb{N}^*_{\tau}$) is $NS(G^*)CS$ in $\mathbb{N}^*_{\mathcal{X}}$. Let $\mathcal{K} \in NSO(\mathbb{N}^*_{\mathcal{X}})$, then $\mathcal{K} \in NS(G)O$ ($\mathbb{N}^*_{\mathcal{X}}$). Since $\mathcal{K} \subseteq \mathcal{K}$ and $\mathcal{K}$ is NSOS in $NS^*_{\mathcal{X}}$, by hypothesis NScl($\mathcal{K}$) $\subseteq \mathcal{K}$. That is $\mathcal{K} \in NS(G)C(\mathbb{N}^*_{\mathcal{X}})$. Hence $NS(G)O(\mathbb{N}^*_{\mathcal{X}})$ $NS(G)C(\mathbb{N}^*_{\mathcal{X}})$. Let $\mathcal{W}^*_1 \in NS(G)C(\mathbb{N}^*_{\mathcal{X}})$ then $\mathcal{W}^{*C}_1$ is a $NS(G)OS$ in $\mathbb{N}^*_{\mathcal{X}}$. But $NS(G)O$ ($\mathbb{N}^*_{\mathcal{X}}$) $\subseteq NSGC(\mathbb{N}^*_{\mathcal{X}})$. Therefore, $\mathcal{W}^{*C}_1 \in NSGC(\mathbb{N}^*_{\mathcal{X}})$. That is $\mathcal{W}^*_1 \in NS(G)O(\mathbb{N}^*_{\mathcal{X}})$. Hence $NS(G)C(\mathbb{N}^*_{\mathcal{X}})$ $NS(G)O(\mathbb{N}^*_{\mathcal{X}})$.
Thus $NS(G)O(\mathbb{N}^*_{\mathcal{X}}) \subseteq NSGC(\mathbb{N}^*_{\mathcal{X}})$.

Theorem 9 *If $\mathcal{W}^*_1$ is NSOS and a $NS(G^*)CS$ in ($\mathbb{N}^*_{\mathcal{X}}$, $\mathbb{N}^*_{\tau}$), then*

(i) *$\mathcal{W}^*_1$ is NS(R)OS in $\mathbb{N}^*_{\mathcal{X}}$*
(ii) *$\mathcal{W}^*_1$ is NS(R)CS in $\mathbb{N}^*_{\mathcal{X}}$*

Proof (i) Let $\mathcal{W}^*_1$ be an NSOS and an $NS(G^*)CS$ in $\mathbb{N}^*_{\mathcal{X}}$. Then NScl($\mathcal{W}^*_1$) $\mathcal{W}^*_1$. That is NS int(NScl($\mathcal{W}^*_1$)) $\mathcal{W}^*_1$. Since $\mathcal{W}^*_1$ is an NSOS, $\mathcal{W}^*_1$ is NS(P)OS in $\mathbb{N}^*_{\mathcal{X}}$. Hence $\mathcal{W}^*_1$ NSint(NScl($\mathcal{W}^*_1$)). Therefore $\mathcal{W}^*_1$ = NSint(NScl($\mathcal{W}^*_1$)). Hence $\mathcal{W}^*_1$ is *NS(R)OS* in $\mathbb{N}^*_{\mathcal{X}}$.
(ii) Let $\mathcal{W}^*_1$ be an NSOS and a $NS(G^*)CS$ in $\mathbb{N}^*_{\mathcal{X}}$. Then NScl($\mathcal{W}^*_1$) $\mathcal{W}^*_1$. That is NScl(NSint($\mathcal{W}^*_1$)) $\mathcal{W}^*_1$. Since $\mathcal{W}^*_1$ is a NSOS, $\mathcal{W}^*_1$ is NSOS in $\mathbb{N}^*_{\mathcal{X}}$. Hence $\mathcal{W}^*_1$ NScl(NSint($\mathcal{W}^*_1$)). Therefore, $\mathcal{W}^*_1$ = NSint(NScl($\mathcal{W}^*_1$)). Hence $\mathcal{W}^*_1$ is NS(R)CS in $\mathbb{N}^*_{\mathcal{X}}$.

3 Neutrosophic g^*-Open Sets

In this section, we introduce Neutrosophic g^*-open sets and studied some of their properties.

Definition 10 A NS $\mathcal{W}^*_1$ is said to be a Neutrosophic g^*-open set ($NS(G^*)OS$ in short) in ($\mathbb{N}^*_{\mathcal{X}}$, $\mathbb{N}^*_{\tau}$) if the complement $\mathcal{W}^{*C}_1$ is $NS(G^*)CS$ in $\mathbb{N}^*_{\mathcal{X}}$. The family of all $NS(G^*)OS$'s of a NTS ($\mathbb{N}^*_{\mathcal{X}}$, $\mathbb{N}^*_{\tau}$) is denoted by $NS(G^*)O$ ($\mathbb{N}^*_{\mathcal{X}}$).

Theorem 10 *A subset $\mathcal{W}^*_1$ of ($\mathbb{N}^*_{\mathcal{X}}$, $\mathbb{N}^*_{\tau}$) is $NS(G^*)OS$ if and only if $\mathcal{W}^*_2 \subseteq$ NSint($\mathcal{W}^*_1$) whenever $\mathcal{W}^*_2$ is $NS(G)CS$ in $\mathbb{N}^*_{\mathcal{X}}$ and $\mathcal{W}^*_2 \subseteq \mathcal{W}^*_1$.*

Proof **Necessity**: Let $\mathcal{W}^*_1$ is NSG*OS in $\mathbb{N}^*_{\mathcal{X}}$. Let $\mathcal{W}^*_2$ be a $NS(G)CS$ in $\mathbb{N}^*_{\mathcal{X}}$ and $\mathcal{W}^*_2 \subseteq \mathcal{W}^*_1$. Then $\mathcal{W}^{*C}_2$ is $NS(G)OS$ in $\mathbb{N}^*_{\mathcal{X}}$ such that $\mathcal{W}^{*C}_1 \subseteq \mathcal{W}^{*C}_2$. Since $\mathcal{W}^{*C}_1$ is $NS(G^*)CS$, we have NScl($\mathcal{W}^{*C}_1$) $\subseteq \mathcal{W}^{*C}_2$. Hence NSint(($\mathcal{W}^*_1$))$^c \subseteq \mathcal{W}^{*C}_2$. Therefore $\mathcal{W}^*_2 \subseteq$ NSint($\mathcal{W}^*_1$).
Sufficiency: Let $\mathcal{W}^*_2 \subseteq$ NSint($\mathcal{W}^*_1$) whenever $\mathcal{W}^*_2$ is $NS(G)CS$ in $\mathbb{N}^*_{\mathcal{X}}$ and $\mathcal{W}^*_2 \subseteq \mathcal{W}^*_1$. Then $\mathcal{W}^{*C}_1 \subseteq \mathcal{W}^{*C}_2$ and $\mathcal{W}^{*C}_2$ is $NS(G)OS$. By hypothesis, NSint(($\mathcal{W}^*_1$))$^c \subseteq$

$\mathcal{W}_2^{*C}$, which implies NScl($\mathcal{W}_1^{*C}$) $\subseteq \mathcal{W}_2^{*C}$. Therefore, $\mathcal{W}_1^{*C}$ is $NS(G^*)CS$ of $\mathbb{N}^*_{\mathcal{X}}$. Hence $\mathcal{W}_1^*$ is $NS(G^*)OS$ in $\mathbb{N}^*_{\mathcal{X}}$.

Theorem 11 *Every NSOS is $NS(G^*)OS$.*

Proof Let $\mathcal{W}_1^*$ be a *NSOS*. Then $\mathcal{W}_1^{*C}$ is NSCS. By Theorem 1, every *NSCS* is $NS(G^*)CS$. Therefore $\mathcal{W}_1^{*C}$ is $NS(G^*)CS$. Hence $\mathcal{W}_1^*$ is $NS(G^*)OS$.

Example 13 Let $\mathbb{N}^*_{\mathcal{X}} = \{w_1, w_2\}$ and let $\mathbb{N}^*_{\tau}$ ={0_{NS}, $\mathcal{K}$, 1_{NS}}is NT on $\mathbb{N}^*_{\mathcal{X}}$, where

$$\mathcal{K} = \left\langle w, \left(\frac{5}{10}, \frac{5}{10}, \frac{7}{10}\right), \left(\frac{3}{10}, \frac{5}{10}, \frac{8}{10}\right)\right\rangle$$

Then,

$$NS\mathcal{W}_1^* = \left\langle w, \left(\frac{2}{10}, \frac{5}{10}, \frac{7}{10}\right), \left(\frac{2}{10}, \frac{5}{10}, \frac{8}{10}\right)\right\rangle$$

is $NS(G^*)OS$ but not *NSOS* in $\mathbb{N}^*_{\mathcal{X}}$.

Theorem 12 *Every NS(R)OS is $NS(G^*)OS$.*

Proof Let $\mathcal{W}_1^*$ be a *NS(R)OS*. Then $\mathcal{W}_1^{*C}$ is *NS(R)CS*. By Theorem 4, every *NS(R)CS* is $NS(G^*)CS$. Therefore $\mathcal{W}_1^{*C}$ is $NS(G^*)CS$. Hence $\mathcal{W}_1^*$ is $NS(G^*)OS$.

Example 14 Let $\mathbb{N}^*_{\mathcal{X}} = \{w_1, w_2\}$ and let $\mathbb{N}^*_{\tau}$ ={0_{NS}, $\mathcal{K}$, 1_{NS}}is NT on $\mathbb{N}^*_{\mathcal{X}}$, where

$$\mathcal{K} = \left\langle w, \left(\frac{3}{10}, \frac{5}{10}, \frac{6}{10}\right), \left(\frac{7}{10}, \frac{5}{10}, \frac{3}{10}\right)\right\rangle.$$

Then

$$NS\mathcal{W}_1^* = \left\langle w, \left(\frac{3}{10}, \frac{5}{10}, \frac{7}{10}\right), \left(\frac{4}{10}, \frac{5}{10}, \frac{6}{10}\right)\right\rangle$$

is $NS(G^*)CS$ but not *NS(R)OS* in $\mathbb{N}^*_{\mathcal{X}}$.

Theorem 13 *Every $NS(G^*)OS$ is $NS(G)OS$.*

Proof Let $\mathcal{W}_1^*$ be a $NS(G^*)OS$ in ($\mathbb{N}^*_{\mathcal{X}}$, $\mathbb{N}^*_{\tau}$). Then $\mathcal{W}_1^{*C}$ is $NS(G^*)CS$. By Theorem 2, every $NS(G^*)CS$ is $NS(G)CS$. Therefore $\mathcal{W}_1^{*C}$ is $NS(G)CS$. Hence $\mathcal{W}_1^*$ is $NS(G)OS$.

Example 15 Let $\mathbb{N}^*_{\mathcal{X}}$ ={w_1, w_2} and let $\mathbb{N}^*_{\tau}$ ={0_{NS}, $\mathcal{K}$, 1_{NS}}is NT on $\mathbb{N}^*_{\mathcal{X}}$, where

$$\mathcal{K} = \left\langle w, \left(\frac{5}{10}, \frac{5}{10}, \frac{4}{10}\right), \left(\frac{2}{10}, \frac{5}{10}, \frac{2}{10}\right)\right\rangle.$$

Then

$$NSW_1^* = \left\langle w, \left(\frac{4}{10}, \frac{5}{10}, \frac{6}{10}\right), \left(\frac{2}{10}, \frac{5}{10}, \frac{6}{10}\right)\right\rangle$$

is $NS(G)OS$ but not $NS(G^*)OS$ in $\mathbb{N}^*_{\mathcal{X}}$.

Theorem 14 *Every $NS(G^*)OS$ is NS(αG)OS.*

Proof Let $\mathcal{W}_1^*$ be a $NS(G^*)OS$ in ($\mathbb{N}^*_{\mathcal{X}}$, $\mathbb{N}^*_{\tau}$). Then $\mathcal{W}_1^{*C}$ is $NS(G^*)CS$. By Theorem 4, every $NS(G^*)CS$ is NSaGCS. Therefore $\mathcal{W}_1^{*C}$ is *NS(aG)CS*. Hence $\mathcal{W}_1^*$ is *NS(aG)OS*.

Example 16 Let $\mathbb{N}^*_{\mathcal{X}} = \{w_1, w_2\}$ and let $\mathbb{N}^*_{\tau} = \{0_{NS}, \mathcal{K}, 1_{NS}\}$ is NT on $\mathbb{N}^*_{\mathcal{X}}$, where

$$\mathcal{K} = \left\langle w, \left(\frac{4}{10}, \frac{5}{10}, \frac{2}{10}\right), \left(\frac{6}{10}, \frac{5}{10}, \frac{7}{10}\right)\right\rangle$$

Then

$$NS\mathcal{W}_1^* = \left\langle w, \left(\frac{6}{10}, \frac{5}{10}, \frac{4}{10}\right), \left(\frac{5}{10}, \frac{5}{10}, \frac{3}{10}\right)\right\rangle$$

is $NS(G^*)OS$ but not $NS(G^*)OS$ in $\mathbb{N}^*_{\mathcal{X}}$.

Theorem 15 *Every $NS(G^*)OS$ is NS(GSP)OS.*

Proof Let $\mathcal{W}_1^*$ be a $NS(G^*)OS$ in ($\mathbb{N}^*_{\mathcal{X}}$, $\mathbb{N}^*_{\tau}$). Then $\mathcal{W}_1^{*C}$ is $NS(G^*)CS$. By Theorem 5, every $NS(G^*)CS$ is *NS(GSP)CS*. Therefore $\mathcal{W}_1^{*C}$ is *NS(GSP)CS*. Hence $\mathcal{W}_1^*$ is NS(GSP)OS.

Example 17 Let $\mathbb{N}^*_{\mathcal{X}} = \{w_1, w_2\}$ and let $\mathbb{N}^*_{\tau} = \{0_{NS}, \mathcal{K}, 1_{NS}\}$is NT on $\mathbb{N}^*_{\mathcal{X}}$, where

$$\mathcal{K} = \left\langle w, \left(\frac{5}{10}, \frac{5}{10}, \frac{5}{10}\right), \left(\frac{6}{10}, \frac{5}{10}, \frac{4}{10}\right)\right\rangle.$$

Then

$$NS\mathcal{W}_1^* = \left\langle w, \left(\frac{5}{10}, \frac{5}{10}, \frac{5}{10}\right), \left(\frac{3}{10}, \frac{5}{10}, \frac{7}{10}\right)\right\rangle$$

is *NS(GSP)OS* however not a $NS(G^*)OS$ in $\mathbb{N}^*_{\mathcal{X}}$.

Theorem 16 *The intersection of two $NS(G^*)OS$ is $NS(G^*)OS$.*

Proof Let $\mathcal{W}_1^*$ and $\mathcal{W}_2^*$ be the two NSG*OS's in $\mathbb{N}^*_{\mathcal{X}}$, $\mathcal{W}_1^{*C}$ and $\mathcal{W}_2^{*C}$ are $NS(G^*)CS$. By Theorem 5, $\mathcal{W}_1^{*C} \cup \mathcal{W}_2^{*C}$ is $NS(G^*)CS$ in $\mathbb{N}^*_{\mathcal{X}}$. Therefore $(\mathcal{W}_1^* \cap \mathcal{W}_2^*)^C$ is $NS(G^*)CS$. Thus $\mathcal{W}_1^* \cap \mathcal{W}_2^*$ is $NS(G^*)OS$ in $\mathbb{N}^*_{\mathcal{X}}$.

Theorem 17 *The union of any two $NS(G^*)OS$'s is not a $NS(G^*)OS$ in general as seen in the subsequent example.*

Example 18 Let $\mathbb{N}^*_{\mathcal{X}} = \{w_1, w_2\}$ and let $\mathbb{N}^*_{\tau}$=$\{0_{NS}, \mathcal{K}, 1_{NS}\}$ is NT on $\mathbb{N}^*_{\mathcal{X}}$, where

$$\mathcal{K} = \left\langle w, \left(\frac{5}{10}, \frac{5}{10}, \frac{1}{10}\right), \left(\frac{0}{10}, \frac{5}{10}, \frac{1}{10}\right)\right\rangle$$

Then,

$$NS'sW^*_1 = \left\langle w, \left(\frac{7}{10}, \frac{5}{10}, \frac{2}{10}\right), \left(\frac{0}{10}, \frac{5}{10}, \frac{1}{10}\right)\right\rangle$$
$$\mathcal{W}^*_2 = \left\langle w, \left(\frac{3}{10}, \frac{5}{10}, \frac{6}{10}\right), \left(\frac{1}{10}, \frac{5}{10}, \frac{0}{10}\right)\right\rangle$$

are $NS(G^*)OS$'s in $\mathbb{N}^*_{\mathcal{X}}$ but $\mathcal{W}^*_1 \cup \mathcal{W}^*_2$ is not a $NS(G^*)CS$ in $\mathbb{N}^*_{\mathcal{X}}$.

Theorem 18 *Let* ($\mathbb{N}^*_{\mathcal{X}}$, $\mathbb{N}^*_{\tau}$) *be a NTS. If* $\mathcal{W}^*_1$ *is NS of* $\mathbb{N}^*_{\mathcal{X}}$ *then the subsequent properties are equal*

(i) $\mathcal{W}^*_1 \in NS(G^*)O(\mathbb{N}^*_{\mathcal{X}})$
(ii) $\mathcal{V} \subseteq NSint(\mathcal{W}^*_1)$ *whenever* $\mathcal{V} \subseteq \mathcal{W}^*_1$ *and* $\mathcal{V}$ *is* $NS(G)CS$ *in* $\mathbb{N}^*_{\mathcal{X}}$
(iii) There exists NOS's G and G_1 *such that* $G_1 \subseteq \mathcal{V} \subseteq G$, *where* $G = NSint(\mathcal{W}^*_1)$, $\mathcal{V} \subseteq \mathcal{W}^*_1$ *and* $\mathcal{V}$ *is NSCS in* $\mathbb{N}^*_{\mathcal{X}}$

Proof (i)$\Rightarrow$(ii): Let $\mathcal{W}^*_1 \in NS(G^*)O(\mathbb{N}^*_{\mathcal{X}})$. Then $\mathcal{W}^{*C}_1$ is $NS(G^*)CS$ in $\mathbb{N}^*_{\mathcal{X}}$. Therefore NScl($\mathcal{W}^{*C}_1$) $\subseteq \mathcal{K}$ whenever $\mathcal{W}^{*C}_1 \subseteq \mathcal{K}$ and $\mathcal{K}$ is $NS(G)OS$ in $\mathbb{N}^*_{\mathcal{X}}$. Taking complement on each sides [NScl($\mathcal{W}^{*C}_1$)]$^C \supseteq \mathcal{U}^C$ whenever $[\mathcal{W}^{*C}_1]^C \supseteq \mathcal{U}^C$. Therefore $\mathcal{U}^C \subseteq NSint(\mathcal{W}^*_1)$ whenever $\mathcal{U}^C \subseteq \mathcal{W}^*_1$ and $\mathcal{U}^C$ is $NS(G)CS$ in $\mathbb{N}^*_{\mathcal{X}}$. Replacing $\mathcal{U}^C$ by $\mathcal{V}$, $\mathcal{V} \subseteq NSint(\mathcal{W}^*_1)$ whenever $\mathcal{V} \subseteq \mathcal{W}^*_1$ and $\mathcal{V}$ is $NS(G)CS$ in $\mathbb{N}^*_{\mathcal{X}}$.
(ii) $\Rightarrow$ (iii): $\mathcal{V} \subseteq NSint(\mathcal{W}^*_1)$ whenever $\mathcal{V} \subseteq \mathcal{W}^*_1$ and $\mathcal{V}$ is $NS(G)CS$ in $\mathbb{N}^*_{\mathcal{X}}$. Hence $NSint(\mathcal{V}) \subseteq \mathcal{V} \subseteq NSint(\mathcal{W}^*_1)$ then there exists NOS's G and G_1 such that $G_1 \subseteq \mathcal{V} \subseteq G$, where $G = NSint(\mathcal{W}^*_1)$ and $G_1 = NSint(\mathcal{V})$.
(iii) $\Rightarrow$ (i): Suppose that there exists NOS's G and G_1 such that $G_1 \subseteq \mathcal{V} \subseteq G$. That is $\mathcal{V} \subseteq NSint(\mathcal{W}^*_1)$. Then NScl($\mathcal{W}^{*C}_1$) $\subseteq \mathcal{V}^C$ whenever $\mathcal{W}^{*C}_1 \subseteq \mathcal{V}^C$ and $\mathcal{V}^C$ is $NS(G)OS$ in $\mathbb{N}^*_{\mathcal{X}}$. Hence $\mathcal{W}^{*C}_1$ is $NS(G^*)CS$ in $\mathbb{N}^*_{\mathcal{X}}$Therefore $\mathcal{W}^*_1 \in NS(G^*)O(\mathbb{N}^*_{\mathcal{X}})$.

Theorem 19 *Let* ($\mathbb{N}^*_{\mathcal{X}}$, $\mathbb{N}^*_{\tau}$) *be an NTS. If* $\mathcal{W}^*_1$ *is NS of* $\mathbb{N}^*_{\mathcal{X}}$. *Then for every* $\mathcal{W}^*_1 \in NS(G^*)O(\mathbb{N}^*_{\mathcal{X}})$ *and every* $\mathcal{W}^*_2 \in NS(\mathbb{N}^*_{\mathcal{X}})$, $NSint(\mathcal{W}^*_1) \subseteq \mathcal{W}^*_2 \subseteq \mathcal{W}^*_1$ *implies* $\mathcal{W}^*_2 \in NS(G^*)O(\mathbb{N}^*_{\mathcal{X}}$.

Proof By hypothesis NSint($\mathcal{W}^*_1$) $\subseteq \mathcal{W}^*_2 \subseteq \mathcal{W}^*_1$. Apply complement on each sides, we get $\mathcal{W}^{*C}_1 \subseteq \mathcal{W}^{*C}_2 \subseteq NScl(\mathcal{W}^{*C}_1)$. Let $\mathcal{W}^{*C}_2 \subseteq \mathcal{K}$ and $\mathcal{K}$ is $NS(G)OS$ in $NS^*_{\mathcal{X}}$. Since $\mathcal{W}^{*C}_1 \subseteq \mathcal{W}^{*C}_2$, $\mathcal{W}^{*C}_1 \subseteq \mathcal{K}$. Since $\mathcal{W}^{*C}_1$ is $NS(G^*)CS$, NScl($\mathcal{W}^{*C}_1$) $\subseteq \mathcal{K}$. Therefore NScl($\mathcal{W}^{*C}_2$) $\subseteq NScl(\mathcal{W}^{*C}_1) \subseteq \mathcal{K}$. Hence $\mathcal{W}^{*C}_2$ is $NS(G^*)CS$ in $\mathbb{N}^*_{\mathcal{X}}$Therefore $\mathcal{W}^*_2$ is $NS(G^*)OS$ in $\mathbb{N}^*_{\mathcal{X}}$. That is $\mathcal{W}^*_2 \in NS(G^*)O(\mathbb{N}^*_{\mathcal{X}})$.

4 Applications of Neutrosophic G^*-Closed Sets

In this section, we introduce Neutrosophic $T^*_{\frac{1}{2}}$ space, Neutrosophic $^*T_{\frac{1}{2}}$ space and applications of Neutrosophic G^*-closed sets.

Definition 11 A NTS $(\mathbb{N}^*_{\mathcal{X}}, \mathbb{N}^*_{\tau})$ is said to be a Neutrosophic $T^*_{\frac{1}{2}}$ space (in short NS$T^*_{\frac{1}{2}}$) if every $NS(G^*)CS$ of $(\mathbb{N}^*_{\mathcal{X}}, \mathbb{N}^*_{\tau})$ is NSCS of $(\mathbb{N}^*_{\mathcal{X}}, \mathbb{N}^*_{\tau})$.

Theorem 20 *Every Neutrosophic $T_{\frac{1}{2}}$ space is Neutrosophic $T^*_{\frac{1}{2}}$ space.*

Proof Let $\mathbb{N}^*_{\mathcal{X}}$be NS$T^*_{\frac{1}{2}}$ a space and let $\mathcal{W}^*_1$ be a $NS(G^*)CS$ of $(\mathbb{N}^*_{\mathcal{X}}, \mathbb{N}^*_{\tau})$. Since every $NS(G^*)CS$ is $NS(G)CS$. By hypothesis $\mathcal{W}^*_1$ is NSCS. Hence $\mathbb{N}^*_{\mathcal{X}}$ is a NS$T^*_{\frac{1}{2}}$ space.

Remark 3 Every Neutrosophic $T^*_{\frac{1}{2}}$space need not be a Neutrosophic $T_{\frac{1}{2}}$ space as seen from the following example.

Example 19 Let $\mathbb{N}^*_{\mathcal{X}} = \{w_1, w_2\}$ and let $\mathbb{N}^*_{\tau} = \{0_{NS}, \mathcal{K}, 1_{NS}\}$ is NT on $\mathbb{N}^*_{\mathcal{X}}$, where

$$\mathcal{K} = \left\langle w, \left(\frac{9}{10}, \frac{5}{10}, \frac{1}{10}\right), \left(\frac{9}{10}, \frac{5}{10}, \frac{1}{10}\right)\right\rangle.$$

Clearly $(\mathbb{N}^*_{\mathcal{X}}, \mathbb{N}^*_{\tau})$ is Neutrosophic $T^*_{\frac{1}{2}}$ space, but not Neutrosophic $T_{\frac{1}{2}}$ space.

Theorem 21 *A NTS $(\mathbb{N}^*_{\mathcal{X}}, \mathbb{N}^*_{\tau})$ is Neutrosophic $T^*_{\frac{1}{2}}$ space if and only if $NS(G^*)OS$ $(\mathbb{N}^*_{\mathcal{X}}) = NSOS(\mathbb{N}^*_{\mathcal{X}})$.*

Proof **Necessity**: Let $\mathcal{W}^*_1$ be a $NS(G^*)OS$ in $\mathbb{N}^*_{\mathcal{X}}$ then $\mathcal{W}^{*C}_1$ is $NS(G^*)CS$ in $\mathbb{N}^*_{\mathcal{X}}$. By hypothesis $\mathcal{W}^{*C}_1$ is NSCS of $\mathbb{N}^*_{\mathcal{X}}$, and therefore $\mathcal{W}^*_1$ is NSOS of $\mathbb{N}^*_{\mathcal{X}}$. Hence $NS(G^*)OS$ of $\mathbb{N}^*_{\mathcal{X}}$ = NSOS$(\mathbb{N}^*_{\mathcal{X}})$.
Sufficiency: Let $\mathcal{W}^*_1$ be a $NS(G^*)CS$ in $\mathbb{N}^*_{\mathcal{X}}$ the $\mathcal{W}^{*C}_1$ is $NS(G^*)OS$ of $\mathbb{N}^*_{\mathcal{X}}$. By our assumption $\mathcal{W}^{*C}_1$ is and NSOS in $\mathbb{N}^*_{\mathcal{X}}$, then $\mathcal{W}^*_1$ is a NSCS in $NS^*_{\mathcal{X}}$. Hence $(\mathbb{N}^*_{\mathcal{X}}, \mathbb{N}^*_{\tau})$ is Neutrosophic $T^*_{\frac{1}{2}}$ space.

Theorem 22 *Let $(\mathbb{N}^*_{\mathcal{X}}, \mathbb{N}^*_{\tau})$ be aNTS and $\mathbb{N}^*_{\mathcal{X}}$ is a NS$T^*_{\frac{1}{2}}$ space then:*

(i) Any union of $NS(G^*)CS$ is $NS(G^*)CS$ in $\mathbb{N}^*_{\mathcal{X}}$
(ii) Any intersection of $NS(G^*)OS$ is $NS(G^*)OS$ in $\mathbb{N}^*_{\mathcal{X}}$.

Proof (i) Let $\{A_i\}i \in j$ is a collection of $NS(G^*)CS$'s in a NS$T^*_{\frac{1}{2}}$. By hypothesis every $NS(G^*)CS$ is an NSCS in $\mathbb{N}^*_{\mathcal{X}}$. But the union of NSCS is an NSCS in $\mathbb{N}^*_{\mathcal{X}}$. Therefore $\{\cup A_i\}_{i\in j}$ is a NSCS in $\mathbb{N}^*_{\mathcal{X}}$. Since every NSCS is an $NS(G^*)CS$, $\{\cup A_i\}_{i\in j}$ is $NS(G^*)CS$ in $\mathbb{N}^*_{\mathcal{X}}$. Hence any union $NS(G^*)CS$'s is $NS(G^*)CS$ in $\mathbb{N}^*_{\mathcal{X}}$.
(ii) It is obvious from (i) by taking the complement.

Definition 12 An NTS $(\mathbb{N}^*_{\mathcal{X}}, \mathbb{N}^*_{\tau})$ is said to be a Neutrosophic $^*T_{\frac{1}{2}}$ space (in short NS$^*T_{\frac{1}{2}}$) if every $NS(G)CS$ of $(\mathbb{N}^*_{\mathcal{X}}, \mathbb{N}^*_{\tau})$ is $NS(G^*)CS$ of $(\mathbb{N}^*_{\mathcal{X}}, \mathbb{N}^*_{\tau})$.

Theorem 23 *Every Neutrosophic $T_{\frac{1}{2}}$ space is a Neutrosophic $^*T_{\frac{1}{2}}$ space.*

Proof Let $\mathbb{N}^*_{\mathcal{X}}$ be a NS$T_{\frac{1}{2}}$ space and let $\mathcal{W}^*_1$ be a $NS(G)CS$ of $(\mathbb{N}^*_{\mathcal{X}}, \mathbb{N}^*_{\tau})$. By hypothesis $\mathcal{W}^*_1$ is a NSCS. Since every NSCS of $(\mathbb{N}^*_{\mathcal{X}}, \mathbb{N}^*_{\tau})$ is a $NS(G^*)CS$ of $(\mathbb{N}^*_{\mathcal{X}}, \mathbb{N}^*_{\tau})$. Hence $\mathbb{N}^*_{\mathcal{X}}$ is a NS$^*T_{\frac{1}{2}}$space.

Theorem 24 *A Space $(\mathbb{N}^*_{\mathcal{X}}, \mathbb{N}^*_{\tau})$ is a NS$T_{\frac{1}{2}}$ space if and only if it is both NS$^*T_{\frac{1}{2}}$ space and NS$T^*_{\frac{1}{2}}$ space.*

Proof **Necessity**: It follows from Theorems 20 and 24.

Sufficiency: $(\mathbb{N}^*_{\mathcal{X}}, \mathbb{N}^*_{\tau})$ is both NS$^*T_{\frac{1}{2}}$ space and NS$T^*_{\frac{1}{2}}$ space. Let $\mathcal{W}^*_1$ be a $NS(G)CS$ of $(\mathbb{N}^*_{\mathcal{X}}, \mathbb{N}^*_{\tau})$. Since $(\mathbb{N}^*_{\mathcal{X}}, \mathbb{N}^*_{\tau})$ is a NS$^*T_{\frac{1}{2}}$ space, then $\mathcal{W}^*_1$ is $NS(G)CS$ of $(\mathbb{N}^*_{\mathcal{X}}, \mathbb{N}^*_{\tau})$. Since $(\mathbb{N}^*_{\mathcal{X}}, \mathbb{N}^*_{\tau})$ is a NS$T^*_{\frac{1}{2}}$ space, then $\mathcal{W}^*_1$ is NSCS of $(\mathbb{N}^*_{\mathcal{X}}, \mathbb{N}^*_{\tau})$. Thus $(\mathbb{N}^*_{\mathcal{X}}, \mathbb{N}^*_{\tau})$ is a NS$T_{\frac{1}{2}}$ space.

Theorem 25 *An NTS $(\mathbb{N}^*_{\mathcal{X}}, \mathbb{N}^*_{\tau})$ is a Neutrosophic $^*T_{\frac{1}{2}}$ space if and only if $NS(G^*)OS$ $(\mathbb{N}^*_{\mathcal{X}}) = NS(G)OS(\mathbb{N}^*_{\mathcal{X}})$.*

Proof **Necessity**: Let $\mathcal{W}^*_1$ be a $NS(G)OS$ $(\mathbb{N}^*_{\mathcal{X}})$ in $NS^*_{\mathcal{X}}$, then $\mathcal{W}^{*C}_1$ is $NS(G)CS$ in $\mathbb{N}^*_{\mathcal{X}}$. By hypothesis $\mathcal{W}^{*C}_1$ is a $NS(G^*)CS$ of $\mathbb{N}^*_{\mathcal{X}}$ and thus $\mathcal{W}^*_1$ is $NS(G^*)OS$ of $\mathbb{N}^*_{\mathcal{X}}$ Hence $NS(G)OS$ $(\mathbb{N}^*_{\mathcal{X}}) = NS(G^*)OS$ $(\mathbb{N}^*_{\mathcal{X}})$.
Sufficiency: Let $\mathcal{W}^*_1$ be an NSCS in $\mathbb{N}^*_{\mathcal{X}}$, the $\mathcal{W}^{*C}_1$ is $NS(G)OS$ of NS^*_{X}. By our assumption $\mathcal{W}^{*C}_1$ is a $NS(G^*)OS$of $\mathbb{N}^*_{\mathcal{X}}$, which implies $\mathcal{W}^*_1$ is a $NS(G^*)CS$ in $\mathbb{N}^*_{\mathcal{X}}$. Hence $(\mathbb{N}^*_{\mathcal{X}}, \mathbb{N}^*_{\tau})$ is a NS$^*T_{\frac{1}{2}}$ space.

References

1. Salama, A.A., Alblowi, S.A.: Neutrosophic set and neutrosophic topological space. ISOR J. Math. **(iii)**(4), 31–35 (2012)
2. Arokiarani, I., Dhavaseelan, R., Jafari, S., Parimala, M.: On some new notions and functions in neutrosophic topological spaces. Neutrosophic Sets Syst. **16**, 16–19 (2017)
3. Ishwarya, P., Bageerathi, K.: On neutrosophic semiopen sets in neutrosophic topological spaces. Int. J. Math. Trends Tech. **214–223**, (2016)
4. Dhavaseelan, R., Jafari, S.: Generalized neutrosophic closed sets. New Trends in Neutrosophic Theory and Applications **II**, 261–273 (2018)
5. Smarandache, F.: Neutrosophic and Neutrosophic Logic. In: First International Conference on Neutrosophic, Neutrosophic Logic, Set, Probability, and Statistics. University of New Mexico, Gallup, NM 87301, USA (2002)
6. Shanthi, V.K., Chandrasekar, S., SafinaBegam, K.: Neutrosophic generalized semi closed sets in neutrosophic topological spaces. Int. J. Res. Advent Technol. **6**(7), 1739–1743 (2018)
7. Venkateswara Rao, V., Srinivasa Rao, Y.: Neutrosophic pre-open sets and pre-closed sets in neutrosophic topology. Int. J. ChemTech Res. **10**(10), 449–458 (2017)

8. Atanassov, K.: Intuitionistic fuzzy sets. Fuzzy Sets Syst. **20**, 87–94 (1986)
9. Dhavaseelan, R., Jafari, S., Hanif Page, Md.: Neutrosophic generalized α-contra-continuity. Creat. Math. Inform. **27**(2), 133–139 (2018)
10. Smaradache, F.: Neutrosophic set: a generalization of intuitionistic fuzzy set. J. Def. Resour. Manag. **1**, 107–114 (2010)
11. Jayanthi, D.: α generalized closed sets in neutrosophic topological spaces. Int. J. Math. Trends Technol. (IJMTT). Special Issue ICRMIT March 2018
12. Salama, A.A., Alblowi, S.A.: Generalized neutrosophic set and generalized neutrosophic topological spaces. J. Comput. Sci. Eng. **(ii)**(7) (2012)

Serving Israeli Queue on Single Product Inventory System with Lead Time for Replenishment

J. Viswanath, C. T. Dorapravina, T. Karthikeyan, and A. Stanley Raj

Abstract Single Non-perishable product stochastic inventory system with maximum capacity of S items is served by a single server, which serves Israeli queue. Reorder is placed only after the level of inventory reaches zero. Customers form an Israeli Queue to get service with at most N different groups with the consideration of unrestricted batch size service. Service time is independent of batch size. Markov structure of the model is identified, and state probabilities are arrived by Numerical approach using MATLAB coding. Also arrived the performance measures like first-order product density of replenishment, mean number of groups in the System, System throughput depends on Inventory level, mean number of groups that are bypassed by the arriving customer. Model validated by numerical illustration.

Keywords Single product inventory system · Lead time for replenishment · Adjustable reorder policy · Israeli queue · Unrestricted batch size service

Mathematics Subject Classification (2010) 60K25 · 90B05 · 90B22

1 Introduction

Stochastic analysis of continuous review inventory systems has been effectively analyzed for the past several decades. The first systematic mathematical theory of inventory and production system is provided in the monograph of Arrow et al. [1],

J. Viswanath (✉) · C. T. Dorapravina
Vel Tech Rangarajan Dr. Sagunthala R&D Institute of Science and Technology, Avadi, Chennai 600062, India
e-mail: mathsviswa@gmail.com

T. Karthikeyan
Department of Mathematics, Ramakrishna Mission Vivekananda College, Mylapore, Chennai 600004, India

A. S. Raj
Department of Physics, Loyola College, Nungambakkam, Chennai 600034, India

R. N. Mohapatra et al. (eds.), *Mathematical Analysis and Computing*, Springer Proceedings in Mathematics & Statistics 344,
https://doi.org/10.1007/978-981-33-4646-8_14

and followed by Hadley and Whitin [2]. Analysis of continuous time inventory models with the consideration of stochastic lead time was done by Zipkin [3]. Exhaustive account on the inventory management has been available in the monographs of Zipkin [4] and Porteus [5]. A new class of single perishable as well as Non-perishable product inventory system with the consideration of Compulsory Waiting Period (CWP) was analyzed see Refs. [6] and [7]. Single product perishable stochastic inventory system with compulsory waiting time for reorder was analyzed by Yadavalli and Udayabaskaran [8]. In inventory maintenance, storing of frozen Vaccines like MMRV, Zoster, and Varicella needs specific temperature conditions. Such inventories never be filled unless all items in the inventory are sold. In such cases, replenishment before the inventory level reaches zero may either harm the existing items which are in the inventory or cause loss to the owner. Once the inventory is filled with its maximum capacity, never be filled again in between the sale unless the items in the inventory reach level zero due to several reasons like air conditioning of items in storage or preservation facilities for quality maintenance of the product. In such cases, the reorder is placed only after the inventory level reaches zero.

In Medical field purchasing, the most expensive Medical devices (MRI scanner for soft tissue imaging, CT scan for bony lesion imaging machines, INUMAC MRI scanner, PET scans, Surgical Robots, Premium CT scanners and Proton Beam Therapy systems) customers who know themselves may club together before purchasing a single item in partnership for their common usage. Service time not be affected to serve a product to a single individual or a group of people who lead by a leader (the one who initiated first). Therefore, service time need not depend on the number of persons in each group. This character among the customer who wait for getting service resembles Israeli queue was introduced by Boxma et al. [9] in their monograph while studying single-server polling system with unlimited batch size. In the literature of queue, unlimited batch service models as an application of end user information system namely videotext, telex network, and Time-division multiple access systems studied by Ammar and Wong [10], Dykeman et al. [11], Liu and Nain [12]. In the past, variety of applications of Queueing-inventory models was analyzed see Refs. [13–24]. But in the past literature, due importance is not given to the stochastic models of single product inventory system serve by single server on Israeli queue disciplined customers. To fill the gap, it is proposed a stochastic Markov model which serves Israeli queue on single product inventory system with lead time for replenishment.

This research article is organized as follows: In Sect. 2, we give model description, Sect. 3 covers Transition rate diagram which represents the dynamism of the model and the governing equations of the model by using stochastic Integral equation technique. In Sect. 4, important performance measures were discussed and Sect. 5 deals with the numerical illustration to validate the model. Section 6 concludes the article.

2 Model Description

A Non-Perishable single product inventory system is considered. Maximum storage capacity of the inventory is S. Reorder is placed when the inventory level falls to zero. Stochastic lead time starts at the time of reordering. Adjustable reorder policy is adopted at the time of replenishment. Lead time follows exponential distribution with rate η. Israeli queue is maintained to serve the customers through the single server facility. Israeli Queue is maintained with batch service, restricted to N groups. Customer arrival follows Poisson process with rate λ and service time follows exponential distribution, which is independent of lead time distribution with parameter μ. Assumed that batch size never affects the service time duration. New arriving customer may find that there are N batches in the system, in which each batch headed by a leader as a representative of their own group to get service for the whole group. The new arriving customer check all the leaders and find any one of the leader who is familiar to him then immediately he joins the group and waiting for the service in the group till his group turn comes for service as a batch mode. The probability for an arriving customer to know any leader who is already waiting in the queue is assumed to be p. When the arriving customer finds that there are $k(1 \leq k < N)$ groups in the system (including the group getting service at the epoch of his arrival), then the arriving customer either joins ith group $(i \leq k)$ with probability $(1-p)^{i-1}p$ or may create a new group with probability $(1-p)^k$. Upon the arrival of the new customer, if there are already N groups and at the same time arriving customer does not know any one of the existing leaders, then he joints Nth group with probability $(1-p)^{N-1}$.

3 Governing Equations

Let $X(t)$ be the number of groups in the Israeli queue at time t and $Y(t)$ be number of items in the inventory at time t. Then, the two-dimensional stochastic processes $\{X(t), Y(t)\}$ is a discrete valued continuous time Markov process with state space $[(i, j), i = 0, 1, 2, ..., N, j = 0, 1, 2, ..., S]$. In Fig. 1, out flow of each state of the system is represented and state probabilities are as follows:

$$Pr(i, j; t) = Pr\{X(t) = i, Y(t) = j | X(0) = 0, Y(0) = S\}, \tag{1}$$

By probability law with the consideration of Mutually exclusive and Exhaustive cases, we get the following Integral equations.

$$Pr(0, S; t) = [1 + \eta Pr(0, 0; t) \Theta e^{-\lambda t}], \tag{2}$$

$$Pr(i, S; t) = [\lambda(1-p)^{i-1} Pr(i-1, S; t) + \eta Pr(i, 0; t)] \Theta e^{-\{\mu + (1-p)^i \lambda\} t}, \tag{3}$$

where $i = 1, 2, ..., N-1$.

$$Pr(N, S; t) = [\lambda(1-p)^{N-1} Pr(N-1, S; t) + \eta Pr(N, 0; t)]\Theta e^{-\mu t} \tag{4}$$

$$Pr(0, j; t) = \mu Pr(1, j+1; t)\Theta e^{-\lambda t}, j = 1, 2, ..., S-1, \tag{5}$$

$$Pr(N, j; t) = (1-p)^{N-1}\lambda Pr(N-1, j; t)\Theta e^{-\mu t}, j = 1, 2, ..., S-1, \tag{6}$$

$$Pr(i, j; t) = [(1-p)^{i-1}\lambda Pr(i-1, j; t) + \mu Pr(i+1, j+1; t)]\Theta e^{-\{\mu+(1-p)^i\lambda\}t}, \tag{7}$$

where $i = 1, 2, ..., N-1$, $j = 1, 2, ..., S-1$.

$$Pr(N, 0; t) = (1-p)^{N-1}\lambda Pr(N-1, 0; t)\Theta e^{-\eta t}, \tag{8}$$

$$Pr(i, 0; t) = [(1-p)^{i-1}\lambda Pr(i-1, 0; t) + \mu Pr(i+1, 1; t)]\Theta e^{-\{\eta+(1-p)^i\lambda\}t}, \tag{9}$$

where $i = 1, 2, ..., N-1$.

$$Pr(0, 0; t) = \mu Pr(1, 1; t)\Theta e^{-(\lambda+\eta)t}, \tag{10}$$

By taking Laplace Stieltjes-Transform on both sides on (2)–(10), we get

$$(\theta + \lambda)Pr^*(0, S; \theta) = [1 + \eta Pr^*(0, 0; \theta)], \tag{11}$$

$$(\theta + \mu + \lambda(1-p)^i)Pr^*(i, S; \theta) = [\lambda(1-p)^{i-1} Pr^*(i-1, S; \theta) + \eta Pr^*(i, 0; \theta)], \tag{12}$$

where $i = 1, 2, ..., N-1$.

$$(\theta + \mu)Pr^*(N, S; \theta) = [\lambda(1-p)^{N-1} Pr^*(N-1, S; \theta) + \eta Pr^*(N, 0; \theta)], \tag{13}$$

$$(\theta + \lambda)Pr^*(0, j; \theta) = \mu Pr^*(1, j+1; \theta), j = 1, 2, ..., S-1, \tag{14}$$

$$(\theta + \mu)Pr^*(N, j; \theta) = \lambda(1-p)^{N-1} Pr^*(N-1, j; \theta), j = 1, 2, ..., S-1, \tag{15}$$

$$(\theta + \mu + \lambda(1-p)^i)Pr^*(i, j; \theta) = [\lambda(1-p)^{i-1} Pr^*(i-1, j; \theta) + \mu Pr^*(i+1, j+1; \theta)], \tag{16}$$

where $i = 1, 2, ..., N-1$, $j = 1, 2, ..., S-1$.

$$(\theta + \eta)Pr^*(N, 0; \theta) = \lambda(1-p)^{N-1} Pr^*(N-1, 0; \theta), \tag{17}$$

$$(\theta + \eta + \lambda(1-p)^i)Pr^*(i, 0; \theta) = [\lambda(1-p)^{i-1} Pr^*(i-1, 0; \theta) + \mu Pr^*(i+1, 1; \theta)], \tag{18}$$

where $i = 1, 2, ..., N-1$.

$$(\theta + \lambda + \eta)Pr^*(0, 0; \theta) = \mu Pr^*(1, 1; \theta), \tag{19}$$

and applying final value theorem of Laplace Stieltjes-Transform in (11)–(19), we get steady-state balance equations as follows:

$$\lambda\pi(0, S) = \eta\pi(0, 0), \tag{20}$$

$$(\mu + \lambda(1-p)^{i})\pi(i, S) = \lambda(1-p)^{i-1}\pi(i-1, S) + \eta\pi(i, 0)], i = 1, 2, ..., N-1, \tag{21}$$

$$\mu\pi(N, S) = \lambda(1-p)^{N-1}\pi(N-1, S) + \eta\pi(N, 0), \tag{22}$$

$$\lambda\pi(0, j) = \mu\pi(1, j+1), j = 1, 2, ..., S-1, \tag{23}$$

$$\mu\pi(N, j) = \lambda(1-p)^{N-1}\pi(N-1, j), j = 1, 2, ..., S-1, \tag{24}$$

$$(\mu + \lambda(1-p)^{i})\pi(i, j) = \lambda(1-p)^{i-1}\pi(i-1, j) + \mu\pi(i+1, j+1), \tag{25}$$

where $i = 1, 2, ..., N-1, j = 1, 2, ..., S-1$

$$\eta\pi(N, 0) = \lambda(1-p)^{N-1}\pi(N-1, 0), \tag{26}$$

$$(\eta + \lambda(1-p)^{i})\pi(i, 0) = \lambda(1-p)^{i-1}\pi(i-1, 0) + \mu\pi(i+1, 1), \tag{27}$$

where $i = 1, 2, ..., N-1$.

$$(\lambda + \eta)\pi(0, 0) = \mu\pi(1, 1), \tag{28}$$

By law of total probability, we get

$$\sum_{i=0}^{N}\sum_{j=0}^{S}\pi(i, j) = 1, \tag{29}$$

Above system of Eqs. (20)–(29) were solved by Matrix method of solving system of non-homogeneous linear equations by using mathematical software MAT LAB.

4 Performance Measures

4.1 Mean Stationary Rate of Events

Let ξ be any one of the events either event of occurrence of Replenishment labeled by r or event of occurrence of demand satisfied labeled by s.
Define $f_{\xi}(t) = lim_{\Delta\to 0} Pr\{\text{an } \xi \text{ event in the interval } (t, t+\Delta)\}/\Delta$. Then, $f_{\xi}(t)$ represents the first-order product density of the ξ event by Srinivasan [25]. Therefore,

$$\text{Stationary mean of } [N(\xi, t)] = \int_0^t f_\xi(t)du, \tag{30}$$

Mean stationary rate of the event ξ is

$$\lim_{t \longrightarrow \infty} \frac{E[N(\xi, t)]}{t} = \lim_{t \longrightarrow \infty} \int_0^t f_\xi(u)du, \tag{31}$$

Let us derive expressions for the first-order product densities of the events to get the mean stationary rate of events.

4.2 First-Order Density of Replenishment

By observation, replenishment can occur in $(t, t + \Delta)$ in the following way: The system is in state $(i, 0)$, $i = 1, 2, ..., N$ at time t and replenishment takes place in the interval $(t, t + \Delta)$.
Hence, we get

$$f_r(t) = \sum_{i=0}^{N} Pr(i, 0; t)\eta, \tag{32}$$

From this, we obtain mean stationary rate of replenishment as

$$E(r) = \sum_{i=0}^{N} \pi(i, 0)\eta. \tag{33}$$

4.3 Mean Number of Groups in the System Depends on Inventory Level

Mean number of groups in the system

$$M(g) = \sum_{i=1}^{N} \sum_{j=0}^{S} i\pi(i, j), \tag{34}$$

4.4 System Throughput

$$T = \left[1 - \sum_{j=1}^{S} \pi(0, j) - \sum_{i=0}^{N} \pi(i, 0) \right] \mu, \tag{35}$$

4.5 Mean Number of Groups that Are Bi-passed by the Arriving Customer

Consider the random variable B_G, where $0 \leq B_G \leq N-1$, which represents the number of groups bypassed by the arriving customer when join for service.

$$Pr(B_G = 0) = \sum_{i=1}^{N} \sum_{j=0}^{S} \pi(i, j) p(1-p)^{i-1} + \sum_{i=0}^{N-1} \sum_{j=0}^{S} \pi(i, j)(1-p)^{i}, \qquad (36)$$

$$Pr(B_G = k) = \sum_{i=k+1}^{N} \sum_{j=0}^{S} \pi(i, j) p(1-p)^{i-k-1}, 1 \leq k \leq N-1, \qquad (37)$$

$$E[B_G = k] = \sum_{k=1}^{N-1} k \sum_{i=k+1}^{N} \sum_{j=0}^{S} \pi(i, j) p(1-p)^{i-k-1}, \qquad (38)$$

5 Numerical Illustration

5.1 Observation of State Probabilities by Maintaining the Number of Items in the Inventory Are Less Than the Number of Groups Allowed in the System for Service

We fix the parameters $\lambda = 8$, $\mu = 15$, $p = 0.4$, $N = 4$, and $S = 3$. The state probabilities are depicted in Table 1. We observe that the sum of probabilities of the states $(i, 0)$, $(i, 1)$, $(i, 2)$ and $(i, 3)$, $i = 0, 1, 2, 3, 4$ are decreasing due to the consideration of batch service, even if the lead time in replenishment with minimum level of inventory for reordering is zero.

Table 1 State probabilities by fixing $N = 4$ and $S = 3$

(i, j)	Pr	(i, j)	Pr	(i, j)	Pr	(i, j)	Pr	(i, j)	Pr
(0, 0)	0.0790	(1, 0)	0.1248	(2, 0)	0.0939	(3, 0)	0.0426	(4, 0)	0.0603
(0, 1)	0.1119	(1, 1)	0.0684	(2, 1)	0.0394	(3, 1)	0.0094	(4, 1)	0.0011
(0, 2)	0.0831	(1, 2)	0.0597	(2, 2)	0.0307	(3, 2)	0.0251	(4, 2)	0.0029
(0, 3)	0.0494	(1, 3)	0.0443	(2, 3)	0.0345	(3, 3)	0.0175	(4, 3)	0.0221

5.2 First-Order Product Density of Replenishment

Control on Arrival Rate: Fixing the parameters $\mu = 15, \eta = 5$ and $p = 0.4$. Maintaining the level of inventory and number of leaders in the queue as $S = 4$ and $N = 4$, respectively. By varying arrival rate λ from 5 to 10 and obtained mean stationary rate $E(r)$ of replenishment. The variation is tabulated in first two columns in Table 2 and depicted in Fig. 2. It is observed that when arrival rate is increased quite natural that the mean stationary rate of Replenishment too increased due to the reason that the inventory reaches zero often and reordering process starts for next replenishment.

Control on Service Rate: Fixing parameters $\lambda = 8, \eta = 5$, and $p = 0.4$. Maintaining the level of inventory and number of leaders in the queue as $S = 4$ and $N = 4$, respectively. By varying service rate μ from 12 to 17 and obtained mean stationary rate $E(r)$ of replenishment. The variation is tabulated in 3rd and 4th columns in Table 2 and depicted in Fig. 2. It is observed that when service rate is increased as an effect the mean stationary rate of Replenishment too increased due to the reason that more customers get service due that the inventory meets reordering point quickly as a result.

Control on Replenishment Rate: Fixing parameters $\lambda = 8, \mu = 15$, and $p = 0.4$. Maintaining the level of inventory and number of leaders in the queue as $S = 4$ and $N = 4$, respectively. By varying replenishment rate η from 2 to 7 and obtained mean stationary rate $E(r)$ of replenishment. The variation is tabulated in 5th and 6th columns in Table 2 and depicted in Fig. 2. It is observed that when replenishment rate is increased as a result the mean stationary rate of Replenishment is too increased due to the reason that when rate of replenishment is increased it means that expected lead time is decreased.

Table 2 First-order product density of replenishment

λ	$E(r)$	μ	$E(r)$	η	$E(r)$
5	1.1452	12	1.5196	2	1.1316
6	1.3142	13	1.5457	3	1.3539
7	1.4608	14	1.5680	4	1.4932
8	1.5869	15	1.5869	5	1.5869
9	1.6948	16	1.6033	6	1.6536
10	1.7867	17	1.6175	7	1.7029

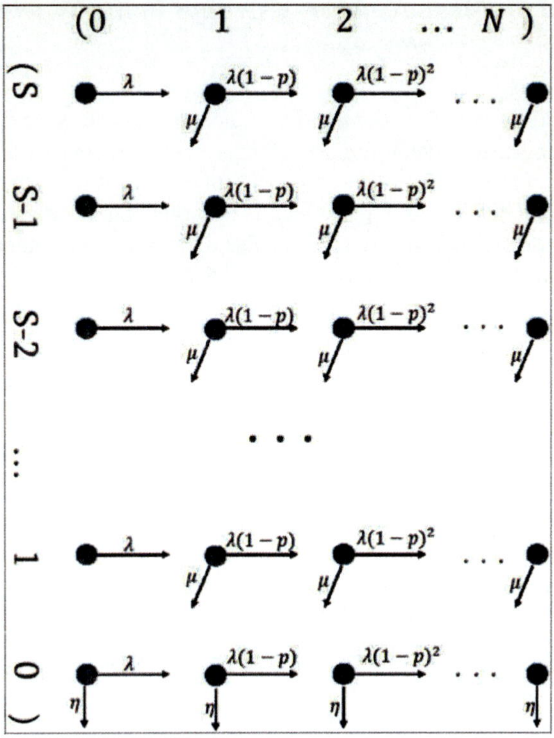

Fig. 1 State out flow transition rate diagram

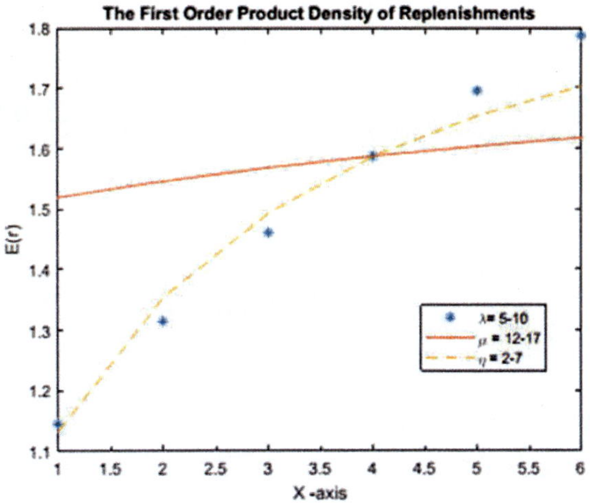

Fig. 2 Mean stationary rate of replenishment

5.3 Mean Number of Groups in the System Depends on Inventory Level

Control on Arrival Rate: By fixing the parameters $\mu = 15, \eta = 5, p = 0.4$ and varying arrival rate λ from 5 to 10. It is arrived that mean number of Groups in the System for three different cases namely for $N = S, N < S$, and $N > S$ and listed in Table 3 and depicted in Fig. 3, and it is observed that mean number of groups in the system is increased while increasing the arrival rate independent of number of groups in the system and number of items in the inventory.

Control on Service Rate: By fixing the parameters $\lambda = 8, \eta = 5, p = 0.4$ and varying arrival rate μ from 12 to 17. It is arrived that mean number of Groups in the System for three different cases namely for $N = S, N < S$, and $N > S$ and listed in Table 4 and depicted in Fig. 4, and it is observed that in general mean number of groups in the system is decreased while increasing the service rate it is quite natural that more people got service as a result number of groups in the system tending to zero.

Control on Replenishment Rate: By fixing the parameters $\lambda = 8, \mu = 15, p = 0.4$ and varying Replenishment rate η from 2 to 7. It is arrived that mean number of groups in the system for three different cases namely $N = S, N < S$, and $N > S$ and listed in Table 5 and depicted in Fig. 5, and it is observed that mean number of groups in the system is decreased while increasing the replenishment rate. Since rate of replenishment is increased it implies that mean lead time is decreased as a result customers will get service without waiting for long period in the system intern mean number of groups in the system decreased. It is quite natural that more people get served and number of groups in the system tending to zero.

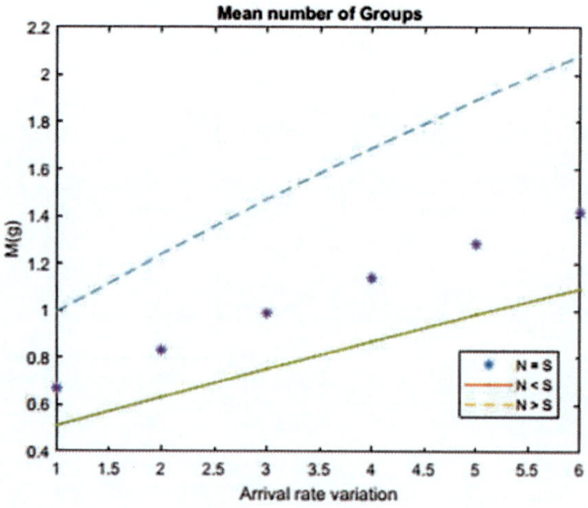

Fig. 3 Mean groups in the system by control on arrival rate

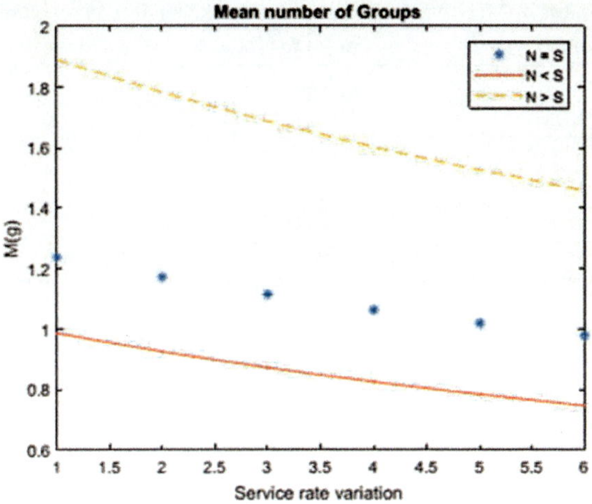

Fig. 4 Mean number of groups in the system by control on service rate

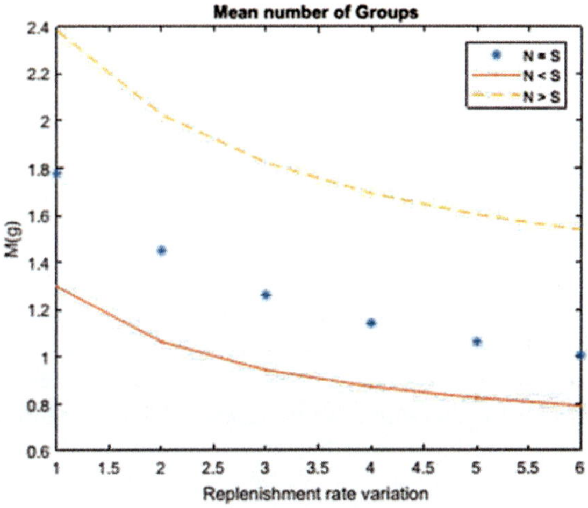

Fig. 5 Mean number of groups in the system by control on replenishment rate

Table 3 Mean number of groups in the system by control on arrival rate

λ	$M(g)$ for $(N = S)$	$M(g)$ for $(N < S)$	$M(g)$ for $(N > S)$
5	0.6727	0.5137	0.9957
6	0.8349	0.6343	1.2400
7	0.9928	0.7542	1.4734
8	1.1439	0.8718	1.6925
9	1.2868	0.9862	1.8960
10	1.4210	1.0966	2.0840

Table 4 Mean number of groups in the system by control on service rate

μ	$M(g)$ for $(N = S)$	$M(g)$ for $(N < S)$	$M(g)$ for $(N > S)$
12	1.2380	0.9855	1.8901
13	1.1724	0.9246	1.7831
14	1.1147	0.8715	1.6881
15	1.0637	0.8249	1.6033
16	1.0182	0.7838	1.5273
17	0.9775	0.7472	1.4589

Table 5 Mean number of groups in the system by control on replenishment rate

η	$M(g)$ for $(N = S)$	$M(g)$ for $(N < S)$	$M(g)$ for $(N > S)$
2	1.7775	1.2981	2.3830
3	1.4501	1.0650	2.0238
4	1.2624	0.9439	1.8216
5	1.1439	0.8718	1.6925
6	1.0637	0.8249	1.6033
7	1.0064	0.7924	1.5382

5.4 System Throughput Depends on Inventory Level

Control on Arrival Rate: Calculated the system throughput by fixing the parameters $\mu = 15$, $\eta = 5$, $p = 0.4$ and and controlling the arrival rate λ for three different cases namely for $N = S$, $N < S$, and $N > S$ and listed in Table 6 and depicted in Fig. 6. It is observed that system throughput is increased irrespective of the level of inventory and the number of groups in the system while arrival rate is increased. It is obvious that since no customer impatience is considered in our model as a result if arrival rate increases then more people get service.

Table 6 System throughput by control on arrival rate

λ	$M(g)$ for $(N = S)$	$M(g)$ for $(N < S)$	$M(g)$ for $(N > S)$
5	4.1687	4.2602	4.3857
6	4.7299	4.8993	4.9716
7	5.2070	5.4706	5.4620
8	5.6104	5.9795	5.8696
9	5.9504	6.4317	6.2069
10	6.2366	6.8332	6.4857

Fig. 6 System throughput by control on arrival rate

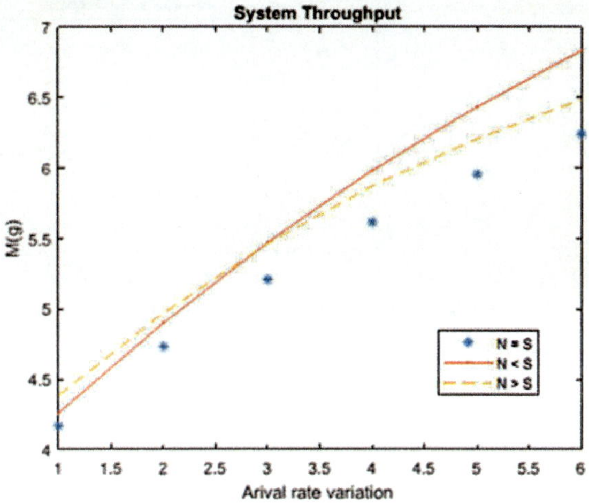

Table 7 System throughput by control on service rate

μ	$M(g)$ for $(N = S)$	$M(g)$ for $(N < S)$	$M(g)$ for $(N > S)$
12	5.2701	5.6292	5.5109
13	5.3986	5.7616	5.6474
14	5.5111	5.8774	5.7660
15	5.6104	5.9795	5.8696
16	5.6986	6.0700	5.9605
17	7.5256	6.1508	6.0408

Control on Service Rate: Calculated the system throughput by fixing the parameters $\lambda = 8, \eta = 5, p = 0.4$ and controlling the service rate μ for three different cases ($N = S, N < S$, and $N > S$) listed in Table 7 also depicted in Fig. 7. It is observed that system throughput is increased while service rate is increased. It is obvious that when service rate is increased (mean service time is decreased) as a result, more number of demands will be satisfied and system throughput is increased.

Control on Replenishment Rate: Calculated the system throughput by fixing the parameters $\lambda = 8, \mu = 15, p = 0.4$ and controlling the replenishment rate η for three different cases ($N = S, N < S$, and $N > S$) and listed in Table 8 and depicted in Fig. 8. It is observed that system throughput is increased while replenishment rate is increased. While increasing the rate of replenishment, mean lead time is reduced. Availability of the product in the inventory is ensured as a result more demand will be satisfied by the server.

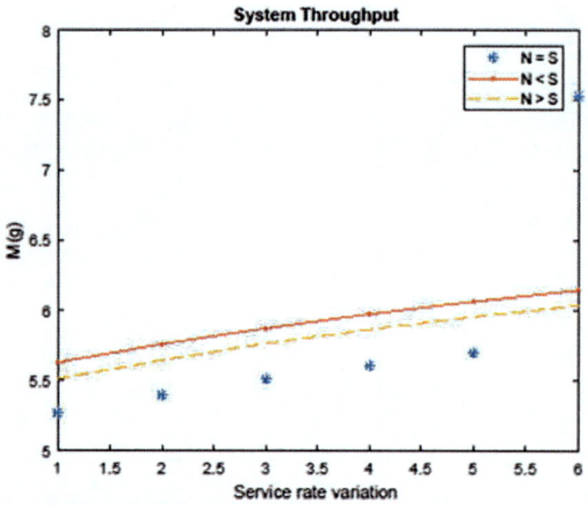

Fig. 7 System throughput by control on service rate

Table 8 System throughput by control on replenishment rate

η	$M(g)$ for $(N = S)$	$M(g)$ for $(N < S)$	$M(g)$ for $(N > S)$
2	4.1155	5.0168	4.2079
3	4.8546	5.5251	5.0131
4	5.3087	5.8052	5.5226
5	5.6104	5.9795	5.8696
6	5.8230	6.0970	6.1190
7	5.9794	6.1809	6.3057

Fig. 8 System throughput by control on replenishment

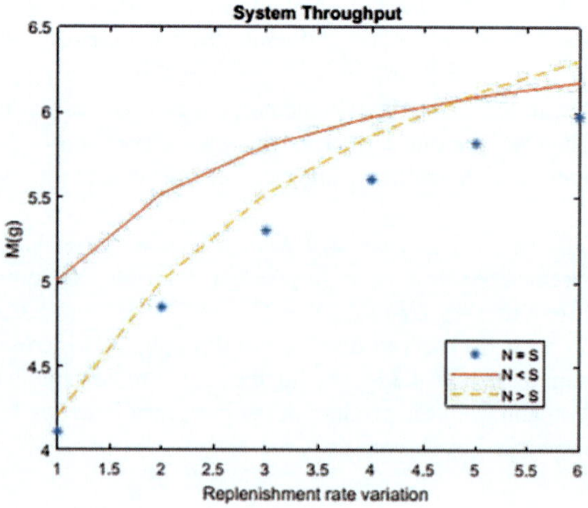

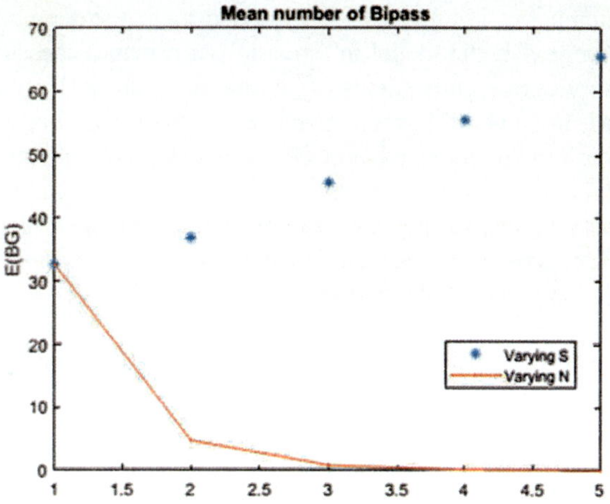

Fig. 9 Mean number of groups that are bypassed by the arriving customer

5.5 *Mean Number of Groups that are Bypassed by the Arriving Customer*

Fixing the parameters $\lambda = 0.1$, $\mu = 1$, $\eta = 0.2$, and $p = 0.4$. First by Maintaining the number of Israeli group in the system as $N = 8$ and varying maximum level of the inventory from $S = 8$ to 3 then it is observed that mean number of bypass by arriving customer is increased, secondly Maintaining maximum level of inventory as $S = 8$ and varying the number of Israeli group in the system for service from $N = 8$ to 3 then it is observed that mean number of bypass by arriving customer is decreased due to the reason that while number of groups allowed is restricted and as a result mean number of bypass by arriving customer will be reduced. Both are depicted in Fig. 9.

6 Conclusion

We considered Single Non-perishable product stochastic inventory system with single server that satisfies Israeli queue with the assumption of independent lead time and service time both follow exponential distribution. Reorder is placed only after the level of inventory reaches zero. Arrival processes followed Poisson. Considered unrestricted batch size service and service time is independent of batch size. By Ito-integral equations, transient state equations arrived. Using Laplace transform technique and final value theorem of Laplace transform, we arrived steady-state equation. By Matrix algebra and the use of MAT LAB software, the system of Non-homogeneous equations was solved. State probabilities were calculated. Effective performance measures like first-order product density of replenishment, mean number of groups in the System, System throughput depends on Inventory level, mean number of groups that are bypassed by the arriving customer were derived. Effectiveness of the model in terms of performance measures based on the arrival rate, service rate, and replenishment rate was analyzed by observing the figures and tables. Mean number of groups bypassed by the arriving customer is varying according to the variation on number of groups in the queue and number of items in the inventory. Number of customers served in a unit of time is increased while increase of arrival, service, and replenishment rates. By varying the arrival, service, and replenishment rates, as a result Mean number of replenishment is increased. Mean number of groups in the system is fluctuated.

Acknowledgements Authors would like to extend their gratitude to the reviewers whose comments and suggestions greatly improved the paper.

References

1. Arrow, M.J., Karlin, S., Scarf, H.: Studies in Mathematical Theory of Inventory and Production. Standford University Press, Stanford, California (1958)
2. Hadley, G., Whitin, T.M.: Analysis of Inventory Systems. Prentice-Hall, Englewood Cliffs, NJ, USA (1963)
3. Zipkin, P.: Stochastic lead-times in continuous-time inventor models. Naval Res. Logist. Quart. **33**, 763–774 (1986)
4. Zipkin, P.: Foundations of Inventory Management. McGraw Hill, Boston, MA, USA (2000)
5. Porteus, E.L.: Foundations of Stochastic Inventory Theory. Stanford University Press, California (2002)
6. Udayabaskaran, S., Viswanath, J., Kothandaraman, M.: A new class of perishable single product inventory systems with compulsory waiting period for reordering. Far East J. Math. Sci. **99**(8), 1245–1260 (2016)
7. Udayabaskaran, S., Viswanath, J., Geetha, S., Kothandaraman, M.: Analysis of s stochastic inventory system with compulsory wait for reorder. Far East J. Math. Sci. **99**(11), 1733–1752 (2016)
8. Yadavalli, V.S.S., Udayabaskaran, S.: A single product inventory system with compulsory waiting period for reordering and with stochastic lead-time. Eur. J. Ind. Eng. **12**, 682–707 (2018)
9. Boxma, O.J., van der Wal, J., Yechiali, U.: Polling with batch service. Stoch. Models **24**(4), 604–625 (2008)
10. Ammar, M.H., Wong, J.W.: On the optimality of cyclic transmission in teletext systems. IEEE Trans. Commun. COM **35**(1), 68–73 (1987)
11. Dykeman, H.D., Ammar, M.H., Wong, J.W.: In scheduling algorithms for videotex systems under broadcast delivery. In: Proceedings of the International Conference on Communications (ICC '86), pp. 1847–1851 (1986)
12. Liu, Z., Nain, P.: Optimal scheduling in some multiqueue single-server systems. IEEE Trans. Automat. Control **37**(2), 247–252 (1992)
13. Berman, O., Kim, E.: Dynamic inventory strategies for profit maximization in a service facility with stochastic service, demand and lead time. Math. Method Oper. Res. **60**, 497–521 (2004)
14. Berman, O., Sapna, K.P.: Inventory management at service facilities for systems with arbitrarily distributed service times. Stochas. Models **16**, 343–360 (2000)
15. Berman, O., Sapna, K.P.: Optimal control of service facilities holding inventory. Comput. Operat. Res. **28**, 429–441 (2001)

16. Sivakumar, B., Arivarignan, G.: A perishable inventory system at service facilities with negative customers. Int. J. Inf. Manage. Sci. **17**, 193–210 (2005)
17. Krishnamoorthy, A., Deepak, T.G., Narayanan, V.C., Vineetha, K.: Control policies for inventory with service time. Stochas. Anal. Appl. **24**, 889–899 (2006)
18. Deepak, T.G., Krishnamoorthy, A., Narayanan, V.C., Vineetha, K.: Inventory with service time and transfer of customers and inventory. Annals Operat. Res. **160**, 191–213 (2008)
19. Yadavalli, V.S.S., Sivakumar, B., Arivarignan, G.: Inventory system with renewal demands at service facility. Int. J. Prod. Econ **114**, 252–264 (2008)
20. Sivakumar, B.: An inventory system with retrial demands and multiple server vacation. Qual. Technol. Quant. Manage. **8**, 125–146 (2011)
21. Saffari, M., Asmussen, S., Haji, R.: The M/M/1 queue with inventory, lost sale, and general lead times. Queue. Syst. **75**, 65–77 (2013)
22. Lawrence, A.S., Sivakumar, B., Arivarignan, G.: A perishable inventory system with service facility and finite source. Appl. Mathe. Model. **37**, 4771–4786 (2013)
23. Zhao, N., Lian, Z.: A queueing-inventory system with two classes of customers. Int. J. Prod. Econ. **129**, 225–231 (2011)
24. Yue, D., Zhao, G., Qin, Y.: An M/M/I queueing-inventory system with geometric batch demands and lost sales. J. Syst. Sci. Complex. **31**, 1024–1041 (2018)
25. Srinivasan, S.K.: Stochastic Point Processes and Their Applications. Griffin, London (1974)

A Multi-criteria Decision-Making Framework Based on the Prospect Theory Under Bipolar Intuitionistic Fuzzy Soft Environment

S. Anita Shanthi and Prathipa Jayapalan

Abstract This paper deals with multi-criteria decision-making (MCDM) problem using prospect theory. The data set of l alternatives and r criteria of MCDM problem is represented in terms of bipolar intuitionistic fuzzy soft set ($BIFSS$). A score function on $BIFSS$ is defined. A procedure of prospect theory based on $BIFSS$ is framed and it generates the $BIFS$ prospect values by which the alternatives are ranked. Moreover, the $BIFS$ value function curve is depicted. Finally, an illustration is given to show the applicability of the procedure.

Keywords Bipolar intuitionistic fuzzy soft set · Prospect theory · MCDM problem · *BIFS* score function

Mathematics Subject Classification (2010) 94 D

1 Introduction

Zadeh [1] introduced fuzzy set theory. Atanassov [2] developed intuitionistic fuzzy set theory. Jana and Pal [3] introduced $BIFSS$.

In a real-life decision situation, decision-makers often face uncertainties and ambiguities. To tackle these, Kahneman and Tversky [4] developed a theory using prospect values. Thillaigovindan et al. [5] proposed score values on $IVIFSSRT$. Tian et al. [6] developed prospect theory on IFS for venture capitalists.

This paper deals with prospect theory based on $BIFSS$. A score function serves as a tool in computing the weighing function of prospect theory. The value function curve is depicted. The applicability of the procedure is explained by an example.

S. Anita Shanthi (✉) · P. Jayapalan
Department of Mathematics, Annamalai University, Annamalai Nagar, Cuddalore, India
e-mail: shanthi.anita@yahoo.com

P. Jayapalan
e-mail: prathipajayapalan@gmail.com

R. N. Mohapatra et al. (eds.), *Mathematical Analysis and Computing*,
Springer Proceedings in Mathematics & Statistics 344,
https://doi.org/10.1007/978-981-33-4646-8_15

2 Score Function of $BIFSS$

Definition of $BIFS$ set is given in [7]. Here, $BIFS$ score function is dealt with.

Definition 1 Score function for $BIFSS$:
$bs_{(F(e)(x))} = \frac{\mu^n_{F(e)}(x)+\mu^p_{F(e)}(x)+(-1-\nu^n_{F(e)}(x))+(1-\nu^p_{F(e)}(x))}{2+\mu^n_{F(e)}(x)+\mu^p_{F(e)}(x)+\nu^n_{F(e)}(x)+\nu^p_{F(e)}(x)+2\pi^n_{F(e)}(x)+2\pi^p_{F(e)}(x)}$,
where $\pi^n_{F(e)}(x) = -1 - \mu^n_{F(e)}(x) - \nu^n_{F(e)}(x)$
and $\pi^p_{F(e)}(x) = 1 - \mu^p_{F(e)}(x) - \nu^p_{F(e)}(x)$.
π^n and π^p denote the negative and positive hesitancy, respectively of $x \in U, e \in A$.

3 MCDM Problem Based on Prospect Theory Under $BIFSS$

In this section, the concepts necessary for the development of prospect theory under $BIFSS$ are defined. Consider U={$y_1, y_2, ..., y_l$}, l alternatives, $E = \{e_1, e_2, ..., e_r\}$, r criteria. Each alternative y_i is represented as $BIFSS$ over U, based on the criterion e_j. $y_{ij} = ((\mu^n_{ij}, \mu^p_{ij}), (\nu^n_{ij}, \nu^p_{ij}))$. By using prospect theory, a MCDM problem is solved under $BIFSS$. Moreover, the ranking depends on the maximum value of $BIFSS$ prospect value from which the desirable alternative is chosen.

Definition 2 Given a $BIFSS$
(F, E)={x, $(\mu^n_{F(e)}(x), \mu^p_{F(e)}(x)), (\nu^n_{F(e)}(x), \nu^p_{F(e)}(x) : x \in X\}$, the $BIFSS$ is aggregated to a single value, by finding the $BIFS$ fuzzy degree,
$\lambda_{ij} = 1 - |\mu^n_{ij} - \nu^n_{ij} + \mu^p_{ij} - \nu^p_{ij}|, \ \forall\, i = 1, 2, ..., l, \ \ j = 1, 2, ..., r.$

Definition 3 For a $BIFSS$, the variable $y^{'}_{ij}$ is defined as

$y^{'}_{ij} = \lambda_{ij} - \frac{1}{l}(\sum\limits_{i=1}^{l} \lambda_{ij}), \ \forall\, j = 1, 2, ..., r$

Definition 4 The $BIFS$ value function is denoted as $Bv(y^{'}_{ij})$ and is defined as follows.
$Bv(y^{'}_{ij}) = \rho(y^{'}_{ij})^s \ \ if \ \ y^{'}_{ij} \geq 0$ and $-\tau(y^{'}_{ij})^t \ \ if \ \ y^{'}_{ij} < 0$.

The $BIFS$ value function is given in the form of a power function. Moreover, if $y^{'}_{ij} \geq 0$ then it is a gain, and if $y^{'}_{ij} < 0$ then it is a loss, where s and t are the $BIFS$ parameters of risk attitude, ρ and τ are the coefficient of $BIFS$ gain and loss aversion, respectively. The parameters are given as $s = 0.93, t = 0.52, \tau = 2.25$, and $\rho = 1.27$ from [6, 8]. From prospect theory, it is proposed that the $BIFS$ value function curve is concave and convex for gains and losses, respectively, and is S shaped.

Definition 5 The $BIFS$ weighting function is defined as follows:
$g^+(bs^+_{ij}) = \frac{\eta(bs^+_{ij}+1)^\gamma}{\eta(bs^+_{ij}+1)^\gamma+(1-bs^+_{ij})^\gamma}$

$g^-(bs_{ij}^-) = \frac{\eta(bs_{ij}^-+1)^\delta}{\eta(bs_{ij}^-+1)^\delta+(1-bs_{ij}^-)^\delta}$.
g^+ and g^- lie in [0, 1]. η, γ and δ are constant parameters such that $\eta = 1.08$, $\gamma = 0.61$ and $\delta = 0.69$, respectively, from [6, 8].

Definition 6 The $BIFS$ positive prospect $Bv(f_{ij}^+)$ and negative prospect $Bv(f_{ij}^-)$ are defined as follows:

$$Bv(f_{ij}^+) = \sum_{j=1}^{r} g^+(bs_{ij}^+)Bv(y_{ij}^{'}) + \beta^+\pi_{ij}^n\pi_{ij}^p Bv(y_{ij}^{'})$$

$$Bv(f_{ij}^-) = \sum_{j=1}^{r} g^-(bs_{ij}^-)Bv(y_{ij}^{'}) + \beta^-\pi_{ij}^n\pi_{ij}^p Bv(y_{ij}^{'}), \quad i = 1, 2, ..., l.$$

where β^+ and β^- are positive and negative prospect parameters, respectively, such that $\beta^+ = 0.62$ and $\beta^- = 0.49$ from [6]. For an alternativ,e y_i corresponding to each e_j the positive prospect and negative prospect is denoted as $Bv(f_{ij}^+)$ and $Bv(f_{ij}^-)$, respectively. Further, $Bv(f_{ij}^+)$ consists of all positive outcomes with gains and $Bv(f_{ij}^-)$ consists of all negative outcomes with losses.

Definition 7 The bipolar intuitionistic fuzzy soft prospect value $Bv(f_i)$ is defined as

$Bv(f_i) = \frac{Z_i^+}{Z_i^+ - Z_i^-}$, where $Z_i^+ = \sum_{j=1}^{r} Bv(f_{ij}^+)$ and

$Z_i^- = \sum_{j=1}^{r} Bv(f_{ij}^-) \quad \forall i = 1, 2, ..., l.$

4 Procedure

Procedure for MCDM problem on $BIFSS$:
Step: 1 Construct a $BIFSS$ for the alternative k_i and for the criterion e_j.
Step: 2 Determine λ_{ij} and $y_{ij}^{'}$ by Definitions 2 and 3.
Step: 3 Calculate the value function $Bv(y_{ij}^{'})$ for each alternative y_i by Definition 4.
Step: 4 Compute weighting function values $g^+(bs_{ij}^+)$ and $g^-(bs_{ij}^-)$, in accordance to the values of bs_{ij}^+ and bs_{ij}^- by Definition 5.
Step: 5 Obtain the $BIFS$ positive and negative prospect value $Bv(f_{ij}^+)$ and $Bv(f_{ij}^-)$ by Definition 6.
Step: 6 Calculate the prospect value $Bv(f_i)$ by Definition 7.
Step: 7 The alternatives are ranked depending on $BIFS$ prospect value. Alternative with maximum prospect value is the best.

Example 1 An expert committee decides to evaluate a group of five companies for selection of best emerging company among the following companies $K = \{k_1, k_2, k_3, k_4, k_5\}$ where k_1, k_2, k_3, k_4 and k_5 are the alternatives. Criteria for deciding the best company are $E = \{e_1, e_2, e_3, e_4\}$, where e_1= Net profit, e_2= Amount of income tax, e_3=Growth rate and e_4= market value, respectively. Depending on these criteria, the best emerging company is selected.

Step 1. $BIFSS(F, E)$ data set:

U	e_1	e_2
k_1	$((-0.2, 0.6)(-0.3, 0.16))$	$((-0.19, 0.7), (-0.24, 0.16))$
k_2	$((-0.22, 0.37), (-0.25, 0.1))$	$((-0.36, 0.56), (-0.4, 0.29))$
k_3	$((-0.38, 0.5), (-0.43, 0.26))$	$((-0.15, 0.23), (-0.17, 0.12))$
k_4	$((-0.3, 0.68), (-0.32, 0.08))$	$((-0.23, 0.64), (-0.33, 0.14))$
k_5	$((-0.21, 0.45), (-0.35, 0.18))$	$((-0.18, 0.37), (-0.21, 0.07))$

U	e_3	e_4
k_1	$((-0.15, 0.46), (-0.35, 0.04))$	$((-0.22, 0.46), (-0.32, 0.2))$
k_2	$((-0.3, 0.62), (-0.29, 0.21))$	$((-0.34, 0.6), (-0.5, 0.32))$
k_3	$((-0.19, 0.5), (-0.21, 0.07))$	$((-0.42, 0.7), (-0.4, 0.3))$
k_4	$((-0.14, 0.22), (-0.15, 0.02))$	$((-0.27, 0.57), (-0.29, 0.19))$
k_5	$((-0.38, 0.67), (-0.4, 0.3))$	$((-0.28, 0.46), (-0.38, 0.1))$

Step 2. The results of y'_{ij}.

U	e_1	e_2	e_3	e_4
k_1	−0.108	−0.198	−0.206	0.048
k_2	0.132	0.082	0.014	−0.032
k_3	0.142	0.262	−0.036	0.028
k_4	−0.188	−0.208	0.204	0.008
k_5	0.022	0.062	0.024	−0.052

Step 3. The BIFS values function $Bv(y'_{ij})$ corresponding to the criteria e_j.

U	e_1	e_2	e_3	e_4
k_1	−0.707	−0.969	−0.989	0.08
k_2	0.193	0.124	0.024	−0.375
k_3	0.207	0.365	−0.399	0.046
k_4	−0.944	−0.994	0.289	0.014
k_5	0.036	0.09	0.039	−0.484

The value function graph is shown in Fig. 1.

Step 4. The weighting function values $g^+(bs^+_{ij})$ and $g^-(bs^-_{ij})$ corresponding to the criteria e_j.

U	e_1	e_2	e_3	e_4
k_1	0.6271	0.6507	0.6269	0.577
k_2	0.5649	0.5687	0.588	0.5984
k_3	0.5623	0.5393	0.604	0.583
k_4	0.6353	0.6366	0.5504	0.5873
k_5	0.5842	0.5709	0.5852	0.5949

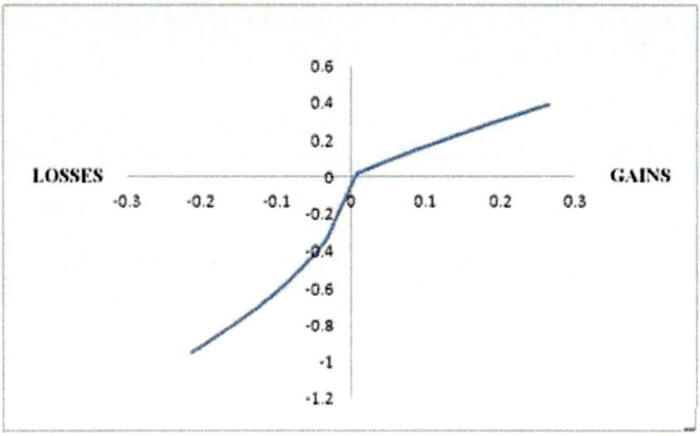

Fig. 1 Value function

Step 5. The $BIFS$ positive and $BIFS$ negative prospect value $Bv(f_{ij}^{+})$ and $Bv(f_{ij}^{-})$ corresponding to the criteria e_j.

U	e_1	e_2	e_3	e_4
k_1	−0.4851	−0.681	−0.7415	0.05
k_2	0.142	0.07	0.015	−0.227
k_3	0.122	0.295	−0.292	0.026
k_4	−0.59	−0.6509	0.256	0.009
k_5	0.025	0.07	0.023	−0.323

Step 6. The $BIFS$ prospect value $Bv(f_i)$ for each alternatives are
$Bv(f_1) = 0.026$
$Bv(f_2) = 0.503$
$Bv(f_3) = 0.897$
$Bv(f_4) = 0.175$
$Bv(f_5) = 0.773.$

Step 7. The alternatives are ranked. From the prospect values $Bv(f_i)$, $Bv(f_3)$ is the maximum and thus the company k_3 is best emerging company.

5 Conclusion

Here, we have dealt with the prospect theory under $BIFSS$ environment. The score function defined plays a vital role in computing the weighing function. The curve depicted reveals the profits and losses in the value function.

Acknowledgements The authors are thankful to the reviewers for their valuable comments, which have greatly helped in improving the quality of the paper.

Conflict of interest The authors declare that they have no conflict of interest.

References

1. Zadeh, L.A.: Fuzzy sets. Inf. Control **8**, 338–353 (1965)
2. Atanassov, K.: Intuitionistic fuzzy sets. Fuzzy Sets Syst. **20**(1), 87–96 (1986)
3. Jana, C., Pal, M.: Application of bipolar intuitionistic fuzzy soft sets in decision making problem. Int. J. Fuzzy Syst. Appl. **7**(3), 33–55 (2018)
4. Kahneman, D., Tversky, A.: Prospect theory: an analysis of decision under risk. Econometrica **47**(2), 263–292 (1979)
5. Thillaigovindan, N., Anita Shanthi, S., Vadivel Naidu, J.: A better score function for multiple criteria decision making in fuzzy environment with criteria choice under risk. Expert. Syst. Appl. **59**, 78–85 (2016)
6. Tian, X., Zeshui, X., Jian, G., Herrera-Viedma, E.: How to select a promising enterprise for venture capitalists with prospect theory under intuitionistic fuzzy circumstance. Appl. Soft Comput. **67**, 756–763 (2018)
7. Anita Shanthi, S., Jayapalan, P.: A multi-criteria decision making problem based on score function of bipolar intuitionistic fuzzy soft set. AIP Conf. Proc. **2177**, 020004(1-6) (2019)
8. Tversky, A., Kahneman, D.: Advances in prospect theory. J. Risk Uncertainity **5**(4), 297–323 (1992)

Equitable Power Domination Number of Generalized Petersen Graph, Balanced Binary Tree, and Subdivision of Certain Graphs

S. Banu Priya and A. Parthiban

Abstract Let G (V, E) be graph. A set $S \subseteq V$ is said to be a power dominating set (PDS) if every vertex $u \in V - S$ is observed by certain vertices in S by the following rules: (i) if a vertex v in G is in PDS, then it dominates itself and all the adjacent vertices of v and (ii) if an observed vertex v in G has $k > 1$ adjacent vertices and if $k - 1$ of these vertices are already observed, then the remaining one non-observed vertex is also observed by v in G. A power dominating set $S \subseteq V$ in $G(V, E)$ is said to be an equitable power dominating set (EPDS), if for every vertex $v \in V - S$ there exists an adjacent vertex $u \in S$ such that the difference between the degree of u and degree of v is less than or equal to 1, i.e., $|d(u) - d(v)| \leq 1$. The minimum cardinality of an equitable power dominating set of G is called the equitable power domination number of G, denoted by $\gamma_{epd}(G)$. An edge is said to be subdivided if the edge xy is replaced by the path: x wy, where w is the new vertex. A graph obtained by subdividing each edge of a graph G is called subdivision of G, and is denoted by $S(G)$. In this paper, we establish the equitable power domination number of subdivision of certain classes of graphs. We also obtain the equitable power domination number of the generalized Petersen graphs and balanced binary tree.

Keywords Power dominating set · Power domination number · Equitable power dominating set · Equitable power domination number · Generalized petersen graphs · Balanced binary tree · And subdivision graph

---

S. Banu Priya
Department of Mathematics, St. Peter's Institute of Higher Education and Research, Avadi, Chennai 600054, Tamil Nadu, India
e-mail: banupriyasadagopan@gmail.com

A. Parthiban (✉)
Department of Mathematics, School of Chemical Engineering and Physical Sciences, Lovely Professional University, Phagwara 144411, Punjab, India
e-mail: mathparthi@gmail.com; lncs@springer.com

R. N. Mohapatra et al. (eds.), *Mathematical Analysis and Computing*,
Springer Proceedings in Mathematics & Statistics 344,
https://doi.org/10.1007/978-981-33-4646-8_16

1 Introduction

Only simple, finite, undirected, and connected graphs are considered in this paper. A dominating set of a graph $G = (V, E)$ is a set S of vertices such that every vertex v in $V - S$ has at least one neighbor in S. The minimum cardinality of a dominating set of G is called the domination number of G, denoted by $\gamma_d(G)$ [1]. A dominating set $S \subseteq V$ in $G(V, E)$ is said to be an equitable dominating set if for every $v \in V - S$ there exists an adjacent vertex u such that the difference between degree of u and degree of v is less than or equal to 1, that is, $|d(u) - d(v)| \leq 1$. The minimum cardinality of an equitable dominating set of G is called the equitable domination number of G, denoted by $\gamma_{ed}(G)$ [2]. A set $S \subseteq V$ is said to be a power dominating set (PDS) of G if every vertex $u \in V - S$ is observed by some vertices in S using the following rules: (a) If a vertex v in G is in PDS, then it dominates itself and all the adjacent vertices of v and (b) if an observed vertex v in G has $k > 1$ adjacent vertices and if $k - 1$ of these vertices are already observed, then the remaining non-observed vertex is also observed by v in G. The minimum cardinality of a power dominating set of G is called the power domination number of G, denoted by $\gamma_{pd}(G)$ [3].

A power dominating set $S \subseteq V$ in $G(V, E)$ is said to be an equitable power dominating set, if for every vertex $v \in V - S$ there exists an adjacent vertex $u \in S$ such that the difference between the degree of u and degree of v is less than or equal to 1, that is, $|d(u) - d(v)| \leq 1$. The minimum cardinality of an equitable power dominating set of G is called the equitable power domination number of G, denoted by γepd (G) [4]. For more results, one can refer to [5, 6]. In this paper, we obtain the equitable power domination number of the generalized Petersen graphs and balanced binary tree (Figs. 1, 2, 3).

Fig. 1 $GP(9, 2)$

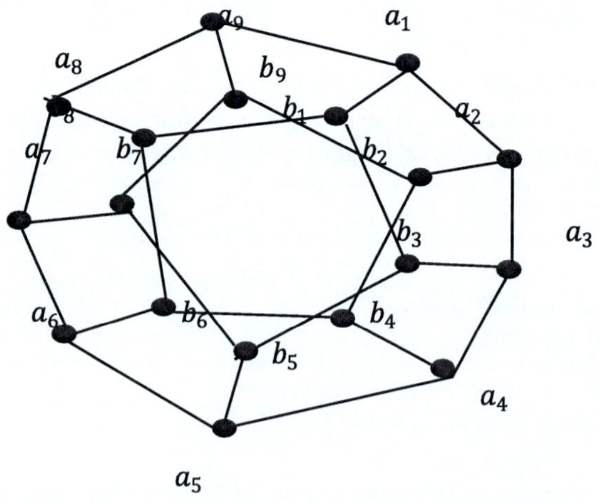

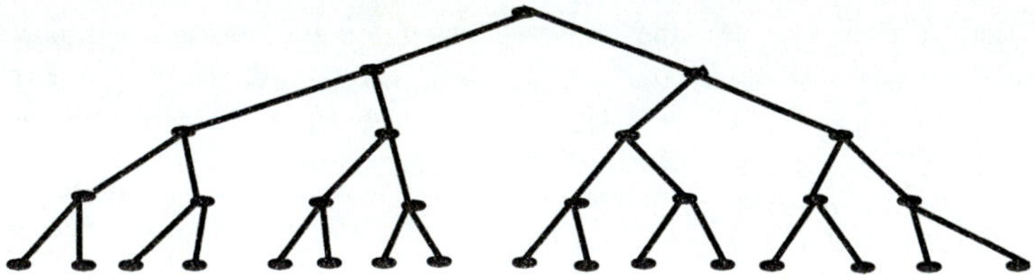

Fig. 2 Balanced binary tree

Fig. 3 A graph G and subdivision of G, S(G)

2 Main Results

For the sake of convenience, by EPDS and EPDN, we mean an equitable power dominating set and the equitable power domination number, respectively.

2.1 Equitable Power Domination Number of the Generalized Petersen Graphs and Balanced Binary Tree

First, we recall the definition of the generalized Petersen graph for the sake of completeness.

Definition 1 [10] The generalized Petersen graph $GP(n, k)$ is defined to be a graph with $V(GP_n, k) = \{a_i, b_i : 0 \leq i \leq n-1\}$ and $E(GP_n, k) = \{a_i\ a_{i+1}, a_i\ b_i, b_i\ b_{i+k} : 0 \leq i \leq n-1$ where the subscripts are expressed as integers modulo $n(n \geq 5)$and$k(k \geq 1)\}$.

Note

1. $GP(n, k)$ is isomorphic to $GP(n, n-k)$.
2. Without restriction of generality, one may consider the generalized Petersen graph $GP(n, k)$ with $k \leq \lceil (n-1)/2 \rceil$.

Theorem 2 *Let $GP(n, k)$ be the generalized Petersen graph.*

$$\text{Then } \gamma_{\text{epd}}(GP(n,k)) = \begin{cases} 2, \text{ for } k = 1, 2 \text{ and } m \geq 4 \\ 3, \text{ for } m \geq 10 \text{ and } k \geq 3 \end{cases}.$$

Proof Let $GP(n,k)$ be the given generalized Petersen graph with $V = \{a_1, a_2, \ldots, a_n, b_1, b_2, \ldots, b_n\}$ and edge set $E(GP(n,k)) = \{a_i a_{i+1}, a_i b_i, b_i b_{i+k} : 0 \leq i \leq n-1\}$. To obtain the equitable power domination number of $GP(n,k)$, we consider the following two cases:

Case 1 For $k = 1, 2$ and $m \geq 4$

Without loss of generality, we choose any one of b_i's, $1 \leq i \leq n$ to be in S, say b_1. Note that b_1 equitably power dominates b_3, a_1, and b_{n-1}. Now the observed vertices b_3, a_1, and b_{n-1} have more than one non-observed vertices and so fail to observe their neighboring vertices which leads to choose another vertex to be in EPDS. Then one can choose either b_2 or b_n to be in S for the sake of minimum cardinality. Now it is easy to see that all the remaining non-observed vertices are observed by their respective neighbors and therefore $|S| = 2$.

Case 2 For m $\geq$ 10 and k $\geq$ 3

Construction of EPDS is similar to Case 1.

3 Equitable Power Domination Number of the Balanced Binary Tree

We recall a few relevant definitions needed for this section for the sake of convenience.

Definition 3 [7] A graph without cycles is called an acyclic graph and a connected acyclic graph is called as a tree.

Definition 4 [7] A binary tree is a tree in which each vertex has at most two pendant vertices.

Definition 5 [7] A balanced binary tree is a binary tree in which the left and right sub-trees of every vertex differ in height by no more than one.

Theorem 6 *Let $B(1,k)$ be a balanced binary tree. Then $\gamma_{\text{epd}}\ (B(1,k)) = \sum_{n=0}^{n=k} 2^n - 2^{n-1}$.*

Proof Let $B(1,k)$ be the given balanced binary tree on k levels with vertex set

$$V = \left\{ \begin{array}{c} a_0, a_1, a_2, a_1^{'}, a_2^{'}, a_3^{'}, a_4^{'}, a_1^{''}, a_2^{''}, a_3^{''}, a_4^{''}, a_5^{''}, a_6^{''}, a_7^{''}, a_8^{''}, \ldots \\ , a_1^n, a_2^n, a_3^n, a_4^n, a_5^n, a_6^n, a_7^n, a_8^n, \ldots, a_n^n \end{array} \right\}$$

where $a_1^n, a_2^n, a_3^n, a_4^n, a_5^n, a_6^n, a_7^n, a_8^n, \ldots, a_n^n$ are the pendant vertices. To obtain an equitable power dominating set S, without loss of generality, we choose a_0 to be in S. The vertex a_0 equitably power dominates a_1 and a_2. Now the vertices a_1 and a_2

have two non-observed vertices $a_1^{'}, a_2^{'}$and$a_3^{'}, a_4^{'}$, respectively. So one has to choose any one between a_1 and a_2, say a_1, then a_2 is observed by a_0. Again as a_2 has two non-observed vertices $a_3^{'}$ and $a_4^{'}$, so one has to choose any one between $a_3^{'}$ and $a_4^{'}$, say $a_3^{'}$. Also a_1 in S observes $a_1^{'}$ and $a_2^{'}$. Proceeding in the same way, finally, we need to choose $a_1^n, a_2^n, a_3^n, a_4^n, a_5^n, a_6^n, a_7^n, a_8^n, \ldots, a_n^n$ as they are the pendent vertices and there are no adjacent vertices satisfying the desired equitable property. Thus, we obtain the sequence of vertices, namely, $a_0, a_1, a_1', a_3', a_1'', a_3'', a_5'', a_7'', \ldots$ and so on. That is,

$\gamma_{\text{epd}}\ (B(1,1)) = 1$
$\gamma_{\text{epd}}\ (B(1,2)) = 1 + 2$
$\gamma_{\text{epd}}\ (B(1,3)) = 1 + 2 + 2^3$
$\gamma_{\text{epd}}\ (B(1,4)) = 1 + 2 + 2^2 + 2^4$
$\gamma_{\text{epd}}\ (B(1,5)) = 1 + 2 + 2^2 + 2^3 + 2^5$

Thus $\gamma_{\text{epd}}\ (B(1,k)) = \sum_{n=0}^{n=k} 2^n - 2^{n-1}$.

3.1 *Equitable Power Domination Number of Subdivision of Certain Classes of Graphs*

The concept of subdivision in graphs was introduced by Trudeau, Richard J in 1993. We recall the definition of subdivision of a graph.

Definition 7 An edge is said to be subdivided if the edge uv is replaced by the path: uwv, where w is the new vertex. A graph obtained by subdividing each edge of a graph G is called subdivision of the graph G, and is denoted by $S(G)$.

Theorem 8 *Let* G *be graph on* n *vertices. Then* $\gamma_{epd}(S(G)) \geq \gamma_{epd}(G)$.

Proof Let G be the given graph with V $= \{v_1, v_2, \ldots, v_n\}$ and edge set E $= \{e_1, e_2, \ldots, e_n\}$. Obtain the subdivision of G, denoted $S(G)$, as follows: $V(S(G)) = V(G) \cup E(G)$ and $E(S(G)) = \{(v_i e_i), (e_i v_j)$: for $1 \leq i \leq n$ and $i+1 \leq j \leq m-1\}$. We consider the following two cases in obtaining an EPDS of S(G):

Case 1 For a vertex v_i incident with e_i for which $|d(v_i) - d(e_i)| \geq 1$ for at least one "i". Then one has to choose e_i to be in S. Thus $\gamma_{epd}(S(G)) \geq \gamma_{epd}(G)$.

Case 2 For a vertex v_i incident with e_i for which $|d(v_i) - d(e_i)| < 1$ for $1 \leq i \leq n$. Then S remains the same. Thus $\gamma_{epd}(S(G)) = \gamma_{epd}(G)$.

Theorem 9 [4] *Let* $C_n, n \geq 3$ *be a cycle. Then* $\gamma_{epd}(C_n)$*= 1.*

Theorem 10 *Let* $C_n, n \geq 3$ *be a cycle. Then* $\gamma_{epd}(S(C_n))$*= 1.*

Proof Let C_n be a cycle with $V(C_n) = \{v_1, v_2, \ldots, v_n\}$. When one performs the subdivision on C_n, the resultant graph is again a cycle on $2n$ vertices. So by Theorem 9, $\gamma_{epd}(S(C_n)) = 1$.

Theorem 11 [4] *Let $P_n, n \geq 1$ be a path. Then $\gamma_{epd}(P_n)$= 1.*

Theorem 12 *Let $P_n, n \geq 3$ be a path. Then $\gamma_{epd}(S(P_n))$= 1.*

Proof Let P_n be a path with $V(P_n) = \{v_1, v_2, \ldots, v_n\}$. An easy check shows that when one performs the subdivision on P_n, the resultant graph is again a path on $2n - 1$ vertices. So by Theorem 11, we deduce that $\gamma_{epd}(S(P_n)) = 1$.

Definition 13 [7] If any two distinct vertices of a graph G are adjacent, then G is said to be complete graph and it is denoted by K_n.

Theorem 14 [4] *For a complete graph K_n, $\gamma_{epd}(K_n) = 1$.*

Theorem 15 *Let $S(K_n)$ be the subdivision of a complete graph K_n. Then $\gamma_{epd}(S(K_n)) = m + n$, for $n \geq 5$.*

Proof Let K_n be a complete graph with $V(K_n) = \{v_1, v_2, \ldots, v_n\}$ and $E(K_n) = \{e_1, e_2, \ldots, e_m\}$. By the definition of a complete graph, the degree of each vertex v_i, $d(v_i) = n-1$ for $1 \leq i \leq n$. Obtain the subdivision of a complete graph K_n, denoted by $S(K_n)$ as follows: $V(S(K_n)) = V_1 \cup V_2$, where $V_1 = V(K_n) = \{v_1, v_2, \ldots, v_n\}$ and $V_2 = E(K_n)$. One can notice that the subdivided graph of a complete graph K_n gives rise to the graph such that no two adjacent vertices with $|d(u) - d(v)| \leq 1$ and violate the equitable property. So to obtain an equitable power dominating set, one has to choose the entire vertex set to be in EPDS. Thus $|S| = m + n$.

Theorem 17 *For a complete bipartite graph $K_{m,n}$, $m, n \geq 4$,*

$$\gamma_{epd}\ (S(K_{m,n})) = \begin{cases} mn + m + n, & if |m - n| \geq 2; \\ 1, & otherwise. \end{cases}$$

Proof Let $K_{m,n}$ be the given complete bipartite graph with $V(K_{m,n}) = V_1 \cup V_2$, where $V_1 = \{u_1, u_2, \ldots, u_m\}$ and $V_2 = \{v_1, v_2, \ldots, v_n\}$ be the two partition sets of $K_{m,n}$. When we construct the subdivision of $K_{m,n}$, the number of vertices of $S(K_{m,n})$ becomes $m + n + mn$. Then the following two cases arise:

Case 1 $|m - n| \geq 2$.

It is clear that $d(u_i) = n$ for every u_i in V_1 and $d(v_i) = m$ for every v_i in V_2. And the degree of newly added vertices is two. Since $|m - n| \geq 2$, the equitable property does not hold well between any two adjacent vertices. Then the entire vertex set of $S(K_{m,n})$ must be chosen to form the equitable power domination set.

Case 2 $|m - n| < 2$

Let $S =\{u_1\}$. Choosing one vertex from one of the partitions with degree more than one is enough to get an equitable power dominating set S. Therefore $\gamma_{epd}\ (K_{m,n}) = 1$, whenever $|m - n| < 2$.

Note γ_{epd} $(S(K_{2,2})) = \gamma_{epd}$ $(S(K_{3,3})) = 1$.

Definition 18 [7] The wheel graph with n spokes, $W_{1,n}$, is the graph that consists of a cycle C_n and one additional vertex, say u, that is, adjacent to all the vertices of the cycle C_n.

Theorem 19 [4] *For a wheel graph* $W_{1,n}$, $n \geq 5$, γ_{epd} $(W_{1,n}) = 2$.

Theorem 20 *Let* $W_{1,n}$, $n \geq 3$ *be a wheel graph. Then* γ_{epd} $(S(W_{1,n})) = n + 1$.

Proof Let $W_{1,n}$ be the given wheel graph on "$n + 1$" vertices with $V(W_{1,n}) = \{v_0, v_1, v_2, \ldots, v_n\}$ where v_0 is the central vertex (the hub) and $v_i : 1 \leq i \leq n$ are the rim vertices, and $E(W_{1,n}) = \{v_1', v_2', \ldots, v_n'\ u_1, u_2, \ldots, u_n\}$, where $u_1, u_2, \ldots, u_n$ represent the spokes of $W_{1,n}$. It is interesting to note that the number of vertices in the subdivision of wheel graph, $S(W_{1,n})$, is $3n + 1$. Now to obtain an equitable power dominating set S, one has to choose the central vertex which is of maximum degree and no adjacent vertices equitably power dominate with it. Also from the remaining vertices, without loss of generality, choose v_1 to be in S as v_1 equitably power dominates v_1', v_n', and u_1. For v_1' and v_n', the only non-observed vertices are v_2 and v_n, respectively, and hence are observed. For v_2 and v_n, there are two non-observed vertices (one in the rim and another in the spoke). So we have to choose any one of the vertices, for the sake of minimum cardinality, we choose u_2 and u_n to be in S. And proceeding thus, one has to choose all the vertices in the spokes of $W_{1,n}$. Thus $S = \{v_0, u_1, u_2, \ldots, u_n\}$ and $|S| = n + 1$.

Definition 21 [7] The gear graph G_n is obtained from a wheel graph $W_{1,n}$ by subdividing each edge of the outer $n-$ cycle of $W_{1,n}$ just once.

Note 1. Gear graph G_n has $2n + 1$ vertices.

2. γ_{epd} $(G_3) = \gamma_{epd}$ $(G_4) = 1$.

Definition 22 [7] Let P_n be a path. Then the $n-$ ladder graph is defined as $P_2\times P_n$.

We label the vertices of the first and second copies of P_n as $\{v_1, v_2, \ldots, v_n\}$.and $\{v_1', v_2', \ldots, v_n'\}$, respectively.
We call a set $W = \{v_1, v_2, v_1', v_2', \ldots v_{n-1}, v_{n-1}', v_n, v_n'\}$.

Theorem 23 *For the* n-*ladder graph*

$$P_2\times P_n, \gamma_{epd}(S(P_2 \times P_n)) = \begin{cases} n-1, when S = \{v \in W\}; \\ n-2, otherwise. \end{cases}$$

Proof Let $G = P_2\times P_n$ be the given $n-$ ladder graph. Note that $|V(G)| = 2n$. Obtain the subdivision of $P_2\times P_n$ with $V(S(P_2 \times P_n)) = \{v_1, v_2, \ldots, v_n, v_1', v_2', \ldots v_{n-1}', v_n, v_n'\} \cup \{u_1, u_2, \ldots, u_{n-1}, v_1', v_2', \ldots, v_{n-1}', w_1, w_2, \ldots, w_n\}$. Now one can see that $|d(u) - d(v)| \leq 1$ for every $u, v \in S(P_2\times P_n)$. Let S denote the required equitable power dominating set of G. Then the following two cases arise:

Case 1 When $S = \{v;\ v \in W\}$

Without loss of generality, let S $= \{v_1\}$. It is easy to see that v_1 equitably power dominates u_1 and w_1 whereas u_1 equitably power dominates v_2 and w_1 equitably power dominates $v_1^{'}$. Moreover, the vertices $v_2$and$v_2^{'}$ have two adjacent vertices that cannot be observed and so one has to include one vertex w_2 to be in S. Proceeding this way we get γ_{epd} (S($P_2 \times P_n$)) $= n - 1$.

Case 2 When $S \subseteq V{-}W$

Consider S $= \{v_3\}$, then v_3 equitably power dominates w_1, u_3, and $u_{\ 2}$. Now there are two non-observed vertices for the already observed vertices w_1, u_3, and $u_{\ 2}$. None of them equitably power dominate any other vertices. Hence, we must choose all the vertices w_1, w_2,..., w_n to be in S. Hence γ_{epd}(S ($P_2 \times P_n$)) $= n - 2$.

Definition 24 The *n*-barbell graph is obtained by connecting two copies of a complete graph K_n by a bridge.

Theorem 25 [4] *Let G be a n− barbell graph. Then* $\gamma_{epd}(G) = 2$, for$n > 2$.

Theorem 26 *Let G be a n− barbell graph.*

Then $\gamma_{epd}(S(G)) = 2(m + n) + 1$, for$n > 6$.

Proof Let G be the given $n-$ barbell graph with $(G) = \{v_1, v_2, \ldots, v_n, v_1', v_2', \ldots, v_n'\}$. Note that when we perform a subdivision on $n-$ barbell graph, the resultant graphs have no adjacent vertices equitably power dominating any of its neighbors. Therefore, we must choose all the vertices to be in S. Hence γ_{epd} $(S(G)) = 2(m + n) + 1$.

4 Applications

The concept of domination helps in computer and in communication to route the information between nodes [8, 9]. Eventually power domination plays a vital role in PMU (phase measurement unit), by minimizing the number of units that are placed in the nodes. It is desirable to minimize the PMU and at the same time it monitors the entire system. Circuits with high voltage may get damaged when connected with very low voltage, and for smooth conduction of entire system, nodes with equal or tend to be equal may have better transmission. We believe that the concept of equitable power domination would play a vital impact in the field of electric power companies and ad hoc networking.

References

1. Narasingh, D.: Graph Theory with Applications to Engineering and Computer Science. New Delhi (1994)
2. Swaminathan, V., Dharmalingam, K.M.: Degree equitable domination on graphs. Kragujevac J. Math. **35**(1), 191–197 (2011)
3. Thenmozhi, B., Prabha, R.: Power domination of middle graph of path, cycle and star. Int. J. Pure Appl. Math. **114**(5), 13–19 (2017)
4. Banu Priya, S., Srinivasan, N.: Equitable power domination number of certain graphs. Int. J. Eng. Technol. **7** (4.10), 349–354 (2018)
5. Banu Priya, S., Parthiban, A., Srinivasan, N.: Equitable power domination number of the central graph of certain graphs. Int. J. Sci. Res. Rev. **7**(3), 2155–2162 (2018)
6. Banu Priya, S., Parthiban, A., Srinivasan, N.: Equitable power domination number of Mycielskian of certain graphs. Int. J. Eng. Technol. 7(4.10), 842–845 (2018)
7. Bondy, J.A., Murthy, U.S.R.: Graph Theory with Applications. Elsevier, North Holland, New York (1986)
8. Coxeter, H.: Self-dual configurations and regular graphs. Bull. Am. Math. Soc **56**, 413–455 (1950)
9. Haynes, T.W., Hedetniemi, S.T., Slater P.J.: Domination in Graphs-Advanced Topics. Marcel Dekker, New York, (1998)
10. Alspach, B.: The classification of Hamiltonian generalized Petersen graphs. J. Comb. Theory, Seri. B **34**(3), 293–312 (1983)

Numerical Solutions of Bivariate Nonlinear Integral Equations with Cardinal Splines

Xiaoyan Liu

Abstract The main objective of this work is to develop an efficient technique for solving bivariate linear or nonlinear Volterra integral equations. The method is based upon the cardinal spline functions on small compact supports. We express the known and unknown functions as linear combinations of translations of the spline functions. The integral equation is reduced to a system of algebra equations. Since the coefficient matrix for the algebraic system is nearly triangular. It is relatively straightforward to solve for the unknowns and an approximation of the original solution with high precision is achieved. Comparisons are made between our schemes and other techniques proposed in recent papers, and the improvement of our method is demonstrated with several numerical examples.

Keywords Bivariate integral equations · Nonlinear integral equations · Spline functions · Numerical methods

1 Introduction

Since the paper [1] by Schoenberg published in 1946, spline functions have gained more and more popularity in a variety of applications. They are piecewise polynomial functions with great approximation properties and excellent flexibility. They are very useful in the construction of wavelet bases and multi-resolution approximations. They are also optimal in the sense that they provide the signal interpolant with the least oscillating energy. Integral equations appear in many fields, including dynamic systems, mathematical applications in economics, communication theory, optimization and optimal control systems, biology and population growth, continuum and quantum mechanics, kinetic theory of gases, electricity and magnetism, potential theory, geophysics, etc. Many differential equations with boundary values

X. Liu (✉)
University of La Verne, La Verne, USA
e-mail: xliu@laverne.edu

R. N. Mohapatra et al. (eds.), *Mathematical Analysis and Computing*,
Springer Proceedings in Mathematics & Statistics 344,
https://doi.org/10.1007/978-981-33-4646-8_17

can be reformulated as integral equations. There are also some problems that can be expressed only in terms of integral equations. Abundant papers have appeared on solving integral equations, for example, Polyanin summarized different solutions of integral equations in [2]. In [3, 4], We discussed numerical methods using cardinal splines in solving systems of linear and nonlinear integral equations of one variable. In this paper, we are going to explore the applications of cardinal splines in solving bivariate integral equations.

We are interested in the Volterra integral equations of the second kind

$$u(x, y) = g(x, y) + \int_a^x \int_c^y K(x, y, s, t, u(s, t))dtds, \; (x, y) \in [a, b] \times [c, d] \tag{1}$$

where the kernel $K(x, y, s, t, u(s, t))$ and $g(x, y)$ are known functions, and $u(x, y)$ is to be determined.

This paper is divided into six sections. In Sects. 2 and 3, bivariate box splines and interpolations are presented. In Sect. 4, the applications of cardinal splines on solving integral equations are explored. The unknown functions are expressed as linear combinations of horizontal translations of a cardinal spline function. Then a system of equations on the coefficients is deducted. We can solve the system and a good approximation of the original solution is obtained. The sufficient condition for the existence of the inverse matrix is discussed and the convergence is investigated. In Sect. 5, the numerical examples are given. The nonlinear system on unknowns is solved and an accurate approximation of the original solution is obtained in each case. Section 6 contains the conclusion remarks.

2 Bivariate Box Spline

Unlike theories of univariate splines, the theory of multivariate splines is far from complete because of the arbitrariness of the region Ω and its partitions. Fortunately, for our purpose, we just need to consider some simple cases. Let us introduce the concept of Box Splines (cf. [5]) first. Box splines are the natural generalization of univariate B-splines on the uniform mesh.

Let

$$\mathbf{X}_n = \{\mathbf{x}^1, \mathbf{x}^2, ..., \mathbf{x}^n\} \subset \mathbf{Z}^s \setminus \{\mathbf{0}\}$$

be a direction set with

$$Span\mathbf{X}_n = \mathbf{R}^s$$

and consider the affine cube

$$[\mathbf{X}_n] = [\mathbf{x}^1, \mathbf{x}^2, \ldots, \mathbf{x}^n] = \{\sum_{i=1}^{n} t_i \mathbf{x}^i : -\frac{1}{2} \le t_i < \frac{1}{2}, i = 1, 2, \ldots, n\},$$

since $Span\mathbf{X}_n = \mathbf{R}^s$, the s-dimensional volume of $[\mathbf{X}_n]$, denoted by $vol_s[\mathbf{X}_n]$ is positive.

Rearrange $\{\mathbf{x}^1, \mathbf{x}^2, \ldots, \mathbf{x}^n\}$ if necessary, so that $vol_s[\mathbf{x}^1, \mathbf{x}^2, \ldots, \mathbf{x}^s] > 0$, and we have the following definition of the box spline $M(\cdot|\mathbf{X}_n)$ with direction set $\mathbf{X}_n$.

Definition 1 Set

$$M(\mathbf{x}|\mathbf{x}^1, \mathbf{x}^2, \ldots, \mathbf{x}^s) = \begin{cases} \frac{1}{vol_s[\mathbf{x}^1, \mathbf{x}^2, \ldots, \mathbf{x}^s]}, & if \ \ \mathbf{x} \in [\mathbf{x}^1, \mathbf{x}^2, \ldots, \mathbf{x}^s] \\ 0 & elsewhere \end{cases} .$$

Then for $m = s + 1, s + 2, ..., n$, define, inductively,

$$M(\mathbf{x}|\mathbf{x}^1, \mathbf{x}^2, \ldots, \mathbf{x}^m) = \int_{-1/2}^{1/2} M(\mathbf{x} - t\mathbf{x}_m|\mathbf{x}^1, \mathbf{x}^2, \ldots, \mathbf{x}^{m-1})\, dt \tag{2}$$

and set $M(\cdot|\mathbf{X}_n) = M(\mathbf{x}|\mathbf{x}^1, \mathbf{x}^2, \ldots, \mathbf{x}^n)$.

Remark. Let $s = 1$ and $\mathbf{x}^1 = \mathbf{x}^2 = \cdots = \mathbf{x}^n = 1$, then $M(x|\mathbf{X}_n) = B_n(x)$ is the n-th degree univariate B-spline. Furthermore, if we choose $\mathbf{x}_1 = \mathbf{x}_2 = \cdots = \mathbf{x}_u = \mathbf{e}_1 = (1, 0)$ and $\mathbf{x}_{u+1} = \mathbf{x}_{u+2} = \cdots = \mathbf{x}_n = \mathbf{e}_2 = (0, 1)$, then $M((x, y)|\mathbf{X}_n) = B_u(x)B_{n-u}(y)$ is the tensor product of two univariate B-splines.

Naturally, Box splines have a lot of similarities as the univariate B-splines. For example, we have the following important propositions (cf. [5]).

Proposition 1 *Let* $\mathbf{X}_n \subset Z^s \setminus \{0\}$ *with* $\langle \mathbf{X}_n \rangle = \mathbf{R}^s$. *Then the box spline* $M(\cdot|\mathbf{X}_n)$ *has the following properties:*

(i) $suppM(\cdot|\mathbf{X}_n) = [\mathbf{X}_n]$.

(ii) $M(\mathbf{x}|\mathbf{X}_n) > 0$ *for* $\mathbf{x}$ *in the interior of* $[\mathbf{X}_n]$.

(iii) Set $B_{\mathbf{X}_n} = \{\sum_{j=1}^{s-1} c_j \mathbf{x}^{i_j} + \sum_j b_j \mathbf{x}^{i'_j} : -\frac{1}{2} \le c_j \le \frac{1}{2}, b_j = \pm\frac{1}{2}, 1 \le i_1 < \cdots < i_{n-1} \le n\}$
where $\{i'_j\}$ *denotes the complementary set of* $\{i_j\}_1^{s-1}$ *with respect to* $\{1, 2, \ldots, n\}$. *(Clearly,* $vol_s B_{\mathbf{X}_n} = 0$ *and* $B_{\mathbf{X}_n} \subset [\mathbf{X}_n]$.*) Then the restriction of* $M(\cdot|\mathbf{X}_n)$ *to each component of the complements of* $B_{\mathbf{X}_n}$ *is a polynomial of total degree n-s. (*$B_{\mathbf{X}_n}$ *is called the grid partition of the box spline* $M(\cdot|\mathbf{X}_n)$.*)*

(iv) Let

$$r(\mathbf{X}_{n)} = \min\{\#\mathbf{Y} : \mathbf{Y} \subset \mathbf{X}_n, (\mathbf{X}_n \backslash \mathbf{Y}) \ne \mathbf{R}^s\} - 2.$$

Then $M(\cdot|\mathbf{X}_n) \in C^{r(\mathbf{X}_n)}(\mathbf{R}^s)$.

Proposition 2 *For any* $\mathbf{x} \in \mathbf{R}^s$, *write*

$$\mathbf{x} = \sum_{i=1}^{n} t_i \mathbf{x}^i$$

where each $t_i = t_i(\mathbf{x})$ *is linear in* $\mathbf{x}$. *Then*

$$(n-s)M(\mathbf{x}|\mathbf{X}_n) = \sum_{i=1}^{n}\{(\frac{1}{2}+t_i)M(\mathbf{x}+\frac{1}{2}\mathbf{x}^i|\mathbf{X}_n\backslash\{\mathbf{x}^i\}) + (\frac{1}{2}-t_i)M(\mathbf{x}-\frac{1}{2}\mathbf{x}^i|\mathbf{X}_n\backslash\{\mathbf{x}^i\})\} \tag{3}$$

whenever each $M(\cdot|\mathbf{X}_n\backslash\{\mathbf{x}^i\})$ *is continuous at* $\mathbf{x}\pm\frac{1}{2}\mathbf{x}^i$, $i = 1, 2, \ldots, n$.

Proposition 3

$$\sum_{\mathbf{j}\in\mathbf{R}^s} M(\cdot - \mathbf{j}|\mathbf{X}_n) \equiv 1 \tag{4}$$

Proposition 4

$$\int_{\mathbf{R}^s} M(\cdot|\mathbf{X}_n)f(\mathbf{x})d\mathbf{x} = \int_{[-1/2,1/2]^n} f(\sum_{i=1}^{n} t_i\mathbf{x}^i)dt_1dt_2...dt_n \tag{5}$$

for all $f \in C(\mathbf{R}^s)$.

Proposition 5 *The Fourier transform of* $M(\cdot|\mathbf{X}_n)$ *is*

$$\hat{M}(y|\mathbf{X}_n) = \hat{M}(\cdot|\mathbf{X}_n)(y) = \prod_{i=1}^{n} \frac{sin(y\cdot\mathbf{x}^i/2)}{y\cdot\mathbf{x}^i/2}. \tag{6}$$

Nevertheless, box splines do not have all the properties the univariate B-splines posses. For example, box splines do not always have minimal supports as the univariate B-splines do. A function in a multivariate spline space is said to have minimal support if there does not exist a nontrivial function in the same space that vanishes identically outside any proper subset of this support. A spline function with minimal support is called a B-spline. A lot of research has been done toward bivariate B-splines.

3 Interpolation by Bivariate Splines

Cardinal interpolation problem by bivariate box splines can be defined as follows. Let ϕ be a compactly supported continuous function in $\mathbf{R}^s$, with $\{\widehat{\phi}(2\pi\mathbf{j})\} \in l^1(Z^s)$. Let $\mathcal{S}(\phi)$ be a vector space of the box spline series

$$\sum_{\mathbf{j}\in\mathbf{Z}^s} c_{\mathbf{j}}\phi(\cdot - \mathbf{j}).$$

The problem of cardinal interpolation from $\mathcal{S}(\phi)$ can be stated as follows (cf. [6]): for a given data sequence $\mathbf{F} = \{f_{\mathbf{j}}\}, \mathbf{j} \in \mathbf{Z}^s$, determine a coefficient sequence $\mathbf{C} = \{c_{\mathbf{j}}\}, \mathbf{j} \in \mathbf{Z}^s$, such that

$$s_{\mathbf{F}}(\mathbf{x}) = \sum_{\mathbf{j}\in \mathbf{Z}^s} c_{\mathbf{j}}\phi(\cdot - \mathbf{j})$$

from $\mathcal{S}(\phi)$ satisfies

$$s_{\mathbf{F}}(\mathbf{k}) = f_{\mathbf{k}}, \ \mathbf{k} \in \mathbf{Z}^s,$$

or equivalently, let $\Phi = \{\phi(\mathbf{j})\}$, using the notation $\mathbf{C} * \Phi$ to represent the convolution of $\mathbf{C}$ with Φ on $\mathbf{Z}^s$, then the above equations can be written as

$$\mathbf{C} * \Phi = \mathbf{F}.$$

When $s = 2$, the interpolation of a function $f(x, y) \in l^p$ by box splines in $S_{3\mu}^{2\mu-1}(\Omega, \Delta_1)$ or $S_{3\mu+1}^{2\mu}(\Omega, \Delta_1)$ can be implemented by the above scheme. However, other than $B_1(x, y) \in S_1^0(\Omega, \Delta_1)$, the amount of calculation involved in finding the inverse of Φ gets bigger as the partition gets finer. As for $B_1(x, y) \in S_1^0(\Omega, \Delta_1)$, since its value is 1 at one grid point and zero at other grid points, we have $\Phi = \mathbf{I}$, a unit matrix and

$$\mathbf{C} = \mathbf{F} * \Phi^{-1} = \mathbf{F}.$$

4 Numerical Methods Solving Integral Equations

We are interested in solving the bivariate linear or nonlinear Volterra integral equations (1). For integers M and N, let $h = \frac{b-a}{M}$, $l = \frac{d-c}{N}$, $x_i = a + ih$ and $y_j = c + jl$. Let $g(x, y) = \sum_{i=0}^{M}\sum_{j=0}^{N} g_{ij} B(\frac{x-x_i}{h}, \frac{y-y_j}{l})$, $u(x, y) = \sum_{i=0}^{M}\sum_{j=0}^{N} c_{ij} B(\frac{x-x_i}{h}, \frac{y-y_j}{l})$,

$$K(x, y, s, t, u(s, t)) = \sum_{i,j=0}^{(M,N)} \sum_{p,q=0}^{(M,N)} K(x_i, y_j, s_p, t_q, c_{pq}) B(\frac{s - x_i}{h}, \frac{t - y_j}{l}) B(\frac{x - x_p}{h}, \frac{y - y_q}{l}),$$

plug into (1):

$$\sum_{i=0}^{M}\sum_{j=0}^{N} c_{ij} B(\frac{x - x_i}{h}, \frac{y - y_j}{l}) = \sum_{i=0}^{M}\sum_{j=0}^{N} g_{ij} B(\frac{x - x_i}{h}, \frac{y - y_j}{l}) + \int_a^y \int_c^x \sum_{i,j=0}^{(M,N)} \sum_{p,q=0}^{(M,N)} B(\frac{s - x_p}{h}, \frac{t - y_q}{l}) K(x_i, y_j, s_p, t_q, c_{pq}) B(\frac{x - x_i}{h}, \frac{y - y_j}{l}) ds dt$$

Let $x = x_i, y = y_j$:

$$c_{ij} = f_{ij} + \int_a^{y_j}\int_c^{x_i}\sum_{p=0}^{M}\sum_{q=0}^{N} B(\frac{s-x_p}{h}, \frac{t-y_q}{l})K(x_i, y_j, s_p, t_q, c_{pq})dsdt \; (S3)$$

$i = 0, 1, ..., N, j = 0, 1, 2, ..., M.$

Remark If the integral equation (1) has a unique solution, then the linear system $(S3)$ is consistent. Furthermore, $u(x, y) = \sum_{i=0}^{M}\sum_{j=0}^{N} c_{ij} B(\frac{x-x_i}{h}, \frac{y-y_j}{l})$ approximate the solution of (1) to a convergence rate of h^2.

5 Numerical Examples

Example 1 (cf. [7]) Let $(x, y) \in [0, 1] \times [0, 1]$

$$u(x, y) = g(x, y) + \int_0^y\int_0^x (x + y - t - s)\, u^2(s, t)\, dsdt$$

$$where\, g(x, y) = x + y - \frac{1}{12}(xy)\left(x^3 + 4x^2y + 4xy^2 + y^3\right)$$

$$K(x, y, s, t, u(s, t)) = (x + y - t - s)u^2(s, t).$$

$$(The\ exact\ Solution\ is\ u(x, y) = (x + y).)$$

For integers $M = N = 8$, let $h = l = \frac{1}{8}$.

$$g(x, y) = \sum_{i=0}^{M}\sum_{j=0}^{N} g_{ij} B(\tfrac{x-x_i}{h}, \tfrac{y-y_j}{l}),$$

$$u(x, y) = \sum_{i=0}^{M}\sum_{j=0}^{N} c_{ij} B(\tfrac{x-x_i}{h}, \tfrac{y-y_j}{l}),$$

$$K(x, y, s, t, u(s, t)) = \sum_{i,j=0}^{(M,N)}\sum_{p,q=0}^{(M,N)} K(x_i, y_j, s_p, t_q, c_{pq})B(\tfrac{s-x_i}{h}, \tfrac{t-y_j}{l})B(\tfrac{x-x_p}{h}, \tfrac{y-y_q}{l})$$

Applying the above method, we get

$$\sum_{i=0}^{M}\sum_{j=0}^{N} c_{ij} B(\frac{x-x_i}{h}, \frac{y-y_j}{l})$$

$$= \sum_{i=0}^{M}\sum_{j=0}^{N} g_{ij} B(\frac{x-x_i}{h}, \frac{y-y_j}{l}) +$$

$$\int_0^y\int_0^x \sum_{i,j=0}^{(M,N)}\sum_{p,q=0}^{(M,N)} (x_i + y_j - s_p - t_q)c_{pq}^2 B(\frac{x-x_i}{h}, \frac{y-y_j}{l})B(\frac{s-x_p}{h}, \frac{t-y_q}{l})dsdt$$

Let $x = x_i, y = y_j$:

$$c_{ij} = g_{ij} + \int_0^{y_j}\int_0^{x_i}\sum_{p=0}^{8}\sum_{q=0}^{8} B(\frac{s-x_p}{h}, \frac{t-y_q}{l})(x_i + y_j - s_p - t_q)c_{pq}^2 dsdt$$

$i = 0, 1, ..., 8, j = 0, 1, 2, ..., 8$. Solving the system, we obtain

$[c_{0,0}, c_{0,1}, c_{0,2}, c_{0,3}, c_{0,4}, c_{0,5}, c_{0,6}, c_{0,7}, c_{0,8};\ c_{1,0}, c_{1,1}, c_{1,2}, c_{1,3}, c_{1,4}, c_{1,5}, c_{1,6}, c_{1,7}, c_{1,8};\ c_{2,0}, c_{2,3}, c_{2,4}, c_{2,5}, c_{2,6}, c_{2,7}, c_{2,8};\ c_{2,1}, c_{2,2}, c_{3,0}, c_{3,1}, c_{3,2}, c_{3,3}, c_{3,4}, c_{3,5}, c_{3,6}, c_{3,7}, c_{3,8};\ c_{4,0}, c_{4,1}, c_{4,2}, c_{4,3}, c_{4,4}, c_{4,5}, c_{4,6}, c_{4,7}, c_{4,8};\ c_{5,0}, c_{5,1}, c_{5,2}, c_{5,3}, c_{5,4}, c_{5,5}, c_{5,6}, c_{5,7}, c_{5,8};\ c_{6,0}, c_{6,1}, c_{6,2}, c_{6,3}, c_{6,4}, c_{6,5}, c_{6,6}, c_{6,7}, c_{6,8};\ c_{7,0}, c_{7,1}, c_{7,2}, c_{7,3}, c_{7,4}, c_{7,5}, c_{7,6}, c_{7,7}, c_{7,8};\ c_{8,0}, c_{8,1}, c_{8,2}, c_{8,3}, c_{8,4}, c_{8,5}, c_{8,6}, c_{8,7}, c_{8,8}] =$

[0, 0.125, 0.25, 0.375, 0.5, 0.625, 0.75, 0.875, 1; 0.125, 0.24999, 0.37497, 0.499945, 0.62490, 0.74985, 0.87479, 0.99971, 1.12463; 0.25, 0.37497, 0.49992, 0.62485, 0.74975, 0.87464, 0.99951, 1.12435, 1.24917; 0.375, 0.49994, 0.624859, 0.74972, 0.87457, 0.99938, 1.12416, 1.24890, 1.37360; 0.5, 0.62490, 0.74975, 0.87457, 0.99934, 1.12406, 1.24874, 1.37336, 1.49792; 0.625, 0.74985, 0.87464, 0.999381, 1.12406, 1.24868, 1.37323, 1.49770, 1.62209; 0.75, 0.87479, 0.99951, 1.12416, 1.24874, 1.37323, 1.49763, 1.62193, 1.74609, 0.875, 0.99971, 1.12435, 1.24890, 1.37336, 1.49770, 1.62193, 1.74600, 1.86990, 1, 1.12463, 1.24917, 1.37360, 1.49792, 1.62209, 1.74609, 1.86990, 1.99346]

Error $< 6.6\times 10^{-3}$.

Example 2 (cf. [7]) Let $(x, y) \in [0, 1] \times [0, 1]$

$$u(x, y) = g(x, y) + \int_0^y\int_0^x (xs^2 + \cos t)u^2(s, t)\, dsdt,$$

where $g(x, y) = x \sin y \left(1 - \frac{1}{9}x^2\left(\sin^2 y\right)\right) + \frac{1}{10}x^6\left(\frac{1}{2}\sin(2y) - y\right)$. (The exact solution is $u(x, y) = x \sin y$.)

Let $M = N = 8, h = l = \frac{1}{8}$, $g(x, y) = \sum_{i=0}^{M}\sum_{j=0}^{N} g_{ij} B(\frac{x-x_i}{h}, \frac{y-y_j}{l})$, $u(x, y) = \sum_{i=0}^{M}\sum_{j=0}^{N} c_{ij} B(\frac{x-x_i}{h}, \frac{y-y_j}{l})$,

$$\left(xs^2 + \cos t\right)u^2(s, t) = \sum_{i,j=0}^{(M,N)}\sum_{p,q=0}^{(M,N)} (x_i s_p^2 + \cos t_q)c_{pq}^2 B(\frac{s-x_i}{h}, \frac{t-y_j}{l})B(\frac{x-x_p}{h}, \frac{y-y_q}{l})$$

$$\sum_{i=0}^{M}\sum_{j=0}^{N} c_{ij} B(\frac{x-x_i}{h}, \frac{y-y_j}{l})$$

$$= \sum_{i=0}^{M}\sum_{j=0}^{N} g_{ij} B(\frac{x-x_i}{h}, \frac{y-y_j}{l}) +$$

$$\int_0^y \int_0^x \sum_{i,j=0}^{(M,N)} \sum_{p,q=0}^{(M,N)} (x_i s_p^2 + \cos t_q) c_{pq}^2 B(\frac{x-x_i}{h}, \frac{y-y_j}{l}) B(\frac{s-x_p}{h}, \frac{t-y_q}{l}) dsdt$$

Let $x = x_i, y = y_j$:

$$c_{ij} = g_{ij} + \int_0^{y_j} \int_0^{x_i} \sum_{p=0}^{8}\sum_{q=0}^{8} B(\frac{s-x_p}{h}, \frac{t-y_q}{l})(x_i s_p^2 + \cos t_q) c_{pq}^2 dsdt$$

$i = 0, 1, ..., N$, $j = 0, 1, 2, ..., M$. Solving the system, we obtain

$[c_{0,0}, c_{0,1}, c_{0,2}, c_{0,3}, c_{0,4}, c_{0,5}, c_{0,6}, c_{0,7}, c_{0,8};\ c_{1,0}, c_{1,1}, c_{1,2}, c_{1,3}, c_{1,4}, c_{1,5}, c_{1,6}, c_{1,7}, c_{1,8};\ c_{2,0}, c_{2,3}, c_{2,4}, c_{2,5}, c_{2,6}, c_{2,7}, c_{2,8};\ c_{2,1}, c_{2,2}, c_{3,0}, c_{3,1}, c_{3,2}, c_{3,3}, c_{3,4}, c_{3,5}, c_{3,6}, c_{3,7}, c_{3,8};\ c_{4,0}, c_{4,1}, c_{4,2}, c_{4,3}, c_{4,4}, c_{4,5}, c_{4,6}, c_{4,7}, c_{4,8};\ c_{5,0}, c_{5,1}, c_{5,2}, c_{5,3}, c_{5,4}, c_{5,5}, c_{5,6}, c_{5,7}, c_{5,8};\ c_{6,0}, c_{6,1}, c_{6,2}, c_{6,3}, c_{6,4}, c_{6,5}, c_{6,6}, c_{6,7}, c_{6,8};\ c_{7,0}, c_{7,1}, c_{7,2}, c_{7,3}, c_{7,4}, c_{7,5}, c_{7,6}, c_{7,7}, c_{7,8};\ c_{8,0}, c_{8,1}, c_{8,2}, c_{8,3}, c_{8,4}, c_{8,5}, c_{8,6}, c_{8,7}, c_{8,8}] =$

[0, 0, 0, 0, 0, 0, 0, 0, 0; 0, 0.015585, 0.03093, 0.04579, 0.05994, 0.07316, 0.08524, 0.09599, 0.10525; 0, 0.03117, 0.06186, 0.09158, 0.11989, 0.14632, 0.17048, 0.19199, 0.21050, 0, 0.04676, 0.09279, 0.13738, 0.17984, 0.21950, 0.25574, 0.28800, 0.31577; 0, 0.06235, 0.12374, 0.18320, 0.23981, 0.29270, 0.34103, 0.38405, 0.42109; 0, 0.07795, 0.15470, 0.22904, 0.29982, 0.36594, 0.42637, 0.48016, 0.52649; 0, 0.09357, 0.18568, 0.27491, 0.35987, 0.43924, 0.51179, 0.57639, 0.63202; 0, 0.10920, 0.21670, 0.32084, 0.42001, 0.51266, 0.59736, 0.67280, 0.73780; 0, 0.12485, 0.24777, 0.36686, 0.48027, 0.58625, 0.68316, 0.76951, 0.84395]

Error $< 2.5\times10^{-3}$

6 Conclusion

The proposed method is a simple and effective procedure for solving bivariate Volterra integral equations of the second kind. The methods can be adapted easily to the Volterra integral equations of the first kind, which have the form $u(x, y) = \int\int_A K(x, y, s, t, u(s, t))dtds$. The methods can also be extended to the Fredholm and Volterra integral equations of the first kind or the second kind, where the integral is on an infinite set. The higher degree cardinal splines could also be applied to nonlinear integral equations; the resulting system of coefficients will be a little more complicated nonlinear systems, which takes more time and effort to solve. Compared with the recent paper [7], our method is more effective.

Acknowledgements The author thanks La Verne Academy and the provost of University of La Verne for their generous support for the research activity.

References

1. Schoenberg, I.J.: On trigonometric spline functions. J. Math. Mech. **13**, 795–825 (1964)
2. Polyanin, A.D.: Handbook of Integral Equations. CRC Press LLC (1998)
3. Liu, X., Xie, J.: Numerical methods for solving systems of Fredholm integral equations with cardinal splines. AIP Conf. Proc. **1637**, 590 (2014)
4. Liu, X., Liu, Z., Xie, J.: Solving systems of volterra integral equations with cardinal splines. J. Appl. Math. Phys. (JAMP) **3**(11) (2015)
5. De Boor, C., Devore, R.: Approximation by smooth multivariate splines. Trans. Am. Math. Soc. **276**, 775–788 (1983)
6. Chui, C.K.: Multivariate Splines. SIAM, Philadelphia (1988)
7. Babolian, E., Bazm, S., Lima, P.: Numerical solution of Nonlinear two-dimensional integral equations using rationalized Haar functions. Commun. Nonlinear Sci. Numer. Simulat. **16**, 1164–1175 (2011)

Bifurcation Analysis and Chaos Control for a Discrete Fractional-Order Prey–Predator System

A. George Maria Selvam, D. Vignesh, and R. Janagaraj

Abstract Allee effect relates to the fitness of an individual and the density of the population in an ecosystem. This type of positive association may lead to a population size below which the persistence of the species is not possible. In this work, we consider a fractional-order discrete-time system representing interactions of predator and prey involving Holling type II response and Allee effect. The existence results of the equilibrium points together with the stability of the system are discussed. The chaotic behavior of the system is analyzed with the bifurcation theory to prove the existence of periodic doubling and Neimark–Sacker bifurcations. The control strategy are employed to the system to study the containment of the chaos and simulations are performed to support the results.

Keywords Population dynamics · Fractional order · Discrete · Equilibrium points · Stability · Allee effect · Holling type II · Bifurcation

Mathematics Subject Classification (2010) 34A08 · 37N25 · 39A28 · 39A30 · 92D25

1 Introduction

The biotic and abiotic factors in the ecosystem have a great impact on the survival of the species population. The presence of these factors also makes the modeling of the system in the ecosystem more complicated. Not all of these factors can be considered

A. George Maria Selvam (✉) · D. Vignesh
Department of Mathematics, Sacred Heart College (Autonomous),
Tirupattur, Vellore, Tamil Nadu, India
e-mail: agmshc@gmail.com

R. Janagaraj
Department of Mathematics, Faculty of Engineering, Karpagam Academy of Higher Education,
Coimbatore, Tamil Nadu, India

R. N. Mohapatra et al. (eds.), *Mathematical Analysis and Computing*,
Springer Proceedings in Mathematics & Statistics 344,
https://doi.org/10.1007/978-981-33-4646-8_18

while modeling an ecosystem but some factors that are crucial, predictable, and have greater impact are taken into consideration. Apart from these factors, there are other aspects that are to be considered are immigration, emigration, hunting, and natural calamities. The construction of models of the ecosystem also takes account of disease spread among the species, spread of virus and parasite, interaction between the same and different species.

The basis of the mathematical modeling of ecosystem revolves around the conservation of the species from extinction. It was in the late eighteenth-century models were developed on the growth of population and extinction [6]. According to Malthus, the population of species grew in a geometrical pattern. One of the most important breakthroughs in population biology was Verhulst's logistic model in 1838. After the origin of the logistic model, different types of models investigating the social and natural processes in the environment have been studied. Various qualitative properties of the system have been analysed employing bifurcation theories and understanding of stability, periodic orbits, attractors, control strategies.

The Allee effect is a phenomenon characterizing the interrelationship between the population density and fitness. Warder Clyde Allee was the first to describe this concept during the 1930s [3]. He observed the growth of goldfish in presence of more individuals in the tank that lead to the origin of Allee effect. Allee effect study has made notable contribution to the field of mathematical biology [2, 9, 11, 13, 17]. Biological facts on the Allee function are studied in [12]. The stability analysis of biological population model of species interaction with or without Allee effect was studied in [16, 20].

The development of the theory of arbitrary order calculus has brought about radical changes in the study of physical, biological and chemical processes due to the memory and hereditary effects considered in models [15]. The behavioral analysis of the system is better understood with a model constructed with fractional order. The discrete version of the factional calculus has recently gained attraction and are widely used in modeling species interactions.

The paper is organized as follows: Discretization of the fractional derivative system is illustrated in Sect. 2. Section 3 analyses the stability results of the system at equilibrium points and the bifurcations of the discretized system is carried out using the bifurcation theory in Sect. 4. Strategies of controlling chaos are implemented in Sect. 5. Examples with Numerical simulations are presented in Sect. 6 followed by a conclusion.

2 Mathematical Description of the Model

The system of two species interaction with Allee effect and functional response of Holling type II is

$$\frac{dx}{dt} = rx(1-x) - \frac{axy}{1+bx}\left[\frac{x}{m+x}\right]$$
$$\frac{dy}{dt} = \frac{dxy}{1+bx} - cy \tag{1}$$

Here $x(t)$ and $y(t)$ denote prey and predator populations. Parameters r, a, b, c, d, m of the system (1) take positive values and represents the growth rate of prey, predation rate, consumption rate, rate at which the predator grows, decline of predator population and Allee constant, respectively.

The non-local property of the fractional order systems have been of great importance as it considers present and past stages of a population. On generalization of (1) to arbitrary order, we have

$$D_t^{\beta}(x) = rx - rx^2 - \frac{axy}{1+bx}\left[\frac{x}{m+x}\right]$$
$$D_t^{\beta}(y) = \frac{dxy}{1+bx} - cy \tag{2}$$

with $x(0) = x_0$ & $y(0) = y_0$, where $\beta \in (0, 1]$ with Caputo derivative of arbitrary order defined by ${}_bD_t^{\beta} p(\ell) = \frac{1}{\Gamma(1-\beta)} \int_b^{\ell} \frac{p^{(\kappa)}(\tau)}{(\ell-\tau)^{\beta-\kappa+1}} d\tau$, for $\kappa - 1 < \beta < \kappa$.

2.1 *Discretization of (2)*

The transformation of continuous models into its discrete counterpart has been carried out in [1, 4, 8, 10, 14, 19]. The discretized prey–predator system of fractional order with Allee effect imposed is

$$x(t+1) = x(t) + \frac{\lambda^{\beta}}{\Gamma(1+\beta)}\left[rx(t)(1-x(t)) - \frac{ax(t)y(t)}{1+bx(t)}\left(\frac{x(t)}{m+x(t)}\right)\right]$$
$$y(t+1) = y(t) + \frac{\lambda^{\beta}}{\Gamma(1+\beta)}\left[\frac{dx(t)y(t)}{1+bx(t)} - cy(t)\right] \tag{3}$$

where $\beta \in (0, 1]$ and $\lambda > 0$ is defined as time interval.

3 Stability Results

This section establishes the existence results of the equilibrium points of the system (3). The Jury conditions are employed to the analysis of stability using Jacobian matrices.

3.1 Equilibrium Points of (3) and Their Existence

Non-negative equilibrium points of the system (3) are obtained from

$$\begin{aligned} rx(1-x) - \frac{axy}{1+bx}\left(\frac{x}{m+x}\right) &= 0 \\ \frac{dxy}{1+bx} - cy &= 0 \end{aligned} \tag{4}$$

The Equilibrium points are

1. $ES_0 = (0, 0)$
2. $ES_1 = (1, 0)$
3. $ES_2 = \left(x^*, \frac{dr(m+(1-m)x^*-(x^*)^2)}{ac}\right)$.

where $x^* = \frac{c}{d-cb}$.

Theorem 1 *The existence of equilibrium points satisfy*

1. *The equilibrium points ES_0 and ES_1 always exists.*
2. *Interior equilibrium point ES_2 exists if $m > \frac{c}{d-cb}$ and $c < \frac{d}{b}$.*

3.2 Stability Analysis

Discussion of stability of (3) is carried out using Jacobian matrices at equilibrium points. The Jacobian matrix of (3) at (x, y) is

$$J(x,y) = \begin{bmatrix} 1 + \mathcal{A}\left[r(1-2x) - B\left(1 + \frac{m-bx^2}{[(m+x)(1+bx)]}\right)\right] & -\mathcal{A}\frac{ax^2}{[(m+x)(1+bx)]} \\ \frac{\mathcal{A}dy}{(1+bx)^2} & 1 + \mathcal{A}\left[\frac{dx}{1+bx} - c\right] \end{bmatrix} \tag{5}$$

where $\mathcal{A} = \frac{\lambda^\beta}{\Gamma(1+\beta)}$, $B = \frac{axy}{(m+x)(1+bx)}$. From (5), we have the characteristic polynomial given by

$$\Theta(\mu) = \mu^2 - T\mu + D = 0 \tag{6}$$

$T = 2 + A\left[r(1-2x) - B\left(1 + \frac{m-bx^2}{[(m+x)(1+bx)]}\right) + \frac{dx}{1+bx} - c\right]$ and $D = \Bigg(1 + A \left[r(1-2x) - B\left(1 + \frac{m-bx^2}{[(m+x)(1+bx)]}\right)\right]\Bigg)\left[1 + A\left(\frac{dx}{1+bx} - c\right)\right] + \frac{A^2adx^2y}{(m+x)(1+bx)^3}$ are the trace and determinant of (5). We make use of the following lemma [18] to provide the relation between roots and quadratic equation.

Lemma 1 *The roots of $\Theta(\mu) = 0$ be $\mu_{1,2}$. Suppose $\Theta(1) > 0$, then equilibrium point (x^*, y^*) is*

1. *a sink if* $|\mu_1| < 1$, $|\mu_2| < 1 \Leftrightarrow \Theta(-1) > 0$, $\Theta(0) < 1$
2. *a saddle point* $|\mu_1| < 1$, $|\mu_2| > 1$ *(or* $|\mu_1| > 1$, $|\mu_2| < 1$*)* $\Leftrightarrow \Theta(-1) < 0$.
3. *a source* $|\mu_1| > 1$, $|\mu_2| > 1 \Leftrightarrow \Theta(-1) > 0$, $\Theta(0) > 1$
4. $|\mu_1| = -1$, $|\mu_2| \neq 1 \Leftrightarrow \Theta(-1) = 0$ $T \neq 0$ *and* 2.
5. *complex with* $|\mu_1| = |\mu_2| \Leftrightarrow T^2 - 4D < 0$, $\Theta(0) = 1$.

Theorem 2 *The equilibrium point* ES_0 *is a*

1. *source, if* $|\mu_2| > 1$, *i.e.,* $\lambda > \left[\frac{2\Gamma(1+\beta)}{c}\right]^{\frac{1}{\beta}}$.
2. *saddle point for* $|\mu_2| < 1$, *i.e.,* $0 < \lambda < \left[\frac{2\Gamma(1+\beta)}{c}\right]^{\frac{1}{\beta}}$.
3. *non-hyperbolic for* $\lambda = \left[\frac{2\Gamma(1+\beta)}{c}\right]^{\frac{1}{\beta}}$.

Proof For ES_0, the Jacobian is

$$J_{ES_0} = J(0,0) = \begin{bmatrix} 1 + \frac{\lambda^\beta}{\Gamma(1+\beta)} r & 0 \\ 0 & 1 - \frac{\lambda^\beta}{\Gamma(1+\beta)} c \end{bmatrix}$$

whose eigenvalues are $\mu_1 = 1 + \frac{\lambda^\beta}{\Gamma(1+\beta)} r$ and $\mu_2 = 1 - \frac{\lambda^\beta}{\Gamma(1+\beta)} c$.
Since $\frac{\lambda^\beta}{\Gamma(1+\beta)} > 0$ for $0 < \beta \le 1$.

(i) Since, $|m_1| > 1$. Then ES_0 is unstable for $\left|1 - \frac{\lambda^\beta}{\Gamma(1+\beta)} c\right| > 1$ which yields

$$\lambda > \left[\frac{2\Gamma(1+\beta)}{c}\right]^{\frac{1}{\beta}}.$$

(ii) ES_0 is unstable (saddle) for $\left|1 - \frac{\lambda^\beta}{\Gamma(1+\beta)} c\right| < 1$, that is

$$0 < \lambda < \left[\frac{2\Gamma(1+\beta)}{c}\right]^{\frac{1}{\beta}}.$$

(iii) Proof follows (i) and (ii).

Theorem 3 *The axial equilibrium point* ES_1 *is*

1. *stable for* $d < c(1+b)$ and $\lambda <$ min $\left\{\left[\frac{2\Gamma(1+\beta)}{r}\right]^{\frac{1}{\beta}}, \left[\frac{2(b+1)\Gamma(1+\beta)}{c(1+b)-d}\right]^{\frac{1}{\beta}}\right\}$;
2. *unstable for* $d > c(1+b)$ and $\lambda >$ max $\left\{\left[\frac{2\Gamma(1+\beta)}{r}\right]^{\frac{1}{\beta}}, \left[\frac{2(b+1)\Gamma(1+\beta)}{c(1+b)-d}\right]^{\frac{1}{\beta}}\right\}$
3. *non-hyperbolic for* $d = c(1+b)$ or $\lambda = \left[\frac{2\Gamma(1+\beta)}{r}\right]^{\frac{1}{\beta}}$ or $\lambda = \left[\frac{2(b+1)\Gamma(1+\beta)}{c(1+b)-d}\right]^{\frac{1}{\beta}}$.

Proof For ES_1, Jacobian matrix is

$$J_{ES_1} = J(1,0) = \begin{bmatrix} 1 - \frac{\lambda^\beta}{\Gamma(1+\beta)} r & -\frac{\lambda^\beta}{\Gamma(1+\beta)} \frac{a}{(1+m+2b)} \\ 0 & 1 + \frac{\lambda^\beta}{\Gamma(1+\beta)} \left(\frac{d}{1+b} - c\right) \end{bmatrix}$$

whose eigen values are $\mu_1 = 1 - \frac{\lambda^\beta}{\Gamma(1+\beta)} r$ and $\mu_2 = 1 + \frac{\lambda^\beta}{\Gamma(1+\beta)} \left(\frac{d}{1+b} - c\right)$. Since $\frac{\lambda^\beta}{\Gamma(1+\beta)} > 0$ for $0 < \beta \le 1$.

(a) ES_1 is stable if $\left|1 - \frac{\lambda^\beta}{\Gamma(1+\beta)} r\right| < 1$ and $\left|1 + \frac{\lambda^\beta}{\Gamma(1+\beta)} \left(\frac{d}{1+b} - c\right)\right| < 1$ which yields

$$d < c(1+b) \text{ and } \lambda < \min \left\{ \left[\frac{2\Gamma(1+\beta)}{r}\right]^{\frac{1}{\beta}}, \left[\frac{2(b+1)\Gamma(1+\beta)}{c(1+b)-d}\right]^{\frac{1}{\beta}} \right\}.$$

(b) ES_1 is unstable if $\left|1 - \frac{\lambda^\beta}{\Gamma(1+\beta)} r\right| > 1$ and $\left|1 + \frac{\lambda^\beta}{\Gamma(1+\beta)} \left(\frac{d}{1+b} - c\right)\right| > 1$, i.e.,

$$d > c(1+b) \text{ and } \lambda > \max \left\{ \left[\frac{2\Gamma(1+\beta)}{r}\right]^{\frac{1}{\beta}}, \left[\frac{2(b+1)\Gamma(1+\beta)}{c(1+b)-b}\right]^{\frac{1}{\beta}} \right\}.$$

(c) Proof is similar to (a) and (b).

$$J_{\text{ES}_2} = \begin{bmatrix} 1 + \mathcal{A}a_{11} & -\mathcal{A}a_{12} \\ \mathcal{A}a_{21} & 1 + \mathcal{A}a_{22} \end{bmatrix} \tag{7}$$

From J_{ES_2}, we get $\Theta(\mu) = \mu^2 - T\mu + D = 0$, with $T = 2 + \mathcal{A}K_1$ and $D = 1 + \mathcal{A}K_1 + \mathcal{A}^2 K_2$. where $\mathcal{A} = \frac{\lambda^\beta}{\Gamma(1+\beta)}$, $K_1 = a_{11} + a_{22}$, $K_2 = a_{11}a_{22} + a_{12}a_{21}$, $B = \frac{ax^*y^*}{(m+x^*)(1+bx^*)}$, $a_{11} = r(1-2x^*) - B\left[1 + \frac{m - bx^{*2}}{[(m+x^*)(1+bx^*)]}\right]$, $a_{12} = \frac{a(x^*)^2}{(m+x^*)(1+bx^*)}$, $a_{21} = \frac{dy^*}{(1+bx^*)^2}$ and $a_{22} = \frac{dx^*}{1+bx^*} - c$. The eigen values are

$$\mu_{1,2} = 1 + \frac{\mathcal{A}K_1}{2} \pm \frac{\mathcal{A}}{2}\sqrt{K_1^2 - 4K_2}.$$

Theorem 4 *The equilibrium point ES_2 is a*

1. *sink if one of the following is satisfied:*

   (a) $\Pi \ge 0$ and $\lambda < \lambda_2$,
   (b) $\Pi < 0$ and $\lambda < \lambda_3$,

2. *source if one of the following is satisfied:*

   (a) $\Pi \geq 0$ and $\lambda > \lambda_1$,
   (b) $\Pi < 0$ and $\lambda > \lambda_3$,

3. ES_2 *is unstable (saddle) if*

   a. $\Pi \geq 0$ and $\lambda_2 < \lambda < \lambda_1$,

4. ES_2 *is non-hyperbolic if one of the following is satisfied:*

   (a) $\Pi > 0$ and $\lambda = \lambda_1$ or $\lambda = \lambda_2$,
   (b) $\Pi < 0$ and $\lambda = \lambda_3$,

$$\Pi = (K_1^2 - 4K_2) \text{ and } \lambda_1 = \left\{\Gamma(1+\beta)\left[\frac{-K_1+\sqrt{K_1^2-4K_2}}{K_2}\right]\right\}^{\frac{1}{\beta}},$$

$$\lambda_2 = \left\{\Gamma(1+\beta)\left[\frac{-K_1-\sqrt{K_1^2-4K_2}}{K_2}\right]\right\}^{\frac{1}{\beta}}, \lambda_3 = \left\{\left[\frac{-\Gamma(1+\beta)K_1}{K_2}\right]\right\}^{\frac{1}{\beta}}$$

4 Bifurcation Theory

The qualitative nature of a model can be investigated by bifurcation analysis. The occurrence of bifurcation is due to a change in the critical value of a parameter based on which bifurcation analysis is carried out. The bifurcation of a parameter reveals the chaotic behavior of the system. The point at which the bifurcation takes place also known as the bifurcation point is identified by standard bifurcation techniques [5].

4.1 Periodic Doubling Bifurcation

The system with change in parameter λ at equilibrium point ES_2 has flip bifurcation if the eigenvalue of system passes through -1 another eigenvalue is neither 1 nor -1 [7]. When the value is varied continuously there will be a cascade of periodic doublings.

Quadratic equation obtained from (7) is

$$\Theta(\mu) = \mu^2 - (2 + \mathcal{A}K_1)\mu + (1 + \mathcal{A}K_1 + \mathcal{A}^2K_2).$$

If $\Pi \geq 0$ and $\lambda = \lambda_1$ or λ_2, then

$$\mu_{1,2} = 1 + \frac{\mathcal{A}K_1}{2} \pm \frac{\mathcal{A}}{2}\sqrt{K_1^2 - 4K_2}.$$

are the eigenvalues of (3) at ES_2.

Theorem 5 *Periodic doubling bifurcation at ES_2 occurs for $\Pi \geq 0$ and $\lambda = \lambda_1$ or λ_2 and $\mu_1 = -1$,*

$$\mu_2 = \frac{K_1\sqrt{K_1^2 - 4K_2} - K_1^2 + 3K_2}{K_2} \neq \pm 1.$$

4.2 Neimark–Sacker Bifurcation

If the system (3) at ES_2 has eigen values that are complex conjugate with absolute value 1, then the occurrence of Neimark–Sacker bifurcation is ensured [11]. From (7), the quadratic equation is given by

$$\Theta(\mu) = \mu^2 - (2 + \mathcal{A}K_1)\mu + (1 + \mathcal{A}K_1 + \mathcal{A}^2 K_2).$$

If $\Pi < 0$ and $\lambda = \lambda_3$, then

$$\mu_{1,2} = 1 - \frac{K_1^2}{2K_2} \pm i\frac{K_1}{2K_2}\sqrt{4K_2 - K_1^2}.$$

are the eigenvalues at ES_2.

Theorem 6 *At ES_2, Neimark–Sacker bifurcation occurs if $\Pi < 0$ and $\lambda = \lambda_3$, and*

$$\left|\mu_{1,2}\right| = \left|1 - \frac{K_1^2}{2K_2} \pm i\frac{K_1}{2K_2}\sqrt{4K_2 - K_1^2}\right| = 1.$$

5 Control Strategies

Control system with control terms introduced in (3) is

$$\begin{aligned} x(t+1) &= x(t) + \frac{\lambda^\beta}{\Gamma(1+\beta)}\left[rx(t)(1-x(t)) - \frac{ax(t)y(t)}{1+bx(t)}\left(\frac{x(t)}{m+x(t)}\right)\right] + S(t) \\ y(t+1) &= y(t) + \frac{\lambda^\beta}{\Gamma(1+\beta)}\left[\frac{dx(t)y(t)}{1+bx(t)} - cy(t)\right] \end{aligned} \tag{8}$$

where Feedback control force is $S(t) = -s_1(x(t) - x^*) - s_2(y(t) - y^*)$ with feedback gains $s_{1,2}$. The Jacobian of system (8) at (x^*, y^*) is

$$J_1\left(x^*, y^*\right) = \begin{bmatrix} 1 + \mathcal{A}a_{11} - s_1 & -\mathcal{A}a_{12} - s_2 \\ \mathcal{A}a_{21} & 1 + \mathcal{A}a_{22} \end{bmatrix}, \tag{9}$$

the variables $\mathcal{A}, a_{11}, a_{12}, a_{21}, a_{22}$ are defined in (7). The quadratic equation of (9) is

$$\mu^2 - (2 + \mathcal{A}K_1 - s_1)\mu + (\mathcal{A}^2 K_2 + \mathcal{A}K_1 + 1 - \mathcal{A}a_{22}s_1 + \mathcal{A}a_{21}s_2 - s_1) = 0. \tag{10}$$

The eigenvalues of (10) be
$\mu_{1,2} = \frac{(2+\mathcal{A}K_1-s_1)\pm\sqrt{(2+\mathcal{A}K_1-s_1)^2-4(\mathcal{A}^2K_2+\mathcal{A}K_1+1-\mathcal{A}a_{22}s_1+\mathcal{A}a_{21}s_2-s_1)}}{2}$ and

$$\mu_1\mu_2 = \mathcal{A}a_{21}(\mathcal{A}a_{12} + s_2) + 1 + \mathcal{A}a_{11}(\mathcal{A}a_{22} + 1) + \mathcal{A}a_{22}(1 - s_1) - s_1 \tag{11}$$

The equations $\mu_1 = \pm 1$ and $\mu_1\mu_2 = 1$ confirms $|\mu_{1,2}| \leq 1$. Suppose $\mu_1\mu_2 = 1$, then (11 yields

$$l_1 : \mathcal{A}K_1 + \mathcal{A}^2 K_2 = \mathcal{A}a_{22}s_1 + s_1 - \mathcal{A}a_{21}s_2$$

Suppose $\mu_1 = 1$ or $\mu_1 = -1$, then equation (10) yields

$$l_2 : \mathcal{A}^2 K_2 = \mathcal{A}a_{22}s_1 - \mathcal{A}a_{21}s_2,\ l_3 : 2\mathcal{A}K_1 + \mathcal{A}^2 K_2 + 4 = 2s_1 + \mathcal{A}a_{22}s_1 - \mathcal{A}a_{21}s_2.$$

The lines l_1, l_2, l_3 bound the triangular region inside which the eigenvalues lie.

The control system with strategy to control Neimark–Sacker bifurcation is

$$\begin{aligned} x(t+1) &= \delta x(t) + \frac{\delta\lambda^\beta}{\Gamma(1+\beta)}\left[rx(t)(1-x(t)) - \left(\frac{ax(t)^2 y(t)}{(1+bx(t))(m+x(t))}\right)\right] \\ &\quad + (1-\delta)x(t) \\ y(t+1) &= \delta y(t) + \frac{\delta\lambda^\beta}{\Gamma(1+\beta)}[dx(t)y(t)/(1+bx(t)) - cy(t)] \\ &\quad + (1-\delta)y(t) \end{aligned} \tag{12}$$

where $0 < \delta < 1$. Neimark–Sacker bifurcation can be eliminated by appropriate choice of λ. The Jacobian of (12) at (x^*, y^*) is

$$J_2\left(x^*, y^*\right) = \begin{bmatrix} 1 + \delta\mathcal{A}a_{11} & -\delta\mathcal{A}a_{12} \\ \mathcal{A}\delta a_{21} & 1 + \delta\mathcal{A}a_{22} \end{bmatrix}. \tag{13}$$

where $\mathcal{A}, a_{11}, a_{12}, a_{21}, a_{22}$ are given in (7). The stability of the system (12) is guaranteed by the presence of the eigen values within unit disk.

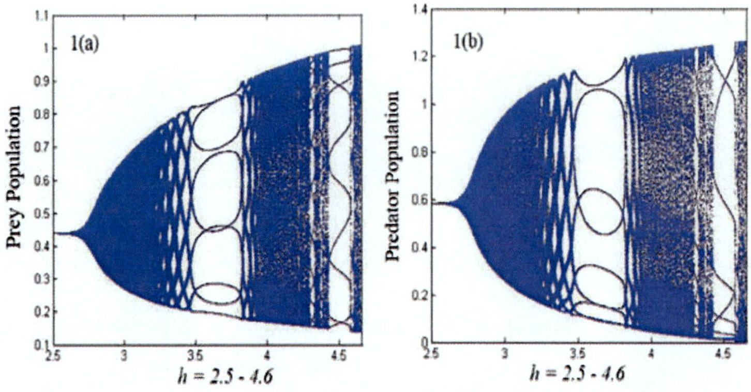

Fig. 1 Neimark–Sacker bifurcation diagram in (λ, x); (λ, y) planes of the system (3)

6 Numerical Examples

This section provides some numerical examples with simulations

Example 1 With $\beta = 0.8, r = 0.35, a = 0.43, b = 0.1, c = 0.4, d = 0.95, m = 0.1$, and $2.5 \leq \lambda \leq 4.6$ in system (3) and initial values $x(0) = 0.5,\ y(0) = 0.4$, equilibrium point is $ES_2 = (x^*, y^*) = (0.4396, 0.5845)$.

Consider $B = 0.1962, K_1 = -0.1820, K_2 = 0.075, \Pi = -0.2677, \lambda_3 = 2.7713, \mathcal{A} = 2.4267$. Eigen values are $\mu_{1,2} = 0.7742 \pm i\,0.6278$ with $|\mu_{1,2}| = 1$. Conditions of Theorem (4) are obtained at ES_2.

Figure 1 illustrates the chaos in system with Neimark–Sacker bifurcation. The transition of the system from stable position to chaos are presented in Figs. 2 and 3 for $\lambda > \lambda_3$. The equilibrium point ES_2 is asymptotically stable for $\lambda < \lambda_3 = 2.7713$ and bifurcates at $\lambda = \lambda_3$.

Example 2 Let $\beta = 0.8, r = 0.35, a = 0.43, b = 0.1, c = 0.4, d = 0.95, m = 0.1, \lambda = 2.78$, and $x(0) = 0.5, y(0) = 0.4$. Example (1) confirms the occurrence of bifurcation varying $\lambda \in [2.5, 4.6]$ and unstable closed orbit at $\lambda = 2.78$ is given in Fig. 4 with $ES_2 = (0.4396, 0.5845)$.

System (12) with control terms for the above chosen values becomes

$$x(t+1) = x(t) + \frac{\delta\lambda^{\beta}}{\Gamma(1+\beta)}\left[rx(t)(1-x(t)) - ax(t)y(t)/(1+bx(t))\left(\frac{x(t)}{m+x(t)}\right)\right]$$
$$y(t+1) = y(t) + \frac{\delta\lambda^{\beta}}{\Gamma(1+\beta)}\left[dx(t)y(t)/(1+bx(t)) - cy(t)\right] \tag{14}$$

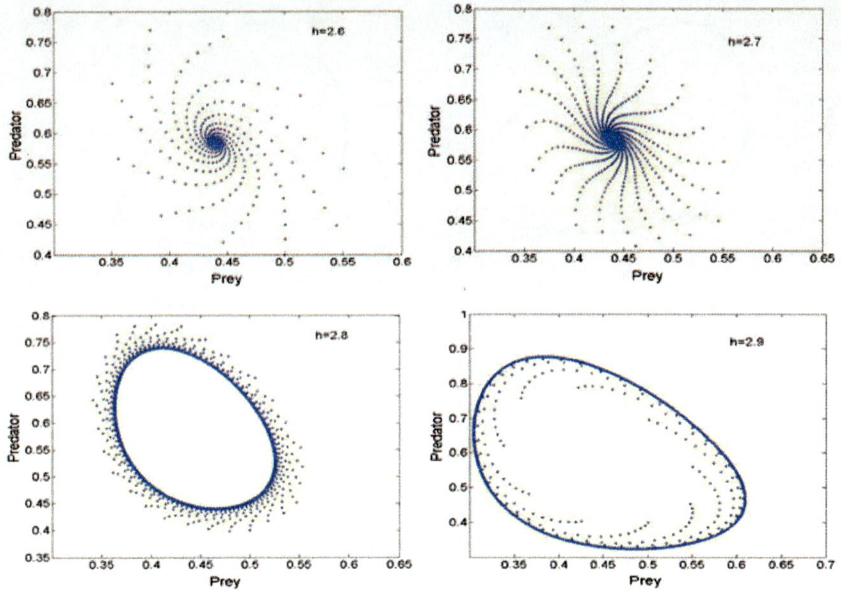

Fig. 2 Different periodic orbits of the bifurcation of the system

with $\beta = 0.8, r = 0.35, a = 0.43, b = 0.1, c = 0.4, d = 0.95, m = 0.1, \lambda = 2.78,$ and $0 < \delta < 1$. Jacobian and characteristic equation of system (14) at ES_2 are

$$J_2\left(x^*, y^*\right) = \begin{pmatrix} 1 - 1.2055\,\delta & -0.3588\,\delta \\ 1.2395\,\delta & 1 \end{pmatrix}.$$

$\mu^2 - (2 - 1.2055\ \delta)\mu + 0.4447\ \delta^2 - 1.2055\ \delta + 1 = 0$. Stability of the system (14) with control terms are illustrated in Fig. 5.

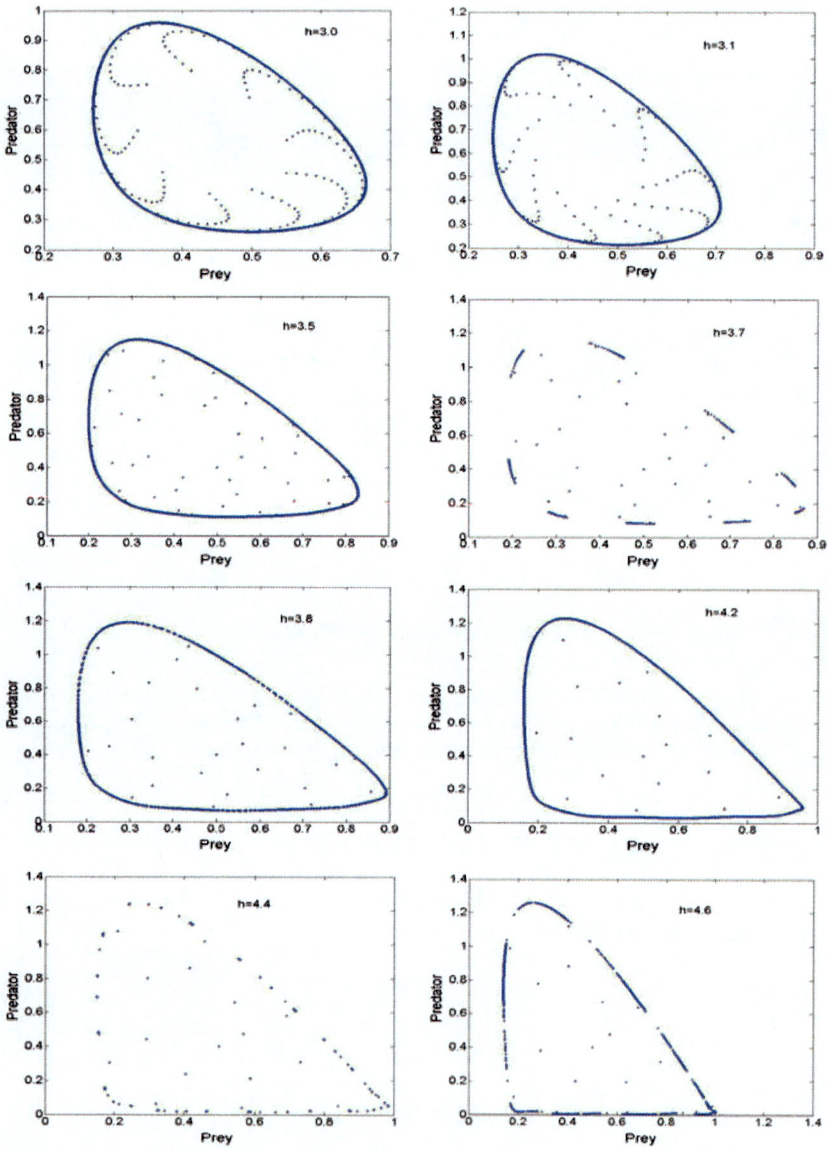

Fig. 3 Different periodic orbits of the bifurcation of the system

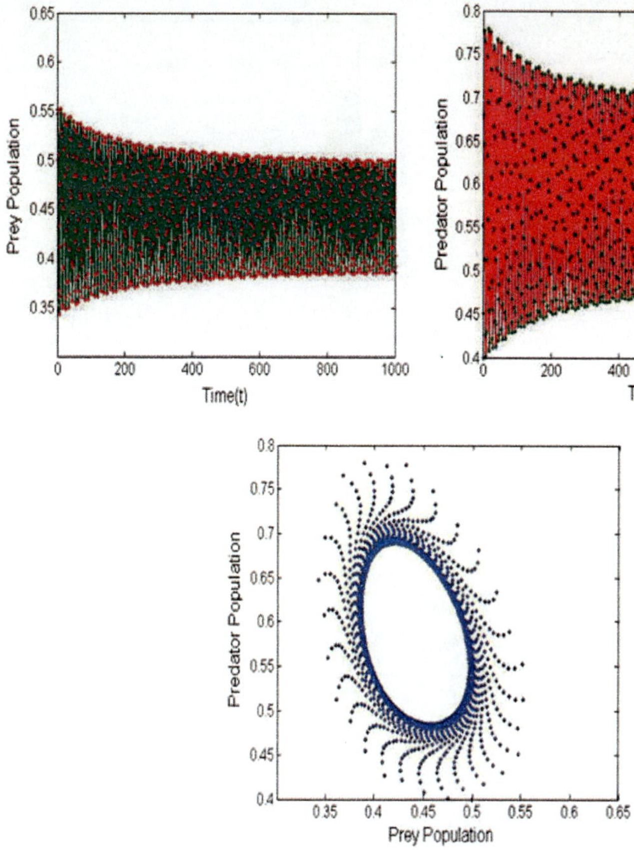

Fig. 4 Phase portrait with time plot for the system (3)

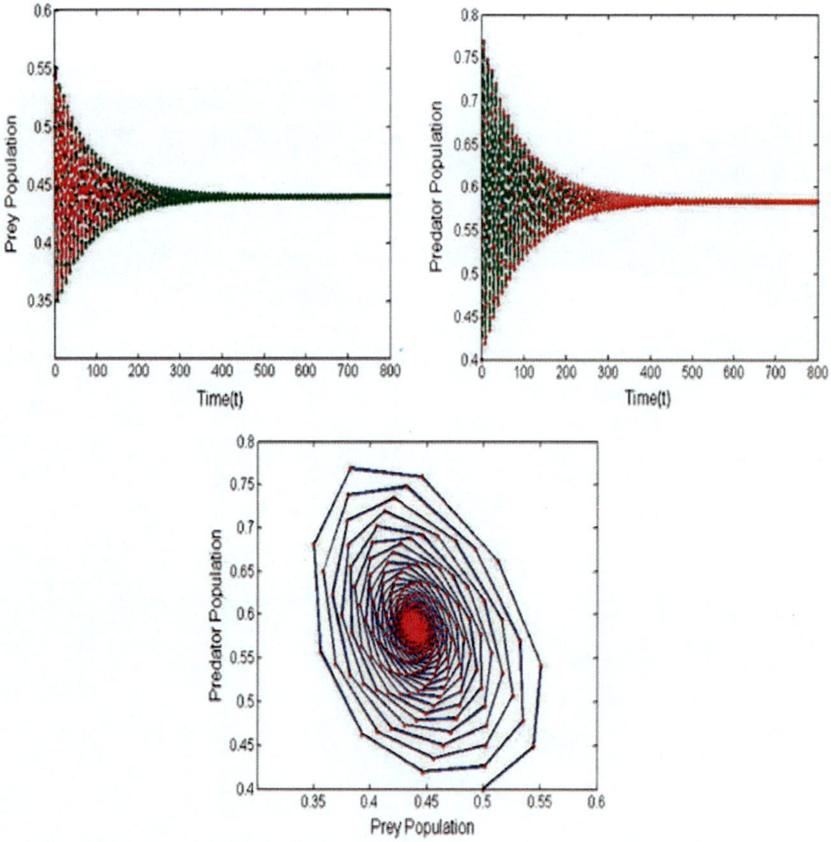

Fig. 5 Phase portrait with time plot for the system (14)

7 Conclusion

The predator–prey interaction model with Allee effect and functional response of Holling type II is considered and stability conditions are provided at all positive equilibrium points. The bifurcation analysis is carried out using traditional techniques and simulations supporting results are presented. The control strategy is implemented and controlling of chaos is numerically confirmed with simulations.

References

1. Agarwal, R.P., El-Sayed, A.M.A., Salman, S.M.: Fractional—order Chua's system: discretization, bifurcation and chaos. Adv. Differ. Equ. **2013**(320) (2013)
2. Gumus, O.A., Kose, H.: On the stability of delay population dynamics related with Allee effects. Math. Comput. Appl. **17**(1), 56–67 (2012)
3. Allee, W.C.: Animal aggressions. University of Chicago press, Chicago (1931)
4. Singh, A., Elsadany, A.A., Elsonbaty, A.: Complex dynamics of a discrete fractional-order leslie-gower predator-prey model. Math. Meth. Appl. Sci. 1–16 (2019)
5. Din, Q.: Complexity and chaos control in a discrete-time prey-predator model. Commun. Nonlinear Sci. Numer. Simul. **49**, 113–134 (2017)
6. Edelstein Keshet, L.: Mathematical Models in Biology. Society for Industrial and Applied Mathematics, New York (2005)
7. Elaydi, S.N.: Discrete Chaos with Applications in Science and Engineering. Chapman and Hall/CRC, Baca Raton (2008)
8. Elsadany, A.A., Matouk, A.E.: Dynamical behaviors of fractional-order Lotka-Volterra predator-prey model and its discretization. Appl. Math. Comput. **49**, 269–283 (2015)
9. Fowler, M.S, Ruxton, G.D.: Population dynamics consequences of Allee effects. J. Theor. Biol. **215**, 39–46 (2002)
10. Selvam, A.G.M., Janagaraj, R.: Numerical analysis of a fractional order discrete prey—predator system with functional response. Int. J. Eng. Technol. **7**(4.10), 681–684 (2018)
11. Selvam, A.G.M., Janagaraj, R., Vignesh, D.: Allee effect and Holling type–II response in a discrete fractional order prey—predator model. IOP Conf. Ser.: J. Phys. **1139**, 1–7 (2018)
12. Kangalgil, F.: The local stability analysis of a nonlinear discrete-time population model with delay and Allee effect. Cumhur. Sci. J. **38**(3), 480–487 (2017)
13. Kangalgil, F., Gumus, O.A.: Allee effect in a new population model and stability analysis. Gen. Math. Notes. **35**(1), 1–6 (2016)
14. Matouk, A.E., Elsadany, A.A.: Dynamical analysis, stabilization and discretization of a chaotic fractional-order GLV model. Nonlinear Dyn. **85**, 1597–1612 (2016)
15. Miller, K.S., Ross, B.: An Introduction to the Fractional Calculus and Fractional Differential Equations. Wiley (1993)
16. Pk, J., Ghorais, S.: (2017) Stability of prey–predator model with holling type response function and selective harvesting. J. Appl. Computat. Math. **6**, 358 (2017)
17. Stephens, P.A., Sutherland, W.J., Freckleton, R.P.: What is the Allee effect. Oikos **87**(1), 185–190
18. Liu, X., Xiao, D.: Complex dynamic behaviors of a discrete-time predator—prey system. Chaos Solitons Fractals **32**(2007), 80–94 (2007)
19. Shi, Y., Ma, Q., Ding, X.: Dynamical behaviors in a discrete fractional-order predator-prey system. Filomat **32**(17), 5857–5874 (2018)
20. Zhou, S., Liu, Y., Wang, G.: The stability of predator-prey systems subject to the Allee effects. Theor. Popul. Biol. **67**, 23–31 (2005)

A Distinct Method for Solving Fuzzy Assignment Problems Using Generalized Quadrilateral Fuzzy Numbers

D. Stephen Dinagar and B. Christopar Raj

Abstract The purpose of this paper is to present a distinct method to solve Fuzzy Assignment Problem by assuming that a decision maker is uncertain about the precise values of the assignment costs. In the proposed method, the assignment costs are represented by Generalized Quadrilateral Fuzzy Numbers. A new method is introduced to find an optimal solution of an Assignment Problem using a new ranking function. An illustration is provided to strengthen the proposed method. The decision makers can easily comprehend this method and it also enables an easier platform for real time applications.

Keywords Fuzzy Assignment Problem(s) (FAP) · Generalized Quadrilateral Fuzzy Number (GQFN) · Ranking function

Mathematics Subject Classification (2010) 03E72 · 90C08

1 Introduction

To assign a task or a job to the most suitable with minimal cost is the real art of Technology. An Assignment Problem is a special case of transportation problem, in which the goal is to assign a number of origins to equal number of destinations at a minimum cost or maximum profit.

Theoretically, the assignment costs involved in any Assignment Problem are taken as crisp values but in real situations, the assignment cost cannot be precise values due to several factors. To overcome this difficulty, fuzzy numbers were introduced.

D. Stephen Dinagar · B. Christopar Raj (✉)
Department of Mathematics, T.B.M.L. College
(Affiliated to Bharathidasan University), Porayar 609307, India
e-mail: christofer2010@gmail.com

D. Stephen Dinagar
e-mail: dsdina@rediffmail.com

R. N. Mohapatra et al. (eds.), *Mathematical Analysis and Computing*,
Springer Proceedings in Mathematics & Statistics 344,
https://doi.org/10.1007/978-981-33-4646-8_19

In this paper, the assignment costs are represented as Generalized Quadrilateral Fuzzy Numbers. A new method is proposed to find an optimal solution for an Assignment Problem using ranking function.

Fuzzy sets and fuzzy theory were introduced by Zadeh [12]. Chanas and Kutcha [2] presented a concept of the optimal solution for transportation problem with fuzzy cost coefficients represented as fuzzy numbers. Amarpreet Kaur and Amit Kumar [1] introduced new methods to find the Initial basic feasible solution and fuzzy optimal solution in which the transportation costs are generalized trapezoidal fuzzy number. Chi-Jen Lin, Ue-Pyng Wen [3] proposed an efficient algorithm based on the labeling method for solving the linear fractional programming. Laxminarayan Sahoo and Santanu Kumar Ghosh [4] used a new defuzzification method based on statistical beta distribution. Thorani and Ravi Shankar [10] developed new classical algorithms using fundamental theorems for Fuzzy Assignment Problems and discussed variations in Fuzzy Assignment Problems also. Nagoor Gani and Mohamed [5] presented a new ranking method fuzzy numbers where the Fuzzy Assignment Problem is transformed into crisp Assignment Problem in the LPP form and solved by using LINGO 9.0.

Stephen Dinagar and Abirami [7] presented new arithmetic operations for interval valued fuzzy numbers. Studies have been done to solve fuzzy transportation problems using Generalized Quadrilateral Fuzzy Numbers [6, 8, 9]. In the present study, an attempt is made to extend these ideas.

In this paper, Sect. 2 gives the basic notions, related to the present study. In Sect. 3, the Generalized Quadrilateral Fuzzy Number(GQFN) is proposed and its arithmetic operations are discussed. A computational procedure is given in Sect. 4. A relevant illustration is provided in Sect. 5. The conclusion of the work is briefed in Sect. 6.

2 Preliminaries

In this section, some basic definitions of fuzzy sets, fuzzy numbers and arithmetic operations of Generalized Quadrilateral Fuzzy Numbers are recalled.

2.1 Fuzzy Set

A fuzzy set $\tilde{A} = \{(x, \mu_A(x), x \in A, \mu_A(x) \in [0, 1]\}$. In this pair $\{(x, \mu_A(x))\}$, the first element x belongs to the classical set A and the second element $\mu_A(x)$ belongs to the interval [0, 1], called membership function.

2.2 Convex Fuzzy Set

A fuzzy set $\tilde{A}$ is convex if $\mu_{\tilde{A}}(\lambda x_1 + (1 - \lambda x_2)) \geq \min(\mu_{\tilde{A}}(x), \mu_{\tilde{A}}(x)), x_1, x_2 \in X$ and $\lambda \in [0, 1]$. Alternatively, a fuzzy set is convex, if all α - level sets are convex.

2.3 Fuzzy Number

A fuzzy set $\tilde{A}$ on $\Re$ must possess at least the following three properties to qualify a fuzzy number:

(i) $\tilde{A}$ must be a normal fuzzy set;
(ii) $\tilde{A}$ must be a convex fuzzy set;
(iii) $\tilde{A}$ must be closed and bounded.

2.4 Trapezoidal Fuzzy Number

A fuzzy number $\tilde{A} = (a, b, c, d)$ is said to be trapezoidal fuzzy number if its membership function is given by

$$\mu_{\tilde{A}}(x) = \begin{cases} \frac{(x-a)}{(b-a)}, & a \leq x < b \\ 1, & b \leq x < c \\ \frac{(x-d)}{(c-d)}, & c \leq x < d \end{cases}$$

3 Generalized Quadrilateral Fuzzy Number (GQFN)

A new fuzzy number $\tilde{A} = (a, b, c, d; \omega_1, \omega_2)$ is defined as Generalized Quadrilateral Fuzzy Number, if its membership function is given by (Fig. 1)

$$\mu_{\tilde{A}}(x) = \begin{cases} \omega_1 \frac{(x-a)}{(b-a)}, & a \leq x < b \\ \frac{(x-b)\omega_2+(c-x)\omega_1}{(c-b)}, & b \leq x < c \\ \omega_2 \frac{(x-d)}{(c-d)}, & c \leq x < d \\ 0, & otherwise \end{cases}$$

Fig. 1 Generalized Quadrilateral Fuzzy Number

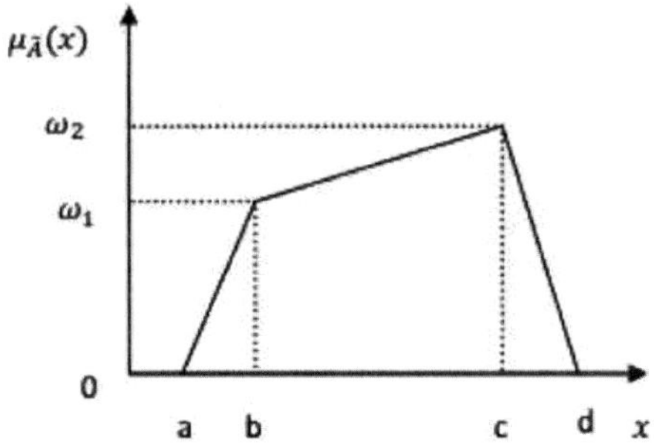

3.1 *Arithmetic Operations on GQFN:*

In this section, the arithmetic operations between two Generalized Quadrilateral Fuzzy Numbers are defined on the universal set of real numbers $\Re$.

Let $\tilde{A_1} = (a_1, b_1, c_1, d_1; \omega_{A_1^1}, \omega_{A_1^2})$ and $\tilde{A_2} = (a_2, b_2, c_2, d_2; \omega_{A_2^1}, \omega_{A_2^2})$

(i) Addition for GQFN:

$$\tilde{A_1} \oplus \tilde{A_2} = (a_1 + a_2, b_1 + b_2, c_1 + c_2, d_1 + d_2; min(\omega_{A_1^1}, \omega_{A_2^1}), min(\omega_{A_1^2}, \omega_{A_2^2})).$$

(ii) Subtraction for GQFN:

$$\tilde{A_1} \ominus \tilde{A_2} = (a_1 - d_2, b_1 - c_2, c_1 - c_2, d_1 - b_2; min(\omega_{A_1^1}, \omega_{A_2^1}), min(\omega_{A_1^2}, \omega_{A_2^2}))$$

(iii) Scalar Multiplication for GQFN:

$$\lambda\tilde{A_1} = \begin{cases} \lambda a_1, \lambda b_1 \lambda c_1 \lambda d_1; (\omega_{A_1^1}, \omega_{A_1^2}) & \lambda \geq 0 \\ \lambda d_1, \lambda c_1 \lambda b_1 \lambda a_1; (\omega_{A_1^1}, \omega_{A_1^2}) & \lambda < 0. \end{cases}$$

3.2 *Ranking Function for GQFN*

We propose a new ranking function $\Re : F(\Re) \to \Re$, which maps each fuzzy number into the real number. Let $\tilde{A}$=$(a, b, c, d; \omega_1, \omega_2)$, then

$$\Re(\tilde{A}) = \left(\frac{a + b + c + d}{4}\right)\left(\frac{\omega_1 + \omega_2}{2}\right)$$

4 Fuzzy Assignment Problem

4.1 Fuzzy Assignment Problem in the Tabulated Form

Workers/Jobs	Job 1	Job 2	Job 3		Job n
Worker 1	$\tilde{c_{11}}$	$\tilde{c_{12}}$	$\tilde{c_{13}}$	...	$\tilde{c_{1n}}$
Worker 2	$\tilde{c_{21}}$	$\tilde{c_{22}}$	$\tilde{c_{23}}$	....	$\tilde{c_{2n}}$
Worker 3	$\tilde{c_{31}}$	$\tilde{c_{32}}$	$\tilde{c_{33}}$	...	$\tilde{c_{3n}}$
⋮	⋮	⋮	⋮	⋮	⋮
Worker n	$\tilde{c_{n1}}$	$\tilde{c_{n2}}$	$\tilde{c_{n3}}$	...	$\tilde{c_{nn}}$

4.2 The Mathematical Formulation of Fuzzy Assignment Problem

$$Min\ \tilde{z} = \sum_{i=1}^{n}\sum_{j=1}^{n} \tilde{c_{ij}} \otimes x_{ij}$$

subject to

$$\sum_{j=1}^{n} x_{ij} = 1, i = 1, 2, 3, \ldots, n$$

$$\sum_{i=1}^{n} x_{ij} = 1, j = 1, 2, 3, \ldots, n$$

$$\text{where } x_{ij} = \begin{cases} 1 & \text{if } i^{th} \text{ person is assigned to } j^{th} \text{ job} \\ 0 & \text{otherwise} \end{cases}$$

c_{ij} = the cost associated with assigning i^{th} resource to j^{th} activity

4.3 Computational Procedure

In this section, a computational procedure is formulated to solve the Fuzzy Assignment Problem with GQFN.

Step 1: Check whether, the number of sources is equal to the number of destinations; if it is not equal, add a dummy rows or dummy columns with zeros in assignment matrix.

Step 2: Convert the fuzzy assignment cost into crisp values.
Step 3: Locate smallest element in each column.
Step 4: If each column and each row have only one assignment, optimal solution is reached. Otherwise go to step 5.
Step 5: If any row has more than one assignment, find the penalty.
Step 6: Find the maximum penalty, choose the corresponding smallest element and make the assignment then omit the corresponding row and column values except assignment value. If there is a tie in maximum penalty, find difference between the next smallest elements.
Step 7: Continue the process until each column and each row have only one assignment.

5 Numerical Illustration

In this section, the assignment is solved with Generalized Quadrilateral Fuzzy Numbers (Table 1).

Solution:

Since, the number of rows is equal to the number of columns, the given problem is balanced Fuzzy Assignment Problem (Tables 1, 2 and 3).

Table 1 Assignment problem

Workers	Jobs			
	J1	J2	J3	J4
A	(3, 5, 6, 7; 0.2, 0.4)	(5, 8, 11, 12; 0.3, 0.5)	(9, 10, 11, 15; 0.2, 0.4)	(5, 8, 10, 11; 0.3, 0.5)
B	(7, 8, 10, 11; 0.3, 0.6)	(3, 5, 6, 7; 0.2, 0.4)	(6, 8, 10, 12; 0.3, 0.6)	(5, 8, 9, 10; 0.2, 0.4)
C	(2, 4, 5, 6; 0.2, 0.4)	(5, 7, 10, 11; 0.3, 0.6)	(8, 11, 13, 15; 0.2, 0.4)	(4, 6, 7, 10; 0.3, 0.6)
D	(6, 8, 10, 12; 0.3, 0.6)	(2, 5, 6, 7; 0.2, 0.5)	(5, 7, 10, 11; 0.3, 0.5)	(2, 4, 5, 7; 0.2, 0.4)

Table 2 Convert the fuzzy assignment cost into crisp values

Workers	Jobs			
	J1	J2	J3	J4
A	1.575	3.6	3.375	3.4
B	4.05	1.575	4.05	2.4
C	1.275	3.7125	3.525	3.0375
D	4.05	1.75	3.3	1.35

Table 3 Locate smallest element in each column

Workers	Jobs			
	J1	J2	J3	J4
A	(3, 5, 6, 7; 0.2, 0.4)	(5, 8, 11, 12; 0.3, 0.5)	(9, 10, 11, 15; 0.2, 0.4)	(5, 8, 10, 11; 0.3, 0.5)
B	(7, 8, 10, 11; 0.3, 0.6)	**(3, 5, 6, 7; 0.2, 0.4)**	(6, 8, 10, 12; 0.3, 0.6)	(5, 8, 9, 10; 0.2, 0.4)
C	**(2, 4, 5, 6; 0.2, 0.4)**	(5, 7, 10, 11; 0.3, 0.6)	(8, 11, 13, 15; 0.2, 0.4)	(4, 6, 7, 10; 0.3, 0.6)
D	(6, 8, 10, 12; 0.3, 0.6)	(2, 5, 6, 7; 0.2, 0.5)	**(5, 7, 10, 11; 0.3, 0.5)**	**(2, 4, 5, 7; 0.2, 0.4)**

Table 4 Make the assignment and delete the other values in the corresponding rows and columns

Workers	Jobs			
	J1	J2	J3	J4
A	(3, 5, 6, 7; 0.2, 0.4)	(5, 8, 11, 12; 0.3, 0.5)	(9, 10, 11, 15; 0.2, 0.4)	*
B	(7, 8, 10, 11; 0.3, 0.6)	(3, 5, 6, 7; 0.2, 0.4)	(6, 8, 10, 12; 0.3, 0.6)	*
C	(2, 4, 5, 6; 0.2, 0.4)	(5, 7, 10, 11; 0.3, 0.6)	(8, 11, 13, 15; 0.2, 0.4)	*
D	*	*	*	**(2, 4, 5, 7; 0.2, 0.4)**

Table 5 Again locate the smallest element in each column

Workers	Jobs			
	J1	J2	J3	J4
A	(3, 5, 6, 7; 0.2, 0.4)	(5, 8, 11, 12; 0.3, 0.5)	**(9, 10, 11, 15; 0.2, 0.4)**	*
B	(7, 8, 10, 11; 0.3, 0.6)	**(3, 5, 6, 7; 0.2, 0.4)**	(6, 8, 10, 12; 0.3, 0.6)	*
C	**(2, 4, 5, 6; 0.2, 0.4)**	(5, 7, 10, 11; 0.3, 0.6)	(8, 11, 13, 15; 0.2, 0.4)	*
D	*	*	*	**(2, 4, 5, 7; 0.2, 0.4)**

Table 6 Delete all values in the rows and columns except assignment values

Workers	Jobs			
	J1	J2	J3	J4
A	*	*	**(9, 10, 11, 15; 0.2, 0.4)**	*
B	*	**(3, 5, 6, 7; 0.2, 0.4)**	*	*
C	**(2, 4, 5, 6; 0.2, 0.4)**	*	*	*
D	*	*	*	**(2, 4, 5, 7; 0.2, 0.4)**

Since 4th row has more than one assignment, optimal solution is not reached. Find the largest penalty. The largest penalty is in the 4th column. The smallest value is (2, 4, 5, 7; 0.2, 0.4) and make the assignment in (D, J4), then omit the other values in the corresponding rows and columns (Tables 4 and 5).

Now, each column and each row have only one assignment. Therefore optimal assignment is reached (Table 6).

The optimal assignment and the corresponding assignment cost is given below

Worker	Job	Assignment cost
A	3	**(9, 10, 11, 15; 0.2, 0.4)**
B	2	**(3, 5, 6, 7; 0.2, 0.4)**
C	1	**(2, 4, 5, 6; 0.2, 0.4)**
D	4	**(2, 4, 5, 7; 0.2, 0.4)**

Minimum Assignment cost = (9, 10, 11, 15; 0.2, 0.4) + (3, 5, 6, 7; 0.2, 0.4) + (2, 4, 5, 6; 0.2, 0.4) + (2, 4, 5, 7; 0.2, 0.4)

Minimum Assignment cost = (16, 23, 27, 35; 0.2, 0.4)

Rank of minimum Assignment cost = 7.575.

6 Conclusion

In this paper, a distinct method is employed to solve Fuzzy Assignment Problem with Generalized Quadrilateral Fuzzy Number(GQFN). The proposed method is a systematic procedure and it consumes less time, less iteration as compared to the existing methods and it provides an optimal solution. This Technique may be extended to solve problems like transportation problems, project scheduling problems and network problems.

References

1. Kaur, A., Kumar, A.: A new approach for solving fuzzy transportation problems using generalized trapezoidal fuzzy numbers. Appl. Soft Comput. **12**, 1201–1213 (2012)
2. Chanas, S., Kutcha, D.: A concept of the optimal solution of the transportation problem with fuzzy cost coefficients. Fuzzy Sets Syst. 299–305 (1996)
3. Lin C.J., Wen, U.P.: A labeling algorithm for the fuzzy assignment problem. Fuzzy Sets Syst. 373–391 (2004)
4. Sahoo, L., Ghosh, S.K.: Solving assignment problem with linguistic costs. J. New Theory 26–37 (2017)
5. Gani, A.N., Mohamed, V.N.: Solution of a fuzzy assignment problem by using a new ranking method. Int. J. Fuzzy Math. Arch. 2, 8–16 (2013)
6. Dinagar, D.S., Palanivel, K.: The transportation problem in fuzzy environment. Int. J. Algorithms, Comput. Math. **2**, 65–71 (2009)
7. Dinagar, D.S., Abirami, D.: A Note on fuzzy critical path in project scheduling using TOPSIS ranking method. Int. J. Appl. Fuzzy Sets Artif. Intell. **6**, 5–15 (2016)
8. Dinagar, D.S., Christopar raj, B.: A novel approach for solving fully fuzzy transportation problems using generalized quadrilateral fuzzy numbers. Am. Int. J. Res. Sci., Technol., Eng. Math. 199–204 (2019)

9. Dinagar, D.S., Christopar raj, B.: A method for solving fully fuzzy transportation problem with generalized quadrilateral fuzzy number. Malaya J. Math. **S**(1), 24–27 (2019)
10. Thorani, Y.L.P., Ravishankar, N.: Fuzzy assignment problem with generalized fuzzy numbers. Appl. Math. Sci. **7**(71), 3511–3537 (2013)
11. Zadeh, L.A.: Fuzzy sets. Inf. Control **8**, 338–353 (1965)
12. Zadeh, L.A.: Fuzzy set as a basis for a theory of possibility. Fuzzy Sets Syst. 3–28 (1978)

Eigen Value Estimates for Fractional Sturm-Liouville Boundary Value Problem

Anil Chavada and Nimisha Pathak

Abstract In this article, we consider fractional Strum-Liouville boundary value problem containing Caputo derivative of order α, $1 < \alpha \leq 2$ with mixed boundary conditions. We establish Cauchy–Schwarz-type inequality to determine a lower bound for the smallest eigenvalues. We give a comparison between the smallest eigenvalues and its lower bounds obtained from the Lyapunov-type and Cauchy-Schwarz-type inequalities. The result shows that Lyapunov-type inequality gives the worse and Cauchy–Schwarz-type inequality gives better lower bound estimates for the smallest eigenvalues. We then use these inequalities to obtain an interval where a linear combination of certain Mittag-Leffler functions has no real zeros.

Keywords Lyapunov inequality · Caputo fractional derivative · Cauchy–Schwarz inequality · Mittag-Leffler function

1 Introduction

The Lyapunov inequality [6] has proved to be very useful in the study of spectral properties of ordinary differential equations. This inequality can be stated as follows [1]:

The nontrivial solution to the boundary value problem $u''(t) + q(t)u(t) = 0, a < t < b, u(a) = u(b) = 0$, exists, where $q : [a, b] \to \mathbb{R}$ is a continuous function, then

$$\int_a^b |q(s)|ds > \frac{4}{b-a}.$$

A. Chavada · N. Pathak (✉)
Department of Applied Mathematics, Faculty of Technology and Engineering,
The Maharaja Sayajirao University of Baroda, Vadodara, India
e-mail: nimisha.pathak-appmath@msubaroda.ac.in

A. Chavada
e-mail: anilc-appmath@msubaroda.ac.in

R. N. Mohapatra et al. (eds.), *Mathematical Analysis and Computing*,
Springer Proceedings in Mathematics & Statistics 344,
https://doi.org/10.1007/978-981-33-4646-8_20

The research on Lyapunov-Type Inequalities (LTIs) for Fractional Boundary Value Problems (FBVPs) has begun since 2013. In [3, 4, 7], the authors have established LTIs for FBVPs of order α, $\alpha \epsilon (1, 2]$ with different boundary conditions. In [9], Pathak obtained the LTI for FBVP of order $2 < \alpha \leq 3$. Jleli and Samet in [4], considered a Caputo fractional differential equation with Sturm–Liouville boundary conditions:

$$({}_{a}^{C}D^{\alpha}u)(t) + q(t)u(t) = 0, a < t < b, 1 < \alpha < 2 \tag{1}$$

$$pu(a) - ru'(a) = u(b) = 0, \tag{2}$$

where $p > 0, \frac{r}{p} > \frac{b-a}{\alpha-1}$ and $q : [a, b] \to \mathbb{R}$ is a continuous function. They established a Lyapunov-type inequality as follows:

$$\int_a^b |q(s)|ds \geq \left(1 + \frac{p}{r}(b-a)\right)\frac{\Gamma\alpha}{(b-a)^{\alpha-1}} \tag{3}$$

In [8], the authors obtained LTI and Cauchy–Schwarz inequality(CSI) for fractional boundary value problem with Hilfer derivative of order $1 < \alpha \leq 2$ and with Caputo derivative of order $2 < \alpha \leq 3$. Furthermore, they applied these inequalities to improve bounds for the smallest eigenvalues to obtain intervals where certain Mittag-Leffler (M-L) functions have no real zeros. Motivated by the above works, we consider the problem (1)–(2) and establish CSI. As an application, we then use this inequality to obtain an interval where a certain combination of M-L functions has no real zeros. Moreover, we give a comparison between the smallest eigenvalues and its lower bounds obtained from LTI and CSI.

2 Preliminaries

In this section, we recall some basic definitions which are further used in this paper.

Definition 2.1 Let $\alpha \geqslant 0$ and f be a real continuous function defined on $[a, b]$. The Riemann–Liouville fractional integral of order α is defined by

$$({}_{a}I^{0}f)(t) = f(t)$$

and

$$({}_{a}I^{\alpha}f)(t) = \frac{1}{\Gamma(\alpha)}\int_a^t (t-s)^{\alpha-1}f(s)ds, \alpha > 0, t \in [a, b].$$

Definition 2.2 The Caputo derivative of fractional order $\alpha \geqslant 0$ is defined by

$$({}_a^C D^0 f)(t) = f(t)$$

and

$$({}_a^C D^\alpha f)(t) = \frac{1}{\Gamma(m-\alpha)} \int_a^t (t-s)^{m-\alpha-1} f^m(s)ds, \alpha > 0, t \in [a, b]$$

where m is the smallest integer greater of equal to α.

Definition 2.3 The one and two-parameter M-L functions are defined, respectively, by

$$E_\alpha(z) := \sum_{k=0}^{\infty} \frac{z^k}{\Gamma(\alpha k + 1)}, \quad (\alpha, z \in R; \alpha > 0)$$

and

$$E_{\alpha,\beta}(z) := \sum_{k=0}^{\infty} \frac{z^k}{\Gamma(\alpha k + \beta)}, \quad (\alpha, \beta, z \in R; \alpha, \beta > 0).$$

Definition 2.4 The Laplace transform of the function

$$\phi(t) = t^{\beta-1} E_{\alpha,\beta}(\pm\lambda t^\alpha)$$

is given by

$$(\mathcal{L}\phi)(s) = \frac{s^{\alpha-\beta}}{s^\alpha \mp \lambda}$$

and its inverse relationship is given as

$$\mathcal{L}^{-1}\left[\frac{s^{\alpha-\beta}}{s^\alpha \mp \lambda}\right] = t^{\beta-1} E_{\alpha,\beta}(\pm\lambda t^\alpha)$$

for more details, refer [5, 10].

3 Main Result

The main result of this paper is given in Theorem 3.1.

Lemma 3.1 $u \in C[a, b]$ *is a solution of* (1)–(2) *if and only if it satisfies the integral equation [4]*

$$u(t) = \int_a^b G(t,s)q(s)u(s)ds, t \in [a,b] \tag{4}$$

where G is the Green's function associated to (1)–(2) is given by

$$G(t,s) = \frac{1}{\Gamma(\alpha)} \begin{cases} \dfrac{\left(\frac{r}{p} + t - a\right)(b-s)^{\alpha-1}}{\left(\frac{r}{p} + b - a\right)} - (t-s)^{\alpha-1}, a \le s \le t \le b, \\ \dfrac{\left(\frac{r}{p} + t - a\right)(b-s)^{\alpha-1}}{\left(\frac{r}{p} + b - a\right)}, a \le t \le s \le b. \end{cases} \tag{5}$$

Lemma 3.2 *[4] Let $1 < \alpha < 2$, $p > 0$, $\frac{r}{p} > \frac{1}{\alpha-1}$, the linear combination of certain M-L functions is given by*

$$pE_{\alpha,2}(x) + qrE_{\alpha,1}(x)$$

have no real zeros for $x \in (-(1 + \frac{p}{r})\Gamma(\alpha), 0]$.

Lemma 3.3 *[8] The Cauchy-Schwarz-type inequality is defined by*

$$1 \le \left\{ \int_a^b \int_a^b |G(t,s)q(s)|^2 dsdt \right\}. \tag{6}$$

Now, we consider a Caputo fractional differential equation with Sturm–Liouville boundary condition by replacing $a = 0$ and $b = 1$ in (1)–(2) as

$$\begin{cases} ({}_0^C D^\alpha u)(t) + q(t)u(t) = 0, \ 0 < t < 1, 1 < \alpha < 2 \\ pu(0) - ru'(0) = u(1) = 0. \end{cases} \tag{7}$$

We are ready to state and prove our main results.

Theorem 3.1 *If a nontrivial continuous solution of the problem (7) exists, then for (7) the Cauchy–Schwarz inequality is*

$$\int_0^1 |q(s)|ds \geqslant \left\{ \frac{1}{(\Gamma(\alpha))^2(2\alpha-1)} \left[\frac{\left(\frac{r}{p}\right)^2 + \left(\frac{r}{p}\right) + \frac{1}{3}}{\left(\frac{r}{p}+1\right)^2} + \frac{1}{2\alpha} \right] - \frac{2}{(\Gamma(\alpha))^2\left(\frac{r}{p}+1\right)} \int_0^1 \left(\frac{r}{p} + t\right) t^\alpha \beta(1,\alpha) {}_2F_1(1-\alpha, 1; \alpha+1, t)dt \right\}^{-\frac{1}{2}}, \tag{8}$$

where, ${}_2F_1(a,b;c;t)$ is a hypergeometric function and $\beta(m,n)$ is a Beta function.

Proof Taking CSI in (4), we get

$$u^2(t) \leq \left[\int_0^1 |G(t,s)q(s)|^2 ds\right]\left[\int_0^1 u^2(s)ds\right].$$

Integrating both the sides with respect to t from 0 to 1, we get

$$\int_0^1 u^2(t)dt \leq \left[\int_0^1\int_0^1 |G(t,s)q(s)|^2 dsdt\right]\left[\int_0^1\int_0^1 u^2(s)dsdt\right],$$

which after some simplifications give

$$\int_0^1 |q(s)|ds \geq \left[\int_0^1\int_0^1 |G(t,s)|^2 dsdt\right]^{-\frac{1}{2}}. \tag{9}$$

Using Eq. (5) with $a = 0$ and $b = 1$ gives

$$G(t,s) = \frac{1}{\Gamma(\alpha)}\begin{cases} \dfrac{\left(\frac{r}{p}+t\right)(1-s)^{\alpha-1}}{\frac{r}{p}+1} - (t-s)^{\alpha-1}, 0 \leq s \leq t \leq 1, \\ \dfrac{\left(\frac{r}{p}+t\right)(1-s)^{\alpha-1}}{\frac{r}{p}+1}, 0 \leq t \leq s \leq 1. \end{cases} \tag{10}$$

Substituting (10) in (9) and simplifying, we get

$$\int_0^1 |q(s)|ds \geqslant \left\{\frac{1}{(\Gamma(\alpha))^2(2\alpha-1)}\left[\frac{\left(\frac{r}{p}\right)^2+\left(\frac{r}{p}\right)+\frac{1}{3}}{\left(\frac{r}{p}+1\right)^2}+\frac{1}{2\alpha}\right]\right.$$
$$\left. - \frac{2}{(\Gamma(\alpha))^2\left(\frac{r}{p}+1\right)}\int_0^1\left(\frac{r}{p}+t\right)t^{\alpha}\beta(1,\alpha)_2F_1(1-\alpha,1;\alpha+1,t)dt\right\}^{-\frac{1}{2}}$$

which concludes the proof.

4 Applications

Taking $q(t) = \lambda, a = 0, b = 1, p = 1$, and $r = 2$ in (1)–(2), we get the following fractional Sturm–Liouville eigenvalue problem:

$$\left({}_0^C D^\alpha u\right)(t) + \lambda u(t) = 0, 0 < t < 1, 1 < \alpha < 2 \tag{11}$$

$$u(0) - 2u'(0) = u(1) = 0. \tag{12}$$

Next, we give three methods to estimate the lower bound for the smallest eigenvalue of problem (11) with boundary conditions (12) by using the following definitions given in [8]

Definition 2.1 A Lyapunov-Type Inequality Lower Bound (LTILB) is defined as a lower bound estimate for the smallest eigenvalue obtained from Lyapunov-type inequality given by (3).

We obtain a lower bound for the smallest eigenvalue of problem (11) with boundary condition (12) is:

$$\lambda \geqslant \frac{3}{2}\Gamma(\alpha). \tag{13}$$

Definition 2.2 A Cauchy–Schwarz Inequality Lower Bound (CSILB) is defined as an estimate of the lower bound for the smallest eigenvalue obtained from the Cauchy–Schwarz inequality of type given in Eq. (8).

We obtain the Cauchy-Schwarz-type inequality of (11) with boundary condition (12) is

$$\begin{aligned}\lambda \geqslant \Bigg\{ & \frac{1}{(\Gamma(\alpha))^2(2\alpha-1)}\left[\frac{19}{27} + \frac{1}{2\alpha}\right] \\ & - \frac{2}{3(\Gamma(\alpha))^2}\int_0^1 (2+t)t^\alpha \beta(1,\alpha)\, {}_2F_1(1-\alpha, 1; \alpha+1, t)dt\Bigg\}^{-\frac{1}{2}}.\end{aligned} \tag{14}$$

In [2], eigenvalues $\lambda \in \mathbb{R}$ of (11)–(12) are the solution of the linear combination of certain M-L functions is:

$$2E_{\alpha,1}(-\lambda) + E_{\alpha,2}(-\lambda) = 0. \tag{15}$$

Now, comparing the non-zero solutions of Eq. (15) for $1.5 < \alpha \leq 2$ with CSILB given by Eq. (14) and LTILB given by the Eq. (13), we get the following comparison figure (See: Fig. 1). This figure clearly demonstrates that between the two estimates considered here, the LTILB provides the worse estimate and the CSILB provide better estimate for the smallest eigenvalues of (11). We use MATHEMATICA and MATLAB code to find the smallest eigenvalue of the M-L functions.

We consider the integer-order case, i.e., $\alpha = 2$. For this case, the LTILB and CSILB for the smallest eigen value λ of (11) are given as 1.5 and 3.3310, respectively. (See Eqs. (13) and (14)). For $\alpha = 2$, the problem (11)–(12) can be solved in closed form using the tools from integer-order calculus. Results show, the smallest λ of (11)–(12) is the root of equation (15), which give the smallest λ as 3.3731. Comparing

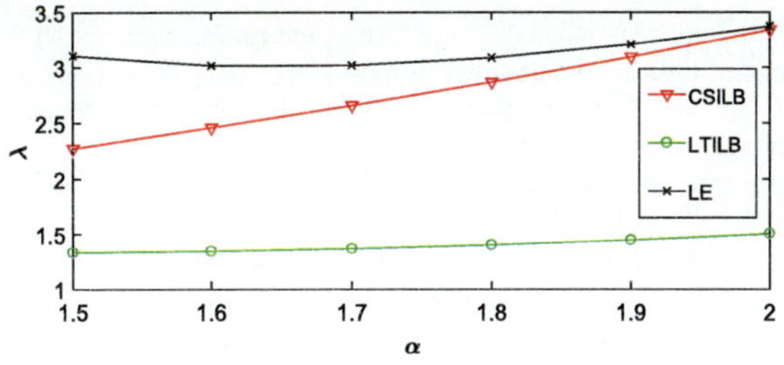

Fig. 1 Comparison of the lower bounds for λ obtained from Lyapunov-type and Cauchy–Schwarz inequalities with the lowest eigenvalue. ($-\circ-$: LTILB; $-\nabla-$: CSILB; $-\times-$:LE-the Lowest Eigenvalue λ)

these λ with its estimate above, it is clear that between LTILB and CSILB for the integer α the CSILB provides the best estimate for the smallest eigenvalue.

5 Conclusion

In this paper, we established Cauchy–Schwarz-type inequality for fractional Strum–Liouville boundary value problem containing Caputo derivative of order α, $1 < \alpha \leq 2$ with mixed boundary conditions to determine a lower bound for the smallest eigenvalues. We give a comparison between the smallest eigenvalues and its lower bounds obtained from the Lyapunov-type and Cauchy–Schwarz-type inequalities. Results showed that Lyapunov-type inequality gives the worse and Cauchy–Schwarz-type inequality gives better lower bound estimates for the smallest eigenvalues. We then used these inequalities to obtain an interval where a linear combination of certain Mittag-Leffler function has no real zeros.

References

1. Brown, R.C., Hinton, D.B.: Lyapunov inequalities and thier applications. In: Survey on Classical Inequalities, pp. 1–25, Math. Appl, 517 Kluwer Acad. Publ., Dordrecht
2. Duan, J.S., Wang, Z., Liu, Y.L., Qiu, X.: Eigenvalue problems for fractional ordinary differential equations. Chaos, Solitons and Fractals **46**, 46–53 (2013)
3. Ferreira, R.A.C.: On a Lyapunov-type inequality and the zeros of a certain Mittag-Leffler function. J. Math. Anal. Appl. **412**, 1058–1063 (2014)
4. Jleli, M., Samet, B.: Lyapunov-type inequalities for fractional boundary-value problems. Electron. J. Differ. Equ. **88**, 1–11 (2015)
5. Kilbas, A.A., Srivastava, H.M., Trujillo, J.J.: Theory and applications of fractional differential equations. North-Holland Math. Stud. **204**. Elsevier, Amsterdam (2006)
6. Lyapunov, A.M.: Problem general de la stabilite du mouvement (French Translation of a Russian paper dated 1893) Ann. Fac. Sci. Univ. Toulouse **2**, 27–247 (1907). Reprinted as Ann. Math. Stud. **17**, Princeton (1947)

7. Pathak, N.: Lyapunov-type inequality and eigenvalue estimates for fractional problems. Dissertation Doctor of Philosophy, Southern Illinois University USA, 2016
8. Pathak, N.: Lyapunov-type inequality for fractional boundary value problems with Hilfer derivative. Math. Inequalities Appl. **21**(1), 179–200 (2018)
9. Pathak, N.: Lyapunov-type inequality and eigenvalue analysis for a fractional problems of order α, IJIRSET **8**(2), 1391–1398 (2019)
10. Podlubny, I.: Fractional differential equation. In: Mathematics in Science and Engineering, vol. 198. Academic Press, San Diego, CA (1999)

Ball Convergence of a Fifth-Order Method for Solving Equations Under Weak Conditions

Ioannis K. Argyros, Santhosh George, and Shobha M. Erappa

Abstract We develop a ball convergence for a fifth-order method to find a solution for an equation. Earlier studies used conditions on the sixth derivative not present in the methods. Moreover, no error estimates are provided. That is why we used conditions up to the second derivative. Numerical experiments validate the theoretical results.

Keywords Steffensen's method · Newton's method

AMS Subject Classification: 49M15 · 65D10

1 Introduction

In this article, we find a solution t_* of equation

$$g(t) = 0, \tag{1}$$

with $g : D \subseteq \mathbb{R} \to \mathbb{R}$ and D is convex.

Chebyshev's, Halley's, Euler's, Super Halley's, [1–28] use the second derivative g'' making them expensive in nature. In the current work, we analyze the local con-

I. K. Argyros
Department of Mathematical Sciences, Cameron University,
Lawton, OK 73505, USA
e-mail: iargyros@cameron.edu

S. George
Department of Mathematical and Computational Sciences, NIT Karnataka,
Mangalore 575025, India
e-mail: sgeorge@nitk.ac.in

S. M. Erappa (✉)
Department of Mathematics, MIT, Manipal Academy of Higher Education,
Manipal 576104, India
e-mail: shobha.me@gmail.com

R. N. Mohapatra et al. (eds.), *Mathematical Analysis and Computing*,
Springer Proceedings in Mathematics & Statistics 344,
https://doi.org/10.1007/978-981-33-4646-8_21

vergence of the Steffensen-type method of order four defined for each $n = 0, 1, 2, \ldots$ by

$$s_n = t_n - \frac{2g(t_n)}{(1+\sqrt{1-2A_n})g(t_n)} \tag{2}$$

$$t_{n+1} = t_n - \frac{\frac{2(g(t_n)+g(s_n))}{g(t_n)}}{1+\sqrt{1-2\frac{g'(t_n)(g(t_n)+g(s_n))}{g(t_n)^2}}},$$

where t_0 is an initial point and $A_n = \frac{g'(t_n)g(t_n)}{|psi'(t_n)^2}$. Method (2) was studied in [17]. It was shown that the order of convergence is five using the derivative of order six. Observe that the iterative procedure (2) uses four functional evaluations per step. Therefore, the efficiency index $EI = p^{\frac{1}{m}}$, where p gives the convergence order and m indicates the number of functional evaluations, and $EI = 5^{\frac{1}{4}} = 1.4953$. Consider an example, for ϕ on $D = [-\frac{1}{2}, \frac{5}{2}]$ like

$$\phi(s) = \begin{cases} s^3 \ln s^2 + s^5 - s^4, & s \neq 0 \\ 0, & s = 0. \end{cases} \tag{3}$$

Choose $t_* = 1$. We have

$$\phi'(s) = 3s^2 \ln s^2 + 5s^4 - 4s^3 + 2s^2, \ \phi'(1) = 3,$$
$$\phi''(s) = 6s \ln s^2 + 20s^3 - 12s^2 + 10s$$
$$\phi'''(s) = 6 \ln s^2 + 60s^2 - 24s + 22.$$

But then function ϕ''' is not bounded on D. In this work, conditions only on the first Fréchet derivative are used, so we can expand method (2). This is done in Sect. 2. The experiments appear in Sect. 3.

2 Ball Convergence

Consider the parameters $l_1 > 0, l_2 > 0, m_1 \geq 1$ and $m_2 \geq 0$, and functions u_1 and $\bar{u}_1$ on $[0, \frac{1}{l_1})$ defined as

$$u(t) = \frac{1}{2(1-l_1 1t)}(l_2 + \frac{4m_1^2 m_2}{(1-l_1 t)^2})t,$$

scalar R_A by

$$R_A = \frac{2}{2l_1 + l_2}$$

and

$$\bar{u}_1(t) = u_1(t) - 1.$$

We have that $\bar{u}_1(0) = -1 < 0$, and

$$\bar{u}_1(R_A) = \frac{4m_1^2 m_2 R_A}{(1 - l_1 r_A)^2} > 0,$$

since $\frac{l_2 R_A}{2(1-l_1 R_A)} = 1$. By the Intermediate Value Theorem (IVT), $\bar{u}_1$ has roots on $(0, R_A)$. Denote by R_1 the minimal such zero. Moreover, let p and $\bar{u}_p$ be on $[0, \frac{1}{l_1})$ as

$$p(t) = 2m_1 m_2 (1 + u_1(t))t - (1 - l_1 t)^2$$

and

$$\bar{u}_p(t) = p(t) - 1.$$

Then, $\bar{u}_p(0) = -1 < 0$ and $\bar{u}_p(t) \to +oo$ as $t \to \frac{1}{l_1}^-$. It follows that function $\bar{u}_p$ has a minimal zero in $(0, \frac{1}{l_1})$ denoted by R_p. Furthermore, let functions u_2 and $\bar{u}_2$ be on $[0, \frac{1}{l_1})$ by

$$u_2(t) = \frac{1}{2(1 - l_1 t)}\{l_2 t + 2[\frac{2m_2 m_1^2 (1 + u_1(t))t}{(1 - l_1 t)^2} + 2m_1 u_1(t)]\}$$

and

$$\bar{u}_2(t) = u_2(t) - 1.$$

Then, we get again that $h_2(0) = -1 < 0$ and $\bar{u}_2(R_A) > 0$. It follows that function $\bar{u}_2$ has a least zero $R_2 \in (0, R_A)$. Set

$$R = \min\{R_1, R_2, R_p\}. \tag{4}$$

Then, we get that for each $t \in [0, R)$

$$0 \le u_1(t) < 1, \tag{5}$$

$$0 \le p(t) < 1, \tag{6}$$

and

$$0 \le u_2(t) < 1. \tag{7}$$

Let the closed and open balls in $\mathbb{R}$ be denoted by $\bar{\mathcal{B}}(v, \rho)$ and $\mathcal{B}(v, \rho)$, respectively.

Theorem 1 *Let $\Psi : D \subseteq \mathbb{R} \to \mathbb{R}$ be twice differentiable. If there exist $t_* \in D$, $l_1 > 0$, $l_2 > 0$, $m_1 \ge 1$, and $m_2 \ge 0$ so that $\forall t, s \in D$,*

$$g(t_*) = 0, \ \ g(t_*) \neq 0, \tag{8}$$

$$|g(t_*)^{-1}(g(t) - g(t_*))| \leq l_1|t - t_*|, \tag{9}$$

$$|g(t_*)^{-1}(g(t) - g(s))| \leq l_2|t - s|, \tag{10}$$

$$|g(t_*)^{-1}g(t)| \leq m_1 \tag{11}$$

$$|g(t_*)^{-1}g'(t)| \leq m_2 \tag{12}$$

and

$$\bar{\mathcal{B}}(t_*, R) \subseteq D, \tag{13}$$

with R *given by (4). Then,* $\lim_{n\to oo} t_n = t_*$ *and for* $e_n = \|t_n - t_*\|$,

$$|s_n - t_*| \leq u_1(e_n)e_n < e_n < R, \tag{14}$$

and

$$|t_{n+1} - t_*| \leq u_2(e_n)e_n < e_n. \tag{15}$$

Moreover, if $S \in [R, \frac{2}{l_1})$ *with* $\bar{\mathcal{B}}(t_*, S) \subset D$, *then* t_* *is the unique solution of (1) in* $\bar{\mathcal{B}}(t_*, S)$.

Proof The estimations (14) and (15) are shown using mathematical induction as follows. Using the conditions $t_0 \in \mathcal{B}(t_*, R) - \{t_*\}$, (4), and (9), we obtain

$$|g(t_*)^{-1}(g(t_0) - g(t_*))| \leq l_1 e_0 < l_1 R < 1. \tag{16}$$

It follows from (16) that $g(t_0) \neq 0$ and

$$|g(t_0)^{-1}g(t_*)| \leq \frac{1}{1 - l_1 e_0}. \tag{17}$$

We can write by (8) that

$$g(t_0) = g(t_0) - g(t_*) = \int_0^1 g(t_* + \gamma(t_0 - t_*)(t_0 - t_*)d\gamma. \tag{18}$$

Notice that $|t_* + \gamma(t_0 - t_*) - t_*| = \gamma e_0 < R$, for each $\gamma \in [0, 1]$. That is $t_* + \gamma(t_0 - t_*) \in \mathcal{B}(t_*, R)$. Then, by (11) and (18), we get

$$\begin{aligned} |g(t_*)^{-1}g(t_0)| &\leq |\int_0^1 g(t_*)^{-1}g(t_* + \gamma(t_0 - t_*)(t_0 - t_*)d\gamma| \\ &\leq m_1 e_0. \end{aligned} \tag{19}$$

We also have by (12)

$$|g(t_*)^{-1}g'(t_0)| \leq m_2. \tag{20}$$

In view of (6), (17), (19), and (20), we have in turn the estimate

$$\begin{aligned} |2A_0| \leq 2|A_0| &\leq 2\,g(t_*)^{-1}g'(t_0)||g(t_*)^{-1}g(t_0)||g(t_*)^{-1}g(t_0)|^{-2} \\ &\leq \frac{2m_1m_2e_0}{(1-l_1e_0)^2} \\ &< \frac{2m_1m_2R}{(1-l_1R)^2} < 1, \end{aligned} \tag{21}$$

so $1-2A_0 > 0$. Hence, s_0 exists. Then, by (2), (4), (5), (8), (10), (17), and (19)–(21), we get in turn

$$\begin{aligned} |s_0 - t_*| &\leq |t_0 - t_* - g(t_0)^{-1}g(t_0)| \\ &\quad + |1 - \frac{2}{1+\sqrt{1-2A_0}}||g(t_0)^{-1}g(t_0)| \\ &\leq |g(t_0)^{-1}g(t_*)||\int_0^1 g(t_*)^{-1}(g(t_* + \gamma(t_0 - t_*)) - g(t_0))(t_0 - t_*)d\gamma| \\ &\quad + \frac{2|A_0|}{|1+\sqrt{1-2A_0}|^2}|g(t_0)^{-1}g(t_*)||g(t_*)^{-1}g(t_0)| \\ &\leq \frac{l_2e_0^2}{2(1-l_1e_0)} + \frac{2|A_0|m_1e_0}{1-l_1e_0}|A_0| \\ &\leq \frac{l_2e_0^2}{2(1-l_1e_0)} + \frac{2m_1^2m_2e_0^2}{(1-l_1e_0)^3} \\ &= u_1(e_0)e_0 < e_0 < R. \end{aligned} \tag{22}$$

Using (6), (17), (19), (20), and (22), we get the estimations

$$\begin{aligned} &2|\frac{g'(t_0)(g(t_0)+g(s_0))}{g(t_0)^2}| \\ &\leq 2|\frac{g(t_*)^{-1}g'(t_0)(g(t_*)^{-1}g(t_0) + g(t_*)^{-1}g(s_0))}{(g(t_*)^{-1}g(t_0))^2}| \\ &\leq \frac{2m_1m_2(e_0 + |s_0 - t_0 * |)}{(1-l_1e_0)^2} \\ &\leq \frac{2m_1m_2(1+u_1(e_0))e_0}{(1-l_1e_0)^2} \\ &< \frac{2m_1m_2(1+u_1(R))R}{(1-l_1R)^2} < 1, \end{aligned}$$

so

$$B_0 = 1 - 2\frac{g'(t_0)(g(t_0) + g(s_0))}{g(t_0)^2} > 0. \tag{23}$$

We need the estimate

$$\begin{aligned}
&|g(t_0) - \frac{2(g(t_0) + g(s_0))}{1 + \sqrt{B_0}}||g(t_*)^{-1}| \\
&= |\frac{g(t_0)(\sqrt{B_0} - 1) - 2g(s_0)}{1 + \sqrt{B_0}}||g(t_*)^{-1}| \\
&\le |\frac{2g'(t_0)g(t_0)(g(t_0) + g(s_0))}{(1 + \sqrt{B_0})^2 g(t_0)^2}||g(t_*)^{-1}| \\
&\quad +2|\frac{g(s_0)}{1 + \sqrt{B_0}}||g(t_*)^{-1}| \\
&\le \frac{2\, g(t_*)^{-1}g'(t_0)||g(t_*)^{-1}g(t_0)|(|g(t_0)| + |g(s_0)||g(t_*)^{-1}|)}{|g(t_*)^{-1}g(t_0)|^2} \\
&\quad +2\, g(s_0)||g(t_*)^{-1}| \\
&\le \frac{2m_1^2 m_2(1 + u_1(e_0))e_0^2}{(1 - l_1 e_0)^2} + 2m_1 u_1(e_0)e_0.
\end{aligned} \tag{24}$$

Moreover, we can write

$$\begin{aligned}
t_1 - t_* &= t_0 - t_* - g(t_0)^{-1}g(t_0) \\
&\quad +(g(t_0) - 2\frac{(g(t_0) + g(s_0))}{1 + \sqrt{B_0}})\frac{1}{g(t_0)}.
\end{aligned} \tag{25}$$

Then, by (4), (7), (17), and (23)–(25), we get

$$\begin{aligned}
|t_1 - t_*| &\le |t_0 - t_* - g(t_0)^{-1}g(t_0)| \\
&\quad +|g(t_0) - 2\frac{(g(t_0) + g(s_0))}{1 + \sqrt{B_0}}||g(t_*)^{-1}||g(t_0)^{-1}g(t_*)| \\
&\le \frac{l_2 e_0^2}{2(1 - l_1 e_0)} \\
&\quad +\frac{1}{1 - l_1 e_0}[\frac{2m_2 m_1^2(1 + u_1(e_0))e_0^2}{(1 - l_1 e_0)^2} \\
&\quad +2m_1 u_1(e_0)e_0] \\
&= u_2(e_0)e_0 < e_0 < R.
\end{aligned}$$

By simply replacing t_0, s_0, t_1 by t_k, s_k, t_{k+1} in the previous calculations, with $u_2(e_0) \in [0, 1)$ we obtain estimates (14) and (15). Using the estimate $|t_{k+1} - t_*| \le |t_k - t_*| < R$, we get that $t_{k+1} \in \mathcal{B}(t_*, R)$ and $\lim_{k\to oo} t_k = t_*$. Let $Q = \int_0^1 g(s_* + \gamma(t_* - s_*)d\gamma$ for $s_* \in \bar{\mathcal{B}}(t_*, S)$ with $g(s_*) = 0$. By (8) we obtain

$$|g(t_*)^{-1}(Q - g(t_*))| \leq \int_0^1 l_1|s_* + \gamma(t_* - s_*) - t_*|d\gamma$$
$$\leq \int_0^1 (1-\gamma)|t_* - s_*|d\gamma \leq \frac{l_1}{2}S < 1, \quad (26)$$

so Q is invertible. Hence form the identity $0 = g(t_*) - g(s_*) = Q(t_* - s_*)$, $t_* = s_*$. □

3 Numerical Examples

Example 1 Set $D = \mathbb{R}$. Define function g on D by

$$g(t) = \sin(t). \quad (27)$$

Hence, for $t_* = 0$, $l_1 = l_2 = m_1 = m_2 = 1$. The parameters are

$$R_A = 0.6667,\ R_1 = 0.1935,\ R_p = 0.3008,\ R_2 = 0.0892 = R.$$

Example 2 Set $D = [-1, 1]$. Consider function g on D by

$$g(t) = e^t - 1. \quad (28)$$

By (28) and $t_* = 0$, we have $l_1 = e - 1 < l_2 = e$, $m_1 = m_2 = 2$. The radii are

$$R_A = 0.3249,\ R_1 = 0.0451,\ R_p = 0.1007,\ R_2 = 0.0104 = R.$$

Example 3 By 3, we get $l_1 = l_2 = 146.6629073$, $m_1 = m_2 = 2$. The scalars are

$$R_A = 0.0045,\ R_1 = 0.0022,\ R_p = 0.0059,\ R_2 = 0.0012 = R.$$

Conflict of interest The researchers have no conflict of interest.

References

1. Amat, S., Hernández, M.A., Romero, N.: A modified Chebyshev's iterative method with at least sixth order of convergence. Appl. Math. Comput. **206**(1), 164–174 (2008)
2. Amat, S., Busquier, S., Plaza, S.: Dynamics of the King's and Jarratt iterations. Aequationes. Math. **69**, 212–213 (2005)
3. Argyros, I.K.: Convergence and Application of Newton-type Iterations. Springer (2008)
4. Argyros, I.K., Hilout, S.: Computational methods in nonlinear Analysis, World Scientific Publ. Co. New Jersey, USA (2013)

5. Argyros, I.K., Chen, D., Quian, Q.: The Jarratt method in Banach space setting. J. Comput. Appl. Math. **51**, 103–106 (1994)
6. Candela, V., Marquina, A.: Recurrence relations for rational cubic methods I: the Halley method. Computing **44**, 169–184 (1990)
7. Chen, J.: Some new iterative methods with three-order convergence. Appl. Math. Comput. **181**, 1519–1522 (2006)
8. Chun, C., Neta, B., Scott, M.: Basins of attraction for optimal eighth order methods to find simple roots of nonlinear equations. Appl. Math. Comput. **227**, 567–592 (2014)
9. Cordero, A., Torregrosa, J.: Variants of Newton's method using fifth order quadrature formulas. Appl. Math. Comput. **190**, 686–698 (2007)
10. Cordero, A., Maimo, J., Torregrosa, J., Vassileva, M.P., Vindel, P.: Chaos in King's iterative family. Appl. Math. Lett. **26**, 842–848 (2013)
11. Cordero, A., Magrenan, A., Quemada, C., Torregrosa, J.R.: Stability study of eight-order iterative methods for solving nonlinear equations. J. Comput. Appl. Math **291**, 348–357 (2016)
12. Cordero, A., Hueso, J.L., Martinez, E., Torregrossa, J.R.: Steffensen type methods for solving non-linear equations. J. Comput. Appl. Math. **236**, 3058–3064 (2012)
13. Ezquerro, J.A., Hernández, M.A.: A uniparametric Halley-type iteration with free second derivative. Int. J. Pure and Appl. Math. **6**(1), 99–110 (2003)
14. Ezquerro, J.A., Hernández, M.A.: New iterations of R-order four with reduced computational cost. BIT Numer. Math. **49**, 325–342 (2009)
15. Frontini, M., Sormani, E.: Some variants of Newton's method with third order convergence. Appl. Math. Comput. **140**, 419–426 (2003)
16. Gutiérrez, J.M., Hernández, M.A.: Recurrence relations for the super-Halley method. Computers Math. Applic. **36**(7), 1–8 (1998)
17. Grau, M., Noguera, N.: A variant of Cauchy's method with accelerated fifth-order convergence. Appl. Math. Lett. **17**, 509–517 (2004)
18. Hernández, M.A., Salanova, M.A.: Sufficient conditions for semilocal convergence of a fourth order multipoint iterative method for solving equations in Banach spaces. Southwest J. Pure Appl. Math **1**, 29–40 (1999)
19. Jaiswal, J.P.: A new third-order derivative free method for solving nonlinear equations. Universal J. Appl. Math. **1**(2), 131–9135 (2013)
20. King, R.F.: A family of fourth-order methods for nonlinear equations. SIAM. J. Numer. Anal. **10**, 876–879 (1973)
21. Maheshwari, A.K.: A fourth order iterative method for solving nonlinear equations. Appl. Math. Comput. **211**, 283–391 (2009)
22. Parhi, S.K., Gupta, D.K.: Semi-local convergence of a Stirling-like method in Banach spaces. Int. J. Comput. Methods **7**(02), 215–228 (2010)
23. Petkovic, M.S., Neta, B., Petkovic, L., Džunič, J.: Multipoint methods for solving nonlinear equations. Elsevier (2013)
24. Potra, F.A., Ptak, V.: Nondiscrete Induction and Iterative Processes. Research Notes in Mathematics, vol. 103. Pitman Publ, Boston, MA (1984)
25. Rall, L.B.: In: Krieger, R.E. (ed.) Computational solution of nonlinear operator equations. New York (1979)
26. Ren, H., Wu, Q., Bi, W.: New variants of Jarratt method with sixth-order convergence. Numer. Algorithms **52**(4), 585–603 (2009)
27. Rheinboldt, W.C.: An adaptive continuation process for solving systems of nonlinear equations. In: Tikhonov, A.N. (ed.) Mathematical models and numerical methods, pub. 3, pp. 129–142. Banach Center, Warsaw Poland (1978)
28. Traub, J.F.: Iterative Methods for the Solution of Equations. Prentice Hall Englewood Cliffs, New Jersey, USA (1964)

Discrete Fractional Sumudu Transform by Inverse Fractional Difference Operator

M. Meganathan, S. Vasuki, B. Chandra Sekar, and G. Britto Antony Xavier

Abstract In this paper, our aim is to develop the discrete Sumudu transform using generalized difference operator. As an application of this transform, the solutions for fractional difference equations with initial conditions are derived. Also, we have obtained discrete Sumudu transform of certain functions and properties are derived.

Keywords Generalized difference operator · Fractional difference · Sumudu transform · Gamma function · Polynomial factorial

1 Introduction

Continuous fractional calculus has been developed by Miller and Ross [20], Oldham and Spanier [23] and Podlubny [24]. Due to its extensive applications in the diverse branches of science and engineering, the discrete fractional calculus has turned out to be the object of many researchers [2–6, 16]. Lately discrete delta fractional calculus has been extended by Atici and Eloe [1], Goodrich [17], and Holm [18]. FFor the modern developments of the theory of discrete fractional calculus, applications of Mittag-Leffler function and fractional integral inequalities can be referred [7–12, 15, 19, 21, 27].

By applying the integral transforms like Mellin, Laplace, Fourier, the solutions of differential equations were acquired. These transforms were made use for the pos-

M. Meganathan (✉) · S. Vasuki · B. Chandra Sekar · G. Britto Antony Xavier
Department of Mathematics, Sacred Heart College (Autonomous),
Tirupattur, Vellore District 635601, Tamil Nadu, India
e-mail: meganathanmath@gmail.com

S. Vasuki
e-mail: vasukishc@gmail.com

B. Chandra Sekar
e-mail: chandrasekarmaths81@gmail.com

G. Britto Antony Xavier
e-mail: brittoshc@gmail.com

R. N. Mohapatra et al. (eds.), *Mathematical Analysis and Computing*,
Springer Proceedings in Mathematics & Statistics 344,
https://doi.org/10.1007/978-981-33-4646-8_22

sible effective changes by a signal in the time domain into a frequency s-domain in the field of Digital Signal Processing (DSP) [26]. Recently in [22, 25], the applications of fractional Fourier transform in X-ray models and simulations are developed.

The article is organized as follows: In Sect. 2, the conceptual idea about delta and its inverse difference operators are presented elaborately. Then we have moved on to derive the discrete Sumudu transform and its functions in Sect. 3. In order to validate, some applications of fractional difference equation with initial conditions are discussed in Sect. 4. Finally in Sect. 5, we have presented the conclusion.

2 Discrete Fractional Calculus

In this section, we present some basic definitions and results on generalized difference operator. Let

$$\mathcal{M}_{r,h} := \{r, r+h, r+2h, ...\}, \quad where\ r \in \mathbb{R}. \tag{1}$$

Definition 1 For the function $u(t)$, the h-difference operator Δ_h is defined by

$$\Delta_h u(t) = \frac{u(t+h) - u(t)}{h}, \tag{2}$$

and its sum is given by

$$\Delta_h^{-1} u(t) = h \sum_{i=0}^{\infty} u(t+ih), \tag{3}$$

Definition 2 [14] (page 5, Definition 2.6) For $h > 0$ and $\alpha \in R$, the falling h-polynomial factorial function is defined by

$$t_h^{(\alpha)} = h^{\alpha} \frac{\Gamma(\frac{t}{h}+1)}{\Gamma(\frac{t}{h}+1-\alpha)}, \tag{4}$$

where $t_h^{(0)} = 1$ and $\frac{t}{h}+1, \frac{t}{h}+1-\alpha, \notin \{0, -1, -2, -3, ...\}$.

Definition 3 Let u be defined on $\mathcal{M}_{r,h}$ and $\alpha > 0$. Then the fractional sum of u is given by

$${}_r\Delta_h^{-\alpha} u(t) = \frac{1}{\Gamma(\alpha)} \sum_{j=r/h}^{t/h-\alpha} (t - \sigma_h(jh))_h^{(\alpha-1)} u(jh)h, \ t \in \mathcal{M}_{r+\alpha,h}. \tag{5}$$

In particular, when $\alpha = 0$, we get

$${}_r\Delta_h^{-0} u(t) = u(t), t \in \mathcal{M}_{r,h}. \tag{6}$$

Definition 4 Let u be the function defined on $\mathcal{M}_{r,h}$, $\alpha \geq 0$, then the Riemann–Liouville fractional difference of u is given by

$$ {}_r\Delta_h^\alpha u(t) := {}_r\Delta_h{}^M \Delta_h^{-(M-\alpha)} u(t), \ t \in \mathcal{M}_{r+M-\alpha,h}. \tag{7}$$

Theorem 1 *Let u be defined on $\mathcal{M}_{r,h}$ and $\alpha, \beta > 0$, then we have*

$$\Delta_{r+\alpha}^\alpha \Delta_r^{-\beta} u(t) = \Delta_r^{\alpha-\beta} u(t), \ t \in \mathcal{M}_{r+\beta+M-\alpha,h}. \tag{8}$$

Definition 5 Let u be the function defined on $\mathcal{M}_{r,h}$, $\alpha \geq 0$, then the αth–order Caputo fractional difference of u is given by

$$ {}_r^C\Delta_h{}^\alpha u(t) := \Delta_h^{-(M-\alpha)} {}_r\Delta_h{}^M u(t), \ t \in \mathcal{M}_{r+M-\alpha,h}. \tag{9}$$

3 The Discrete Sumudu Transform

Definition 6 The generalized Sumudu transform of the function u is given by

$$\mathcal{S}_{r,h}\{u\}(\eta) := \frac{h}{\eta}\int_r^\infty e_{\ominus(h/\eta)}(\sigma(t), r)u(t)\Delta_h t, r \in R, \tag{10}$$

In particular case for $\mathcal{M}_{r,h}$, the discrete Sumudu transform is defined by

$$\mathcal{S}_{r,h}\{u\}(\eta) = \frac{h}{\eta}\sum_{j=0}^{\infty}\left(\frac{\eta}{\eta+h}\right)^{j+1} u(j+r). \tag{11}$$

Definition 7 A function u on $\mathcal{M}_{r,h}$ is of exponential order $\alpha (\alpha > 0)$ if $\exists\, A > 0$ such that

$$|u(t)| \leq A\alpha^t \quad \text{for large } t. \tag{12}$$

Remark 1 Throughout this paper, we use the following sufficient condition in all results, $\forall \eta \in C\ \{-1, 0\}$ such that $|(\eta + h/\eta)| > \alpha$.

Lemma 1 *Let u and v are the two functions on $\mathcal{M}_{r-n,h}$ and $\mathcal{M}_{r,h}$ respectively. Then we have*

$$\mathcal{S}_{r-n,h}\{u(t)\}(\eta) = \left(\frac{\eta}{\eta+h}\right)^{n/h} \mathcal{S}_{r,h}\{u(t)\}(\eta) + \frac{h}{\eta}\sum_{j=0}^{n/h-1} u(jh+r-n)\left(\frac{\eta}{\eta+h}\right)^{j+1} \tag{13}$$

$$\mathcal{S}_{r+n,h}\{v(t)\}(\eta) = \left(\frac{\eta+h}{\eta}\right)^{n/h} \mathcal{S}_{r,h}\{v(t)\}(\eta) - \frac{h}{\eta}\sum_{j=0}^{n/h-1} v(jh+r)\left(\frac{\eta+h}{\eta}\right)^{n/h-1-j} \tag{14}$$

Proof Since

$$
\begin{aligned}
\mathcal{S}_{r-n,h}\{u(t)\}(\eta) &= \frac{h}{\eta}\sum_{j=0}^{\infty} u(jh+r-n)\left(\frac{\eta}{\eta+h}\right)^{j+1} \\
&= \frac{h}{\eta}\sum_{j=n/h}^{\infty} u(jh+r-n)\left(\frac{\eta}{\eta+h}\right)^{j+1} + \frac{h}{\eta}\sum_{j=0}^{n/h-1} u(jh+r-n)\left(\frac{\eta}{\eta+h}\right)^{j+1} \\
&= \frac{h}{\eta}\sum_{j=0}^{\infty} u(jh+r)\left(\frac{\eta}{\eta+h}\right)^{j+n/h+1} + \frac{h}{\eta}\sum_{j=0}^{n/h-1} u(jh+r-n)\left(\frac{\eta}{\eta+h}\right)^{j+1}
\end{aligned}
$$

$$
\mathcal{S}_{r-n,h}\{u(t)\}(\eta) = \left(\frac{\eta}{\eta+h}\right)^{n/h} \mathcal{S}_{r,h}\{u(t)\}(\eta) + \frac{h}{\eta}\sum_{j=0}^{n/h-1} u(jh+r-n)\left(\frac{\eta}{\eta+h}\right)^{j+1}
$$

Now,

$$
\begin{aligned}
\mathcal{S}_{r+n,h}\{v(t)\}(\eta) &= \frac{h}{\eta}\sum_{j=0}^{\infty} u(jh+r-n)\left(\frac{\eta}{\eta+h}\right)^{j+1} \\
&= \frac{h}{\eta}\sum_{j=n/h}^{\infty} v(jh+r)\left(\frac{\eta}{\eta+h}\right)^{j-n/h+1} \\
&= \frac{h}{\eta}\sum_{j=0}^{\infty} v(jh+r)\left(\frac{\eta}{\eta+h}\right)^{j-n/h+1} \\
&\quad - \frac{h}{\eta}\sum_{j=0}^{n/h-1} v(jh+r)\left(\frac{\eta}{\eta+h}\right)^{j-n/h+1} \\
&= \left(\frac{\eta+h}{\eta}\right)^{n/h} \frac{h}{\eta}\sum_{j=0}^{\infty} v(jh+r)\left(\frac{\eta}{\eta+h}\right)^{j+1}
\end{aligned}
$$

$$
\begin{aligned}
&\quad - \frac{h}{\eta}\sum_{j=0}^{n/h-1} v(jh+r)\left(\frac{\eta+h}{\eta}\right)^{n/h-1-j} \\
\mathcal{S}_{r+n,h}\{v(t)\}(\eta) &= \left(\frac{\eta+h}{\eta}\right)^{n/h} \mathcal{S}_{r,h}\{v(t)\}(\eta) \\
&\quad - \frac{h}{\eta}\sum_{j=0}^{n/h-1} v(jh+r)\left(\frac{\eta+h}{\eta}\right)^{n/h-1-j}
\end{aligned}
$$

Definition 8 The βth-order Taylor monomial is defined by

$$h_\beta(t,r) := \frac{(t-r)_h^{(\beta)}}{\Gamma(\beta+1)},\ t \in \mathcal{M}_r. \tag{15}$$

Lemma 2 *For $r_1, r_2 \in R$ and $r_2 - r_1 = \beta$, then we have*

$$h^{\beta+1}\mathcal{S}_{r_2,h}\left\{h_\beta(.,r_1)\right\}(\eta) = (\eta+h)^\beta. \tag{16}$$

Proof From the binomial relation, we have

$$(a+b)^\alpha = \sum_{j=0}^{\infty}\binom{\alpha}{j} a^j b^{\alpha-j} \tag{17}$$

$$\binom{\alpha}{j} := \frac{\alpha^{(j)}}{j!}. \tag{18}$$

From the relation (17), we have

$$\binom{-\alpha}{j} = (-1)^j\binom{j+\alpha-1}{\alpha-1} \tag{19}$$

$$\frac{1}{(1-z)^\alpha} = ((-z)+1)^{-\alpha} = \sum_{j=0}^{\infty}\binom{j+\alpha-1}{\alpha-1} z^j \tag{20}$$

for $\alpha \in \mathbb{R}$ and $|z| < 1$.

Since $r_2 - r_1 = \beta$, then we have

$$\begin{aligned}
(\eta+h)^\beta &= \frac{h^{\beta+1}}{\eta+h}\frac{1}{\left(1-\left(\frac{\eta}{\eta+h}\right)\right)^{\beta+1}} \\
&= \frac{h^{\beta+1}}{\eta+h}\sum_{j=0}^{\infty}\binom{j+\beta}{\beta}\left(\frac{\eta}{\eta+h}\right)^j \\
&= \frac{h^{\beta+1}}{\eta}\sum_{j=0}^{\infty}\binom{j+\beta}{\beta}\left(\frac{\eta}{\eta+h}\right)^{j+1} \\
&= \frac{h^{\beta+1}}{\eta}\sum_{j=0}^{\infty}\frac{(j+\beta)^{(\beta)}}{\Gamma(\beta+1)}\left(\frac{\eta}{\eta+h}\right)^{j+1}
\end{aligned}$$

$$= \frac{h^{\beta+1}}{\eta} \sum_{j=0}^{\infty} h_{\beta}(j + r_2, r_1) \left(\frac{\eta}{\eta + h} \right)^{j+1}$$

$$(\eta + h)^{\beta} = h^{\beta+1} \mathcal{S}_{r_2,h} \left\{ h_{\beta}(.,r_1) \right\} (\eta)$$

Definition 9 Define the convolution of two functions u, v on $\mathcal{M}_{r,h}$ by

$$(u * v)(t) = h \sum_{j=r/h}^{t/h-1} u(jh)v(t - \sigma_h(jh) + r), t \in \mathcal{M}_{r,h} \tag{21}$$

Lemma 3 *Let* u, v *be defined on* $\mathcal{M}_{r,h}$, *then*

$$\mathcal{S}_{r,h} \{(u * v)\} (\eta) = (\eta + h) \mathcal{S}_{r,h} \{u\} \mathcal{S}_{r,h} \{v\} (\eta) \tag{22}$$

Proof Since

$$\mathcal{S}_{r,h} \{(u * v)\} (\eta) = \frac{h}{\eta} \sum_{j=0}^{\infty} (u * v)(jh + a) \left(\frac{\eta}{\eta + h} \right)^{j+1},$$

$$= h\frac{h}{\eta} \sum_{j=0}^{\infty} \left(\frac{\eta}{\eta + h} \right)^{j+1} \sum_{k=r/h}^{j+r/h-1} u(kh)v(jh + r - \sigma_h(kh) + r),$$

$$\mathcal{S}_{r,h} \{(u * v)\} (\eta) = h\frac{h}{\eta} \sum_{j=0}^{\infty} \sum_{k=0}^{j-1} \left(\frac{\eta}{\eta + h} \right)^{j+1} u(kh + r)v(jh - \sigma_h(kh) + r).$$

Substitution $jh - \sigma_h(kh) = \alpha$ yields
$\Rightarrow jh - kh - h = \alpha \Rightarrow jh = \alpha + kh + h \Rightarrow j = \frac{\alpha}{h} + k + 1.$

$$\mathcal{S}_{r,h} \{(u * v)\} (\eta) = h\frac{h}{\eta} \sum_{\alpha=0}^{\infty} \sum_{k=0}^{\infty} \left(\frac{\eta}{\eta + h} \right)^{\alpha/h+k+1} u(kh + r)v(\alpha + r),$$

$$= (\eta + h) \left(\frac{h}{\eta} \sum_{j=0}^{\infty} u(kh + r) \left(\frac{\eta}{\eta + h} \right)^{k+1} \right)$$

$$\left(\frac{h}{\eta} \sum_{\alpha=0}^{\infty} v(\alpha + r) \left(\frac{\eta}{\eta + h} \right)^{\alpha/h+1} \right),$$

$$\mathcal{S}_{r,h} \{(u * v)\} (\eta) = (\eta + h) \mathcal{S}_{r,h} \{u\} \mathcal{S}_{r,h} \{v\} (\eta).$$

Theorem 2 *For the function* u *and* v, *we have*

$$\mathcal{S}_{r+\alpha,h}\left\{{}_r\Delta_h^{-\alpha}u\right\}(\eta) = \frac{(\eta+h)^{\alpha}}{h^{\alpha-1}}\mathcal{S}_{r,h}\left\{u\right\}(\eta) \tag{23}$$

$$\mathcal{S}_{r+\alpha-M,h}\left\{{}_r\Delta_h^{-\alpha}u\right\}(\eta) = \frac{\eta^{M}}{h^{\alpha-1}(\eta+h)^{M-\alpha}}\mathcal{S}_{r,h}\left\{u\right\}(\eta) \tag{24}$$

Proof Consider, $\mathcal{S}_{r+\alpha-M,h}\left\{{}_r\Delta_h^{-\alpha}u\right\}(\eta) = \frac{h}{\eta}\sum_{j=0}^{\infty}\left(\frac{\eta}{\eta+h}\right)^{j+1} {}_r\Delta_h{}^{-\alpha}u(jh+r+\alpha-M)$

$$\begin{aligned}
&= \left(\frac{\eta}{\eta+h}\right)^{M} \mathcal{S}_{r+\alpha,h}\left\{{}_r\Delta_h^{-\alpha}u\right\}(\eta) + \\
&\quad \frac{h}{\eta}\sum_{j=0}^{M-1}\left(\frac{\eta}{\eta+h}\right)^{j+1} {}_r\Delta_h{}^{-\alpha}u(jh+r+\alpha-M)
\end{aligned} \tag{25}$$

$$= \left(\frac{\eta}{\eta+h}\right)^{M} \mathcal{S}_{r+\alpha,h}\left\{{}_r\overset{-\alpha}{\underset{h}{\Delta}}u\right\}(\eta)$$

taking M zeros of ${}_r\Delta_h^{-\alpha}u$ into account. Furthermore, by (16), (21), and (22)

$$\begin{aligned}
\mathcal{S}_{r+\alpha}\left\{{}_r\Delta_h^{-\alpha}u\right\}(\eta) &= \frac{h}{\eta}\sum_{j=0}^{\infty}\left(\frac{\eta}{\eta+h}\right)^{j+1} {}_r\Delta_h{}^{-\alpha}u(jh+r+\alpha) \\
&= \frac{h}{\eta}\sum_{j=0}^{\infty}\left(\frac{\eta}{\eta+h}\right)^{j+1}\sum_{k=r/h}^{j+r/h}\frac{(jh+r+\alpha-\sigma_h(kh))_h^{(\alpha-1)}}{\Gamma(\alpha-h+1)}u(kh)h \\
&= \frac{h}{\eta}\sum_{j=0}^{\infty}\left(\frac{\eta}{\eta+h}\right)^{j+1}\sum_{k=r/h}^{j+r/h} hu(kh)h_{\alpha-1}(jh+r-kh+r, r-(\alpha-1)) \\
&= \frac{h^2}{\eta}\sum_{j=0}^{\infty}\left(\frac{\eta}{\eta+h}\right)^{j+1}(u * h_{\alpha-1}(.,r-(\alpha-1)))(jh+r) \\
&= h\mathcal{S}_{r,h}\left\{u * h_{\alpha-1}(.,r-(\alpha-1))\right\}(\eta) \\
&= h(\eta+h)\mathcal{S}_{r,h}\left\{u\right\}(\eta)\mathcal{S}_{r,h}\left\{h_{\alpha-1}(.,r-(\alpha-1))\right\} \\
&= \frac{h}{h^{\alpha}}(\eta+h)(\eta+h)^{\alpha-1}\mathcal{S}_{r,h}\left\{u\right\}(\eta) \\
\mathcal{S}_{r+\alpha,h}\left\{{}_r\Delta_h^{-\alpha}u\right\}(\eta) &= \frac{(\eta+h)^{\alpha}}{h^{\alpha-1}}\mathcal{S}_{r,h}\left\{u\right\}(\eta).
\end{aligned}$$

Then we obtain

$$\mathcal{S}_{r+\alpha-M,h}\left\{{}_r\Delta_h^{-\alpha}u\right\}(\eta) = \left(\frac{\eta}{\eta+h}\right)^M \frac{(\eta+h)^\alpha}{h^{\alpha-1}}\mathcal{S}_{r,h}\{u\}(\eta)$$
$$= \frac{\eta^M}{h^{\alpha-1}(\eta+h)^{M-\alpha}}\mathcal{S}_{r,h}\{u\}(\eta).$$

Theorem 3 *Let u be a function defined on $\mathcal{M}_{r,h}$, then*

$$\mathcal{S}_{r+M-\alpha,h}\left\{{}_r\Delta_h^\alpha u\right\}(\eta) = \frac{h^\alpha(\eta+h)^{M-\alpha}}{\eta^M}\mathcal{S}_{r,h}\{u\}(\eta) - \sum_{j=0}^{M-1}\eta^{j-M}{}_r\Delta_h{}^{\alpha-M+j}u(r+M-\alpha) \tag{26}$$

Proof Since $\alpha = M$, the result is true. Therefore for $\alpha \neq M$, hence it follows from (23) that

$$\mathcal{S}_{r+M-\alpha,h}\left\{{}_r\Delta_h^\alpha u\right\}(\eta) = \mathcal{S}_{r+M-\alpha,h}\left\{{}_r\Delta_h^M \Delta_h^{-(M-\alpha)}u\right\}(\eta)$$
$$= \left(\frac{h}{\eta}\right)^M \mathcal{S}_{r+M-\alpha,h}\left\{\Delta_h^{-(M-\alpha)}u\right\}(\eta)$$
$$-\sum_{j=0}^{M-1}\eta^{j-M}\Delta_h^j\Delta_r^{-(M-\alpha)}u(r+M-\alpha)$$
$$\mathcal{S}_{r+M-\alpha,h}\left\{{}_r\Delta_h^\alpha u\right\}(\eta) = \frac{h^\alpha}{\eta^M}(\eta+h)^{M-\alpha}\mathcal{S}_{r,h}\{u\}(\eta)$$
$$-\sum_{j=0}^{M-1}\eta_r^{j-M}\Delta_h^{\alpha-M+j}u(r+M-\alpha).$$

Theorem 4 *Let u defined on $\mathcal{M}_{r,h}$, then we have*

$$\mathcal{S}_{r+M-\alpha,h}\left\{{}_r^c\Delta_h{}^\alpha u\right\}(\eta) = \frac{h^{\alpha+1}}{\eta^M}(\eta+h)^{M-\alpha}\left[\mathcal{S}_{r,h}\{u\}(\eta) - \frac{1}{h^M}\sum_{j=0}^{M-1}\eta^j\Delta_h^j u(r)\right] \tag{27}$$

Proof Since $\alpha = M$, the result is true. If $\alpha \neq M$ and hence it follows from (23) that

$$\mathcal{S}_{r+M-\alpha,h}\left\{{}_r^c\Delta_h^\alpha u\right\}(\eta) = \mathcal{S}_{r+M-\alpha,h}\left\{\Delta_h^{-(M-\alpha)}\Delta_h^M u\right\}(\eta)$$
$$= \frac{(\eta+h)^{M-\alpha}}{h^{M-\alpha-1}}\mathcal{S}_{r,h}\left\{\Delta_h^M u\right\}(\eta)$$
$$\mathcal{S}_{r+M-\alpha,h}\left\{{}_r^C\Delta_h^\alpha u\right\}(\eta) = \frac{h^{\alpha+1}}{\eta^M}(\eta+h)^{M-\alpha}\left[\mathcal{S}_{r,h}\{u\}(\eta) - \frac{1}{h^M}\sum_{j=0}^{M-1}\eta^{j-M}\Delta_h^j u(r)\right]$$

Lemma 4 *Let u on $\mathcal{M}_{r,h}$ be given. For any $\xi \in \mathcal{M}_0$ and $\alpha > 0$, we have*

$$ {}_r^C\Delta_h^{\alpha+\xi} u(t) = {}_r^C \Delta_h^{\alpha} \Delta^{\xi} u(t) \tag{28}$$

Proof Since, we have

$$\begin{aligned}
{}_r^C\Delta_h{}^{\alpha+\xi} u(t) &= \Delta_h^{-(M+\xi-\alpha-\xi)}{}_r\Delta_h{}^{M+\xi} u(t) \\
&= \Delta_h^{-(M-\alpha)} \Delta_r^M \Delta_r^{\xi} u(t) \\
&= {}_r\Delta_h{}^{-(M-\alpha)+M+\xi} u(t) \\
&= {}_r^C\Delta_h{}^{\alpha+\xi} u(t) \\
{}_r^C\Delta_h{}^{\alpha+\xi} u(t) &= {}_r^C\Delta_h{}^{\alpha} \Delta^{\xi} u(t)
\end{aligned}$$

Corollary 1 *For the function u, we have*

$$\mathcal{S}_{r+M-\alpha}\left\{{}^C\Delta_r^{\alpha+\xi} u\right\}(\eta) = \frac{h^{\alpha+1}(\eta+h)^{M-\alpha}}{\eta^{M+\xi}}$$

$$\left[h^{\xi}\mathcal{S}_{r,h}\{u\}(\eta) - \frac{1}{h^M}\sum_{j=0}^{M+\xi-1} \eta^{j-M}\Delta^j u(r)\right] \tag{29}$$

Proof Since $\alpha = M$, (27) is true. If $\alpha \neq M$, then we have

$$\begin{aligned}
\mathcal{S}_{r+M-\alpha,h}\left\{{}_r^C\Delta_h{}^{\alpha+\xi} u\right\}(\eta) &= \mathcal{S}_{r+M-\alpha,h}\left\{\Delta_h^{-(M+\xi-\alpha-\xi)}{}_r\Delta_h{}^{M+\xi} u(t)\right\}(\eta) \\
&= \mathcal{S}_{r+M-\alpha,h}\left\{\Delta_h^{-(M-\alpha)}{}_r\Delta_h{}^{M+\xi} u(t)\right\}(\eta) \\
&= \frac{(\eta+h)^{M-\alpha}}{h^{M-\alpha-1}}\mathcal{S}_{r,h}\left\{{}_r\Delta_h{}^{M+\xi} u(t)\right\}(\eta) \\
\mathcal{S}_{r+M-\alpha,h}\left\{{}_r^C\Delta_h{}^{\alpha+\xi} u\right\}(\eta) &= \frac{h^{\alpha+1}(\eta+h)^{M-\alpha}}{\eta^{M+\xi}} \\
&\quad \left[h^{\xi}\mathcal{S}_{r,h}\{u\}(\eta) - \frac{1}{h^M}\sum_{j=0}^{M+\xi-1}\eta^{j-M}\Delta_h^j u(r)\right]
\end{aligned}$$

4 Applications

Here, we present the solutions of certain initial value problems using the Sumudu transform.

Example 1 For the function u, the unique solution to the IVP

$$ {}_h\Delta^{\alpha}_{r+\alpha-M} Z(t) = u(t), t \in \mathcal{M}_r \tag{30} $$

$\Delta^j z(r+\alpha-M) = A_j$, $j \in \{0, 1, ..., M-1\}$, $A_j \in \mathbb{R}$ is given by

$$ z(t) = \sum_{j=0}^{M-1} \varsigma_j (t-r)^{(\alpha+j-M)} + {}_r\Delta_h{}^{-\alpha} u(t), \ t \in \mathcal{M}_{r+\alpha-M}, \tag{31} $$

where

$$ \varsigma_j = \frac{\Delta^{\alpha-M+j}_{r+\alpha-M} z(r)}{\Gamma(\alpha+j-M+1)} = \sum_{\xi=0}^{j} \sum_{k=0}^{j-\xi} \frac{(-1)^k}{j!} (j-k)^{(M-\alpha)} \binom{j}{\xi} \binom{j-\xi}{k} A_\xi. \tag{32} $$

Proof Since $\alpha = M$, the result holds. If $\alpha \neq M$, we have

$$ \mathcal{S}_{r,h} \left\{ {}_h\Delta^{\alpha}_{r+\alpha} z \right\} (\eta) = \mathcal{S}_{r,h} \{u\} (\eta). \tag{33} $$

Then from (26), it follows

$$ \frac{h^{\alpha}(\eta+h)^{M-\alpha}}{\eta^M} \mathcal{S}_{r+\alpha-M,h} \{z\} (\eta) - \sum_{j=0}^{M-1} \eta^{j-M} \Delta^{\alpha-M+j}_{r+\alpha-M} z(r) = \mathcal{S}_{r,h} \{u\} (\eta) \tag{34} $$

and hence

$$ \mathcal{S}_{r+\alpha-M,h} \{z\} (\eta) = \frac{\eta^M}{h^{\alpha}(\eta+h)^{M-\alpha}} \mathcal{S}_{r,h} \{u\} (\eta) + \sum_{j=0}^{M-1} \frac{\eta^j}{h^{\alpha}(\eta+h)^{M-\alpha}} \Delta^{\alpha-M+j}_{r+\alpha-M} z(r) \tag{35} $$

By (24), we have

$$ \frac{\eta^M}{h^{\alpha}(\eta+h)^{M-\alpha}} \mathcal{S}_{r,h} \{u\} (\eta) = \mathcal{S}_{r+\alpha-M,h} \left\{ {}_r\Delta_h^{-\alpha} u \right\} (\eta). \tag{36} $$

Using (13), we get

$$ \begin{aligned} \frac{\eta^j}{h^{\alpha}(\eta+h)^{M-\alpha}} &= \frac{1}{h^{\alpha}} \left(\frac{\eta}{\eta+h} \right)^j (\eta+h)^{j-M+\alpha} \\ &= \left(\frac{\eta}{\eta+h} \right)^j h^{j-M+1} \mathcal{S}_{r+\alpha+j-M,h} \left\{ h_{\alpha+j-M}(.,r) \right\} (\eta) \\ &= h^{j-M+1} (\mathcal{S}_{r+\alpha-M,h} \left\{ h_{j-M+\alpha}(.,r) \right\} (\eta) \end{aligned} $$

$$-\frac{h}{\eta}\sum_{k=0}^{j-1}\left(\frac{\eta}{\eta+h}\right)^{k+1} h_{\alpha+j-M}(kh+r+\alpha-M,r)(\eta))$$
$$= h^{j-M+1}\mathcal{S}_{r+\alpha-M,h}\left\{h_{\alpha+j-M}(.,r)\right\}(\eta)$$

since $h_{\alpha+j-M}(kh+r+\alpha-M,r)$

$$= \frac{(kh+\alpha-M)_h^{(\alpha+j-M)}}{\Gamma(\alpha+j-M+1)}$$
$$= h^{\alpha+j-M}\frac{\Gamma\left(\dfrac{kh+\alpha-M}{h}+1\right)}{\Gamma\left(\dfrac{kh+\alpha-M}{h}+1-(\alpha+j-M)\right)\Gamma(\alpha+j-M+1)}$$
$$= h^{\alpha+j-M}\frac{\Gamma(kh+\alpha-M+h)}{\Gamma((k-j+1)h+(\alpha-M)(1-h))\Gamma(\alpha+j-M+1)}$$
$$= 0$$

Consequently, we have $\mathcal{S}_{r+\alpha-M}\left\{z\right\}(\eta)$

$$= \mathcal{S}_{r+\alpha-M}\left\{\Delta_h^{-\alpha}u\right\}(\eta) + \sum_{j=0}^{M-1}\Delta_{r+\alpha-M}^{\alpha-M+j}z(r)h^{j-M+1}\mathcal{S}_{r+\alpha-M}\left\{h_{\alpha+j-M}(.,r)\right\}(\eta)$$
$$= \mathcal{S}_{r+\alpha-M}\left\{\sum_{j=0}^{M-1}\Delta_{r+\alpha-M}^{\alpha-M+j}h^{j-M+1}z(r)h_{\alpha+j-M}(.,r) + {}_r\Delta_h{}^{-\alpha}u\right\}(\eta) \tag{37}$$

Since Sumudu transform is a one-to-one operator, we conclude that for $t \in \mathcal{M}_{r+\alpha-M}$,

$$z(t) = \sum_{j=0}^{M-1}\left(\frac{\Delta_{r+\alpha-M}^{\alpha-M+j}z(r)}{\Gamma(\alpha+j-M+1)}\right)(t-r)^{(\alpha+j-M)} + {}_r\Delta_h{}^{-\alpha}u(t), \tag{38}$$

where

$$\frac{\Delta_{r+\alpha-M}^{\alpha-M+j}z(r)}{\Gamma(\alpha+j-M+1)} = \sum_{\xi=0}^{j}\sum_{k=0}^{j-\xi}\frac{(-1)^k}{j!}(j-k)^{(M-\alpha)}\binom{j}{\xi}\binom{j-\xi}{k}{}_h\Delta^j z(r+\alpha-M), \tag{39}$$

Example 2 Consider the IVP, Replacing Riemann–Liouville fractional difference by the Caputo fractional difference as

$${}_h^C\Delta_{r+\alpha-M}^{\alpha}Z(t) = u(t),\ t \in \mathcal{M}_{r,h}, \tag{40}$$

$\Delta_h^j z(r+\alpha-M) = A_j,\ j \in \{0, 1, ..., M-1\},\ A_j \in \mathbb{R}.$

Applying the Sumudu transform to both sides, we get

$$\mathcal{S}_{r,h}\left\{{}_h\Delta^{\alpha}_{r+\alpha}z\right\}(\eta) = \mathcal{S}_{r,h}\left\{u\right\}(\eta). \tag{41}$$

Then from (27) it follows

$$\frac{h^{\alpha+1}}{\eta^M}(\eta+h)^{M-\alpha}\left[\mathcal{S}_{r+\alpha-M}\left\{z\right\}(\eta) - \frac{1}{h^M}\sum_{j=0}^{M-1}\eta^j A_j\right] = \mathcal{S}_{r,h}\left\{u\right\}(\eta). \tag{42}$$

By (24), we have

$$\begin{aligned}\mathcal{S}_{r+\alpha-M}\left\{z\right\}(\eta) &= \frac{1}{h^M}\sum_{j=0}^{M-1}\eta^j A_j + \frac{\eta^M}{h^{\alpha+1}(\eta+h)^{M-\alpha}}\mathcal{S}_{r,h}\left\{u\right\}(\eta)\\ &= \frac{1}{h^M}\sum_{j=0}^{M-1}\eta^j A_j \mathcal{S}_{r+\alpha-M}\left\{{}_r\Delta_h^{-\alpha}u\right\}(\eta),\end{aligned}$$

and also we have

$$\mathcal{S}_0\left\{t_h^{(n)}\right\}(\eta) = n!\eta^n, \quad n \in \mathcal{M}_0, \tag{43}$$

hence

$$z(t) = \frac{1}{h^M}\sum_{j=0}^{M-1}A_j\frac{(t-r-\alpha+M)^{(j)}}{j!} + {}_r\Delta_h{}^{-\alpha}u(t) \tag{44}$$

5 Conclusion

In this work, we proved some results with the discrete fractional Sumudu transform using inverse difference operator. We obtained discrete fractional Sumudu transform of Taylor's monomial and fractional sums for fractional order. The advantage of our findings is we present some examples for solving the initial value problem using discrete fractional Sumudu transform.

Acknowledgements Catalyzed and financially supported by Tamilnadu State Council for Science and Technology, Dept. of Higher Education, Government of Tamilnadu.

References

1. Atici, F.M., Eloe, P.W.: Two-point boundary value problems for finite fractional difference equations. J. Differ. Equ. Appl. https://doi.org/10.1080/10236190903029241
2. Atıcı, F.M., Eloe, P.W.: Electr. J. Qual. Theor. Differ. Equ. Spec. Ed. I **2009**, 1 (2009)

3. Abdeljawad, T., Atici, F.: Abstr. Appl. Anal. **2012**, 406757 (2012)
4. Abdeljawad, T.: Discr. Dyn. Nat. Soc. **2013**, 406910 (2013)
5. Abdeljawad, T.: Adv. Differ. Equ. **2013**, 36 (2013)
6. Abdeljawad, T., Jarad, F., Baleanu, D.: Adv. Differ. Equ. **2012**, 72 (2012)
7. Abdeljawad, T., Baleanu, D.: J. Nonlinear Sci. Appl. **10**, 1098 (2017)
8. Abdeljawad, T., Baleanu, D.: Adv. Differ. Equ. **2017**, 78 (2017)
9. Abdeljawad, T., Baleanu, D.: Choas Solitons Fractals **102**, 106 (2017)
10. Agarwal, P., Al-Mdallal, Q., Cho, Y.J., Jain, S.: Fractional differential equations for the generalized Mittag-Leffer function. Adv. Differ. Equ. **2018**(1), 58 (2018)
11. Agarwal, P., El-Sayed, A.A.: Non-standard finite difference and Chebyshev collocation methods for solving fractional diffusion equation. Phys. A Stat. Mech. Appl. **500**, 40–49 (2018)
12. Baltaeva, U., Agarwal, P.: Boundary value problems for the third order loaded equation with non characteristic type change boundaries. Math. Methods Appl. Sci. **41**(9), 3307–3315 (2018)
13. Britanak, V., Rao, K.R.: The fast generalized discrete Fourier transforms: a unified approach to the discrete sinusoidal transforms computation. Signal Process. **79**, 135–150 (1999). Dec
14. Bastos, N.R.O., Ferreira, R.A.C., Torres, D.F.M.: Discrete-time fractional variational problems. Signal Process. **91**(3), 513–524 (2011)
15. Choi, J., Agarwal, P.: Certain fractional integral inequalities involving hypergeometric operators. East Asian Math. **30**(3), 283–291 (2014)
16. Goodrich, C., Peterson, A.: Discrete Fractional Calculus. Springer, New York (2015)
17. Goodrich, C.S.: Solutions to a discrete right-focal boundary value problem. Int. J. Differ. Equ. **5**, 195–216 (2010)
18. Holm, M.: Sum and difference compositions in discrete fractional calculus. CUBO Math. J. **13**(3) (2011)
19. Jain, S., Agarwal, P., Kilicman, A.: Pathway fractional integral operator associated with 3m-Parametric Mittag-Leffer functions. Int. J. Appl. Comput. Math. **4**(5) (2018)
20. Miller, K.S., Ross, B.: An Introduction to the Fractional Calculus and Fractional Differential Equations. Wiley, New York (1993)
21. Mehrez, K., Agarwal, P.: New Hermite Hadamard type integral inequalities for convex functions and their applications. J. Comput. Appl. Math. **350**, 274–285 (2018)
22. Ni, L., Da, X., Hu, H., Liang, Y., Xu, R.: PHY-aided secure communication via weighted fractional Fourier transform. Wirel. Commun. Mob. Comput. **2018**, 10 pages (2018). Article ID 7963451
23. Oldham, K., Spanier, J.: The Fractional Calculus: Theory and Applications of Differentiation and Integration to Arbitrary Order. Dover Publications Inc, Mineola, New York (2002)
24. Podlubny, I.: Fractional Differential Equations. Academic Press, New York (1999)
25. Pedersen, A.F., Simons, H., Detlefs, C., Poulsen, H.F.: The fractional Fourier transform as a simulation tool for lens-based X-ray microscopy. J. Synchrotron Radiat. **25**, 717–728 (2018)
26. Smith, S.W.: The Scientist and Engineer's Guide to Digital Signal Processing, 2nd edn. California Technical Publishing San Diego, California (1999)
27. Sitho, S., Ntouyas, S.K., Agarwal, P., Tariboon, J.: Noninstantaneous impulsive inequalities via conformable fractional calculus. J. Inequalities Appl. **1**, 261 (2018)

Generalized Mittag-Leffler Factorial Function with Sums

Jaraldpushparaj Simon and Britto Antony Xavier Gnanaprakasam

Abstract This research aims to express several functions, like trigonometric functions, Mittag-Leffler (ML) function. For that we have introduced Generalized Mittag-Leffler function factorial by which one can express certain functions as well as series by sum of polynomial factorials. This model will be used to obtain solutions of fractional difference equations also.

Keywords Difference equation · Mittag-Leffler function · Discrete Maclaurin series

AMS classification: 39A10 · 33E12

1 Introduction

The ML functions plays a vital role during the last twenty years. Mittag-Leffler functions are applicable in solving the problems in mathematical sciences, physical sciences, life sciences, scientific engineering, and earth sciences [1]. The general form of Mittag-Leffler function (ML function) of power series form is given by

$$E_{\alpha,\beta}(z) = \sum_{k=0}^{\infty} \frac{z^k}{\Gamma(\beta + \alpha k)}, \quad \alpha, \beta \in C, \mathbb{R}(\alpha) > 0, \mathbb{R}(\beta) > 0, z \in C. \tag{1}$$

In (1), the polynomial z^k can be replaced by polynomial factorial to obtain solution of difference equations. For more details on ML function, one can refer Lang [2], [3], Hilfer [4, 5], and Saxena [6].

J. Simon (✉) · B. A. X. Gnanaprakasam
Department of Mathematics, Sacred Heart College (Autonomous),
Tirupattur, 635601 Vellore, Tirupattur District, Tamil Nadu, India
e-mail: jaraldss@gmail.com

B. A. X. Gnanaprakasam
e-mail: brittoshc@gmail.com

R. N. Mohapatra et al. (eds.), *Mathematical Analysis and Computing*,
Springer Proceedings in Mathematics & Statistics 344,
https://doi.org/10.1007/978-981-33-4646-8_23

Recently, due to the vast potential of the applications in several applied fields in practical problems such as Rheology, Fluid flow, Electrical, Diffusive transport akin to Diffusion, Probability, and Statistical Distribution theory, we have introduced the EMLF function (3) as

$$e_\nu(\lambda, k, c) = \sum_{j=0}^{\infty} \lambda^j \frac{(k + (a-1) + cj\nu)^{(j\nu)}}{\Gamma(j\nu + 1)}, \tag{2}$$

where λ and c are constants [7].

In this paper, we introduce new Generalized Mittag-Leffler Factorial (GMLF) functions arrived from (2) with replacing the constants λ and c by functions $\bar{\lambda}$ and $\bar{c}$ defined on $\mathbb{N}(0)$. By GMLF functions, we are able to express the Maclaurin series as GMLF functions. This GMLF function satisfies certain type of fractional difference equation, hence we are able to obtain solutions of fractional difference equations.

2 Basic Definitions and Related Theorems

In this section, we deals with the preliminaries which will be used in the subsequent sections. Here we use the definition $\Delta_\ell u(k) = u(k+\ell) - u(k)$.

Definition 1 [8] Let $\ell, k, \nu \in (-\infty, \infty)$, if $k/\ell + 1 - \nu \notin \{0, -1, -2, \ldots\}$ then the ℓ-polynomial factorial is defined as

$$k_\ell^{(\nu)} = \ell^\nu \frac{\Gamma(k/\ell + 1)}{\Gamma(k/\ell + 1 - \nu)}, \tag{3}$$

where Γ is the gamma function and $k_\ell^{(n)} = k(k-\ell)(k-2\ell)\cdots(k-(n-1)\ell)$ if $n \in N$.

Definition 2 [8] For $-1 < \ell < 1, \ell \neq 0$ and $k, \nu \in \mathbb{R}$, the ℓ-extorial function, denoted as $e_\nu(k_\ell)$, is defined as

$$e_\nu(k_\ell) = 1 + \frac{k_\ell^{(\nu)}}{\Gamma(1\nu + 1)} + \frac{k_\ell^{(2\nu)}}{\Gamma(2\nu + 1)} + \frac{k_\ell^{(3\nu)}}{\Gamma(3\nu + 1)} + \cdots + \infty. \tag{4}$$

If $\ell \in (-\infty, \infty)$, $\ell \neq 0$ and k is a multiple of ℓ and $\nu \in N$, then (9) is defined, and which case all except finite terms of $e_\nu(k_\ell)$ are zero.

Definition 3 [7] For $c \in [0, 1]$, $|\lambda| < 1$, $|\ell| < 1$, $j\nu + 1 \notin N(0)$, $k \in \mathbb{R}$. The Extended Mittag-Leffler Factorial (EMLF) function is defined as

$$e_\nu(\lambda, k_\ell, c) = \sum_{j=0}^{\infty} \frac{\lambda^j}{\Gamma(j\nu + 1)} \Big(k - (a+\ell) + cj\ell\nu\Big)_\ell^{(j\nu)}. \tag{5}$$

The function $e_\nu(\lambda, k_{-\ell}, 1)$ is defined for $\lambda \in (-\infty, \infty)$, $\ell > 0$, if $k - (a+\ell)$ is positive and multiple of ℓ and also $e_\nu(\lambda, k_\ell, 1)$ is defined if $k - (a+\ell)$ is negative and multiple of ℓ.

Special Cases:

(i) When $\ell = 0$, $a = -1$, $\lambda = \nu = 1$, $e_1(1, k_0, c) = e^k$.
(ii) $e_\nu(k_\ell, 0) =$ Extorial Function. (Newly defined function [7])
(iii) $e_\nu(\lambda, k_0, 1)$ is the existing MLF function.

Theorem 4 *[7] The inverse difference of product of EMLF function $e_\nu(\lambda, k_\ell, c)$ and the polynomial factorial function $k_\ell^{(n)}$ is given by*

$$\Delta_\ell^{-1}[k_\ell^{(n)} e_\nu(\lambda, k_\ell, c)] = k_\ell^{(n)} \Delta_\ell^{-1} e_\nu(\lambda, k_\ell, c) - \Delta_\ell^{-1}[\Delta_\ell^{-1} e_\nu(\lambda, (k+\ell)_\ell, c) \Delta_\ell k_\ell^{(n)}]. \tag{6}$$

Also (6) is a solution of $\Delta_\ell u(k) = k_\ell^{(n)} e_\nu(\lambda, k_\ell, c)$.

Theorem 5 *[7] If $m \in N(1)$, $\ell \in (0, \infty)$, $k \in [0, \infty)$ and $c \in -N(1)$, then*

$$\Delta_\ell^{-m} e_\nu(\lambda, k_\ell, c)\Big|\Big|_{(m-1)\ell+j}^{k} = \sum_{r=m}^{[\frac{k}{\ell}]} \frac{(r-1)^{(m-1)}}{(m-1)!} e_\nu(\lambda, (k - r\ell)_\ell, c),\ \ j = k - [\frac{k}{\ell}]\ell, \tag{7}$$

which is a solution of the difference equation $\Delta_\ell^m u(k) = e_\nu(\lambda, k_\ell, c)$.

Definition 6 (*Generalized Mittag-Leffler Factorial function*) Let $\bar{\lambda}$ and $\bar{c}$ be any two real valued functions defined on $\mathbb{N}(0)$. Let ν be the real number such that $j\nu + 1 \notin \mathbb{N}(0)$ for all $j \in \mathbb{N}(0)$ and $\ell \neq 0$. Then the GMLF function is defined by

$$e_\nu(\bar{\lambda}, k_\ell, \bar{c}) = \sum_{j=0}^{\infty} \frac{\bar{\lambda}(j)}{\Gamma(j\nu + 1)} (k + \bar{c}j)^{(j\nu)}. \tag{8}$$

Remark 7 (8) becomes EMLF function if $\bar{\lambda}(j) = \lambda^j$, $|\lambda| < 1$ and $\bar{c}(j) = cj\ell\nu - (a+\ell)$, $c \in [0, 1]$. The GMLF function $e_\nu(\bar{\lambda}, k_\ell, \bar{c})$ becomes discrete Maclaurin series when $\nu = 1$, $\bar{\lambda}(j) = \dfrac{\Delta^j f(0)}{\ell^j}$, $\bar{c}(j) = 0$.

3 Summation of EMLF Function

In this section, we obtain numerical solution of certain type of higher order difference equations.

Theorem 8 *Let $m \in N(1)$ and $k \in (m\ell, \infty)$. Then we have*

$$\sum_{r=m}^{[\frac{k}{\ell}]} \frac{(r-1)^{(m-1)}}{(m-1)!}(k-r\ell)_\ell^{(n)}.e(\lambda,(k-r\ell)_\ell,c) = F_m^{(n)}(k) - F_m^{(n)}((m-1)\ell+j), \tag{9}$$

where $F_m^{(n)}(k) = \Delta_\ell^{-m}\big(F_{(m-1)}^{(n)}(k) - F_m^{(n)}((m-2)\ell+j)\big)$, $m = 2, 3, \cdots$ *and* $F_1^{(n)}(k) = \Delta_\ell^{-1}\big(k_\ell^{(n)}e(\lambda, k_\ell, c)\big)$. *Also (9) is a numerical solution of the* m^{th} *order* ℓ*-difference equation* $\Delta_\ell^m u(k) = k_\ell^{(n)} e_\nu(\lambda, k_\ell, c)$.

Proof From (5) and $\Delta_\ell^{-1}e(\lambda, k_\ell, c) = \frac{1}{\lambda}e(\lambda,(k-c)_\ell,c)$,

$\Delta_\ell k_\ell^{(n)} = n\ell(k\ell)^{(n-1)}$, $\Delta_\ell^{-1}k_\ell^{(\nu)} = \dfrac{k_\ell^{(\nu+1)}}{(\nu+1)\ell}$ and also from Theorem (4) we have

$$\Delta_\ell^{-1}[k_\ell^{(n)}e(\lambda,k_\ell,c)] = k_\ell^{(n)}\frac{1}{\lambda}e(\lambda,(k-c)_\ell,c) - \Delta_\ell^{-1}\Big[\frac{1}{\lambda}e(\lambda,(k+\ell-c)_\ell,c)(n\ell)k_\ell^{(n-1)}\Big]. \tag{10}$$

Taking $m = 1$ in Theorem (5) yields $\Delta_\ell^{-1}e(\lambda,k_\ell,c)\Big|_j^k = \sum_{r=1}^{[\frac{k}{\ell}]} e(\lambda,(k-r\ell)_\ell,c)$.

Applying the Theorem (5) to (10),

$$\sum_{r=1}^{[\frac{k}{\ell}]}(k-r\ell)_\ell^{(n)}e(\lambda,(k-r\ell)_\ell,c) = F_1^{(n)}(k) - F_1^{(n)}(j), \tag{11}$$

where $F_1^{(n)}(k) = \Delta_\ell^{-1}[k_\ell^{(n)}e(\lambda,k_\ell,c)]$ again operating Δ_ℓ^{-1} on both sides of (11)

$$\sum_{r=2}^{[\frac{k}{\ell}]}\frac{(r-1)^{(1)}}{1!}(k-r\ell)_\ell^{(n)}.e(\lambda,(k-r\ell)_\ell,c) = F_2^{(n)}(k) - F_2^{(n)}(\ell+j).$$

Similarly, if we proceed we get (9).

4 Discrete Maclaurin Series

The Maclaurin series for successively differentiable functions is given by

$f(k) = f(0) + \frac{f'(0)}{1!}k + \frac{f''(0)}{2!}k^2 + \frac{f'''(0)}{3!}k^3 + \cdots$. Here k belongs to certain open intervals. It is possible to obtain discrete Maclaurin series for $k \in (-\infty, \infty)$.

Definition 9 The discrete Maclaurin series with polynomial factorial function is defined by

$$f(k) = f(0) + \frac{\Delta_\ell f(0)}{\ell 1!}k_\ell^{(1)} + \frac{\Delta_\ell^2 f(0)}{\ell^2 2!}k_\ell^{(2)} + \frac{\Delta_\ell^3 f(0)}{\ell^3 3!}k_\ell^{(3)} + \cdots. \tag{12}$$

Example 10 For the function $f(k) = a^k$, we have $\Delta_\ell a^k = a^k(a_\ell - 1)$, $\Delta_\ell^2 a^k = a^k(a_\ell - 1)^2, \ldots$, and (12) becomes $a^k = a^0 + \frac{a^0(a^\ell - 1)}{\ell}\frac{k_\ell^{(1)}}{1!} + \frac{a^0(a^\ell - 1)^2}{\ell}\frac{k_\ell^{(2)}}{2!} + \frac{a^0(a^\ell - 1)^3}{\ell}\frac{k_\ell^{(3)}}{3!} + \cdots$ which is the Discrete Maclaurin Series for a^k. When k is a multiple of ℓ, (12) becomes finite series. If $k = 8, \ell = 4, a = 2$, then we get $2^8 = 2^0 + \frac{2^0(2^4 - 1)}{4}\frac{8_4^{(1)}}{1!} + \frac{2^0(2^4 - 1)^2}{16}\frac{8_4^{(2)}}{2!} = 256.$

Theorem 11 *The fractional difference of sine and cosine functions are*

$$\mathop{\Delta}_{\ell}^{\nu} \sin k = 2^\nu \sin^\nu(\ell/2) \sin\big((\nu\pi)/2 + k + \nu\ell/2\big). \tag{13}$$

$$\textit{and } \mathop{\Delta}_{\ell}^{\nu} \cos k = 2^\nu \cos^\nu(\pi/2 - \ell/2) \cos\big((\nu\pi)/2 + k + \nu\ell/2\big). \tag{14}$$

Proof The trigonometric formulas of sin and cos functions are
$\sin(A + B) = \sin A \cos B + \cos A \sin B$, $\sin(A - B) = \sin A \cos B - \cos A \sin B$.
Simplifying the above formulas we get

$$\sin(k + \ell) - \sin k = 2\cos(k + \ell/2)\sin(\ell/2)$$
$$\mathop{\Delta}_{\ell} \sin k = 2\sin(\ell/2)\sin(\pi/2 + k + \ell/2)$$
$$\mathop{\Delta}_{\ell}^{2} \sin k = \mathop{\Delta}_{\ell}[\mathop{\Delta}_{\ell} \sin k] = 2\sin(\ell/2)\mathop{\Delta}_{\ell}[\sin(\pi/2 + k + \ell/2)]$$
$$\mathop{\Delta}_{\ell}^{2} \sin k = 2^2 \sin^2(\ell/2)\sin(2\pi/2 + k + 2\ell/2).$$

Similarly, if we proceed we get (13) and in the same way, we can derive (14).

Theorem 12 *The GMLF function of sine series is given by*

$$\sin(k) = \sin(0) + \frac{\Delta_\ell \sin(0)}{\ell 1!}k_\ell^{(1)} + \frac{\Delta_\ell^2 \sin(0)}{\ell^2 2!}k_\ell^{(2)} + \cdots. \tag{15}$$

Proof From the definition of the GMLF function,

$$e_\nu(\bar{\lambda}, k_\ell, \bar{c}) = \sum_{j=0}^{\infty} \frac{\bar{\lambda}(j)}{\Gamma(j\nu + 1)}(k + \bar{c}j)^{(j\nu)}.$$

Now (15) is obtained by $\bar{\lambda}(j) = \frac{\Delta_\ell^j \sin 0}{\ell^j}$, $\bar{c}(j) = 0$ and $\nu = 1$.

Theorem 13 *The GMLF function of cosine series is given by*

$$\cos(k) = \cos(0) + \frac{\Delta_\ell \cos(0)}{\ell 1!}k_\ell^{(1)} + \frac{\Delta_\ell^2 \cos(0)}{\ell^2 2!}k_\ell^{(2)} + \cdots. \tag{16}$$

Proof From the definition of the GMLF function,

$$e_\nu(\bar{\lambda}, k_\ell, \bar{c}) = \sum_{j=0}^{\infty} \frac{\bar{\lambda}(j)}{\Gamma(j\nu+1)} (k+\bar{c}j)^{(j\nu)}.$$

Now taking the function $\bar{\lambda}(j) = \overset{j}{\underset{\ell}{\Delta}} \cos 0$, $\bar{c}(j) = 0$ and $\nu = 1$, gives (16).

Example 14 If the function $f(k) = \sin k$ and $k = 9, \ell = 3, \pi = 22/7$, we have
$\sin k = \sin 0 + \frac{2\sin(\ell/2)\sin(\pi/2+\ell/2)}{\ell}\frac{k_\ell^{(1)}}{1!} + \frac{2^2\sin^2(\ell/2)\sin((2\pi)/2+(2\ell)/2)}{\ell^2}\frac{k_\ell^{(2)}}{2!} +$
$\frac{2^3\sin^3(\ell/2)\sin((3\pi)/2+(3\ell)/2)}{\ell^2}\frac{k_\ell^{(3)}}{3!}$
$0.41 = 0 + 0.43 - 1.70 + 1.68 \Rightarrow 0.41 = 0.41.$

Example 15 If the function $f(k) = \cos k$ and $k = 6, \ell = 3, \pi = 22/7$
$\cos k = \cos 0 + \frac{2\cos(\pi/2-\ell/2)\cos(\pi/2+\ell/2)}{\ell}\frac{k_\ell^{(1)}}{1!} + \frac{2^2\cos^2(\pi/2-\ell/2)\cos((2\pi)/2+(2\ell)/2)}{\ell^2}\frac{k_\ell^{(2)}}{2!}$
$0.96 = 1 - 3.98 + 3.94 \Rightarrow 0.96 = 0.96.$

Corollary 16 *The inverse fractional difference of the sine and cosine functions are obtained by*

$$\overset{-\nu}{\underset{\ell}{\Delta}} \sin k = \frac{1}{2^\nu \sin^\nu(\ell/2)} \sin(k - \nu\ell/2 - (\nu\pi)/2). \tag{17}$$

RHS of (17) is a closed solution of the fractional order ℓ-difference equation $\overset{\nu}{\underset{\ell}{\Delta}} u(k) = \sin k$.

$$\overset{-\nu}{\underset{\ell}{\Delta}} \cos k = \frac{1}{2^\nu \cos^\nu(\pi/2-\ell/2)} \cos(k - \nu\ell/2 - (\nu\pi)/2). \tag{18}$$

Similarly, RHS of (18) is a exact solution of fractional order ℓ-difference equation $\overset{\nu}{\underset{\ell}{\Delta}} u(k) = \cos k$.

Proof The proof of the corollary follows from (13) and (14).

The following theorem shows that the inverse difference of product of polynomial factorial and trigonometric function takes GMLF form.

Theorem 17 *Let the cosine real function and the polynomial factorial real function* $k_\ell^{(m)}$ *be any two functions. Then we have*

$$\Delta_\ell^{-1}[k_\ell^{(m)} \cos nk] = \sum_{t=0}^{m} m_1^{(t)} \ell^{(t)} k_\ell^{(m-t)} \times \frac{\cos(n(k+t\ell) - (t+1)\pi/2 - (t+1)n\ell/2)}{(-1)^t 2^{t+1} \cos^{t+1}(\pi/2 - n\ell/2)}, \tag{19}$$

which is a exact solution of the ℓ-difference equation $\underset{\ell}{\Delta} u(k) = k_\ell^{(m)} \cos nk$.

Proof Taking $n = 1$, and $e_\nu(\lambda, k_\ell, c) = \cos nk$ in (6) we get,
$\Delta_\ell^{-1}[k_\ell^{(1)} \cos nk] = k_\ell^{(1)} \Delta_\ell^{-1} \cos nk - \Delta_\ell^{-1}[\Delta_\ell^{-1} \cos n(k+\ell)\Delta_\ell k_\ell^{(1)}]$.
From equation (18), we arrive

$$\Delta_\ell^{-1}[k_\ell^{(1)} \cos nk] = k_\ell^{(1)} \frac{\cos(nk - n\ell/2 - \pi/2}{2\cos(\pi/2 - n\ell/2)} - \Delta_\ell^{-1}\Big[\frac{\cos(n(k+\ell) - n\ell/2 - \pi/2}{2\cos(\pi/2 - n\ell/2)}.\ell\Big]$$

$$\Delta_\ell^{-1}[k_\ell^{(1)} \cos nk] = \frac{k_\ell^{(1)}\Big[\cos(nk - n\ell/2 - \pi/2\Big]}{2\cos(\pi/2 - n\ell/2)} - \frac{\ell\Big[\cos(n(k+\ell) - 2n\ell/2 - 2\pi/2\Big]}{2^2\cos^2(\pi/2 - n\ell/2)}.$$

By product formula as given in (6) and iteration method we get (17).

Theorem 18 *Let the sine function and the polynomial factorial function $k_\ell^{(m)}$ be two real valued functions. Then we have*

$$\Delta_\ell^{-1}[k_\ell^{(m)} \sin nk] = \sum_{t=0}^{m} m_1^{(t)} \ell^{(t)} k_\ell^{(m-t)} \times \frac{\sin(n(k+t\ell) - (t+1)\pi/2 - (t+1)n\ell/2)}{(-1)^t 2^{t+1} \sin^{t+1}(\pi/2 - n\ell/2)}, \tag{20}$$

which is an exact solution of the ℓ-difference equation $\underset{\ell}{\Delta} u(k) = k_\ell^{(m)} \sin nk$.

Proof The proof follows by proceeding as the Theorem (17).

5 Fractional Difference of Trigonometric Functions

In this section, we derive the difference and inverse difference equations of the sine and cosine functions. By using $(x^n + \frac{1}{x^n}) = 2\cos n\theta$ and the binomial function $(x + \frac{1}{x})^n = x^n + n_{c_1} x^{n-1}\frac{1}{x} + n_{c_2} x^{n-2}\frac{1}{x^2} + \cdots$, we arrive the following theorems.

Theorem 19 *The GMLF function of $\sin^2 k$ series is given by*

$$\sin^2(k) = \sin^2(0) + \frac{\Delta_\ell \sin^2(0)}{\ell 1!} k_\ell^{(1)} + \frac{\Delta_\ell^2 \sin^2(0)}{\ell^2 2!} k_\ell^{(2)} + \cdots. \tag{21}$$

Proof From the definition of the GMLF function,

$$e_\nu(\bar\lambda, k_\ell, \bar c) = \sum_{j=0}^{\infty} \frac{\bar\lambda(j)}{\Gamma(j\nu + 1)} (k + \bar c(j))^{(j\nu)}.$$

Now taking the function $\bar\lambda(j) = \dfrac{\Delta_\ell^j \sin^2 0}{\ell^j}$, $\bar c(j) = 0$ and $\nu = 1$, gives (21).

Theorem 20 *The GMLF function of $\cos^2$ series is given by*

$$\cos^2(k) = \cos^2(0) + \frac{\Delta_\ell \cos^2(0)}{\ell 1!} k_\ell^{(1)} + \frac{\Delta_\ell^2 \cos^2(0)}{\ell^2 2!} k_\ell^{(2)} + \cdots. \tag{22}$$

Proof From the definition of the GMLF function,

$$e_\nu(\bar{\lambda}, k_\ell, \bar{c}) = \sum_{j=0}^{\infty} \frac{\bar{\lambda}(j)}{\Gamma(j\nu+1)}(k+\bar{c}j)^{(j\nu)}.$$

Now taking the function $\bar{\lambda}(j) = \dfrac{\Delta_\ell^j \cos^2 0}{\ell^j}$, $\bar{c}(j) = 0$ and $\nu = 1$, gives (22).

Theorem 21 *The fractional difference and the inverse fractional difference of the cosine function are*

$$\underset{\ell}{\overset{\nu}{\Delta}} \cos^2 k = \frac{1}{2}\Big(2^\nu \cos^\nu(\pi/2-\ell)\cos(\nu\pi/2+2k+\nu\ell)\Big). \tag{23}$$

Also (23) is an exact solution of fractional ℓ-difference equation $\underset{\ell}{\overset{\nu}{\Delta}} u(k) = \cos^2 k$ and

$$\underset{\ell}{\overset{-\nu}{\Delta}} \cos^2 k = \frac{1}{2}\Big[\frac{\cos(\nu\pi/2-2k+\nu\ell)}{2^\nu\cos^\nu(\pi/2-\ell)}\Big]. \tag{24}$$

Proof We know that $(x+\frac{1}{x})^2 = x^2 + 2_{c_1}x^1.\frac{1}{x} + 2_{c_2}x^0.\frac{1}{x^2} = (x^2+\frac{1}{x^2})+2$

$$(x+\tfrac{1}{x})^2 = 2\cos 2\theta + 2$$

$$(2\cos\theta)^2 = 2\cos 2\theta + 2 = \frac{\cos 2\theta + 1}{2} = \frac{1}{2^1}\big(\cos 2\theta + 1\cos 0\theta\big)$$

$$\underset{\ell}{\Delta}(\cos^2 k) = \Delta_\ell\big(\frac{1}{2^1}\cos 2\theta + \frac{1}{2^1}1\cos 0\theta\big)$$

$$\underset{\ell}{\Delta}(\cos^2 k) = \frac{1}{2}\big[2\cos(\pi/2-\ell)\cos(\pi/2+2k+\ell)\big].$$

Similarly we can proceed up to $\underset{\ell}{\overset{\nu}{\Delta}}(\cos^2 k)$ and we get the result. Replacing ν by $-\nu$, we get (24).

6 Conclusion

From our findings it is possible to express functions as GMLF function. The GMLF function contains polynomial factorial from which one can find higher order difference and its inverse which will be used to obtain solutions of several types of fractional difference equations. When $\ell \to 0$, the difference equation goes to differential equation and this model is useful for solving differential equations also.

References

1. Haubold, H.J., Mathai, A.M., Saxena, R.K.: Mittag-leffler functions and their applications. J. Appl. Math. https://doi.org/10.1155/2011/298628
2. Lang, K.R.: Astrophysical Formulae in Gas Processes and High-Energy Astrophysics, vol. 1. Springer, New York (1999)
3. Lang, K.R.: Astrophysical Formulae in Space, Time, Matter and Cosmology, vol. 2. Springer, New York (1999)
4. Hilfer, R.: Fractional diffusion based on Riemann-Liouville fractional derivatives. J. Phys. Chem. **104**(3), 914–924 (2000)
5. Hilfer, R.: Applications of Fractional Calculus in Physics. World Scientific, Singapore (2000)
6. Saxena, R.K.: Certain properties of generalized Mittag-Leffler function. In: Proceedings of the 3rd Annual Conference of the Society for Special Functions and Their Applications. Chennai, India, pp. 77–81 (2002)
7. Xavier, G.B.A., Jaraldpushparaj, S.: Extended Mittag-Leffler functions, series and its sum. J. Phys.: Conf. Ser. **1597**, 012005. https://doi.org/10.1088/1742-6596/1597/1/012005
8. Xavier, G.B.A., John Borg, S., Jaraldpushparaj, S.: Extorial solutions for fractional and partial difference equations with applications. In: AIP Conference Proceedings, **2095**, 030004 (2019). https://doi.org/10.1063/1.5097515

Performance Analysis of an M/M/1 Queue with Single Working Vacation with Customer Impatience Subject to Catastrophe

B. Thilaka, B. Poorani, and S. Udayabaskaran

Abstract A single server queue with Poisson arrival and exponential service times subject to a policy of single working vacation with customer impatience is considered. The service times are different for an active phase and a working vacation phase. The customer is allowed to leave the system during the working vacation phase. Catastrophes, when they occur, wipe out the system which results in the system being inactive for a random period of time. Explicit expressions for the transient probabilities of the close-down period, maintenance state, active state, working vacation state and system size for active phase and working vacation phase have been obtained. The corresponding steady-state analysis and performance measures are also obtained. The effects of various parameters on the system performance measures are studied using numerical examples.

Keywords Catastrophe · Close-down state · Maintenance state · Transient and steady-state probabilities · Working vacation · Customer impatience/Packet loss

Mathematics Subject Classification (2010) MSC 68M20

B. Thilaka
Department of Applied Mathematics, Sri Venkateswra College of Engineering, Kancheepuram 602117, India
e-mail: bthilaka@svce.ac.in

B. Poorani (✉)
Department of Mathematics, KCG college of Technology, Chennai 600097, India
e-mail: poorani.b1980@gmail.com

S. Udayabaskaran
Department of Mathematics, Vel Tech Rangarajan Dr. Sagunthala R&D Institute of Science and Technology, Chennai 600062, India
e-mail: udayabaskaran.s@gmail.com

R. N. Mohapatra et al. (eds.), *Mathematical Analysis and Computing*, Springer Proceedings in Mathematics & Statistics 344,
https://doi.org/10.1007/978-981-33-4646-8_24

1 Introduction

Queueing systems have been analysed on account of their wide range of applications in computer systems, telecommunication systems, wireless networks, etc. An excellent overview of the fundamental methods and results on queuing theory is given in the monographs of [5, 8, 11, 14–16, 23]. Wireless Sensor Network has many successful applications like automated factories, remote telemedicine, military area protection and so on. Energy consumption is a very significant issue for battery-powered mobile stations. The IEEE 802.16 e standard outlines sleep mode and idle mode operations on the MAC layer to save the energy of the MSs. The performance of the sleep mode and idle mode operations have been analysed by various researchers. A few of them are [4, 13, 17–21, 24–27].

Catastrophic events are an important aspect of the dynamics of computer networks, which are known as phase transitions. Dabrowski [10] gives a survey of the research of phase transition in communication networks and discusses the characteristics. The lifetime of a sensor can be increased by changing the sensor from the active mode to the working vacation mode as it has a lower service rate than the active mode. Boutoumi and Gharbi [6] proposed a policy for improving energy consumption and a latency efficiency technique in wireless sensor nodes based on a combination of normal vacation and working vacation policies.

Packet loss occurs when one or more packets of data travelling across a computer network fail to reach their destination. It is caused by errors in data transmission, typically across wireless networks. It is measured as a percentage of packets lost with respect to packets sent. Wireless networks are susceptible to a number of factors that can corrupt or lose packets in transit. In WSN, packet delivery performance is of high importance in energy-constrained networks as it translates into a network lifetime indicator. Packet losses are highly correlated over short time periods, but are independent over longer periods. Packet delivery performance in WSN has been studied in various papers [9, 22, 28].

Customer impatience (though voluntary) may be associated with packet loss in WSN. Even though packet loss occurs during transmission, the fact that they are not transmitted is considered as being lost from the system. Now, customers reneging from a queueing system leave before being served. In effect, this means that though the customers have been in the system for a period of time, their purpose of being served is not achieved. Loss of data also effectively means that the packets have been lost from the system. Hence, we look upon customer impatience as packet loss and analyse the system subject to packet loss.

Queueing systems with customer impatience have been studied very extensively. The concept of customer's impatience is that a customer reneges the queue after having waited for a sufficient time in the queue. The impact of customer's impatience has been analysed in [2, 3, 7, 12] for various queueing models.

This has motivated us to analyse the performances of an M/M/1 queueing system with the server under maintenance, sleep mode, single working vacation with customer impatience subject to catastrophe. This paper is structured as follows: Sect.

2 contains the mathematical model description and transient state probabilities of the system. Steady-state analysis is described in Sect. 3. In Sect. 4, some key performance measures under steady-state conditions are obtained. Numerical examples are explained to illustrate the effects of system parameters on the performance measures in Sect. 5. Finally, Sect. 6 concludes the paper.

2 Model Description and Analysis

We consider an M/M/1 queueing system subject to a policy of single working vacation with customer impatience. The server is in busy phase (normal busy mode), in slow phase (working vacation mode), in sleep mode and in maintenance mode. The server is said to be in state 1 if it is in busy mode, in state 2 if it is in working vacation mode, state 0 if it is in sleep mode and state M if it is in maintenance state. Customers arrive according to a Poisson process only to the busy phase with arrival rate λ. During the normal busy mode and working vacation mode, the service rates are, respectively, μ_1 and μ_2 such that $\mu_2 < \mu_1$. The server can switch from busy mode to working vacation mode with a rate of η but not vice verse.

When there is no customer in the system during the busy period and in the working vacation period, the server moves to sleep mode in order to save power. Once the customer arrives, the system switches automatically to the busy phase. Customer impatience is assumed to occur only in the working vacation phase with a rate ξ. Catastrophes are assumed to arrive as a Poisson process with the rate γ. Once a catastrophe occurs, all the customers are wiped out from the entire system and it enters the maintenance state. During the maintenance period, no customer is allowed to enter the system. At the end of the maintenance period, the system moves to sleep mode with the rate α.

Let $X(t)$ denote the number of customers in the system at time t when the server is in a busy state and $J(t)$ denote the state of the server at time t. The joint process $\{X(t), J(t), t \geq 0\}$ is Markov.

The state space of the system is given by
$\Omega = \{(0,0), (0, M)\} \bigcup \{(1,1), (2,1), (3,1), ...\} \bigcup \{(1,2), (2,2), (3,2), ...\}$

From the state transition diagram, by using probability laws (Fig. 1)

$$p(i, j, t) = P[X(t) = i, Y(t) = j; t/i = 0, 1, 2, ...\infty, j = 0, 1, 2]$$

$$p(0, 0, t) = e^{-\lambda t} + [p(1, 1, t)\mu_1 + p(1, 2, t)\mu_2 + p(M, t)\alpha]©e^{-\lambda t} \quad (1)$$

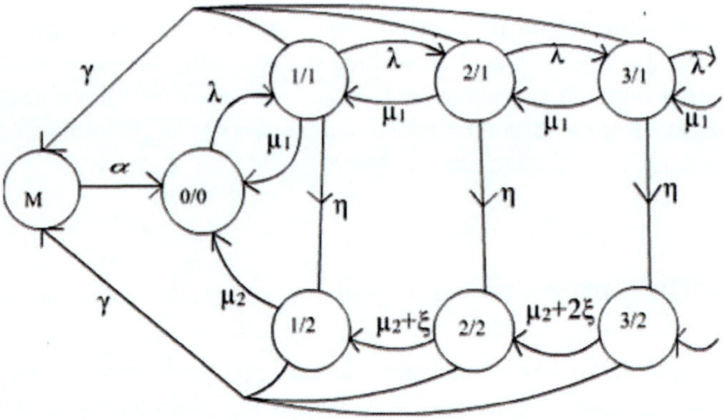

Fig. 1 Transition diagram

$$p\,(M,t) = \sum_{i=1}^{\infty}[p\,(i,1,t)\,\gamma + p\,(i,2,t)\,\gamma]©e^{-\alpha t} \tag{2}$$

$$p\,(1,1,t) = [p\,(0,0,t)\,\lambda + p\,(2,1,t)\,\mu_1]©e^{-(\lambda+\mu_1+\eta+\gamma)t} \tag{3}$$

$$p\,(i,1,t) = [p\,(i-1,0,t)\,\lambda + p\,(i+1,1,t)\,\mu_1]©e^{-(\lambda+\mu_1+\eta+\gamma)t}, i \geq 2 \tag{4}$$

$$p\,(i,2,t) = [p\,(i,1,t)\,\eta + p\,(i+1,2,t)\,(\mu_2 + i\xi)]©e^{-(\mu_2+(i-1)\xi+\gamma)t}, i \geq 1 \tag{5}$$

Taking the Laplace transform of (1)–(5) with respect to time and denoting the transform variable by s

$$(s+\lambda)\,p^*\,(0,0,s) = 1 + p^*\,(1,1,s)\,\mu_1 + p^*\,(1,2,s)\,\mu_2 + p^*\,(M,s)\,\alpha \tag{6}$$

$$(s+\alpha)\,p^*\,(M,s) = \sum_{i=1}^{\infty}[p^*\,(i,1,s)\,\gamma + p^*\,(i,2,s)\,\gamma] \tag{7}$$

$$(s+\lambda+\mu_1+\eta+\gamma)\,p^*\,(1,1,s) = p^*\,(0,0,s)\,\lambda + p^*\,(2,1,s)\,\mu_1 \tag{8}$$

$$(s+\lambda+\mu_1+\eta+\gamma)\,p^*(i,1,s)=p^*(i-1,0,s)\,\lambda+p^*(i+1,1,s)\,\mu_1,\quad i\geq 2 \tag{9}$$

$$(s+\mu_2+(i-1)\xi+\gamma)\,p^*(i,2,s)=p^*(i,1,s)\,\eta+p^*(i+1,2,s)\,(\mu_2+i\xi),\quad i\geq 1 \tag{10}$$

Defining the Probability Generating Functions as

$$G_1^*(u,s)=\sum_{n=1}^{\infty}p^*(i,1,s)\,u^i \tag{11}$$

$$G_2^*(u,s)=\sum_{n=1}^{\infty}p^*(i,2,s)\,u^i \tag{12}$$

$$G^*(u,s)=p^*(0,0,s)+p^*(M,s)+G_1^*(u,s)+G_2^*(u,s) \tag{13}$$

The system of equations (9), after some manipulation, yields

$$(s+\lambda+\mu_1+\gamma+\eta)\left[G_1^*(u,s)-p^*(1,1,s)\right]=\lambda u G_1^*(u,s)+\frac{\mu_1}{u}\left[G_1^*(u,s)-p^*(1,1,s)\,u-p^*(1,1,s)\,u^2\right]$$

$$G_1^*(u,s)=\frac{\lambda u^2 p^*(0,0,s)-u\mu_1 p^*(1,1,s)}{-\lambda u^2+u(s+\lambda+\mu_1+\gamma+\eta)-\mu_1} \tag{14}$$

The system of equations (10), after some simplification, yields

$$(s+\mu_2+\gamma-\xi)\,G_2^*(u,s)+\xi u\frac{\partial G_2^*(u,s)}{\partial u}=\eta G_1^*(u,s)+\frac{\mu_2}{u}G_2^*(u,s)-\mu_2 p^*(1,2,s)+\xi\left[\frac{\partial G_2^*(u,s)}{\partial u}-p^*(1,2,s)\right]-\frac{\xi}{u}\left[G_2^*(u,s)-p^*(1,2,s)\,u\right]$$

$$\frac{\partial G_2^*(u,s)}{\partial u}(\xi u-\xi)+G_2^*(u,s)\left[s+\mu_2+\gamma-\xi-\frac{\mu_2}{u}+\frac{\xi}{u}\right]=\eta G_1^*(u,s)-\mu_2 p^*(1,2,s) \tag{15}$$

$$\frac{\partial G_2^*(u,s)}{\partial u} - G_2^*(u,s)\left(\frac{s+\gamma}{\xi(1-u)} - \left(\frac{\mu_2-\xi}{\xi u}\right)\right)$$
$$= \frac{\mu_2}{\xi(1-u)}p^*(1,2,s) - \frac{\eta}{\xi(1-u)}G_1^*(u,s) \tag{16}$$

At u = 1, (14) and (15) reduce to

$$G_1^*(1,s) = \frac{p^*(0,0,s)\lambda - \mu_1 p^*(1,1,s)}{s+\gamma+\eta} \tag{17}$$

$$G_2^*(1,s) = \frac{\eta G_1^*(1,s) - \mu_2 p^*(1,2,s)}{s+\gamma} \tag{18}$$

Adding (17) and (18), we have

$$(s+\gamma)\left[G_1^*(1,s) + G_2^*(1,s)\right] = p^*(0,0,s)\lambda_1 - \mu_1 p^*(1,1,s)$$
$$-\mu_2 p^*(1,2,s) \tag{19}$$

Using (6) and (7), the total probability law is confirmed.

$$Thatis, G_1^*(1,s) + G_2^*(1,s) + p^*(M,s) + p^*(0,0,s) = \frac{1}{s} \tag{20}$$

The zeros of $-\lambda u^2 + u(s+\lambda+\mu_1+\gamma+\eta) - \mu_1 = 0$ are given by

$$u_1 = \frac{(s+\lambda+\mu_1+\gamma+\eta) - \sqrt{(s+\lambda+\mu_1+\gamma+\eta)^2 - 4\lambda\mu_1}}{2\lambda}$$
$$u_2 = \frac{(s+\lambda+\mu_1+\gamma+\eta) + \sqrt{(s+\lambda+\mu_1+\gamma+\eta)^2 - 4\lambda\mu_1}}{2\lambda}$$

These roots satisfy the conditions

$$u_1 + u_2 = \frac{s+\lambda+\mu_1+\gamma+\eta}{\lambda} \tag{21}$$

$$u_1 u_2 = \frac{\mu_1}{\lambda} \tag{22}$$

$$|u_1| < 1, |u_2| > 1 \tag{23}$$

$$-\lambda u^2 + u\left(s+\lambda+\mu_1+\gamma+\eta\right) - \mu_1 = \frac{(u-u_1)\left(\mu_1 - uu_1\lambda\right)}{u_1} \tag{24}$$

where $0 < u_1 < 1 < u_2$.

Invoking the analyticity of $G_1^*(u,s)$, we have

$$p^*(1,1,s) = \frac{\lambda u_1}{\mu_1} p^*(0,0,s) \tag{25}$$

Substituting (25) in (14), we obtain

$$G_1^*(u,s) = \frac{\lambda u u_1}{\mu_1 - uu_1\lambda} p^*(0,0,s)$$

$$G_1^*(u,s) = \sum_{n=1}^{\infty}\left(\frac{\lambda u u_1}{\mu_1}\right)^i p^*(0,0,s) \tag{26}$$

Equating the coefficients in (26), we get

$$p^*(i,1,s) = \frac{\lambda^i u_1^i}{\mu_1^i} p^*(0,0,s),\ i \geq 1 \tag{27}$$

Using (26) in (16) yields

$$\frac{\partial G_2^*(u,s)}{\partial u} - G_2^*(u,s)\left(\frac{s+\gamma}{\xi(1-u)} - \left(\frac{\mu_2-\xi}{\xi u}\right)\right)$$
$$= \frac{\mu_2}{\xi(1-u)} p^*(1,2,s) - \frac{\eta}{\xi(1-u)} \sum_{n=1}^{\infty}\left(\frac{\lambda u u_1}{\mu_1}\right)^i p^*(0,0,s) \tag{28}$$

The integrating factor (IF) of (28) is

$$IF = (1-u)^{\frac{(s+\gamma)}{\xi}} u^{\frac{(\mu_2-\xi)}{\xi}}$$

Multiplying both sides of (28) by IF yields

$$\frac{\partial}{\partial u}\left[G_2^*(u,s)(1-u)^{\frac{(s+\gamma)}{\xi}} u^{\frac{(\mu_2-\xi)}{\xi}}\right] = \frac{\mu_2}{\xi}(1-u)^{\frac{(s+\gamma)}{\xi}-1} u^{\frac{\mu_2}{\xi}-1} p^*(1,2,s)$$
$$-\frac{\eta}{\xi}\sum_{i=1}^{\infty}\left(\frac{\lambda u_1}{\mu_1}\right)^i u^{\frac{\mu_2}{\xi}+i-1}(1-u)^{\frac{(s+\gamma)}{\xi}-1} p^*(0,0,s) \tag{29}$$

On integration, we have

$$G_2^*(u,s)(1-u)^{\frac{(s+\gamma)}{\xi}}u^{\frac{(\mu_2-\xi)}{\xi}} = \frac{\mu_2}{\xi}\int_0^u (1-v)^{\frac{(s+\gamma)}{\xi}-1}v^{\frac{\mu_2}{\xi}-1}p^*(1,2,s)\,dv$$
$$-\frac{\eta}{\xi}\sum_{i=1}^{\infty}\left(\frac{\lambda u_1}{\mu_1}\right)^i\int_0^u v^{\frac{\mu_2}{\xi}+i-1}(1-v)^{\frac{(s+\gamma)}{\xi}-1}p^*(0,0,s)\,dv$$

$$G_2^*(u,s) = \left[\frac{\mu_2}{\xi}\int_0^u (1-v)^{\frac{(s+\gamma)}{\xi}-1}v^{\frac{\mu_2}{\xi}-1}p^*(1,2,s)\,dv\right.$$
$$\left.-\frac{\eta}{\xi}\sum_{i=1}^{\infty}\left(\frac{\lambda u_1}{\mu_1}\right)^i\int_0^u v^{\frac{\mu_2}{\xi}+i-1}(1-v)^{\frac{(s+\gamma)}{\xi}-1}p^*(0,0,s)\,dv\right](1-u)^{\frac{-(s+\gamma)}{\xi}}u^{\frac{(\xi-\mu_2)}{\xi}} \tag{30}$$

In the limiting case as $u \to 1$, we get

$$G_2^*(1,s) = \left[\frac{\mu_2}{\xi}\int_0^1 (1-v)^{\frac{(s+\gamma)}{\xi}-1}v^{\frac{\mu_2}{\xi}-1}p^*(1,2,s)\,dv\right.$$
$$\left.-\frac{\eta}{\xi}\sum_{i=1}^{\infty}\left(\frac{\lambda u_1}{\mu_1}\right)^i\int_0^1 v^{\frac{\mu_2}{\xi}+i-1}(1-v)^{\frac{(s+\gamma)}{\xi}-1}p^*(0,0,s)\,dv\right]\lim_{u\to 1}(1-u)^{\frac{-(s+\gamma)}{\xi}} \tag{31}$$

Since $G_2^*(1,s) = \sum_{n=1}^{\infty} p^*(n,2,s)$ is well defined and

$$\lim_{u\to 1}(1-u)^{\frac{-(s+\gamma)}{\xi}} = \infty$$

(31) reduces to

$$\frac{\mu_2}{\xi}\int_0^1 (1-v)^{\frac{(s+\gamma)}{\xi}-1}v^{\frac{\mu_2}{\xi}-1}p^*(1,2,s)\,dv - \frac{\eta}{\xi}\sum_{i=1}^{\infty}\kappa^i$$
$$\int_0^1 v^{\frac{\mu_2}{\xi}+i-1}(1-v)^{\frac{(s+\gamma)}{\xi}-1}p^*(0,0,s)\,dv = 0$$

By using Beta function, this can be written as

$$\frac{\mu_2}{\xi}p^*(1,2,s)\,\beta\left(\frac{\mu_2}{\xi},\frac{s+\gamma}{\xi}\right) = \frac{\eta}{\xi}\sum_{i=1}^{\infty}\left(\frac{\lambda u_1}{\mu_1}\right)^i\beta\left(\frac{\mu_2}{\xi}+i,\frac{s+\gamma}{\xi}\right)p^*(0,0,s)$$

$$p^*(1,2,s) = \frac{\frac{\eta}{\xi}\sum_{i=1}^{\infty}\left(\frac{\lambda u_1}{\mu_1}\right)^i \beta\left(\frac{\mu_2}{\xi}+i, \frac{s+\gamma}{\xi}\right)}{\frac{\mu_2}{\xi}\beta\left(\frac{\mu_2}{\xi}, \frac{s+\gamma}{\xi}\right)} p^*(0,0,s)$$

where

$$\beta(a,b) = \int_0^1 u^{a-1}(1-u)^{b-1}du$$

On simplification, the above equation becomes

$$p^*(1,2,s) = \frac{\eta}{\mu_2}\sum_{n=1}^{\infty}\left(\frac{\lambda u_1}{\mu_1}\right)^n \prod_{j=1}^{n}\frac{\mu_2+j\xi}{(s+\gamma+\mu_2+j\xi)} p^*(0,0,s) \tag{32}$$

Substituting $i = 1$ in Eq. (10) gives

$$p^*(2,2,s) = \frac{1}{\mu_2+\xi}\left[(s+\mu_2+\gamma)\, p^*(1,2,s) - \eta p^*(1,1,s)\right]$$

Substituting Eqs. (32) and (25) in the above equation, we obtain

$$p^*(2,2,s) = \frac{\eta}{\mu_2+\xi}\left[\frac{1}{\mu_2}\sum_{n=1}^{\infty}\left(\frac{\lambda u_1}{\mu_1}\right)^n \prod_{j=1}^{n}\frac{(\mu_2+j\xi)(s+\gamma+\mu_2)}{(s+\gamma+\mu_2+j\xi)} - \frac{\lambda u_1}{\mu_1}\right] p^*(0,0,s) \tag{33}$$

Substituting $i = 2$ in Eq. (10) yields

$$p^*(3,2,s) = \frac{1}{\mu_2+2\xi}\left[(s+\mu_2+\gamma+\xi)\, p^*(2,2,s) - \eta p^*(2,1,s)\right]$$

Substituting Eqs. (33) and (27) in the above equation gives

$$p^*(3,2,s) = \frac{\eta}{\mu_2+2\xi}\left[\sum_{n=1}^{\infty}\left(\frac{\lambda u_1}{\mu_1}\right)^n \prod_{j=1}^{n}\frac{(\mu_2+j\xi)(s+\gamma+\mu_2)(s+\gamma+\mu_2+\xi)}{(s+\gamma+\mu_2+j\xi)(\mu_2+\xi)} \frac{1}{\mu_2} - \frac{\lambda u_1}{\mu_1}\frac{(s+\gamma+\mu_2+\xi)}{\mu_2+\xi} - \left(\frac{\lambda u_1}{\mu_1}\right)^2\right] p^*(0,0,s)$$

In general

$$p^{*}(i,2,s)=\frac{\eta}{\mu_2+(i-1)\xi}\left[\sum_{n=1}^{\infty}\left(\frac{\lambda u_1}{\mu_1}\right)^{n}\prod_{j=1}^{n}\frac{\mu_2+j\xi}{(s+\gamma+\mu_2+j\xi)}\right.$$

$$\prod_{l=0}^{i-2}\frac{(s+\gamma+\mu_2+l\xi)}{(\mu_2+l\xi)}-\sum_{d=1}^{i-2}\left(\frac{\lambda u_1}{\mu_1}\right)^{d}\prod_{r=d}^{i-2}\frac{(s+\gamma+\mu_2+r\xi)}{(\mu_2+r\xi)}$$

$$\left.-\left(\frac{\lambda u_1}{\mu_1}\right)^{i-1}\right]p^{*}(0,0,s)\,,\,i\geq 3 \quad (34)$$

Substitute Eqs. (27) and (34) in (7) to get

$$p^{*}(M,s)=\frac{\gamma}{s+\alpha}\left[\sum_{i=1}^{\infty}\left(\frac{\lambda u_1}{\mu_1}\right)^{i}+\frac{\eta}{\mu_2}\sum_{n=1}^{\infty}\left(\frac{\lambda u_1}{\mu_1}\right)^{n}\prod_{j=1}^{n}\frac{\mu_2+j\xi}{(s+\gamma+\mu_2+j\xi)}\right.$$

$$+\frac{\eta}{\mu_2+\xi}\left[\frac{1}{\mu_2}\sum_{n=1}^{\infty}\left(\frac{\lambda u_1}{\mu_1}\right)^{n}\prod_{j=1}^{n}\frac{(\mu_2+j\xi)(s+\gamma+\mu_2)}{(s+\gamma+\mu_2+j\xi)}-\frac{\lambda u_1}{\mu_1}\right]$$

$$+\sum_{i=3}^{\infty}\left\{\frac{\eta}{\mu_2+(i-1)\xi}\left(\sum_{n=1}^{\infty}\left(\frac{\lambda u_1}{\mu_1}\right)^{n}\prod_{j=1}^{n}\frac{\mu_2+j\xi}{(s+\gamma+\mu_2+j\xi)}\right.\right.$$

$$\prod_{l=0}^{i-2}\frac{(s+\gamma+\mu_2+l\xi)}{(\mu_2+l\xi)}-\sum_{d=1}^{i-2}\left(\frac{\lambda u_1}{\mu_1}\right)^{d}\prod_{r=d}^{i-2}\frac{(s+\gamma+\mu_2+r\xi)}{(\mu_2+r\xi)}$$

$$\left.\left.\left.-\left(\frac{\lambda u_1}{\mu_1}\right)^{i-1}\right)\right\}\right]p^{*}(0,0,s) \quad (35)$$

Using (27), (34) and (35), Eq. (6) becomes

$$p^{*}(0,0,s)\left[1-H^{*}(s)\right]=\frac{1}{s+\lambda}$$

$$=\frac{1}{(s+\lambda)(1-H^{*}(s))}=\frac{1}{(s+\lambda)}\sum_{l=0}^{\infty}\left(H^{*}(s)\right)^{l} \quad (36)$$

where

$$H^*(s) = \frac{\lambda u_1}{s+\lambda} + \frac{\eta}{s+\lambda}\sum_{n=1}^{\infty}\left(\frac{\lambda u_1}{\mu_1}\right)^n \prod_{j=1}^{n}\frac{\mu_2 + j\xi}{(s+\gamma+\mu_2+j\xi)} + \frac{\alpha}{s+\lambda}\frac{\gamma}{s+\alpha}$$
$$\left[\sum_{i=1}^{\infty}\left(\frac{\lambda u_1}{\mu_1}\right)^i + \frac{\eta}{\mu_2}\sum_{n=1}^{\infty}\left(\frac{\lambda u_1}{\mu_1}\right)^n \prod_{j=1}^{n}\frac{\mu_2 + j\xi}{(s+\gamma+\mu_2+j\xi)}\right.$$
$$+\frac{\eta}{\mu_2+\xi}\left[\frac{1}{\mu_2}\sum_{n=1}^{\infty}\left(\frac{\lambda u_1}{\mu_1}\right)^n \prod_{j=1}^{n}\frac{(\mu_2 + j\xi)(s+\gamma+\mu_2)}{(s+\gamma+\mu_2+j\xi)} - \frac{\lambda u_1}{\mu_1}\right]$$
$$+\sum_{i=3}^{\infty}\left\{\frac{\eta}{\mu_2+(i-1)\xi}\left(\sum_{n=1}^{\infty}\left(\frac{\lambda u_1}{\mu_1}\right)^n \prod_{j=1}^{n}\frac{\mu_2 + j\xi}{(s+\gamma+\mu_2+j\xi)}\right.\right.$$
$$\prod_{l=0}^{i-2}\frac{(s+\gamma+\mu_2+l\xi)}{(\mu_2+l\xi)} - \sum_{d=1}^{i-2}\left(\frac{\lambda u_1}{\mu_1}\right)^d \prod_{r=d}^{i-2}\frac{(s+\gamma+\mu_2+r\xi)}{(\mu_2+r\xi)} -$$
$$\left.\left.\left.\left(\frac{\lambda u_1}{\mu_1}\right)^{i-1}\right)\right\}\right] \tag{37}$$

Taking the inverse Laplace transform [1] yields

$$L^{-1}\left[\left(s-\sqrt{s^2-a^2}\right)^k\right] = \frac{ka^k}{t}I_k(at). \tag{38}$$

Hence

$$L^{-1}(u_1) = e^{-(\lambda+\mu_1+\gamma+\eta)t}\left(\frac{\mu_1}{\lambda}\right)^{1/2}\frac{I_1\left(2\sqrt{\lambda\mu_1}t\right)}{t}$$
$$L^{-1}\left(u_1^r\right) = e^{-(\lambda+\mu_1+\gamma+\eta)t}\left(\frac{\mu_1}{\lambda}\right)^{r/2}\frac{rI_r\left(2\sqrt{\lambda\mu_1}t\right)}{t}$$

By using the notation, we have

$$\Phi_{1;r}(t) = L^{-1}\left(u_1^r\right),$$

Inversion of (27) gives

$$p(i,1,t) = \left(\frac{\lambda}{\mu_1}\right)^i \Phi_{1;i}(t) \copyright p(0,0,t),\ i \geq 1 \tag{39}$$

Inversion of (32), (33) and (34) gives

$$p(1,2,t) = \frac{\eta}{\mu_2}\sum_{n=1}^{\infty}\left(\frac{\lambda}{\mu_1}\right)^n \Phi_{1;n}(t)\prod_{j=1}^{n}(\mu_2+j\xi)\copyright\Psi_1(t)\copyright p(0,0,t) \tag{40}$$

where

$$\Psi_1(t) = L^{-1}\left(\prod_{j=1}^{n}\frac{1}{s+\mu_2+\gamma+j\xi}\right)$$

$$= e^{-(\mu_2+\gamma+\xi)t} © e^{-(\mu_2+\gamma+2\xi)t} © e^{-(\mu_2+\gamma+3\xi)t} ©....© e^{-(\mu_2+\gamma+j\xi)t}$$

$$p(2,2,t) = \frac{\eta}{\mu_2+\xi}\left[\frac{1}{\mu_2}\sum_{n=1}^{\infty}\left(\frac{\lambda}{\mu_1}\right)^n \Phi_{1;n}(t) © \prod_{j=1}^{n}(\mu_2+j\xi)\right.$$
$$\left.\Psi_1(t)©\Psi_2(t) - \frac{\lambda}{\mu_1}\Phi_{1;1}(t)\right] © p(0,0,t) \quad (41)$$

$$p(i,2,t) = \frac{\eta}{\mu_2+(i-1)\xi}\left[\sum_{n=1}^{\infty}\left(\frac{\lambda}{\mu_1}\right)^n \Phi_{1;n-i+1}(t) © \prod_{j=1}^{n}(\mu_2+j\xi)\right.$$
$$\prod_{l=0}^{i-2}(\Psi_1(t)©\Psi_2(t)©...\Psi_i(t)) - \sum_{d=1}^{i-2}\prod_{r=d}^{i-2}\frac{1}{\mu_2+r\xi}\left(\frac{\lambda}{\mu_1}\right)^d \Phi_{1;d-r+1}(t)$$
$$\left.©\Psi_{i-d+r}(t) - \left(\frac{\lambda}{\mu_1}\right)^{i-1}\Phi_{1;i-1}(t)\right] © p(0,0,t), i \geq 3 \quad (42)$$

where

$$\Psi_i(t) = L^{-1}\left[u_1(s+\mu_2+\gamma+(i-2)\xi)\right] = e^{-(\lambda+\mu_1+\gamma+\eta)t}\left\{\left[\frac{d}{dt}\left(\frac{\mu_1}{\lambda}\right)^{1/2}\right.\right.$$
$$\left.\left.\frac{I_1(2\sqrt{\lambda\mu_1}t)}{t} + \delta(t)\right] + (\mu_2+(i-2)\xi-\lambda-\mu_1-\eta)\left(\frac{\mu_1}{\lambda}\right)^{1/2}\frac{I_1(2\sqrt{\lambda\mu_1}t)}{t}\right\}$$
$$, i \geq 2$$

Inversion of (35) gives

$$
\begin{aligned}
p(M,t) = \gamma e^{-\alpha t} © & \left[\sum_{i=1}^{\infty} \left(\frac{\lambda}{\mu_1}\right)^i \Phi_{1;i}(t) + \frac{\eta}{\mu_2} \sum_{n=1}^{\infty} \left(\frac{\lambda}{\mu_1}\right)^n \Phi_{1;n}(t) \prod_{j=1}^{n} (\mu_2 + j\xi) \right. \\
& © \Psi_1(t) + \frac{\eta}{\mu_2+\xi} \left[\frac{1}{\mu_2} \sum_{n=1}^{\infty} \left(\frac{\lambda}{\mu_1}\right)^n \Phi_{1;n}(t) © \prod_{j=1}^{n} (\mu_2 + j\xi) \Psi_1(t) \right. \\
& \left. © \Psi_2(t) - \frac{\lambda}{\mu_1} \Phi_{1;1}(t) \right] + \sum_{i=3}^{\infty} \left\{ \frac{\eta}{\mu_2 + (i-1)\xi} \left(\sum_{n=1}^{\infty} \left(\frac{\lambda}{\mu_1}\right)^n \Phi_{1;n-i+1}(t) \right. \right. \\
& © \prod_{j=1}^{n} (\mu_2 + j\xi) \prod_{l=0}^{i-2} (\Psi_1(t) © \Psi_2(t) © \ldots \Psi_i(t)) - \sum_{d=1}^{i-2} \prod_{r=d}^{i-2} \frac{1}{\mu_2 + r\xi} \left(\frac{\lambda}{\mu_1}\right)^d \\
& \left. \left. \left. \Phi_{1;d-r+1}(t) © \Psi_{i-d+r}(t) - \left(\frac{\lambda}{\mu_1}\right)^{i-1} \Phi_{1;i-1}(t) \right) \right\} \right] © p(0,0,t)
\end{aligned} \tag{43}
$$

Inversion of Eq. (36) yields

$$
p(0,0,t) = e^{-\lambda t} \sum_{l=0}^{\infty} \left(H^{©(n)} t\right)^l \tag{44}
$$

where

$$
\begin{aligned}
H(t) = & \lambda e^{-\lambda t} © \Phi_{1;1}(t) + \eta e^{-\lambda t} © \sum_{n=1}^{\infty} \left(\frac{\lambda}{\mu_1}\right)^n \Phi_{1;n}(t) \prod_{j=1}^{n} (\mu_2 + j\xi) © \Psi_1(t) + \\
& \alpha e^{-\lambda t} © \gamma e^{-\alpha t} © \left[\sum_{i=1}^{\infty} \left(\frac{\lambda}{\mu_1}\right)^i \Phi_{1;i}(t) + \frac{\eta}{\mu_2} \sum_{n=1}^{\infty} \left(\frac{\lambda}{\mu_1}\right)^n \Phi_{1;n}(t) \prod_{j=1}^{n} (\mu_2 + j\xi) \right. \\
& © \Psi_1(t) + \frac{\eta}{\mu_2+\xi} \left[\frac{1}{\mu_2} \sum_{n=1}^{\infty} \left(\frac{\lambda}{\mu_1}\right)^n \Phi_{1;n}(t) © \prod_{j=1}^{n} (\mu_2 + j\xi) \Psi_1(t) \right. \\
& \left. © \Psi_2(t) - \frac{\lambda}{\mu_1} \Phi_{1;1}(t) \right] + \sum_{i=3}^{\infty} \left\{ \frac{\eta}{\mu_2 + (i-1)\xi} \left(\sum_{n=1}^{\infty} \left(\frac{\lambda}{\mu_1}\right)^n \Phi_{1;n-i+1}(t) \right. \right. \\
& © \prod_{j=1}^{n} (\mu_2 + j\xi) \prod_{l=0}^{i-2} (\Psi_1(t) © \Psi_2(t) © \ldots \Psi_i(t)) - \sum_{d=1}^{i-2} \prod_{r=d}^{i-2} \frac{1}{\mu_2 + r\xi} \left(\frac{\lambda}{\mu_1}\right)^d \\
& \left. \left. \left. \Phi_{1;d-r+1}(t) © \Psi_{i-d+r}(t) - \left(\frac{\lambda}{\mu_1}\right)^{i-1} \Phi_{1;i-1}(t) \right) \right\} \right]
\end{aligned} \tag{45}
$$

3 Steady-State Distribution

The steady state probabilities are given by

$$\pi\,(i, j) = \lim_{t\to\infty} p\,(i, j, t)\,, i = 1, 2, ...; j = 0, M, 1, 2$$

From the final value theorem for Laplace transform, we get

$$\pi\,(i, j) = \lim_{s\to 0} sp^{*}\,(i, j, s)\,, i = 1, 2, ...; j = 0, M, 1, 2$$

From (27), we have

$$\pi\,(i, 1) = \left(\frac{\lambda z_1}{\mu_1}\right)^{i} \pi\,(0, 0)\,, i \geq 1 \tag{46}$$

where

$$z_1 = \lim_{s\to 0} u_1 = \frac{(\lambda + \mu_1 + \gamma + \eta) - \sqrt{(\lambda + \mu_1 + \gamma + \eta)^2 - 4\lambda\mu_1}}{2\lambda}$$

From (32), we obtain

$$\pi\,(1, 2) = \frac{\eta}{\mu_2} \sum_{n=1}^{\infty} \left(\frac{\lambda z_1}{\mu_1}\right)^{n} \prod_{j=1}^{n} \frac{\mu_2 + j\xi}{\gamma + \mu_2 + j\xi} \pi\,(0, 0) \tag{47}$$

From (33), we get

$$\pi\,(2, 2) = \frac{\eta}{\mu_2 + \xi} \left[\frac{1}{\mu_2} \sum_{n=1}^{\infty} \left(\frac{\lambda z_1}{\mu_1}\right)^{n} \prod_{j=1}^{n} \frac{(\mu_2 + j\xi)\,(\gamma + \mu_2)}{\gamma + \mu_2 + j\xi} - \frac{\lambda z_1}{\mu_1} \right] \pi\,(0, 0) \tag{48}$$

From (34), we obtain

$$\pi\,(i, 2) = \frac{\eta}{\mu_2 + (i-1)\xi} \left[\sum_{n=1}^{\infty} \left(\frac{\lambda z_1}{\mu_1}\right)^{n} \prod_{j=1}^{n} \frac{\mu_2 + j\xi}{\gamma + \mu_2 + j\xi} \prod_{l=0}^{i-2} \frac{\gamma + \mu_2 + l\xi}{\mu_2 + l\xi} - \sum_{d=1}^{i-2} \left(\frac{\lambda z_1}{\mu_1}\right)^{d} \prod_{r=d}^{i-2} \frac{\gamma + \mu_2 + r\xi}{\mu_2 + r\xi} - \left(\frac{\lambda z_1}{\mu_1}\right)^{i-1} \right] \pi\,(0, 0)\,, i \geq 3 \tag{49}$$

From (35), we have

$$\pi\,(M)=\frac{\gamma}{\alpha}\left[\sum_{i=1}^{\infty}\left(\frac{\lambda z_1}{\mu_1}\right)^i+\sum_{n=1}^{\infty}\left(\frac{\lambda z_1}{\mu_1}\right)^n\prod_{j=1}^{n}\frac{\mu_2+j\xi}{\gamma+\mu_2+j\xi}\left(\frac{\eta}{\mu_2}+\frac{\eta(\gamma+\mu_2)}{\mu_2(\mu_2+\xi)}\right)\right.$$
$$-\frac{\eta}{\mu_2+\xi}\left(\frac{\lambda z_1}{\mu_1}\right)+\sum_{i=3}^{\infty}\left\{\frac{\eta}{\mu_2+(i-1)\xi}\left(\sum_{n=1}^{\infty}\left(\frac{\lambda z_1}{\mu_1}\right)^n\prod_{j=1}^{n}\frac{\mu_2+j\xi}{\gamma+\mu_2+j\xi}\right.\right.$$
$$\left.\left.\left.\prod_{l=0}^{i-2}\frac{\gamma+\mu_2+l\xi}{\mu_2+l\xi}-\sum_{d=1}^{i-2}\left(\frac{\lambda z_1}{\mu_1}\right)^d\prod_{r=d}^{i-2}\frac{\gamma+\mu_2+r\xi}{\mu_2+r\xi}-\left(\frac{\lambda z_1}{\mu_1}\right)^{i-1}\right)\right\}\right]\pi\,(0,0) \tag{50}$$

By using the total probability, we get

$$\pi\,(M)+\pi\,(0,0)+\sum_{i=1}^{\infty}\pi\,(i,1)+\sum_{i=1}^{\infty}\pi\,(i,2)=1$$

$$\pi\,(0,0)=\left[1+\left(1+\frac{\gamma}{\alpha}\right)\left[\sum_{i=1}^{\infty}\left(\frac{\lambda z_1}{\mu_1}\right)^i+\sum_{n=1}^{\infty}\left(\frac{\lambda z_1}{\mu_1}\right)^n\prod_{j=1}^{n}\frac{\mu_2+j\xi}{\gamma+\mu_2+j\xi}\right.\right.$$
$$\left(\frac{\eta}{\mu_2}+\frac{\eta(\gamma+\mu_2)}{\mu_2(\mu_2+\xi)}\right)-\frac{\eta}{\mu_2+\xi}\left(\frac{\lambda z_1}{\mu_1}\right)+\sum_{i=3}^{\infty}\left\{\frac{\eta}{\mu_2+(i-1)\xi}\right.$$
$$\left(\sum_{n=1}^{\infty}\left(\frac{\lambda z_1}{\mu_1}\right)^n\prod_{j=1}^{n}\frac{\mu_2+j\xi}{\gamma+\mu_2+j\xi}\prod_{l=0}^{i-2}\frac{\gamma+\mu_2+l\xi}{\mu_2+l\xi}\right.$$
$$\left.\left.\left.\left.-\sum_{d=1}^{i-2}\left(\frac{\lambda z_1}{\mu_1}\right)^d\prod_{r=d}^{i-2}\frac{\gamma+\mu_2+r\xi}{\mu_2+r\xi}-\left(\frac{\lambda z_1}{\mu_1}\right)^{i-1}\right)\right\}\right]\right]^{-1} \tag{51}$$

4 Steady-State Performance Measures

4.1 Expected Number of Customers in Busy State

The mean number of customers in the busy state is denoted by $E(B)$ as

$$E(B)=\sum_{i=1}^{\infty}i\pi\,(i,1)=\sum_{i=1}^{\infty}i\frac{\lambda_1^i z_1^i}{\mu_1^i}\pi\,(0,0)$$

$$= \frac{\lambda_1 z_1}{\mu_1}\left(1 - \frac{\lambda_1 z_1}{\mu_1}\right)^{-2} \pi\,(0,0)$$

$$= \frac{\lambda_1 z_1 \mu_1}{(\mu_1 - \lambda_1 z_1)^2}\pi\,(0,0) \tag{52}$$

4.2 Expected Number of Customers in Working Vacation State

The mean number of customers in the Working vacation state is denoted by $E(WV)$ as

$$E(WV) = \sum_{i=1}^{\infty} i\pi\,(i,2) = \pi\,(1,2) + 2\pi\,(2,2) + \sum_{i=3}^{\infty} i\pi\,(i,2)$$

$$= \sum_{n=1}^{\infty}\left(\frac{\lambda z_1}{\mu_1}\right)^n \prod_{j=1}^{n} \frac{\mu_2 + j\xi}{(\gamma + \mu_2 + j\xi)}\pi\,(0,0)\left[\frac{\eta}{\mu_2} + \frac{2\eta\,(\gamma+\mu_2)}{\mu_2\,(\mu_2+\xi)}\right.$$
$$\left. + \frac{3\eta\,(\gamma+\mu_2)\,(\gamma+\mu_2+\xi)}{\mu_2\,(\mu_2+\xi)\,(\mu_2+2\xi)} + ...\right] - \left(\frac{\lambda z_1}{\mu_1}\right)\pi\,(0,0)\left[\frac{2\eta}{\mu_2+\xi} + \right.$$
$$\left.\frac{3\eta(\gamma+\mu_2+\xi)}{(\mu_2+\xi)(\mu_2+2\xi)} + \frac{4\eta(\gamma+\mu_2+\xi)(\gamma+\mu_2+2\xi)}{(\mu_2+\xi)(\mu_2+2\xi)(\mu_2+3\xi)} + ...\right] -$$
$$\left(\frac{\lambda z_1}{\mu_1}\right)^2 \pi\,(0,0)\left[\frac{3\eta}{\mu_2+2\xi} + \frac{4\eta(\gamma+\mu_2+2\xi)}{(\mu_2+2\xi)(\mu_2+3\xi)} + \right.$$
$$\left.\frac{5\eta(\gamma+\mu_2+2\xi)(\gamma+\mu_2+3\xi)}{(\mu_2+2\xi)(\mu_2+3\xi)(\mu_2+4\xi)} + ...\right] - ...$$

$$= \sum_{n=1}^{\infty}\left(\frac{\lambda z_1}{\mu_1}\right)^n \prod_{j=1}^{n} \frac{\mu_2 + j\xi}{(\gamma+\mu_2+j\xi)}\frac{\eta}{\mu_2}\left[1 + \sum_{k=1}^{\infty}(k+1)\right.$$
$$\left.\prod_{l=1}^{k}\frac{\gamma+\mu_2+(l-1)\xi}{\mu_2+l\xi}\right]\pi\,(0,0) - \sum_{m=1}^{\infty}\left(\frac{\lambda z_1}{\mu_1}\right)^m \frac{\eta}{\mu_2+m\xi}\left[(m+1)\right.$$
$$\left. + \sum_{k=m+1}^{\infty}(k+1)\prod_{l=m+1}^{k}\frac{\gamma+\mu_2+(l-1)\xi}{\mu_2+l\xi}\right]\pi\,(0,0) \tag{53}$$

4.3 Expected Number of Customers in the System

The mean number of customers in the system is denoted by $E(X)$ as

$$E(X) = \sum_{i=1}^{\infty} i\pi\,(i, 1) + \sum_{i=1}^{\infty} i\pi\,(i, 2) \tag{54}$$

$$E(X) = E(B) + E(WV)$$

where $E(B)$ and $E(WV)$ are obtained from (52) and (53).

5 Numerical Illustration

For numerical illustrations, the values $\lambda = 0.3$, $\mu_1 = 0.7$ and $\mu_2 = 0.3$ such that $\mu_2 < \mu_1$ and $\gamma = 0.5$, $\eta = 4$ and $\alpha = 2$ $\xi = 1$ are chosen. The steady-state distribution is tabulated in Table 1.

Using steady-state distribution, the values of some performance measures are given in Table 2.

Figure 2 indicates that the values of ρ_1 decrease when the values of η increase.

Table 1 Steady-state distribution

$\pi(M) = 0.1052$	$\pi(0, 0) = 0.4741$	$\pi(i, 1)$	$\pi(i, 2)$
		$\pi(1, 1) = 0.0260$	$\pi(1, 2) = 0.1550$
		$\pi(2, 1) = 0.0014$	$\pi(2, 2) = 0.0152$
		$\pi(3, 1) = 0.0000786$	$\pi(3, 2) = 0.0094$
		$\pi(4, 1) = 0.00000432$	$\pi(4, 2) = 0.0017$
		$\pi(i, 1) = 0, \mathrm{i} \geq 5$	$\pi(5, 2) = 0.000057$
			$\pi(i, 2) = 0, \mathrm{i} \geq 6$

Table 2 Performance measures

Probability that the server is in Maintenance mode	0.1052
Probability that the server is in Sleep mode	0.4741
Probability that the server is in Busy mode	0.0276
Probability that the server is in Working vacation mode	0.3931

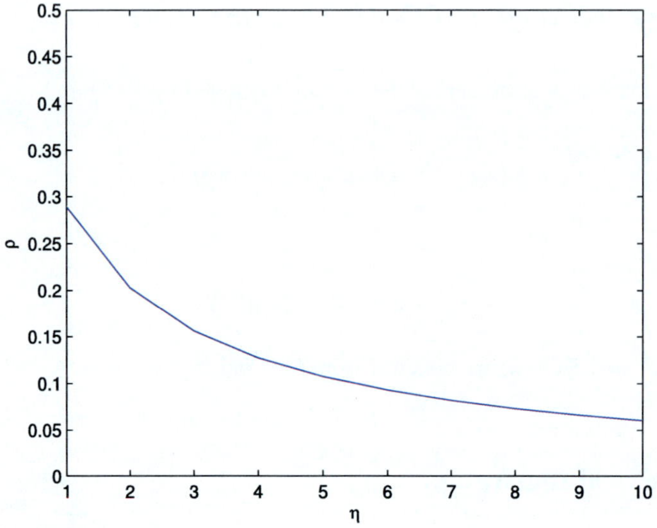

Fig. 2 Variation of ρ_1 versus η

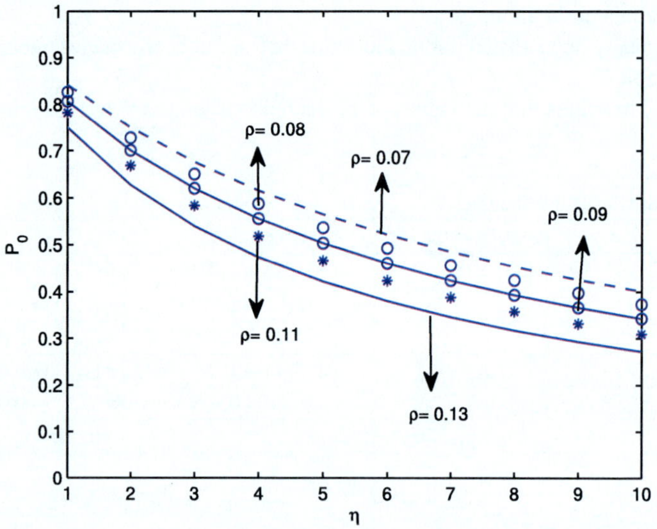

Fig. 3 Probability that the server is in sleep mode versus η

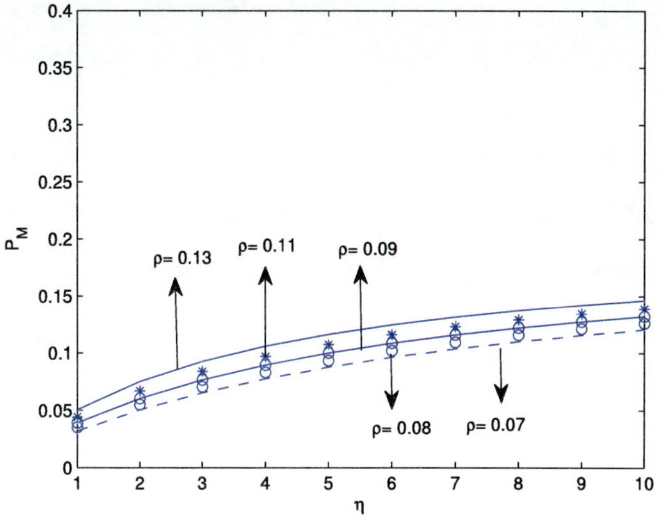

Fig. 4 Probability that the server is in Maintenance mode versus η

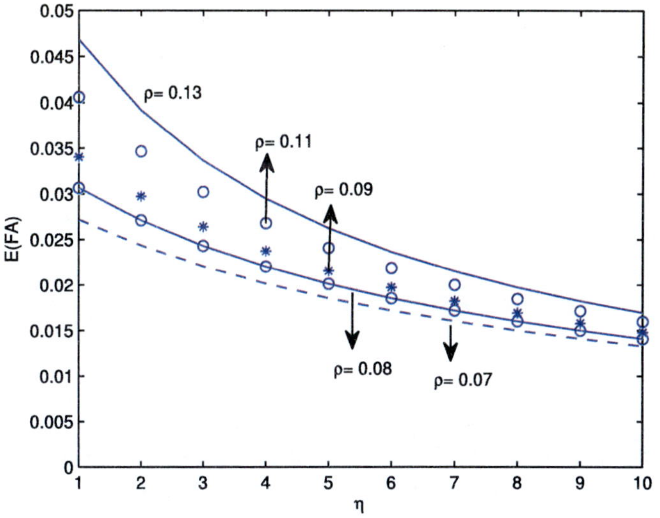

Fig. 5 Mean number of customers in busy state versus η

Figure 3 reveals that the probability $\pi(0, 0)$ that the system is in sleep mode decreases gradually when η increases for various values of ρ_1; also the sleep mode probability $\pi(0, 0)$ increases when ρ_1 decreases.

Figure 4 illustrates that the probability $\pi(M)$ that the system is in maintenance mode of the system increases when the transition rate η increases for various values of ρ_1; also the maintenance mode probability $\pi(M)$ decreases when ρ_1 decreases.

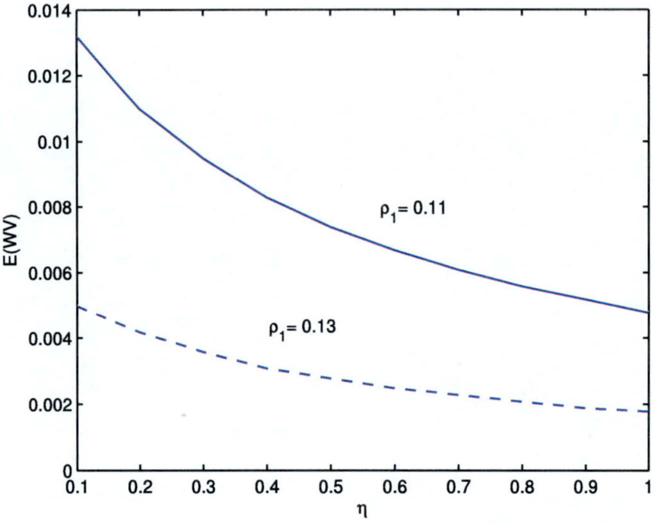

Fig. 6 Mean number of customers in working vacation state versus η

From Fig. 5, it is observed that the average number of customers in the busy state decreases when η increases as expected for different values of ρ_1; also the mean number of customers in the busy state increases when ρ_1 increases.

Figure 6 shows that the average number of customers in the working vacation state decreases initially before approaching a steady state when η increases for different values of ρ_1; also the mean number of customers in the working vacation state increases when ρ_1 increases.

6 Conclusion

In this chapter, an M/M/1 queueing system with the server operating in three modes—maintenance mode, sleep mode and active mode (busy state and working vacation state where the working vacation state incorporates customer impatience)—subject to catastrophes is considered. Explicit expressions are obtained for the transient probabilities of the system in the three different modes. The steady-state probabilities and some steady-state performance measures are obtained. Finally, graphical illustrations are presented and the effects of various parameters on the system performance measures are discussed.

References

1. Abramowitz, M., Stegun, I.A.: Handbook of Mathematical Functions: With Formulas, Graphs, and Mathematical Tables, vol. 55. Courier Corporation (1965)
2. Altman, E., Borovkov, A.A.: On the stability of retrial queues. Queueing Syst. **26**(3–4), 343–363 (1997)
3. Baccelli, F., Boyer, P., Hebuterne, G.: Single-server queues with impatient customers. Adv. Appl. Probab. **16**(4), 887–905 (1984)
4. Baek, S., Son, J.J., Choi, B.D.: Performance analysis of push-to-talk over IEEE 802.16 e with sleep mode and idle mode. Telecommun. Syst. **47**(3–4), 291–302 (2011)
5. Bertsekas, D.P., Gallager, R.G.: Data networks (1986)
6. Boutoumi, B., Gharbi, N.: Two thresholds working vacation policy for improving energy consumption and latency in WSNS. In: International Conference on Queueing Theory and Network Applications, pp. 168–181. Springer (2018)
7. Boxma, O.J., de Waal, P.R.: Multiserver queues with impatient customers. In: ITC, vol. 14, p. 8 (1994)
8. Chan, W.C.: Performance Analysis of Telecommunications and Local Area Networks, vol. 533. Springer Science & Business Media (2006)
9. Cirstea, C., Cernaianu, M., Gontean, A.: Packet loss analysis in wireless sensor networks routing protocols. In: 2012 35th International Conference on Telecommunications and Signal Processing (TSP), pp. 37–41. IEEE (2012)
10. Dabrowski, C.: Catastrophic event phenomena in communication networks: a survey. Comput. Sci. Rev. **18**, 10–45 (2015)
11. Daigle, J.N.: Queueing theory for telecommunications. In: Addison-Wesley Series in Telecommunications (1991)
12. Daley, D.: General customer impatience in the queue Gi/G/1. J. Appl. Probab. **2**(1), 186–205 (1965)
13. De Turck, K., De Vuyst, S., Fiems, D., Wittevrongel, S.: Performance analysis of the IEEE 802.16 e sleep mode for correlated downlink traffic. Telecommun. Syst. **39**(2), 145 (2008)
14. Gelenbe, E., Pujolle, G.: Introduction to queueing networks (1987)
15. Giambene, G.: Queuing theory and telecommunications: networks and applications (2005)
16. Hayes, J.F., Babu, T.V.G.: Modeling and Analysis of Telecommunications Networks, vol. 6. Wiley Online Library (2004)
17. Huo, Z., Jin, S., Wang, Z.: Performance analysis and evaluation for sleep-mode with uplink/downlink traffics in IEEE 802.16 e network. J. Netw. **6**(8), 1145 (2011)
18. Hwang, E., Kim, K.J., Son, J.J., Choi, B.D.: The power-saving mechanism with periodic traffic indications in the IEEE 802.16 e/m. IEEE Trans. Veh. Technol. **59**(1), 319–334 (2009)
19. Hwang, E., Lee, Y.H., Kim, K.J., Son, J.J., Choi, B.D.: Performance analysis of power saving mechanism employing both sleep mode and idle mode in IEEE 802.16 e. IEICE Trans. Commun. **92**(9), 2809–2822 (2009)
20. Lei, H., Nilsson, A.A.: Queuing analysis of power management in the IEEE 802.11 based wireless lans. IEEE Trans. Wirel. Commun. **6**(4), 1286–1294 (2007)
21. Seo, J.B., Lee, S.Q., Park, N.H., Lee, H.W., Cho, C.H.: Performance analysis of sleep mode operation in IEEE 802.16 e. In: IEEE 60th Vehicular Technology Conference, 2004. VTC2004-Fall. 2004, vol. 2, pp. 1169–1173. IEEE (2004)
22. Srinivasan, K., Dutta, P., Tavakoli, A., Levis, P.: Understanding the causes of packet delivery success and failure in dense wireless sensor networks. In: Proceedings of the 4th International Conference on Embedded Networked Sensor Systems, pp. 419–420. ACM (2006)
23. Takagi, H.: Queueing analysis: a foundation of performance evaluation. Vac. Priority Syst. **1**(1) (1991)
24. Xiao, Y.: Energy saving mechanism in the IEEE 802.16 e wireless man. IEEE Commun. Lett. **9**(7), 595–597 (2005)

25. Zhang, Y.: Performance modeling of energy management mechanism in IEEE 802.16 e mobile wimax. In: 2007 IEEE Wireless Communications and Networking Conference, pp. 3205–3209. IEEE (2007)
26. Zhang, Y., Fujise, M.: Energy management in the IEEE 802.16 e MAC. IEEE Commun. Lett. **10**(4), 311–313 (2006)
27. Zhang, Y., Zheng, C., Zhang, H., Dai, J.: Power saving class I sleep mode in IEEE802.16e system. In: The 9th International Conference on Advanced Communication Technology, vol. 3, pp. 1487–1491 (2007)
28. Zhao, J., Govindan, R.: Understanding packet delivery performance in dense wireless sensor networks. In: Proceedings of the 1st International Conference on Embedded Networked Sensor Systems, pp. 1–13. ACM (2003)

Generalization of Permutation on Infinite Kolakoski Sequence with Finite Number of Strings

L. Vigneswaran, N. Jansirani, and V. R. Dare

Abstract The binary alphabet $\Sigma = \{1, 2\}$ can generate an infinite Kolakoski sequence $[K]$ by the concept of σ_0 and σ_1. A linear mapping $K : R^3 \rightarrow R^3$ over the standard basis generates a Kolakoski array is established along with the possibilities of string w, further the permutation of Kolakoski array with the string of length $|w| = 3$ is discussed. Many communication networks require secure transfer of Information, permutations are frequently used in communication networks and parallel and distributed systems. Finally the properties of permutation Kolakoski array in abstract algebra over the composition of function as operation are obtained.

Keywords Basis · Dimension · Kolakoski · Permutation · String

Mathematics Subject Classification (2000) 15A04 · 11Y55 · 68R15

1 Introduction

An infinite Kolakoski sequence [1, 2] over the binary alphabet $\Sigma = \{1, 2\}$ also known as Oldenburger-Kolakoski sequence consisting of blocks and positions was recreated by William Kolakoski who discussed about it in 1965. An one-sided infinite Kolakoski sequence $K = 1221121221221121122121121221121 \ldots$ can be generated in array [3] with the string of length $|w| = 3$ along with standard basis of R^3 over

L. Vigneswaran (✉) · N. Jansirani
Department of Mathematics, Queen Mary's College, University of Madras, Chennai, India
e-mail: ace25vkas@gmail.com

N. Jansirani
e-mail: njansirani@gmail.com

V. R. Dare
Department of Mathematics, Madras Christian College, University of Madras, Chennai, India
e-mail: rajkumardare@yahoo.com

R. N. Mohapatra et al. (eds.), *Mathematical Analysis and Computing*, Springer Proceedings in Mathematics & Statistics 344,
https://doi.org/10.1007/978-981-33-4646-8_25

the binary alphabet $\Sigma = \{1, 2\}$. In this paper we generate the permutation Kolakoski array from a group of w_n with degree n [4, 5]. Also the effect of transformation of an infinite Kolakoski sequence of string into permutation array with degree n in algebraic structure is discussed [6, 7].

2 Basic Definitions

Definition 1 A non-empty set Σ of symbols, called the alphabet. The strings are finite sequence of symbols from the alphabet. If the alphabet $\Sigma = \{1,2\}$, then $w = abab$ and $w = aaabbba$ are strings on Σ.

Definition 2 An infinite Kolakoski sequence K over the binary alphabet $\Sigma = \{1, 2\}$ under two iterating operations σ_0 and σ_1 are [1, 2], $\sigma_0(\text{Even}) = \begin{cases} 1 \to 1 \\ 2 \to 11 \end{cases}$ $\sigma_1(\text{Odd}) = \begin{cases} 1 \to 2 \\ 2 \to 22 \end{cases}$. If an infinite Kolakoski sequence with seed value as 1 under two iterating operations σ_0 and σ_1, then the classical Kolakoski sequence is $K = 122112122122112112212112122112112122122112122121\ldots$

Definition 3 A regular palindrome is a string of numbers or letters that is the same forward as backward [8]. For an example, the string $w = \{121, 212\}$ and $w = abba$ are palindrome.

Definition 4 A set $S = u_1, u_2, \ldots u_n$ of vectors is a basis of V if it has the following two properties: (a) S is linearly independent (b) S spans V.

Definition 5 Let S be a non-empty set. A bijective function $f : S \to S$ is called a permutation [7]. If S has n elements, then the permutation is said to be of degree n. It is also known as symmetric group of degree n and $O(S_n) = n!$.

3 Basis Vectors of R^3 Over Standard Basis on Kolakoski Strings

Suppose $K : R^3 \to R^3$ is a linear mapping, then the basis vectors of Kolakoski strings of array with dimension $n = 3$ will be $K(u_1, u_2, u_3) = (u_1, 2u_1, u_1)$ and $K(u_1, u_2, u_3) = (u_1 + 2u_2, 2u_1 + u_2, u_1 + 2u_2)$. Let's start with an initial string 122 of length $|w| = 3$ from on an infinite Kolakoski sequence, then the basis vectors will be $K(u_1, u_2, u_3) = (u_1 + 2u_2 + u_3, 2u_1 + u_2 + 2u_3, u_1 + 2u_2 + 2u_3)$. Further the standard basis of R^3 is $K = (e_1, e_2, e_3) = \{(1, 0, 0), (0, 1, 0), (0, 0, 1)\}$. Then the image of basis vectors is $K(1, 0, 0) = (1, 2, 1)$, $K(0, 1, 0) = (2, 1, 2)$ and $K(0, 0, 1) = (1, 2, 2)$.Hence,

$$K_1 = \begin{pmatrix} 1 & 2 & 1 \\ 2 & 1 & 2 \\ 1 & 2 & 2 \end{pmatrix}$$

Therefore for every string of length $|w| = 3$ of basis vectors in Kolakoski sequence can be transformed into an array over the binary alphabet $\Sigma = \{1, 2\}$. Also the basis vectors of Kolakoski array from the next string of sequence, $K(u_1, u_2, u_3) = (2u_1, u_1, 2u_1)$, $K(u_1, u_2, u_3) = (2u_1 + u_2, u_1 + 2u_2, 2u_1 + u_2)$. The next string of an infinite Kolakoksi sequence is 112 then the basis vectors will be

$$K(u_1, u_2, u_3) = (2u_1 + u_2 + u_3, u_1 + 2u_2 + u_3, 2u_1 + u_2 + 2u_3)$$

The standard basis of R^3 is

$$K(1, 0, 0) = (2, 1, 2)$$
$$K(0, 1, 0) = (1, 2, 1)$$
$$K(0, 0, 1) = (1, 1, 2)$$

Hence,

$$K_2 = \begin{pmatrix} 2 & 1 & 1 \\ 1 & 2 & 1 \\ 2 & 1 & 2 \end{pmatrix}$$

Therefore for every string of length $|w| = 3$ of Kolakoski sequence can be transformed into an array over the binary alphabet $\Sigma = \{1, 2\}$.

Theorem 1 *If* $[a_{ij}]_{i=1,j=n} = [a_{ij}]_{i=j=n}$ *then every* $[a_{ij}]$ *of* K *has* $n - 1$ *dimensional image.*

Proof Let, the basis vectors of $K : R^3 \to R^3$ as $K(u_1, u_2, u_3) = (u_1, 2u_1, u_1)$, $K(u_1, u_2, u_3) = (u_1 + 2u_2, 2u_1 + u_2, u_1 + 2u_2)$. The palindrome string over the binary alphabet $\Sigma = \{1, 2\}$ will be 121 and 212. If the basis vectors of R^3 are $K(u_1, u_2, u_3) = (u_1 + 2u_2 + u_3, 2u_1 + u_2 + 2u_3, u_1 + 2u_2 + u_3)$ then the image of basis vector is

$$K(1, 0, 0) = (1, 2, 1)$$
$$K(0, 1, 0) = (2, 1, 2)$$
$$K(0, 0, 1) = (1, 2, 1)$$

$$K = \begin{pmatrix} 1 & 2 & 1 \\ 2 & 1 & 2 \\ 1 & 2 & 1 \end{pmatrix} \quad (1)$$

$$\sim \begin{pmatrix} 1 & 0 & 1 \\ 0 & 1 & 0 \\ 0 & 0 & 0 \end{pmatrix} \tag{2}$$

Hence $dim(Im(k)) = 2 = n - 1$.
If $K(u_1, u_2, u_3) = (u_1, 2u_1, u_1)$, $K(u_1, u_2, u_3) = (u_1 + 2u_2, 2u_1 + u_2, u_1 + 2u_2)$ then the basis vectors of R^3 is $K(u_1, u_2, u_3)$=$(u_1 + 2u_2 + 2u_3, 2u_1 + u_2 + u_3, u_1 + 2u_2 + 2u_3)$. Hence the image of basis vector is

$$K(1, 0, 0) = (1, 2, 1)$$
$$K(0, 1, 0) = (2, 1, 2)$$
$$K(0, 0, 1) = (2, 1, 2)$$

$$K = \begin{pmatrix} 1 & 2 & 2 \\ 2 & 1 & 1 \\ 1 & 2 & 2 \end{pmatrix} \tag{3}$$

$$\sim \begin{pmatrix} 1 & 0 & 0 \\ 0 & 1 & 1 \\ 0 & 0 & 0 \end{pmatrix} \tag{4}$$

Hence, $dim(Im(k)) = 2 = n - 1$. □

Theorem 2 *The inverse of every non-palindrome Kolakoski array is similar to the change of basis of K.*

Proof Let's consider the Kolakoski string of length $|w| = 3$. If $K : R^3 \to R^3$ be a linear mapping, then the standard basis is $E = (e_1, e_2, e_3) = \{(1, 0, 0), (0, 1, 0), (0, 0, 1)\}$. The inverse is exists on Kolakoski array only with non-palindrome string. Hence the possibilities of string are $\{122,\ 112,\ 221,\ 211\}$. Let, $K(u_1, u_2, u_3) = \{(1, 2, 1), (2, 1, 2), (1, 2, 2)\}$. Then the change of basis is $u_1 = (1, 2, 1) = e_1 + 2e_2 + e_3, u_2 = (2, 1, 2) = 2e_1 + e_2 + 2e_3, u_3 = (1, 2, 2) = e_1 + 2e_2 + 2e_3$. Hence,

$$K = \begin{pmatrix} 1 & 2 & 1 \\ 2 & 1 & 2 \\ 1 & 2 & 2 \end{pmatrix} \tag{5}$$

$$K^{-1} = \begin{pmatrix} 1 & 1 & -1 \\ 1 & 0 & 0 \\ -1 & 0 & 1 \end{pmatrix} \tag{6}$$

From K with standard basis of R^3,

$$u_1 + 2u_2 + u_3 = 1, u_1 + 2u_2 + u_3 = 0, u_1 + 2u_2 + u_3 = 0$$
$$2u_1 + u_2 + 2u_3 = 0, 2u_1 + u_2 + 2u_3 = 1, 2u_1 + u_2 + 2u_3 = 0$$
$$u_1 + 2u_2 + 2u_3 = 0, u_1 + 2u_2 + 2u_3 = 0, u_1 + 2u_2 + 2u_3 = 1$$

Solving all these equations we get $u_1 = 1, u_2 = 1, u_3 = -1, u_1 = 1, u_2 = 0, u_3 = 0, u_1 = -1, u_2 = 0, u_3 = 1$. Therefore,

$$\begin{aligned} e_1 &= (1, 0, 0) = u_1 + u_2 - u_3 \\ e_1 &= (0, 1, 0) = u_1 + 0u_2 + 0u_3 \\ e_1 &= (0, 0, 1) = -u_1 + 0u_2 + u_3 \end{aligned}$$

Hence,

$$K = \begin{pmatrix} 1 & 1 & -1 \\ 1 & 0 & 0 \\ -1 & 0 & 1 \end{pmatrix} \tag{7}$$

Equation (6) and (7) holds the proof of the statement. □

4 Permutation on Kolakoski Strings

Let S be a non-empty set. A bijective function $f{:}S{\to}S$ is called a permutation [9, 10]. If S has n elements, then the permutation is said to be of degree n. S_n is a group under composition of functions as operation. The group S_n is called the permutation group on n symbols [11]. It is also known as symmetric group of degree n and $O(S_n)$=$n!$. Let's take the binary alphabet $\Sigma = \{1, 2\}$ in which we use σ_0 and σ_1 alternatively. To generate an infinite Kolakoski sequence, let σ_0(Even)= $\begin{cases} 1 \to 1 \\ 2 \to 11 \end{cases}$ and σ_1(Odd)= $\begin{cases} 1 \to 2 \\ 2 \to 22. \end{cases}$ Hence the classical Kolakoski sequence is $K = 12211212212211211\ldots$

The total number of possibilities of strings to generate Kolakoski sequence over the binary the alphabet $\Sigma = \{1, 2\}$ is $2^n - (n-1)$ for $n = 3$. Let the strings are $w = \{122, 112, 121, 212, 211, 221\} = S$. Hence the permutations of Kolakoski strings are

$$P_1 = \begin{pmatrix} 1\,2\,1\,2\,1\,2 \\ 2\,1\,2\,1\,2\,1 \end{pmatrix} \; P_2 = \begin{pmatrix} 1\,2\,1\,2\,1\,2 \\ 1\,2\,2\,1\,1\,2 \end{pmatrix} \quad P_3 = \begin{pmatrix} 1\,2\,1\,2\,1\,2 \\ 1\,2\,1\,2\,1\,2 \end{pmatrix}$$

$$P_4 = \begin{pmatrix} 1\,2\,1\,2\,1\,2 \\ 2\,1\,1\,2\,2\,1 \end{pmatrix}$$

Since we use binary alphabet $\Sigma = \{1, 2\}$ each of them repeating for twice so $O(S_n) = 2n! = 4$.

Theorem 3 *If $w = \{122, 112, 121, 212, 221, 211\}$ over the binary alphabet $\Sigma = \{1, 2\}$, then $(w_2, \bullet)$ is an abelian group.*

Proof The permutation of Kolakoski strings are

$$P_1 = \begin{pmatrix} 1 & 2 & 1 & 2 & 1 & 2 \\ 2 & 1 & 2 & 1 & 2 & 1 \end{pmatrix} P_2 = \begin{pmatrix} 1 & 2 & 1 & 2 & 1 & 2 \\ 1 & 2 & 2 & 1 & 1 & 2 \end{pmatrix} \quad P_3 = \begin{pmatrix} 1 & 2 & 1 & 2 & 1 & 2 \\ 1 & 2 & 1 & 2 & 1 & 2 \end{pmatrix}$$

$$P_4 = \begin{pmatrix} 1 & 2 & 1 & 2 & 1 & 2 \\ 2 & 1 & 1 & 2 & 2 & 1 \end{pmatrix}$$

Case (i)

$$P_1 P_1 = \begin{pmatrix} 1 & 2 & 1 & 2 & 1 & 2 \\ 2 & 1 & 2 & 1 & 2 & 1 \end{pmatrix} \begin{pmatrix} 1 & 2 & 1 & 2 & 1 & 2 \\ 2 & 1 & 2 & 1 & 2 & 1 \end{pmatrix} = \begin{pmatrix} 1 & 2 & 1 & 2 & 1 & 2 \\ 1 & 2 & 1 & 2 & 1 & 2 \end{pmatrix} = P_3$$

$$P_1 P_2 = \begin{pmatrix} 1 & 2 & 1 & 2 & 1 & 2 \\ 2 & 1 & 2 & 1 & 2 & 1 \end{pmatrix} \begin{pmatrix} 1 & 2 & 1 & 2 & 1 & 2 \\ 1 & 2 & 2 & 1 & 1 & 2 \end{pmatrix} = \begin{pmatrix} 1 & 2 & 1 & 2 & 1 & 2 \\ 2 & 1 & 1 & 2 & 2 & 1 \end{pmatrix} = P_4$$

$$P_1 P_3 = \begin{pmatrix} 1 & 2 & 1 & 2 & 1 & 2 \\ 2 & 1 & 2 & 1 & 2 & 1 \end{pmatrix} \begin{pmatrix} 1 & 2 & 1 & 2 & 1 & 2 \\ 1 & 2 & 1 & 2 & 1 & 2 \end{pmatrix} = \begin{pmatrix} 1 & 2 & 1 & 2 & 1 & 2 \\ 2 & 1 & 2 & 1 & 2 & 1 \end{pmatrix} = P_1$$

$$P_1 P_4 = \begin{pmatrix} 1 & 2 & 1 & 2 & 1 & 2 \\ 2 & 1 & 2 & 1 & 2 & 1 \end{pmatrix} \begin{pmatrix} 1 & 2 & 1 & 2 & 1 & 2 \\ 2 & 1 & 1 & 2 & 2 & 1 \end{pmatrix} = \begin{pmatrix} 1 & 2 & 1 & 2 & 1 & 2 \\ 1 & 2 & 2 & 1 & 1 & 2 \end{pmatrix} = P_2$$

Case (ii)

$$P_2 P_1 = \begin{pmatrix} 1 & 2 & 1 & 2 & 1 & 2 \\ 1 & 2 & 2 & 1 & 1 & 2 \end{pmatrix} \begin{pmatrix} 1 & 2 & 1 & 2 & 1 & 2 \\ 2 & 1 & 2 & 1 & 2 & 1 \end{pmatrix} = \begin{pmatrix} 1 & 2 & 1 & 2 & 1 & 2 \\ 2 & 1 & 1 & 2 & 2 & 1 \end{pmatrix} = P_4$$

$$P_2 P_2 = \begin{pmatrix} 1 & 2 & 1 & 2 & 1 & 2 \\ 1 & 2 & 2 & 1 & 1 & 2 \end{pmatrix} \begin{pmatrix} 1 & 2 & 1 & 2 & 1 & 2 \\ 1 & 2 & 2 & 1 & 1 & 2 \end{pmatrix} = \begin{pmatrix} 1 & 2 & 1 & 2 & 1 & 2 \\ 1 & 2 & 1 & 2 & 1 & 2 \end{pmatrix} = P_3$$

$$P_2 P_3 = \begin{pmatrix} 1 & 2 & 1 & 2 & 1 & 2 \\ 1 & 2 & 2 & 1 & 1 & 2 \end{pmatrix} \begin{pmatrix} 1 & 2 & 1 & 2 & 1 & 2 \\ 1 & 2 & 1 & 2 & 1 & 2 \end{pmatrix} = \begin{pmatrix} 1 & 2 & 1 & 2 & 1 & 2 \\ 1 & 2 & 2 & 1 & 1 & 2 \end{pmatrix} = P_2$$

$$P_2 P_4 = \begin{pmatrix} 1 & 2 & 1 & 2 & 1 & 2 \\ 1 & 2 & 2 & 1 & 1 & 2 \end{pmatrix} \begin{pmatrix} 1 & 2 & 1 & 2 & 1 & 2 \\ 2 & 1 & 1 & 2 & 2 & 1 \end{pmatrix} = \begin{pmatrix} 1 & 2 & 1 & 2 & 1 & 2 \\ 2 & 1 & 2 & 1 & 2 & 1 \end{pmatrix} = P_1$$

Case (iii)

$$P_3 P_1 = \begin{pmatrix} 1 & 2 & 1 & 2 & 1 & 2 \\ 1 & 2 & 1 & 2 & 1 & 2 \end{pmatrix} \begin{pmatrix} 1 & 2 & 1 & 2 & 1 & 2 \\ 2 & 1 & 2 & 1 & 2 & 1 \end{pmatrix} = \begin{pmatrix} 1 & 2 & 1 & 2 & 1 & 2 \\ 2 & 1 & 2 & 1 & 2 & 1 \end{pmatrix} = P_1$$

$$P_3 P_2 = \begin{pmatrix} 1 & 2 & 1 & 2 & 1 & 2 \\ 1 & 2 & 1 & 2 & 1 & 2 \end{pmatrix} \begin{pmatrix} 1 & 2 & 1 & 2 & 1 & 2 \\ 1 & 2 & 2 & 1 & 1 & 2 \end{pmatrix} = \begin{pmatrix} 1 & 2 & 1 & 2 & 1 & 2 \\ 1 & 2 & 2 & 1 & 1 & 2 \end{pmatrix} = P_2$$

$$P_3 P_3 = \begin{pmatrix} 1 & 2 & 1 & 2 & 1 & 2 \\ 1 & 2 & 1 & 2 & 1 & 2 \end{pmatrix} \begin{pmatrix} 1 & 2 & 1 & 2 & 1 & 2 \\ 1 & 2 & 1 & 2 & 1 & 2 \end{pmatrix} = \begin{pmatrix} 1 & 2 & 1 & 2 & 1 & 2 \\ 1 & 2 & 1 & 2 & 1 & 2 \end{pmatrix} = P_3$$

$$P_3 P_4 = \begin{pmatrix} 1 & 2 & 1 & 2 & 1 & 2 \\ 1 & 2 & 1 & 2 & 1 & 2 \end{pmatrix} \begin{pmatrix} 1 & 2 & 1 & 2 & 1 & 2 \\ 2 & 1 & 1 & 2 & 2 & 1 \end{pmatrix} = \begin{pmatrix} 1 & 2 & 1 & 2 & 1 & 2 \\ 2 & 1 & 1 & 2 & 2 & 1 \end{pmatrix} = P_4$$

Case (iv)

$$P_4P_1 = \begin{pmatrix} 1\,2\,1\,2\,1\,2 \\ 2\,1\,1\,2\,2\,1 \end{pmatrix} \begin{pmatrix} 1\,2\,1\,2\,1\,2 \\ 2\,1\,2\,1\,2\,1 \end{pmatrix} = \begin{pmatrix} 1\,2\,1\,2\,1\,2 \\ 1\,2\,2\,1\,1\,2 \end{pmatrix} = P_2$$

$$P_4P_2 = \begin{pmatrix} 1\,2\,1\,2\,1\,2 \\ 2\,1\,1\,2\,2\,1 \end{pmatrix} \begin{pmatrix} 1\,2\,1\,2\,1\,2 \\ 1\,2\,2\,1\,1\,2 \end{pmatrix} = \begin{pmatrix} 1\,2\,1\,2\,1\,2 \\ 2\,1\,2\,1\,2\,1 \end{pmatrix} = P_1$$

$$P_4P_3 = \begin{pmatrix} 1\,2\,1\,2\,1\,2 \\ 2\,1\,1\,2\,2\,1 \end{pmatrix} \begin{pmatrix} 1\,2\,1\,2\,1\,2 \\ 1\,2\,1\,2\,1\,2 \end{pmatrix} = \begin{pmatrix} 1\,2\,1\,2\,1\,2 \\ 2\,1\,1\,2\,2\,1 \end{pmatrix} = P_4$$

$$P_4P_4 = \begin{pmatrix} 1\,2\,1\,2\,1\,2 \\ 2\,1\,1\,2\,2\,1 \end{pmatrix} \begin{pmatrix} 1\,2\,1\,2\,1\,2 \\ 2\,1\,1\,2\,2\,1 \end{pmatrix} = \begin{pmatrix} 1\,2\,1\,2\,1\,2 \\ 1\,2\,1\,2\,1\,2 \end{pmatrix} = P_3$$

Table 1 Cayley table for Kolakoski strings of permutation array

•	P_1	P_2	P_3	P_4
P_1	P_3	P_4	P_1	P_2
P_2	P_4	P_3	P_2	P_1
P_3	P_1	P_2	P_3	P_4
P_4	P_2	P_1	P_4	P_3

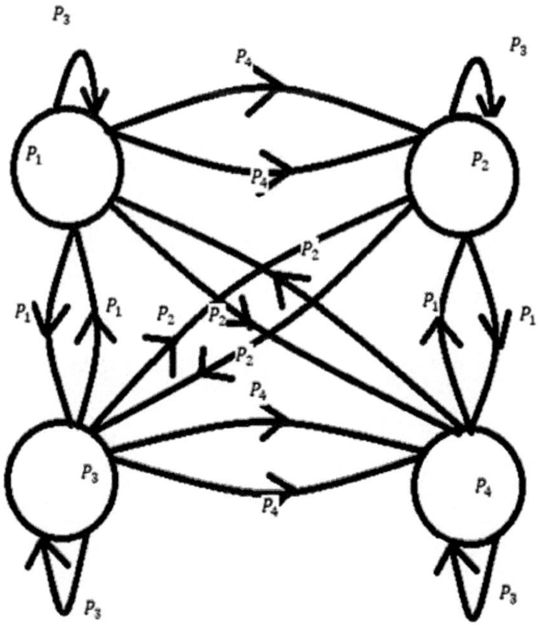

Fig. 1 Cayley graph for Kolakoski strings

∗ Closure: The body of the Cayley table contains only the element of w, w is closed with respect to •.
∗ Associativity: Composition of function is associative, w is associative.
∗ Identity: P_3 is the identity element of w_2.
∗ Inverse: $P_1^{-1} = P_1$, $P_2^{-1} = P_2$, $P_3^{-1} = P_3$, $P_4^{-1} = P_4$. The inverse exists for every element. $(w_2, \bullet)$ is a group.
∗ Commutative: From the Cayley table $P_1.P_2 = P_2.P_1$, etc., all the elements are commutate to each other. Hence $(w_2, \bullet)$ is an abelian group. □

Theorem 4 *If* Σ*={1,2} then every non-palindrome string of length* $|w| = 3$ *generates a permutation array.*

Proof Let, $K(u_1, u_2, u_3)$=$(u_1 + 2u_2, 2u_1 + u_2, u_1 + 2u_2)$. An invertible Kolakoski array can be occured from non-palindrome strings $\{122, 112, 211, 221\}$. Then the basis vectors of R^3 is

$$K(u_1, u_2, u_3) = (u_1 + 2u_2 + u_3, 2u_1 + u_2 + 2u_3, u_1 + 2u_2 + 2u_3)$$

Then the image of basis vector is

$$K(1, 0, 0) = (1, 2, 1)$$
$$K(0, 1, 0) = (2, 1, 2)$$
$$K(0, 0, 1) = (1, 2, 2)$$

Hence

$$K = \begin{pmatrix} 1 & 2 & 1 \\ 2 & 1 & 2 \\ 1 & 2 & 2 \end{pmatrix} \sim \begin{pmatrix} 1 & 0 & 0 \\ 0 & 1 & 0 \\ 0 & 0 & 1 \end{pmatrix}$$

Here the image of basis vectors are generates standard basis of R^3 with $dim(Im(K)) = 3$. Similarly, if

$$K(u_1, u_2, u_3) = (2u_1 + u_2 + 2u_3, u_1 + 2u_2 + u_3, 2u_1 + u_2 + u_3)$$

Then the image of basis vector is

$$K(1, 0, 0) = (2, 1, 2)$$
$$K(0, 1, 0) = (1, 2, 1)$$
$$K(0, 0, 1) = (2, 1, 1)$$

Hence

$$K = \begin{pmatrix} 2 & 1 & 2 \\ 1 & 2 & 1 \\ 2 & 1 & 1 \end{pmatrix} \sim \begin{pmatrix} 1 & 0 & 0 \\ 0 & 1 & 0 \\ 0 & 0 & 1 \end{pmatrix}$$

Then the image of basis vectors of non-palindrome string of length $|w| = 3$ generates a permutation array. □

5 Conclusion

An infinite Kolakoski sequence over the binary alphabet $\Sigma = \{1, 2\}$ of string with length $|w| = 3$ can be transformed into an array of size (3×3) over the basis vectors of R^3 is shown and its generating permutation array with standard basis of R^3 is obtained. Furthermore, well-known properties of abstract algebra are extended to permutation Kolakoski array. Future work focuses to find the applications of Kolakoski array in matrix algebra, namely, Data compression, Image analysis, and security in transfer of information.

References

1. Sing, B.: More Kolakoski sequences: [Math.CO] (2010)
2. Sing, Bernd: Kolakoski-(2m, 2n) are limit-periodic model sets. J. Math. Phys. **44**(2), 899–912 (2003)
3. Jansirani, N., Vigneswaran, L., Dare, V.R.: A new recursive two dimensional pattern On Kolakoski sequence. IJCSE **07**(05), 2347–2693 (2019)
4. Vigneswaran, L., Jansirani, N., Dare, V.R.: A tree structure of Kolakoski two dimensional word. Int. J. Anal. Exp. Model. Anal. **11**(10), 246–252 (2019)
5. Vigneswaran, L., Jansirani, N., Dare, V.R.: A non binary tree with random walk on Kolakoski sequence. ICGTA (2019) (Presented and Submitted)
6. Dobson, E., Spiga, P., Verret, G.: Cayley graphs on abelian groups. Combinatorica **36**(4), 371–393 (2016)
7. Grech, M., Kisielewicz, A.: 2-closed abelian permutation groups. Electron. Notes Discret. Math. **68**, 83–88 (2018)
8. Kupin, E.J., Rowland, E.S.: Bounds on the frequency of 1 in the Kolakoski word (2008). arXiv:0809.2776
9. Kisielewicz, A.: Supergraphs and graphical complexity of permutation groups. Ars Comb. **101**, 193–207 (2011)
10. Mohanty, S.P., Sridharan, M.R., Shukla, S.K.: Graphical cyclic permutation groups: combinatorics and graph theory. Lect. Notes Math. **885**, 339–346 (1981)
11. Imrich, W., Watkins, M.A.: On automorphism groups of cayley graphs. Period. Math. Hung. **7**, 243–258 (1976)

A Stabilized Numerical Algorithm for Singularly Perturbed Delay Differential Equations via Exponential Fitting

N. Sathya Kumar and R. Nageshwar Rao

Abstract In this paper, we present a stabilized numerical algorithm based on central differences for a class of boundary value problems of singularly perturbed differential equation with a small delay. The numerical algorithm is analyzed for convergence and numerical examples are solved to demonstrate the applicability of the method.

Keywords Fitting parameter · Central differences · Singular perturbation problem · Delay · Numerical methods

Mathematics Subject Classification (2010) 65L10 · 65L11

1 Introduction

We consider a class of boundary value problems of singularly perturbed differential equation with a small delay in the convection term

$$\epsilon y''(x) + a(x)y'(x-\delta) + b(x)y(x) = f(x), \ 0 < x < 1, \tag{1}$$

subject to the interval and boundary conditions

$$\begin{aligned} y(x) &= \phi(x), \ -\delta \le x \le 0, \\ y(1) &= \beta, \end{aligned} \tag{2}$$

N. S. Kumar · R. N. Rao (✉)
Department of Mathematics, School of Advanced Sciences, Vellore Institute of Technology, Vellore 632014, Tamil Nadu, India
e-mail: nrao_ragi@yahoo.co.in

N. S. Kumar
e-mail: sathya.maths001@gmail.com

R. N. Mohapatra et al. (eds.), *Mathematical Analysis and Computing*, Springer Proceedings in Mathematics & Statistics 344,
https://doi.org/10.1007/978-981-33-4646-8_26

where $0 < \epsilon \ll 1$ is the perturbation parameter and δ is the delay parameter. Let $(\epsilon - \delta a(x)) > 0 \; \forall \; x \in [0, 1]$, $a(x) \geq M > 0$, $b(x) \leq -\theta < 0$ where M and θ are positive constants, $f(x)$, $\phi(x)$ are sufficiently smooth functions and β is a constant.

The numerical analysis of singular perturbation problems has always been far from trivial because of the boundary layer behavior of the solution. The theory and applications of singularly perturbed differential equations can be found in the monographs [1–8]. Boundary value problems of delay differential equations occur in the study of signal transmission with time delays in control theory [9], first exit problems in neurobiology [10, 11], the study of optically bistable devices [12], in describing the human pupil-light reflex [13], and in variety of models for physiological processes or diseases [14, 15]. Lange and Miura [16, 17] gave an asymptotic approach to solve boundary value problems for second-order singularly perturbed differential-difference equations with small shifts. Extensive numerical work had been initiated by Kadalbajoo and Sharma in their papers [18–21]. Some numerical aspects of this type of problems with small shifts were considered in [22, 23]. Rao and Chakravarthy [24] presented a tridiagonal finite difference method for singularly perturbed differential-difference equations with small shift. Mohapatra and Natesan [25] proposed finite difference method on a adaptively generated grid for singularly perturbed delay difference equation with a small delay.

The brief overview of the present paper is as follows: In Sect. 2, some properties of analytical solutions of the continuous problem are listed. In Sect. 3, we carry out the discretization of the continuous problem whose solution exhibits a boundary layer at the left end of the underlying interval, and thus obtain the finite difference scheme. In Sect. 4, the convergence of numerical method is discussed. In Sect. 5, the numerical method for problems with right end boundary layer is discussed. The efficiency of our presented work is shown by carrying out numerical investigations on several test problems which are stated in Sect. 6. Finally the conclusions follow in the last section.

2 The Continuous Problem

We consider that the shift parameter (δ) is smaller than singular perturbation parameter (ϵ). Now, to tackle the term containing delay, we use Taylor's series as pointed out by Cunningham ([26], p. 222) and Tian [27] in his thesis work. From the Taylor's series expansion of the term $y'(x - \delta)$, we have

$$y'(x - \delta) \approx y'(x) - \delta y''(x).$$

Thus we have from Eq. (1) the approximating equation

$$(\epsilon - \delta a(x))y''(x) + a(x)y'(x) + b(x)y(x) = f(x), \; 0 \leq x \leq 1, \tag{3}$$

$$\begin{aligned} y(0) &= \phi(0) = \phi_0(say), \\ y(1) &= \beta. \end{aligned} \tag{4}$$

Some properties of the solution of (3) and (4) are shown below:
Let L be the operator corresponding to Eq. (3), i.e.,

$$L \equiv (\epsilon - \delta a(x))\frac{d^2}{dx^2} + a(x)\frac{d}{dx} + b(x)I.$$

The operator L satisfies the following continuous minimum principle and stability estimate:

Lemma 1 *Suppose $\pi(x)$ be any sufficiently smooth function satisfying $\pi(0) \geq 0$ and $\pi(1) \geq 0$. Then $L\pi(x) \leq 0$ for all $x \in (0, 1)$ implies that $\pi(x) \geq 0$ for all $x \in [0, 1]$.*

Proof Let $z \in [0, 1]$ be such that $\pi(z) = \min_{x\in[0,1]} \pi(x)$ and assume that $\pi(z) < 0$. Clearly $z \notin \{0, 1\}$, therefore $\pi'(z) = 0$ and $\pi''(z) \geq 0$. Now we have

$$L\pi(z) = (\epsilon - \delta a(x))\pi''(z) + a(x)\pi'(z) + b(x)\pi(z) > 0,$$

which contradicts our assumption, therefore we must have $\pi(z) \geq 0$ and thus $\pi(x) \geq 0\ \forall x \in [0, 1]$. Now we are able to show the stability of solutions of the continuous problem (3, 4).

Lemma 2 *Let $y(x)$ be the solution of the problem (3, 4), then we have*

$$||y|| \leq \theta^{-1}||f|| + \max(|\phi_0|, |\beta|),$$

where $||.||$ is the l_∞ norm given by $||y|| = \max_{x\in[0,1]} |y(x)|$.

Proof Let us construct the two barrier functions $\psi^{\pm}$ defined by

$$\psi^{\pm}(x) = \theta^{-1}||f|| + \max(|\phi_0|, |\beta|) \pm y(x).$$

Then we have

$$\begin{aligned} \psi^{\pm}(0) &= \theta^{-1}||f|| + \max(|\phi_0|, |\beta|) \pm y(0) \\ &= \theta^{-1}||f|| + \max(|\phi_0|, |\beta|) \pm \phi_0, \text{ since } y(0) = \phi_0 \geq 0, \end{aligned}$$

$$\begin{aligned} \psi^{\pm}(1) &= \theta^{-1}||f|| + \max(|\phi_0|, |\beta|) \pm y(1) \\ &= \theta^{-1}||f|| + \max(|\phi_0|, |\beta|) \pm \beta, \text{ since } y(1) = \beta \geq 0, \end{aligned}$$

and we have

$$\begin{aligned} L_\epsilon \psi^{\pm} &= (\epsilon - \delta a(x))(\psi^{\pm}(x))' + a(x)(\psi^{\pm}(x))' + b(x)(\psi^{\pm}(x)) \\ &= b(x)(\theta^{-1}||f|| + \max(|\phi_0|, |\beta|)) \pm L_\epsilon y(x) \\ &= b(x)(\theta^{-1}||f|| + \max(|\phi_0|, |\beta|)) \pm f(x). \end{aligned}$$

We have $b(x)\theta^{-1} \leq -1$, since $b(x) \leq -\theta < 0$.
Using this inequality in the above inequality, we get

$$\begin{aligned} L_\epsilon \psi^{\pm}(x) &\leq (-||f|| \pm f(x)) + b(x)\max(|\phi_0|, |\beta|) \\ &\leq 0 \forall x \in (0, 1), \text{ since } ||f|| \geq f(x). \end{aligned}$$

Therefore, by the minimum principle [5], we obtain $\psi^{\pm}(x) \geq 0$ for all $x \in [0, 1]$, which gives the required estimate.

Lemma 1 implies that the solution is unique and since the problem under consideration is linear, the existence of the solution is implied by its uniqueness. Further, the boundedness of the solution is implied by Lemma 2.

Lemma 3 *Let $y(x) = y_0 + z_0$ be the zeroth-order approximation to the solution of (3) and (4), where y_0 represents the zeroth-order approximate outer solution (i.e., the solution of the reduced problem) and z_0 represents the zeroth-order approximate solution in the boundary layer region. Then for a fixed positive integer i,*

$$\lim_{h\to 0} y(ih) \approx y_0 + (\phi(0) - y_0(0)) \exp\{-a(0)i\rho\},$$

$$\textit{where } \rho = \frac{h}{\epsilon - \delta a(0)}.$$

Proof Let $y_0(x)$ be the solution of reduced problem

$$a(x){y_0}'(x) + b(x)y_0(x) = f(x),$$

$y_0(1) = \beta$ and $z_0(t)$ is the solution of the boundary value problem

$${z_0}''(t) + a(0){z_0}'(t) = 0, z_0(0) = \phi(0) - y_0(0), z_0(\infty) = 0,$$

$$\text{where } t = \frac{x}{\epsilon - \delta M}.$$

From the theory of singular perturbations it is well-known that the zeroth-order asymptotic approximation to the solution (3) and (4) is (cf. [4], pp. 22–26)

$$y(x) = y_0(x) + \frac{a(0)}{a(x)}(\phi(0) - y_0(0)) \times \exp\left\{-\int_0^x \left(\frac{a(x)}{\epsilon - \delta a(x)}\right) dx\right\}.$$

As we are considering the differential equations on sufficiently small sub-intervals, the coefficients could be assumed to be locally constant. Hence

$$y(x) \approx y_0(x) + (\phi(0) - y_0(0)) \times \exp\left\{-\left(\frac{a(0)}{\epsilon - \delta a(0)}\right)x\right\}.$$

So, at the nodal points we have

$$y(x_i) \approx y_0(x_i) + (\phi(0) - y_0(0)) \exp\left\{-\left(\frac{a(0)}{\epsilon - \delta a(0)}\right)x_i\right\}, i = 0, 1, 2, \ldots, N.$$

$$i.e., \; y(ih) \approx y_0(ih) + (\phi(0) - y_0(0)) \exp\left\{-\left(\frac{a(0)}{\epsilon - \delta a(0)}\right)ih\right\}.$$

Therefore,

$$\lim_{h \to 0} y(ih) \approx y_0 + (\phi(0) - y_0(0)) \exp\{-a(0)i\rho\} \text{ for } i = 0, 1, 2, \ldots, N,$$

$$\text{where } \rho = \frac{h}{\epsilon - \delta a(0)} \text{(cf. [5]; pp. 93–94)}.$$

3 Exponentially Fitted Second-Order Central Difference Method

Classical methods are not expected to perform well in the overall range of h (mesh parameter) and ϵ (perturbation parameter) values, since the numerical schemes may contain exponentials as coefficients. We propose here an exponentially fitted second-order central difference method for the boundary value problem (3, 4).

We divide the interval [0, 1] into N equal parts with constant mesh length h. Let $0 = x_0, x_1, x_2, \ldots, x_n = 1$ be the mesh points. Then we have $x_i = ih$, $i = 0, 1, 2, \ldots, N$. Using central difference formulae, the finite difference representation of Eq. (3) may be written at a typical mesh point $x_i, i = 0, 1, 2, \ldots, N$, according to

$$(\epsilon - \delta a_i)\left(\frac{y_{i-1} - 2y_i + y_{i+1}}{h^2} - \frac{h^2}{12}{y_i}^4 - \frac{h^4}{360}{y_i}^6 - \cdots\right) + \\ a_i\left(\frac{y_{i+1} - y_{i-1}}{2h} - \frac{h^2}{6}{y_i}^3 - \frac{h^4}{120}{y_i}^5 - \cdots\right) + b_i y_i = f_i, \tag{5}$$

and $a(x_i) = a_i$; $b(x_i) = b_i$; $f(x_i) = f_i$; $y(x_i) = y_i$.

The boundary conditions become

$$y_0 = \phi(x_0),\ y_N = \beta. \tag{6}$$

Equation (5) provides the basis for a second-order method; however, at this stage, the left-hand side of Eq. (5) is not tridiagonal, because it involves the difference ${y_i}^{(3)}$. Evidently tridiagonal estimate of ${y_i}^{(3)}$ correct to $O(h^5)$ is required and this estimate is obtained as follows (cf. [28]):

Differentiating (3) once with respect to x, then using central difference formulae, gives a tridiagonal $O(h^5)$ approximation for ${y_i}^{(3)}$ as follows:

$${y_i}^{(3)} = \frac{1}{\epsilon - \delta a_i}\left((\delta a_i{}' - a_i)y_i{}'' - (a_i{}' + b_i)y_i{}' - b_i{}'y_i + f_i{}'\right) \tag{7}$$

and $a'(x_i) = a_i{}'$; $b'(x_i) = b_i{}'$; $f'(x_i) = f_i{}'$; $y'(x_i) = y_i{}'$.
By substituting (7) in (5) and simplifying, we obtain

$$\begin{aligned}
\frac{1}{\rho}(y_{i-1} - 2y_i + y_{i+1}) = {} & \left(\frac{a_i}{2} + \frac{a_i\delta a_i{}'\rho}{6} - \frac{{a_i}^2\rho}{6} + \frac{a_i a_i{}'\rho h}{12} + \frac{a_i b_i \rho h}{12}\right) y_{i-1} \\
& + \left(-\frac{a_i\delta a_i{}'\rho}{3} + \frac{{a_i}^2\rho}{3} - \frac{a_i b_i{}'\rho h}{12} - b_i h\right) y_i \\
& + \left(-\frac{a_i}{2} + \frac{a_i\delta a_i{}'\rho}{6} - \frac{{a_i}^2\rho}{6} - \frac{a_i a_i{}'\rho h}{12} - \frac{a_i b_i \rho h}{12}\right) y_{i+1} \\
& + \frac{a_i f_i{}'\rho h^2}{6} + hf_i + R_i, \text{ where } R_i = O(h^3).
\end{aligned}$$

Here we introduced a fitting parameter $\sigma(\rho)$ in a finite difference scheme and it is required to find $\sigma(\rho)$ in such a way that the solution of (5) converges uniformly in ϵ to the solution of (3, 4). Hence we obtain a finite difference scheme as follows:

$$E_i y_{i-1} - F_i y_i + G_i y_{i+1} = H_i + T_i; \quad i = 1, 2, \ldots, N-1, \tag{8}$$

where

$$\begin{aligned}
E_i &= \frac{\sigma}{\rho} - \frac{a_i}{2} - \frac{a_i\delta a_i{}'\rho}{6} + \frac{{a_i}^2\rho}{6} - \frac{a_i a_i{}'\rho h}{12} - \frac{a_i b_i \rho h}{12}, \\
F_i &= \frac{2\sigma}{\rho} - \frac{a_i\delta a_i{}'\rho}{3} + \frac{{a_i}^2\rho}{3} - \frac{a_i b_i{}'\rho h^2}{12} - b_i h, \\
G_i &= \frac{\sigma}{\rho} + \frac{a_i}{2} + \frac{a_i\delta a_i{}'\rho}{6} + \frac{{a_i}^2\rho}{6} + \frac{a_i a_i{}'\rho h}{12} + \frac{a_i b_i \rho h}{12}, \\
H_i &= \frac{a_i f_i{}'\rho h^2}{6} + hf_i \text{ and } T_i \text{ is the truncation error of } O(h^3).
\end{aligned}$$

3.1 Calculation of the Fitting Parameter

We consider $h = O(\epsilon - \delta a_i)$ such that $\frac{h}{\epsilon-\delta a_i}$ is finite. Now, taking limit as $h \to 0$ in (8), using Lemma 3 and simplifying, we get the constant fitting parameter as

$$\sigma(\rho) = \frac{a_0\rho}{2}\coth\left(\frac{a(0)\rho}{2}\right) - \frac{\rho^2 {a_0}^2}{6}, \tag{9}$$

where $\rho = \frac{h}{\epsilon-\delta a(0)}$.
To obtain the solution of the original problem, we solve the tridiagonal system (8) where σ is given by (9) subject to the boundary conditions (6). We use Thomas algorithm to solve the tridiagonal system.

4 Stability Consideration

The Numerical scheme (8) can be rewritten as

$$\left(-\frac{\sigma}{\rho} + u_i\right) y_{i-1} + \left(\frac{2\sigma}{\rho} + v_i\right) y_i + \left(-\frac{\sigma}{\rho} + w_i\right) y_{i+1} + g_i + T_i = 0, \tag{10}$$

where

$$\begin{aligned}
u_i &= \frac{a_i}{2} + \frac{a_i\delta {a_i}'\rho}{6} - \frac{{a_i}^2\rho}{6} + \frac{a_i {a_i}'\rho h}{12} + \frac{a_i b_i \rho h}{12}, \\
v_i &= \frac{a_i\delta {a_i}'\rho}{3} - \frac{{a_i}^2\rho}{3} + \frac{a_i {b_i}'\rho h^2}{6} + b_i h, \\
w_i &= -\frac{a_i}{2} - \frac{a_i\delta {a_i}'\rho}{6} - \frac{{a_i}^2\rho}{6} - \frac{a_i {a_i}'\rho h}{12} - \frac{a_i b_i \rho h}{12}, \\
g_i &= \frac{a_i {f_i}'\rho h^2}{6} + h f_i, \; T_i = O(h^3).
\end{aligned}$$

Incorporating the boundary conditions $y_0 = \phi(x_0) = \phi(0)$, $y_N = \beta$ we obtain the system of equations in the matrix form as

$$(D + P)y + M + T(h) = 0, \tag{11}$$

where

$$D = \left[-\frac{\sigma}{\rho}, \frac{2\sigma}{\rho}, \frac{-\sigma}{\rho}\right] = \begin{bmatrix} \frac{2\sigma}{\rho} & \frac{-\sigma}{\rho} & 0 & \dots & 0 \\ \frac{-\sigma}{\rho} & \frac{2\sigma}{\rho} & \frac{-\sigma}{\rho} & \dots & 0 \\ 0 & . & . & \dots & . \\ . & . & . & \dots & . \\ 0 & \dots & 0 & \frac{-\sigma}{\rho} & \frac{2\sigma}{\rho} \end{bmatrix}$$

and

$$P = [u_i, v_i, w_i] = \begin{bmatrix} v_1 & w_1 & 0 & \dots & 0 \\ u_2 & v_2 & w_2 & \dots & 0 \\ 0 & . & . & \dots & . \\ . & . & . & \dots & . \\ 0 & \dots & 0 & u_{N-1} & v_{N-1} \end{bmatrix}$$

are tridiagonal matrices of order $N-1$, and

$$M = \left[g_1 + \left(-\frac{\sigma}{\rho} + u_1\right)\phi_0, g_2, g_3, \dots, g_{N-2}, g_{N-1} + \left(-\frac{\sigma}{\rho} + w_{N-1}\right)\beta\right]^T,$$

and $y = [y_1, y_2, \dots, y_{N-1}]^T$, $T(h) = [T_1, T_2, \dots, T_{N-1}]^T$, $O = [0, 0, \dots, 0]^T$ are the associated vectors of Eq. (11).

Let $Y = [Y_1, Y_2, \dots, Y_{N-1}]^T \cong y$ which satisfies the equation

$$(D + P)Y + M = 0. \tag{12}$$

Let $e_i = Y_i - y_i$, $i = 1, 2, \dots, N-1$ be the discretization error so that $E = [e_1, e_2, \dots, e_{N-1}]^T = Y - y$.

Subtracting Eq. (11) from Eq. (12) we get

$$(D + P)E = T(h). \tag{13}$$

Let $|a_i| \le C_1$, $|b_i| \le C_2$

Let $p_{i,j}$ be the (i, j)th element of the matrix P, then

$$|P_{i,i+1}| = |w_i| \le \frac{C_1}{2} + \frac{C_1C_2}{12}\rho h + \frac{{C_1}^2}{6}\rho;\ i = 1, 2, \dots, N-2,$$
$$|P_{i,i-1}| = |u_i| \le \frac{C_1}{2} + \frac{C_1C_2}{12}\rho h + \frac{{C_1}^2}{6}\rho;\ i = 2, 3, \dots, N-1.$$

Thus for sufficiently small h (i.e., as $h \to 0$), we have

$$\frac{\sigma}{\rho} + |P_{i,i+1}| < 0,\ i = 1, 2, \dots, N-2,$$
$$\frac{\sigma}{\rho} + |P_{i,i-1}| < 0,\ i = 2, 3, \dots, N-1.$$

Hence the matrix $(D + P)$ is irreducible [29].

Let S_i be the sum of the elements of the ith row of the matrix $(D + P)$, then

$$S_i = \frac{\sigma}{\rho} + \frac{{a_i}^2}{6\epsilon}h - \frac{a_i b_{i+1}}{12\epsilon}h^2 + \frac{a_i b_{i-1}}{12\epsilon}h^2 - b_i h$$
$$- \frac{a_i}{2} - \frac{a_i a_{i+1}}{24\epsilon}h + \frac{a_i a_{i-1}}{24\epsilon}h - \frac{a_i b_i}{12\epsilon}h^2, \text{ for } i = 1,$$
$$S_i = -\frac{a_i b_{i+1}}{12\epsilon}h^2 + \frac{a_i b_{i-1}}{12\epsilon}h^2 - b_i h, \text{ for } i = 2, 3, \ldots, N-2,$$
$$S_i = \frac{\sigma}{\rho} + \frac{a_i}{2} + \frac{a_i a_{i+1}}{24\epsilon}h - \frac{a_i a_{i-1}}{24\epsilon}h + \frac{a_i b_i}{12\epsilon}h^2$$
$$- \frac{{a_i}^2}{6\epsilon}h - \frac{a_i b_{i+1}}{12\epsilon}h^2 + \frac{a_i b_{i-1}}{12\epsilon}h^2 - b_i h, \text{ for } i = N-1.$$

Let $C_{1*} = \min_{1 \le i \le N-1} |a_i|$, $C_1^* = \max_{1 \le i \le N-1} |a_i|$, $C_{2*} = \min_{1 \le i \le N-1} |b_i|$, $C_2^* = \max_{1 \le i \le N-1} |b_i|$.
Then $0 < C_{1*} \le C_1 \le C_1^*$, $0 < C_{2*} \le C_2 \le C_2^*$.
For sufficiently small h, $(D + P)$ is monotone [29, 30].

Hence $(D + P)^{-1}$ exists and $(D + P)^{-1} \ge 0$.

From the error Eq. (13) we have

$$||E|| \le ||(D + P)^{-}1|| \cdot ||T||.$$

For sufficiently small h, we have

$$S_i > hQ_1 \text{ for } i = 1, \text{ where } Q_1 = -\frac{C_1 C_2}{12}\rho - C_2, \tag{14}$$

$$S_i > hQ_2 \text{ for } i = 2, 3, \ldots, N-2, \text{ where } Q_2 = -C_2, \tag{15}$$

$$S_i > hQ_3 \text{ for } i = N-1, \text{ where } Q_3 = -C_2 + \frac{C_1 C_2}{12}\rho. \tag{16}$$

Let $(D + P)^{-1}_{i,k}$ be the (i, k)th element of $(D + P)^{-1}$ and we define

$$||(D + P)^{-1}|| = \max_{1 \le i \le N-1} \sum_{k=1}^{N-1} (D + P)^{-1}_{i,k}$$

and

$$||T(h)|| = \max_{1 \le i \le N-1} |T_i|.$$

Since $(D + P)^{-1} \ge 0$ and $\sum_{k=1}^{N-1} (D + P)^{-1}_{i,k} \cdot S_k = 1$ for $i = 1, 2, ..., N-1$.

we have

$$(D+P)^{-1}_{i,k} \leq \frac{1}{S_k} < \frac{1}{h} Q_1 \;\; \text{for}\, k = 1,$$
$$(D+P)^{-1}_{i,k} \leq \frac{1}{S_k} < \frac{1}{h} Q_3 \;\; \text{for}\, k = N-1.$$

Further

$$\sum_{n=2}^{N-2} (D+P)^{-1}_{i,k} \leq \frac{1}{\min\limits_{2\leq k\leq N-2} S_k} \leq \frac{1}{h} Q_2 \;\; \text{for}\; i = 1, 2, \ldots, N-1.$$

From the error Eq. (13), using Eqs. (14) and (15), we get

$$||E|| = \frac{1}{h} |Q_1 + Q_2 + Q_3| \times |T(h)| = O(h^2).$$

This establishes the convergence of the finite difference scheme (8).

5 Right-End Boundary Layer Problems

Finally, we discuss our method for singularly perturbed two point boundary value problems with right-end boundary layer of the underlying interval. When $0 < (\epsilon - \delta a(x)) \ll 1$, $a(x) \leq M < 0$, $b(x) < 0$ throughout the interval [0, 1], where M is some negative constant, the boundary value problem (3) and (4) displays a boundary layer at $x = 1$.

Lemma 4 *Let $y(x) = y_0 + z_0$ be the zeroth-order approximation to the solution of (3) and (4), where y_0 represents the zeroth-order approximate outer solution (i.e., the solution of the reduced problem) and z_0 represents the zeroth-order approximate solution in the boundary layer region. Then for a fixed positive integer i,*

$$\lim_{h\to 0} y(ih) = y_0(0) + (\beta - y_0(1)) \exp\left[a(1)\left(\frac{1}{\epsilon - \delta a(1)} - i\rho\right)\right], \textit{where}\, \rho = \frac{h}{\epsilon - \delta a(1)}.$$

Proof The proof is based on asymptotic analysis (cf. [4], pp. 22–26), and similar to the proof of Lemma 3.

Applying the same procedure as in Sect. 2 and using Lemma 4, we will get the tridiagonal system (8) with fitting parameter as

$$\sigma(\rho) = \frac{a_0\rho}{2} \coth\left(\frac{a(0)\rho}{2}\right) - \frac{\rho^2 {a_0}^2}{6}$$

and it can be solved easily by Thomas Algorithm.

6 Numerical Results

To elucidate the importance and applicability of the present method, we have considered three boundary value problems for singularly perturbed linear differential-difference equations where the boundary layer is to the left end of the interval [0, 1] and also one problem in which the boundary layer lies in the right end of the interval. Since the exact solutions of problems are not known for various δ values, we use the double mesh principle as stated below to discuss the maximum absolute errors for the examples:

$$E_N = \max_{0 \le i \le N} |{y_i}^N - {y_{2i}}^{2N}|.$$

The numerical rates of convergence for the presented scheme for the solutions of the problems is given by

$$R_N = \frac{\log|E_N/E_{2N}|}{\log 2}.$$

Example 1 (17, pp. 254). $\epsilon y''(x) + y'(x - \delta) + y(x) = 0$, subject to the interval and boundary conditions $y(x) = 1; -\delta \le x \le 0, y(1) = 1$.

Example 2 $\epsilon y''(x) + (1 + x)y'(x - \delta) - e^{-x}y(x) = 1$, subject to the interval and boundary conditions $y(x) = 0; -\delta \le x \le 0, y(1) = 1$.

Example 3 (20, pp. 699). $\epsilon y''(x) + 0.25y'(x - \delta) - y(x) = 0$, subject to the interval and boundary conditions $y(x) = 1; -\delta \le x \le 0, y(1) = 0$.

Example 4 (20, pp. 707). $\epsilon y''(x) - y'(x - \delta) + y(x) = 0$, subject to the interval and boundary conditions $y(x) = 1; -\delta \le x \le 0, y(1) = -1$.

Table 1 The maximum absolute errors for Example 1 when $\delta = 0.5\epsilon$

$\epsilon \downarrow N \rightarrow$	100	200	300	400	500
2^{-1}	1.1574e−05	2.8917e−06	1.2851e−06	7.2287e−07	4.6263e−07
2^{-2}	4.8309e−05	1.2084e−05	5.3710e−06	3.0213e−06	1.9336e−06
2^{-3}	1.7367e−04	4.3442e−05	1.9308e−05	1.0861e−05	6.9512e−06
2^{-4}	4.2672e−04	1.0663e−04	4.7389e−05	2.6655e−05	1.7059e−05
2^{-5}	9.5159e−04	2.3696e−04	1.0525e−04	5.9183e−05	3.7874e−05
2^{-6}	2.0626e−03	5.0567e−04	2.2387e−04	1.2575e−04	8.0429e−05
2^{-7}	4.5963e−03	1.0701e−03	4.6795e−04	2.6181e−04	1.6711e−04
2^{-8}	1.1106e−02	2.3516e−03	9.9313e−04	5.4641e−04	3.4610e−04
2^{-9}	2.6720e−02	5.6488e−03	2.2549e−03	1.1945e−03	7.3871e−04
2^{-10}	5.8092e−02	1.3695e−02	5.5310e−03	2.8534e−03	1.7161e−03
E_N	5.8092e−02	1.3695e−02	5.5310e−03	2.8534e−03	1.7161e−03
R_N	1.9992	1.9999	1.9999	1.9999	1.9984

7 Conclusions

In this paper, solution to the boundary value problems is presented with an exponentially fitted tridiagonal finite difference method for singularly perturbed differential-difference equations containing a small negative shift. The method is best suited for problems with shift parameter smaller than the perturbation parameter. In order to demonstrate the proposed method and to show the effect of shift parameter on the boundary layer behavior of the solution, an extensive amount of computational work has been carried out. The maximum absolute error is shown in the form of Tables 1, 2, 3, and 4 for the considered examples. The effect of shift on the boundary layer

Table 2 The maximum absolute errors for Example 2 when $\delta = 0.5\epsilon$

$\epsilon \downarrow N \rightarrow$	100	200	300	400	500
2^{-1}	3.9728e−07	8.9797e−08	3.8582e−08	2.1340e−08	1.3520e−08
2^{-2}	1.8068e−06	4.2227e−07	1.8436e−07	1.0289e−07	6.5571e−08
2^{-3}	6.9193e−06	1.5513e−06	6.7116e−07	3.7347e−07	2.3771e−07
2^{-4}	2.3408e−05	4.6241e−06	1.9298e−06	1.0580e−06	6.6833e−07
2^{-5}	8.6029e−05	1.3524e−05	5.1451e−06	2.7013e−06	1.6670e−06
2^{-6}	3.6776e−04	4.7216e−05	1.5343e−05	7.3168e−06	4.2531e−06
2^{-7}	1.4411e−03	2.0786e−04	6.0108e−05	2.5351e−05	1.3350e−05
2^{-8}	4.0018e−03	8.4262e−04	2.7598e−04	1.1582e−04	5.7693e−05
2^{-9}	8.3937e−03	2.4117e−03	9.8937e−04	4.8293e−04	2.6267e−04
2^{-10}	1.4429e−02	5.2338e−03	2.5278e−03	1.4156e−03	8.6852e−04
E_N	1.4429e−02	5.2338e−03	2.5278e−03	1.4156e−03	8.6852e−04
R_N	1.463	1.886	2.0213	2.0144	2.0108

Table 3 The maximum absolute errors for Example 3 when $\delta = 0.5\epsilon$

$\epsilon \downarrow N \rightarrow$	100	200	300	400	500
2^{-1}	2.7948e−06	6.9878e−07	3.1057e−07	1.7469e−07	1.1182e−07
2^{-2}	8.2231e−06	2.0558e−06	9.1371e−07	5.1396e−07	3.2893e−07
2^{-3}	2.1202e−05	5.3006e−06	2.3558e−06	1.3251e−06	8.4809e−07
2^{-4}	5.0721e−05	1.2675e−05	5.6329e−06	3.1684e−06	2.0278e−06
2^{-5}	1.1830e−04	2.9529e−05	1.3124e−05	7.3823e−06	4.7244e−06
2^{-6}	2.6957e−04	6.7189e−05	2.9837e−05	1.6772e−05	1.0736e−05
2^{-7}	5.9574e−04	1.4835e−04	6.5530e−05	3.6880e−05	2.3573e−05
2^{-8}	1.3553e−03	3.1834e−04	1.4069e−04	7.8658e−05	5.0106e−05
2^{-9}	3.5552e−03	6.9622e−04	3.0240e−04	1.6523e−04	1.0489e−04
2^{-10}	8.3805e−03	1.8159e−03	7.0948e−04	3.5322e−04	2.2308e−04
E_N	8.3805e−03	1.8159e−03	7.0948e−04	3.5322e−04	2.2308e−04
R_N	1.9999	2.0000	1.9998	1.9992	1.9999

Table 4 The maximum absolute errors for Example 4 when $\delta = 0.5\epsilon$

$\epsilon \downarrow N \rightarrow$	100	200	300	400	500
2^{-1}	6.6237e−06	1.6559e−06	7.3594e−07	4.1397e−07	2.6495e−07
2^{-2}	2.7058e−05	6.7646e−06	3.0064e−06	1.6911e−06	1.0823e−06
2^{-3}	7.3514e−05	1.8368e−05	8.1627e−06	4.5914e−06	2.9385e−06
2^{-4}	1.1141e−04	2.7805e−05	1.2354e−05	6.9470e−06	4.4461e−06
2^{-5}	2.3101e−04	5.7765e−05	2.5674e−05	1.4442e−05	9.2429e−06
2^{-6}	5.7268e−04	1.4300e−04	6.3542e−05	3.5740e−05	2.2873e−05
2^{-7}	1.2856e−03	3.1881e−04	1.4147e−04	7.9536e−05	5.0894e−05
2^{-8}	2.8233e−03	6.8255e−04	3.0121e−04	1.6899e−04	1.0805e−04
2^{-9}	6.5312e−03	1.4654e−03	6.3334e−04	3.5285e−04	2.2473e−04
2^{-10}	1.6202e−02	3.3468e−03	1.3750e−03	7.4714e−04	4.6989e−04
E_N	1.6202e−02	3.3468e−03	1.3750e−03	7.4714e−04	4.6989e−04
R_N	1.9997	1.9999	2.0000	1.9999	1.9999

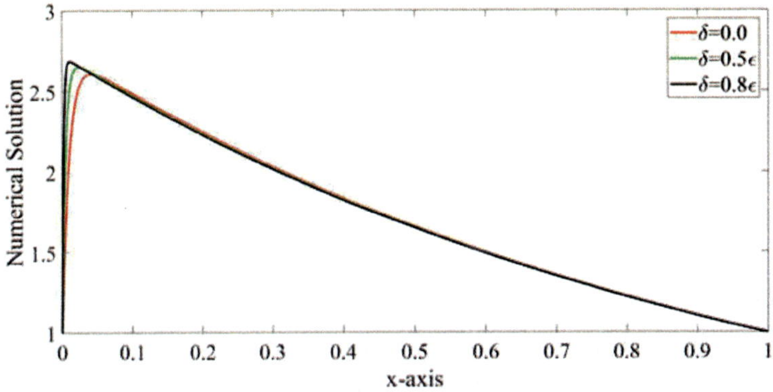

Fig. 1 Graph of numerical solution of Example 1 for $\epsilon = 10^{-2}$

of the solution has been analyzed from solution of the considered examples for different values of delay and are plotted as Figs. 1, 2, 3, and 4. From the figures, we observed that as the shift parameter increases, thickness of the layer decreases in the case where the solution exhibits layer behavior on the left side. Whereas thickness of the layer increases in the case where the solution exhibits boundary layer behavior on the right side of the interval. On the basis of the various numerical solutions of a variety of illustrations, it is concluded that the present method provides significant advantage for the linear singularly perturbed differential-difference equations.

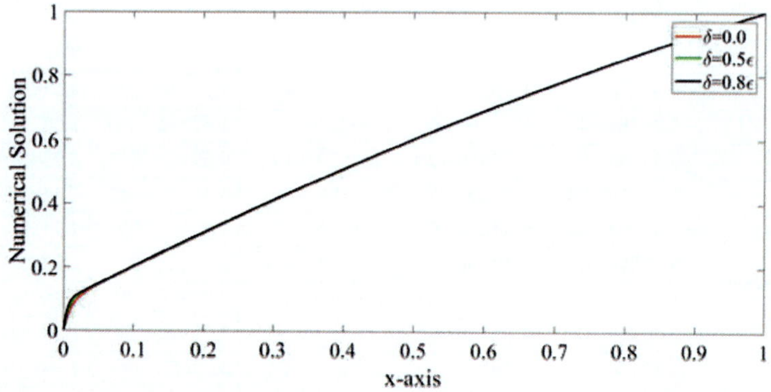

Fig. 2 Graph of numerical solution of Example 2 for $\epsilon = 10^{-2}$

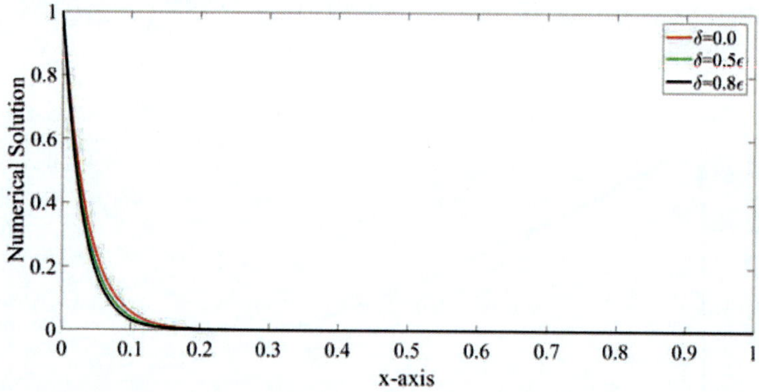

Fig. 3 Graph of numerical solution of Example 3 for $\epsilon = 10^{-2}$

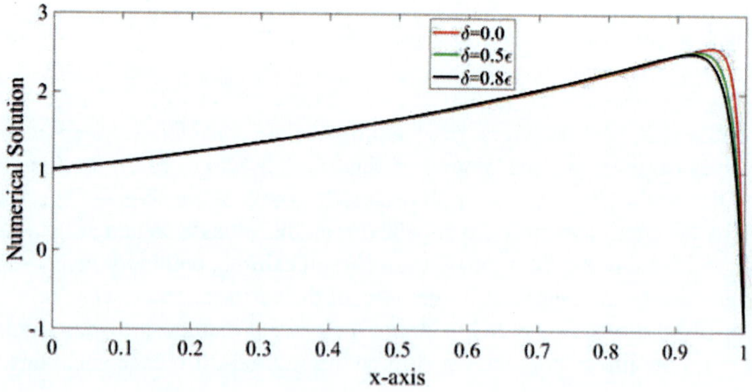

Fig. 4 Graph of numerical solution of Example 4 for $\epsilon = 10^{-2}$

References

1. Bender, C.M., Orszag, S.A.: Advanced Mathematical Methods for Scientists and Engineers. Mc Graw-Hill, New York (1978)
2. Kevorkian, J., Cole, J.D.: Perturbation Methods in Applied Mathematics. Springer, New York (1981)
3. Nayfeh, A.H.: Perturbation Methods. Wiley, New York (1973)
4. O' Malley R.E.: Introduction to Singular Perturbations. New York, Academic Press (1974)
5. Doolan, E.P., Miller, J.J.H., Schilders, W.H.A.: Uniform Numerical Methods for Problems with Initial and Boundary Layers. Boole Press, Dublin (1980)
6. Farrel, P.A., Hegarty, A.F., Miller, J.J.H., O'Riordan, E., Shishkin, G.I.: Robust Computational Techniques for Boundary Layers. New York, Chapman & Hall/CRC (2000)
7. Miller, J.J.H., O'Riordan, E., Shishkin, G.I.: Fitted Numerical Methods for Singular Perturbation Problems. Singapore, World Scientific (1996)
8. Roos, H.G., Stynes, M., Tobiska, L.: Numerical Methods for Singularly Perturbed Differential Equations. Springer, Berlin (1996)
9. Glizer, V.Y.: Asymptotic solution of a boundary-value problem for linear singularly-perturbed functional differential equations arising in optimal control theory. J. Optimizat. Theory Appl. **106**(2), 309–335 (2000)
10. Stein, R.B.: A theoretical analysis of neuronal variability. Biophys. J. **5**(2), 173–194 (1965)
11. Tuckwell, H.C.: On the first-exit time problem for temporally homogeneous Markov processes. J. Appl. Prob. **13**, 39–48 (1976)
12. Derstine, M.W., Gibbs, H.M., Hopf, F.A., Kaplan, D.L.: Bifurcation gap in a hybrid optical system. Physical Rev. A **26**, 3720(R) (1982)
13. Longtin, A., Milton, J.: Complex oscillations in the human pupil light reflex with mixed and delayed feedback. Math. Biosci. **90**(1–2), 183–199 (1988)
14. Mackey, M.C., Glass, L.: Oscillations and chaos in physiological control systems. Science 287–289 (1977)
15. Wazewska-Czyzewska, M., Lasota, A.: Mathematical models of the red cell system. Mat. Stos. **6**, 25–40 (1976)
16. Lange, C.G., Miura, R.M.: Singular perturbation analysis of boundary-value problems for differential difference equations: V. Small shifts with layer behavior. SIAM J. Appl. Math. **54**(1), 249–272 (1994)
17. Lange, C.G., Miura, R.M.: Singular perturbation analysis of boundary-value problems for differential difference equations: VI. Small shifts with rapid oscillations. SIAM J. Appl. Math. **54**(1), 273–283 (1994)
18. Kadalbajoo, M.K., Sharma, K.K.: Numerical treatment of a mathematical model arising from a model of neuronal variability. J. Math. Anal. Appl. **307**(2), 606–627 (2005)
19. Kadalbajoo, M.K., Sharma, K.K.: A numerical method based on finite difference for boundary value problems for singularly perturbed delay differential equations. Appl. Math. Comput. **197**(2), 692–707 (2008)
20. Kadalbajoo, M.K., Sharma, K.K.: Numerical analysis of singularly perturbed delay differential equations with layer behavior. Appl. Math. Comput. **157**(1), 11–28 (2004)
21. Kadalbajoo, M.K., Sharma, K.K.: Parameter-uniform fitted mesh method for singularly perturbed delay differential equations with layer behavior. Electron. Trans. Numer. Anal. **23**, 180–201 (2006)
22. Kadalbajoo, M.K., Ramesh, V.P.: Hybrid method for numerical solution of singularly perturbed delay differential equations. Appl. Math. Comput. **187**(2), 797–814 (2007)
23. Ramos, J.I.: Exponential methods for singularly perturbed ordinary differential-difference equations. Appl. Math. Comput. **182**(2), 1528–1541 (2006)
24. Rao, R.N., Chakravarthy, P.P.: An exponentially fitted tridiagonal finite difference method for singularly perturbed differential-difference equations with small shift. Ain Shams Eng. J. **5**, 1351–1360 (2014)

25. Mohapatra, J., Natesan, S.: Uniform convergence analysis of finite difference scheme for singularly perturbed delay differential equation on an adaptively generated grid. Numer. Math. Theory Methods Appl. **3**, 1–22 (2010)
26. Cunningham, W.J.: Introduction to Nonlinear Analysis. McGraw-Hill Book Company Inc, New York (1958)
27. Tian H.: Numerical treatment of singularly perturbed delay differential equations. Ph.D. thesis, University of Manchester (2000)
28. Boqucz, E.A., Walker, J.D.A.: Fourth order finite difference methods for two-point boundary value problems. IMA J. Numer. Anal. **4**(1), 69–82 (1984)
29. Varga, R.S.: Matrix Iterative Analysis. Prentice-Hall, Englewood Cliffs, New Jersey (1962)
30. Young, D.M.: Iterative Solution of Large Linear Systems. Academic Press, New York (1971)

Generalized Inverse of Special Infinite Matrix

V. Subharani, N. Jansirani, and V. R. Dare

Abstract In this article, Special Infinite Matrix (SIM) and Complementary Special Infinite Matrix (CSIM) are introduced and its basic properties are studied. Various types of Symmetric Properties for SIM and CSIM are discussed. The existence of Generalized Inverse and Special Inverses for SIM and CSIM are derived. The complexity of SIM and CSIM are analyzed and suitable examples are provided.

Keywords Circulant · Drazin inverse · Generalized inverse · Symmetry

Mathematics Subject Classification (2010) 15A09 · 05A05

1 Introduction

In Digital Image Processing, when recovering the original image from the degraded image generalized inverse is used [4]. Robotic research the concept of generalized inverse are contributed a lot [12]. Statistics, Numerical Linear Algebra are rapidly handled the concepts via Generalized Inverse. The concept of Drazin inverse was shown to be very useful in various applied mathematical setting [6, 10, 11]. This has a lot of applications in Singular Differential Equation or Difference Equations, Markov Chains, Cryptography, Iterative Method or Multibody System Dynamics. In 1920, Moore was first introduced the notion of the Generalized Inverse of a matrix for any dimension. In 1955, unaware of Moore's work, Penrose defined the Generalized

V. Subharani (✉) · N. Jansirani
Department of Mathematics, Queen Mary's College, University of Madras, Chennai, India
e-mail: subhashreethiyag@gmail.com

N. Jansirani
e-mail: njansirani@gmail.com

V. R. Dare
Department of Mathematics, Madras Christian College, University of Madras, Chennai, India
e-mail: rajkumardare@yahoo.com

R. N. Mohapatra et al. (eds.), *Mathematical Analysis and Computing*,
Springer Proceedings in Mathematics & Statistics 344,
https://doi.org/10.1007/978-981-33-4646-8_27

Inverse. Both the definition is equivalent [1–3]. Since the Generalized Inverse called as a Moore-Penrose inverse. If A is a square non-singular matrix then there exists a unique matrix B, such that $AB = BA = I$, then B is called as an inverse of A. If A is a singular or a rectangle matrix, no such inverse exists. If inverse of A exists, then the system of linear equation has a unique solution. If the linear equations are inconsistent, then least-square solutions are used. Generalized Inverse possesses the property that it must reduce to inverse of A, when it is a non-singular matrix and it should exists for a larger class. It should satisfy the properties of inverse such as $(A^{-1})^{-1} = A$, $(A^{-1})^* = (A^*)^{-1}$, etc. and should provide the answer to the question such as consistency or least-square solutions. The second section contains the basic definitions and preliminaries for this article. In third section the SIM and CSIM are examined with the property of Circulant, Symmetry and Pascal symmetry. In fourth section, the existence of Generalized Inverses of SIM and CSIM are analyzed and examples are provided. In fifth section, the Special Inverses such as Drazin, Group Inverse of SIM and CSIM are investigated and verified through the examples.

2 Basic Definitions and Preliminaries

Let A be a matrix over the field of real numbers. The dimension of the matrix is the total number of rows and columns of a matrix, defined as $m \times n$, where m is a number of rows and n is a number of columns, which are positive integer. The elements of the matrix are defined as a_{ij} is the ith row and jth column a matrix, where $0 < i \leq m, 0 < j \leq n$ [7]. Throughout this paper the bottom-most row is the first row and left-most column is the first column. The dimension $(m \times n)$ is represent the first m rows and n columns of a given matrix.

$$\text{Structure of The Matrix is } A = \begin{pmatrix} a_{m1} & a_{m2} & \cdots & a_{m(n-1)} & a_{mn} \\ a_{(m-1)1} & a_{(m-1)2} & \cdots & a_{(m-1)(n-1)} & a_{(m-1)n} \\ \cdots & \cdots & \ddots & \vdots & \vdots \\ a_{21} & a_{22} & \cdots & a_{2(n-1)} & a_{2n} \\ a_{11} & a_{12} & \cdots & a_{1(n-1)} & a_{1n}. \end{pmatrix}$$

If $A = (a_{ij}) \in R^{m \times n}$, then

- The transpose of A is defined as $A^T = (a_{ij})^T = a_{ji}$
- The secondary transpose of A is defined as $A^S = (a_{ij})^S = a_{n-j+1,m-i+1}$

$A = (a_{ij}) \in R^{n \times n}$ is

- Symmetric if $A = A^T$
- Secondary symmetric if $A = A^S$
- N-symmetric if $AA^T = A^T A$
- SN-symmetric if $AA^S = A^S A$
- Orthogonal if $AA^T = A^T A = I$
- Secondary Orthogonal if $AA^S = A^S A = I$
- Pascal symmetric if $a_{ij} = a_{rs}$, for $i + j = r + s$ [8].

If A is a Bi-symmetric then A is both symmetric and secondary symmetric matrix and is a BN-symmetric if A is both N-symmetric and BN-symmetric matrix. A is bi-orthogonal if A is both orthogonal and secondary orthogonal matrix. The left boundary of A is defined as $a_{11}a_{21}a_{31}...a_{m1}a_{m2}a_{m3}...a_{mn}$ and the right boundary of A is defined as $a_{11}a_{12}a_{13}...a_{1n}a_{2n}a_{3n}...a_{mn}$. If A is a primary diagonal symmetric matrix then the right boundary and the left boundary are equal. The secondary left boundary of A is defined as $a_{1n}a_{1(n-1)}a_{1(n-2)}...a_{12}a_{11}a_{21}a_{31}...a_{(m-1)1}a_{m1}$ and the secondary right boundary of A is defined as $a_{1n}a_{2n}a_{3n}...a_{(m-1)n}a_{mn}a_{m(n-1)}a_{m(n-2)}...a_{m2}a_{m1}$. If the secondary right boundary and secondary left boundary of A is equal then it is said to be a secondary diagonal symmetric matrix. If $A = (a_{ij}) \in R^{n\times n}$, is a tripotent matrix then $A^3 = A$ and k-tripotent matrix if $A^k = (A^3)^k$ where $k \geq 2$ [13]. Any A have the rank r if it has atleast one submatrix of order r which is non-singular [1].

3 Complexity of SIM and CSIM

In this section, SIM and CSIM are introduced and analyzed with the properties of Circulant, Symmetry and Pascal symmetry. Various types of Palindromes are defined and examined with the examples.

Definition 1 The Special Infinite Matrix (SIM) $A = (a_{ij})_{i,j\geq 1} \in R^{m\times n}$, is defined as $a_{ij} = \left\{\begin{matrix} 1 \; (i+j-1) \;\; mod \;\; 3 = 0 \\ 0 \qquad\quad otherwise. \end{matrix}\right\}$.

Then the corresponding SIM of A is $\begin{pmatrix} \vdots & \vdots & \vdots & \vdots & \vdots & \iddots \\ 0 & 1 & 0 & 0 & 1 & \dots \\ 0 & 0 & 1 & 0 & 0 & \dots \\ 1 & 0 & 0 & 1 & 0 & \dots \\ 0 & 1 & 0 & 0 & 1 & \dots \\ 0 & 0 & 1 & 0 & 0 & \dots \end{pmatrix}$.

Definition 2 The Complementary Special Infinite Matrix (CSIM) $A = (a_{ij})_{i,j\geq 1} \in R^{m\times n}$, is defined as $a_{ij} = \left\{\begin{matrix} 0 \; (i+j-1) \;\; mod \;\; 3 \neq 0 \\ 1 \qquad\quad otherwise. \end{matrix}\right\}$.

Then the corresponding CSIM Of A is $\begin{pmatrix} \vdots & \vdots & \vdots & \vdots & \vdots & \iddots \\ 1 & 0 & 1 & 1 & 0 & \dots \\ 1 & 1 & 0 & 1 & 1 & \dots \\ 0 & 1 & 1 & 0 & 1 & \dots \\ 1 & 0 & 1 & 1 & 0 & \dots \\ 1 & 1 & 0 & 1 & 1 & \dots \end{pmatrix}$.

Definition 3 $A = (a_{ij})_{i,j\geq 1} \in R^{m\times n}$, is Circulant if

$$a_{ij} = \left\{\begin{matrix} a_{i+j-1} \;\; if \;\; (i+j) \leq n+1 \\ a_{i+j-n-1} \; if \;\; (i+j) > n+1, \end{matrix}\right\} \text{ where } i, j, n, m \in N.$$

Structure of Circulant Matrix

$$\mathfrak{Cir}(A) = \begin{pmatrix} a_n & a_1 & \dots & a_{n-2} & a_{n-1} \\ a_{n-1} & a_n & \dots & a_{n-3} & a_{n-2} \\ \vdots & \vdots & \ddots & \vdots & \vdots \\ a_2 & a_3 & \dots & a_n & a_1 \\ a_1 & a_2 & \dots & a_{n-1} & a_n \end{pmatrix}$$

Example 1 $\mathfrak{Cir}(A) = \begin{pmatrix} e & d & b & a & c \\ c & e & d & b & a \\ a & c & e & d & b \\ b & a & c & e & d \end{pmatrix}_{4\times5}$ is a circulant matrix of dimension 4×5.

In this example $a_1 = b, a_2 = a, a_3 = c, a_4 = e, a_5 = d$, consider a_{34} be the arbitary position of A, which comes under the condition $(i + j) > n + 1$, $(3 + 4) > 6$ then $a_{34} = a_{(i+j-n-1)}, a_{(7-6)} = a_1 = b$.

Definition 4 $A = (a_{ij})_{i,j\geq1} \in R^{m\times n}$, is said to be a right (or left) boundary palindrome matrix if the right (or left) boundary of A and it's reverse are equal. A is said to be a secondary right (or secondary left) boundary palindrome matrix if the secondary right (or left boundary) and it's reverse are equal.

If the boundary palindrome of a matrix and the secondary boundary palindrome of a matrix are equal then it is called as a palindrome matrix.

Definition 5 The primary left diagonal boundary of a matrix is the elements from bottom left corner to top right corner. The primary left diagonal boundary is defined as

$$a_{11}a_{21}a_{12}a_{31}a_{22}a_{13}\dots a_{n1}a_{(n-1)2}\dots a_{2(n-1)}a_{2n}\dots a_{n(n-1)}a_{(n-1)n}a_{nn}.$$

The primary right diagonal boundary of a matrix is the element from top right corner to bottom left corner. The primary right diagonal boundary is defined as

$$a_{nn}a_{(n-1)n}a_{n(n-1)}\dots a_{2n}a_{2(n-1)}\dots a_{(n-1)2}a_{n1}\dots a_{13}a_{22}a_{31}a_{12}a_{21}a_{11}.$$

Definition 6 The secondary right diagonal boundary of a matrix is the elements from bottom right corner to top left corner. The secondary right diagonal boundary is defined as

$$a_{1n}a_{1(n-1)}a_{2n}a_{1(n-2)}a_{2(n-1)}a_{3n}\dots a_{11}a_{22}\dots$$
$$a_{nn}a_{21}a_{32}\dots a_{n(n-1)}\dots a_{(n-1)1}a_{n2}a_{n1}.$$

The secondary left diagonal boundary of a matrix is the elements from top left corner to bottom right corner. The secondary left diagonal boundary is defined as

$$a_{n1}a_{n2}a_{(n-1)1}\dots a_{n(n-1)}\dots a_{32}a_{21}a_{nn}\dots a_{22}a_{11}\dots$$
$$a_{3n}a_{2(n-1)}a_{1(n-2)}a_{2n}a_{1(n-1)}a_{1n}.$$

Definition 7 If the primary right diagonal boundary and left diagonal boundary of a matrix are same then the matrix is called as a palindrome primary diagonal matrix. If the secondary right diagonal boundary and left diagonal boundary of a matrix are same then the matrix is called as a palindrome secondary diagonal matrix. Every palindrome primary diagonal need not be a palindrome secondary diagonal and vice versa.

Example 2 Let $A=\begin{pmatrix} a & b & d & a & a \\ b & a & c & a & b \\ c & b & a & d & d \\ b & a & b & c & a \\ a & a & c & a & b \end{pmatrix}$. Then the right boundary of a matrix is $aacabadba$. The left boundary of a matrix is $abcbabdaa$. The secondary right boundary of a matrix is $bacaabcba$. The secondary left boundary of a matrix is $badbaadba$. The secondary right boundary of a matrix is $bacaabcba$.

Example 3 Let $A = \begin{pmatrix} a & a & b & a & a \\ a & a & a & b & a \\ b & a & a & a & b \\ a & b & a & a & a \\ a & a & b & a & a \end{pmatrix}$ is a Palindrome and Pascal matrix.

Example 4 Let $A = \begin{pmatrix} 0 & 1 & 0 \\ 0 & 1 & 1 \\ 1 & 0 & 0 \end{pmatrix}$ be a symmetric matrix but not a secondary symmetric matrix and $A = \begin{pmatrix} 1 & 0 & 1 \\ 0 & 0 & 1 \\ 1 & 1 & 0 \end{pmatrix}$ be a secondary symmetric matrix but not a symmetric matrix. But $A = \begin{pmatrix} 0 & 1 & 1 \\ 1 & 0 & 1 \\ 1 & 1 & 0 \end{pmatrix}$ is a BN-symmetric matrix.

Theorem 1 *Any (m, n) dimension of SIM and CSIM are Circulant matrices, where $m, n \geq 3$.*

Proof By the Method of Mathematical Induction, to prove this theorem is true for 4×4. Let $A = \begin{pmatrix} 0 & 0 & 1 & 0 \\ 1 & 0 & 0 & 1 \\ 0 & 1 & 0 & 0 \\ 0 & 0 & 1 & 0 \end{pmatrix}$ then the possible submatrix are $\begin{pmatrix} 1 & 0 & 0 \\ 0 & 1 & 0 \\ 0 & 0 & 1 \end{pmatrix}$ $\begin{pmatrix} 0 & 0 & 1 \\ 1 & 0 & 0 \\ 0 & 1 & 0 \end{pmatrix}$ $\begin{pmatrix} 0 & 1 & 0 \\ 0 & 0 & 1 \\ 1 & 0 & 0 \end{pmatrix}$ by the definition of circulant, the SIM is a circulant type. Furthermore the right boundary and left boundary are equal then the submatrices are secondary diagonal symmetric matrix. Hence the theorem is true for $m, n = 4$. Then this theorem is true for some r. To prove this theorem is true for $r + 1$. In SIM, the possibility of extension of dimensions are $(m \times n + 1)$, $(m + 1 \times n)$, $(m + 1 \times n + 1)$. In all the possibilities the right boundary of SIM decides whether it is a circulant or not. But any dimension of SIM, the right boundary and left boundary are same. Hence the theorem.

Theorem 2 *If A is a submatrix of SIM of dimension* (3×3), *then the following statements exist*

1. *A is a Bi-orthogonal matrix and tripotent matrix.*
2. *A is a k-tripotent and* A^T, A^S *is a tripotent and k-tripotent matrix.*

Proof The possible submatrices of SIM of dimension 3×3 are $\begin{pmatrix} 1 & 0 & 0 \\ 0 & 1 & 0 \\ 0 & 0 & 1 \end{pmatrix}$, $\begin{pmatrix} 0 & 0 & 1 \\ 1 & 0 & 0 \\ 0 & 1 & 0 \end{pmatrix}$, $\begin{pmatrix} 0 & 1 & 0 \\ 0 & 0 & 1 \\ 1 & 0 & 0 \end{pmatrix}$. Furthermore the possible submatrices of SIM of dimension 3×3 are permutation matrix, secondary diagonal symmetric matrix. The permutation matrix are orthonormal, orthogonal matrix and $A^2 = I$, $A = A^T = A^{-1}$. Hence they are a symmetric, N-symmetric and N-unitary matrix. Since the submatrices are also a secondary orthogonal matrix then $A^S = A^{-1} = I$. Hence they are a secondary symmetric and SN-symmetric matrix. Finally, the submatrices of SIM of dimension 3×3 are Bi-symmetric, BN-symmetric, Bi-orthogonal matrix. Since $A^2 = I$, they are tripotent and k-tripotent.

Theorem 3 *If A is a SIM (or CSIM) of dimension* $3n \times 3n$. *then the following statements are hold.*

1. $A^T, A^S, A^{T^S}, A^{S^T}$, $(A + A^T)$, $(A^T + A)$, $(A + A^T)^k$, $(A^T + A)^k$ *are circulant matrix.*
2. *Any submatrix of any dimension of SIM ((or CSIM)) are circulant matrix.*

Proof Consider A is a SIM. The proof is by the method of Mathematical Induction. To prove this theorem is true for $n = 3$, then $A = \begin{pmatrix} 1 & 0 & 0 \\ 0 & 1 & 0 \\ 0 & 0 & 1 \end{pmatrix}$, then $A^T, A^S, A^{T^S}, A^{S^T}$, $(A + A^T)$, $(A^T + A)$, $(A + A^T)^k$, $(A^T + A)^k$ are circulant matrices. Hence this is true for $n = r$ for some positive integer. Consider the dimension $n = r + 1$, furthermore the SIM is a palindrome matrix if the dimension is $3n$. Hence the theorem. The submatrix of right boundary and left boundary are equal, hence it is circulant. Similar proof holds for CSIM.

Theorem 4 *If* $A \in R^{(3n \times 3n)}$ *is SIM (or CSIM), then A is a BN-symmetric matrix and palindrome matrix.*

Proof The proof follows by the Mathematical Induction Method. Let A be a SIM. To prove this theorem is true for $n = 2$, then $A = \begin{pmatrix} 1 & 0 & 0 & 1 & 0 & 0 \\ 0 & 1 & 0 & 0 & 1 & 0 \\ 0 & 0 & 1 & 0 & 0 & 1 \\ 1 & 0 & 0 & 1 & 0 & 0 \\ 0 & 1 & 0 & 0 & 1 & 0 \\ 0 & 0 & 1 & 0 & 0 & 1 \end{pmatrix}$,

$$A^T = \begin{pmatrix} 1&0&0&1&0&0 \\ 0&1&0&0&1&0 \\ 0&0&1&0&0&1 \\ 1&0&0&1&0&0 \\ 0&1&0&0&1&0 \\ 0&0&1&0&0&1 \end{pmatrix} = A^S, \text{ Hence it is a BN-Symmetric matrix.}$$

The primary (or secondary) right diagonal of a matrix and the primary (or secondary) left diagonal of a matrix are same. Hence it is a palindrome matrix. Hence the theorem is true for $n = 2$. Then assume that this true for $n = r$ for some positive integer r.

Consider $n = r + 1$, $(A)_{r+1} = \begin{pmatrix} 1\,0\,0 \ldots 1\,0\,0 \\ 0\,1\,0 \ldots 0\,1\,0 \\ 0\,0\,1 \ldots 0\,0\,1 \\ \vdots\;\vdots\;\vdots\;\ddots\;\vdots\;\vdots\;\vdots \\ 1\,0\,0 \ldots 1\,0\,0 \\ 0\,1\,0 \ldots 0\,1\,0 \\ 0\,0\,1 \ldots 0\,0\,1 \end{pmatrix}_{3(r+1)} = A^T = A^S$, hence the theorem. Similar proof is true for CSIM of dimension $3n \times 3n$.

4 Existence of Generalized Inverses for SIM and CSIM

In this section, Generalized Inverse of SIM are derived, the existence of Semi-Inverse, Least-Square G-Inverse, Minimum Norm G-Inverse of SIM are also studied and the concepts are explained with the examples.

Definition 8 $A \in R^{(m\times n)}$ is a SIM (or CSIM). Then $A^{\ddagger} \in R^{(n\times m)}$ is a G-Inverse of A if $AA^{\ddagger}A = A$. The G-Inverse is not unique.

Definition 9 $A \in R^{(m\times n)}$ is a SIM (or CSIM). Then $A^{\ddagger} \in R^{(n\times m)}$ is a $\{1, 2\}$-inverse or Semi-Inverse of A if $AA^{\ddagger}A = A$ and $A^{\ddagger}AA^{\ddagger} = A^{\ddagger}$.

Definition 10 $A \in R^{(m\times n)}$ is a SIM (or CSIM). Then $A^{\ddagger} \in R^{(n\times m)}$ is a $\{1, 3\}$-inverse or a Least-Square G-Inverse of A if $AA^{\ddagger}A = A$ and $(AA^{\ddagger})^T = AA^{\ddagger}$.

Definition 11 $A \in R^{(m\times n)}$ is a SIM (or CSIM). Then $A^{\ddagger} \in R^{(n\times m)}$ is a $\{1, 4\}$-inverse or a Minimum Norm G-Inverse of A if $AA^{\ddagger}A = A$ and $(A^{\ddagger}A)^T = A^{\ddagger}A$.

Definition 12 $A \in R^{(m\times n)}$ is a SIM (or CSIM). Moore-Penrose $A^{\ddagger} \in R^{(n\times m)}$ of $A \in R^{(m\times n)}$ is the unique matrix which satisfies the following 4-conditions

$$AA^{\ddagger}A = A \tag{1}$$

$$A^{\ddagger}AA^{\ddagger} = A^{\ddagger} \tag{2}$$

$$(AA^{\ddagger})^T = AA^{\ddagger} \tag{3}$$

$$(A^{\ddagger}A)^T = A^{\ddagger}A \tag{4}$$

Theorem 5 *$A \in R^{(m\times n)}$ is a SIM (or CSIM). If $A^{\ddagger}$ exists then it is unique.*

Proof If $A^{\ddagger}$ exists, then $A\{1,2,3,4\}$ exist. If $M, N \in A\{1,2,3,4\}$ then

$$\begin{aligned} M &= MAM = M(AM)^T = MM^TA^T = MM^T(ANA)^T \\ &= M(AM)^T(AN)^T = MAMAN = MAN = (MA)^T(NA)^TN \\ &= A^TM^TA^TN^TN = A^TN^TN = NAN = N \end{aligned}$$

Hence $A^{\ddagger}$ is unique.

Example 5 Consider 4×4 dimension of CSIM A is $\begin{pmatrix} 0&0&1&0\\ 1&0&0&1\\ 0&1&0&0\\ 0&0&1&0 \end{pmatrix}$,

then $A^{\ddagger} = \begin{pmatrix} 0&0&0.5&0\\ 0.5&0&0&0.5\\ 0&1&0&0\\ 0&0&0.5&0 \end{pmatrix}$. $AA^{\ddagger}A = A$, $A^{\ddagger}AA^{\ddagger} = A^{\ddagger}$ and

$(AA^{\ddagger})^T = \begin{pmatrix} 0.5&0&0&0.5\\ 0&0&1&0\\ 0&1&0&0\\ 0.5&0&0&0.5 \end{pmatrix} = AA^{\ddagger}$, $(A^{\ddagger}A)^T = \begin{pmatrix} 0.5&0&0&0.5\\ 0&0&1&0\\ 0&1&0&0\\ 0.5&0&0&0.5 \end{pmatrix} = A^{\ddagger}A.$

Property 1 If $A \in R^{(m\times n)}$ is a SIM (or CSIM). Then $(A^{\ddagger})^{\ddagger} = A$, $(A^T)^{\ddagger} = (A^{\ddagger})^T$, $A^{\ddagger} = (A^TA)^{\ddagger}A^T = A^T(A^TA)^{\ddagger}$.

Proof Consider 4×4 dimension of CSIM A is $\begin{pmatrix} 1&1&0&1\\ 0&1&1&0\\ 1&0&1&1\\ 1&1&0&1 \end{pmatrix}$, then

$A^{\ddagger} = \begin{pmatrix} 0.125&0.25&-0.25&0.125\\ -0.25&0.5&0.5&-0.25\\ 0.25&-0.5&0.5&0.25\\ 0.125&0.25&-0.25&0.125 \end{pmatrix}$. $A^{\ddagger\ddagger} = \begin{pmatrix} 0.125&0.25&-0.25&0.125\\ -0.25&0.5&0.5&-0.25\\ 0.25&-0.5&0.5&0.25\\ 0.125&0.25&-0.25&0.125 \end{pmatrix}^{\ddagger} = A,$

$(A^T)^{\ddagger} = \begin{pmatrix} 0.125&0.25&-0.25&0.125\\ -0.25&0.5&0.5&-0.25\\ 0.25&-0.5&0.5&0.25\\ 0.125&0.25&-0.25&0.125 \end{pmatrix} = (A^{\ddagger})^T$

and $A^{\ddagger} = (A^TA)^{\ddagger}A^T = A^T(A^TA)^{\ddagger}$.

From this, SIM and CSIM have satisfied the properties.

Theorem 6 *$A \in R^{m\times n}$ is a SIM (or CSIM). If $A^{\ddagger}$ is a $A\{1\}$-inverse of A and $\lambda \in R$, then*

1. $(A^{\ddagger})^T \in A^T\{1\}$
2. *If A is non-singular, then $A^{(1)} = A^{-1}$ is unique.*

3. $\lambda^{\ddagger} A^{(1)} \in (\lambda A)\{1\}$
4. $rank(A^{\ddagger}) \geq rank(A)$
5. *If S and T are permutation matrices, then* $T^{-1}A^{\ddagger}S^{-1} \in SAT\{1\}$.
6. $AA^{\ddagger}$ *and* $A^{\ddagger}A$ *are idempotent and have the same rank as* A.

Proof (1) $A^{\ddagger}$ is a $A\{1\}$-inverse of A. Then
$AA^{\ddagger}A = A \Rightarrow (AA^{\ddagger}A)^T = A^T \Rightarrow A^T A^{\ddagger^T} A^T = A^T$. Hence, $A^{\ddagger^T}\{1\} \in A^T\{1\}$.
(2) If A is non-singular, then $AA^{-1} = I = A^{-1}A$.
$A = AA^{\ddagger}A \Rightarrow A^{-1}AA^{-1} = A^{-1}AA^{\ddagger}AA^{-1} \Rightarrow A^{-1} = A^{\ddagger}$.
(3) λ is a scalar. It is defined by
$\lambda^{\ddagger} = \begin{Bmatrix} \lambda^{-1}\ if & \lambda \neq 0 \\ 0\ \ if & \lambda = 0 \end{Bmatrix}$. If $\lambda^{\ddagger}A^{(1)} \in (\lambda A)\{1\}$,
then $\lambda A \lambda^{\ddagger} A^{(1)} \lambda A = \lambda\lambda^{\ddagger}\lambda A = \lambda A$.
(4) $AA^{\ddagger}A = A$, $rank(AA^{\ddagger}A) \leq rank(AA^{\ddagger}) \leq rank(A)$.
(5) Since S and T are permutation matrices, S and T are invertible,
$S^{(-1)} = S^T$, $P^{(-1)} = P^T$. $AA^{\ddagger}A = A$
$(SAT)T^{-1}A^{\ddagger}S^{-1}(SAT) = SA(TT^{-1})A^{\ddagger}(S^{-1}S)AT = SAA^{\ddagger}AT = SAT$.
Then $T^{-1}A^{\ddagger}S^{-1} \in SAT\{1\}$
(6) $(AA^{\ddagger})^2 = AA^{\ddagger}AA^{\ddagger} = AA^{\ddagger}$. Similarly $(A^{\ddagger}A)^2 = A^{\ddagger}AA^{\ddagger}A = A^{\ddagger}A$.
Hence, $AA^{\ddagger}$ and $A^{\ddagger}A$ are idempotent.
$rank AA^{\ddagger} \leq rank(A) = rank(AA^{\ddagger}A) \leq rank(A^{\ddagger}A)$.

Example 6 Consider A is a SIM of dimension 5×5.

$$\text{Let } A = \begin{pmatrix} 0&1&0&0&1 \\ 0&0&1&0&0 \\ 1&0&0&1&0 \\ 0&1&0&0&1 \\ 0&0&1&0&0 \end{pmatrix} \text{ then } A^{\ddagger} = \begin{pmatrix} 0&0.25&0&0&0.25 \\ 0&0&0.5&0&0 \\ 0.5&0&0&0.5&0 \\ 0&0.25&0&0&0.25 \\ 0&0&0.5&0&0 \end{pmatrix}$$

It satisfies the following $AA^{\ddagger}A = A$, $A^T A^{\ddagger^T} A^T = A^T$

$$AA^{\ddagger} = \begin{pmatrix} 0&0.5&0&0&0.5 \\ 0.5&0&0&0.5&0 \\ 0&0&1&0&0 \\ 0&0.5&0&0&0.5 \\ 0.5&0&0&0.5&0 \end{pmatrix} = A^{\ddagger}A,\ (AA^{\ddagger})^T = \begin{pmatrix} 0&0.5&0&0&0.5 \\ 0.5&0&0&0.5&0 \\ 0&0&1&0&0 \\ 0&0.5&0&0&0.5 \\ 0.5&0&0&0.5&0 \end{pmatrix} = A^{\ddagger}A \text{ and}$$

$$(A^{\ddagger}A)^T = \begin{pmatrix} 0&0.5&0&0&0.5 \\ 0.5&0&0&0.5&0 \\ 0&0&1&0&0 \\ 0&0.5&0&0&0.5 \\ 0.5&0&0&0.5&0 \end{pmatrix} = A^{\ddagger}A. \quad (AA^{\ddagger})^2 = \begin{pmatrix} 0&0.5&0&0&0.5 \\ 0.5&0&0&0.5&0 \\ 0&0&1&0&0 \\ 0&0.5&0&0&0.5 \\ 0.5&0&0&0.5&0 \end{pmatrix} = AA^{\ddagger}$$

$$\text{and } (A^{\ddagger}A)^2 = \begin{pmatrix} 0&0.5&0&0&0.5 \\ 0.5&0&0&0.5&0 \\ 0&0&1&0&0 \\ 0&0.5&0&0&0.5 \\ 0.5&0&0&0.5&0 \end{pmatrix} = A^{\ddagger}A.$$

Hence $AA^{\ddagger}$, $A^{\ddagger}A$ are idempotent and $rank(AA^{\ddagger}) = rank(A^{\ddagger}A) = rank(A)$.

Theorem 7 *$A \in R^{(m\times n)}$ is a SIM (or CSIM). Let $R, S \in A\{1\}$, and let $P = RAS$, then $P \in A\{1, 2\}$.*

Proof $R, S \in A\{1\}$, then $ARA = A$, $ASA = A$
$APA = A(RAS)A = ASA = A$.
$PAP = (RAS)A(RAS) = R(ASA)(RAS) = (RAR)AS = RAS = P$.
Hence, $P \in \{1, 2\}$.

Theorem 8 *$A \in R^{(m\times n)}$ is a SIM (or CSIM). $A^{\ddagger} \in A\{1\}$, then $A^{\ddagger} \in A\{2\}$ iff $rank(AA^{\ddagger}) = rank(A^{\ddagger})$.*

Proof $A^{\ddagger} \in A\{2\}$, then $A^{\ddagger}AA^{\ddagger} = A^{\ddagger} \Rightarrow A \in A^{\ddagger}\{1\} \Rightarrow rank(A^{\ddagger}) = rank(AA^{\ddagger})$.
Conversely, consider $rank(A^{\ddagger}) = rank(AA^{\ddagger}) \Rightarrow A^{\ddagger} = YAA^{\ddagger}$ for some Y matrix.
$A^{\ddagger}AA^{\ddagger} = YAA^{\ddagger}AA^{\ddagger}$
$A^{\ddagger}AA^{\ddagger} = YAA^{\ddagger} = A^{\ddagger}$, then $A^{\ddagger} \in A\{2\}$. Hence the theorem.

Theorem 9 *$A \in R^{(m\times n)}$ is a SIM (or CSIM), if $A^{\ddagger} \in A\{1, 2\}$ then $rank(A) = rank(A^{\ddagger})$.*

Proof Since $A^{\ddagger} \in A\{1\}$ then $AA^{\ddagger}A = A$, furthermore $AA^{\ddagger}$ is idempotent and the *rank* is equal to A.
$A^{\ddagger} \in A\{2\}$ then $A^{\ddagger}AA^{\ddagger} = A^{\ddagger}$ and $AA^{\ddagger}$ is idempotent and the *rank* is equal to A.
Hence, $rank(A) = rank(A^{\ddagger})$.

Theorem 10 *If $A \in R^{(m\times n)}$ is a SIM (or CSIM) then $A^{\ddagger} = A^{(1,4)}AA^{(1,3)}$.*

Proof From $A^{\ddagger} = A^{(1,4)}AA^{(1,3)}$, consider $A^{\ddagger} = X$ then $X = XAX$, hence $A^{\ddagger} \in \{2\}$. And $A^{\ddagger}A$, $AA^{\ddagger}$ are symmetric because of $\{3, 4\}$-inverse.

Hence, $A^{\ddagger} \in A\{1, 2, 3, 4\}$. Furthermore $A\{1, 2\}$-inverse of a SIM (or CSIM) is a $\{2\}$-inverse. Similarly $A\{1, 2, 3\}$-inverse of a SIM (or CSIM) is a $\{1, 3\}$-inverse and a $\{2, 3\}$-inverse. The existence of a $\{1, 2, 3, 4\}$-inverse, demonstrated the existence of $\{i, j, ..., k\}$-inverse for all possible choices of one, two, three integers i, j, âŁ¦, k from the set $\{1, 2, 3, 4\}$. If $\{1, 2, 3, 4\}$-inverse exists then it is unique.

Theorem 11 *$A \in R^{(m\times n)}$ is a SIM (or CSIM). The set $A\{1, 3\}$ consists of solution for P of $AP = AH$, H is a $\{1, 3\}$ inverse of A.*

Proof $H \in A\{1, 3\}$, then $AHA = A, (AH)^T = AH$ and $P \in A\{1, 3\}$, then $APA = A, (AP)^T = AP$.
$AH = APAH = (AP)^T(AH)^T = P^TA^T = (AP)^T = AP$.

Hence P is a solution of $AP = AH$. Conversely, consider $AH = AP$ with $H \in A\{1, 3\}$. $AHA = A$ and $APA = A$ which implies $P \in A\{1\}$.

Since $AH = AP \Rightarrow (AH)^T = (AP)^T \Rightarrow AH = (AP)^T \Rightarrow AP = (AP)^T \Rightarrow P \in A\{3\}$. Hence $P \in A\{1, 3\}$.

Example 7 Let A be a submatrix of CSIM of dimension 4×4.

$$A = \begin{pmatrix} 0 & 1 & 1 & 0 \\ 1 & 0 & 1 & 1 \\ 1 & 1 & 0 & 1 \\ 0 & 1 & 1 & 0 \end{pmatrix}, \text{ then } A^{\ddagger} = \begin{pmatrix} -0.125 & 0.25 & 0.25 & -0.125 \\ 0.25 & -0.5 & 0.5 & 0.25 \\ 0.25 & 0.5 & -0.5 & 0.25 \\ -0.125 & 0.25 & 0.25 & -0.125 \end{pmatrix}$$

$$(AA^{\ddagger})^T = \begin{pmatrix} 0.5 & 0 & 0 & 0.5 \\ 0 & 0 & 1 & 0 \\ 0 & 1 & 0 & 0 \\ 0.5 & 0 & 0 & 0.5 \end{pmatrix} = AA^{\ddagger} \text{ and } (A^{\ddagger}A)^T = \begin{pmatrix} 0.5 & 0 & 0 & 0.5 \\ 0 & 0 & 1 & 0 \\ 0 & 1 & 0 & 0 \\ 0.5 & 0 & 0 & 0.5 \end{pmatrix} = A^{\ddagger}A$$

5 Existence of Special Inverses

In this section the Special Inverses like Drazin inverse and Group inverse of SIM and CSIM are studied. Examples are provided.

Definition 13 The following equations applicable only to square matrices

$$A^k A^{\ddagger} A = A^k \qquad (1^k)$$
$$AA^{\ddagger} = A^{\ddagger}A \qquad (5)$$
$$A^k A^{\ddagger} = A^{\ddagger} A^k \qquad (5^k)$$
$$A(A^{\ddagger})^k = (A^{\ddagger})^k A \qquad (6^k)$$

where k is an $index$ of A which is the smallest positive integer satisfies the condition that $rank A^k = rank A^{k+1}$.

Definition 14 $A \in R^{(n \times n)}$ is a SIM (or CSIM). The Drazin inverse of A is a unique matrix $A^{\ddagger} \in R^{(n \times n)}$ satisfying the relation $A^{k+1}A^{\ddagger} = A^k$, $AA^{\ddagger} = A^{\ddagger}A$, $A^{\ddagger}AA^{\ddagger} = A^{\ddagger}$, where k is the index of A. The unique $\{1^k, 2, 5\}$-inverse is called Drazin inverse of A, and is denoted by A^D.

Example 8 Let A be a submatrix of CSIM of dimension 4×4. $A = \begin{pmatrix} 1 & 0 & 1 & 1 \\ 1 & 1 & 0 & 1 \\ 0 & 1 & 1 & 0 \\ 1 & 0 & 1 & 1 \end{pmatrix}$,

$$A^{\ddagger} = \begin{pmatrix} 0.125 & -0.25 & 0.25 & 0.125 \\ 0.125 & 0.5 & -0.5 & -0.25 \\ -0.125 & 0.5 & 0.5 & 0.25 \\ 0.125 & -0.25 & 0.25 & 0.125 \end{pmatrix}. \text{ Then } AA^{\ddagger} = \begin{pmatrix} 0.5 & 0 & 0 & 0.5 \\ 0 & 0 & 1 & 0 \\ 0 & 1 & 0 & 0 \\ 0.5 & 0 & 0 & 0.5 \end{pmatrix} = A^{\ddagger}A.$$

Definition 15 $A \in R^{(n \times n)}$ is a SIM (or CSIM). The Drazin inverse A of index 0 or 1 is called the Group inverse of A or $\{1, 2, 5\}$-inverse and is denoted by $A^{\#}$. When the Group inverse exists, it is unique.

The Group inverse does not exist for all square matrices, but only those of $index$ 1.

Theorem 12 $A \in R^{(n\times n)}$ *is a SIM (or CSIM). If Group inverse exists then it is unique.*

Proof If $A^{\#}$ exists, then $A\{1,2,5\}$ exist.
If $P, Q \in A\{1,2,5\}$, $E = AP = PA$ and $F = AQ = QA$
$P = AP = AQAP = FE. \quad Q = QA = QAPA = FE.$
Furthermore, $P = PAP = EP = FP = QAP = QE = QF = QAQ = Q.$
Hence $A^{\#}$ is unique.

Property 2 If $A \in R^{(n\times n)}$ is a SIM (or CSIM), then $A^{D^{D^{D}}} = A^{D}$, $A^{T^{D}} = A^{D^{T}}$, $A^{l^{D}} = A^{D^{l}}, l = 1, 2 \ldots.$

Example 9 Let A be a CSIM of dimension 6×6. $A = \begin{pmatrix} 0&1&0&0&1&1\\ 1&0&1&1&0&1\\ 1&1&0&1&1&0\\ 0&1&1&0&1&1\\ 1&0&1&1&0&1\\ 1&1&0&1&1&0 \end{pmatrix}$, then $A^{D} =$

$$\begin{pmatrix} -0.125 & 0.125 & 0.125 & -0.125 & 0.125 & 0.125\\ 0.125 & -0.125 & 0.125 & 0.125 & -0.125 & 0.125\\ 0.125 & 0.125 & -0.125 & 0.125 & 0.125 & -0.125\\ -0.125 & 0.125 & 0.125 & -0.125 & 0.125 & 0.125\\ 0.125 & -0.125 & 0.125 & 0.125 & -0.125 & 0.125\\ 0.125 & 0.125 & -0.125 & 0.125 & 0.125 & -0.125 \end{pmatrix}.$$

It satisfies the following conditions $A^{D^{D^{D}}} = A^{D}$, $\quad A^{T^{D}} = A^{D^{T}}$, $A^{l^{D}} = A^{D^{l}}, l = 1, 2 \ldots.$

Property 3 If $A \in R^{(n\times n)}$ is a SIM (or CSIM). Then $A^{\#^{\#}} = A$, $A^{T^{\#}} = A^{\#^{T}}$, $A^{l^{\#}} = A^{\#^{l}}$, for every positive integer l.

Example 10 Let A be a submatrix of dimension of SIM.

$$A = \begin{pmatrix} 0&1&0&0\\ 0&0&1&0\\ 1&0&0&1\\ 0&1&0&0 \end{pmatrix}, \quad A^{\#} = \begin{pmatrix} 0 & 0.5 & 0 & 0\\ 0 & 0 & 1 & 0\\ 0.5 & 0 & 0 & 0.5\\ 0 & 0.5 & 0 & 0 \end{pmatrix}.$$

$$A^{\#^{\#}} = \begin{pmatrix} 0 & 0.5 & 0 & 0\\ 0 & 0 & 1 & 0\\ 0.5 & 0 & 0 & 0.5\\ 0 & 0.5 & 0 & 0 \end{pmatrix}^{\#} = A, \qquad A^{T^{\#}} = \begin{pmatrix} 0 & 0.5 & 0 & 0\\ 0 & 0 & 1 & 0\\ 0.5 & 0 & 0 & 0.5\\ 0 & 0.5 & 0 & 0 \end{pmatrix}^{\#} = A^{\#^{T}}.$$

Theorem 13 $A \in R^{(n\times n)}$ *is a SIM (or CSIM). A have a index k. Then* $\{A^{l} : l \geq k\}$ *have the same rank, the same range and same null space and their transpose* $\{A^{l^{T}} : l \geq k\}$ *have the same rank, the same range and same null space.*

Proof $A \in R^{(n\times n)}$ is a SIM (or CSIM) and have *index* k, then $rank(A^{k}) = rank(A^{k+1})$. $A^{k+1}A^{D} = A^{k} \Rightarrow A^{l-k}A^{k+1}A^{D} = A^{l-k}A^{k} \Rightarrow A^{l+1} = A^{l} (l \geq k)$.

Hence, $A^l : l \geq k$ have the same rank, the same range and same null space. Since A is a symmetric matrix, then their transpose $\{A^{l^T} : l \geq k\}$ have the same rank, the same range and same null space.

Example 11 A is a submatrix of SIM of dimension 4×4.

$$A = \begin{pmatrix} 1\,0\,0\,1 \\ 0\,1\,0\,0 \\ 0\,0\,1\,0 \\ 1\,0\,0\,1 \end{pmatrix}, \text{ then } AA^{\ddagger} = \begin{pmatrix} 0.5\;0\;0\;0.5 \\ 0\;\;0\;1\;\;0 \\ 0\;\;1\;0\;\;0 \\ 0.5\;0\;0\;0.5 \end{pmatrix} = A^{\ddagger}A$$

$$A^2A^{\ddagger} = \begin{pmatrix} 1\,0\,0\,1 \\ 0\,1\,0\,0 \\ 0\,0\,1\,0 \\ 1\,0\,0\,1 \end{pmatrix} = A^{\ddagger}A^2. \quad A^{\ddagger^2}A = \begin{pmatrix} 0.25\;0\;0\;0.25 \\ 0\;\;1\;0\;\;0 \\ 0\;\;0\;1\;\;0 \\ 0.25\;0\;0\;0.25 \end{pmatrix} = AA^{\ddagger^2}.$$

$$A^3A^{\ddagger} = \begin{pmatrix} 2\,0\,0\,2 \\ 0\,0\,1\,0 \\ 0\,1\,0\,0 \\ 2\,0\,0\,2 \end{pmatrix} = A^{\ddagger}A^3. \quad A^{\ddagger^3}A = \begin{pmatrix} 0.125\;0\;0\;0.125 \\ 0\;\;0\;1\;\;0 \\ 0\;\;1\;0\;\;0 \\ 0.125\;0\;0\;0.125 \end{pmatrix} = AA^{\ddagger^3}.$$

This result can be generalized to any k.

6 Conclusions

Various types of Symmetric properties are discussed for Special Infinite Matrix (SIM) and Complement of Special Infinite Matrix (CSIM). Also all types of Inverse exist and unique for both SIM and Complement of CSIM is established. Generally a unique generalized inverse exists for any singular matrix but the result need not be true for Drazin and Group inverse. But a remarkable property which states that all inverse exist and unique for both SIM and CSIM is found. In future, investigation on range symmetric and decomposition of SIM will be focused.

References

1. Ben-Israel, A., Greville, T.N.E.: Generalized Inverses: Theory and Applications, 2nd ed. Rutgers University (2001)
2. Campbell, S.L.: The Drazin inverse and systems of second order linear differential equations. Linear Multilinear Algebra **14**, 195–198 (1983)
3. Campbell, S.L., Meyer, C.D.: Generalized Inverses of Linear Transformations. Dover, New York (1991)
4. Chountasis, S., Katsikis, V.N., Pappas, D.: Applications of the Moore-Penrose Inverse in Digital Image Restoration, vol. 2009, Article ID 170724, 12 pp. Hindawi Publishing Corporation Mathematical Problems in Engineering. https://doi.org/10.1155/2009/170724
5. Cline, R.E.: An application of representation of a matrix. MRC Technical Report, # 592 (1965)
6. Drazin, M.P.: Pseudo-inverse in associative rings and semigroups. Am. Math. Monthly **65**, 506–514 (1958)

7. Horn, R.A., Johnson, C.R.: Matrix Analysis. Cambridge University Press (1985)
8. Jansirani, N., Rajkumar Dare, V., Subramanian, K.G.: Sturmian Arrays, Advances in Image Analysis and Applications (Chapter 9). Research Publishing, Printed in Singapore (2011). ISBN-13:978-08-7923-5, ISBN-10:381-08-7923-7
9. Lee, A.: Secondary symmetric secondary skew symmetric secondary orthogonal matrices. Period. Math. Hungary **7**, 63–70 (1976)
10. Li, X., Wei, Y.: A note on the representations for the Drazin inverse of 2x2 block matrices. Linear Algebra Appl. **423**, 332–338 (2007)
11. Meyer, C.D., Rose, N.J.: The index and the Drazin inverse of block triangular matrices. SIAM J. Appl. Math. **33**, 1–7 (1977)
12. Schwartz, E.M., Doty, K.L.: Application of the Weighted Generalized-Inverse in Control Optimization and Robotics (1992)
13. Subharani, V., Jansirani, N., Dare, V.R.: Special properties of circulant infinite array. Int. J. Anal. Exp. Modal Anal. **XI**(X), 219–227 (2019)

A New Approach for Non-linear Fractional Heat Transfer Model by Adomian Decomposition Method

Gunvant A. Birajdar

Abstract In this paper, the attempt has been made to obtain the dual solution of heat conduction and mass transfer model which is in the form of non-linear fractional differential equation. For modelling this, the Caputo fractional derivative is used, and Adomian decomposition method is employed for getting its solution. Also, for each model, the test problems are solved by using MATLAB Software.

Keywords Adomian decomposition method · Caputo fractional derivative · Heat transfer model

Mathematics Subject Classification (2010) 34A08

1 Introduction

Since last two decades or more than that the new branch of mathematics called "Fractional Calculus" came in picture with so many applications in different fields. Particularly, fractional differential equations have applications in material science, biochemistry, engineering, etc. It has been found that so many day today real life problems such as frequency dependent damping behaviour of substance, viscoelasticity, heat and mass transfer and Newtonian fluid expressed by fractional differential equations [1, 2]. The relevance of such physical process motivate us to build easy, coherent and accurate technique for solving linear and non-linear fractional differential equations.

The well-known classical methods are mathematical method [1, 2], numerical methods [3–7], iterative method [8, 9] and Adomian decomposition method (ADM) [10–14]. Adomian decomposition method is introduced by Adomian [10, 11] in 1980. Moreover, Adomian has been shown the importance and usefulness of the

G. A. Birajdar (✉)
School of Rural Development, Tata Institute of Social Sciences,
Tuljapur, Osmanabad (MS) 413601, India
e-mail: gabirajdar11@gmail.com

R. N. Mohapatra et al. (eds.), *Mathematical Analysis and Computing*,
Springer Proceedings in Mathematics & Statistics 344,
https://doi.org/10.1007/978-981-33-4646-8_28

method while dealing with non-linear equations. The researcher Wazwaz [15] has applied ADM to obtain the solution of different kind of differential equations. On this line, Dhaigude and Birajdar [16] employed the discrete ADM to system of initial value problem for fractional partial differential equations.

The number of researchers discussed the area in material science like of heat and mass transfer phenomena which are governed by non-linear boundary value problems and has been inquired because of its dual form. It has been reported that it's having potential applications in a variety of chemical and physical models. Also, it is challenging task to solve such models by using analytical methods. Therefore, an attempt is made to obtain the solution of such problems by ADM. The aim of this article to investigate the non-linear fractional BVP which is characterized by the existence of the dual solution.

Consider the fractional differential equation

$$D_t^{\alpha} v(t) + v^n(t) = 0, \quad 1 < \alpha \leq 2 \ \ n \in Z \tag{1.1}$$

with boundary condition

$$v'(a) = 0, v(b) = 1 \quad a \leq t \leq b. \tag{1.2}$$

fractional differential equation (1.1) and boundary conditions (1.2) are known as fractional boundary value problem (BVP).

2 Basic Definitions and Notations

This section is dealing with basic definitions, notations. Also, the properties of fractional derivative operator viz Riemann-Liouville and Caputo and their relation.

Definition 2.1 [1] A real-valued function $g(t), t > 0$ is called in space $C_{\alpha}, \alpha \in \Re$ if $\exists\ p > \alpha$ a real number such that $g(t) = t^p g_1(t)$ where $g_1(t) \in C[0, \infty)$.

Definition 2.2 [1] A function $g(t), t > 0$ is called to be in space $C_{\alpha}^m, m \in N \bigcup \{0\}$ if $g^m \in C_{\alpha}$.

Definition 2.3 [1] Suppose that $g \in C_{\alpha}$ & $\alpha \geq -1$, then Riemann-Liouville (R-L) integral operator of real order of function $g(t)$ of fractional power α is denoted by $J^{\alpha} g(t)$ and is defined as

$$J^{\alpha} g(t) = \frac{1}{\Gamma(\alpha)} \int_0^t (t - \tau)^{\alpha - 1} g(\tau) d\tau, \quad t > 0, \ \alpha > 0$$

The R-L integral operator has beautiful property [1] as

$$J^{\alpha} t^{\gamma} = \frac{\Gamma(\gamma + 1) t^{\gamma + \alpha}}{\Gamma(\gamma + \alpha + 1)}$$

Definition 2.4 [17] For $m \in Z$ integer of power $\alpha > 0$, the Caputo fractional derivative of function $v(t)$ with respect to t of order $\alpha > 0$ is defined as

$$D_t^{\alpha} f(t) = \begin{cases} \frac{1}{\Gamma(m-\alpha)} \int_0^t (t-\tau)^{m-\alpha-1} f^m d\tau, & \text{for } m-1 < \alpha < m; \\ f^m(t), & \text{for } \alpha = m \in N. \end{cases}$$

The following is the important relation between Caputo an R-L fractional operator is as follows

$$J^{\alpha}(D_t^{\alpha} g(t)) = J^{\alpha}(J^{m-\alpha} g^{(m)}(t)) = J^m g^{(m)}(t) = g(t) - \sum_{k=0}^{m-1} g^{(k)}(0) \frac{t^k}{k!}.$$

3 Adomian Decomposition Method

This section is devoted for development of ADM for fractional BVP.

Consider the equation

$$L^{\alpha} v(t) + M v(t) = 0 \tag{3.1}$$

where $L^{\alpha}(.)$ is linear invertible fractional operator and $M(.)$ is a non-linear operator. Operating both sides, the R-L integral operator J^{α} on Eq. (3.1) and using boundary conditions, we have

$$v(t) = f(t) - J^{\alpha}[M(v)] \tag{3.2}$$

We assume that the Eqs. (1.1)–(1.2) has series solution according to ADM procedure,

$$v(t) = \sum_{n=0}^{\infty} v_n(t) \tag{3.3}$$

for the solution $v(t)$, and the infinite series for the non-linear term $M(v)$ as

$$M(v) = \sum_{n=0}^{\infty} B_n(v_0, v_1, \ldots, v_n), \tag{3.4}$$

where the each term of $v_n(t)$ is lead to the compact solution of $v(t)$ which will find our recursively. The term B_n are the Adomian polynomials [15, 18], which are obtained from the following formula

$$B_n = \frac{1}{n!}\left[\frac{d^n}{d\lambda^n} g\left(\sum_{k=0}^{\infty} \lambda^k v_k(t)\right)\right]_{\lambda=0}, \quad n \geq 0 \tag{3.5}$$

Based on the above formula (3.5), we listed first few terms of Adomian polynomial from $B_0, B_1, \ldots, B_4$, we have

$$
\begin{aligned}
B_0 &= g(v_0),\\
B_1 &= v_1 g^{'}(v_0),\\
B_2 &= v_2 g^{'}(v_0) + \frac{v_1^2}{2!} g^{''}(v_0)\\
B_3 &= v_3 g^{'}(v_0) + u_1 v_2 g^{''}(v_0) + \frac{v_1^3}{3!} g^{'''}(v_0)\\
B_4 &= v_4 g^{'}(v_0) + (\frac{v_2^2}{2!} + v_1 v_3) g^{''}(v_0) + \frac{v_1^2 v_2}{2!} g^{'''}(v_0) + \frac{v_1^4}{4!} g^{iv}(v_0)
\end{aligned}
\tag{3.6}
$$

where $g(v(t))$ is the non-linear function. Substituting Eqs. (3.3) and (3.4) into Eq. (3.2), we obtain

$$\sum_{n=0}^{\infty} v_n(t) = \xi - J^{\alpha} B_n = \xi - J^{\alpha}\left\{\frac{1}{n!}\left[\frac{d^n}{d\lambda^n} g\left(\sum_{k=0}^{\infty} \lambda^k v_k(t)\right)\right]_{\lambda=0}\right\} \tag{3.7}$$

Identifying $v_0(t) = \varphi$, the recursive scheme

$$v_0(t) = \xi \tag{3.8}$$

$$v_{k+1}(t) = -J^{\alpha} B_k, k \geq 0. \tag{3.9}$$

in this way all the terms can determine $v_n(t)$. The series solution of $v(t)$ is obtained, where the Adomian approximates $\varphi_{m+1} = \sum_{n=0}^{m} v_n(x)$.

By using the boundary conditions, it is easy to determine the unknown constant. As per above discussion, we will select a BVP that hold dual solution.

The aim of the ADM is to construct an approximate analytical solution to non-linear BVPs.

4 Modelling of the Physical Problems

Cooling fins are significantly used to intensify the heat transfer between a solid surface and its radiative, convective or convective-radiative surface [19]. These smooth surfaces are extensively utilize for illustration, for cooling electric transformers, the cylinders of aircraft engines, as well as other heat transfer devises [19, 20]. The temperature distribution of a straight rectangular fin with a set power-law temperature reliant surface heat flux can be obtained by the solutions of a one-dimensional steady-state heat conduction equation which, in dimensionless form as follows

$$D^{\alpha} v(t) - \gamma^2 v^{i+1} = 0, \quad 1 < \alpha \leq 2 \tag{4.1}$$

$$v'(0) = 0, v(1) = 1, \ 0 \leq t \leq 1 \tag{4.2}$$

where γ is the convective-conductive parameter. It is reported that if $-4 \leq i \leq -2$, then the BVP (4.1) either exists dual solution or no solution for a said convective-conductive constant [19, 20].

4.1 Heat Transfer Model

If i takes value $i = -4$, the non-linear BVP Eq. (4.1) portray a heat transfer model which represents in the following form

$$D^{\alpha} v(t) - \gamma^2 v^{-3} = 0, \quad 1 < \alpha \leq 2 \tag{4.3}$$

$$v'(0) = 0, v(1) = 1. \tag{4.4}$$

Eq. (4.3) writes in the following operator form

$$D^{\alpha} v(t) = \gamma^2 v^{-3}. \tag{4.5}$$

Apply R-L integral operator J^{α} on the both sides of Eq. (4.5). Also, using Adomian polynomial, we got the following recursive formula

$$v_0(t) = \theta \tag{4.6}$$

$$v_{k+1}(t) = J^{\alpha}\left[\gamma^2 B_k\right], k \geq 0 \tag{4.7}$$

where B_k, for $k \geq 0$, be the Adomian polynomials construct for the non-linear term $\dfrac{1}{v^3}$ as follows

$$\begin{aligned}
B_0 &= \frac{1}{v_0^3}, \\
B_1 &= -3v_1 \frac{1}{v_0^4} \\
B_2 &= -3v_2 \frac{1}{v_0^4} + 6v_1^2 \frac{1}{v_0^5} \\
B_3 &= -3v_3 \frac{1}{v_0^4} + 12v_1 v_2 \frac{1}{v_0^5} - 10v_1^3 \frac{1}{v_0^6} \\
B_4 &= -3v_4 \frac{1}{v_0^4} + 12(\frac{1}{2}v_2^2 + v_1 v_3) \frac{1}{v_0^5} - 30v_1^2 v_2 \frac{1}{v_0^6} + 15v_1^4 \frac{1}{v_0^7} \\
\ldots &= \ldots \\
\ldots &= \ldots
\end{aligned} \tag{4.8}$$

which obtained from the formula (3.5). The following are terms of the solution,

$$
\begin{aligned}
v_0(t) &= \theta \\
v_1(t) &= \frac{\gamma^2}{\theta^3}\frac{t^{2\alpha}}{\Gamma(2\alpha+1)} \\
v_2(t) &= \frac{-3\gamma^4}{\theta^7}\frac{t^{4\alpha}}{\Gamma(4\alpha+1)} \\
v_3(t) &= \frac{\gamma^6}{\theta^{11}}\left\{\frac{\Gamma(4\alpha+1)}{\Gamma(6\alpha+1)} + \frac{6\Gamma(4\alpha+1)}{\Gamma(2\alpha+1)^2\Gamma(6\alpha+1)}\right\}t^{4\alpha}
\end{aligned} \tag{4.9}
$$

Therefore, the solution of above boundary value problem is

$$
\begin{aligned}
v(t) &= v_0 + v_1 + v_2 + v_3 + \ldots \\
v(t) &= \theta + \frac{\gamma^2}{\theta^3}\frac{t^{2\alpha}}{\Gamma(2\alpha+1)} - \frac{3\gamma^4}{\theta^7}\frac{t^{4\alpha}}{\Gamma(4\alpha+1)} + \\
&\quad \frac{\gamma^6}{\theta^{11}}\left\{\frac{\Gamma(4\alpha+1)}{\Gamma(6\alpha+1)} + \frac{6\Gamma(4\alpha+1)}{\Gamma(2\alpha+1)^2\Gamma(6\alpha+1)}\right\}t^{4\alpha}
\end{aligned} \tag{4.10}
$$

To find unknown θ, we use boundary condition $v(1) = 1$. The comparison of Eq. (4.10) with the results in [19–21], we considered the $\gamma = \frac{2}{5}$ in (4.10) and after manipulations we get two different values of θ as follows: $\theta = 0.560584612$ and $\theta = 0.8944216050$ (Figs. 1 and 2).

Using these two values in Eq. (4.10), we obtained dual solution and its graphical representation as follows for distinct values of α.

4.2 Reaction-Diffusion Model

For $i = -2$, in the Eq. (4.1) express the following BVP which govern the reaction-diffusion model in porous catalysts,

$$D^\alpha v(t) - \gamma^2 v^{-1} = 0, \quad 1 < \alpha \le 2 \tag{4.11}$$

$$v^{'}(0) = 0, v(1) = 1, \quad 0 \le t \le 1. \tag{4.12}$$

The above equation can be write in operator form, we have

$$D^\alpha v(t) = \gamma^2 v^{-1}, \quad 1 < \alpha \le 2 \tag{4.13}$$

Similarly apply the ADM procedure, we have

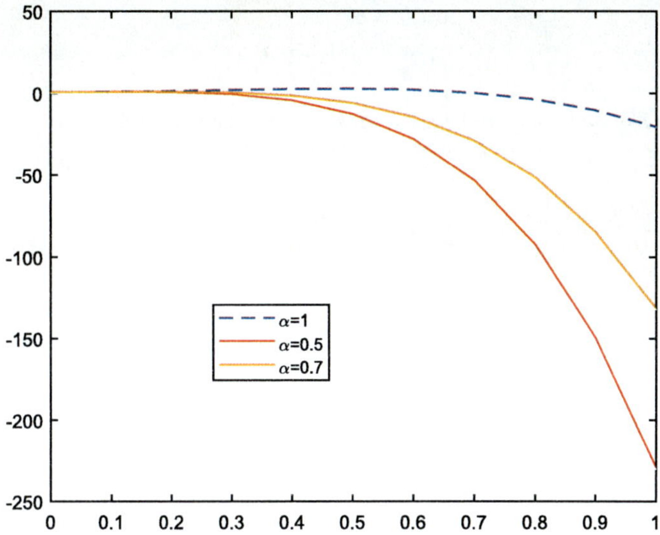

Fig. 1 Solution of $v(t)$ for $\theta = 0.560584612$

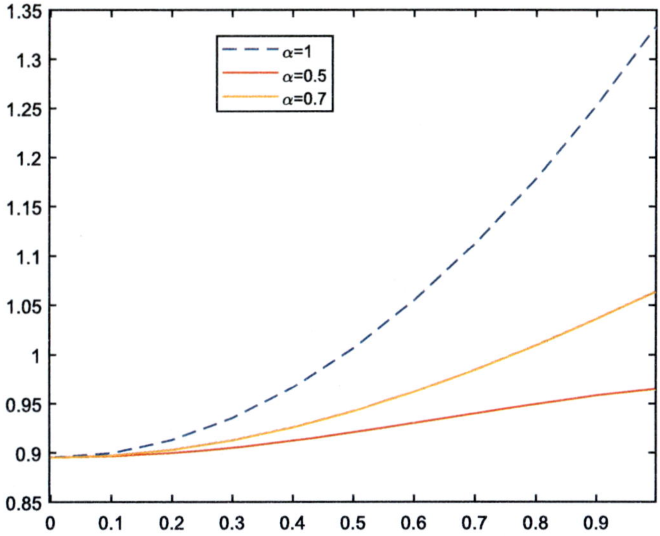

Fig. 2 Solution of $v(t)$ for $\theta = 0.8944216050$

$$v_0(t) = \theta \tag{4.14}$$
$$v_{k+1}(t) = J^{\alpha}\left[\gamma^2 B_k\right], k \geq 0 \tag{4.15}$$

where the Adomian polynomials B_k, for $k \geq 0$, for the non-linear term v^{-3} are given as

$$\begin{aligned}
B_0 &= \frac{1}{v_0},\\
B_1 &= -v_1\frac{1}{v_0^2}\\
B_2 &= -v_2\frac{1}{v_0^2} + v_1^2\frac{1}{v_0^3}\\
B_3 &= -v_3\frac{1}{v_0^2} + 2v_1v_2\frac{1}{v_0^3} - v_1^3\frac{1}{v_0^4},\\
B_4 &= -v_4\frac{1}{v_0^2} + 2(\frac{1}{2}v_2^2 + v_1v_3)\frac{1}{v_0^3} - 3v_1^2v_2\frac{1}{v_0^4} + 15v_1^4\frac{1}{v_0^5},\\
\ldots &= \ldots\\
\ldots &= \ldots
\end{aligned} \tag{4.16}$$

which obtained from the definitional formula [10, 11]. This in turn gives the following solution components,

$$\begin{aligned}
v_0(t) &= \theta\\
v_1(t) &= \frac{\gamma^2}{\theta}\frac{t^{2\alpha}}{\Gamma(2\alpha+1)}\\
v_2(t) &= \frac{\gamma^4}{\theta^3}\frac{t^{4\alpha}}{\Gamma(4\alpha+1)}\\
v_3(t) &= \frac{\gamma^6}{\theta^5}\left\{1 + \frac{\Gamma(4\alpha+1)}{\Gamma(2\alpha+1)}\right\}\frac{t^{6\alpha}}{\Gamma(6\alpha+1)}\\
\ldots &= \ldots\\
\ldots &= \ldots
\end{aligned} \tag{4.17}$$

Therefore, the solution of above BVP is

$$v(t) = v_0 + v_1 + v_2 + v_3 + \ldots \tag{4.18}$$
$$v(t) = \theta + \frac{\gamma^2}{\theta}\frac{t^{2\alpha}}{\Gamma(2\alpha+1)} + \frac{\gamma^4}{\theta^3}\frac{t^{4\alpha}}{\Gamma(4\alpha+1)} + \frac{\gamma^6}{\theta^5}\left\{1 + \frac{\Gamma(4\alpha+1)}{\Gamma(2\alpha+1)}\right\}\frac{t^{6\alpha}}{\Gamma(6\alpha+1)}$$

To find unknown constant θ by using boundary condition $v(1) = 1$, as well as to compare Eq. (4.18) with the results in Semary and Hassan [22] (2015),

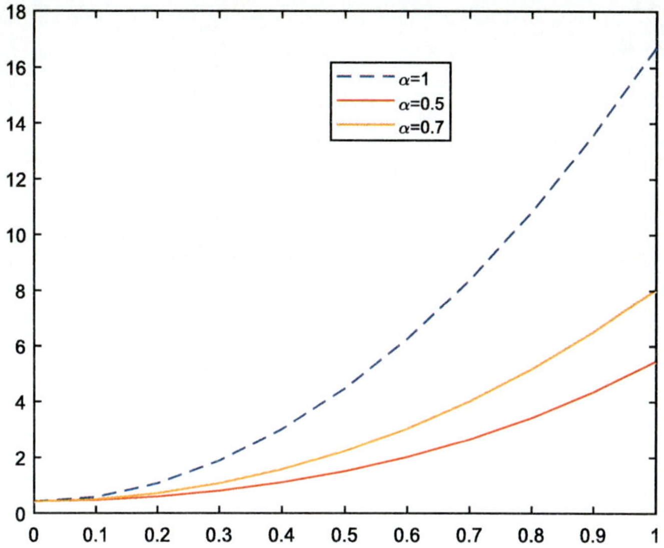

Fig. 3 Solution of $v(t)$ for $\theta = 0.3931282510$

we put $\gamma = \frac{7}{10}$ in (4.10) and after simplifying we get two distinct values of θ as follows: $\theta = 0.3931282510$ and $\theta = 0.6532152418$. Using these two values in equation (4.18), we obtained the dual solution. Its graphical representation is as follows for distinct values of α. The following is the graphical solution for $\alpha = 1,\ \alpha = 0.5\ and\ \ \alpha = 0.7$ (Figs. 3 and 4).

5 Existence and Uniqueness

Theorem 5.1 *[18] Suppose that the non-linear operator N satisfies a uniform Lipschitz condition*

$$| Nu_1 - Nu_2 | \leq K \mid u_1 - u_2 \mid, \quad a \leq s, \quad t \leq b, u_i \in \Re, i = 1, 2$$

Furthermore, we suppose that

$$K = sup \int_b^a \mid G(s,t) \mid dt < 1$$

then the BVP (1.1) and (1.2) has unique solution.

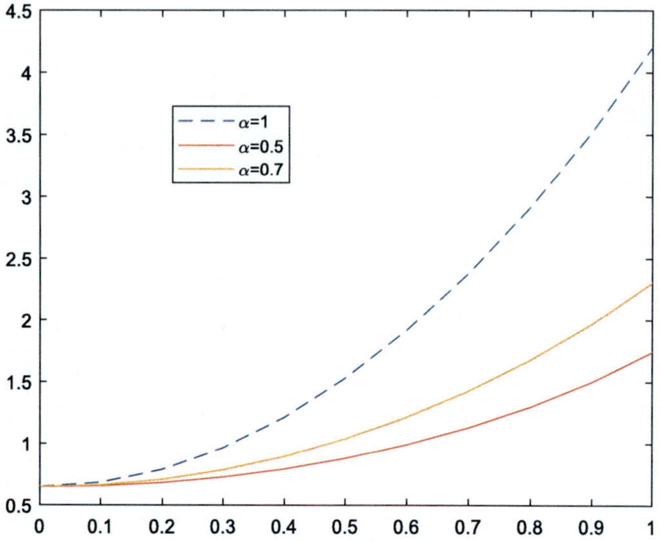

Fig. 4 Solution of $v(t)$ for $\theta = 0.6532152418$

6 Conclusion

This paper presented a new approach using the ADM to achieve a dual solution of fractional non-linear BVPs. It exhibits that ADM validates the solutions of these models by easily, accurately and systematically by using the Adomian polynomials. This method can be extended for solving variety of numerical problems including scientific model for fractional non-linear heat and mass transfer, heat transfer of cooling fins, reaction and diffusion in porous catalysts and for reaction transport. The test problems shows not only the efficiency of the method but also smooth and stable solution.

References

1. Podlubny, I.: Fractional Differential Equations. Academic Press, San Diego (1999)
2. Samko, S.G., Kilbas, A.A., Marichev, O.I.: Fractional Integral and Derivatives: Theory and Applications. Gordon and Breach, Yverdon (1993)
3. Birajdar, G.A.: Numerical solution of time fractional Navier-Stokes equation by discrete Adomian decomposition method. Nonlinear Eng. **3**(1), 1–6 (2014)
4. Birajdar, G.A.: An implicit numerical method for semilinear space-time fractional diffusion equation. Walailak J. Sci. Tech. **12**(11), 983–994 (2015)
5. Birajdar, G.A., Dhaigude, D.B.: An implicit numerical method for semilinear fractional diffusion equation. In: International Conference on Mathematical Sciences, pp. 674–678. Elsevier (2014)

6. Birajdar, G.A.: Stability of nonlinear fractional diffusion equation. Lib. Math. (New Series) **36**(1), 1–12 (2016)
7. Birajdar, G.A.: An implicit finite difference method for semilinear time-fractional diffusion equation. Dyn. Contin. Discret. Impuls. Syst. Ser. A Math. Anal. **27**(3a), 209–217 (2020)
8. Daftardar-Gejji, V., Jafari, H.: An iterative method for solving nonlinear functional equations. J. Math. Anal. Appl. **316**, 753–763 (2006)
9. Dhaigude, C.D., Nikam, V.R.: Solution of fractional partial differential equations using iterative method. Fract. Calc. Appl. Anal. **15**(4), 684–699 (2012)
10. Adomian, G.: Solving Frontier Problems of Physics: The Decomposition Method. Kluwer, Boston (1994)
11. Adomian, G.: A review of the decomposition method in applied mathematics. J. Math. Anal. Appl. **135**, 501–544 (1988)
12. Daftardar-Gejji, V., Bhalekar, S.: Solving multi-term linear and nonlinear diffusion-wave equations of fractional order by Adomian decomposition. Appl. Math. Comput. **202**, 113–120 (2008)
13. Dhaigude, D.B., Birajdar, G.A., Nikam, V.R.: Adomain decomposition method for fractional Benjamin-Bona-Mahony-Burger's equations. Int. J. Appl. Math. Mech. **8**(12), 42–51 (2012)
14. Dhaigude, D.B., Birajdar, G.A.: Numerical solution of fractional partial differential equations by discrete Adomian decomposition method. Adv. Appl. Math. Mech. **6**(1), 107–119 (2014)
15. Wazwaz, A.M.: Partial Differential Equations and Solitary Waves Theory. Higher Education Press and Springer, Beijing and Berlin (2009)
16. Dhaigude, D.B., Birajdar, G.A.: Numerical solution of system of fractional partial differential equations by discrete Adomian decomposition method. J. Fract. Calc. Appl. **3**(12), 1–11 (2012)
17. Caputo, M.: Linear models of dissipation whose Q is almost independent. II. Geo-phys. J. Roy. Astron. **13**, 529–5397 (1967)
18. Wazwaz, A.M., Rach, R., Bougoffa, L.: Dual solutions for nonlinear boundary value problems by the Adomian decomposition method. Int. J. Numer. Methods Heat Fluid Flow **26**(8), 2393–2409 (2016)
19. Abbasbandy, S., Shivanian, E.: Predictor homotopy analysis method and its application to some nonlinear problems. Commun. Nonlinear Sci. Numer. Simulat. **16**, 2456–2468 (2011)
20. Barletta, A., Magyari, E., Keller, B.: Dual mixed convection flows in a vertical channel. Int. J. Heat Mass Transf. **48**, 4835–4845 (2005)
21. Ellery, A., Simpson, M.: An analytical method to solve a general class of nonlinear reactive transport models. Chem. Eng. J. **169**, 313–318 (2011)
22. Semary, M., Hassan, H.: A new approach for a class of nonlinear boundary value problems with multiple solutions. J. Assoc. Arab. Univ. Basic Appl. Sci. **17**, 27–35 (2015)

Bifurcation and Chaos Control for Discrete Fractional-Order Prey–Predator Model with Square Root Interaction

A. George Maria Selvam, S. Britto Jacob, and R. Dhineshbabu

Abstract This research article is related with the local stability of the steady states of fractional-order prey–predator system with square root interaction term. A discrete form of the system is proposed for discussion. The interaction period is proportionate to the square root of the prey species density. The existence of steady states, stability of the steady states, flip or period-doubling bifurcation (PDB), Neimark–Sacker bifurcation (NSB), controlling chaos are investigated. To verify the analytical results, we plot time series, phase plane,s and bifurcations diagrams for biologically feasible parameter values. From the numerical illustrations we understand that the discrete model exhibits rich dynamic behavior and also expresses the complicated dynamical behavior of the system in the presence of square root term interaction.

Keywords Prey–Predator System · Stability · Square root Interaction term and Bifurcation.

Mathematics Subject Classification (2010) 34C23 · 37C25 · 39A30 · 92D30.

1 Introduction

In mathematical biology, the study of population dynamics is classified into single-species model, two-species model, and multi-species model. Many researchers used different approaches to study the dynamical behaviors of population models with differential equations in ordinary, partial, and stochastic forms. Also they represented

A. G. M. Selvam (✉) · S. B. Jacob
Department of Mathematics, Sacred Heart College (Autonomous), Tirupattur 635 601, Tamil Nadu, India
e-mail: agmshc@gmail.com

R. Dhineshbabu
Department of Mathematics, Sri Venkateswara College of Engineering and Technology (Autonomous), Chittoor 517 127, Andhra Pradesh, India

R. N. Mohapatra et al. (eds.), *Mathematical Analysis and Computing*,
Springer Proceedings in Mathematics & Statistics 344,
https://doi.org/10.1007/978-981-33-4646-8_29

the systems by difference equations, fractional-order differential equations (FODEs). In particular, FODEs are providing an exact or approximate description of different nonlinear phenomena. The superior position of fractional-order system is that they allow higher number of degrees of freedom than an integer-order system. FODEs are preferably used for the reason that they are evidently associated to the systems with memory effects which existent is in utmost every biological phenomena. In mathematical biology, numerous applications are available because FODE has close relations to fractals.

The relation between predator and their prey are modeled by Lotka (1925) and Volterra (1926) independently. This model gained a great importance due to its broad range of applications in real life. Many academics provided a considerable contribution on continuous-time and discrete-time domain [1–5]. Certain nonlinear and linear functional responses are used to form the phenomena of interaction among prey and predator, like Beddington-DeAngelies type, Lesile-Gower type, Crowley-Martin type, Holling type, and Kolmogorov type.

In recent times, some researchers analyzed the dynamical behavior of nonlinear interactions of population models and some other studies investigate the approximate solution of nonlinear FODE models [6]. Elsadany [7] derived the results for fractional-order Lotka–Volterra prey-predator model with discretization. The dynamic behavior and structure of discrete-domain prey–predator model in two-dimensional spaces with different assumptions and conditions have been analyzed and discussed by researchers [8–10]. The main purpose of this paper is to present the discrete-time fractional-order prey–predator model incorporating with square root interaction term and to analyze the non-local properties induced by of fractional-order derivatives.

The rest of the paper investigates and presents the results in the following order: Discrete version of the prey–predator system with fractional order is modeled in Sect. 2. Existence of steady states and analysis of the local stability of the proposed model is dealt in Sect. 3. In Sect. 4, the bifurcation theory for the corresponding system is derived through PDB and NSB. Chaos control procedure is presented in Sect. 5. Section 6 deals with numerical illustrations where the analytical results are verified. Section 7 gives the conclusion of the paper.

2 Predator–Prey Model with Square Root Interaction

In population dynamics, the interaction among species has been analyzed with different set of conditions and assumptions. As a significance of the herd behavior, in [11–13], authors choose the two-species competition model, where the interaction of prey species is represented by square root term instead of simple prey population. The attention of prey–predator model subjected to square root interaction term is more appropriate and has fascinated interest, which inspires us to model the following system.

$$\frac{du}{dt} = ru(t)[1 - u(t)] - a\sqrt{u(t)}v(t),$$
$$\frac{dv}{dt} = c\sqrt{u(t)}v(t) - bv(t), \tag{1}$$

where all the parameters are positive. Here, the prey population represents the herd behavior. Hence the interaction among the predator and prey occurs at outer corridor of prey.

The corresponding nonlinear fractional-order two-dimensional prey–predator system with square root interaction of system (1) is

$$D^{\alpha}u(t) = ru(t)[1 - u(t)] - a\sqrt{u(t)}v(t),$$
$$D^{\alpha}v(t) = c\sqrt{u(t)}v(t) - bv(t), \tag{2}$$

where $0 \leq \alpha < 1$, particularly if $\alpha = 1$, then the system (2) is a regular integer-order system and D^{α} is the Caputo fractional-order derivative. In the proposed model, $u(t)$ and $v(t)$ represents prey density and predator density over time t. The intrinsic per capita growth rate of prey is r, mortality rate of prey in the presence of predator is a, consumption rate is c , when there is no prey the death rate of predator is b.

In mathematical modeling, discretization process is employed to convert the continuous-time domain to discrete-time domain. Now, utilizing the discretization process to a fractional-order system (2), we get the discrete fractional-order prey–predator model with square root interaction as follows:

$$u_{t+1} = u_t + \frac{s^{\alpha}}{\alpha\Gamma(\alpha)}\left(ru_t[1 - u_t] - a\sqrt{u_t}v_t\right),$$
$$v_{t+1} = v_t + \frac{s^{\alpha}}{\alpha\Gamma(\alpha)}\left(c\sqrt{u_t}v_t - bv_t\right). \tag{3}$$

3 Existence and Stability Analysis of Steady States

To find the steady states of (3), it is enough to solve the following system of equations:

$$ru(t)[1 - u(t)] - a\sqrt{u(t)}v(t) = 0,$$
$$c\sqrt{u(t)}v(t) - bv(t) = 0. \tag{4}$$

After simple computation, we summarize the result as the following lemma

Lemma 1 *The proposed system (3) has the extinction steady state* $E_0 = (0, 0)$ *and the boundary steady state* $E_1 = (1, 0)$, *if* $u \neq 0$. *System (3) has a unique positive steady state* $E_2 = \left(\frac{b^2}{c^2}, \frac{rb[c^2-b^2]}{ac^3}\right)$, *provided* $c > b$.

Here, we analyze the stability properties of the steady states of discretized fractional-order system (3) which is determined by the parameters α, s, r, a, b, and c. The variation matrix V of (3) at any steady states is

$$V(u,v) = \begin{bmatrix} 1+\frac{s^\alpha}{\alpha\Gamma(\alpha)}\left(r-2ru-\frac{av}{2\sqrt{u}}\right) & -\frac{s^\alpha}{\alpha\Gamma(\alpha)}\left(a\sqrt{u}\right) \\ \frac{s^\alpha}{\alpha\Gamma(\alpha)}\left(\frac{cv}{2\sqrt{u}}\right) & 1+\frac{s^\alpha}{\alpha\Gamma(\alpha)}\left(-b+c\sqrt{u}\right) \end{bmatrix}.$$

Analysis of stability of the steady states of (3) is determined by the modules of eigenvalues.

3.1 Extinction Steady State

Now, in system (3) we have square root interaction term, then the extinction point has singularity. Hence the variation matrix is in-determinant at this point. To discuss the stability of the extinction steady state, we replace the variables u_t by U_t^2 and v_t by V_t in system (1), respectively. So the modified form is

$$\begin{aligned} U_{t+1} &= U_t + \frac{s^\alpha}{\alpha\Gamma(\alpha)}\left(\frac{r}{2}U_t(1-U_t^2)-\frac{a}{2}V_t\right), \\ V_{t+1} &= V_t + \frac{s^\alpha}{\alpha\Gamma(\alpha)}\left(V_t\left(cU_t-b\right)\right). \end{aligned} \tag{5}$$

Now the variation matrix for (5) is

$$V(u*,v*) = \begin{bmatrix} 1+\frac{A}{2}\left(r-3rU^2\right) & -\frac{Aa}{2} \\ \frac{AcV}{2} & 1+A\left(cU-b\right) \end{bmatrix},$$

where $A = \frac{s^\alpha}{\alpha\Gamma(\alpha)}$.

In the system of equations (5), if $U=0$ then the extinction steady state $E_0=(0,0)$ always exists.

Theorem 1 *If $r>0$ and $|\lambda_1|>1$ then E_0 is saddle, if $0<s<\sqrt[\alpha]{\frac{2\alpha\Gamma(\alpha)}{b}}$, source if $s>\sqrt[\alpha]{\frac{2\alpha\Gamma(\alpha)}{b}}$ and non-hyperbolic if $s=\sqrt[\alpha]{\frac{2\alpha\Gamma(\alpha)}{b}}$.*

Proof The variation matrix at E_0 is given by

$$V(E_0) = \begin{bmatrix} 1+\frac{Ar}{2} & -\frac{Aa}{2} \\ 0 & 1-Ab \end{bmatrix}.$$

The eigenvalues are $\lambda_1 = 1+\frac{Ar}{2}$ and $\lambda_2 = 1-Ab$. Since $r>0$ and $|\lambda_1|>1$ then E_0 is saddle, if $0<s<\sqrt[\alpha]{\frac{2\alpha\Gamma(\alpha)}{b}}$, source if $s>\sqrt[\alpha]{\frac{2\alpha\Gamma(\alpha)}{b}}$ and non-hyperbolic if $s=\sqrt[\alpha]{\frac{2\alpha\Gamma(\alpha)}{b}}$.

3.2 Exclusion Steady State

Theorem 2 *The steady state $E_1 = (1, 0)$ of system (3) has the following properties for each admissible value of corresponding parameters:*

R-1. If either (i) or (ii) is satisfied, then E_1 is stable (sink).

(i) $\nabla \geq 0$ and $r_1 < r < r_2$,
(ii) $\nabla < 0$ and $r < r_2$.

R-2. If either (i) or (ii) is satisfied, then E_1 is unstable (source).

(i) $\nabla \geq 0$ and $r > \max\{r_1, r_2\}$,
(ii) $\nabla < 0$ and $r > r_2$.

R-3. If either (i) or (ii) is satisfied, then E_1 is non-hyperbolic.

(i) $\nabla \geq 0$ and $r = r_1$,
(ii) $\nabla < 0$ and $r = r_2$.

R-4. For further parameter values excluding those values in $(R-1)$ to $(R-3)$, it is called saddle point,

where $r_1 = \frac{2A(b-c)-4}{A[A(b-c)-2]}$ *and* $r_2 = \frac{c-b}{1+A(c-b)}$.

Proof Variation matrix of (3) at E_1 is

$$V(E_1) = \begin{bmatrix} 1 - Ar & -Aa \\ 0 & 1 + A(c-b) \end{bmatrix}. \tag{6}$$

The characteristic equation of $V(E_1)$ is

$$F(\lambda) = \lambda^2 - (2 - A(r + b - c))\,\lambda + 1 - A\,(r + b - c) + A^2 r(b - c) = 0.$$

After the computation, we get the eigenvalues of $V(E_1)$ as

$$\lambda_{1,2} = \frac{[2 - A(r + b - c)] \pm A\sqrt{\nabla}}{2},$$

where $\nabla = (r + c - b)^2$.

Hence the conditions $(R-1)$ to $(R-3)$ are satisfied, as the exclusion steady-state E_1 satisfies the conditions for local stability.

3.3 Coexistence Steady State

Theorem 3 *The coexistence steady state E_2 of (3) has the following properties for each admissible value of corresponding parameters:*

S-1. If either (i) *or* (ii) *is satisfied, then* E_2 *is sink.*

(i) $\nabla_1 \geq 0$ *and* $c_1 < c^2 < c_2$,
(ii) $\nabla_1 < 0$ *and* $c^2 < c_2$.

S-2. If either (i) *or* (ii) *is satisfied, then* E_2 *is source.*

(i) $\nabla_1 \geq 0$ *and* $c^2 > max\{c_1, c_2\}$,
(ii) $\nabla_1 < 0$ *and* $c^2 > c_2$.

S-3. If either (i) *or* (ii) *is satisfied, then* E_2 *is non-hyperbolic.*

(i) $\nabla_1 \geq 0$ *and* $c^2 = c_1$,
(ii) $\nabla_1 < 0$ *and* $c^2 = c_2$.

S-4. For further parameter values excluding those values in $(S-1)$ *to* $(S-3)$, *it is called a saddle point,*

where $c_1 = \frac{Arb^2(6+Ab)}{[Ar(2+Ab)+8]}$ *and* $c_2 = \frac{b^2(3+Ab)}{(1+Ab)}$ *and* $\nabla_1 = a_{11}^2 - 4a_{12}a_{21}$.

Proof Variation matrix of (3) at E_2 is

$$V(E_2) = \begin{bmatrix} 1 + Aa_{11} & -Aa_{12} \\ Aa_{21} & 1 \end{bmatrix}, \tag{7}$$

where $a_{11} = \frac{rc^2-3rb^2}{2c^2}$, $a_{12} = \frac{ba}{c}$ and $a_{21} = \frac{r(c^2-b^2)}{2ac}$.
The characteristic equation of $V(E_2)$ is

$$F(\lambda) = \lambda^2 - \delta_1\lambda + \delta_2 = 0,$$

where $\delta_1 = 2 + Aa_{11}$ and $\delta_2 = 1 + Aa_{11} + A^2a_{12}a_{21}$.

Hence the conditions $(S-1)$ to $(S-3)$ holds, as the coexistence (interior) steady-state E_2 satisfies all the sufficient conditions for local stability.

4 Bifurcation Analysis

This section presents the existence and analysis of NSB and PDB at the exclusion and coexistence steady-states $E_{1,2}$ of the corresponding system (3). In dynamical system, there exists different types of bifurcations from the steady states, when a particular parameter passes through the critical value. When it occurs the stability of the system differs.

4.1 Periodic-Doubling Bifurcation of (3)

To discuss the PDB, consider r as the specific bifurcation parameter. When any one of the eigenvalues of Variation matrix is -1 at a steady state and another eigenvalue is either 1 or -1, then the exclusion steady-state E_1 sustains PDB. The variation matrix of (3) at E_1 is given in (6). The characteristic polynomial of (6) is

$$F(\lambda) = \lambda^2 - (2 - A(r + b - c))\,\lambda + 1 - A(r + b - c) + A^2 r(b - c).$$

From Theorem 2, if $\nabla \geq 0$ and $r = r_1$, then the eigenvalues of E_1 are

$$\lambda_{1,2} = \frac{[2 - A(r + b - c)] \pm A\sqrt{\nabla}}{2}.$$

Hence, the following theorem is subsequent of the above analysis.

Theorem 4 *In PDB, the exclusion steady-state E_1 loses its stability whenever $\nabla \geq 0$, $r = r_1$, $\lambda_1 = -1$ and $\lambda_2 = 1 - A(b - c) \neq \pm 1$.*

4.2 Neimark–Sacker Bifurcation of (3)

Here c is considered as a specific bifurcation parameter to analyze the NSB. Corresponding to the coexistence steady-state E_2, NSB happens only when two eigenvalues of the variation matrix at a steady-state E_2 are a pair of complex conjugate numbers with component 1. The variation matrix of (3) at E_2 is in (7).

Then The characteristic polynomial of (7) is

$$F(\lambda) = \lambda^2 - (2 + Aa_{11})\,\lambda + 1 + Aa_{11} + A^2 a_{12} a_{21}.$$

By Theorem 3, if $\nabla_1 < 0$ and $c^2 = c_2$, then the eigenvalues E_2 are

$$\lambda_{1,2} = \frac{(Aa_{11} + 2)}{2} \pm \frac{Ai\sqrt{4a_{12}a_{21} - a_{11}^2}}{2}.$$

This analysis resulting into the succeeding theorem.

Theorem 5 *In NSB, the coexistence steady-state E_2 loses its stability, whenever $\nabla_1 < 0$ and $c^2 = c_2$ and*

$$|\lambda_{1,2}| = \left| \frac{(Aa_{11} + 2)}{2} \pm \frac{Ai\sqrt{4a_{12}a_{21} - a_{11}^2}}{2} \right| = 1.$$

5 Chaos Control of the Proposed System (3)

Now, two different control methods are employed to the proposed system, to attain the stable fractional chaotic paths from unstable fractional periodic paths, employing the linear feedback control method to system (3).

Hence, the system (3) in the controlled form is

$$\begin{aligned} u_{t+1} &= u_t + \frac{s^\alpha}{\alpha\Gamma(\alpha)}\left(ru_t[1-u_t] - a\sqrt{u_t}v_t\right) + S_t, \\ v_{t+1} &= v_t + \frac{s^\alpha}{\alpha\Gamma(\alpha)}\left(c\sqrt{u_t}v_t - bv_t\right), \end{aligned} \tag{8}$$

where the feedback controlling force is presented by $S_t = -p_1(u_n - u^*) - p_2(v_n - v^*)$ and feedback gains denoted by $p_{1,2}$.

The variation matrix of (8) at E_1 is

$$V_1(E_1) = \begin{bmatrix} 1 - Ar - p_1 & -Aa - p_2 \\ 0 & 1 + A(c-b) \end{bmatrix},$$

where $A = \frac{s^\alpha}{\alpha\Gamma(\alpha)}$.

The characteristic equation of $V_1\,(E_1)$ is

$$\lambda^2 - (2 - A(r+b-c) - p_1)\,\lambda + 1 - A(r+b-c) + A^2(rb - rc) - p_1(1 - Ab + Ac) = 0. \tag{9}$$

Assumed that the eigenvalues of (9) are λ_1, λ_2, then we have

$$\lambda_1\lambda_2 = 1 - A(r+b-c) + p_1[A(b-c) - 1] + A^2 r(b-c). \tag{10}$$

The outlines of marginal stability are estimated by the calculation $\lambda_1 = \pm 1$ and $\lambda_1\lambda_2 = 1$. From these restrictions, we assure that the eigenvalues λ_1 and λ_2 consist absolute value which is less than 1.

Assume $\lambda_1\lambda_2 = 1$, then (10) gives

$$\ell_1 : p_1(Ab - Ac - 1) = A[r + (1 - Ar)(b - c)].$$

Suppose that $\lambda_1 = 1$, then equation (9) gives

$$\ell_2 : (p_1 + Ar)(b - c) = 0$$

or $\lambda_1 = -1$, then equation (9) gives

$$\ell_3 : p_1[A(b-c) - 2] = (Ab - Ac - 2)(2 - Ar).$$

The stable eigenvalues lies inside the trilateral field which is bounded by the outlines ℓ_1, ℓ_2, and ℓ_3.

Now introducing the hybrid control strategy, to control the chaos formed by NSB in the proposed system (3), considering that system (3) exhibits NSB at E_2, then the respective controlled system is given by

$$
\begin{aligned}
u_{t+1} &= \phi u_t + \phi \frac{s^\alpha}{\alpha\Gamma(\alpha)} \left(r u_t(1-u_t) - a\sqrt{u_t} v_t\right) + (1-\phi)u_t, \\
v_{t+1} &= \phi v_t + \phi \frac{s^\alpha}{\alpha\Gamma(\alpha)} \left(c\sqrt{u_t} v_t - b v_t\right) + (1-\phi)v_n,
\end{aligned} \tag{11}
$$

where $0 < \phi < 1$.

Both parameter perturbation and feedback control are used as a controlling strategy in (11). Also, by the appropriate selection of controlled parameter ϕ, the NSB of E_2 for (11) can be progressive (overdue) or entirely removed.

The variation matrix of (11) evaluated at E_2 is

$$
V_2(E_2) = \begin{bmatrix} 1 + A\phi a_{11} & -A\phi a_{12} \\ A\phi a_{21} & 1 \end{bmatrix}. \tag{12}
$$

Here, the characteristic equation of $V_2(E_2)$ is

$$
\lambda^2 - (2 + A\phi a_{11})\lambda + 1 + A\phi a_{11} + A^2\phi^2 a_{12}a_{21} = 0.
$$

Hence, the conditions for stability of coexistence steady-state E_2 for controlled system (11) is summarized in the below theorem.

Theorem 6 *The steady-state E_2 of the system (11) is locally asymptotically stable if*

$$
|2 + A\phi a_{11}| < 2 + A\phi a_{11} + A^2\phi^2 a_{12}a_{21} < 2.
$$

6 Numerical Illustrations

To ensure the analytical results presented in the previous sections, we plot the bifurcation diagrams, time graphs, and phase portraits for the proposed system (3).

Example 1 To discuss the PDB, we choose different biologically feasible parametric values, $r \in [2.5, 3.71]$, $a = 3.99$, $b = 2.79$, $c = 1.93$, $s = 0.37$, $\alpha = 0.39$ with initial values $u_0 = 0.4$, $v_0 = 0.3$ for the system (3). Already we know that system (3) have an exclusion steady state $E_1 = (1, 0)$. Here, from Theorem 4 it follows as $\nabla = 3.0864 \geq 0$ and $r_1 = 2.6168$. Also the characteristic polynomial evaluated at E_1 with $r_1 = 2.6168$ is

$$
F(\lambda) = \lambda^2 + 0.6573\lambda - 0.3427 = 0. \tag{13}
$$

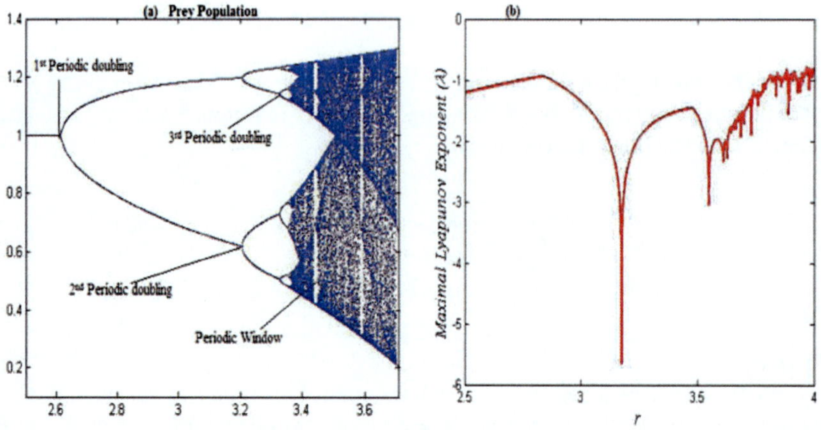

Fig. 1 **a** Flip bifurcation diagram for system (3) around E_1 in $(r-u)$ plane, **b** Maximum Lyapunov Exponent corresponding to (a)

The roots of (13) are $\lambda_1 = -1$ and $\lambda_2 = 0.3427 \neq \pm 1$. Hence, from Theorem 4, the conditions for PDB are attained near extinction steady-state E_1 at r_1 which is the critical value. From Fig. 1a, we note that the exclusion steady-state E_1 of system (3) is stable for $r < r_1$, and lose its stability when $r = r_1$. For $r > r_1$, we see that the periodic curves of periods 2, 4, 8, 16, 32 and non-periodic oscillations which is known as chaos appears. The maximum Lyapunov exponent (LEs) associated with Fig. 1a is figured in 1(b) which confirms the presence of the chaos and aperiodic paths in the specified parametric space. Therefore, from the observations, we have some positive LE values and some negative LE values, which indicates that there are some stable steady states or stable periodic windows in the chaotic area. Moreover, the presence of chaos is confirmed when the Lyapunov exponent characteristics are implying in the positive region.

Example 2 To verify the NSB analytical results, we consider $c \in [2.5, 2.71]$, $a = 1.99$, $b = 1.79$, $r = 2.89$, $\alpha = 0.99$ and $s = 0.55$ with initial values $u_0 = 0.4$, $v_0 = 0.3$, then we get $\nabla_1 = -4.6626 < 0$ and $c = c_2 = 2.5332$. From Theorem 5, the coefficients of system (3) are satisfied. From the calculation, we obtain, at $c = c_2 = 2.5332$, the system (3) have coexistence steady state $E_2 = (0.4993, 0.5138)$. The characteristic polynomial evaluated at E_2 is

$$F(\lambda) = \lambda^2 - 1.6002\lambda + 1 = 0. \tag{14}$$

The roots of (14) are $\lambda_{1,2} = 0.8001 \pm i0.5999$ with $|\lambda_{1,2}| = 1$. Hence, from Theorem 5, the conditions of NSB are attained near the coexistence steady-state E_2 at c_2 which is known as a critical value. In Fig. 2a, b, we observe that coexistence steady-state E_2 of (3) is stable for $c < c_2 = 2.5332$, loses its stability at $c = c_2 = 2.5332$

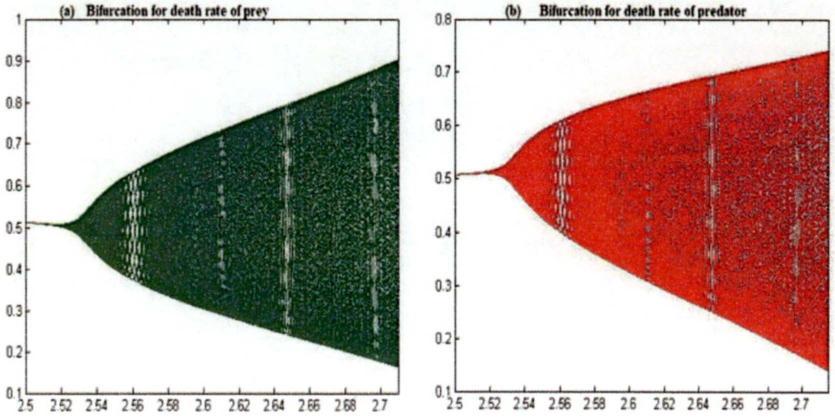

Fig. 2 NSB graphs of system (3) in $(c - u)$ and $(c - v)$ planes

and inviting new invariant cycle appears for c in the range of [2.5332, 2.55], which are plotted in Fig. 3b, c. To verify these observations, different phase portraits of c related to Fig. 2 are in Fig. 3. The quasi-periodic orbits appears when $c = 2.56$, when the value of c is increased, the curves collapsed and also observe that the inviting chaotic paths have occurred which are shown in Fig. 3g–i.

Example 3 Next, we take $a = 3.99$, $b = 2.79$, $c = 1.93$, $s = 0.37$, $\alpha = 0.39$, and $r = r = 2.6168$ with initial values $u_0 = 0.4$, $v_0 = 0.3$. In this case, the exclusion steady-state $E_1 = (1, 0)$ of system (3) is unstable. The plots of u_t and v_t are shown in Fig. 4a for system (3). Linear feedback control strategy is employed to attain the asymptotic stability of steady-state E_1. Let us consider the controlled system (8) where the feedback controlling force is considered as $Sn = -p_1(u_t - 1) - p_2(v_t - 0)$ with gains $p_1 = -0.01$ and $p_2 = 0.315$. The graphs of u_t and v_t are shown in Fig. 4b for system (8). Our analytical results are confirmed from these graphs.

Example 4 Finally, consider the parametric values as $a = 1.99$, $b = 1.79$, $r = 2.89$, $\alpha = 0.99$, $s = 0.55$ and with initial values $(u_0, v_0) = (0.4, 0.3)$. From Example 4, we see that the system (3) undergoes NSB as c varies in [2.5, 2.71]. In Figure 5 we see a closed invariant curve appears at $c = 2.534$ enfolding this unstable coexistence steady state $E_2 = (0.4990, 0.5140)$. Hence, the controlled system (11) is

$$
\begin{aligned}
u_{t+1} &= u_t + \phi \frac{s^{\alpha}}{\alpha\Gamma(\alpha)} \left(ru_t[1 - u_t] - a\sqrt{u_t}v_t \right), \\
v_{t+1} &= v_t + \phi \frac{s^{\alpha}}{\alpha\Gamma(\alpha)} \left(c\sqrt{u_t}v_t - bv_t \right),
\end{aligned} \tag{15}
$$

where $a = 1.99$, $b = 1.79$, $r = 2.89$, $\alpha = 0.99$, $s = 0.55$, $c = 2.534$ and $0 < \phi < 1$. Then variation matrix of (15) at E_2 is

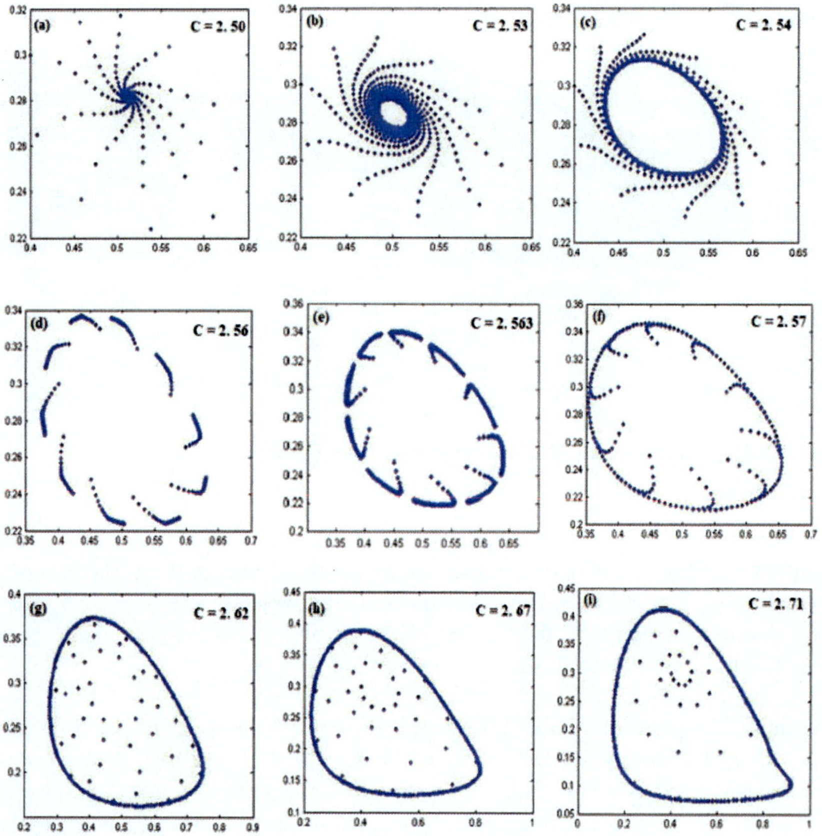

Fig. 3 Phase plane graphs for different values of c corresponding to Fig. 2

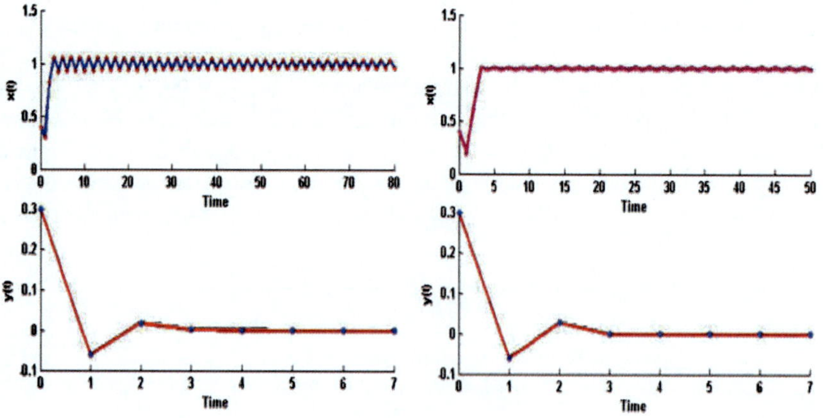

Fig. 4 **a** Plots of u_t and v_t for system (3), **b** Plots of u_t and v_t for controlled system (8)

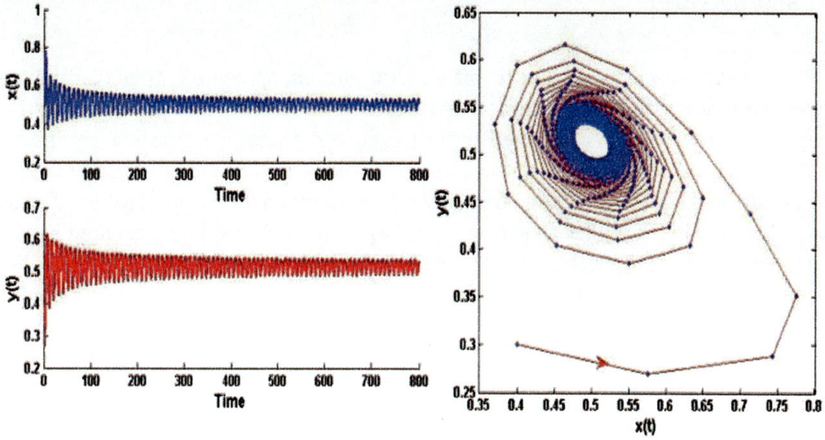

Fig. 5 **a** Plots of u_t and v_t for system (3), **b** Phase portrait for system (3)

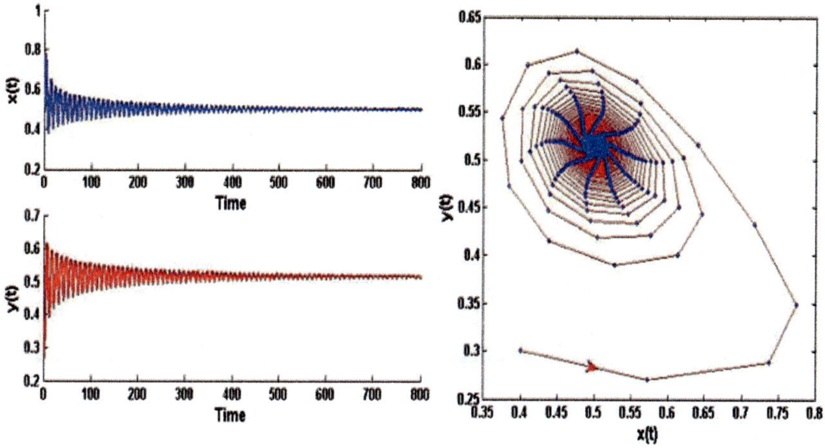

Fig. 6 **a** Plots of u_t and v_t for system (11), **b** Phase plane graphs for system (11)

$$\begin{bmatrix} 1 - 0.3990\phi & -0.7811\phi \\ 0.5122\phi & 1 \end{bmatrix}. \tag{16}$$

The characteristic polynomial of (16) is

$$F(\lambda) = \lambda^2 - (2 - 0.3990\phi)\lambda + 0.4001\phi^2 - 0.3990\phi + 1. \tag{17}$$

When $0 < \phi < 0.99990$, the solutions of (17) lies in the unit open disk. Also, the time graphs for u_t, v_t of (17) are in Fig. 6 with $\phi = 0.98$. From the graphs in Fig. 6a, b, we see that the coexistence steady-state E_2 is stable.

7 Conclusion

This research article is related to bifurcation and chaos control of prey–predator model with square root interaction term defined by the discrete-time FODEs. Discrete version of the system is obtained by utilization of piecewise constant arguments method. The variation matrix of the extinction steady state is indeterminate because the system had square root term. To discuss the stability of the extinction steady state, we re-arrange the variables as $u(t) = U^2(t)$ and $v(t) = V(t)$. The presented results demonstrates the herd behavior of prey, which plays a vital role for the stability of the proposed system. In particular, the corresponding model is highly sensitive to bifurcation parameter. Also we employ the feedback and perturbation control strategies to control the chaos. The numerical illustrations validate our analytical results. The cascades of PDB with chaotic sets are shown. Numerical simulations and graphical representation exhibit the rich dynamics of the proposed system.

References

1. Berryman, A.A.: The origins and evolution of predator prey theory. Ecology **73**, 1530–1535 (1992)
2. Chen, B.-S., Chen, J.-J.: Bifurcation and chaotic behavior of a discrete singular biological economic system. Appl. Math. Comput. **219**, 2371–2386 (2012)
3. Elsadany, A.A., Matouk, A.E.: Dynamical Behaviors of Fractional order Lotka - Volterra predator prey model and its discretization. J. Appl. Math. Comput. (2014). https://doi.org/10.1007/s12190-014-0838-6
4. George Maria Selvam, A., Dhineshbabu, R., Britto Jacob, S.: Dynamical behaviors of fractional order prey predator interactions. Int. J. Adv. Res. **V**(VII), 2184–2189 (2017)
5. Elaydi, S.: An Introduction to Difference Equations, 3rd edn. Springer International Edition, First Indian Reprint (2008)
6. El-Sayed, A.M.A., El-Mesiry, A., El-Saka, H.: On the fractional-order logistic equation. Appl. Math. Lett. **20**, 817–823 (2007)
7. Elsadany, A.-E. A., EL-Metwally, H. A., Elabbasy, E. M., Agiza, H. N.: Chaos and bifurcation of a nonlinear discrete prey-predator system. Comput. Ecol. Software **2**(3):169–180 (2012
8. George Maria Selvam, A., Britto Jacob, S., Dhineshbabu, R.: Bifurcation and chaos control for prey predator model with step size in discrete time. J. Phys. Conf. Ser., **1543**, 012010 (2020)
9. Petras, I.: Fractional order nonlinear systems—modeling. Analysis and simulation. Higher Education Press, Springer (2010)
10. Agarwal, R.P., El-Sayed, A.M.A., Salman, S.M.: Fractional-order Chuas system: discretization,bifurcation and chaos. Adv. Difference Equations **320** (2013)
11. Braza, P.: Predator-prey dynamics with square root functional responses. Nonlinear Analysis: Real World Applications, 13(4): 1837-1843, 2012
12. Zhu, X., Dai, Y., Li, Q., Zhao, K.: Stability and Hopf bifurcation of a modified predator-prey model with a time delay and square root response function. Adv. Difference Equations **235** (2017)
13. George Maria Selvam, A., Ragavan, C., Britto Jacob, S.: Analysis of discrete time prey—predator model with square root interaction. J. Appl. Sci. Comput. **VI**(III), 126–131 (2019)

Factorization of *EP* Operators in Krein Spaces

A. Vinoth and P. Sam Johnson

Abstract A closed range bounded operator on a Hilbert space is said to be an EP operator if the operator commutes with its Moore-Penrose inverse. In this paper, we characterize EP operators through factorization in the Krein space settings.

Keywords Moore-Penrose inverse · EP operator · Krein space

Mathematics Subject Classification (2010) 47B50 · 15A09 · 46C20

1 Introduction

The notion of EP matrix was introduced by Schwerdtfeger [16]. A square matrix is said to be an EP matrix if $R(A) = R(A^*)$. One of the main reasons for studying EP matrix is that it commutes with its Moore-Penrose inverse [15]. Many authors studied about EP matrices [1, 5, 10, 12, 14, 17]. The notion of EP was extended to Hilbert spaces, C*-algebras, Banach algebras [3, 4, 6–9]. The study EP matrices on finite dimensional Krein spaces was done by Jayaraman [11]. In this paper, we give the characterization of EP operators in the Krein space settings.

Definition 1 A Krein space is an indefinite inner product space $(\mathcal{K}, [\cdot, \cdot])$ (the form $[\cdot, \cdot]$ is sesquilinear and Hermitian) such that there exists an automorphism J of

The present work of the second author was partially supported by the National Board for Higher Mathematics (NBHM), Ministry of Atomic Energy, Government of India (Reference No.2/48(16)/2012/NBHM(R.P.)/R&D 11/9133).

A. Vinoth (✉)
Department of Mathematics, St. Xavier's College, Palayamkottai 627002, India
e-mail: vinoth.antony1729@gmail.com

P. Sam Johnson
Department of Mathematical and Computational Sciences, National Institute of Technology Karnataka, Surathkal 575025, Mangaluru, India
e-mail: sam@nitk.edu.in

R. N. Mohapatra et al. (eds.), *Mathematical Analysis and Computing*,
Springer Proceedings in Mathematics & Statistics 344,
https://doi.org/10.1007/978-981-33-4646-8_30

$\mathcal{K}$ which squares to the identity (called fundamental symmetry), $\langle x, y\rangle \equiv [Jx, y]$ defines a positive definite inner product and $(\mathcal{K}, \langle\cdot,\cdot\rangle)$ is a Hilbert space.

The following notations are useful for viewing Krien space in another form.

1. $\beta^+ \equiv \{x \in \mathcal{K} : [x, x] > 0\}$ is called the "positive cone"
2. $\beta^- \equiv \{x \in \mathcal{K} : [x, x] < 0\}$ is called the "negative cone"
3. $\beta^0 \equiv \{x \in \mathcal{K} : [x, x] = 0\}$ is called the "neutral cone."

Definition 2 If the inner product space $\mathcal{K}$ admits a fundamental decomposition of the form $\mathcal{K} = \mathcal{K}^+ \oplus \mathcal{K}^-$; $\mathcal{K}^+ \subset \beta^+ \cup \{0\}$, $\mathcal{K}^- \subset \beta^- \cup \{0\}$, where the subspaces $\mathcal{K}^+, \mathcal{K}^-$ are complete with respect to the norm $\|x\| = \left|[x, x]\right|^{1/2}$, then we say that $\mathcal{K}$ is a Krein space.

Let A be a bounded linear operator from a Hilbert space $\mathcal{H}_1$ to a Hilbert space $\mathcal{H}_2$. It is known that the Moore-Penrose inverse of A exists if and only if $\mathcal{R}(A)$ is a closed subspace of $\mathcal{H}_2$. As every closed subspace of a Krein space is not necessarily a Krein space, closed range condition on A in Krein space settings is not sufficient for the existence of Moore-Penrose inverse.

Example 1 [2] Consider $\mathcal{K} = \{(x_i)_{i=1}^{\infty} : \sum_{i=1}^{\infty} |x_i|^2 < \infty, x_i \in \mathbb{C}\}$ with the inner product

$$[(x_i)_{i=1}^{\infty}, (y_i)_{i=1}^{\infty}] = \sum_{i=1}^{\infty}(-1)^i x_i \overline{y_i}.$$

Let $L = \{(x_i)_{i=1}^{\infty} : x_{2i} = \frac{2i}{2i-1}x_{2i-1}, i = 1, 2, 3, \ldots\}$. Here L is a closed subspace of $\mathcal{K}$, but L is not complete with respect to the given inner product.

In [13], the author has given necessary and sufficient conditions for the existence of Moore-Penrose inverse in Krein spaces. For Krein spaces $\mathcal{K}_1$ and $\mathcal{K}_2$, $\mathcal{B}(\mathcal{K}_1, \mathcal{K}_2)$ denotes the algebra of all bounded linear operators from $\mathcal{K}_1$ to $\mathcal{K}_2$ and $\mathcal{B}(\mathcal{K}, \mathcal{K}) = \mathcal{B}(\mathcal{K})$. For an operator $A \in \mathcal{B}(\mathcal{K})$, we denote the adjoint of an operator with respect to indefinite inner product by A^*, its kernel by $\mathcal{N}(A)$, and its range by $\mathcal{R}(A)$. If $A \in \mathcal{B}(\mathcal{K}_1)$ and $B \in \mathcal{B}(\mathcal{K}_2)$, then $A \oplus B$ denotes their direct sum acting on $\mathcal{K}_1 \oplus \mathcal{K}_2$, here $\mathcal{K}_1 \oplus \mathcal{K}_2$ is the exterior 2-sum of $\mathcal{K}_1$ and $\mathcal{K}_2$. The inner product on $\mathcal{K}_1 \oplus \mathcal{K}_2$ is defined as $[(k_1, k_2), (k_1', k_2')]_{\mathcal{K}_1 \oplus \mathcal{K}_2} = [k_1, k_1']_{\mathcal{K}_1} + [k_2, k_2']_{\mathcal{K}_2}$. At the same time, for any two orthogonal subspaces $\mathcal{M}_1$ and $\mathcal{M}_2$ of $\mathcal{K}$, the internal orthogonal direct sum is denoted by $\mathcal{M}_1 \oplus^{\perp} \mathcal{M}_2$.

Definition 3 [13] A subspace $\mathcal{L}$ of a Krein space $\mathcal{K}$ is said to be regular if $\mathcal{L} \oplus^{\perp} \mathcal{L}^{\perp} = \mathcal{K}$. An operator $A \in \mathcal{B}(\mathcal{K}_1, \mathcal{K}_2)$ is regular if both $\mathcal{R}(A)$ and $\mathcal{N}(A)$ are regular.

Theorem 1 *[13] Let $A \in \mathcal{B}(\mathcal{K}_1, \mathcal{K}_2)$. Then A admits unique Moore-Penrose inverse if and only if A is regular.*

Theorem 2 *[13] Let $A \in \mathcal{B}(\mathcal{K})$ be regular. Then the following are equivalent:*

1. $AA^{\dagger} = A^{\dagger}A$.

2. $\mathcal{N}(A)^{\perp} = \mathcal{R}(A)$.
3. $\mathcal{N}(A) = \mathcal{N}(A^*)$.
4. $\mathcal{R}(A) = \mathcal{R}(A^*)$.

Definition 4 [13] An operator $A \in \mathcal{B}(\mathcal{K})$ is called an EP operator if A is regular and $AA^{\dagger} = A^{\dagger}A$.

2 Main Results

Motivated by the seminal paper [9], we give some characterizations of EP operators on Krein spaces through factorization in this section.

Lemma 1 *Let $A_1 \in \mathcal{B}(\mathcal{K}_1)$ and $A_2 \in \mathcal{B}(\mathcal{K}_2)$ be regular. Then $A_1 \oplus A_2$ is EP if and only if A_1 and A_2 are EP.*

Proof Suppose $A_1 \oplus A_2$ is EP and $x \in \mathcal{N}(A_1)$. Then $(x, 0) \in \mathcal{N}(A_1 \oplus A_2) = \mathcal{N}(A_1^* \oplus A_2^*)$ and $x \in \mathcal{N}(A_1^*)$. On the other hand, if $x \in \mathcal{N}(A_1^*)$, then we have $x \in \mathcal{N}(A_1)$. Hence, A_1 is EP. Similarly, A_2 is also EP. Conversely, suppose A_1, A_2 are EP and $(x, y) \in \mathcal{N}(A_1 \oplus A_2)$, then $A_1x = 0$ and $A_2y = 0$. This implies $x \in \mathcal{N}(A_1) = \mathcal{N}(A_1^*)$ and $y \in \mathcal{N}(A_2) = \mathcal{N}(A_2^*)$. Hence $(x, y) \in \mathcal{N}(A_1^* \oplus A_2^*)$. Therefore, $A_1 \oplus A_2$ is EP.

Lemma 2 *Let $A_1 \in \mathcal{B}(\mathcal{K}_1)$ and $A_2 \in \mathcal{B}(\mathcal{K}_2)$ be regular and $U \in \mathcal{B}(\mathcal{K}_2, \mathcal{K}_1)$ be injective such that $A_1 = UA_2U^*$. Then A_1 is EP if and only if A_2 is EP.*

Proof Suppose A_2 is EP and $x \in \mathcal{N}(A_1)$. Then $UA_2U^*x = 0$. Since U is injective, $A_2U^*x = 0$ implies that $U^*x \in \mathcal{N}(A_2) = \mathcal{N}(A_2^*)$, which in turn, implies that $UA_2^*U^*x = 0$, equivalently $x \in \mathcal{N}(A_1^*)$. The other implication follows in a similar way. Hence, A_1 is EP. Conversely, suppose A_1 is EP and $x \in \mathcal{N}(A_2)$. Therefore, $A_2x = 0$. Since U is injective, U^* is surjective. Hence, for $x \in \mathcal{K}_2$ there exists $y \in \mathcal{K}_1$ such that $U^*y = x$. Therefore, $A_2U^*y = 0$ implies that $UA_2U^*y = A_1y = 0$. Since A_1 is EP, $A_1^*y = UA_2^*U^*y = 0$. Using injectivity of U and $U^*y = x$, we get $x \in \mathcal{N}(A_2^*)$. The other implication follows in a similar way. Hence, A_2 is EP.

Remark 1 The previous lemma is not true if U is not injective. Consider the Krein space $\mathcal{K}$ in Example 1. Let A_1, U, A_2 be operators on $\mathcal{K}$ defined by $A_1(x_1, x_2, \ldots,) = (x_1, 0, x_3, 0, \ldots)$, $U(x_1, x_2, \ldots) = (x_1 - x_2, x_3 - x_4, \ldots)$ and $A_2(x_1, x_2, \ldots) = (x_2, 0, x_4, \ldots)$, respectively. Then we have $U^*(x_1, x_2, \ldots,) = (x_1, x_1, x_2, x_2, \ldots)$. Here $A_1 = UA_2U^*$ and A_1 is EP, but U is not injective and A_2 is not EP.

Theorem 3 *Let $A \in \mathcal{B}(\mathcal{K})$ be regular. Then the following are equivalent:*

1. *A is EP.*
2. *There exist Krein spaces $\mathcal{H}_1, \mathcal{L}_1, U_1 \in \mathcal{B}(\mathcal{H}_1 \oplus \mathcal{L}_1, \mathcal{K})$ unitary and $B_1 \in \mathcal{B}(\mathcal{H}_1)$ isomorphism such that $A = U_1(B_1 \oplus 0)U_1^*$.*

3. *There exist Krein spaces* $\mathcal{H}_2, \mathcal{L}_2, U_2 \in \mathcal{B}(\mathcal{H}_2 \oplus \mathcal{L}_2, \mathcal{K})$ *isomorphism and* $B_2 \in \mathcal{B}(\mathcal{H}_2)$ *isomorphism such that* $A = U_2(B_2 \oplus 0)U_2^*$.
4. *There exist Krein spaces* $\mathcal{H}_3, \mathcal{L}_3, U_3 \in \mathcal{B}(\mathcal{H}_3 \oplus \mathcal{L}_3, \mathcal{K})$ *injective and* $B_3 \in \mathcal{B}(\mathcal{H}_3)$ *isomorphism such that* $A = U_3(B_3 \oplus 0)U_3^*$.

Proof Assume that A is EP. Let $\mathcal{H}_1 = \mathcal{R}(A), \mathcal{L}_1 = \mathcal{N}(A)$. Since $\mathcal{R}(A)$ and $\mathcal{N}(A)$ are regular, they are Krein spaces. Then the map $U_1 : \mathcal{H}_1 \oplus \mathcal{L}_1 \to \mathcal{K}$ is defined by $U_1(x, y) = x + y$ for all $x \in \mathcal{R}(A)$, $y \in \mathcal{N}(A)$. To say U_1 is unitary we have to show U_1 is surjective and $[U_1(y_1, z_1), U_1(y_2, z_2)] = [(y_1, z_1), (y_2, z_2)]$. This can be done since $\mathcal{R}(A) \oplus^{\perp} \mathcal{N}(A) = \mathcal{K}$. In fact, we can explicitly say $U_1^* k = (P_{\mathcal{R}(A)}k, P_{\mathcal{N}(A)}k)$, $k \in \mathcal{K}$. Next $B_1 = A\,|_{\mathcal{R}(A)}: \mathcal{R}(A) \to \mathcal{R}(A)$ is isomorphism, since $\mathcal{R}(A^*) = \mathcal{R}(A)$. Hence, $A = U_1(B_1 \oplus 0)U_1^*$. This proves (1) $\Rightarrow$ (2). (2) $\Rightarrow$ (3), (3) $\Rightarrow$ (4) are obvious. (4) $\Rightarrow$ (1) follows from Lemmas 1 and 2.

Remark 2 Theorem 3 gives a key idea to construct Moore-Penrose inverse of an EP operator. If $A = U_1(B_1 \oplus 0)U_1^*$, then $A^{\dagger} = U_1(B_1^{-1} \oplus 0)U_1^*$. Also, if we do not assume U_3 is injective, then A is not necessarily EP.

In Theorem 3, if we assume B_1 is injective with closed range, then A is not necessarily EP. The next characterization is given through simultaneous factorization of A and A^* of the form $A = U(B \oplus 0)U^*$ and $A^* = U(C \oplus 0)U^*$ with U, B and C injective.

Theorem 4 *Let* $A \in \mathcal{B}(\mathcal{K})$ *be regular. Then the following are equivalent:*

1. *A is* EP.
2. (a) *There exist Krein spaces* $\mathcal{H}_1$ *and* $\mathcal{L}_1$, $V_1 \in \mathcal{B}(\mathcal{H}_1 \oplus \mathcal{L}_1, \mathcal{K})$ *injective,* $W_1 \in \mathcal{B}(\mathcal{H}_1 \oplus \mathcal{L}_1, \mathcal{K})$, $S_1 \in \mathcal{B}(\mathcal{K}, \mathcal{H}_1 \oplus \mathcal{L}_1)$, $B_1 \in \mathcal{B}(\mathcal{H}_1)$ *injective and* $C_1 \in \mathcal{B}(\mathcal{H}_1)$ *such that* $A = V_1(B_1 \oplus 0)S_1$ *and* $A^* = W_1(C_1 \oplus 0)S_1$.
   (b) *There exist Krein spaces* $\mathcal{H}_2$ *and* $\mathcal{L}_2$, $V_2 \in \mathcal{B}(\mathcal{H}_2 \oplus \mathcal{L}_2, \mathcal{K})$, $W_2 \in \mathcal{B}(\mathcal{H}_2 \oplus \mathcal{L}_2, \mathcal{K})$ *injective,* $S_2 \in \mathcal{B}(\mathcal{K}, \mathcal{H}_2 \oplus \mathcal{L}_2)$, $B_2 \in \mathcal{B}(\mathcal{H}_2)$ *and* $C_2 \in \mathcal{B}(\mathcal{H}_2)$ *injective such that* $A = V_2(B_2 \oplus 0)S_2$ *and* $A^* = W_2(C_2 \oplus 0)S_2$.

Proof (1) $\Rightarrow$ (2): The proof follows from Theorem 3.
(2) $\Rightarrow$ (1): Assume (a) holds. $A = V_1(B_1 \oplus 0)S_1$ and V_1 and B_1 are injective, we get $\mathcal{N}(A) = S_1^{-1}(\{0\} \oplus \mathcal{L}_1)$ and $A^* = W_1(C_1 \oplus 0)S_1$ gives $S_1^{-1}(\{0\} \oplus L_1) \subseteq \mathcal{N}(A^*)$. Therefore, $\mathcal{N}(A) \subseteq \mathcal{N}(A^*)$. By (b) we get $\mathcal{N}(A^*) \subseteq \mathcal{N}(A)$. Hence, A is EP.

Theorem 5 *Let* $A \in \mathcal{B}(\mathcal{K})$ *be regular. Then the following are equivalent:*

1. *A is* EP.
2. *There exist Krein spaces* $\mathcal{H}_1, \mathcal{L}_1, U \in \mathcal{B}(\mathcal{H}_1 \oplus \mathcal{L}_1, \mathcal{K})$ *isomorphism and* $B \in \mathcal{B}(\mathcal{H}_1)$ *isomorphism and* $C \in \mathcal{B}(\mathcal{H}_1)$ *such that* $A = U(B \oplus 0)U^{-1}$ *and* $A^* = U(C \oplus 0)U^{-1}$.

Proof (1) $\Rightarrow$ (2): The proof follows from Theorem 3.
(2) $\Rightarrow$ (1): From the proof of (2) $\Rightarrow$ (1) in Theorem 4 we get $\mathcal{N}(A) \subseteq \mathcal{N}(A^*)$. Taking adjoint in given A and A^*, then $A^* = (U^*)^{-1}(B^* \oplus 0)U^*$, $A = (U^*)^{-1}(C^* \oplus 0)U^*$. In the same argument, we get $\mathcal{N}(A^*) \subseteq \mathcal{N}(A)$. Hence A is EP.

In above statement of the theorem, we can replace A^* by $A^\dagger$, since $\mathcal{N}(A^*) = \mathcal{N}(A^\dagger)$. The next theorem uses simultaneous factorization of AA^* and A^*A.

Theorem 6 *Let $A \in \mathcal{B}(\mathcal{K})$ be regular. Then the following are equivalent:*

1. *A is EP.*
2. *There exist Krein spaces $\mathcal{H}_1, \mathcal{L}_1$, $U_1 \in \mathcal{B}(\mathcal{H}_1 \oplus \mathcal{L}_1, \mathcal{K})$ unitary and $B_1 \in \mathcal{B}(\mathcal{H}_1)$ isomorphism such that $A^*A = U_1(B_1^*B_1 \oplus 0)U_1^*$ and $AA^* = U_1(B_1B_1^* \oplus 0)U_1^*$.*
3. *(a) There exist Krein spaces $\mathcal{H}_2$ and $\mathcal{L}_2$, $V_2 \in \mathcal{B}(\mathcal{H}_2 \oplus \mathcal{L}_2, \mathcal{K})$ injective, $W_2 \in \mathcal{B}(\mathcal{H}_2 \oplus \mathcal{L}_2, \mathcal{K})$, $S_2 \in \mathcal{B}(\mathcal{K}, \mathcal{H}_2 \oplus \mathcal{L}_2)$, $B_2 \in \mathcal{B}(\mathcal{H}_2)$ injective and $C_2 \in \mathcal{B}(\mathcal{H}_2)$ such that $A^*A = V_2(B_2 \oplus 0)S_2$ and $AA^* = W_2(C_2 \oplus 0)S_2$.*
   (b) There exist Krein spaces $\mathcal{H}_3$ and $\mathcal{L}_3$, $V_3 \in \mathcal{B}(\mathcal{H}_3 \oplus \mathcal{L}_3, \mathcal{K})$, $W_3 \in \mathcal{B}(\mathcal{H}_3 \oplus \mathcal{L}_3, \mathcal{K})$ injective, $S_3 \in \mathcal{B}(\mathcal{K}, \mathcal{H}_3 \oplus \mathcal{L}_3)$, $B_3 \in \mathcal{B}(\mathcal{H}_3)$ and $C_3 \in \mathcal{B}(\mathcal{H}_3)$ injective such that $AA^* = V_3(B_3 \oplus 0)S_3$ and $AA^* = W_3(C_3 \oplus 0)S_3$.
4. *There exist Krein spaces $\mathcal{H}_4, \mathcal{L}_4$, $U_4 \in \mathcal{B}(\mathcal{H}_4 \oplus \mathcal{L}_4, \mathcal{K})$ isomorphism and $B_4 \in \mathcal{B}(\mathcal{H}_4)$ isomorphism and $C_4 \in \mathcal{B}(\mathcal{H}_4)$ such that $A^*A = U_4(B_4 \oplus 0)U_4^{-1}$ and $AA^* = U_4(C_4 \oplus 0)U_4^{-1}$.*

Proof (1) $\Rightarrow$ (2): By Theorem 3, (2) $\Rightarrow$ (3), (2) $\Rightarrow$ (4) and (3) $\Rightarrow$ (1) are obvious. (2) $\Rightarrow$ (1): As in proof of (2) $\Rightarrow$ (1) in Theorem 4, we get $\mathcal{N}(AA^*) = \mathcal{N}(A^*A)$. But we know that $\mathcal{N}(AA^*) = \mathcal{N}(A^*)$ and $\mathcal{N}(A^*A) = \mathcal{N}(A)$. Therefore $\mathcal{N}(A) = \mathcal{N}(A^*)$. Hence, A is EP.

Theorem 7 *Let $A \in \mathcal{B}(\mathcal{K})$ be regular. Then the following are equivalent:*

1. *A is EP.*
2. *There exists an isomorphism $N_1 \in \mathcal{B}(\mathcal{K})$ such that $A^* = N_1A$.*
3. *There exists $N_2 \in \mathcal{B}(\mathcal{K})$ injective such that $A^* = N_2A$.*
4. *There exist $S_1, S_2 \in \mathcal{B}(\mathcal{K})$ such that $A^* = S_1A$ and $A = S_2A^*$.*

Proof (1) $\Rightarrow$ (2): By Theorem 3, we have $A = U(B_1 \oplus 0)U^*$ with $U \in \mathcal{B}(\mathcal{H} \oplus \mathcal{L}, \mathcal{K})$ unitary and $B \in \mathcal{B}(\mathcal{H})$ an isomorphism. If we take $N_1 = U(B^*B^{-1} \oplus I)U^* : \mathcal{K} \to \mathcal{K}$, then N_1 is an isomorphism with $A^* = N_1A$. (2) $\Rightarrow$ (3) is direct and (2) $\Rightarrow$ (4) follows from $A = N_1^{-1}A^*$. (3) $\Rightarrow$ (1) : As $A^* = N_1A$, we get $\mathcal{N}(A) \subseteq \mathcal{N}(A^*)$. But N_1 is injective which implies that $\mathcal{N}(A^*) \subseteq \mathcal{N}(A)$. Hence A is EP. (4) $\Rightarrow$ (1) : By $A^* = S_1A$, we have $\mathcal{N}(A) \subseteq \mathcal{N}(A^*)$ and by $A = S_2A^*$ we get that $\mathcal{N}(A^*) \subseteq \mathcal{N}(A)$.

Theorem 8 *Let $A \in \mathcal{B}(\mathcal{K})$ be regular. Then the following are equivalent:*

1. *A is EP.*
2. *There exists an isomorphism $N_1 \in \mathcal{B}(\mathcal{K})$ such that $A^\dagger = N_1A = AN_1$.*
3. *There exists $N_2 \in \mathcal{B}(\mathcal{K})$ injective such that $A^\dagger = N_2A$.*
4. *There exist $S_1, S_2 \in \mathcal{B}(\mathcal{K})$ such that $A^\dagger = S_1A$ and $A = S_2A^\dagger$.*

Proof (1) $\Rightarrow$ (2): By Theorem 3, we have $A = U(B_1 \oplus 0)U^*$ with $U \in \mathcal{B}(\mathcal{H} \oplus \mathcal{L}, \mathcal{K})$ unitary and $B \in \mathcal{B}(\mathcal{H})$ an isomorphism. If we take $N_1 = U(B^{-2} \oplus I)U^* : \mathcal{K} \to \mathcal{K}$, then N_1 is an isomorphism with $A^\dagger = N_1A = AN_1$. As $\mathcal{N}(A^\dagger) = \mathcal{N}(A^*)$, the rest follows from the proof of Theorem 7.

Theorem 9 *Let $A \in \mathcal{B}(\mathcal{K})$ be regular. Then the following are equivalent:*

1. *A is EP.*
2. *There exists an isomorphism $N_1 \in \mathcal{B}(\mathcal{K})$ such that $A^*A = N_1AA^*$.*
3. *There exists $N_2 \in \mathcal{B}(\mathcal{K})$ injective such that $A^*A = N_2AA^*$.*
4. *There exist $S_1, S_2 \in \mathcal{B}(\mathcal{K})$ such that $A^*A = S_1AA^*$ and $AA^* = S_2A^*A$.*

Proof (1) $\Rightarrow$ (2): By Theorem 3, we have $A = U(B_1 \oplus 0)U^*$ with $U \in \mathcal{B}(\mathcal{H} \oplus \mathcal{L}, \mathcal{K})$ unitary and $B \in \mathcal{B}(\mathcal{H})$ an isomorphism. If we take $N_1 = U(B^*B(B^*)^{-1}B^{-1} \oplus I)U^* : \mathcal{K} \to \mathcal{K}$, then N_1 is an isomorphism with $A^*A = N_1AA^*$. As $\mathcal{N}(AA^*) = \mathcal{N}(A^*)$ and $\mathcal{N}(A^*A) = \mathcal{N}(A)$. The rest follows from the proof of Theorem 7.

References

1. Baksalary, O.M., Trenkler, G.: Characterizations of EP, normal, and Hermitian matrices. Linear Multilinear Algebra **56**(3), 299–304 (2008). https://doi.org/10.1080/03081080600872616
2. Bognár, J.: Indefinite Inner Product Spaces. Springer, New York-Heidelberg (1974). Ergebnisse der Mathematik und ihrer Grenzgebiete, Band 78
3. Brock, K.G.: A note on commutativity of a linear operator and its Moore-Penrose inverse. Numer. Funct. Anal. Optim. **11**(7–8), 673–678 (1990). https://doi.org/10.1080/01630569008816396
4. Campbell, S.L., Meyer, C.D.: EP operators and generalized inverses. Can. Math. Bull **18**(3), 327–333 (1975)
5. Cheng, S., Tian, Y.: Two sets of new characterizations for normal and EP matrices. Linear Algebra Appl. **375**, 181–195 (2003). https://doi.org/10.1016/S0024-3795(03)00650-5
6. Djordjević, D.S.: Products of EP operators on Hilbert spaces. Proc. Am. Math. Soc. **129**(6), 1727–1731 (electronic) (2001). https://doi.org/10.1090/S0002-9939-00-05701-4
7. Djordjević, D.S.: Characterizations of normal, hyponormal and EP operators. J. Math. Anal. Appl. **329**(2), 1181–1190 (2007). https://doi.org/10.1016/j.jmaa.2006.07.008
8. Djordjević, D.S., Koliha, J.J.: Characterizing Hermitian, normal and EP operators. Filomat **21**(1), 39–54 (2007). https://doi.org/10.2298/FIL0701039D
9. Drivaliaris, D., Karanasios, S., Pappas, D.: Factorizations of EP operators. Linear Algebra Appl. **429**(7), 1555–1567 (2008). https://doi.org/10.1016/j.laa.2008.04.026
10. Hartwig, R.E., Katz, I.J.: On products of EP matrices. Linear Algebra Appl. **252**, 339–345 (1997). https://doi.org/10.1016/0024-3795(95)00693-1
11. Jayaraman, S.: EP matrices in indefinite inner product spaces. Funct. Anal. Approx. Comput. **4**, 23–31 (2012)
12. Katz, I.J., Pearl, M.H.: On EPr and normal EPr matrices. J. Res. Nat. Bur. Standards Sect. B **70B**, 47–77 (1966)
13. Mary, X.: Moore-Penrose inverse in Kreĭn spaces. Integral Equ. Oper. Theory **60**(3), 419–433 (2008). https://doi.org/10.1007/s00020-008-1562-0
14. Meyer Jr., C.D.: Some remarks on EP_r matrices, and generalized inverses. Linear Algebra Appl. **3**, 275–278 (1970)
15. Pearl, M.H.: On generalized inverses of matrices. Proc. Cambridge Philos. Soc. **62**, 673–677 (1966)
16. Schwerdtfeger, H.: Introduction to Linear Algebra and the Theory of Matrices. P. Noordhoff, Groningen (1950)
17. Tian, Y., Wang, H.: Characterizations of EP matrices and weighted-EP matrices. Linear Algebra Appl. **434**(5), 1295–1318 (2011). https://doi.org/10.1016/j.laa.2010.11.014

On the Spectral Parameters of Certain Cartesian Products of Graphs with P_2

S. Sarah Surya and P. Subbulakshmi

Abstract A structure descriptor that is largely studied in the context of spectral graph theory is the energy of a graph. It is defined as the sum of the absolute values of the eigenvalues of the adjacency matrix of the graph. Spectral Graph Theory is the study of properties of graphs and the matrices associated with them such as its adjacency matrix or Laplacian matrix. It has enormous applications in diverse areas such as chemistry, coding theory, information theory, geographic studies, etc. In this paper, we have obtained the values of energy, spectral radius, second-largest eigenvalue, least eigenvalue, spread and separator of book graph, ladder graph and prism graph and supplement our results using the MATLAB programs.

Keywords Energy · Eigenvalues · Spectral radius · Cartesian product of graphs

Mathematics Subject Classification (2010) MSC 05C50 · MSC 34L16

1 Introduction

Algebraic graph theory is the branch of mathematics that deals with the algebraic properties of matrices associated with the graphs. In particular, Spectral Graph Theory is the study of properties of graphs and the matrices associated with them such as its adjacency matrix or Laplacian Matrix. The adjacency matrix of a graph is always a real symmetric matrix and hence the eigenvalues are always real numbers. By the Perron-Frobenius theory of nonnegative matrices, the adjacency matrix of any connected graph is irreducible and its spectral radius has multiplicity one. The spectrum of the graph G, denoted by $Spec(G)$, is the multiset of eigenvalues of G. We write

S. Sarah Surya · P. Subbulakshmi (✉)
Department of Mathematics, Stella Maris College (Affiliated to the University of Madras), Chennai 600086, India
e-mail: subbulakshmi216@gmail.com

S. Sarah Surya
e-mail: sara24solomon@gmail.com

R. N. Mohapatra et al. (eds.), *Mathematical Analysis and Computing*, Springer Proceedings in Mathematics & Statistics 344,
https://doi.org/10.1007/978-981-33-4646-8_31

$Spec(G) = \begin{pmatrix} \lambda_1 & \lambda_2 & \cdots & \lambda_p \\ m_1 & m_2 & \cdots & m_p \end{pmatrix}$, where $\lambda_1, \lambda_2, \ldots, \lambda_p$ are the eigenvalues and m_i is the multiplicity of λ_i, $1 \leq i \leq p$ [21]. This is a graph invariant which characterizes the topological structure of the graph.

Though the motivation for the definition of energy of a graph came in 1930s, the formal definition was first given by Ivan Gutman in 1978. Initially, very few mathematicians were attracted by this definition. But, in the past few decades, the interest on this graph invariant has considerably increased and different variants of graph energy have been widely studied. Numerous results on bounds of energy and other spectral parameters have been obtained. Also, a lower bound on the least eigenvalue in terms of the chromatic number and an upper bound on the chromatic number in terms of the least eigenvalue of a graph has been obtained.

There are several results on the values of energy and other spectral parameters for various classes of graphs such as circulant graphs [16], hypercube and its complement [4], fibonomial graphs [1], complement of regular line graphs [2], non commuting graphs and conjugacy class graph of dihedral groups [18, 19], Cayley graph over a finite chain ring [22], unitary one-matching bi-Cayley graph over a finite commutative ring [15], unitary Cayley graph and its complement [12] and gcd-graphs [22].

In this paper, we have obtained the values of energy, spectral radius, second-largest eigenvalue, least eigenvalue, spread and separator of book graph, ladder graph and prism graph and supplement our results using the MATLAB programs.

2 Motivation

Spectral graph theory is an extremely beneficial and utilitarian subject. In the context of large complex networks, there is a strong correlation between graph energy and the betweenness and eigencentrality of vertices. As the exact computation of these centrality measures is quite expensive and requires global processing of the network, our research opens up possibilities for the estimation of these centrality measures based only on local information [20].

The spectral radius has the potential to describe the distribution of a single dominant element when a transportation network is represented through a suitable matrix [23]. In chemistry, the spectra of a molecular graph and the corresponding eigenvalues are closely linked to the molecular stability and related chemical properties. The upper bound for the capacity of a channel was obtained in terms of the largest real eigenvalue of the channel graph. The maximum entropy of a specific information source can be expressed in terms of the largest eigenvalue of its connection matrix [5, 13, 24].

The second-largest eigenvalue is employed in some of the most successful methods for bounding the expansion of a graph in coding theory [7]. In general, the second-largest eigenvalue of a graph gives information about expansion and randomness properties. The smallest eigenvalue gives information about independence number and chromatic number and interlacing gives information about substructures [3].

3 Preliminaries

Definition 1 [9] The *energy* of a graph is defined as the sum of the absolute values of the eigenvalues of its adjacency matrix. i.e., if A is a $n \times n$ matrix with n eigenvalues $\lambda_1, \lambda_2, \ldots, \lambda_n$, then

$$E(G) = \sum_{i=1}^{n} |\lambda_i|$$

Definition 2 [6, 8, 14] If the eigenvalues of the adjacency matrix of a graph G are written in a nonincreasing manner, i.e., $\lambda_1 \geq \lambda_2 \geq \cdots \geq \lambda_n$, λ_1 is referred to as the *spectral radius* or *index*, λ_2 is the *second largest eigenvalue*, and λ_n is the *least eigenvalue* of G.

Definition 3 [10] The *spread* and the *separator* of a graph G are defined as $S(G) = \lambda_1(G) - \lambda_n(G)$, $S_A(G) = \lambda_1(G) - \lambda_2(G)$, respectively.

Definition 4 [17] The *Book* B_m is the graph $S_m \times P_2$ where S_m is the star with m edges. See Fig. 1.

Definition 5 [11] *Ladder Graphs* F_m are defined as $P_2 \times P_m$. See Fig. 2.

Definition 6 [11] *Prism graphs* or (*cyclic ladder graphs*) H_m are defined as $P_2 \times C_m$. See Fig. 3.

Fig. 1 B_{11}

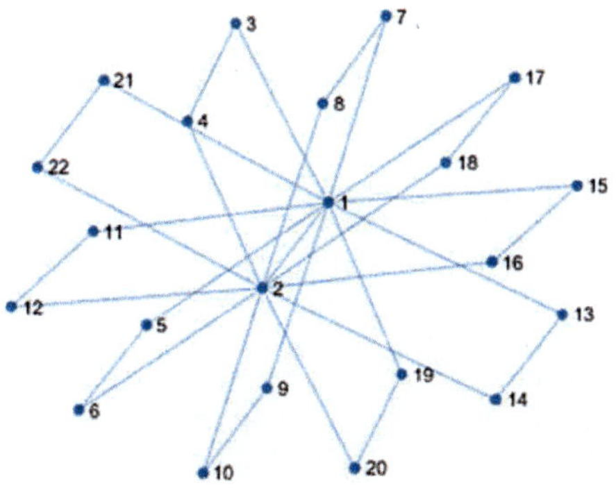

Fig. 2 F_7

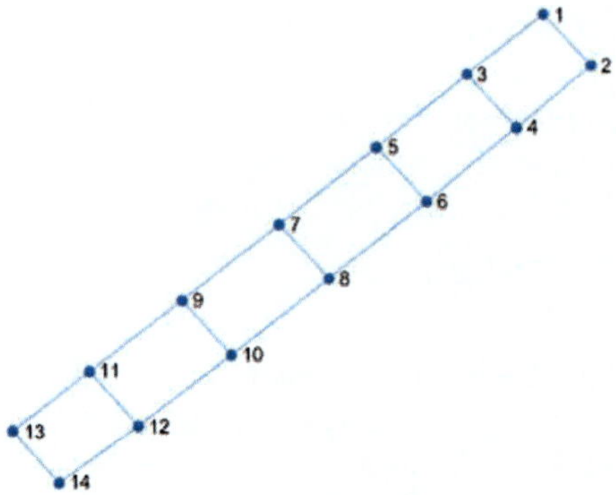

Fig. 3 H_8

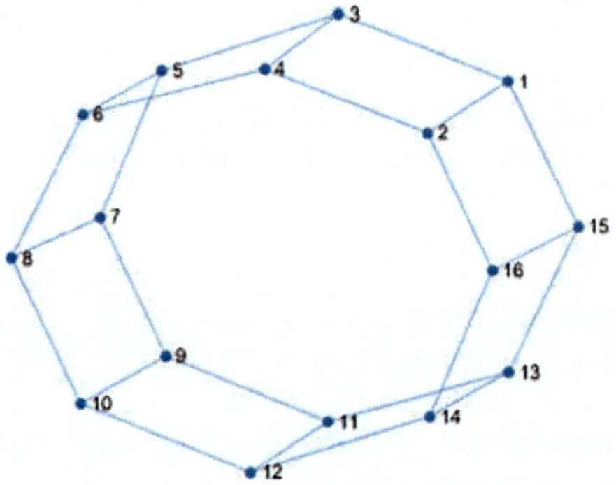

4 Main Results

Theorem 1 *The Book Graph B_m has the following parameters:*

(i) *Energy* = $2(m-1)+4\sqrt{m+1}$
(ii) *Spectral Radius* = $\sqrt{m+1}+1$
(iii) *Second-Largest Eigenvalue* = $\sqrt{m+1}-1$
(iv) *Least Eigenvalue* = $-\sqrt{m+1}-1$
(v) *Spread* = $2\sqrt{m+1}+2$
(vi) *Separator* = 2

Proof The adjacency matrix of the Book graph B_m is of the form

$$A(B_m)=\begin{pmatrix} 0 & J \\ J^T & 0 \end{pmatrix}$$

where

$$J = \begin{pmatrix} 1 & 1 & 1 & \cdots & 1 \\ 1 & 1 & 0 & \cdots & 0 \\ 1 & 0 & 1 & \cdots & 0 \\ \vdots & \vdots & & \ddots & \\ 1 & 0 & 0 & \cdots & 1 \end{pmatrix}$$

and the characteristic equation is of the form

$$(\lambda - 1)^{m-1}(\lambda + 1)^{m-1}(\lambda^2 - (\sqrt{m+1} + 1)^2)(\lambda^2 - (\sqrt{m+1} - 1)^2)$$

Hence, $Spec(G) = \begin{pmatrix} 1 & -1 & (\sqrt{m+1}+1) & (\sqrt{m+1}-1) & (-\sqrt{m+1}+1) & (-\sqrt{m+1}-1) \\ m-1 & m-1 & 1 & 1 & 1 & 1 \end{pmatrix}$

Thus, Energy of the Book Graph is

$$\begin{aligned} E(B_m) &= (m-1)|1| + (m-1)|-1| + 1|\sqrt{m+1}+1| + 1|\sqrt{m+1}-1| \\ &\quad + 1|-\sqrt{m+1}+1| + 1|-\sqrt{m+1}-1| \end{aligned}$$

$$\begin{aligned} &= (m-1) + (m-1) + 2(\sqrt{m+1}+1) + 2(\sqrt{m+1}-1) \\ &= 2(m-1) + 4\sqrt{m+1} \end{aligned}$$

From $Spec(G)$, (ii) to (vi) can be easily verified.

Theorem 2 *The Ladder Graph F_m has the following parameters:*

(i) *Energy* $\approx 3m - 2$

(ii)

$$\textit{Spectral Radius} \approx \begin{cases} 2 & \text{if } m = 2 \\ \sqrt{2}+1 & \text{if } m = 3 \\ e+1 & \text{if } m = 4 \\ \sqrt{3}+1 & \text{if } m = 5 \\ 3 & \text{if } m \geq 6 \end{cases}$$

(iii)

$$\textit{Second-Largest Eigenvalue} \approx \begin{cases} 0 & \text{if } m = 2 \\ 1 & \text{if } m = 3 \\ e & \text{if } m = 4 \\ 2 & \text{if } 5 \leq m \leq 7 \\ 3 & \text{if } m \geq 8 \end{cases}$$

(iv)

$$\text{Least Eigenvalue} \approx \begin{cases} -2 & \text{if } m = 2 \\ -(\sqrt{2}+1) & \text{if } m = 3 \\ -(e+1) & \text{if } m = 4 \\ -(\sqrt{3}+1) & \text{if } m = 5 \\ -3 & \text{if } m \geq 6 \end{cases}$$

(v)

$$Spread \approx \begin{cases} 0 & \text{if } m = 2 \\ 2(\sqrt{2}+1) & \text{if } m = 3 \\ 2(e+1) & \text{if } m = 4 \\ 2(\sqrt{3}+1) & \text{if } m = 5 \\ 6 & \text{if } m \geq 6 \end{cases}$$

(vi)

$$Separator \approx \begin{cases} 2 & \text{if } m = 2 \\ \sqrt{2} & \text{if } m = 3 \\ 1 & \text{if } m = 4 \\ \sqrt{3}-1 & \text{if } m = 5 \\ 0 & \text{if } m \geq 6 \end{cases}$$

Proof For $m = 3$, i.e., if $G = F_3$, the spectrum and eigenvalues of its 6×6 adjacency matrix are obtained as $Spec(G) = \begin{pmatrix} (-\sqrt{2}-1) & -1 & (-\sqrt{2}+1) & (\sqrt{2}-1) & 1 & (\sqrt{2}+1) \\ 1 & 1 & 1 & 1 & 1 & 1 \end{pmatrix}$ and hence $E(F_3) = 7.657$. Similarly, for $m = 4$, i.e., if $G = F_4$, we can find that $E(F_4)$= 10.472.

Thus, using mathematical induction, when the dimension of the ladder graph is increased by 1, the number of vertices increases by two and the number of edges increases by three. Further, the $2m \times 2m$ adjacency matrix becomes a $(2m + 2) \times (2m + 2)$ adjacency matrix which increases the value of energy by 2 (approx.).

Therefore, it follows that $E(F_m) \approx 3m - 2$.

Also, (ii) to (vi) can be verified easily.

Working along similar lines, we can prove the following.

Theorem 3 *The Prism Graph H_m has the following parameters:*

(i) *Energy* $\approx 3m - 3$
(ii) *Spectral Radius* ≈ 3

(iii)

$$\textit{Second-Largest Eigenvalue} \approx \begin{cases} 1 & \textit{if } m = 3, 4 \\ e & \textit{if } m = 5 \\ 2 & \textit{if } 6 \le m \le 8 \\ 3 & \textit{if } m \ge 9 \end{cases}$$

(iv) *Least Eigenvalue* ≈ -3
(v) *Spread* ≈ 6
(vi)

$$\textit{Separator} \approx \begin{cases} 2 & \textit{if } m = 3, 4 \\ 1 & \textit{if } 5 \le m \le 8 \\ 0 & \textit{if } m \ge 9 \end{cases}$$

5 MATLAB Programs

The program to obtain the eigenvalues and energy of the book graph is as follows:

```
m=input('Enter the value of m:\n');
v=1+zeros(m,1);
G1=graph([v], [2:m+1]);
A1 = adjacency(G1);
I1 = eye(m+1)
G2=graph([1], [2]);
A2 = adjacency(G2);
I2 = eye(2)
K1 = kron(A1,I2);
K2 = kron(I1,A2);
Aprod = K1 + K2;
Gprod = graph(Aprod);
plot(Gprod)
A = adjacency(Gprod);
B = full(A);
K = eig(B);
E = sum(abs(K))
```

Following is the program to obtain the eigenvalues and energy of the ladder graph:

```
m=input('Enter the value of m:\n');
G1=graph([1:m-1], [2:m]);
A1 = adjacency(G1);
```

```
I1 = eye(m)
G2=graph([1], [2]);
A2 = adjacency(G2);
I2 = eye(2)
K1 = kron(A1,I2);
K2 = kron(I1,A2);
Aprod = K1 + K2;
Gprod = graph(Aprod);
plot(Gprod)
A = adjacency(Gprod);
B = full(A);
K = eig(B);
E = sum(abs(K))
```

The program to obtain the eigenvalues and energy of the prism graph is given below:

```
m=input('Enter the value of m:\n');
G1=graph([1:m], [2:m 1]);
A1 = adjacency(G1);
I1 = eye(m)
G2=graph([1], [2]);
A2 = adjacency(G2);
I2 = eye(2)
K1 = kron(A1,I2);
K2 = kron(I1,A2);
Aprod = K1 + K2;
Gprod = graph(Aprod);
plot(Gprod)
A = adjacency(Gprod);
B = full(A);
K = eig(B);
E = sum(abs(K))
```

6 Conclusion

The values of energy, spectral radius, second-largest eigenvalue, least eigenvalue, spread and separator of the book graph, ladder graph and prism graph in terms of their dimensions are obtained. Further, we have also given the MATLAB programs for calculating the energy and eigenvalues of these graphs. We hope that the formulae obtained for these spectral parameters will help in an economical way for studying properties of larger dimensions of these graphs. The spectral parameters of variants of hypercube is under investigation.

Acknowledgements We thank DST (FIST 2015) MATLAB R2017b which was used for computation.
Conflict of interest The authors declare that they have no conflict of interest.

References

1. Akbulak, M., Kale, A., Öteleş, A.: On the connectivity properties and energy of Fibonomial graphs. Discrete Appl. Math. **169**, 1–8 (2014)
2. Alinaghipour, F., Ahmadi, B.: On the energy of complement of regular line graph. MATCH Commun. Math. Comput. Chem. **60**, 427–434 (2008)
3. Brouwer, A.E., Haemers, W.H.: Spectra of Graphs, 250 pp. Springer Science and Business Media, New York (2011)
4. Chen, X., Xie, W.: Energy of a hypercube and its complement. Int. J. Algebra **6**, 799–805 (2012)
5. Cohn, M.: On the channel capacity of read/write isolated memory. Discrete Appl. Math. **56**, 1–8 (1995)
6. Cvetković, D., Simić, S.: The second largest eigenvalue of a graph (a survey). Filomat **9**, 449–472 (1995)
7. Dowling Jr, M.C.: Expander graphs and coding theory. Ph.D. Dissertation, Clemson University (2016)
8. Fan, Y.Z., Yu, G.D., Wang, Y.: The chromatic number and the least eigenvalue of a graph. Electron. J. Comb. **19**, 1–8 (2012)
9. Gutman, I.: The energy of a graph. Ber. Math-Statist. Sekt. Forschungszentrum Graz **103**, 1–22 (1978)
10. Hong, W.X., You, L.H.: On the eigenvalues of firefly graphs. Trans. Combi. **3**, 1–9 (2014)
11. Hosaya, H., Harary, F.: On the matching properties of three fence graphs. J. Math. Chem. **12**, 211–218 (1993)
12. Ilić, A.: The energy of unitary Cayley graphs. Linear Algebra Appl. **431**, 1881–1889 (2009)
13. Immink, K.A.S.: Codes for Mass Data Storage Systems, 349 pp. Shannon Foundation Publishers, The Netherlands (2004)
14. Li, X., Shi, Y., Gutman, I.: Graph Energy, 268 pp. Springer Science and Business Media, New York (2012)
15. Liu, X.: Energies of unitary one-matching bi-Cayley graphs over finite commutative rings. Discrete Appl. Math. **259**, 170–179 (2019)
16. Louis, J.: A formula for the energy of circulant graphs with two generators. J. Appl. Math. **2016**, 1–4 (2016)
17. Maheo, M.: Strongly graceful graphs. Discrete Math. **29**, 39–46 (1980)
18. Mahmoud, R., Sarmin, N. H., Erfanian, A.: On the energy of non-commuting graph of dihedral groups. AIP Conf. Proc. **1830**, 070011-1–070011-5 (2017)
19. Mahmoud, R., Sarmin, N.H., Erfanian, A.: The conjugacy class graph of some finite groups and its energy. Malays. J. Fundam. Appl. Sci. **13**, 659–665 (2017)
20. Morzy, M., Kajdanowicz, T.: Graph energies of egocentric networks and their correlation with vertex centrality measures. Entropy **20**, 1–18 (2018)
21. Sampathkumar, E., Bharati, R., Sathish, K., Stephen, S.: On certain energies of a one-point union of Complete Graphs K_n. Proc. Jangjeon Math. Soc. **21**, 59–67 (2018)
22. Suntornpoch, B., Meemark, Y.: Cayley graphs over a finite chain ring and GCD-graphs. Bull. Aust. Math. Soc. **93**, 353–363 (2016)
23. Tinkler, K.J.: The physical interpretation of eigenvalues of dichotomous matrix. Trans. Inst. Br. Geogr. **55**, 17–46 (1972)
24. Wang, T., Jia, L., Sun, F.: Bounds of the spectral radius and the Nordhaus-Gaddum type of the graphs. Sci. World J. **2013**, 1–7 (2013)

Asymptotic Series of a General Symbol of Pseudo-Differential Operators Involving Linear Canonical Transform

Manish Kumar, Tusharakanta Pradhan, and Ram N. Mohapatra

Abstract With the help of linear canonical transform, we characterize a new class of pseudo-differential operators involving a general symbol $\eta(x, y)$. Asymptotic series expansion of pseudo-differential operators associated with the general symbol is derived by using the theory of linear canonical transform.

Keywords Pseudo-differential operators · Asymptotic series · Linear canonical transform

Mathematics Subject Classification (2000) 47G30

1 Introduction

The linear canonical transform (LCT) is a very important tool in time-frequency domain [2–4]. The LCT is a generalization of well-known transforms, such as Fourier transform, Fractional Fourier transform [1, 9, 10] and many more. It consists of the three-parameter family of linear transform and can be parameterized by a 2×2 matrix with a determinant one. For any matrix $A = \begin{bmatrix} \alpha & \beta \\ \gamma & \delta \end{bmatrix} = \begin{bmatrix} \delta & -\beta \\ -\gamma & \alpha \end{bmatrix}^{-1}$ with $\alpha\delta - \beta\gamma = 1$.

One can define LCT by the above matrix A applied over a Lebesgue integrable function $\varphi(x)$ as follows:

M. Kumar (✉) · T. Pradhan
Department of Mathematics, Birla Institute of Technology and Sciences-Pilani, Hyderabad Campus, Hyderabad 500078, Telangana, India
e-mail: manishkumar@hyderabad.bits-pilani.ac.in

R. N. Mohapatra
Department of Mathematics, University of Central Florida, Orlando, FL 32816, USA
e-mail: ram.mohapatra@ucf.edu

R. N. Mohapatra et al. (eds.), *Mathematical Analysis and Computing*,
Springer Proceedings in Mathematics & Statistics 344,
https://doi.org/10.1007/978-981-33-4646-8_32

$$\varphi_A(y) = (\mathscr{C}_A \varphi)(y) = \int_{-\infty}^{\infty} K_A(x, y)\varphi(x)dx, \tag{1}$$

where

$$\begin{aligned} K_A(x, y) &= \frac{1}{\sqrt{2\pi i\beta}} e^{i\left[\frac{\alpha x^2}{2\beta} - \frac{xy}{\beta} + \frac{\delta y^2}{2\beta}\right]} \\ &= M_A e^{i\left[\frac{\alpha x^2}{2\beta} - \frac{xy}{\beta} + \frac{\delta y^2}{2\beta}\right]}, \end{aligned}$$

is the kernel and $\beta \neq 0$. The corresponding inversion formula exists for non-zero β and it is defined by

$$\varphi(x) = (\mathscr{C}_{A^{-1}} \varphi_A)(x) = \int_{-\infty}^{\infty} K_{A^{-1}}(x, y)\varphi_A(y)dy, \tag{2}$$

where

$$\overline{K_A(x, y)} = K_{A^{-1}}(x, y),$$

and

$$K_{A^{-1}}(x, y) = \overline{M_A} e^{i\left[\frac{\alpha x^2}{2\beta} - \frac{xy}{\beta} + \frac{\delta y^2}{2\beta}\right]}.$$

Further, LCT satisfies the additivity and reversibility conditions, and it can be shown as follows:

$$\mathscr{C}_A \mathscr{C}_B = \mathscr{C}_{AB},$$

and

$$\mathscr{C}_{A^{-1}} = (\mathscr{C}_A)^{-1}$$

respectively. Let $\varphi_1(x)$ and $\varphi_2(x)$ be two functions belongs to $L^p(\mathbb{R})$ and there are two constants λ_1 and λ_2 such that

$$\mathscr{C}_A[\lambda_1\varphi_1(x) + \lambda_2\varphi_2(x)] = \lambda_1\mathscr{C}_A(\varphi_1(x)) + \lambda_2\mathscr{C}_A(\varphi_2(x)),$$

called linearity property of LCT. Now we recall some important definition and properties to investigate the proposed work. The space $L^p(\mathbb{R})$ denote the class of measurable functions g defined on real line such that

$$||g||_{L^p(\mathbb{R})} = \left[\int_{-\infty}^{\infty} |g(x)|^p dx\right]^{\frac{1}{p}} < \infty. \tag{3}$$

where $1 \leq p < \infty$. With the help of the work proposed in [8], we can define a new generalized Sobolev space $\mathbb{G}_A^{s,p}(\mathbb{R})$ associated with LCT as follows:

$$||\varphi||_{s,p}^{A} = \left[\int_{-\infty}^{\infty} (1 + |y|^2)^{\frac{sp}{2}} |\varphi_A(y)|^p dy \right]^{\frac{1}{p}}, \tag{4}$$

where $s \in \mathbb{R}$ and $\varphi \in S'(\mathbb{R})$. Let $\mathscr{G}$ be a class of all complex-valued measurable functions $\eta(x, y)$ such that the map is defined by the function $\eta(x, y)$ from $\mathbb{R} \times \mathbb{R} - \{0\}$ to $\mathbb{C}$. Then, the following properties can be seen

$$(i)\ \lim_{x \to \infty} \eta(x, y) = \eta(\infty, y),$$

exists for all $y \in \mathbb{R} - \{0\}$ and bounded by a measurable function.

(ii) We describe $\eta'(x, y) = \eta(x, y) - \eta(\infty, y)$, then

$$\eta'(x, y) = \overline{M}_A \int_{-\infty}^{\infty} K_{A^{-1}}(x, y)(\mathscr{C}_A\eta')(\lambda, y)d\lambda,\ \forall\ (x, y) \in \mathbb{R} \times \mathbb{R} - \{0\}, \tag{5}$$

where $(\mathscr{C}_A\eta')(\lambda, y)$ is complex-valued measurable function in λ and y $\forall(\lambda, y) \in \mathbb{R} \times \mathbb{R} - \{0\}$ and satisfies the estimate

$$|(\mathscr{C}_A\eta')(\lambda, y)| \leq \kappa(\lambda),\ \forall\ \lambda \in \mathbb{R}, \tag{6}$$

where $(1 + |\lambda|^2)^l \kappa(\lambda) \in L^1(\mathbb{R})$, and $l \in \mathbb{N}$. Suppose a sequence $\{r_j\}_{j=0}^{\infty}$ is of the form $r_0 > r_1 > r_2 > \cdots > r_j \to -\infty$ as $j \to \infty$ and $\psi \in C^{\infty}(\mathbb{R})$ such that $0 \leq \psi(y) \leq \infty, \forall\ y \in \mathbb{R} - \{0\}$,

$$\psi(y) = \begin{cases} 0 & \text{if } 0 \leq y \leq \frac{1}{2} \\ 1 & \text{if } y \geq 1 \end{cases}.$$

Assume an infinite sequence of function $\{\eta_j(x, y)\}_{j=0}^{\infty}$ defined on $\mathbb{R} \times \mathbb{R} - \{0\}$, then we can express a function as follows:

$$\eta(x, y) = \sum_{j=0}^{\infty} \psi\left(\frac{y}{t_j}\right) |y|^{r_j} \eta_j(x, y), \tag{7}$$

where $\{t_j\}_{j=0}^{\infty}$ is a sequence of positive real numbers such that $t_j \to \infty$ as $j \to \infty$.

From (7), it is obvious that $\eta(x, y) = 0$, for $|y| \leq \frac{1}{2}, x \in \mathbb{R}$, and $t \in [1, \infty)$,

The global estimate for (7) defined by $\eta(x, y)$ and of remainder of order N is given by

$$\zeta_N(x, y) = \eta(x, y) - \sum_{j=0}^{N-1} |\psi|^{r_j} \eta_j(x, y) \tag{8}$$

$$= \sum_{j=N}^{\infty} \psi\left(\frac{y}{t_j}\right) |y|^{r_j} \eta_j(x, y), \quad \forall y \in \mathbb{R} - \{0\}. \tag{9}$$

Theorem 1 *The following inequalities hold good for the sequence* $\{r_j\}_{j=0}^{\infty}$ *as defined before:*

$$|\eta(x, y)| \leq B|y|^{r_o},$$
$$|\zeta_N(x, y)| \leq B|y|^{r_N}, \quad \forall (x, y) \in \mathbb{R} \times \mathbb{R} - \{0\}, \tag{10}$$

for $N \in \mathbb{N}$. *In particular, the estimates are as follows:*

$$|\eta(\infty, y)| \leq B|y|^{r_0}, \quad |\zeta_N(\infty, y)| \leq B|y|^{r_N}, \quad \forall y \in \mathbb{R} - \{0\}. \tag{11}$$

Proof From [7, pp. 233–234], the proof of this theorem can easily be derived.

Theorem 2 *The following inequalities hold good for the sequence* $\{r_j\}_{j=0}^{\infty}$ *as defined before:*

$$|\mathscr{C}_A \eta^{'}(\lambda, y)| \leq \kappa(\lambda)|y|^{r_o},$$
$$|\mathscr{C}_A \zeta_N^{'}(\lambda, y)| \leq \kappa(\lambda)|y|^{r_N}, \quad \forall (\lambda, y) \in \mathbb{R} \times \mathbb{R} - \{0\}, \tag{12}$$

for $N \in \mathbb{N}$. *In particular the estimates are as follows:*

$$|(\mathscr{C}_A \eta^{'})(\infty, y)| \leq B|y|^{r_0}, \quad |(\mathscr{C}_A \zeta_N^{'})(\infty, y)| \leq B|y|^{r_N}, \quad \forall y \in \mathbb{R} - \{0\}. \tag{13}$$

where $(1 + |\lambda|^2)^p \kappa(\lambda) \in L^1(\mathbb{R})$, $\forall p = 0, 1, 2, 3, \ldots,$ *and* $N \in \mathbb{N}$.

Proof From [7, pp. 133–135], the proof of this theorem can easily be proved.

2 Asymptotic Series Expansion for Pseudo-Differential Operators

In this section, we are developing important theorems and results of [8] with the help of LCT.

Definition 1 Suppose $\eta(x, y) \in \mathscr{G}$. Then the pseudo-differential operator $\eta(x, D) = P_{\eta,A}$ associated with symbol $\eta(x, y)$ using linear canonical transform is given by

$$\eta(x, D) = (\eta(x, y)\varphi_A)(x) = \int_{-\infty}^{\infty} \overline{K_A(x, y)}\eta(x, y)\varphi_A(y)dy, \tag{14}$$

where $\varphi_A(y)$ is defined in (1) $\forall (x, y) \in \mathbb{R} \times \mathbb{R} - \{0\}$.

From [6], we can recall the following Definitions and Lemmas as given below.

Definition 2 A function $\varphi \in \mathbb{C}^\infty(\mathbb{R})$ is member of $\mathbb{S}(\mathbb{R})$ if and only if it satisfies

$$\gamma_{\mu,\nu}(\varphi) = \sup_{x\in\mathbb{R}} |x^\mu D^\nu \varphi(x)| < \infty, \tag{15}$$

for every choice of non-negative integers μ and ν.

The main propose is to extend some results of [7] with the help of a linear canonical transform.

Definition 3 The convolution of two function $f,\ g \in L^1(\mathbb{R})$ is given by

$$(g \star f)(x) = \int_{-\infty}^{\infty} g(x) f(x - y)dy, \tag{16}$$

provided the right-hand side of Eq. (16) exists.

Proof One can find the proof of this lemma in [5].

Lemma 1 *An infinitely differentiable complex-valued function φ satisfies (15) if and only if*

$$\tau_{\mu,\nu}(\varphi) = \sup_{x\in\mathbb{R}} |(1 + |x|^2)^{\frac{m}{2}} D^\beta \varphi(x)| < \infty, \quad \forall m, \beta \in N_0. \tag{17}$$

Lemma 2 *(Peetre) Suppose t, y, and $\lambda \in \mathbb{R}$, then the estimate*

$$\left(\frac{(1 + |y|^2)}{(1 + |\lambda|^2)}\right)^t \leq 2^{|t|}(1 + |y - \lambda|)^{|t|}, \tag{18}$$

hold good.

Proof One can find the proof of this lemma in [5].

Theorem 3 *Suppose $\eta'(x, y) \in \mathcal{G}$; then one can get the following relation:*

$$\mathscr{C}_A\left[e^{ix^2\frac{\alpha}{2\beta}}\eta'(x, D)\varphi(x)\right](y) = \overline{M}_A \int_{-\infty}^{\infty} e^{-i(\lambda^2 - y\lambda)\frac{\delta}{\beta}}\eta_A'(y - \lambda, y)\varphi_A(\lambda)d\lambda, \tag{19}$$

where $\varphi \in \mathbb{S}(\mathbb{R})$ and $x \in \mathbb{R}$.

Proof By the Definition of linear canonical transform (1), one can have

$$\begin{aligned}
&\mathscr{C}_A\left[e^{ix^2\frac{\alpha}{2\beta}}\eta^{'}(x,D)\varphi(x)\right](y)\\
&= M_A\int_{-\infty}^{\infty} e^{i\left[\frac{\alpha x^2}{2\beta}-\frac{xy}{\beta}+\frac{\delta y^2}{2\beta}\right]}\times\eta^{'}(x,y)e^{ix^2\frac{\alpha}{2\beta}}\varphi(x)dx\\
&= M_A\int_{-\infty}^{\infty} e^{i\left[\frac{\alpha x^2}{2\beta}-\frac{xy}{\beta}+\frac{\delta y^2}{2\beta}\right]}\eta^{'}(x,y)e^{ix^2\frac{\alpha}{2\beta}}\times\left(\overline{M}_A\int_{-\infty}^{\infty} e^{-i\left[\frac{\alpha x^2}{2\beta}-\frac{x\lambda}{\beta}+\frac{\delta\lambda^2}{2\beta}\right]}\varphi_A(\lambda)d\lambda\right)dx\\
&= M_A\overline{M}_A\int_{-\infty}^{\infty}\int_{-\infty}^{\infty} e^{i\left[\frac{\alpha x^2}{2\beta}+\frac{\delta y^2}{2\beta}-\frac{\delta\lambda^2}{2\beta}-x(y-\lambda)\frac{1}{\beta}\right]}\eta^{'}(x,y)\varphi_A(\lambda)d\lambda dx\\
&= M_A\overline{M}_A\int_{-\infty}^{\infty}\int_{-\infty}^{\infty} e^{i\left[\frac{\alpha x^2}{2\beta}+\frac{\delta y^2}{2\beta}-\frac{\delta\lambda^2}{2\beta}-x(y-\lambda)\frac{1}{\beta}\right]}\eta^{'}(x,y)\varphi_A(\lambda)d\lambda dx\\
&= M_A\overline{M}_A\int_{-\infty}^{\infty}\int_{-\infty}^{\infty} e^{i\left[\frac{\alpha x^2}{2\beta}+\frac{\delta y^2}{2\beta}+\frac{\delta\lambda^2}{2\beta}-\frac{2y\lambda}{2\beta}-x(y-\lambda)\frac{1}{\beta}\right]}\times e^{i(y\lambda-\lambda^2)\frac{\delta}{\beta}}\eta^{'}(x,y)\varphi_A(\lambda)d\lambda dx\\
&= M_A\overline{M}_A\int_{-\infty}^{\infty}\int_{-\infty}^{\infty} e^{i\left[\frac{\alpha x^2}{2\beta}+(y-\lambda)^2\frac{\delta}{2\beta}-x(y-\lambda)\frac{1}{\beta}\right]}\times e^{i(y\lambda-\lambda^2)\frac{\delta}{\beta}}\eta^{'}(x,y)\varphi_A(\lambda)d\lambda dx\\
&= \overline{M}_A\int_{-\infty}^{\infty} e^{i(y\lambda-\lambda^2)\frac{\delta}{\beta}}\left(M_A\int_{-\infty}^{\infty} e^{i\left[\frac{\alpha x^2}{2\beta}+(y-\lambda)^2\frac{\delta}{\beta}-x(y-\lambda)\frac{1}{\beta}\right]}\eta^{'}(x,y)dx\right)\varphi_A(\lambda)d\lambda\\
&= \overline{M}_A\int_{-\infty}^{\infty} e^{i(y\lambda-\lambda^2)\frac{\delta}{\beta}}\eta^{'}_A(y-\lambda,y)\varphi_A d\lambda.
\end{aligned}$$

Theorem 4 *Assume a symbol $\eta(x,y)\in\mathscr{G}$ and $\eta(x,D)=\eta(\infty,D)+\eta^{'}(x,D)$ is the associated operator, then one can find the following relation:*

$$\begin{aligned}
\mathscr{C}_A\left[e^{ix^2\frac{\alpha}{2\beta}}\eta^{'}(x,D)\varphi(x)\right](y) &= \eta(\infty,y)\mathscr{C}_A\left[e^{ix^2\frac{\alpha}{2\beta}}\varphi(x)\right](y)\\
&\quad+\overline{M}_A\int_{-\infty}^{\infty} e^{-i(\lambda^2-y\lambda)\frac{\delta}{\beta}}\eta^{'}_A(y-\lambda,y)\varphi_A(\lambda)d\lambda,
\end{aligned}$$

where $\varphi\in\mathbb{S}(\mathbb{R})$, $x\in\mathbb{R}$.

Proof We know that

$$\mathscr{C}_A\left[e^{ix^2\frac{\alpha}{2\beta}}\eta^{'}(x,D)\varphi(x)\right](y)=\mathscr{C}_A\left[e^{ix^2\frac{\alpha}{2\beta}}\{\eta(\infty,D)+\eta^{'}(x,D)\}\varphi(x)\right](y).\quad(20)$$

By linearity of linear canonical transform, we get

$$\mathscr{C}_A\left[e^{ix^2\frac{\alpha}{2\beta}}\eta^{'}(x,D)\varphi(x)\right](y)=\mathscr{C}_A\left[e^{ix^2\frac{\alpha}{2\beta}}\eta(\infty,D)\varphi(x)\right](y)+\mathscr{C}_A\left[e^{ix^2\frac{\alpha}{2\beta}}\eta^{'}(x,D)\varphi(x)\right](y).$$

Now using Definition (1) and Theorem 3, we get the required result.

Theorem 5 *Suppose a symbol* $\eta(x, y)$ *belongs to* $\mathcal{G}$ *and* $\eta_A(x, D)$ *associated operator, then one find the following relation:*

$$||\eta(x, D)\varphi||_{s,p}^{A} \leq B_{s,p}||\varphi||_{s+r_0,p}^{A}, \tag{21}$$

where $\varphi \in \mathbb{S}(\mathbb{R})$, *and* $s \in \mathbb{R}$.

Proof From equation (4), we have

$$||e^{ix^2\frac{\alpha}{2\beta}}(\eta(\infty, D)\varphi)||_{s,p} = \left[\int_{-\infty}^{\infty}(1+|y|^2)^{\frac{sp}{2}}|\mathcal{C}_A\left[e^{ix^2\frac{\alpha}{2\beta}}\eta(\infty, D)\varphi(x)\right](y)|^p dy\right]^{\frac{1}{p}}$$
$$= \left[\int_{-\infty}^{\infty}(1+|y|^2)^{\frac{sp}{2}}||\eta(\infty, y)|^p\mathcal{C}_A\left[e^{ix^2\frac{\alpha}{2\beta}}\varphi(x)\right](y)|^p dy\right]^{\frac{1}{p}}.$$

Now the argument of Theorem 1 yields

$$||e^{ix^2\frac{\alpha}{2\beta}}(\eta(\infty, D)\varphi)||_{s,p}^{A} \leq B\left[\int_{-\infty}^{\infty}(1+|y|^2)^{\frac{sp}{2}}(1+|y|^2)^{\frac{r_0 p}{2}} \times \mathcal{C}_A\left[e^{ix^2\frac{\alpha}{2\beta}}\varphi(x)\right](y)|^p dy\right]^{\frac{1}{p}}$$
$$= B\left[\int_{-\infty}^{\infty}(1+|y|^2)^{\frac{(s+r_0)p}{2}} \times |\mathcal{C}_A\left[e^{ix^2\frac{\alpha}{2\beta}}\varphi(x)\right](y)|^p dy\right]^{\frac{1}{p}}$$
$$= B||e^{ix^2\frac{\alpha}{2\beta}}\varphi||_{s+r_0,p}^{A}.$$

Thus, we have

$$||e^{ix^2\frac{\alpha}{2\beta}}(\eta(\infty, D)\varphi)||_{s,p}^{A} \leq B||\varphi||_{s+r_0,p}^{A} \;\; \forall\varphi \in \mathbb{S}(\mathbb{R}) \; and \; s \in \mathbb{R}, \tag{22}$$

now we consider the function

$$U_{s,A}(y) = (1+|y|^2)^{\frac{s}{2}}G_A(y),$$

where $U_{s,A}(y) = 0$ for $|y| \leq \frac{1}{2}$

$$G_A(y) = \overline{M}_A\int_{-\infty}^{\infty}e^{-i(\lambda^2-y\lambda)\frac{\alpha}{\delta}}\eta_A'(y-\lambda, y)\varphi_A(\lambda)d\lambda, \;\; for \; |y| \geq \frac{1}{2}. \tag{23}$$

Therefore

$$U_{s,A}(y) = \overline{M}_A(1+|y|^2)^{\frac{s}{2}}\int_{-\infty}^{\infty}e^{-i(\lambda^2-y\lambda)\frac{\alpha}{\delta}}\eta_A'(y-\lambda, y)\varphi_A(\lambda)d\lambda.$$
$$= \overline{M}_A\int_{-\infty}^{\infty}\left(\frac{1+|y|^2}{1+|\lambda|^2}\right)^{\frac{s}{2}}e^{-i(\lambda^2-y\lambda)\frac{\alpha}{\delta}} \times (1+|\lambda|^2)^{\frac{s}{2}}\eta_A'(y-\lambda, y)\varphi_A(\lambda)d\lambda,$$

Using Lemma 2, we can write

$$U_{s,A}(y) \leq 2^{\frac{|s|}{2}} \overline{M}_A \int_{-\infty}^{\infty} (1+|y-\lambda|^2)^{\frac{|s|}{2}} (1+|\lambda|^2)^{\frac{s}{2}} \times |\eta_A^{'}(y-\lambda, y)||\varphi_A(\lambda)|d\lambda. \tag{24}$$

From Theorem 2, we get

$$\begin{aligned} U_{s,A}(y) &\leq 2^{\frac{|s|}{2}} \overline{M}_A \int_{-\infty}^{\infty} (1+|y-\lambda|^2)^{\frac{|s|}{2}} (1+|\lambda|^2)^{\frac{s}{2}} \times \kappa(y-\lambda)(1+|y|^2)^{\frac{r_0}{2}} |\varphi_A(\lambda)|d\lambda \\ &= 2^{\frac{|s|}{2}} \overline{M}_A \int_{-\infty}^{\infty} (1+|y-\lambda|^2)^{\frac{|s|}{2}} (1+|\lambda|^2)^{\frac{s+r_0}{2}} \kappa(y-\lambda) \left(\frac{1+|y|^2}{1+|\lambda|^2}\right)^{\frac{r_0}{2}} |\varphi_A(\lambda)|d\lambda \\ &= 2^{\frac{|s|+|r_0|}{2}} \overline{M}_A \int_{-\infty}^{\infty} (1+|y-\lambda|^2)^{\frac{|s|}{2}} (1+|y-\lambda|^2)^{\frac{|r_0|}{2}} (1+|\lambda|^2)^{\frac{s+r_0}{2}} \kappa(y-\lambda)|\varphi_A(\lambda)|d\lambda \\ &= 2^{\frac{|s|+|r_0|}{2}} \overline{M}_A \int_{-\infty}^{\infty} (1+|y-\lambda|^2)^{\frac{(|s|+|r_0|)}{2}} (1+|\lambda|^2)^{\frac{s+r_0}{2}} \kappa(y-\lambda)|\varphi_A(\lambda)|d\lambda \\ &= \int_{-\infty}^{\infty} \phi(y-\lambda)h_A(\lambda)d\lambda \\ &= (\phi \star h_A)(y). \end{aligned}$$

Here $\phi(y-\lambda) \in L^1(\mathbb{R})$ by Eq. (6) $\forall s \in \mathbb{R}$ and since $\varphi_A(\lambda) \in \mathbb{S}(\mathbb{R})$, therefore, $h_A(\lambda) \in L^p(\mathbb{R})$ thus, we have $(\phi \star h_A)(y) \in L^p(\mathbb{R})$ and the inequality

$$||\phi \star h_A||_{L^p(\mathbb{R})} \leq ||\varphi||_{L^p(\mathbb{R})}, \tag{25}$$

this implies that

$$||U_{s,A}(y)||_p \leq B_A ||\varphi||^A_{s+r_0,p} \; \forall \varphi \in \mathbb{S}(\mathbb{R}), \; s \in \mathbb{R}, \tag{26}$$

so that

$$||e^{ix^2\frac{\alpha}{2\beta}}(\eta^{'}(x,D)\varphi)||^A_{s,p} = \left\|(1+|y|^2)^{\frac{s}{2}} \mathscr{C}_A\left[e^{ix^2\frac{\alpha}{2\beta}(\eta^{'}(x,D)\varphi)(x)}\right]\right\|_p \tag{27}$$

$$= ||U_{s,A}(y)||_p, \tag{28}$$

using (26), we get

$$||e^{ix^2\frac{\alpha}{2\beta}(\eta^{'}(x,D)\varphi)}||^A_{s,p} \leq B_A ||\varphi||^A_{s+r_0,p}. \tag{29}$$

Therefore

$$||e^{ix^2\frac{\alpha}{2\beta}}(\eta(x,D)\varphi)||^A_{s,p} = ||e^{ix^2\frac{\alpha}{2\beta}}\eta(\infty,D)\varphi + \eta^{'}(x,D)\varphi||^A_{s,p} \tag{30}$$

$$\leq ||e^{ix^2\frac{\alpha}{2\beta}}\eta(\infty,D)\varphi||^A_{s,p} + ||e^{ix^2\frac{\alpha}{2\beta}}\eta^{'}(x,D)\varphi||^A_{s,p}, \tag{31}$$

now using (22) and (29), we have

$$||e^{ix^2\frac{\alpha}{2\beta}}(\eta(x, D)\varphi)||^A_{s,p} \leq (B + B_A)||\varphi||^A_{s+r_0,p}, \quad (32)$$

and hence

$$||e^{ix^2\frac{\alpha}{2\beta}}(\eta(x, D)\varphi)||^A_{s,p} \leq B_{s,p}||\varphi||^A_{s+r_0,p}. \quad (33)$$

Definition 4 Suppose L is a linear operator with $r \in \mathbb{R}$ and for all $s \in \mathbb{R}$, there exist a constant $B_s > 0$ such that

$$||L\varphi|| \leq B_s||\varphi||_{s+r,p}, \ \forall\varphi \in \mathcal{G}^\infty_A, \quad (34)$$

the infimum of all orders r of L is called true order of L.

Definition 5 Suppose $\psi_r((1/t)D\,\eta(x, D))$ be a linear operator for all $\varphi \in \mathcal{G}^\infty_A$ in to itself and satisfy the following inequality:

$$||\psi_r((1/t)D\,\eta(x, D))||_{s,p} \leq B||\varphi||_{s+r,p}, \ \forall\varphi \in \mathcal{G}^\infty_A, \quad (35)$$

then $\psi_r((1/t)D\,\eta(x, D))$ is said to be a canonical operator of degree r, where r is a real number.

Definition 6 Suppose $\psi_{r_j}((1/t)D)\,\eta_j(x, D)$ is a sequence of canonical operators of degree r_j, where $\{r_j\}^\infty_{j=0}$ is defined as before. Then, corresponding sequence $\{\eta_j(x, y)\}^\infty_{j=0} \in \mathcal{G}$, a linear operator $\mathcal{M} : \mathcal{G}^\infty_0 \to \mathcal{G}^\infty_0$ is asymptotically expanded into the series $\psi_{r_j}((1/t)D)\eta_j(x, D)$, if it satisfies the following inequality:

$$t.o.\left[M - \sum_{j=0}^{N} \psi_{r_j}\left(\frac{1}{t_j}D\right)\eta_j(x, D)\right] < r_N, \quad (36)$$

$t.o.(L) = \inf \vartheta(L)$, the greatest lower bounded of $\vartheta(L)$ where $\vartheta(L)$ is defined by $r \in \mathbb{R}$ such that, $\forall s \in \mathbb{R}, \exists B_s \in \mathbb{R}^+$ with property that

$$||L\varphi||_{H^s} \leq B_s||\varphi||_{H^{s+r}}, \forall\varphi \in H^\infty. \quad (37)$$

Theorem 6 *Suppose a sequence of symbols $\{\eta_j(x, y)\}^\infty_{j=0} \in \mathcal{G}$ and a sequence $\{r_j\}^\infty_{j=0}$ as defined before. Then, there exists a sequence of canonical operators of degree r_j, $K_{j,A}$ and a linear operator $P_{\eta,A}$ in $\mathcal{G}^\infty_A$ such that*

(i) $t.o.(P_{\eta,A}) \leq r_0$;

(ii) $P_{\eta,A} \sim \sum_{j=0}^{\infty} K_{j,A}$, *that is*, $t.o.[P_{\eta,A} - \sum_{j=0}^{N} K_{j,A}] < r_N$.

Proof With the help of Theorem 5, one can easily prove part (i). To prove part (ii), we use arguments of [7, pp. 241–242]. We can define canonical operator by the following way:

$$\begin{aligned}(K_{j,A}\varphi)(x) &= e^{-ix^2\frac{\alpha}{2\beta}}\mathscr{C}_{A^{-1}} \\ &\times \left[\psi_{r_j}\left(\frac{y}{t_j}\right)\left[\eta_j(\infty,y)\mathscr{C}_A\left(e^{ix^2\frac{\alpha}{2\beta}}\varphi(x)\right)(y)\right.\right. \\ &\left.\left.+\overline{M}_A\int_{-\infty}^{\infty}e^{-i(\lambda^2-y\lambda)\frac{\alpha}{\delta}}\eta'_{j,A}(y-\lambda,y)\varphi_A(\lambda)d\lambda\right]\right],\end{aligned}$$

where

$$\eta(x,y)=\sum_{j=0}^{\infty}\psi_{r_j}\left(\frac{y}{t}\right)[\eta_j(\infty,y)+\eta'_j(x,y)],$$

is a member of the class of symbol $\mathscr{G}$. Also, we get

$$\begin{aligned}(P_{\eta,A}\varphi)(x)=e^{-ix^2\frac{\alpha}{2\beta}}\mathscr{C}_{A^{-1}}&\left[\eta(\infty,y)\overline{M}_A\left(e^{ix^2\frac{\alpha}{2\beta}}\varphi(x)\right)(y)\right. \\ &\left.+\overline{M}_A\int_{-\infty}^{\infty}e^{-i(\lambda^2-y\lambda)\frac{\alpha}{\delta}}\eta'_A(y-\lambda)\varphi_A(\lambda)d\lambda\right].\end{aligned}$$

Therefore

$$\begin{aligned}&\mathscr{C}_A\left[e^{ix^2\frac{\alpha}{2\beta}}\left(P_{\eta,A}\phi-\sum_{j=0}^{\infty}K_{j,A}\varphi\right)\right](y) \\ &\quad=\left(\eta(\infty,y)-\sum_{j=0}^{N}\psi_{r_j}\left(\frac{y}{t_j}\right)\eta_j(\infty,y)\right) \\ &\quad\times\mathscr{C}_A\left(e^{ix^2\frac{\alpha}{2\beta}}\varphi(x)\right)(y) \\ &\quad+\overline{M}_A\int_{-\infty}^{\infty}e^{-i(\lambda^2-y\lambda)\frac{\alpha}{\delta}} \\ &\quad\times\left(\eta'_A(y-\lambda,y)-\sum_{j=0}^{N}\sum_{j=0}^{N}\psi_{r_j}\left(\frac{y}{t_j}\right)\eta'_{j,A}(y-\lambda,y)\right) \\ &\quad\times\varphi_A(\lambda)d\lambda,\end{aligned}$$

using Definitions 5 and 6, we have

$$\eta(\infty,y)-\sum_{j=0}^{N}\psi_{r_j}\left(\frac{y}{t_j}\right)\eta_j(\infty,y)=\zeta_{N+1}(\infty,y). \tag{38}$$

From Theorem 1 and Eq. (11), we have

$$|\zeta_{N+1}(\infty, y)| \leq B(1+|y|^2)^{\frac{r_{N+1}}{2}},$$

where $y \in \mathbb{R} - \{0\}$ and

$$\eta_A'(y-\lambda, y) - \sum_{j=0}^{N} \psi_{r_j}\left(\frac{y}{t_j}\right) \eta_{j,A}'(y-\lambda, y) = \zeta_{A,N+1}'(y-\lambda, y).$$

Also, $|\zeta_A'(y-\lambda, y)| \leq BK_{N+1}(y-\lambda)(1+|\lambda|^2)^{\frac{r_{N+1}}{2}}$.
Now

$$\begin{aligned} \mathscr{C}_A\left[e^{ix^2\frac{\alpha}{2\beta}}\left(P_{\eta,A}\phi - \sum_{j=0}^{\infty} K_{j,A}\varphi\right)\right](y) \\ = \zeta_{N+1}(\infty, y)\mathscr{C}_A\left(e^{ix^2\frac{\alpha}{2\beta}}\varphi(x)\right)(y) \\ + \overline{M}_A \int_{-\infty}^{\infty} e^{-i(\lambda^2 - y\lambda)\frac{\alpha}{2\beta}} \zeta_A'(y-\lambda, y)\varphi_A(\lambda)d\lambda, \end{aligned}$$

using Theorem 5, we get

$$||P_{\eta,A}\varphi - \sum_{j=0}^{k} K_j\varphi||_{s,p} \leq M||\varphi||_{s+r_{N+1},p}, \quad \forall\varphi \in \mathbb{S}(\mathbb{R}), \quad s \in \mathbb{R}.$$

Hence, we get

$$t.o.\left[P_{\eta,A} - \sum_{j=0}^{k} K_{j,A}\right] \leq r_{N+1} \leq r_N. \tag{39}$$

References

1. Almeida, L.B.: The fractional fourier transform and time-frequency representations. IEEE Trans. Signal Process. **42**(11), 3084–3091 (1994)
2. Collins, S.A.: Lens-system diffraction integral written in terms of matrix optics. JOSA **60**(9), 1168–1177 (1970)
3. Healy, J.J., Alper Kutay, M., Ozaktas, H.M., Sheridan, J.T.: Linear Canonical Transforms: Theory and Applications, vol. 198. Springer (2015)
4. Moshinsky, M., Quesne, C.: Linear canonical transformations and their unitary representations. J. Math. Phys. **12**(8), 1772–1780 (1971)
5. Pathak, R.S.: A Course in Distribution Theory and Applications. Alpha Science Int'l Ltd. (2001)
6. Pathak, R.S., Prasad, A., Kumar, M.: Fractional fourier transform of tempered distributions and generalized pseudo-differential operator. J. Pseudo-Differ. Oper. Appl. **3**(2), 239–254 (2012)

7. Zaidman, S.: On asymptotic series of symbols and of general pseudo-differential operators. Rendiconti del Seminario matematico della Università di Padova **63**, 231–246 (1980)
8. Zaidman, S.: Distributions and Pseudo-Differential Operators, vol. 248. Longman (1991)
9. Zayed, A.I.: A convolution and product theorem for the fractional fourier transform. IEEE Signal Process. Lett. **5**(4), 101–103 (1998)
10. Zayed, A.I.: A class of fractional integral transforms: a generalization of the fractional fourier transform. IEEE Trans. Signal process. **50**(3), 619–627 (2002)

Under (ψ, α, β) Conditions Coupled Coincidence Points with Ordered Generalized Metric Spaces

Richa Sharma, Ruchika Mehta, and Virendra Singh Chouhan

Abstract The objective of this manuscript is to substantiate coupled coincidence point for maps having a mixed j-monotone property under the structure of derived metric spaces equipped under partial order. Also, we demonstrate the oneness of coupled common fixed points of the assumed maps. Some of our outcomes modify and extend comparable outcomes in the literature. The results presented in this paper have been utilized to get the solution of an integral equation.

Keywords Mixed j-monotone property · Coupled coincidence point · Generalized metric space · Partial order

1 Introduction

In the analysis, the Principle of Banach Contraction is considered as the core foundation of fixed point to study the in metric Space. This is widely used fixed point result in abstract branches of mathematics. In many different directions, the theorem has been generalized. In fact, more literature dealing with extension of this historical result.

Alber and Gurre-Delabriere [3] generalized the contraction principle of Banach's in Hilbert spaces known as the Weak contraction principle and demonstrated the presence of fixed points. Rhoades [31] has represented the result given by Alber and Gurre-Delabriere [3] is as well effective in metric spaces which is complete.

R. Sharma
Department of Mathematics, Chandigarh University, Mohali, India
e-mail: richa.tuknait@yahoo.in

R. Mehta
Department of Mathematics and Statistics, Manipal University Jaipur, Jaipur, India
e-mail: ruchika.mehta1981@gmail.com

V. Singh Chouhan (✉)
Department of Mathematics and Statistics, Manipal University Jaipur, Jaipur, India
e-mail: darbarvsingh@yahoo.com

R. N. Mohapatra et al. (eds.), *Mathematical Analysis and Computing*,
Springer Proceedings in Mathematics & Statistics 344,
https://doi.org/10.1007/978-981-33-4646-8_33

Various works on weakly contractions have been discussed, and a number of these are counted in [9, 11, 18, 30, 35].

Khan with others [20] presented the utilization of control type function in problems of metric fixed point. Such type known as 'altering distance function'. Its extension have been utilized in numerous work of fixed point results, few of them are identifies in [5, 15, 16, 19, 26, 34].

In recent, theory of fixed point has progressed quickly with metric spaces under partially ordered (see [9, 19, 27–29]). Firstly Bhaskar and Lakshmikantham [6] studied the presence of coupled fixed points with mixed monotone maps in metric space which is partially ordered. After that, many group of researchers have given contributions on such concept. Ciric and Lakshmikantham [13] acquaint with a new conception of coupled coincidence point. For further works in coupled and coincidence point theorems see, [8, 12, 14, 21, 30].

In 2006, another concept G-metric spaces of generalized metric spaces introduced by Mustafa and Sims [23] is known as generalization of metric spaces. Later, a lot of theorems in G-metric spaces for fixed point have been achieved. To see more outcomes in this spaces, we elude the pursuer to [24, 25, 32]. Voluminous researcher have considered in G-metric space idea of coupled fixed point (see [4, 5, 7, 10, 17, 22, 33]).

The notion of compatible mappings w and w^* type initially proposed by Abbas et al. [1]. Abbas et al. [2] used this idea to prove an oneness theorem under G-metric spaces for coupled fixed point, for other aftereffect of coupled fixed point under G-metric spaces, more details in [5, 10].

The first motive of this manuscript is to get result for nonlinear contraction maps with coupled coincidence point type of results with generalized metric spaces under the framework of partially ordered set having property of mixed p-monotone.

2 Preliminaries

For detailed information on the subsequent definitions and ideas, see Mustafa and Sims [23].

Definition 2.1 [23] Suppose $G : S \times S \times S \to R^+$ be a function and suppose S be a non-empty set that satisfies the following axioms for all $p_1, p_2, p_3, a \in S$:

(G_1) $G(p_1, p_2, p_3) = 0$ if $p_1 = p_2 = p_3$,
(G_2) $0 < G(p_1, p_1, p_2)$ with $p_1 \neq p_2$,
(G_3) $G(p_1, p_1, p_2) \leq G(p_1, p_2, p_3)$ with $p_3 \neq p_2$,
(G_4) $G(p_1, p_2, p_3) = G(p_1, p_3, p_2) = G(p_2, p_3, p_1) = \dots$
(G_5) $G(p_1, p_2, p_3) \leq G(p_1, a, a) + G(a, p_2, p_3)$.

Then function G is called generalized metric, precisely a G-metric on S and pair (S, G) is called G-metric space.

For more insights about the accompanying definitions, we allude the peruser [6, 13].

Definition 2.2 [6] An element $(p, z) \in S \times S$ is known as coupled fixed point for the map $Y : S \times S \to S$ if $Y(p, z) = p,\ Y(z, p) = z$.

Definition 2.3 [6] Suppose $(S, \leq)$ be partially ordered set and a map $Y : S \times S \to S$. Then Y hold the property of mixed monotone if $Y(p, v)$ is monotone increasing in p and monotone decreasing in z, then, for any $p, z \in S$,

$$p_1, p_2 \in S,\ p_1 \leq p_2 \implies Y(p_1, z) \leq Y(p_2, z),$$
$$z_1, z_2 \in S,\ z_1 \leq z_2 \implies Y(p, z_1) \geq Y(p, z_2).$$

Definition 2.4 [13] Presume $(S, \leq)$ be partially ordered set, $Y : S \times S \to S$, $J : S \to S$ be a couple of maps. Then Y has the property of mixed J-monotone if $Y(p, v)$ is monotone J-increasing in p and is monotone J-decreasing in v, then for any $p, z \in S$,

$$p_1, p_2 \in S,\ Jp_1 \leq Jp_2 \implies Y(p_1, z) \leq Y(p_2, z),$$
$$z_1, z_2 \in S,\ Jz_1 \leq Jz_2 \implies Y(p, z_2) \leq Y(p, z_1).$$

Definition 2.5 [13] An element $(p, z) \in S \times S$ is said to be coupled coincidence point of the maps $Y : S \times S \to S$ and $J : S \to S$ if $Y(p, z) = Jp,\ Y(z, p) = Jz$.

Definition 2.6 [13] Suppose S be a non-empty set and we say that the mappings $Y : S \times S \to S$ and $J : S \to S$ are commutative if $JY(p, z) = Y(Jp, Jz)$, for all $p, z \in S$.

An altering distance type function has been presented in [20] as

Definition 2.7 [20]. Υ is a function of altering distance type if $\Upsilon : [0, \infty) \to [0, \infty)$ such that

1. Υ is non-decreasing and smooth.
2. $\Upsilon(t) = 0$ iff $t = 0$.

Abbas et al. [1] proposed the concept of compatible mappings of type w and w^* and used this idea to show a remarkable preciously one theorem of a on maps Y and J in cone metric spaces for coupled fixed point.

Definition 2.8 [1] Mappings $Y : S \times S \to S$ and $J : S \to S$ known as

(1) w-compatible if $J(Y(p, z)) = Y(Jp, Jz)$ whenever $Jp = Y(p, z)$ and $Jz = Y(z, p)$;
(2) w^*-compatible if $J(Y(p, p)) = Y(Jp, Jp)$ whenever $Jp = Y(p, p)$.

3 Main Theorem

Theorem 3.1 *Let $(S, \leq, G)$ be G-metric space which is ordered complete. Presume $J : S \to S$ and $Y : S \times S \to S$ be smooth such that Y has the property of mixed J-monotone and J commutes along Y, such that*

$$\psi(G(Y(p, y), Y(u, v), Y(r, z))) \leq \alpha(N((p, y), (u, v), (r, z))) - \beta(N((p, y), (u, v), (r, z))), \tag{1}$$

where

$$\begin{aligned} N((p, y), (u, v), (r, z)) = max\{&G(Jp, Ju, Jr), G(Jy, Jv, Jz), G(Jp, Y(p, y), Y(p, y)),\\ &G(Ju, Y(u, v), Y(u, v)), G(Jr, Y(r, z), Y(r, z)),\\ &G(Jy, Y(y, p), Y(y, p)), G(Jv, Y(v, u), Y(v, u)),\\ &G(Jz, Y(z, r), Y(z, r))\},\end{aligned}$$

for all $p, y, u, v, r, z \in S$ with $Jr \leq Ju \leq Jp$ and $Jy \leq Jv \leq Jz$, where $\psi, \alpha, \beta : [0, \infty) \to [0, \infty)$ α is continuous, β is lower semicontinuous, and ψ an altering distance function such that

$$\alpha(0) = 0 = \beta(0) \tag{2}$$

and

$$\psi(t) - \alpha(t) + \beta(t) > 0 \quad \textit{for every} \quad t > 0. \tag{3}$$

Presume that $Y(S \times S) \subseteq J(S)$. If there exists $p_0, y_0 \in S$ such that $Jp_0 \leq Y(p_0, y_0)$ and $Jy_0 \geq Y(y_0, p_0)$, then coupled coincidence point preserved by Y and J, that is, $\exists\ p, y \in S,\ J(p) = Y(p, y)$ and $J(y) = Y(y, p)$.

Proof Assume $p_0, y_0 \in S$, $Jp_0 \leq Y(p_0, y_0)$ and $Jy_0 \geq Y(y_0, p_0)$. Utilizing the way that $Y(S \times S) \subseteq J(S)$, take $p_1, y_1 \in S$ such that $Jp_1 = Y(p_0, y_0)$ and $Jy_1 = Y(y_0, p_0)$.

By similar contentions, $Y(S \times S) \subseteq J(S)$, take $p_2, y_2 \in S$ thus $Jp_2 = Y(p_1, y_1)$ and $Jy_2 = Y(y_1, p_1)$. As Y has property of mixed J-monotone, we have $Jp_0 \leq Jp_1 \leq Jp_2$ and $Jy_2 \leq Jy_1 \leq Jy_0$. Proceeding in this way, we have two sequences $\{p_s\}$ and $\{y_s\}$ in S such that

$$Jp_s = Y(p_{s-1}, y_{s-1}) \leq Jp_{s+1} = Y(p_s, y_s)$$

and

$$Jy_{s+1} = Y(y_s, p_s) \leq Jy_s = Y(y_{s-1}, p_{s-1}).$$

If, for some integer s, we have $(Jp_{s+1}, Jy_{s+1}) = (Jp_s, Jy_s)$, then $Y(p_s, y_s) = Jp_s$ and $Y(y_s, p_s) = py_s$, that is, (p_s, y_s) is a coincidence point of Y and J. Thus, we presume that $(Jp_{s+1}, Jy_{s+1}) \neq (Jp_s, Jy_s)\ \forall\ s \in \mathbb{N}$, that is, we presuppose that either $Jp_{s+1} \neq Jp_s$ or $Jy_{s+1} \neq Jy_s$.

For every $s \in \mathbb{N}$, we achieve from (1) that

$$\begin{aligned}\psi(G(Jp_s, Jp_{s+1}, Jp_{s+1}) &= \psi(G(Y(p_{s-1}, y_{s-1}), Y(p_s, y_s), Y(p_s, y_s)))\\ &\leq \alpha(N((p_{s-1}, y_{s-1}), (p_s, y_s), (p_s, y_s)))\\ &\quad - \beta(N((p_{s-1}, y_{s-1}), (p_s, y_s), (p_s, y_s))),\end{aligned}$$

and

$$\begin{aligned}\psi(G(Jy_s, Jy_{s+1}, Jy_{s+1}) &= \psi(G(Y(y_{s-1}, p_{s-1}), Y(y_s, p_s), Y(y_s, p_s)))\\ &\leq \alpha(N((y_{s-1}, p_{s-1}), (y_s, p_s), (y_s, p_s)))\\ &\quad - \beta(N((y_{s-1}, p_{s-1}), (y_s, p_s), (y_s, p_s))),\end{aligned}$$

where

$$\begin{aligned}N((p_{s-1}, y_{s-1}), (p_s, y_s), (p_s, y_s)) &= N((y_{s-1}, p_{s-1}), (y_s, p_s), (y_s, p_s))\\ &= max\{G(Jp_{s-1}, Jp_s, Jp_s), G(Jy_{s-1}, Jy_s, Jy_s),\\ &\qquad G(Jp_s, Jp_{s+1}, Jp_{s+1}), G(Jy_s, Jy_{s+1}, Jy_{s+1})\}.\end{aligned}$$

Now, let us consider three cases.

Case I: If $N((p_{s-1}, y_{s-1}), (p_s, y_s), (p_s, y_s)) = G(Jp_s, Jp_{s+1}, Jp_{s+1})$.
We claim that

$$N((p_{s-1}, y_{s-1}), (p_s, y_s), (p_s, y_s)) = G(Jp_s, Jp_{s+1}, Jp_{s+1}) = 0.$$

In fact, if $G(Jp_s, Jp_{s+1}, Jp_{s+1}) \neq 0$, then

$$\psi(G(Jp_s, Jp_{s+1}, Jp_{s+1}) \leq \alpha(G(Jp_s, Jp_{s+1}, Jp_{s+1})) - \beta(G(Jp_s, Jp_{s+1}, Jp_{s+1})),$$

By (3), which is contradiction.

Case II: If $N((p_{s-1}, y_{s-1}), (p_s, y_s), (p_s, y_s)) = G(Jy_s, Jy_{s+1}, Jy_{s+1})$.
Similarly, we can show that $G(Jy_s, Jy_{s+1}, Jy_{s+1}) = 0$.

Case III: If $N((p_{s-1}, y_{s-1}), (p_s, y_s), (p_s, y_s)) = max\{G(Jp_{s-1}, Jp_s, Jp_s), G(Jy_{s-1}, Jy_s, Jy_s)\}$.

We have

$$\begin{aligned}\psi(G(Jp_s, Jp_{s+1}, Jp_{s+1}) &\leq \alpha(max\{G(Jp_{s-1}, Jp_s, Jp_s), G(Jy_{s-1}, Jy_s, Jy_s)\})\\ &\quad - \beta(max\{G(Jp_{s-1}, Jp_s, Jp_s), G(Jy_{s-1}, Jy_s, Jy_s)\}),\end{aligned} \tag{4}$$

and

$$\psi(G(Jy_s, Jy_{s+1}, Jy_{s+1}) \leq \alpha(max\{G(Jp_{s-1}, Jp_s, Jp_s), G(Jy_{s-1}, Jy_s, Jy_s)\}) \\ - \beta(max\{G(Jp_{s-1}, Jp_s, Jp_s), G(Jy_{s-1}, Jy_s, Jy_s)\}). \tag{5}$$

Let

$$\delta_s = max\{G(Jp_{s-1}, Jp_s, Jp_s), G(Jy_{s-1}, Jy_s, Jy_s)\}.$$

If for an $s \geq 1$, $\delta_s = 0$, thus the result of the theorem follows. So, we presume that

$$\delta_s \neq 0, \ \text{for every } s \geq 1. \tag{6}$$

For some s, $\delta_{s-1} < \delta_s$. From (4) and (5), as ψ is non-decreasing, and

$$\begin{aligned} &\psi(max\{G(Jp_{s-1}, Jp_s, Jp_s), G(Jy_{s-1}, Jy_s, Jy_s)\}) \\ &\quad < \psi(max\{G(Jp_s, Jp_{s+1}, Jp_{s+1}), G(Jy_s, Jy_{s+1}, Jy_{s+1})\}) \\ &\quad = max\{\psi(G(Jp_s, Jp_{s+1}, Jp_{s+1}), \psi(G(Jy_s, Jy_{s+1}, Jy_{s+1}))\} \\ &\quad \leq \alpha(max\{G(Jp_{s-1}, Jp_s, Jp_s), G(Jy_{s-1}, Jy_s, Jy_s)\}) \\ &\quad\quad - \beta(max\{G(Jp_{s-1}, Jp_s, Jp_s), G(Jy_{s-1}, Jy_s, Jy_s)\}), \end{aligned} \tag{7}$$

thus, $\psi(\delta_s) - \alpha(\delta_s) + \beta(\delta_s) \leq 0$. From the presumptions, we have $\delta_s = 0$, which contradicts (6). Thus, $\forall\ s \geq 1$ we deduce that

$$\delta_{s+1} \leq \delta_s, \tag{8}$$

that is, the sequence $\{\delta_s\}$ is non-increasing of real numbers of positive type and thus there exists an $a \geq 0$ thus $\lim_{s\to\infty} \delta_s = a$.

Taking $s \to \infty$ in Eq. (7) and apply the continuity of ψ and α, the lower semi-continuity of β, after that we obtain $\psi(a) \leq \alpha(a) - \beta(a)$, which gives $a = 0$, from our presumptions about ψ, α, β. Thus,

$$\lim_{s\to\infty} max\{G(Jp_{s-1}, Jp_s, Jp_s), G(Jy_{s-1}, Jy_s, Jy_s)\} = 0. \tag{9}$$

Thus,

$$\lim_{s\to\infty} G(Jp_{s-1}, Jp_s, Jp_s) = 0, \ \lim_{s\to\infty} G(Jy_{s-1}, Jy_s, Jy_s) = 0.$$

Next, we claim that $\{Jp_s\}$ and $\{Jy_s\}$ are G-Cauchy sequences.

Presume the contrary, $\{Jp_s\}$ or $\{Jy_s\}$ is not a G-Cauchy sequences, thus

$$\lim_{n,s\to\infty} G(Jp_n, Jp_s, Jp_s) \neq 0 \text{ or } \lim_{n,s\to\infty} G(Jy_n, Jy_s, Jy_s) \neq 0.$$

Then existence of $\varepsilon > 0$ and subsequences of integers $\{n(h)\}$ and $\{s(h)\}$ with $s(h) > n(h) > h$ such that

$$max\{G(Jp_{n(h)}, Jp_{s(h)}, Jp_{s(h)}), G(Jy_{n(h)}, Jy_{s(h)}, Jy_{s(h)})\} \geq \varepsilon. \tag{10}$$

Corresponding to $\{n(h)\}$, we can take $\{s(h)\}$ with $s(h) > n(h)$ which is smallest integer and satisfying (10). So

$$max\{G(Jp_{n(h)}, Jp_{s(h)-1}, Jp_{s(h)-1}), G(Jy_{n(h)}, Jy_{s(h)-1}, Jy_{s(h)-1})\} < \varepsilon. \tag{11}$$

Now, from (G_5), we have

$$G(Jp_{n(h)}, Jp_{s(h)}, Jp_{s(h)}) \leq G(Jp_{n(h)}, Jp_{s(h)-1}, Jp_{s(h)-1}) + G(Jp_{s(h)-1}, Jp_{s(h)}, Jp_{s(h)}).$$

Then, from (11), we get

$$G(Jp_{n(h)}, Jp_{s(h)}, Jp_{s(h)}) < G(Jp_{s(h)-1}, Jp_{s(h)}, Jp_{s(h)}) + \varepsilon. \tag{12}$$

Similarly, from (G_5) and (11), we have

$$G(Jy_{n(h)}, Jy_{s(h)}, Jy_{s(h)}) < G(Jy_{s(h)-1}, Jy_{s(h)}, Jy_{s(h)}) + \varepsilon. \tag{13}$$

Equations (10), (12), and (13) give

$$\begin{aligned} \varepsilon &\leq max\{G(Jp_{n(h)}, Jp_{s(h)}, Jp_{s(h)}), G(Jy_{n(h)}, Jy_{s(h)}, Jy_{s(h)})\} \\ &< max\{G(Jp_{s(h)-1}, Jp_{s(h)}, Jp_{s(h)}), G(Jy_{s(h)-1}, Jy_{s(h)}, Jy_{s(h)})\} + \varepsilon. \end{aligned} \tag{14}$$

Letting, $h \to \infty$ in (14) and using (9), we get

$$\lim_{k\to\infty} max\{G(Jp_{n(h)}, Jp_{s(h)}, Jp_{s(h)}), G(Jy_{n(h)}, Jy_{s(h)}, Jy_{s(h)})\} = \varepsilon. \tag{15}$$

Again, from (G_5) and (11), we get

$$\begin{aligned} G(Jp_{n(h)-1}, Jp_{s(h)-1}, Jp_{s(h)-1}) &\leq G(Jp_{n(h)-1}, Jp_{n(h)}, Jp_{n(h)}) + G(Jp_{n(h)}, Jp_{s(h)-1}, Jp_{s(h)-1}) \\ &< G(Jp_{n(h)-1}, Jp_{n(h)}, Jp_{n(h)}) + \varepsilon, \end{aligned} \tag{16}$$

and

$$\begin{aligned} G(Jy_{n(h)-1}, Jy_{s(h)-1}, Jy_{s(h)-1}) &\leq G(Jy_{n(h)-1}, Jy_{n(h)}, Jy_{n(h)}) + G(Jy_{n(h)}, Jy_{s(h)-1}, Jy_{s(h)-1}) \\ &< G(Jy_{n(h)-1}, Jy_{n(h)}, Jy_{n(h)}) + \varepsilon. \end{aligned} \tag{17}$$

From (16) and (17), we have

$$\begin{aligned} &max\{G(Jp_{n(h)-1}, Jp_{s(h)-1}, Jp_{s(h)-1}), G(Jy_{n(h)-1}, Jy_{s(h)-1}, Jy_{s(h)-1})\} \\ &\quad < max\{G(Jp_{n(h)-1}, Jp_{n(h)}, Jp_{n(h)}), G(Jy_{n(h)-1}, Jy_{n(h)}, Jy_{n(h)}) + \varepsilon. \quad (18) \end{aligned}$$

Using (G_5), we get

$$\begin{aligned} G(Jp_{n(h)}, Jp_{s(h)}, Jp_{s(h)}) &\le G(Jp_{n(h)}, Jp_{s(h)-1}, Jp_{s(h)-1}) + G(Jp_{s(h)-1}, Jp_{s(h)}, Jp_{s(h)}) \\ &\le G(Jp_{n(h)}, Jp_{n(h)-1}, Jp_{n(h)-1}) + G(Jp_{n(h)-1}, Jp_{s(h)-1}, Jp_{s(h)-1}) \\ &\quad + G(Jp_{s(h)-1}, Jp_{s(h)}, Jp_{s(h)}). \end{aligned}$$

Using $G(u, v, v) \le 2G(v, u, u)$, for any $u, v \in S$, we have

$$\begin{aligned} G(Jp_{n(h)}, Jp_{s(h)}, Jp_{s(h)}) &\le 2G(Jp_{n(h)-1}, Jp_{n(h)}, Jp_{n(h)}) + G(Jp_{n(h)-1}, Jp_{s(h)-1}, Jp_{s(h)-1}) \\ &\quad + G(Jp_{s(h)-1}, Jp_{s(h)}, Jp_{s(h)}). \quad (19) \end{aligned}$$

Similarly, we get

$$\begin{aligned} G(Jy_{n(h)}, Jy_{s(h)}, Jy_{s(h)}) &\le 2G(Jy_{n(h)-1}, Jy_{n(h)}, Jy_{n(h)}) + G(Jy_{n(h)-1}, Jy_{s(h)-1}, Jy_{s(h)-1}) \\ &\quad + G(Jy_{s(h)-1}, Jy_{s(h)}, Jy_{s(h)}). \quad (20) \end{aligned}$$

So, from (10), (19) and (20), we obtain

$$\begin{aligned} \varepsilon &\le max\{G(Jp_{n(h)}, Jp_{s(h)}, Jp_{s(h)}), G(Jy_{n(h)}, Jy_{s(h)}, Jy_{s(h)})\} \\ &\le 2max\{G(Jp_{n(h)-1}, Jp_{n(h)}, Jp_{n(h)}), G(Jy_{n(h)-1}, Jy_{n(h)}, Jy_{n(h)})\} \\ &\quad + max\{G(Jp_{n(h)-1}, Jp_{s(h)-1}, Jp_{s(h)-1}), G(Jy_{n(h)-1}, Jy_{s(h)-1}, Jy_{s(h)-1})\} \\ &\quad + max\{G(Jp_{s(h)-1}, Jp_{s(h)}, Jp_{s(h)})), G(Jy_{s(h)-1}, Jy_{s(h)}, Jy_{s(h)}))\}. \quad (21) \end{aligned}$$

From (18) and (21), we get

$$\begin{aligned} &\varepsilon - 2max\{G(Jp_{n(h)-1}, Jp_{n(h)}, Jp_{n(h)}), G(Jy_{n(h)-1}, Jy_{n(h)}, Jy_{n(h)})\} \\ &\quad - max\{G(Jp_{s(h)-1}, Jp_{s(h)}, Jp_{s(h)})), G(Jy_{s(h)-1}, Jy_{s(h)}, Jy_{s(h)}))\} \\ &\quad \le max\{G(Jp_{n(h)-1}, Jp_{s(h)-1}, Jp_{s(h)-1}), G(Jy_{n(h)-1}, Jy_{s(h)-1}, Jy_{s(h)-1})\} \\ &\quad < max\{G(Jp_{n(h)-1}, Jp_{n(h)}, Jp_{n(h)}), G(Jy_{n(h)-1}, Jy_{n(h)}, Jy_{n(h)}) + \varepsilon. \end{aligned} \quad (22)$$

Letting $h \to \infty$ in (22) and using (9), we get

$$\lim_{h\to\infty} max\{G(Jp_{n(h)-1}, Jp_{s(h)-1}, Jp_{s(h)-1}), G(Jy_{n(h)-1}, Jy_{s(h)-1}, Jy_{s(h)-1})\} = \varepsilon. \quad (23)$$

From using the inequality (1), we get

$$\begin{aligned} \psi(G(Jp_{n(h)}, Jp_{s(h)}, Jp_{s(h)})) &= \psi(G(Y(p_{n(h)-1}, y_{n(h)-1}), Y(p_{s(h)-1}, y_{s(h)-1}), Y(p_{s(h)-1}, y_{s(h)-1}))) \\ &\le \alpha(N((p_{n(h)-1}, y_{n(h)-1}), (p_{s(h)-1}, y_{s(h)-1}), (p_{s(h)-1}, y_{s(h)-1}))) \quad (24) \\ &\quad - \beta(N((p_{n(h)-1}, y_{n(h)-1}), (p_{s(h)-1}, y_{s(h)-1}), (p_{s(h)-1}, y_{s(h)-1}))), \end{aligned}$$

and

$$\begin{aligned}\psi(G(Jy_{n(h)}, Jy_{s(h)}, Jy_{s(h)})) &= \psi(G(Y(y_{n(h)-1}, p_{n(h)-1}), Y(y_{s(h)-1}, p_{s(h)-1}), Y(y_{s(h)-1}, p_{s(h)-1}))) \\ &\leq \alpha(N((y_{n(h)-1}, p_{n(h)-1}), (y_{s(h)-1}, p_{s(h)-1}), (y_{s(h)-1}, p_{s(h)-1}))) \\ &\quad - \beta(N((y_{n(h)-1}, p_{n(h)-1}), (y_{s(h)-1}, p_{s(h)-1}), (y_{s(h)-1}, p_{s(h)-1}))).\end{aligned} \tag{25}$$

where

$$\begin{aligned}&N((y_{n(h)-1}, p_{n(h)-1}), (y_{s(h)-1}, p_{s(h)-1}), (y_{s(h)-1}, p_{s(h)-1})) \\ &\quad = N((p_{n(h)-1}, y_{n(h)-1}), (p_{s(h)-1}, y_{s(h)-1}), (p_{s(h)-1}, y_{s(h)-1})) \\ &\quad = max\{G(Jp_{n(h)-1}, Jp_{s(h)-1}, Jp_{s(h)-1}), G(Jy_{n(h)-1}, Jy_{s(h)-1}, Jy_{s(h)-1}), \\ &\qquad G(Jp_{n(h)-1}, Jp_{n(h)}, Jp_{n(h)}), G(Jy_{n(h)-1}, Jy_{n(h)}, Jy_{n(h)}), \\ &\qquad G(Jp_{s(h)-1}, Jp_{s(h)}, Jp_{s(h)}), G(Jy_{s(h)-1}, Jy_{s(h)}, Jy_{s(h)})\}.\end{aligned}$$

Now, by (24) and (25), we have

$$\begin{aligned}&\psi(max\{G(Jp_{n(h)}, Jp_{s(h)}, Jp_{s(h)}), G(Jy_{n(h)}, Jy_{s(h)}, Jy_{s(h)})\}) \\ &\quad = max\{\psi(G(Jp_{n(h)}, Jp_{s(h)}, Jp_{s(h)})), \psi(G(Jy_{n(h)}, Jy_{s(h)}, Jy_{s(h)}))\} \\ &\quad \leq \alpha(Z_s) - \beta(Z_s),\end{aligned} \tag{26}$$

where

$$\begin{aligned}Z_s = max\{&G(Jp_{n(h)-1}, Jp_{s(h)-1}, Jp_{s(h)-1}), G(Jy_{n(h)-1}, Jy_{s(h)-1}, Jy_{s(h)-1}), \\ &G(Jp_{n(h)-1}, Jp_{n(h)}, Jp_{n(h)}), G(Jy_{n(h)-1}, Jy_{n(h)}, Jy_{n(h)}), \\ &G(Jp_{s(h)-1}, Jp_{s(h)}, Jp_{s(h)}), G(Jy_{s(h)-1}, Jy_{s(h)}, Jy_{s(h)}).\end{aligned} \tag{27}$$

Finally, letting $h \to \infty$ in (26) (27) and using (9), (15), and (23), we get

$$\psi(\varepsilon) \leq \alpha(max(\varepsilon, 0, 0)) - \beta(max(\varepsilon, 0, 0)). \tag{28}$$

Therefore, $\psi(\varepsilon) - \alpha(\varepsilon) + \beta(\varepsilon) \leq 0$ and then $\varepsilon = 0$ represent a conflict. Therefore, $\{Jp_s\}$ and $\{Jy_s\}$ are G-Cauchy sequences.

Subsequently, $\{Jp_s\}$ and $\{Jy_s\}$ in complete G-metric space(S, G) are G-Cauchy sequences. Then, there are $p, y \in S$ such that $\{Jp_s\}$ and $\{Jy_s\}$ are, respectively, G-converges to p and y, respectively, we have

$$\lim_{s\to\infty} G(Jp_s, Jp_s, p) = \lim_{s\to\infty} G(Jp_s, p, p) = 0, \tag{29}$$

$$\lim_{s\to\infty} G(Jy_s, Jy_s, y) = \lim_{s\to\infty} G(Jy_s, y, y) = 0. \tag{30}$$

From (29) and (30), the continuity of J, we have

$$\lim_{s\to\infty} G(J(Jp_s), J(Jp_s), Jp) = \lim_{s\to\infty} G(J(Jp_s), Jp, Jp) = 0, \tag{31}$$

$$\lim_{s\to\infty} G(J(Jy_s), J(Jy_s), Jy) = \lim_{s\to\infty} G(J(Jy_s), Jy, Jy) = 0. \tag{32}$$

Thus, $\{J(Jp_s)\}$ is G-convergent to Jp and $\{J(Jp_s)\}$ is G-converges to Jy. Since Y and J commute, so

$$J(Jp_{s+1}) = J(Y(p_s, y_s)) = Y(Jp_s, Jy_s), \tag{33}$$

$$J(Jy_{s+1}) = J(Y(y_s, p_s)) = Y(Jy_s, Jp_s). \tag{34}$$

Using (33) and (34) and the continuity of Y, we get $\{J(Jp_{s+1})\}$ is G-converges to $Y(p, y)$ and $\{J(Jy_{s+1})\}$ is G-converges to $Y(y, p)$. Using uniqueness of the limit, $Y(p, y) = J(p)$ and $Y(y, p) = J(y)$.

Theorem 3.2 *Assume hypothesis of Theorem 3.1 are satisfied. Moreover, presume that S has the properties*

(a) if an increasing sequence $p_s \to p$, then $p_s \leq p$, $\forall$ s,
(b) if a decreasing sequence $y_s \to y$, then $y_s \geq y$, $\forall$ s.

Then, the conclusion of Theorem 3.1 also holds.

Proof On the way of the proof of Theorem 3.1, we have $\{Jp_s\}$ and $\{Jy_s\}$ are G-Cauchy sequences in $J(S)$ with $Jp_s \leq Jp_{s+1}$ and $Jy_s \geq Jy_{s+1}$ for each $s \in \mathbb{N}$. Since $(J(S), G)$ is complete G-metric, soo existence of $p, y \in S$ such that $Jp_s \to Jp$ and $Jy_s \to Jy$.

Since $\{Jp_s\}$ is non-decreasing, $\{Jy_s\}$ is non-increasing, applying the consistency of $(S, G, \leq)$ we get $Jp_s \leq Jp$ and $Jy_s \geq Jy$ for each $s \in \mathbb{N}$. By inequality (1), we obtain

$$\begin{aligned}\psi(G(Jp_{s+1}, Y(p, y), Y(p, y))) &= \psi(G(Y(p_s, y_s), Y(p, y), Y(p, y)))\\ &\leq \alpha(N((p_s, y_s), (p, y), (p, y))) - \beta(N((p_s, y_s), (p, y), (p, y))),\end{aligned} \tag{35}$$

and

$$\begin{aligned}\psi(G(Jy_{s+1}, Y(y, p), Y(y, p))) &= \psi(G(Y(y_s, p_s), Y(y, p), Y(y, p)))\\ &\leq \alpha(N((y_s, p_s), (y, p), (y, p))) - \beta(N((y_s, p_s), (y, p), (y, p))),\end{aligned} \tag{36}$$

where

$$
\begin{aligned}
N((p_s, y_s), (p, y), (p, y)) &= N((y_s, p_s), (y, p), (y, p)) \\
&= max\{G(Jp_s, Jp, Jp), G(Jy_s, Jy, Jy), G(Jp_s, Jp_{s+1}, Jp_{s+1}), \\
&\qquad G(Jp, Y(p, y), Y(p, y)), G(Jy_s, Jy_{s+1}, Jy_{s+1}), \\
&\qquad G(Jy, Y(y, p), Y(y, p))\}.
\end{aligned}
$$

Now, we claim that

$$
max\{G(Jp, Y(p, y), Y(p, y)), G(Jy, Y(y, p), Y(y, p))\} = 0. \tag{37}
$$

If this not true, then $max\{G(Jp, Y(p, y), Y(p, y)), G(Jy, Y(y, p), Y(y, p))\} > 0$. Since $Jp_s \to Jp$ and $Jy_s \to Jy$ for each $s \in \mathbb{N}$,

$$
\begin{aligned}
N((p_s, y_s), (p, y), (p, y)) &= N((y_s, p_s), (y, p), (y, p)) \\
&= max\{G(Jp, Y(p, y), Y(p, y)), G(Jy, Y(y, p), Y(y, p))\}.
\end{aligned}
$$

Combining this with (35) and (36), we get

$$
\begin{aligned}
&\psi(max\{G(Jp_{s+1}, Y(p, y), Y(p, y)), G(Jy_{s+1}, Y(y, p), Y(y, p)))\}) \\
&\quad = max\{\psi(G(Jp_{s+1}, Y(p, y), Y(p, y))), \psi(G(Jy_{s+1}, Y(y, p), Y(y, p))))\} \\
&\quad \leq \alpha(max\{G(Jp, Y(p, y), Y(p, y)), G(Jy, Y(y, p), Y(y, p))\}) \\
&\qquad - \beta(max\{G(Jp, Y(p, y), Y(p, y)), G(Jy, Y(y, p), Y(y, p))\}).
\end{aligned}
$$

Letting $s \to \infty$ it follows that

$$
\begin{aligned}
&\psi(max\{G(Jp, Y(p, y), Y(p, y)), G(Jy, Y(y, p), Y(y, p)))\}) \\
&\quad \leq \alpha(max\{G(Jp, Y(p, y), Y(p, y)), G(Jy, Y(y, p), Y(y, p))\}) \\
&\qquad - \beta(max\{G(Jp, Y(p, y), Y(p, y)), G(Jy, Y(y, p), Y(y, p))\}).
\end{aligned}
$$

From our assumptions about ψ, α, β, which is contradiction. Therefore, (37) holds. After this, it follows $J(p) = Y(p, y)$ and $J(y) = Y(y, p)$.

Example 3.1 Let $S = [0, 1]$ and $G : S \times S \times S \to R^+$ be defined as

$$
G(u, v, r) = |u - v| + |v - r| + |r - u|, \forall u, v, r \in S.
$$

Then (S, G) shows complete G-metric space.

Consider the mapping $Y : S \times S \to S$ defined by $Y(p, y) = \frac{1}{5}p - \frac{1}{3}y^2$ if $p \geq y$ for all $p, y \in S$. Also define $J : S \to S$ by $Jp = p$ for $p \in S$.

Let $\psi, \alpha, \beta : [0, \infty) \to [0, \infty)$ such that $\psi(p) = \alpha(p) = p$ and $\beta(p) = \frac{1}{5}p$. Clearly, ψ is a function of altering distance type, α and β are continuous and lower semicontinuous, respectively, $\alpha(0) = \beta(0) = 0$, also $\psi(t) - \alpha(t) + \beta(t) = \frac{t}{5} > 0$ for all $t > 0$.

Now, for value of (p, y), (u, v), and (r, z), we have the following possibility such that $p \geq u \geq r$, $y \leq v \leq z$.

$$\begin{aligned} G(Y(p,y), Y(u,v), Y(r,z)) &= |(\frac{1}{5}p - \frac{1}{3}y^2) - (\frac{1}{5}u - \frac{1}{3}v^2)| + |(\frac{1}{5}u - \frac{1}{3}v^2) - (\frac{1}{5}r - \frac{1}{3}z^2)| \\ &+ |(\frac{1}{5}r - \frac{1}{3}z^2) - (\frac{1}{5}p - \frac{1}{3}y^2)| \\ &\leq \frac{1}{5}[|(p-u)| + |u-r| + |r-p|] + \frac{1}{3}[|(y^2 - v^2)| + |v^2 - z^2| + |z^2 - y^2|]. \end{aligned}$$

$(y^2 - v^2) \leq (y - v)$.
Similarly, $(v^2 - z^2) \leq (v - z)$ and $(z^2 - y^2) \leq (z - y)$. Thus, we have

$$\begin{aligned} G(Y(p,y), Y(u,v), Y(r,z)) &\leq \frac{1}{5}[|(p-u)| + |u-r| + |r-p|] + \frac{1}{3}[|(y-v)| + |v-z| + |z-y|] \\ &\leq \frac{1}{5}(G(Jp, Ju, Jr)) + \frac{1}{3}(G(Jy, Jv, Jz)) \\ &\leq \frac{8}{15} max\{G(Jp, Ju, Jr), G(Jy, Jv, Jz)\} \\ &\leq \frac{4}{5} N((p,y),(u,v),(r,z)) \\ &= N((p,y),(u,v),(r,z)) - \frac{1}{5} N((p,y),(u,v),(r,z)) \\ &= \alpha(N((p,y),(u,v),(r,z))) - \beta(N((p,y),(u,v),(r,z))), \end{aligned}$$

where

$$\begin{aligned} N((p,y),(u,v),(r,z)) = max\{&G(Jp, Ju, Jr), G(Jy, Jv, Jz), G(Jp, Y(p,y), Y(p,y)), \\ &G(Ju, Y(u,v), Y(u,v)), G(Jr, Y(r,z), Y(r,z)), \\ &G(Jy, Y(y,p), Y(y,p)), G(Jv, Y(v,u), Y(v,u)), \\ &G(Jz, Y(z,r), Y(z,r))\}. \end{aligned}$$

Therefore, every conditions of Theorem (3.1) are satisfied. Furthermore, (0, 0) is the unique coupled coincidence point of Y and J.

Corollary 3.1 *Assume that $(X, \leq)$ be a partially ordered set and (X, G) is complete G-metric space. $M : X \times X \to X$ is a map satisfying (1) (with $p = I_X$) for all $u, v, w, x, y, z \in X$ with $x \geq u \geq w$ and $y \leq v \leq z$. and M has the property of mixed monotone. Moreover, suppose either*

1. *F is continous or*
2. *X possesses:*

(a) if increasing sequence $x_s \to x$, then $x_s \leq x$, $\forall$ s,
(b) if decreasing sequence $y_s \to y$, then $y_s \geq y$, $\forall$ s.

If existence of $x_0, y_0 \in X$ such type $x_0 \leq M(x_0, y_0)$ and $y_0 \geq M(y_0, x_0)$, then M possesses a coupled fixed point.

Theorem 3.3 *Under the assumption of Theorem 3.2, presume that $Jy_0 \leq Jp_0$. Next, it follows $Jp = Y(p, y) = Y(y, p) = Jy$. Furthermore, if Y and J are w-compatible, then Y and J admits (u, u)type coupled coincidence point.*

Proof If $Jy_0 \leq Jp_0$, then $Jy \leq Jy_s \leq Jy_0 \leq Jp_0 \leq Jp_s \leq Jp$ for all $n \in \mathbb{N}$. Thus, if $Jp \neq Jy$(and then $G(Jp, Jp, Jy) \neq 0$ and $G(Jy, Jy, Jp) \neq 0$), so by inequality (1), we get

$$\psi(G(Jy, Jp, Jp)) = \psi(G(Y(y, p), Y(p, y), Y(p, y)))$$
$$\leq \alpha(N((y, p), (p, y), (p, y))) - \beta(N((y, p), (p, y), (p, y))),$$

where

$$N((y, p), (p, y), (p, y)) = max\{G(Jy, Jp, Jp), G(Jp, Jy, Jy)\}.$$

Hence,

$$\psi(G(Jy, Jp, Jp)) \leq \alpha(max\{G(Jy, Jp, Jp), G(Jp, Jy, Jy)\})$$
$$- \beta(max\{G(Jy, Jp, Jp), G(Jp, Jy, Jy)\}). \tag{38}$$

Since $Jy \leq Jp$, therefore applying the same concept we have

$$\psi(G(Jp, Jy, Jy)) \leq \alpha(max\{G(Jy, Jp, Jp), G(Jp, Jy, Jy)\})$$
$$- \beta(max\{G(Jy, Jp, Jp), G(Jp, Jy, Jy)\}). \tag{39}$$

From (38) and (39), we have

$$\psi(max\{G(Jy, Jp, Jp), G(Jp, Jy, Jy)) = max\{\psi(G(Jy, Jp, Jp)), \psi(G(Jp, Jy, Jy))\}$$
$$\leq \alpha(max\{G(Jy, Jp, Jp), G(Jp, Jy, Jy)\})$$
$$- \beta(max\{G(Jy, Jp, Jp), G(Jp, Jy, Jy)\}).$$

Thus, by properties of ψ, α, β functions, we obtain $G(Jy, Jp, Jp) = 0$ and $G(Jp, Jy, Jy) = 0$, a contradiction. Hence $Jp = Jy$, that is, $Jp = Y(p, y) = Y(y, p) = Jy$. Now, let $u = Jp = Jy$. Since Y and J are w-compatible, then

$$Ju = J(Jp) = J(Y(p, y)) = Y(Jp, Jy) = Y(u, u).$$

Thus, Y and J admit a coupled coincidence point of the type (u, u).

If $(S, \leq)$ is a partially ordered set, we enhance the product $S \times S$. We need the following idea of the partial order relation:

$$\text{for each } \ (u, v), (r, z) \in S \times S, \ \ (u, v) \leq (r, z) \iff u \leq r, \ v \geq z. \tag{40}$$

Theorem 3.4 *Under the axioms of Theorem 3.1, assume that, for each (p, y) and (ṕ, ý) in S, there exists (u, v)∈ S × S as (u, v) is comparable to (p, y) and (ṕ, ý). Then, Y and J admit preciously unique common coupled fixed point.*

Proof By Theorem (3.1), set of coupled coincidence point is non-empty. Our aim to show if (p, y) and (ṕ, ý) are coupled coincidence points, that is,

$$J(p) = Y(p, y), \quad J(y) = Y(y, p)$$
$$\text{and} \quad J(\acute{p}) = Y(\acute{p}, \acute{y}), \quad J(\acute{y}) = Y(\acute{y}, \acute{p}),$$

then

$$Jp = J\acute{p} \text{ and } Jy = J\acute{y}. \tag{41}$$

Take an element $(u, v) \in S \times S$ comparable with both of them.

Let $u_0 = u, v_0 = v$ and take $u_1, v_1 \in S$ so that $Ju_1 = Y(u_0, v_0)$ and $Jv_1 = Y(v_0, u_0)$.

Then, similarly by Theorem (3.1), we can take sequences $\{Ju_s\}$ and $\{Jv_s\}$ inductively as follows

$$Ju_{s+1} = Y(u_s, v_s) \textit{ and } Jv_{s+1} = Y(v_s, u_s).$$

Further, set $p_0 = p, y_0 = y, \acute{p}_0 = \acute{p}, \acute{y}_0 = \acute{y}$ and similar way, consider the sequences $\{Jp_s\}, \{Jy_s\}$ and $\{J\acute{p}_s\}, \{J\acute{y}_s\}$.

Since $(Jp, Jy) = (Y(p, y), Y(y, p)) = (Jp_1, Jy_1)$ and $(Y(u, v), Y(v, u)) = (Ju_1, Jv_1)$ are comparable, then $Jp \leq Ju_1$ and $Jy \geq Jv_1$. Applying the mathematical induction, simply we can prove

$$Jp \leq Ju_s \;\; Jy \geq Jv_s \;\; \forall s \in \mathbb{N}.$$

Let $\gamma_s = max\{G(Jp, Jp, Ju_s), G(Jy, Jy, Jv_s)\}$. We will show that $\lim_{s\to\infty} \gamma_s = 0$. First, assume that $\gamma_s = 0$, for an $s \geq 1$.

By inequality (1), we have

$$\begin{aligned}\psi(G(Jp, Jp, Ju_{s+1})) &= \psi(G(Y(p, y), Y(p, y), Y(u_s, v_s))) \\ &\leq \alpha(N((p, y), (p, y), (u_s, v_s))) - \beta(N((p, y), (p, y), (u_s, v_s))),\end{aligned}$$

where

$$N((p,\ y),\ (p,\ y),\ (us,\ vs)) = \max\{G(Jp;\ Jp;\ Ju),\ G(Jy;\ Jy;\ Jv)\}$$

Therefore, we obtain

$$\psi(G(Jp, Jp, Ju_{s+1})) \leq \alpha(max\{G(Jp, Jp, Ju_s), G(Jy, Jy, Jv_s)\}) \\ - \beta(max\{G(Jp, Jp, Ju_s), G(Jy, Jy, Jv_s)\}). \quad (42)$$

Similarly, we have

$$\psi(G(Jy, Jy, Jv_{s+1})) \leq \alpha(max\{G(Jp, Jp, Ju_s), G(Jy, Jy, Jv_s)\}) \\ - \beta(max\{G(Jp, Jp, Ju_s), G(Jy, Jy, Jv_s)\}). \quad (43)$$

Therefore, from (42) and (43), we get

$$\begin{aligned}\psi(\gamma_{s+1}) &= \psi(max\{G(Jp, Jp, Ju_{s+1})), G(Jy, Jy, Jv_{s+1})\} \\ &= max\{\psi(G(Jp, Jp, Ju_{s+1}))), \psi(G(Jy, Jy, Jv_{s+1}))\} \\ &\leq \alpha(max\{G(Jp, Jp, Ju_s), G(Jy, Jy, Jv_s)\}) \\ &\quad - \beta(max\{G(Jp, Jp, Ju_s), G(Jy, Jy, Jv_s)\}) \\ &= \alpha(\gamma_s) - \beta(\gamma_s) \\ &= \alpha(0) - \beta(0). \end{aligned} \quad (44)$$

So, from behavior of ψ, α and β, we deduce $\gamma_{s+1} = 0$. Reciting this procedure, we can prove that $\gamma_p = 0$, for all $p \geq s$. Thus, $\lim_{s\to\infty} \gamma_s = 0$.

Now, let $\gamma_s \neq 0$, for all s and let $\gamma_s < \gamma_{s+1}$, for some s.

As ψ is a function of altering distance type, from (44)

$$\begin{aligned}\psi(\gamma_s) &= \psi(max\{G(Jp, Jp, Ju_s), G(Jy, Jy, Jv_s)\}) \\ &< \psi(\gamma_{s+1}) \\ &= \psi(max\{G(Jp, Jp, Ju_{s+1})), G(Jy, Jy, Jv_{s+1})\} \\ &= max\{\psi(G(Jp, Jp, Ju_{s+1}))), \psi(G(Jy, Jy, Jv_{s+1}))\} \\ &\leq \alpha(max\{G(Jp, Jp, Ju_s), G(Jy, Jy, Jv_s)\}) \\ &\quad - \beta(max\{G(Jp, Jp, Ju_s), G(Jy, Jy, Jv_s)\}) \\ &= \alpha(\gamma_s) - \beta(\gamma_s). \end{aligned}$$

This shows that $\gamma_s = 0$, shows as a contradiction.

Hence, $\gamma_{s+1} \leq \gamma_s$, for all $s \geq 1$. Next, if we go according to Theorem (3.1), we can prove that

$$\lim_{s\to\infty} max\{G(Jp, Jp, Ju_s), G(Jy, Jy, Jv_s)\} = 0. \quad (45)$$

So, $\{Ju_s\} \to Jp$ and $\{Jv_s\} \to Jy$.

In similar way, we can prove that

$$\lim_{s\to\infty} max\{G(J\acute{p}, J\acute{p}, Ju_s), G(J\acute{y}, J\acute{y}, Jv_s)\} = 0. \quad (46)$$

That is, $\{Ju_s\} \to J\acute{t}$ and $\{Jv_s\} \to J\acute{y}$. Finally, since the limit is unique, $Jp = J\acute{p}$ and $Jy = J\acute{y}$.

As $Jp = Y(p, y)$ and $Jy = Y(y, p)$, by commutativity of Y and J, we have

$$J(Jp) = J(Y(p, y)) = Y(Jp, Jy) \text{ and } J(Jy) = J(Y(y, p)) = Y(Jy, Jp). \tag{47}$$

Denote $Jp = z$ and $Jy = r$. Then, from(47), it follows that

$$Jz = Y(z, r) \text{ and } Jr = Y(r, z). \tag{48}$$

Thus (z, r) is a coupled coincidence type of point of Y and J. From (41) with $\acute{p} = z$ and $\acute{y} = r$, it follows $Jz = Jp$ and $Jr = Jy$, that is,

$$Jz = z \text{ and } Jr = r. \tag{49}$$

Thus, from (48) and (49), we get $z = Jz = Y(z, r)$ and $r = Jr = Y(r, z)$. Therefore, (z, r) shows coupled common fixed point of Y and J.

To show the uniqueness, accept that (h, g) is distinct coupled common fixed point of Y and J. Then $h = Jh = Y(h, g)$ and $g = Jg = Y(g, h)$. The pair (h, g) is a coupled coincidence point of Y and J, then we get $Jh = Jp = z$ and $Jg = Jy = r$. Thus, $h = Jh = Jz = z$ and $g = Jg = Jr = r$. This shows preciously unique coupled fixed point, hence the proof.

Theorem 3.5 *With the hypothesis of Theorem (3.2), further assume that each (p, y) and $(\acute{p}, \acute{y})$ in S, existence of $(u, v) \in S \times S$ such type $(F(u, v), F(v, u))$ is comparable to $(Y(p, y), Y(y, p))$ and $(Y(\acute{p}, \acute{y}), Y(\acute{y}, \acute{p}))$. If Y and J are w-compatible, then Y and J admit (u, u) type a unique common coupled fixed point.*

Proof Applying Theorem (3.2), coupled fixed points of Y and J is non-empty. Let (p, y) and $(\acute{p}, \acute{y})$ be two coupled coincidence points of Y and J. Succeeding the proof of Theorem (3.4), we can find

$$Jp = J\acute{p} \text{ and } Jy = J\acute{y}. \tag{50}$$

If (p, y) is coupled coincidence point of Y and J, then (y, p) also a coupled coincidence point of Y and J. By (50), $Jp = Jy$. Put $u = Jp = Jy$. Since $Jp = Y(p, y)$, $Jy = Y(y, p)$ and Y and J are w-compatible, we have $Ju = J(Jp) = J(Y(p, y)) = Y(Jp, Jy) = Y(u, u)$. Thus (u, u) is a common coupled fixed point of Y and J. So, $Ju = Jp = Jy = u$ and therefore we have $u = Ju = Y(u, u)$. Therefore, (u, u) represent common coupled fixed point for Y and J.

For uniqueness of common coupled fixed point of Y and J, let (v, r) be other coupled fixed point of Y and J, that is, $v = Jv = Y(v, r)$ and $r = Jr = Y(r, v)$. Clearly, we have $Jp = Ju = Jv$ and $Jy = Ju = Jr$. Therefore $u = v = r$. Thus, Y and J admit unique common coupled fixed point of the type (u, u).

4 Application to Integral Equations

In this segment, we contemplate the existence and uniqueness of solutions of a nonlinear integral equation utilizing the outcomes demonstrated in Sect. 3.

Consider the accompanying integral equation:

$$p(w) = \int_0^1 (k_1(w, \varsigma) + k_2(w, \varsigma))(f_1(\varsigma, p(\varsigma)) + f_2(\varsigma, p(\varsigma)))d\varsigma + a(w), \quad w \in [0, 1]. \tag{51}$$

We will analyze eq. (51) under the accompanying presumptions:

(i) $k_i : [0, 1] \times [0, 1] \to \mathbb{R} (i = 1, 2)$ are continuous and $k_1(w, \varsigma) \geq 0$ and $k_2(w, \varsigma) \leq 0$.

(ii) $a \in C[0, 1]$.

(iii) $f_1, f_2 : [0, 1] \times \mathbb{R} \to \mathbb{R}$ are continuous functions.

(iv) Existence of constants $\lambda, \mu > 0$ such that for all $w \in [0, 1], t, y \in \mathbb{R}$ and $p \geq y$,

$$0 \leq f_1(w, p) - f_1(w, y) \leq \lambda(p - y)$$

and

$$-\mu(p - y) \leq f_2(w, p) - f_2(w, y) \leq 0.$$

(v) There exist $\gamma, \delta \in C[0, 1]$ such that

$$\begin{aligned} \gamma(w) \leq & \int_0^1 k_1(w, \varsigma)(f_1(\varsigma, \gamma(\varsigma)) + f_2(\varsigma, \delta(\varsigma)))d\varsigma \\ & + \int_0^1 k_2(w, \varsigma)(f_1(\varsigma, \delta(\varsigma)) + f_2(\varsigma, \gamma(\varsigma)))d\varsigma + a(w), \\ \delta(w) \geq & \int_0^1 k_1(w, \varsigma)(f_1(\varsigma, \delta(\varsigma)) + f_2(\varsigma, \gamma(\varsigma)))d\varsigma \\ & + \int_0^1 k_2(w, \varsigma)(f_1(\varsigma, \gamma(\varsigma))) + f_2(\varsigma, \delta(\varsigma))d\varsigma + a(w). \end{aligned}$$

(vi) $3.max(\lambda, \mu) \parallel k_1 - k_2 \parallel_\infty \leq \frac{1}{2}$, where

$$\parallel k_1 - k_2 \parallel_\infty = sup\{(k_1(w, \varsigma) - k_2(w, \varsigma)) : w, \varsigma \in [0, 1]\}.$$

Let the space $S = C[0, 1]$ of continuous functions defined on $[0, 1]$. Define $G : S \times S \times S \to R^+$ by

$$G(u, v, r) = \sup_{w \in [0,1]} \mid u(w) - v(w) \mid + \sup_{w \in [0,1]} \mid v(w) - r(w) \mid$$
$$+ \sup_{t \in [0,1]} \mid r(w) - u(w) \mid, \text{ for all } u, v, r \in S.$$

Then, (S, G) is a G-complete metric space.

This space can also be endowed with a partial order given by

$$u, v \in C[0, 1], u \leq v \iff u(w) \leq v(w), \text{for any } w \in [0, 1].$$

Clearly, if in $S \times S$, we consider the order given by

$$(u, v), (r, z) \in S \times S, (u, v) \leq (r, z) \iff u \leq r \text{ and } v \geq z,$$

Thus, for any $u, v \in S$, we have that $max(u, v), min(u, v) \in S$, condition (40) is fulfilled.

Furthermore, in [27], it is shown that $(C[0, 1], \leq)$ fulfills presumption (1).

Now, we formulate our result.

Theorem 4.1 *Under presumptions (i)–(vi), Eq.* (51) *has a unique solution in* $C[0, 1]$.

Proof We consider the operator $Y : S \times S \to S$ defined by

$$Y(p, y)(w) = \int_0^1 k_1(w, \varsigma)(f_1(\varsigma, p(\varsigma)) + f_2(\varsigma, y(\varsigma)))d\varsigma$$
$$+ \int_0^1 k_2(w, \varsigma)(f_1(\varsigma, y(\varsigma)) + f_2(\varsigma, p(\varsigma)))d\varsigma + a(w), \quad \text{for } w \in [0, 1].$$

By virtue of our suppositions, Y is well defined (this means that for $p, y \in S$ then $Y(p, y) \in S$).

Primarily, we demonstrate that Y has the mixed monotone property.

In fact, for $p_1 \leq p_2$ and $w \in [0, 1]$, we have

$$\begin{aligned}
Y(p_1, y)(w) - Y(p_2, y)(w) &= \int_0^1 k_1(w, \varsigma)(f_1(\varsigma, p_1(\varsigma)) + f_2(\varsigma, y(\varsigma)))d\varsigma \\
&+ \int_0^1 k_2(w, \varsigma)(f_1(\varsigma, y(\varsigma)) + f_2(\varsigma, p_1(\varsigma)))d\varsigma + a(w) \\
&- \int_0^1 k_1(w, \varsigma)(f_1(\varsigma, p_2(\varsigma)) + f_2(\varsigma, y(\varsigma)))d\varsigma \\
&- \int_0^1 k_2(w, \varsigma)(f_1(\varsigma, y(\varsigma)) + f_2(\varsigma, p_2(\varsigma)))d\varsigma - a(w) \\
&= \int_0^1 k_1(w, \varsigma)(f_1(\varsigma, p_1(\varsigma)) - f_1(\varsigma, p_2(\varsigma)))d\varsigma \\
&+ \int_0^1 k_2(w, \varsigma)(f_2(\varsigma, p_1(\varsigma)) - f_2(\varsigma, p_2(\varsigma)))d\varsigma.
\end{aligned} \tag{52}$$

Taking into account that $p_1 \leq p_2$ and our assumptions,

$$f_1(\varsigma, p_1(\varsigma)) - f_1(\varsigma, p_2(\varsigma)) \leq 0,$$
$$f_2(\varsigma, p_1(\varsigma)) - f_2(\varsigma, p_2(\varsigma)) \geq 0,$$

and from (52) we obtain

$$Y(p_1, y)(w) - Y(p_2, y)(w) \leq 0$$

and this proves that $Y(p_1, y) \leq Y(p_2, y)$. Similarly, if $y_1 \geq y_2$ and $w \in [0, 1]$, we have

$$\begin{aligned} Y(p, y_1)(w) - Y(p, y_2)(w) &= \int_0^1 k_1(w, \varsigma)(f_1(\varsigma, p(\varsigma)) + f_2(\varsigma, y_1(\varsigma)))d\varsigma \\ &+ \int_0^1 k_2(w, \varsigma)(f_1(\varsigma, y_1(\varsigma)) + f_2(\varsigma, p(\varsigma)))d\varsigma + a(w) \\ &- \int_0^1 k_1(w, \varsigma)(f_1(\varsigma, p(\varsigma)) + f_2(\varsigma, y_2(\varsigma)))d\varsigma \\ &- \int_0^1 k_2(w, \varsigma)(f_1(\varsigma, y_2(\varsigma)) + f_2(\varsigma, p(\varsigma)))d\varsigma - a(w) \\ &= \int_0^1 k_1(w, \varsigma)(f_2(\varsigma, y_1(\varsigma)) - f_2(\varsigma, y_2(\varsigma)))d\varsigma \\ &+ \int_0^1 k_2(w, \varsigma)(f_1(\varsigma, y_1(\varsigma)) - f_1(\varsigma, y_2(\varsigma)))d\varsigma \end{aligned}$$

and by our assumptions, as $y_1 \geq y_2$,

$$f_2(\varsigma, y_1(\varsigma)) - f_2(\varsigma, y_2(\varsigma)) \leq 0,$$
$$f_1(\varsigma, y_1(\varsigma)) - f_1(\varsigma, y_2(\varsigma)) \geq 0,$$

and thus,

$$Y(p, y_1)(w) - Y(p, y_2)(w) \leq 0,$$

or, equivalently,

$$Y(p, y_1) \leq Y(p, y_2).$$

Therefore, Y has mixed monotone property.
In what follows, we estimate the quantity $G(Y(p, y), Y(u, v), Y(r, z))$ for all $p, y, u, v, r, z \in S$, with $p \geq u \geq r$ and $y \leq v \leq z$.
Indeed, as Y has the mixed monotone property, $Y(p, y) \geq Y(u, v) \geq Y(r, z)$ and we can acquire

$$
\begin{aligned}
G(Y(p,y),Y(u,v),Y(r,z)) = {} & \sup_{w\in[0,1]} \mid Y(p,y)(w) - Y(u,v)(w) \mid \\
& + \sup_{t\in[0,1]} \mid Y(u,v)(w) - Y(r,z)(w) \mid \\
& + \sup_{w\in[0,1]} \mid Y(r,z)(w) - Y(p,y)(w) \mid \\
= {} & \sup_{t\in[0,1]} (Y(p,y)(w) - Y(u,v)(w)) + \sup_{t\in[0,1]} (Y(u,v)(w) - Y(r,z)(w)) \\
& + \sup_{w\in[0,1]} (Y(p,y)(w) - Y(r,z)(w)) \\
= {} & \sup_{w\in[0,1]} \Big[\int_0^1 k_1(w,\varsigma)(f_1(\varsigma,p(\varsigma)) + f_2(\varsigma,y(\varsigma)))d\varsigma \\
& + \int_0^1 k_2(w,\varsigma)(f_1(\varsigma,y(\varsigma)) + f_2(\varsigma,p(\varsigma)))d\varsigma + a(w) \\
& - \int_0^1 k_1(w,\varsigma)(f_1(\varsigma,u(\varsigma)) + f_2(\varsigma,v(\varsigma)))d\varsigma \\
& - \int_0^1 k_2(w,\varsigma)(f_1(\varsigma,v(\varsigma)) + f_2(\varsigma,u(\varsigma)))ds - a(w) \Big] \\
& + \sup_{w\in[0,1]} \Big[\int_0^1 k_1(w,\varsigma)(f_1(\varsigma,u(\varsigma)) + f_2(\varsigma,v(\varsigma)))d\varsigma \\
& + \int_0^1 k_2(w,\varsigma)(f_1(\varsigma,v(\varsigma)) + f_2(\varsigma,u(\varsigma)))d\varsigma + a(w) \\
& - \int_0^1 k_1(w,\varsigma)(f_1(r,r(\varsigma)) + f_2(\varsigma,z(\varsigma)))d\varsigma \\
& - \int_0^1 k_2(w,\varsigma)(f_1(\varsigma,z(\varsigma)) + f_2(\varsigma,r(\varsigma)))d\varsigma - a(w) \Big] \\
& + \sup_{w\in[0,1]} \Big[\int_0^1 k_1(w,\varsigma)(f_1(\varsigma,p(\varsigma)) + f_2(\varsigma,y(\varsigma)))d\varsigma \\
& + \int_0^1 k_2(w,\varsigma)(f_1(\varsigma,y(\varsigma)) + f_2(\varsigma,p(\varsigma)))d\varsigma + a(w) \\
& - \int_0^1 k_1(w,\varsigma)(f_1(\varsigma,r(\varsigma)) + f_2(\varsigma,z(\varsigma)))d\varsigma \Big] \\
& - \int_0^1 k_2(w,\varsigma)(f_1(\varsigma,z(\varsigma)) + f_2(\varsigma,r(\varsigma)))d\varsigma - a(w) \Big]
\end{aligned}
$$

$$
\begin{aligned}
= {} & \sup_{t\in[0,1]} \Big[\int_0^1 k_1(w,\varsigma)[(f_1(\varsigma,p(\varsigma)) - f_1(\varsigma,u(\varsigma))) - (f_2(\varsigma,v(\varsigma)) - f_2(\varsigma,y(\varsigma)))]d\varsigma \\
& - \int_0^1 k_2(w,\varsigma)[(f_1(\varsigma,v(\varsigma)) - f_1(\varsigma,y(\varsigma))) - (f_2(\varsigma,p(\varsigma)) - f_2(\varsigma,u(q)))]dq \Big] \\
& + \sup_{w\in[0,1]} \Big[\int_0^1 k_1(w,\varsigma)[(f_1(\varsigma,u(\varsigma)) - f_1(\varsigma,r(\varsigma))) - (f_2(\varsigma,z(\varsigma)) - f_2(\varsigma,v(\varsigma)))]d\varsigma \\
& - \int_0^1 k_2(w,\varsigma)[(f_1(\varsigma,z(\varsigma)) - f_1(\varsigma,v(\varsigma))) - (f_2(\varsigma,u(\varsigma)) - f_2(\varsigma,r(\varsigma)))]d\varsigma \Big]
\end{aligned}
$$

$$+\sup_{t\in[0,1]}\Big[\int_0^1 k_1(w,\varsigma)[(f_1(\varsigma,p(\varsigma))-f_1(\varsigma,r(\varsigma)))-(f_2(\varsigma,z(\varsigma))-f_2(\varsigma,y(\varsigma)))]d\varsigma$$

$$-\int_0^1 k_2(w,\varsigma)[(f_1(\varsigma,z(\varsigma))-f_1(\varsigma,y(\varsigma)))-(f_2(\varsigma,p(\varsigma))-f_2(\varsigma,r(\varsigma)))]d\varsigma\Big]. \quad (53)$$

By our assumptions(notice that $p \geq u \geq r$ and $y \leq v \leq z$)

$$\begin{aligned}
&f_1(\varsigma,p(\varsigma))-f_1(\varsigma,u(\varsigma))\leq\lambda(p(\varsigma)-u(\varsigma)), && f_2(\varsigma,v(\varsigma))-f_2(\varsigma,y(\varsigma))\geq-\mu(v(\varsigma)-y(\varsigma)),\\
&f_1(\varsigma,v(\varsigma))-f_1(\varsigma,y(\varsigma))\leq\lambda(v(\varsigma)-y(\varsigma)), && f_2(\varsigma,p(\varsigma))-f_2(\varsigma,u(\varsigma))\geq-\mu(p(\varsigma)-u(\varsigma)),\\
&f_1(\varsigma,u(\varsigma))-f_1(\varsigma,r(\varsigma))\leq\lambda(u(\varsigma)-r(\varsigma)), && f_2(\varsigma,z(\varsigma))-f_2(\varsigma,v(\varsigma))\geq-\mu(z(\varsigma)-v(\varsigma)),\\
&f_1(\varsigma,z(\varsigma))-f_1(\varsigma,v(\varsigma))\leq\lambda(z(\varsigma)-v(\varsigma)), && f_2(\varsigma,u(\varsigma))-f_2(\varsigma,r(\varsigma))\geq-\mu(u(\varsigma)-r(\varsigma)),\\
&f_1(\varsigma,p(\varsigma))-f_1(\varsigma,r(\varsigma))\leq\lambda(p(\varsigma)-r(\varsigma)), && f_2(\varsigma,z(\varsigma))-f_2(\varsigma,y(\varsigma))\geq-\mu(z(\varsigma)-y(\varsigma)),\\
&f_1(\varsigma,z(\varsigma))-f_1(\varsigma,y(\varsigma))\leq\lambda(z(\varsigma)-y(\varsigma)), && f_2(\varsigma,p(\varsigma))-f_2(\varsigma,r(\varsigma))\geq-\mu(p(\varsigma)-r(\varsigma)).
\end{aligned}$$

Taking into account these last inequalities and (53), we get

$$\begin{aligned}
G(F(p,y),F(u,v),F(r,z)) \leq\ & \sup_{w\in[0,1]}\Big[\int_0^1 k_1(w,\varsigma)[\lambda(p(\varsigma)-u(\varsigma))+\mu(v(\varsigma)-y(\varsigma))]d\varsigma\\
&+\int_0^1(-k_2(w,\varsigma))[\lambda(v(\varsigma)-y(\varsigma))+\mu(p(\varsigma)-u(\varsigma))]d\varsigma\Big]\\
&+\sup_{w\in[0,1]}\Big[\int_0^1 k_1(w,\varsigma)[\lambda(u(\varsigma)-w(\varsigma))+\mu(z(\varsigma)-v(\varsigma))]d\varsigma\\
&+\int_0^1(-k_2(w,\varsigma))[\lambda(z(\varsigma)-v(\varsigma))+\mu(u(\varsigma)-r(\varsigma))]d\varsigma\Big]\\
&+\sup_{w\in[0,1]}\Big[\int_0^1 k_1(w,\varsigma)[\lambda(p(\varsigma)-r(\varsigma))+\mu(z(\varsigma)-y(\varsigma))]d\varsigma\\
&+\int_0^1(-k_2(w,\varsigma))[\lambda(z(\varsigma)-y(\varsigma))+\mu(p(\varsigma)-r(\varsigma))]d\varsigma\Big]\\
=\ & max(\lambda,\mu)\sup_{w\in[0,1]}\Big[\int_0^1(k_1(w,\varsigma)-k_2(w,\varsigma))(p(\varsigma)-u(\varsigma))d\varsigma\\
&+\int_0^1(k_1(w,\varsigma)-k_2(w,\varsigma))(v(\varsigma)-y(\varsigma))d\varsigma
\end{aligned}$$

$$\begin{aligned}
&+\int_0^1(k_1(w,\varsigma)-k_2(w,\varsigma))(u(\varsigma)-r(\varsigma))d\varsigma\\
&+\int_0^1(k_1(w,\varsigma)-k_2(w,\varsigma))(z(\varsigma)-v(\varsigma))d\varsigma\\
&+\int_0^1(k_1(w,\varsigma)-k_2(w,\varsigma))(p(\varsigma)-r(\varsigma))d\varsigma\\
&+\int_0^1(k_1(w,\varsigma)-k_2(w,\varsigma))(z(\varsigma)-y(\varsigma))d\varsigma\Big].
\end{aligned} \quad (54)$$

Defining

$$(I)=\int_0^1(k_1(w,\varsigma)-k_2(w,\varsigma))(p(\varsigma)-u(\varsigma))d\varsigma,\quad (II)=\int_0^1(k_1(w,\varsigma)-k_2(w,\varsigma))(v(\varsigma)-y(\varsigma))d\varsigma,$$
$$(III)=\int_0^1(k_1(w,\varsigma)-k_2(w,\varsigma))(u(\varsigma)-r(\varsigma))d\varsigma,\quad (IV)=\int_0^1(k_1(w,\varsigma)-k_2(w,\varsigma))(z(\varsigma)-v(\varsigma))d\varsigma,$$
$$(V)=\int_0^1(k_1(w,\varsigma)-k_2(w,\varsigma))(p(\varsigma)-r(\varsigma))d\varsigma,\quad (VI)=\int_0^1(k_1(w,\varsigma)-k_2(w,\varsigma))(z(\varsigma)-y(\varsigma))d\varsigma.$$

and using the Cauchy–Schwartz inequality in (I), we obtain

$$\begin{aligned}(I)&\leq\Big(\int_0^1(k_1(w,\varsigma)-k_2(w,\varsigma))^2d\varsigma\Big)^{\frac{1}{2}}.\Big(\int_0^1(p(\varsigma)-u(\varsigma))^2d\varsigma\Big)^{\frac{1}{2}}\\&\leq\parallel k_1-k_2\parallel_\infty .G(p,u,r).\end{aligned}\tag{55}$$

With similar procedure, we can get the accompanying estimate for

$$(II)\leq\parallel k_1-k_2\parallel_\infty .G(y,v,z),\tag{56}$$
$$(III)\leq\parallel k_1-k_2\parallel_\infty .G(p,u,r),\tag{57}$$
$$(IV)\leq\parallel k_1-k_2\parallel_\infty .G(y,v,z),\tag{58}$$
$$(V)\leq\parallel k_1-k_2\parallel_\infty .G(p,u,r),\tag{59}$$
$$(VI)\leq\parallel k_1-k_2\parallel_\infty .G(y,v,z).\tag{60}$$

from (54)–(60), we have

$$\begin{aligned}G(Y(p,y),Y(u,v),Y(r,z))&\leq max(\lambda,\mu)\parallel k_1-k_2\parallel_\infty[G(p,u,r)+G(y,v,r)\\&\quad+G(p,u,r)+G(y,v,r)\\&\quad+G(p,u,r)+G(y,v,r)]\\&=3.max(\lambda,\mu)\parallel k_1-k_2\parallel_\infty[G(p,u,r)+G(y,v,r)]\\&\leq 3.max(\lambda,\mu)\parallel k_1-k_2\parallel_\infty[(N((p,y),(u,v),(r,z))].\end{aligned}\tag{61}$$

From (61) and assumption (vi) give us

$$G(Y(p,y),Y(u,v),Y(r,z))\leq\frac{1}{2}(N((p,y),(u,v),(r,z))),$$

where

$$\begin{aligned}N((p,y),(u,v),(r,z))=max(&G(p,u,r),G(y,v,z),G(p,Y(p,y),Y(p,y)),\\&G(u,Y(u,v),Y(u,v)),G(r,Y(r,z),Y(r,z)),\\&G(y,Y(y,p),Y(y,p)),G(v,Y(v,u),Y(v,u)),\\&G(z,Y(z,r),Y(z,r))\end{aligned}$$

or, equivalently,

$$G(Y(p, y), Y(u, v), Y(r, z)) \leq (N((p, y), (u, v), (r, z))^2 - [(N((p, y), (u, v), (r, z))^2 - \frac{1}{2}(N((p, y), (u, v), (r, z)))]. \tag{62}$$

Put $\psi(p) = p$, $\alpha(p) = p^2$ and $\beta(p) = p^2 - \frac{1}{2}p$. Obviously, ψ, α, and β are satisfying conditions of Corollary (3.1) and from (62) we get

$$\psi(G(Y(p, y), Y(u, v), Y(r, z))) \leq \alpha(N((p, y), (u, v), (r, z))) - \beta(N((p, y), (u, v), (r, z)))),$$

where

$$N((p, y), (u, v), (r, z)) = max(G(p, u, r), G(y, v, z), G(p, Y(p, y), Y(p, y)), G(u, Y(u, v), Y(u, v)), G(r, Y(r, z), Y(r, z)), G(y, Y(y, p), Y(y, p)), G(v, Y(v, u), Y(v, u)), G(z, Y(z, r), Y(z, r)).$$

This demonstrates that the operator Y fulfills the contractive condition showing up in Corollary (3.1).

Lastly, let γ, δ be the functions appearing in assumption (v); then, by (v), we get

$$\gamma \leq Y(\gamma, \delta), \delta \geq Y(\delta, \gamma).$$

Applying Corollary (3.1), we deduce the existence of $(p, y) \in S \times S$ such that $p = Y(p, y)$ and $y = Y(y, p)$, that is, (p, y) is a solution of the system (51).

5 Results and Discussion

The main result of this paper is on coupled coincidence point for maps having a mixed J-monotone property under the framework of generalized metric spaces equipped with a partial order. After this, we have demonstrated the uniqueness of coupled common fixed points of the assumed maps. Some of our outcomes modify and extend comparable outcomes in the literature. The results presented in this paper have been enforced to get the solution of an integral equation.

6 Conclusions

We established some coupled coincidence point results for maps having a property of mixed J-monotone fulfilling a condition of nonlinear contractive type in the framework of ordered generalized metric spaces.

References

1. Abbas, M., Khan, M.A., Radenovic, S.: Common coupled fixed point theorem in cone metric space for w-compatible mappings. Appl. Math. Comput. **217**, 195–202 (2010)
2. Abbas, M., Khan, A.R., Nazir, T.: Coupled common fixed point results in two generalized metric spaces. Appl. Math. Comput. **217**, 6328–6336 (2011)
3. Alber, I.Y., Guerre-Delabriere, S.: Principle of weakly contractive maps in Hilbert spaces. In: Gohberg, I., Lyubich, Y. (eds.) New Results in Operator Theory, Advanced and Applications, vol. 98, pp. 7–22 (1997)
4. Aydi, H., Damjanovic, B., Samet, B., Shatanawi, W.: Coupled fixed point theorems for nonlinear contractions in partially ordered G-metric spaces. Math. Comput. Model. **54**, 2443–2450 (2011)
5. Aydi, H., Postolache, M., Shatanawi, W.: Coupled fixed point results for (ψ, ϕ)-weakly contractive mappings in ordered G-metric spaces. Comput. Math. Appl. **63**, 298–309 (2012)
6. Bhaskar, T.G., Lakshmikantham, V.: Fixed point theorems in partially ordered metric spaces and applications. Nonlinear Anal. Theory, Methods Appl. **65**(7), 1379–1393 (2006)
7. Cho, Y.J., Rhoades, B.E., Saadati, R., Samet, B., Shatanawi, W.: Nonlinear coupled fixed point theorems in ordered generalized metric spaces with integral type. Fixed Point Theory Appl. **8**:1–14 (2012)
8. Choudhury, B.S., Kundu, A.: A coupled coincidence point result in partially ordered metric spaces for compatible mappings. Nonlinear Anal.: Theory, Methods Appl. **73**, 2524–2531 (2010)
9. Choudhury, B.S., Kundu, A.: (ψ, α, β)-weak contractions in partially ordered metric spaces. Appl. Math. Lett. **25**(1), 6–10 (2012)
10. Choudhury, B.S., Maity, P.: Coupled fixed point results in generalized metric spaces. Math. Comput. Model. **54**, 73–79 (2011)
11. Choudhury, B.S., Konar, P., Rhoades, B.E., Metiya, N.: Fixed point theorems for generalized weakly contractive mappings. Nonlinear Anal.: Theory, Methods Appl. **74**, 2116–2126 (2011)
12. Ciric, L., Lakshmikantham, V.: Coupled random fixed point theorems for nonlinear contractions in partially ordered metric spaces. Stoch. Appl. **27**, 1246–1259 (2009)
13. Ciric, L., Lakshmikantham, V.: Coupled fixed point theorems for nonlinear contractions in partially ordered metric spaces. Nonlinear Anal. Theory, Methods Appl. **70**(12), 4341–4349 (2009)
14. Ding, H.S., Li, L., Radenovic, S.: Coupled coincidence point theorems for generalized nonlinear contraction in partially ordered metric spaces. Fixed Point Theory Appl. **170**, 1–9 (2012)
15. Doric, D.: Common fixed point for generalized (ψ, φ) -weak contractions. Appl. Math. Lett. **22**, 1896–1900 (2009)
16. Dutta, P.N., Choudhury, B.S.: A generalization of contraction principle in metric spaces. Fixed Point Theory Appl. (2008). Article ID 406368
17. Gu, F., Yin, Y.: A new common coupled fixed point theorem in generalized metric space and applications to integral equations. Fixed Point Theory Appl. (2013). Article ID 266
18. Harjani, J., Sadarangani, K.: Fixed point theorems for weakly contractive mappings in partially ordered sets. Nonlinear Anal.: Theory, Methods Appl **71**, 3403–3410 (2009)
19. Harjani, J., Sadarangani, K.: Generalized contractions in partially ordered metric spaces and applications to ordinary differential equations. Nonlinear Anal.: Theory, Methods Appl. **72**, 1188–1197 (2010)

20. Khan, M.S., Swaleh, M., Sessa, S.: Fixed point theorems by altering distances between the points. Bull. Austral. Soc. **30**(1), 1–9 (1984)
21. Luong, N.V., Thuan, N.X.: Coupled fixed points in partially ordered metric spaces and application. Nonlinear Anal. Theory, Methods Appl. **74**, 983–992 (2011)
22. Luong, N.V., Thuan, N.X.: Coupled fixed point theorems in partially ordered G-metric spaces. Math. Comput. Model. **55**, 1601–1609 (2012)
23. Mustafa, Z., Sims, B.: A new approach to generalized metric spaces. J. Nonlinear Convex Anal. **7**(2):289–296 (2006)
24. Mustafa, Z., Sims., B.: Fixed point theorems for contractive mapping in complete G-metric spaces. Fixed Point Theory Appl. **10** (2009). Article ID 917175
25. Mustafa, Z., Obiedat, H., Awawdeh, F.: Some fixed point theorem for mapping on complete G-metric spaces. Fixed Point Theory Appl. (2008) Article ID189870
26. Naidu, S.V.R.: Some fixed point theorems in metric spaces by altering distances. Czech. Math. J. **53**(1), 205–212 (2003)
27. Nieto, J.J., Rodriguez-Lopez, R.: Contractive mapping theorems in partially ordered sets and applications to ordinary differential equations. Order **22**, 223–239 (2005)
28. Nieto, J.J., Lopez, R.R.: Existence and uniqueness of fixed point in partially ordered sets and applications to ordinary differential equations. Acta. Math. Sinica **23**, 2205–2212 (2007)
29. Ran, A.C.M., Reuring, M.C.B.: A fixed point theorem in partially ordered sets and some applications to matrix equations. Proc. Amer. Math. Soc. **132**, 1435–1443 (2004)
30. Razani, A., Parvaneh, V.: Coupled coincidence point results for ψ, α, β-weak contractions in partially ordered metric spaces. J. Appl. Math. (2012). Article ID 496103
31. Rhoades, B.E.: Some theorems on weakly contractive maps. Nonlinear Anal. Theory, Methods Appl. **47**(4), 2683–2693 (2001)
32. Saadati, R., Vaezpour, S.M., Vetro, P., Rhoades, B.E.: Fixed point theorems in generalized partially ordered G-metric spaces. Math. Comput. Model. **52**, 797–801 (2010)
33. Shatanawi, W.: Coupled fixed point theorems in generalized metric spaces. Hacet. J. Math. Stat. **40**(3), 441–447 (2011)
34. Shatanawi, W., Samet, B.: On (ψ, ϕ)-weakly contractive condition in partially ordered metric spaces. Comput. Math. Appl. **62**, 3204–3214 (2011)
35. Zhang, Q., Song, Y.: Fixed point theory for generalized φ-weak contractions. Appl. Math. Lett. **22**, 75–78 (2009)

Dynamic Pricing of Products Based on Visual Quality and E-Commerce Factors

S. Poornima, S. Mohanavalli, S. Swarnalatha, and I. Kesavarthini

Abstract Dynamic pricing is an approach in setting the real-time price for a product or service that is highly flexible. The goal of dynamic pricing is to allow a company that sells goods or services over the Internet to change prices in response to market demands. Dynamic pricing has become critical in e-commerce, mostly due to frequent change in demands and automation. The proposed pricing method attempts to scale up the profits of the firm by calculating an optimal price for the goods by considering numerous factors such as the supply and demand rates for a particular item, the availability of a particular item, competitor rates for the item, and the deviation from ideal image quality of the item. This pricing will help an e-commerce firm to have an unparalleled advantage over their peers and will also allow them to reap high profits. In the proposed method, visual aspects of the goods are included for better categorization and classification of the various items for sale into classes. This classification is done based on their supply and demand rate, availability of the item, the deviation from ideal image quality, etc. The algorithm used for classifying is label encoder and gradient boosting regression algorithm has been used for predicting the optimal prices. The proposed method has been validated and has a prediction accuracy rate of 97%.

Keywords Face · Gender · Feature extraction · Local binary pattern · Moments · Classification · SVM · AdaBoost · Rainforest

S. Poornima (✉) · S. Mohanavalli · S. Swarnalatha · I. Kesavarthini
Department of Information Technology, Sri Sivasubramaniya Nadar College of Engineering, Chennai, India
e-mail: poornimas@ssn.edu.in

S. Mohanavalli
e-mail: mohanas@ssn.edu.in

R. N. Mohapatra et al. (eds.), *Mathematical Analysis and Computing*,
Springer Proceedings in Mathematics & Statistics 344,
https://doi.org/10.1007/978-981-33-4646-8_34

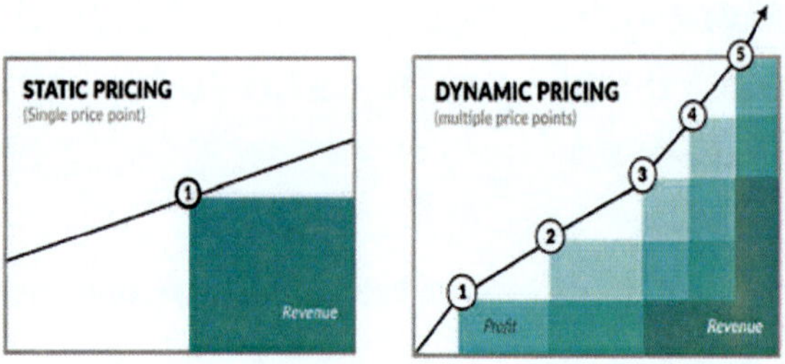

Fig. 1 Graphical representation of static and dynamic pricing. *Source* https://competitoor.com/dynamic-pricing-how-works

1 Introduction

Dynamic pricing is the latest pricing trend that has taken the e-commerce industry by storm. While it is not an entirely new concept and companies have been using it randomly over the years, it is certainly something that is much more relevant in the new age of e-commerce companies.

Simply put, dynamic pricing is a strategy in which product prices continuously adjust, sometimes in a matter of minutes, in response to real-time supply and demand. For example, Amazon, the global ecommerce giant, is one of the largest retailers to have adopted dynamic pricing and updates prices every 10 min.

There are two broad categories of pricing strategies available for online stores—static pricing and dynamic pricing (Fig. 1). In static pricing, price point is maintained for an extended time period, but it does not allow for adjustments during product or service delivery. The customer cannot realize that cost basis is higher than expected, instead pays the established price regardless of changes in your time or costs. Also, it does not allow for adjustments over time to sell off extra inventory or available seats for entertainment types of events. Whereas dynamic pricing (discriminatory pricing) allows to maximize profits with each customer and commonly used in event promotions. Also, dynamic pricing has the ability to adjust prices for fluctuating demand product or services like seafood distributors and restaurants.

2 Related Works

Dynamic pricing enables a firm to increase revenue by better matching supply with demand, responding to shifting demand patterns, and achieving customer segmentation. Throughout the history of commerce, charging different prices for identical goods has been a common practice in markets separated by geography or defined by distinguishable customer types [1]. Among many works on dynamic pricing, a classical work can be found in [2, 3] and an overview in [4]. In [1], it classifies the pricing strategy into three categories: skimming, neutral, and penetration, whereas in [5] the high value/high price and low value/low price are categorized as premium and economy, respectively. Samuel B. Hwang and Sungho Kim [6] used this classification in the development of their dynamic pricing algorithm. They focused their work on real-time adjustment of price by comparing the prices of our competitors, and not in charging different prices to different buyers. This system is based on a mathematical model of sales rate and competitor prices. It includes selection of a suitable pricing policy for optimal pricing of commodities in addressing the future demands better than random sampling [7]. The factors that impact the pricing as concluded from [8, 9, 10, 4] are: (i) item condition, (ii) shipping condition, (iii) inventory of commodity (demand, supply), and (iv) item type. The abovementioned factors have been incorporated in our model for dynamic pricing. However, all the above research failed to take into consideration how the visual quality of a product affected its pricing. In E-commerce, goods promised are not always the same as the goods delivered. There might be a minor deviation in color or a change is visual appearance. This might cause dissatisfaction among the customers. To overcome this, our pricing model considers the visual quality of the commodity as a parameter for optimal pricing.

3 Proposed System

The architecture of the pricing system has mainly five components, namely, data collection, preprocessing, exploration, feature engineering, and modeling as shown in Fig. 2.

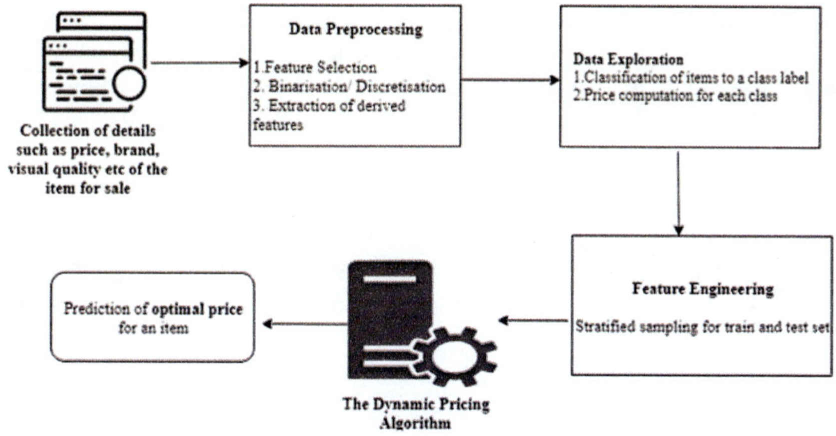

Fig. 2 Proposed system architecture

Table 1 Features in Mercari dataset

Feature	Description	Type
Id	Id of the listing	Numeric
Name	Name of the Product	Categorical
Item_condition_id	The conditions of the item provided by the seller	Numeric
Category_name	Category of the product	Categorical
Brand_name	Brand name of the product	Categorical
Price	The price of the product in USD	Numeric
Shipping	An indicator of whether the shipping is paid by the seller or buyer	Categorical
Item_description	The full description of the product	Categorical
Visual quality	Percentage difference of the visual image of the current product and the ideal product visual image	Numeric
Color	Color of the product	Categorical
Manufacturer	Name of the manufacturer	Categorical
Manufacturer_id	Id of the manufacturer	Numeric
Merchants	Names of the merchants to whom the manufacturers sold their products	Categorical
Prices.DateAdded	The date on which the prices were last updated	Date
Prices.Currency	The currency in which the product is priced, e.g., USD	Categorical
Weight	Weight of the product in kilograms	Numeric
Sizes	The sizes available for a particular product such as large, medium, small, etc	Categorical
Reviews	The reviews given about a particular product	Categorical
Prices.ReturnPolicy	Indicates whether returning the product involves additional payment from the customer's side or not	Numeric
Supply_demand	Indicates whether supply meets the demand for a particular product	Category

3.1 Data Collection

The data for this work was collected from the dataset provided by the Kaggle Competition—"Mercari Price Suggestion Challenge." Table 1 depicts the features provided by the Mercari dataset.

3.2 Data Preprocessing

Data preprocessing involves cleaning and structuring of data, so that the algorithm can predict the optimal prices of the products. As shown in Fig. 2, data preprocessing

Table 2 Features in dataset after feature selection

Feature	Description	Type
Id	Id of the listing	Numeric
Name	Name of the product	Categorical
Item_condition_id	The conditions of the item provided by the seller	Numeric
Category_name	Category of the product	Categorical
Brand_name	Brand name of the product	Categorical
Price	The price of the product in USD	Numeric
Shipping	An indicator of whether the shipping is paid by the seller or buyer	Categorical
Item_description	The full description of the product	Categorical
Visual quality	Percentage difference of the visual image of the current product and the ideal product visual image	Numeric
Supply_demand	Indicates whether supply meets the demand for a particular product	Category

has three functions which are being carried out—feature selection, binarization or discretization, and feature extraction.

Feature Selection

Feature selection is the process of selecting a subset of relevant features from the original set. The three main criteria for selection of a feature are: informativeness, relevance, and non-redundancy. Table 2 shows few essential features selected from the dataset.

The feature visual quality in the above table is created by scraping the image of the product from the image URL provided in the dataset and finding the average pixel difference of the fetched image from the ideal image of the product provided previously by vendor. The image comparison algorithm is used to compare the scraped image and the ideal image. It computes the average pixel difference of the two images. The sample ideal image of the product and the image of the same product which the vendor wants to sell are shown in Fig. 3.

Fig. 3 Ideal and scraped image of the product to be sold

Binarization and Discretization

Discrete values have important roles in data mining and knowledge discovery, since more concise to represent and specify, easy to use, and comprehend than continuous values. Discretization can lead to improved predictive accuracy and also many induction algorithms found in the literature require discrete features. As cited in [10], a typical discretization process broadly consists of four steps, namely, (i) sorting the continuous values of feature to be discretized, (ii) evaluating a cut-point for splitting/adjacent intervals for merging, (iii) splitting or merging intervals of continuous value based on criterion, and (iv) stopping at some point.

Pricing algorithm required the categorical feature "Item_condition_id" and "visual quality" to be discretized. The following distinction has been arrived by the methods described above: for feature "item_condition_id," we obtain 0 for the no description, 1 for the description "new," 2 for the description "new with tags," 3 for the description "new with boxes." For feature "visual quality," we obtain 0 for a percentage difference in visual quality of the product from the ideal image which is equal to 0, we obtain 1 for a difference less than 2%, we obtain 2 for a difference less than or equal to 5%, and we obtain 3 for a difference greater than 5%. Binarization is done for two features, namely, "shipping" and "Supply_demand." When the shipping cost is taken into account by the seller, the value of this feature for that item is 0 and it is made 1 when the shipping cost is not taken up by the seller. In the case of the feature "supply_demand," the item is inputted 1 for this feature when the supply meets the demand and 0 otherwise.

Feature Extraction

Feature extraction is the process of deriving new features either as simple combinations of original features or as a more complex mapping from the original set to the new set. These are the features which are finally present in in the dataset after preprocessing: int_name, cond, int_desc, new, was_described, cat_lenn, shoes, item_condition_id, brand_name, shipping, des_len, name_len, mean_des, word_count, word_name, sub1, sub2, supply_demand, img_score, category name, and sale.

3.3 *Data Exploration*

According to definition in [11], data exploration is the first and foremost step in data analysis, commonly conducted by visual analytics tools for summarizing the main characteristics of a data set such as size, accuracy, initial patterns in the data, and other attributes. Thereafter, the relationships between the different variables are uncovered and organizations can continue the mining process by creating and deploying data models to take action on the insights gained [11]. This proposed work employs classification using label encoders and computes the price range, the average, and median price of top brands.

3.4 *Feature Engineering*

Dataset is split into two separate sets, Training Set and Test Set, in the ratio of 70:30. Here it is the algorithm model that is going to learn from the data to make predictions. There are many ways to select a sample—simple random sampling, stratified random sampling, cluster random sampling, and systematic random sampling. As mentioned in [12], stratified sample guarantees that members from each group will be represented in the sample, so this sampling method is good when we want some members from every group. Thus, this project utilizes stratified sampling for feature engineering process.

3.5 *Modeling*

The prediction of prices of products is done by taking a number of factors into consideration such as supply rate, demand rate, visual quality, item sale conditions, item type, brand name, and sale. Regression is a technique from statistics that is used to predict values of a desired target quantity when the target quantity is "continuous." It is obviously the case that the target quantity can be complex and typically multi-dimensional in contrast to this simple example. There are different types of regressors, namely, linear regression, logistic regression, polynomial regression, stepwise regression, ridge regression, lasso regression, ElasticNet regression, and gradient regression.

Gradient Regression (GR) is used to build an additive model in a forward stage-wise fashion, allowing optimization of arbitrary differentiable loss functions. In each stage, a regression tree is fit on the negative gradient of the given loss function [13]. The main focus of GR is to minimize the loss by adding weak learners using a gradient descent. In this stage-wise additive GR model, one new weak learner is added at a time and existing weak learners are left unchanged. Gradient boosting involves three different functions, namely, loss function, weak learner, and additive model. Loss function is differentiable based on the type of problem being solved and regression may use squared error. Decision trees are used as the weak learner in gradient boosting and the residuals can be corrected from the subsequent outputs obtained in the predictions. Trees are constructed in a greedy manner, choosing the best split points based on purity scores. In this additive model, trees are added one at a time using gradient descent-tree parameterization procedure in order to minimize the loss, but the existing trees remain unchanged. In this proposed work, gradient boosting regressor [14] is used for predicting the prices of the products in the test dataset.

4 Process Flow

This section deals with implementation of the system using various algorithms and discusses about various performance metrics of the results predicted by the system. The process flow of the proposed system is shown in Fig. 4.

The proposed system gets the details of the products to be sold and their various parameters such as previous rates, shipping conditions, item description, item condition, etc. Around 4,00,000 product details were collected from kaggle. Collected data was preprocessed by binarization and discretization methods. Additional feature, "visual quality," is computed by taking the average pixel difference of the given image from the ideal image of the same product. The feature "visual quality" is also discretized by preprocessing methods. Preprocessed data is divided into training datasets and testing datasets. Around 1000 datasets were used as training data. Varied range of data (100, 200, 500, 1000) were used as testing datasets. Training and testing data were given to label encoder algorithms for categorizing the products into various classes. The learning is done by gradient boosting ensemble and the prices are predicted optimally.

The prices of the products are computed in such a manner so as to bring maximum profits to the retailers and at the same time ensuring to increase the sale of the product.

In reference to [15], gradient boosting is a machine learning technique for regression and classification problems producing a prediction model [14] in the form of an ensemble of weak prediction models. The objective of any supervised learning algorithm is to define a loss function and minimize it. Considering mean squared

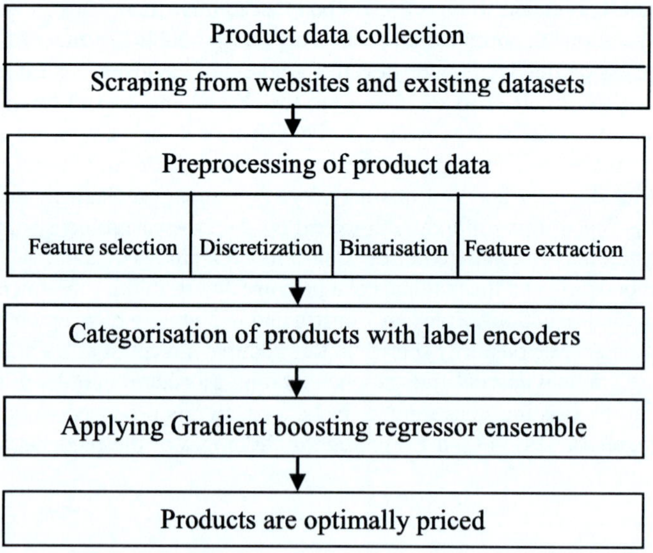

Fig. 4 Flowchart of the proposed model

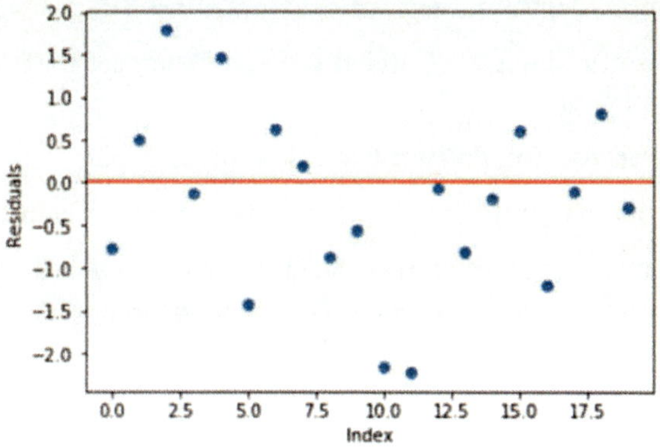

Fig. 5 Sample residuals with mean around 0. *Source* https://medium.com/mlreview/gradient-boosting-from-scratch-1e317ae4587d

error (MSE) as loss which is defined as [15] follows:

$$Loss = MSE = \sum \left(y_i - y_i^p\right)^2 \tag{1}$$

where, $y_i = i$th target value, $y_i^p = i$th prediction, $L\left(y_i - y_i^p\right)$ is Loss function.

Using **gradient descent** and updating our predictions based on a learning rate, we can find the values where MSE is minimum [15].

$$y_i^p = y_i^p + a * \delta \sum \left(y_i - y_i^p\right)^2 / \delta y_i^p \tag{2}$$

which becomes,$y_p^i = y_p^i - a * 2 * \sum \left(y_i - y_i^p\right)$.

where, α is learning rate and $\sum \left(y_i - y_i^p\right)$ is sum of residuals.

So, updating the predictions, viz., the sum of our residuals is close to 0 (or minimum) and predicted values are sufficiently close to actual values.

Intuition Behind Gradient Boosting

Sum of its residuals is assumed initially to be 0 in linear regression (Fig. 5). The intuition behind gradient boosting algorithm is to repetitively leverage the patterns in residuals and strengthen a model with weak predictions and make it better. Once we reach a stage that residuals do not have any pattern that could be modeled, we can stop modeling residuals (otherwise it might lead to overfitting). Algorithmically, minimizing the loss function, such that test loss reaches its minima. The gradient regression function is used to predict prices with the following parameters [15]:

clf = ensemble.GradientBoostRegressor (learning_rate = 0.7, n_estimators = 700, max_depth = 4, warm_start = True, verbose = 1, random_state = 45, max_features = 0.8).

learning_rate: float, optional (default = 0.1.

estimators: int (default = 100).

max_depth: integer, optional (default = 3).

warm_start: bool, default: False When set to True, reuse the solution of the previous.

verbose: int, default: 0 Enable verbose output.

random_state: int, RandomState instance or None, optional (default = None).

max_features: int, float, string or None, optional (default = None)—Number of features to consider when looking for the best split.

5 Implementatıon

The pricing method is implemented using HTML, CSS, and Python packages. The input data must be specified as shown in Table 2. The input data is preprocessed using the data preprocessing methods. The preprocessed data is then fed to the gradient boosting regressor ensemble to compute the predicted price.

5.1 Results and Discussions

The system is designed to classify and predict the price for desired categories like shoes, clothing, and accessories. The system is trained using the features listed in above sections. GBR model is used to train the system with 500 observations and 150 observations to the test holdout. The next step is to train a GBR model with 100 trees assumption and ensemble library and two more additional arguments, namely, *max_depth* and *learning rate*. Max Depth specifies the maximum depth of each tree (i.e., highest level of variable interactions allowed while training the model). Learning rate is used for reducing/shrinking the impact of each additional fitted base learner (tree). It reduces the size of incremental steps and thus penalizes the importance of consecutive iteration.

Prediction

For a sample training and testing dataset of size 100 and 30, respectively, the predicted results are as given in Table 3. The predict function of GBR is used for predicting the prices.

Average percentage difference in price for the above test set is **15.09 approx** and visualization of the price between actual and predicted is shown in Fig. 6.

Analyzing The Predictions

The most important features that contribute in affecting the predicted price of a product are as follows:

Table 3 Predicted prices of products

Id	Name	Price predicted	Percentage difference
0	Miles Kimball Adjustable Memory Foam Slippers	13.81	18.74
1	Bedroom Athletics Bale Slippers Men Round Toe Canvas Black Slipper	27.67	30.83
2	Isotoner Women's Brushed Sweater Knit Clog Slippers	27.51	5.81
3	Isotoner Women's Brushed Sweater Knit Clog Slippers	20.46	3.74
4	Easycomforts Easy Comforts Style Memory Foam Slippers	21.52	5.55
5	Saucony Women's Shadow Original Ankle-high Synthetic Fashion Sneaker	35.01	28.43
6	Muk Luks Womens Jane Suede Moccasin	46.19	2.79
7	Muk Luks Womens Jane Suede Moccasin	47.84	1.78
8	Reebok Skyscape Chase Women Us 9.5 Black Sneakers	22.99	0
9	Lacoste Womens Gazon W5 Sneakers In Navy	87.03	2.45
10	Lacoste Womens Gazon W5 Sneakers In Navy	84.95	0
11	Guess Hadly Women Us Green Winter Boot	46.99	0
12	Guess Hadly Women Us Green Winter Boot	63.97	8.43
13	Zoot Tt Trainer 2.0 Round Toe Synthetic Sneakers	73.16	1.63
14	Zoot Tt Trainer 2.0 Round Toe Synthetic Sneakers	250	0
15	Ugg Australia Airehart Women Us 9 Brown Snow Boot	137.24	14.69
16	Madeline Taken For A Ride Women Round Toe Synthetic Knee High Boot	56.54	24.61
17	Style & Co Andoraa Open-toe Synthetic Heels	66.22	12.23
18	Bearpaw Womens Knit Tall Sheepskin Fold Over Knee-high Boot	74.45	6.36
19	Bearpaw Womens Knit Tall Sheepskin Fold Over Knee-high Boot	49.99	0
20	Baretraps Mirabella Open-toe Leather Slingback Sandal	43.99	0
21	Baretraps Mirabella Open-toe Leather Slingback Sandal	15.76	12.42
22	Unique Bargains Women's Open Toe Cutout Chunky Heel Lace-up Sandals Brown (size 8)	127.09	76.54
23	Coloriffics Danica Peep-toe Canvas Heels	2.99	0

(continued)

Table 3 (continued)

Id	Name	Price predicted	Percentage difference
24	Naot Women Sheryl Sandals	132.3	8.76
25	Naot Women Sheryl Sandals	157.89	0
26	Cesare Paciotti Womens Kid Suede Pumps Heels Shoes	530.15	23.17
27	Michael Michael Kors Mavis Back Zip Women Open-toe Leather Gold Heels	225	0
28	Michael Michael Kors Mavis Back Zip Women Open-toe Leather Gold Heels	91.87	20.9
29	Jessica Simpson Mariani Women Open-toe Canvas Heels	79.94	135.18
30	Jessica Simpson Apple Pumps Hi Heel Women's Shoes,black Color, Size: 10 m,new	19.34	22.6

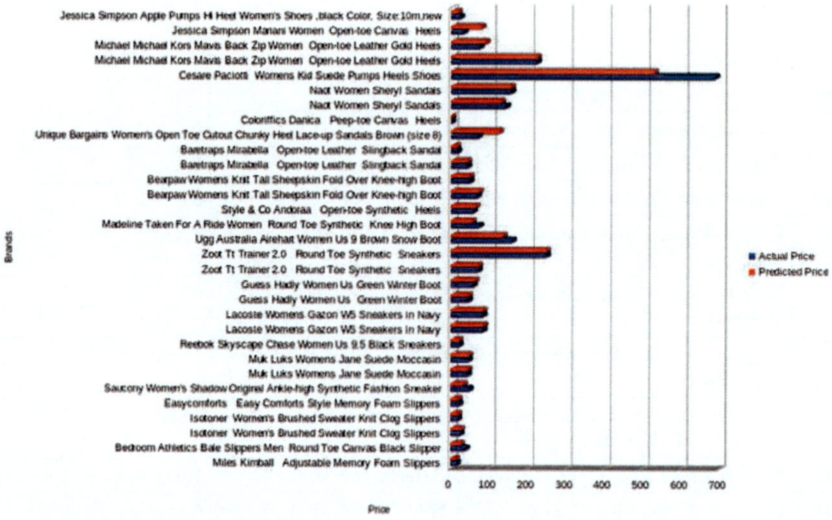

Fig. 6 Visualization of sample results

1. Sale,
2. Img_score,
3. Supply_demand,
4. Shipping, and
5. Item_condition_id.

A few samples from the dataset are considered for analyzing how these features affect the price (Tables 4, 5, 6) for different products.

Table 4 Analysis of effect of features on prices for product 1

Product name	Jessica Simpson Mariani women open-toe canvas heels
Sale	1 (Item on sale)
Img_score	1 (Minor visual deviation)
Supply_demand	0 (Supply meets demand)
Shipping	1 (Shipping cost paid by seller)
Item_condition_id	1 (New item)
Actual price	79.93
Predicted price	33.99 approx
Analysis	The predicted price has a difference of almost 57% as the item is new and is on sale, and has a small deviation visually from the ideal product and the shipping cost is to be paid by the retailer or seller. All these factors contribute to the extreme spike in the price

Table 5 Analysis of effect of features on prices for product 2

Product name	Isotoner women's brushed sweater knit clog slippers
Sale	0 (Item is not on sale)
Img_score	0 (No visual deviation)
Supply_demand	1 (Demand surpasses supply)
Shipping	0 (Shipping cost paid by customer)
Item_condition_id	1 (New item)
Actual Price	26
Predicted price	27.51 approx
Analysis	The predicted price is 5.8% higher than the actual price as the item is not on sale, the image of the item is ideal, the demand for the item is high, shipping cost is paid by customer, and the item is new. The demand for the product has caused this spike in price

Table 6 Analysis of effect of features on prices for product 3

Product name	Baretraps mirabella open-toe leather slingback sandal
Sale	1 (Item is on sale)
Img_score	3 (Major visual deviation)
Supply_demand	0 (Supply meets demand)
Shipping	1 (Shipping cost paid by seller)
Item_condition_id	1 (New item)
Actual price	17.99
Predicted price	15.76 approx
Analysis	The predicted price is 12.4% lesser than the actual price as the item is on sale, there is a major visual deviation, the demand meets supply, shipping cost is paid by seller, and the item is new

6 Conclusion and Future Work

This work uses different features such as sale, shipping, supply, demand, visual quality to dynamically price products of different category by using existing machine learning algorithm, namely, label encoder to categorize the products and gradient boosting regression method to predict the prices. The strengths of the proposed work are: (i) usage of large datasets, (ii) pricing for various categories of items, (iii) pricing of items without compromising the brand value of the user, (iv) prices predicted by algorithm will ensure significant improvement in sales, and (v) significant improvements over the performance metrics of the existing machine learning algorithms. Deviation of the predicted price from the actual prices is less than 10%. The critical features that affect the pricing are identified as Sale, Img_score, Supply_demand, Shipping, and Item_condition_id.

This work can be extended in several dimensions. Incorporating more training data would improve the relevance and categorization of products. Feature extraction can be done in more enhanced way and additional variants of the ensemble classifier such as artificial neural networks or decision trees can be explored and more detailed classification can be done. Implementation of associative rule mining techniques could also be incorporated in making the pricing more efficient.

References

1. Weiss, R.M., Mehrotra, A.K.: Online dynamic pricing: efficiency, equity and the future of E-commerce. Va. J. Law Technol., Summer (2001)
2. Engelson, M.: Pricing Strategy: An Interdisciplinary Approach. Joint Management Strategy, London, England (1995)
3. Nagle, T.T., Holden, R.K., Holden, R.: The Strategy and Tactics of Pricing: A Guide to Profitable Decision Making, 3rd edn. Prentice Hall, Upper Saddle River, New Jersey (2002)
4. Dynamic Pricing Overview. https://www.managingchange.com/dynamic/overview.html
5. Pricing strategies. https://www.marketingteacher.com/Lessons/lesson_pricing.htm
6. Hwang, S.B, Kim, S., Dynamic pricing algorithm for E-commerce. Advances in Systems, Computing Sciences and Software Engineering
7. Lobel, R., Perakis, G.: Dynamic pricing through sampling based optimization. Available SSRN 1748426 (2010)
8. Chen, Y., Farias, V.F.: Robust dynamic pricing with strategic customers. Math. Oper. Res. **43**(4), 1119–1142 (2018)
9. Elmaghraby, W., Keskinocak, P.: Dynamic pricing in the presence of inventory considerations: research overview, current practices, and future directions. Manag. Sci. **49**(10), 1287–1309 (2003)
10. Liu, H., et al.: Discretization: an enabling technique. Data Min. Knowl. Discov. **6**(4), 393–423 (2002)
11. Rouse, M.: Data exploration. https://searchbusinessanalytics.techtarget.com/definition/data-exploration
12. Marshall, M.N.: Sampling for qualitative qualitative research. Fam. Pract. **13**(6), 522–526 (1996)
13. Friedman, J.: Greedy function approximation: a gradient boosting machine. Ann. Stat. **29**(5) (2001)

14. Elith, J., Leathwick, J.R., Hastie, T.: A working guide to boosted regression trees. J. Anim. Ecol. **77**(4), 802–813 (2008)
15. Grover, P.: Gradient boosting from scratch. https://medium.com/mlreview/gradient-boosting-from-scratch-1e317ae4587d

Analysis of Audio-Based News Classification Using Machine Learning Techniques

S. Divya, R. Raghavi, N. Sripriya, S. Mohanavalli, and S. Poornima

Abstract Text classifiers can automatically analyse text using Natural Language Processing (NLP) techniques and then assign categories based on its content. Applying machine learning techniques in the field of NLP has achieved appreciable results. In this work, a system for analysing and classifying news videos based on the audio content using machine learning techniques has been presented. It assists the user to find the genre of a news video without watching it. In the proposed work, NLP techniques are utilized to identify the most correlated unigrams and bigrams, TF-IDF which are the features used to train the model using the machine learning techniques such as Multinomial Naïve-Bayes classifier, Logistic Regression and Support Vector Machines. The performance of various classifiers in classifying the news videos are analysed and presented here. For this purpose, a dataset has been collected, which consists of 25 News videos of CNN news channel which covers almost five categories. However, the classifier models are trained using text news data obtained from BBC news articles. The accuracy of the classifiers is tested for both BBC text news and also for the text news extracted from news video. The experimental results convey that the multinomial Naive-Bayes classifier outperforms the other classifier models for both the noisy and noiseless text input.

S. Divya (✉) · R. Raghavi · N. Sripriya · S. Mohanavalli · S. Poornima
Department of Information Technology, Sri Sivasubramaniya Nadar College of Engineering, Kalavakkam, Chennai, India
e-mail: divya_stephenson@yahoo.com

R. Raghavi
e-mail: raghaviraman7@gmail.com

N. Sripriya
e-mail: sripriyan@ssn.edu.in

S. Mohanavalli
e-mail: mohanas@ssn.edu.in

S. Poornima
e-mail: poornimas@ssn.edu.in

R. N. Mohapatra et al. (eds.), *Mathematical Analysis and Computing*,
Springer Proceedings in Mathematics & Statistics 344,
https://doi.org/10.1007/978-981-33-4646-8_35

Keywords News · Text processing · Text classification · Support vectors · Multinomial Naïve-Bayes · Logistic regression · MLP classifier

1 Introduction

Textual data representation is the common source of information, in which around 80% of those are unstructured data. For the purpose of making text data understandable, text classification is vital to structure the textual data for further processing.

Text classification assigns predefined categories to the available text data, which will help in structuring and arranging it. Text classification can be done either manually or automatically. Manual text classification utilizes a human who interprets the text content and categorizes it accordingly. This method usually can provide quality results but it takes more time and cost that mandates the need for automatic classifiers. Following are a few applications of text classification: Understanding the review of users from social media, identification of spam and non-spam emails, automatically tagging customer queries, and categorizing news articles.

Text classification algorithm is vital among many software systems that perform text processing. The major task is topic classification which categorizes a text document into one of a predefined set of topics. Topic classification is done by mainly identifying the keywords in the text data. Yet another type of classification is sentiment analysis, which aims at the detection of the conflict in the text content, the contextual view it conveys.

Various automatic text classification systems are Rule-based systems, Machine Learning-based systems and Hybrid systems.

Rule-based systems utilize a set of manually generated rhetoric rules, which instructs the model to practice verbally relevant features of a text for the purpose of identifying appropriate categories based on its context. Each rule consists of an antecedent or pattern and a predicted category. This system is human-understandable and gets updated later. The drawback of this system: in-depth domain knowledge is required which lags for the beginners to work on it. Formulating rules for a complicated system is difficult and should be analysed deeply. This system has complexity in maintenance and also has the possibility of creating some negative impact on the results generated from the pre-existing rules, while new rules are added.

Machine Learning-based systems categorize the text on the basis of its experience gained from the past data. The training data is labelled which is utilized by the machine learning algorithm to analyse the relationship between each term and the label assigned for that input. Classification done using machine learning is generally accurate than manually designed rule systems.

Hybrid system is a combination of machine learning-based system and rule-based system, which further improves the classification outcome. Fine-tuning is smoothly applied by including certain rules for the contradictory labels which are not rightly stated by the base classifier.

Text classification is generally stated as an instance of a supervised machine learning model. A labelled dataset (comprising of input text and label) is fed as an input to the classifier. The steps involved in the classification task are as follows: 1. Preparing the dataset: Dataset is identified with the supporting format and few primary pre-processing is performed. 2. Feature extraction: Splitting of dataset as training data and test data is performed which is followed by the transformation of training data into flat features that may be extracted. 3. Training the model: The model is trained using the training data features. 4. Testing the model: Evaluation of the model for test data is performed. 5. Performance Analysis: Performance metrics are used to evaluate the classification accuracy.

The primary objective of this work is to provide users the genre of a news video. This categorized news can be further used for processing like information retrieval, summarization of long news video, etc. In this work, the classifier model is built using a text news dataset. The news videos for which genre have to be identified is converted to text and is given to the model during the testing phase. The model categorizes the converted text content into any one of the identified categories. Further, the news video classification was carried out using various machine learning classifiers and their results are compared and analysed.

2 Literature Survey

In the recent trend, text classification is a vital application of machine learning in the area of identifying the news genre. Generally, texts are represented as feature vectors before it is applied with the machine learning methods. This representation of a term as vector is termed as embedding, which is the main technique for managing and arranging the input text data. This strategy is basically used for feature selection which takes a role in improving the classification performance.

As an extension of Inverse Document Frequency (IDF), Weighted Inverse Document Frequency (WIDF) [1], which is a term weighting method, is used to calculate the frequency of each term in the collection of text. WIDF is proven to be advanced among the various other methods. The selection of feature based on a training set is applied for the purpose of dimensionality reduction and to increase the performance of classification. Chi-Square (CHI) and Information Gain (IG) are some of the various feature selection methods [2] used for classification purposes. This portraits the association between class labels related to the document and to the absence/presence of a term within the input document on the basis of statistical and information theory.

Conditional Mutual Information max–min (CMIM) [3], is a novel feature selection method, which can choose the individually discerning and independent features. This method surpasses the traditional methods used for feature selection. Extensive works to compare a variety of term weighting scheme with SVM [4] is presented and a new scheme known as tf-rf is derived to increase the discerning power of each word. This scheme outperforms the other term weighting schemes.

Latent Dirichlet Allocation (LDA) [5] reduces the dimensions of text to extract features with the help of topic model. Softmax regression algorithm is incorporated in this method for obtaining the solution for multi-class problems. Probability-based models work with an assumption that each article to be a collection of plurality in the mixture of topics and in the identification of hidden information topic with large document. Every text has the topic being selected from topic distribution and a word being chosen from word distribution until traversing every word in the document. Evaluation of classification is done using the metrics such as precision and F1-Measure. Bayesian classification method is applied [6] to choose the indicative feature sub-set for each class with the help of label-specific features to automatically classify the text. Baggenstoss's Probability Density Function (PDF) projection theorem is finally applied. Binary valued feature model has either 0 or 1 as its label to specify whether the term is available in the document or not. The other model is the real-valued feature model in which the feature indicates the Term Frequency (TF), which is the count of the term available in the document. Study projects that the real-valued feature model provides better performance than binary-valued feature model.

After the specific representation and the feature selection, the ML algorithm can be applied to the input. Various text classifiers are developed using machine learning techniques, probabilistic models, etc. Automated text classification is a major field of research, even after the invention of a variety of approaches because of the faulty classifier which needs further improvement.

The performance of Naïve-Bayes classifier [7] is increased by estimating the parameter for the document model and normalizing the length of the document to ignore the problems in the traditional approaches. As an added method, the development of mutual information-weighted Naïve-Bayes classifier is stated to increase the impact of most informative words. Generally, Naïve-Bayes is used in text classification applications due to its simple procedure and effective outcome [8]. But its performance is low since the text modelling is not done well. On the basis of tree-like Bayesian networks [9], the evaluation of Bayesian multinet classifier produces the results which suggest the ability to handle classification in large variables, with appreciable speed and accuracy. Support Vector Machine (SVM) is applied to classification task results in good precision but bad recall. The way to improve recall is to customize the SVM by adjusting its threshold [10]. This gives a better result which describes a fast Decision Tree construction algorithm that takes advantage of the sparsity of text data. In addition to it, a rule simplification method is applied that transforms the decision tree into a logically equivalent rule set [11].

The news articles are deeply analysed [12] and an approach is presented that helps the user to identify the news articles related to a particular category. Results of various algorithms such as Naïve-Bayes, Decision Tree, K-Nearest Neighbour algorithms are stated [13] and the performance is also evaluated. Comparatively, Decision Tree algorithm produces accurately classifies the text.

Many text classification algorithms [14–16] are surveyed and a novel keyword extraction method [17, 18] is proposed which concludes by examining various factors. Categorization experiments are done on noisy texts [19] and are tested and presented. Any erroneous text obtained from the extraction process from some other source like media or recognition system is stated as noisy text. When the classification performance is compared among the clean and noisy (Word Error Rate between ~10 and ~50%) version of the same document, the acceptable performance loss is shown.

3 Machine Learning Classifiers

Classifier which works with the help of past experience is the machine learning-based system for classification. This does not depend on any hand-crafted rules for performing classification, instead learns from experience to perform classification. This system utilizes the pre-labelled dataset which includes the text and its relevant label. The label states the association between the text and the label assigned to it. Initially, the extraction of feature is done, in which, the transformation of text into its representative numerical form is done in the form of a vector. Now, the machine learning algorithm is trained using the training data that includes a pair of feature sets and categories to produce a classification model.

After the training is done, the model can initiate its prediction. The same feature selection is applied for the transformation of new text into feature sets, which is given into the model for prediction.

Specifically, on complicated classification tasks, machine learning classifies accurately than the rule-based system. Few familiar machine leaning algorithms for classification model creation are Naïve-Bayes algorithm, Support Vector Machines, Logistic Regression, Random Forest and Multi-Layer Perceptron (MLP). The following will describe the overview of the classification algorithms.

Naïve-Bayes classifier: Based on the Bayes theorem, Naïve-Bayes classifiers hold a collection of classification algorithms. This family of algorithms shares a common feature, which states that 'every pair of features being classified is independent to each other'. This produces better results, while data availability is less and has insufficient computational resources. Bayes theorem computes the conditional probabilities of two events to occur on the basis of the occurrence of each individual event to occur. This describes that a text, which is being represented as vector contains certain information about the probabilities of the occurrence of that text within the text of a given category. This supports the algorithm to calculate the likelihood of that text's belonging to the category.

Multinomial Naïve-Bayes Classifier: Multinomial event model allows the feature vectors to represent the frequencies with which certain events are generated by a multinomial (P1, P2, ... PN), where P(I) is the probability that event 1 occurs. The feature vector holds the count of event 1 that is being observed in a particular instance. This event model is used for classification, where events represent the occurrence of a word in the input document. The multinomial Naïve-Bayes classifier is shown as a linear classifier in log space. If a specific class and the feature value do not occur together in the training set, there will be zero probability, which is based on its frequency. This leads to a problem of wiping out all the information in the other probability. Small sample correction called pseudo-count is used, in order to eliminate the probability set to be zero. Regularization of Naïve-Bayes is known as Laplace smoothing while the value of pseudo-count is one.

Support Vector Machines: SVM does not require huge training data to produce accurate results. SVM draws a 'line' or hyperplane for the purpose of dividing it into two, in which one sub-space holds vectors that belong to a particular group and the other sub-spaces hold vectors that do not belong to it. These vectors are the representation of training data and a group holds the tag in which the input tag is tagged in.

Logistic Regression: In this supervised classification algorithm, the output value can be discrete for any input, which is a set of features. This regression model models the data with the help of the sigmoid function. Based on the decision threshold, this algorithm becomes a classification technique. Setting the threshold value is an important aspect and it is purely dependent on the classification task itself. Following two arguments affect the decision for the value of the threshold value.

Low Precision/High Recall: While the number of false negatives has to be reduced without reducing the false positive count, a decision value with low precision and high recall value is chosen.

High Precision/Low Recall: During the necessity of reduction of false positive without reducing the count of false negative, a decision value with a high value of precision and low value of recall is chosen.

Random Forest Classifier: This is an ensemble algorithm, which combines more than one algorithm. A huge number of uncorrelated Decision Trees are built on the basis of averaging random selection of predictor variables. For nonlinear regression, Decision Trees are proven to be less effective. To improve its capability, various techniques like bootstrap aggregation or bagging are performed. Bagging improves the prediction accuracy of Decision Trees, it results in the possibility of interoperability.

An improvement over bagged trees is introduced by de-correlating the trees, in which the variance is decreased, while the trees are averaged. Each time, while constructing Decision Trees, a split in the tree is considered and m predictors, which are chosen at random are used to split candidates from the complete set of predictors.

Since the m predictors are generated at each split (m), and when choosing $m \approx p$, this states that the number of predictors chosen at each split is approximately equal to the square root of the total number of predictors, p. Random Forest trees are insensitive to skewed distributions. Due to the fact that the RF trees do not need to be mapped into normal value domains, outliers and the methodology for imputations are less required; this is considered as one of the most efficient predictive machine learning techniques.

Multi-layer Perceptron: A deep neural network composed of more than one perceptron. This incorporates an input layer, which receives the signal, an output layer, which predicts the output and any number of hidden layers in between those two layers, which do the computation of the MLP. Multi-layer Perceptron is normally used o supervised learning problems, which trains on input–output pairs and trains itself to model the correlation between those inputs and outputs. During training, weights and bias of the model are adjusted in-order to reduce the error. Backpropagation is done to make the weights and bias being adjusted according to the error. The error can be measured using a variety of methods like Mean Squared Error (MSE). The forward pass moves the input layer to the output layer through the hidden layers, and the output layer prediction is done against the ground truth labels. Backward pass does backpropagation and the chain rule of calculus, with respect to various weights and bias, the error function is partial derivate. This differentiation gives a gradient or the error landscape along which the parameter can be adjusted as they move the MLP one step closer to the error minimum. Any gradient-related optimization algorithm can be used to do the gradient descent. This is repeated until the error value is negligible and that state is called convergence.

4 Proposed Work

The main focus of the proposed audio-based news (converted to text) classification task is to accurately classify the news video articles into their respective category. Here, the category of the given news is predicted based on the high occurrence of correlated unigrams and bigrams of a specific class that were identified in training. The text classification module comprises of mainly three tasks:

i. Pre-processing, Feature extraction with TF-IDF for classification.
ii. Training the classification model using Multinomial NB, Logistic Regression and MLP Classifier excusing benchmark dataset is as shown in Fig. 1.
iii. Testing the audio-based converted News using the trained models such as Multinomial NB, Logistic Regression and MLP Classifier is as shown in Fig. 2.

Tokenization: A pre-processing method that breaks a sentence into words or other meaningful elements called tokens. This aims at investigating the words in a sentence.

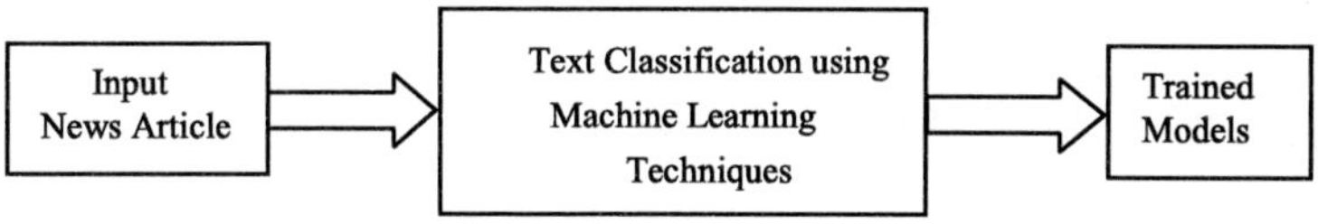

Fig. 1 Training of Classifier Models

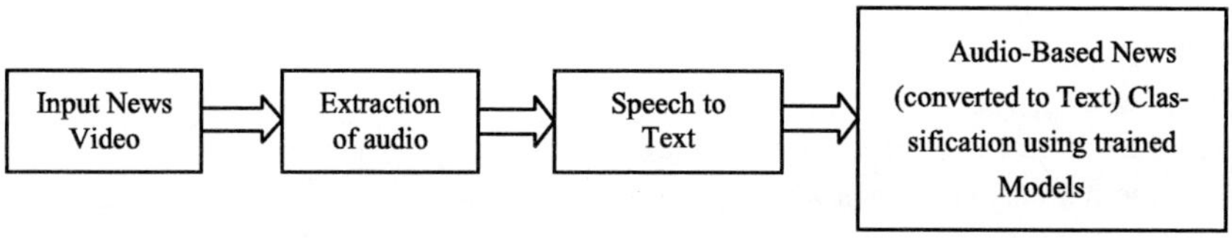

Fig. 2 Classification of converted news using trained models

Stop Words removal: Document holds huge words which do not provide any important significance to be used in classification algorithms, such as {'a', 'about', 'above', 'across', 'after', 'again'}. These words are removed from the texts.

N-Gram technique: This technique is used for extracting features from the set of n-words that occur in that order in a text. This cannot represent a text but can be used as a feature to represent a text: 1-gram is single word representation of a text using its words, which misses its order, 2-gram and 3-gram show the specified number of connected words to detect more information.

Term Frequency-Inverse Document Frequency: This is used along with the term frequency to minimize the effect of common words in the dataset. IDF holds the higher weight of words with the min/max frequencies term in the document. TF-IDF is represented in the equation given below:

$$W(d, t) = TF(d, t) * \log(\frac{N}{df(t)})$$

Here, N represents the number of documents and $df(t)$ is the number of documents that contain the term t in the corpus.

The above-mentioned preprocessing steps are carried out and the features are extracted for categorization. During training, features are extracted using TF-IDF, which is a measure that uses two statistical methods, Term Frequency and Inverse Document Frequency. Term frequency is the count of each term appearing in document d against the count of all words in the document, using which the most correlated unigrams and bigrams are identified for each category from the given passage. After the training phase, the category of the news article is predicted in the testing phase based on the high occurrence of correlated unigrams and bigrams of a specific class that were identified. The model is then tested with the news text which is converted from the collected news video.

In the training phase, machine learning models using Multinomial NB, Logistic Regression, SVM, Random Forest classifier and MLP Classifier, are trained with the BBC text dataset for the classification task. *BBC News articles* of total 1900 of 5 categories labelled as Politics, Business, Sports, Technology and Entertainment are available in the dataset. Train test split percentage is (67, 33), i.e. 67% of the dataset is used for training and 33% is used for testing.

Further, manually collected news videos, which are converted into text, are also tested to identify the category using the above models. The collected 25 *CNN News videos* (converted into text) which consist of 5 categories labelled as Politics, Business, Sports, Technology and Entertainment are used.

Using Multinomial NB, Logistic Regression, SVM, Random Forest classifier and MLP Classifier, the category of the news is identified. The performance of each classifier is measured by evaluating how accurately the news is categorized. The accurate prediction of news category of each classifier is measured.

In another instance, the news video is considered, from which the audio is extracted and is converted into text. Now, this text is given as input to the various classifier models. This categorized news text extracted from news video will be structured and can provide much more information to the user. The accuracy measure for the classification of news text extracted from video is evaluated.

5 Metrics and Experimental Results

The performance of each classifier is measured using evaluation metrics. Some of the metrics used to evaluate the performance include recall, precision, accuracy, F-measure, etc. These metrics are calculated based on the confusion matrix, which is shown in Fig. 3. Confusion matrix comprises of true positives (TP), false positives (FP), false negatives (FN) and true negatives (TN). The implication of these elements may vary based on the application for which it evaluates.

Accuracy: The fraction of correct prediction over all the possible corrections made.

$$\text{accuracy} = \frac{(TP + TN)}{(TP + FP + FN + TN)}$$

Sensitivity: The fraction of positives that are predicted correctly

$$\text{sensitivity} = \frac{TP}{(TP + FN)}$$

Specificity: The fraction of negatives that are predicted right.

$$\text{specificity} = \frac{TN}{(TN + FP)}$$

Precision: Correctly predicted positive to the proportion of all positives (i.e. positive predictive value).

Fig. 3 Confusion matrix

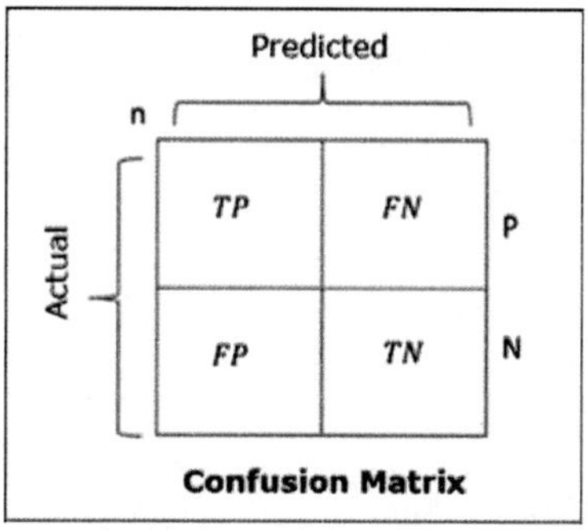

$$\text{precision} = \frac{\sum_{i=1}^{L} TP_i}{\sum_{i=1}^{L} TP_i + FP_i}$$

Recall: Number of the relevant document retrieved to the fraction of total number of document retrieved.

$$\text{recall} = \frac{\sum_{i=1}^{L} TP_i}{\sum_{i=1}^{L} TP_i + FN_i}$$

F1 Score: Difference between the precision and the recall

$$\text{F1 - Score} = \frac{\sum_{i=1}^{L} 2TP_i}{2TP_i + FP_i + FN_i}$$

The classifiers such as MLP Classifier, Multinomial NB, Logistic Regression, SVM and Random Forest Classifier trained on the news articles of 1900 records in train–test split percentage (67, 33%) are evaluated and used. The trained models such as MLP Classifier, Multinomial NB and Logistic Regression are used for testing the given dataset of five news videos belonging to each category. The output of each classifier individually is evaluated using the confusion matrix based on the observed true positives, true negatives, false positives and false negatives. MLP Classifier and the Multinomial NB Classifier validated across the news videos resulted in 0.80% of true positives and 0.20% of false positives, whereas the Logistic Regression model validated across the news videos resulted in 0.84% of true positives and 0.16% of false positives; hence, the performance measure observed is high in Logistic Regression comparatively to other models. As shown in Fig. 4, the number of true positives for the category business and politics is less than the other category because the correlated unigrams and bigrams are used as the feature come under multi-labels. For example, some words may belong to more than one category. Thus, by improving the threshold level of the correlated n-grams, the number of true positives can be increased. Also as shown in Fig. 5, the number of false positives in Logistic Regression is less than the other models which is also another factor in the increase in performance.

Fig. 4 True positives for different classifiers

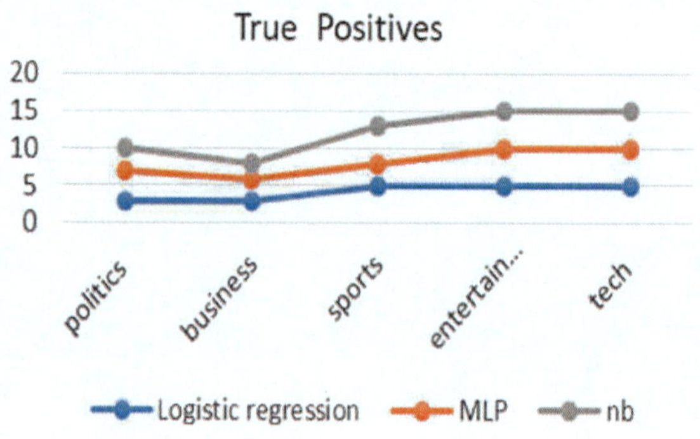

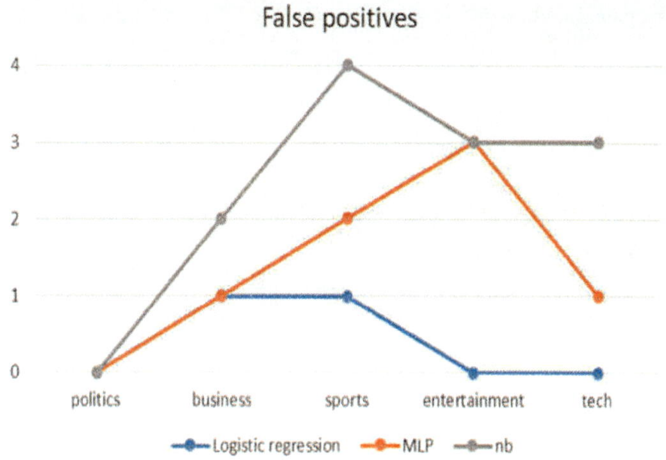

Fig. 5 False positives for different classifiers

Accuracy measure of different classifiers trained on News articles computed using the confusion matrix is shown in Table 1 and the accuracy measure of classification of audio-based News (converted to text) using trained models is shown in Table 2. MLP Classifier and Logistic Regression models validated across the news videos resulted in 0.94% of overall accuracy, whereas the Multinomial Naïve Bayes model validated across the news videos resulted in 0.96% of overall accuracy; hence, the performance measure observed is high in Multinomial Naïve-Bayes model comparatively to the other models.

Since the classification is done to the news text converted from the news video, it is important to evaluate the performance of the speech recognition task too. Word Error Rate (WER) is a common metric used to calculate the recognition accuracy.

Table 1 Accuracy of different classifiers on BBC News articles (Text)

Classifier	Accuracy
MLP classifier	98.2
Multinomial Naïve-Bayes	97.4
Logistic regression	97.3
Support vector machine	97.0
Random forest classifier	84.0

Table 2 Accuracy of classifiers on audio-based News videos

Classifier	Accuracy
MLP classifier	94
Multinomial Naive-Bayes	96
Logistic regression	94

This can be calculated as

$$WER = \frac{S + D + 1}{N},$$

where
S—Number of substitutions,
D—Number of Deletions,
I—Number of Insertions,
C—Number of correct words,
N—Number of words in the reference ($N = S + D + C$).

Lower *WER* often indicates that the ASR software is more accurate in recognizing speech, while a higher WER often indicates lower ASR accuracy. Accuracy in the classification of the news text extracted from the video using various classifiers is shown below.

The evaluation of the accuracy of various classification models concludes that Multinomial Naïve-Bayes model outperforms the other classifiers on BBC Text News articles. In the case of the classification of news text extracted from news video, Multinomial Naïve-Bayes outperforms the other classifiers. It has been concluded that Multinomial Naïve-Bayes model performs well for both noisy and noiseless text input.

6 Conclusion and Future Work

A system for classifying news videos based on the audio content using machine learning techniques has been presented. In this work, features are extracted using NLP techniques and machine learning classifier model is built to perform the classification task. A dataset consisting of 25 News videos of CNN news channel which covers almost 5 categories has been collected to carry out the experiments. News videos for which genre have to be identified is converted to text and is given to the model during the testing phase. The model categorizes the converted text content into any one of the identified categories. Further, the news video classification was carried out using various machine learning classifiers and their results are compared and analysed. The analysis concludes that Multinomial Naive-Bayes classifier outperforms the other classifier models for both the noisy and noiseless text input. In future, the semantic analysis of audio-based news can be incorporated to enhance classification accuracy.

References

1. Tokunaga, T.: Text categorization based on weighted inverse document frequency. In: Special Interest Groups and Information Process Society of Japan (SIG-IPSJ) (1994)
2. Rogati, M., Yang, Y.: High-performing feature selection for text classification. In: International Conference on Information and Knowledge Management (2003)
3. Wang, L.: Feature selection withconditionalmutual information Maxi Min in text categorization. In: International Conference on Information and Knowledge Management (2004)
4. Lan, M., Sung, H.-B., Tan, C.L.: A comparative study on term weighting schemes for text categorization. In: International Joint Conference on Neural Networks (2005)
5. Li, Z., Shang, W.: News text classification model based on topic model. In: IEEE/ACIS 15th International Conference on Computer and Information Science (ICIS) (2016)
6. Tang, B., He, H.: A Bayesian classification approach using class-specific features. IEEE Trans. Know. Data Eng. (2016)
7. Kim, S.B., Rim, H.C., Yook, D.S., Lim, H.S.: Effective methods for improving Naïve Bayes text classifiers. LNAI 2417 (2002)
8. Schneider: Techniques for improving the performance of Naive-Bayes for text classification, vol. 3406. LNCS (2005)
9. Klopotek, M., Woch, M.: Very large Bayesian networks in text classification. ICCS 2003, LNCS 2657 (2003)
10. Shanahan, J., Roma, N.: Improving SVM text classification performance through threshold adjustment. LNAI 2837 (2003)
11. Johnson, D.E., Oles, F.J., Zhang, T., Goetz, T.: A decision-tree-based symbolic rule induction system for text categorization. IBM Syst. J. (2002)
12. Nowshed, C.A., Seddiqui, M.H., Das, S.: Bangla news classification using Naive-Bayes classifier. In: 16th International Conference on Computer and Information Technology (ICCIT). IEEE (2014)
13. Menaka, S., Radha, N.: Text classification using keyword extraction technique. Int. J. Adv. Res. Comput. Sci. Software Eng. (2013)
14. Aggarwal, C.C., Cheng, X.Z.: A survey of text classification algorithms. Mining text data. Springer US (2012)
15. Schapire, R.E., Singer, Y.: Boostexter: a boosting-based system for text categorization. Mach. Learn. **39**(2), 135–168 (2000)
16. Deng, X., Li, Y., Weng, J., Zhang, J.: Feature selection for text classification: a review. Multimed Tools Appl. (2018)
17. Gong, Y.L., Zhu, J., Tang, S.P.: Keywords extraction based on text classification. Adv. Mater. Res. (2013)
18. Joachims, T.: Optimizing search engines using clickthrough data. In: International Conference on Knowledge Discovery and Data Mining (2002)
19. Vinciarelli, A.: Noisy text categorization, pattern recognition. In: 17th International Conference on (ICPR '04) (2004)
20. Mirończuk, M.M., Protasiewicz, J.: A recent overview of the state-of-the-art elements of text classification. J. Expert Syst. Appl. (2018)

On Generalized Vector Variational-Like Inequalities and Nonsmooth Multiobjective Programming Problems Using Limiting Subdifferential

B. B. Upadhyay and Priyanka Mishra

Abstract In this article, we study generalized Minty and Stampacchia vector variational-like inequalities with their weaker forms and nonsmooth multiobjective programming problem. Under strong invexity hypothesis, we establish equivalence among the solutions of generalized vector variational-like inequalities and efficient minimizers of order s of the nonsmooth multiobjective programming problem. Furthermore, with the help of KKM-Fan theorem, we derive certain conditions under which the solutions of vector variational-like inequalities exist. The results presented in this paper generalize, unify, and sharpen the works of Li and Yu [21], Upadhyay et al. [31] and Upadhyay and Mishra [32, 33].

Keywords Multiobjective programming problems · Strong invexity · Efficient minimizers

1 Introduction

The assumption of convexity is often too strong for the applications in optimization theory, economics, and probability theory, see [12, 28, 30]. Due to wider applications of convex function, many researchers have shown their interest in the study of the special class of functions, which possesses several properties of convex functions. In this aspect, Hanson [14] introduced the class of nonconvex functions, which was later named as invex functions by Craven [6]. Invex function has various applications in the field of variational inequality problems and nonlinear optimization, for more exposition, see [4, 24, 36]. The notion of strongly η-invex functions is introduced by Jeyakumar and Mond [17]. The concept of invexity was extended by Reiland [29] for nonsmooth functions.

B. B. Upadhyay (✉) · P. Mishra
Department of Mathematics, Indian Institute of Technology Patna,
Patna, Bihar 801106, India
e-mail: bhooshan@iitp.ac.in

R. N. Mohapatra et al. (eds.), *Mathematical Analysis and Computing*,
Springer Proceedings in Mathematics & Statistics 344,
https://doi.org/10.1007/978-981-33-4646-8_36

In many real-world problems, the algorithms for solving nonsmooth multiobjective programming problems (for short, (NMPP)) terminate after finite steps and we get approximate solutions only. Therefore, in computational as well as analytical point of view, the study of approximate solutions becomes very useful. Many authors have been studied several variants of approximate efficient solution, see [8, 13, 31, 33, 35]. During the study of the convergence of numerical techniques, Cromme [7] defined the notion of strict local minimizers. The notion of strict local minimizer was extended by Auslender [3] to higher order strict local minimizer. Jiménez [18] gave the concept of strict minimizer of order s, for multiobjective programming problems.

For finite dimensional spaces, Giannessi [10] defined the concept of vector variational inequality. In literature, several researchers studied the applications of Minty vector variational-like inequality (for short (MVVLI)) and the Stampacchia vector variational-like inequality (for short (SVVLI)) for (NMPP), for more details, we refer to [2, 5, 9, 11, 15, 20, 23, 34]. Li and Yu [21] showed the equivalence among the solutions of (MVVLI), (SVVLI), and multiobjective programming problem (for short (MPP)) for directionally differentiable invex functions. Al-Homidan and Ansari [1] establish the equivalence among the solutions of (MVVLI), (SVVLI), and (NMPP) involving nonsmooth invex functions. Oveisiha and Zafarani [27] showed the equivalence between the solution of (MVVLI) and (NMPP) using η-invex functions with limiting subdifferential. Upadhyay and Mishra [32] proved the relationship among the solutions of (GMVVLI), (GSVVLI), and (NMPP) using Clarke subdifferential.

Motivated by the works of [22, 31–33], we consider a class of (GMVVLI), (GSVVLI) with its weaker form (WGMVVLI), (WGSVVLI), respectively, in terms of limiting subdifferentials and a class of (NMPP). We showed the equivalence among the solutions of (GMVVLI), (GSVVLI), and efficient minimizer of order s with respect to (for abbreviation w.r.t.) ϑ of (NMPP). Furthermore, we obtain existence results for the solutions of (GMVVLI) and (GSVVLI).

The organization of this paper is given as follows: in Sect. 2, some basic definitions and preliminaries are given; in Sect. 3, we proved the equivalence between the solutions of (GMVVLI), (GSVVLI), and the efficient minimizer of order s w.r.t. ϑ of (NMPP); in Sect. 4, we showed the equivalence between the solutions of (WGMVVLI), (WGSVVLI), and strict minimizer of order s w.r.t. ϑ of (NMPP); in Sect. 5, we derive existence results for the solutions of (GMVVLI), (GSVVLI).

2 Definitions and Preliminaries

Let $\emptyset \neq \Omega \subseteq \mathbb{R}^n$ and $\vartheta : \Omega \times \Omega \to \mathbb{R}^n$ be a vector valued function. Let $\Psi : \Omega \to \mathbb{R}$ be lower semicontinuous function. Let $C := -\mathbb{R}^p_+ \setminus \{0\}$ and $\text{int}C := -\text{int}\mathbb{R}^p_+$ be its interior.

For all $y_1, y_2 \in \mathbb{R}^p$, we use the following ordering relation:

$$y_1 \leq_C y_2 \iff y_1 - y_2 \in C; \quad y_1 \nleq_C y_2 \iff y_1 - y_2 \notin C.$$

Now, we recall the following concepts related with limiting subdifferential from [25].

Definition 1 The Fréchet subdifferential of Ψ at $y \in \Omega$ is defined as

$$\hat{\partial}\Psi(y) := \left\{ \xi \in \mathbb{R}^n : \liminf_{y' \to y} \frac{\Psi(y') - \Psi(y) - \langle \xi, y' - y \rangle}{\|y' - y\|} \geq 0 \right\}.$$

Definition 2 The limiting subdifferential of Ψ at $\tilde{y} \in \Omega$ is defined as

$$\partial_M \Psi(\tilde{y}) := \limsup_{y \to \tilde{y}} \hat{\partial}\Psi(y),$$

where lim sup is the Painlevé-Kuratowski outer limit.

Definition 3 [24] $\Psi : \Omega \to \mathbb{R}$ is said to be locally Lipschitz on Ω, if for each $y \in \Omega$, there exist constants $L, \delta > 0$, s.t. for every $y_1, y_2 \in B(y; \delta) \cap \Omega$, one has

$$|\Psi(y_1) - \Psi(y_2)| \leq L\|y_1 - y_2\|.$$

Definition 4 [24] Ω is said to be an invex set w.r.t. ϑ, if for every $y_1, y_2 \in \Omega$, one has

$$y_2 + \mu\vartheta(y_1, y_2) \in \Omega, \quad \forall \mu \in [0, 1].$$

From now onwards, Ω is an invex set w.r.t. ϑ, unless otherwise specified.

Definition 5 $\Psi : \Omega \to \mathbb{R}$ is said to be strongly preinvex of order s, if we can get a scalar $\beta > 0$, such that, for all $y_2, y_1 \in \Omega$ and $\mu \in [0, 1]$, one has

$$\Psi(y_2 + \mu\vartheta(y_1, y_2)) \leq (1 - \mu)\Psi(y_2) + \mu\Psi(y_1) - \mu(1 - \mu)\beta\|\vartheta(y_1, y_2)\|^s.$$

Definition 6 $\Psi : \Omega \to \mathbb{R}$ is said to be strongly invex of order s, if we can get a scalar $\beta > 0$, such that, for all $y_2, y_1 \in \Omega$, one has

$$\Psi(y_2) - \Psi(y_1) \geq \langle \xi, \vartheta(y_2, y_1) \rangle + \beta||\vartheta(y_2, y_1)||^s, \quad \forall \xi \in \partial_M \Psi(y_1).$$

Definition 7 [19] $\Gamma : \Omega \to 2^{\Omega}$ is said to be strongly invariant monotone of order s, if we can get a scalar $\beta > 0$, such that

$$\langle \xi, \vartheta(y_2, y_1) \rangle + \langle \zeta, \vartheta(y_1, y_2) \rangle \leq -\beta\{||\vartheta(y_2, y_1)||^s + ||\vartheta(y_1, y_2)||^s\},$$

for any $y_1, y_2 \in \Omega$, and any $\xi \in \Gamma(y_1), \zeta \in \Gamma(y_2)$.

Condition A [37] Let $\Psi : \Omega \to \mathbb{R}$, then

$$\Psi(y_2 + \vartheta(y_1, y_2)) \leq \Psi(y_1), \quad \forall y_1, y_2 \in \Omega.$$

Condition C [26] Let $\vartheta : \Omega \times \Omega \to \mathbb{R}^n$, then for all $y_2, y_1 \in \Omega,\ \mu \in [0, 1]$, one has

(i) $\vartheta(y_1, y_1 + \mu\vartheta(y_2, y_1)) = -\mu\vartheta(y_2, y_1)$,
(ii) $\vartheta(y_2, y_1 + \mu\vartheta(y_2, y_1)) = (1 - \mu)\vartheta(y_2, y_1)$.

Remark 1 Yang et al. [37] have shown that, for all $\mu_1,\ \mu_2 \in [0, 1]$, one has

$$\vartheta(y_1 + \mu_2\vartheta(y_2, y_1), y_1 + \mu_1\vartheta(y_2, y_1)) = (\mu_2 - \mu_1)\vartheta(y_2, y_1).$$

The following Lemma can be proved in a similar way as the proof of Lemma 3.1 in [16].

Lemma 1 *Let Ψ be strongly invex of order s, then $\partial_M\Psi$ is strongly invariant monotone of order s, that is*

$$\langle \xi, \vartheta(y_1, y_2)\rangle + \langle \zeta, \vartheta(y_2, y_1)\rangle \leq -\beta\{||\vartheta(y_1, y_2)||^s + ||\vartheta(y_2, y_1)||^s\},$$

for all $y_2, y_1 \in \Omega$, $\xi \in \partial_M\Psi(y_2)$ and *$\zeta \in \partial_M\Psi(y_1)$.*

Theorem 1 *[25] Let Ψ be Lipschitz on an open set containing $[y_2, y_1]$ in Ω. Then, one has*

$$\Psi(y_1) - \Psi(y_2) \leq \langle \xi, y_1 - y_2\rangle, \quad \text{for some } \xi \in \partial_M\Psi(u);\ u \in [y_2, y_1[.$$

Lemma 2 *[32] Let Ψ is strongly invex of order s on Ω, such that ϑ satisfy the Condition C, then Ψ is strongly preinvex of order s on Ω.*

We study the following nonsmooth multiobjective programming problem:

$$\text{(NMPP)} \qquad \begin{aligned} &\text{Minimize } \ \Psi(y) = (\Psi_1(y), \ldots, \Psi_p(y)) \\ &\text{subject to } \ y \in \Omega, \end{aligned}$$

where $\Psi_k : \Omega \to \mathbb{R},\ k \in \mathcal{K} := \{1, 2, \ldots, p\}$ are lower semicontinuous functions on Ω.

For $\Psi = (\Psi_1, \ldots, \Psi_p)$, limiting subdifferential of Ψ at $y \in \Omega$ is defined as

$$\partial_M\Psi(y) := \partial_M\Psi_1(y)\times, \ldots, \partial_M\Psi_p(y).$$

Definition 8 [32] $\tilde{y} \in \Omega$ is said to be efficient minimizer of order s w.r.t. ϑ of (NMPP), if for some $\beta \in \text{int}\mathbb{R}^p_+$, one has

$$\Psi(y) - \Psi(\tilde{y}) - \beta||\vartheta(y, \tilde{y})||^s \notin C,\ \forall y \in \Omega.$$

Definition 9 [32] $\tilde{y} \in \Omega$ is said to be strict minimizer of order s w.r.t. ϑ of (NMPP), if for some $\beta \in \text{int}\mathbb{R}^p_+$, one has

$$\Psi(y) - \Psi(\tilde{y}) - \beta||\vartheta(y, \tilde{y})||^s \notin \text{int}C, \ \forall y \in \Omega.$$

Now, we study the following generalized Minty and Stampacchia type vector variational-like inequalities in terms of limiting subdifferential:

(GMVVLI) To find $\tilde{y} \in \Omega$, such that

$$\langle \partial_M \Psi(y), \vartheta(y, \tilde{y}) \rangle \not\subseteq C, \ \forall y \in \Omega.$$

(GSVVLI) To find $\tilde{y} \in \Omega$, such that, for any $y \in \Omega$, we can get $\zeta \in \partial_M \Psi(\tilde{y})$, one has

$$\langle \zeta, \vartheta(y, \tilde{y}) \rangle \notin C.$$

(WGMVVLI) To find $\tilde{y} \in \Omega$, such that

$$\langle \partial_M \Psi(y), \vartheta(y, \tilde{y}) \rangle \not\subseteq \text{int}C, \ \forall y \in \Omega.$$

(WGSVVLI) To find $\tilde{y} \in \Omega$, such that, for any $y \in \Omega$, we can get $\zeta \in \partial_M \Psi(\tilde{y})$, one has

$$\langle \zeta, \vartheta(y, \tilde{y}) \rangle \notin \text{int}C.$$

3 Relationship Between (GMVVLI), (GSVVLI), and (NMPP)

In this section, using the powerful tool of limiting subdifferential, we establish certain relations among the solutions of (GMVVLI), (GSVVLI), and efficient minimizer of order s w.r.t. ϑ of (NMPP).

Theorem 2 *Let each Ψ_k, $k \in \mathcal{K}$ be strongly invex of order s, ϑ is skew and $\tilde{y}$ is an efficient minimizer of order s of (NMPP) w.r.t. ϑ, then $\tilde{y}$ solves (GMVVLI). Furthermore, let ϑ satisfy the Condition C, each Ψ_k, $k \in \mathcal{K}$ be locally Lipschitz on Ω and satisfy the Condition A. Let $\tilde{y}$ solves (GMVVLI), then $\tilde{y} \in \Omega$ is an efficient minimizer of order s w.r.t. ϑ of (NMPP).*

Proof Assume that $\tilde{y} \in \Omega$ be an efficient minimizer of order s w.r.t. ϑ of (NMPP). Therefore, for all $y \in \Omega$, we can get a vector $\beta \in \text{int}\mathbb{R}^p_+$, such that

$$\Psi(y) - \Psi(\tilde{y}) - \beta||\vartheta(y, \tilde{y})||^s \notin C. \tag{1}$$

From strong invexity of order s of Ψ_k, $k \in \mathcal{K}$, for any $y \in \Omega$, we get

$$\Psi_k(\tilde{y}) - \Psi_k(y) \geq \langle \xi_k, \vartheta(\tilde{y}, y) \rangle + \beta_k||\vartheta(\tilde{y}, y)||^s, \ \forall \xi_k \in \partial_M \Psi_k(y), \ k \in \mathcal{K}. \tag{2}$$

Since ϑ is skew and $\beta_k > 0$, from (2), we get

$$\Psi_k(y) - \Psi_k(\tilde{y}) - \beta_k||\vartheta(y, \tilde{y})||^s \leq \langle \xi_k, \vartheta(y, \tilde{y})\rangle - 2\beta_k||\vartheta(y, \tilde{y})||^s, \quad \forall k \in \mathcal{K}. \quad (3)$$

From (1) and (3), we have

$$\langle \partial_M \Psi(y), \vartheta(y, \tilde{y})\rangle \nsubseteq C, \ \forall y \in \Omega.$$

Therefore, $\tilde{y}$ solves (GMVVLI).
Conversely, let $\tilde{y} \in \Omega$ solves (GMVVLI), but not efficient minimizer of order s w.r.t. ϑ of (NMPP). Then, for some $y \in \Omega$, such that for all $\beta \in \text{int}\mathbb{R}^p_+$, we have

$$\Psi(y) - \Psi(\tilde{y}) - \beta||\vartheta(y, \tilde{y})||^s \in C. \quad (4)$$

Let $y(\mu) := \tilde{y} + \mu\vartheta(y, \tilde{y})$ for all $\mu \in [0, 1]$. Let $\mu' \in (0, 1)$, then from Theorem 1, we can get $\mu_k \in]0, \mu']$ and $\xi_k \in \partial_M \Psi_k(y(\mu_k))$, such that

$$\mu'\langle \xi_k, \vartheta(y, \tilde{y})\rangle \leq \Psi_k(\tilde{y} + \mu'\vartheta(y, \tilde{y})) - \Psi_k(\tilde{y}), \quad \forall k \in \mathcal{K}. \quad (5)$$

From Lemma 2, each Ψ_k, $k \in \mathcal{K}$ is strongly preinvex of order s. Therefore, we can get a $\beta \in \text{int}\mathbb{R}^p_+$, such that

$$\Psi_k(y(\mu')) \leq (1 - \mu')\Psi_k(\tilde{y}) + \mu'\Psi_k(y) - \mu'(1 - \mu')\beta_k||\vartheta(y, \tilde{y})||^s, \ \forall k \in \mathcal{K}. \quad (6)$$

From (5) and (6), we get

$$\langle \xi_k, \vartheta(y, \tilde{y})\rangle \leq \Psi_k(y) - \Psi_k(\tilde{y}) - (1 - \mu')\beta_k||\vartheta(y, \tilde{y})||^s, \ \forall k \in \mathcal{K}. \quad (7)$$

From (7), we get

$$\langle \xi_k, \vartheta(y, \tilde{y})\rangle \leq \Psi_k(y) - \Psi_k(\tilde{y}), \quad \forall k \in \mathcal{K}. \quad (8)$$

Let $\mu^* < \min\{\mu_1, \mu_2, \ldots, \mu_p\}$. From Lemma 1, each $\partial_M \Psi_k$, $k \in \mathcal{K}$ is strongly invariant monotone of order s. Therefore, for every $\xi_k \in \partial_M \Psi_k(y(\mu_k))$ and $\xi_k^* \in \partial_M \Psi_k(y(\mu^*))$, we get

$$\langle \xi_k, \vartheta(y(\mu^*), y(\mu_k))\rangle + \langle \xi_k^*, \vartheta(y(\mu_k), y(\mu^*))\rangle \leq -\beta_k[||\vartheta(y(\mu_k), y(\mu^*))||^s + ||\vartheta(y(\mu^*), y(\mu_k))||^s], \quad \forall k \in \mathcal{K}. \quad (9)$$

Since ϑ is skew, from (9), we get

$$\langle \xi_k - \xi_k^*, \vartheta(y(\mu_k), y(\mu^*))\rangle \geq \beta_k||\vartheta(y(\mu_k), y(\mu^*))||^s, \quad \forall k \in \mathcal{K}. \quad (10)$$

From (10) and Remark 1, we have

$$\bar{\beta}_k||\vartheta(y, \tilde{y})||^s \leq \langle \xi_k - \xi_k^*, \vartheta(y, \tilde{y})\rangle, \text{ where } \bar{\beta}_k = (\mu_k - \mu^*)^{s-1}, \ \forall k \in \mathcal{K}. \quad (11)$$

From (11), we get

$$\langle \xi_k, \vartheta(y, \tilde{y}) \rangle \geq \langle \xi_k^*, \vartheta(y, \tilde{y}) \rangle + \bar{\beta}_k ||\vartheta(y, \tilde{y})||^s, \quad \forall k \in \mathcal{K}. \tag{12}$$

From (8) and (12), we get

$$\Psi_k(y) - \Psi_k(\tilde{y}) \geq \langle \xi_k^*, \vartheta(y, \tilde{y}) \rangle + \bar{\beta}_k ||\vartheta(y, \tilde{y})||^s, \quad \forall k \in \mathcal{K}. \tag{13}$$

From (13), we get

$$\langle \xi_k^*, \vartheta(y, \tilde{y}) \rangle \leq \Psi_k(y) - \Psi_k(\tilde{y}) - \bar{\beta}_k ||\vartheta(y, \tilde{y})||^s, \quad \forall k \in \mathcal{K}. \tag{14}$$

From (4) and (14), we have

$$\langle \partial_M \Psi(y(\mu^*)), \vartheta(y, \tilde{y}) \rangle \subseteq C. \tag{15}$$

Multiplying (15) by $-\mu^*$ and from Remark 1, we have

$$\langle \partial_M \Psi(y(\mu^*)), \vartheta(y(\mu^*), \tilde{y}) \rangle \subseteq C,$$

which contradicts our assumption.

Theorem 3 *Let each Ψ_k, $k \in \mathcal{K}$ be strongly invex function of order s on Ω and $\tilde{y} \in \Omega$ solves (GSVVLI), then $\tilde{y}$ is an efficient minimizer of order s w.r.t. ϑ of (NMPP).*

Proof Let $\tilde{y} \in \Omega$ solves (GSVVLI), then for each $y \in \Omega$, we can get $\zeta \in \partial_M \Psi(\tilde{y})$, such that

$$\langle \zeta, \vartheta(y, \tilde{y}) \rangle \notin C. \tag{16}$$

From strong invexity of order s of Ψ_k, $k \in \mathcal{K}$, we can get a constant $\beta \in \text{int}\mathbb{R}_+^p$, such that

$$\langle \zeta_k, \vartheta(y, \tilde{y}) \rangle + \beta_k ||\vartheta(y, \tilde{y})||^s \leq \Psi_k(y) - \Psi_k(\tilde{y}), \quad \forall y \in \Omega. \tag{17}$$

From (16) and (17), we get

$$\Psi(y) - \Psi(\tilde{y}) - \beta ||\vartheta(y, \tilde{y})||^s \notin C.$$

Hence, $\tilde{y} \in \Omega$ is efficient minimizer of order s w.r.t. ϑ of (NMPP).

Remark 2 However, (GSVVLI) is not necessary condition for a point to be the efficient minimizer of order s w.r.t. ϑ of (NMPP). To justify this fact, consider the following nonsmooth multiobjective programming problem

$$\text{(P2)} \qquad \begin{aligned} &\text{Minimize } \Psi(y) = (\Psi_1(y), \Psi_2(y)) \\ &\text{subject to } y \in \Omega, \end{aligned}$$

where $\Omega = [-1, 1]$, $\Psi_1, \Psi_2 : \Omega \to \mathbb{R}$ and $\vartheta : \Omega \times \Omega \to \mathbb{R}$ be defined as follows:

$$\Psi_1 = \begin{cases} y^2 - y, & y \geq 0, \\ y^2 - 2y, & y < 0, \end{cases} \quad \Psi_2 = \begin{cases} y^2 + 1, & y \geq 0, \\ y^2 + e^{-y}, & y < 0, \end{cases}$$

$$\text{and} \quad \vartheta(y_1, y_2) = \begin{cases} 1 - y_2, & y_1 \geq 0 \text{ and } y_2 < 0, \\ y_1 - y_2, & \text{elsewhere.} \end{cases}$$

Now, we can evaluate that

$$\partial_M \Psi(y) = \begin{cases} (2y - 1, 2y), & y > 0, \\ \{(t, k) : t \in [-2, -1],\ k \in [-1, 0]\}, & y = 0, \\ (2y - 2, 2y - e^{-y}), & y < 0. \end{cases}$$

Evidently, for $\beta_k = 1,\ k = 1, 2$, Ψ_1 and Ψ_2 are strongly invex of order $s = 2$.

Moreover, for $\beta = (1, 1)$, $\tilde{y} = 0$ is efficient minimizer of order 2 w.r.t. ϑ, as for any $y \in \Omega$, we get

$$\Psi(y) - \Psi(\tilde{y}) - \beta||\vartheta(y, \tilde{y})||^s \notin \bar{C}, \text{ where } \bar{C} := -\mathbb{R}^2_+ \setminus \{0\}.$$

But $\tilde{y} = 0$ does not solves (GSVVLI), as for any $y > 0$, we get

$$\langle \zeta, \vartheta(y_1, \tilde{y}) \rangle \in \bar{C},\ \forall \zeta \in \partial_M \Psi(0) \text{ where } \bar{C} := \mathbb{R}^2_+ \setminus \{0\}.$$

We conclude the following relation between the solutions of (GSVVLI) and (GMVVLI), from Theorems 2 and 3.

Corollary 1 *Let each $\Psi_k,\ k \in \mathcal{K}$ be strongly invex function of order s and ϑ is skew. Let ϑ and $\Psi_k,\ k \in \mathcal{K}$ satisfy the Condition C and Condition A, respectively. If $\tilde{y} \in \Omega$ solves (GSVVLI), then $\tilde{y}$ also solves (GMVVLI).*

4 Relationship Between (WGMVVLI), (WGSVVLI), and (NMPP)

In this section, we showed the equivalence between the solutions of (WGMVVLI), (WGSVVLI), and strict minimizer of order s w.r.t. ϑ of (NMPP).

Proposition 1 *Let each $\Psi_k,\ k \in \mathcal{K}$ be strongly invex of order s and $\tilde{y}$ solves (WGSVVLI), then $\tilde{y}$ also solves (WGMVVLI).*

Proof Assume that $\tilde{y} \in \Omega$ solves (WGSVVLI), then we can get $\zeta \in \partial_M \Psi(\tilde{y})$, such that for any $y \in \Omega$, we get

$$\langle \zeta, \vartheta(y, \tilde{y})\rangle \notin \text{int}C. \tag{18}$$

From Lemma 1, $\partial_M \Psi_k$ is strongly invariant monotone of order s. Therefore, we can get a constant $\beta \in \text{int}\mathbb{R}^p_+$, such that

$$\langle \xi_k - \zeta_k, \vartheta(y, \tilde{y})\rangle \geq \beta_k[||\vartheta(y, \tilde{y})||^s + ||\vartheta(\tilde{y}, y)||^s], \quad \forall k \in \mathcal{K}, \tag{19}$$

for all $\zeta_k \in \partial_M \Psi_k(\tilde{y}),\ \xi_k \in \partial_M \Psi_k(y),$ and $y \in \Omega$.

From (18) and (19), we get

$$\langle \partial_M \Psi(y), \vartheta(y, \tilde{y})\rangle \nsubseteq \text{int}C, \ \forall y \in \Omega.$$

Hence, $\tilde{y} \in \Omega$ solves (WGMVVLI).

Proposition 2 *Let each $\Psi_k,\ k \in \mathcal{K}$ be locally Lipschitz, ϑ is skew and affine in first argument. If $\tilde{y} \in \Omega$ solves (WGMVVLI), then $\tilde{y}$ also solves (WGSVVLI).*

Proof Assume that $\tilde{y}$ solves (WGMVVLI), then for each $y \in \Omega$, we have

$$\langle \partial_M \Psi(y), \vartheta(y, \tilde{y})\rangle \nsubseteq \text{int}C.$$

Let $y(\mu_n) = \tilde{y} + \mu_n \vartheta(y, \tilde{y})$, where $\mu_n \downarrow 0$. Since $\tilde{y}$ solves (WGMVVLI), then

$$\langle \partial_M \Psi(y(\mu_n)), \vartheta(y(\mu_n), \tilde{y})\rangle \nsubseteq \text{int}C. \tag{20}$$

Since ϑ is skew and affine in first argument, we get

$$\begin{aligned} 0 &= \langle \xi_n, \vartheta(y(\mu_n), y(\mu_n))\rangle \\ &= \mu_n\langle \xi_n, \vartheta(y, y(\mu_n))\rangle + (1-\mu_n)\langle \xi_n, \vartheta(\tilde{y}, y(\mu_n))\rangle, \ \forall \xi_n \in \partial_M \Psi(y(\mu_n)). \end{aligned} \tag{21}$$

From (20) and (21), we get

$$\langle \xi_n, \vartheta(y, y(\mu_n))\rangle \notin \text{int}C. \tag{22}$$

Since $\partial_M \Psi_k,\ k \in \mathcal{K}$ is compact, $\xi_n \in \partial_M \Psi(y(\mu_n))$, $\xi_n \to \zeta$, and $y(\mu_n) \to \tilde{y}$ as $n \to \infty$, we have $\zeta \in \partial_M \Psi(\tilde{y})$.

Therefore, for all $y \in \Omega$, we can get $\zeta \in \partial_M \Psi(\tilde{y})$, such that

$$\langle \zeta, \vartheta(y, \tilde{y})\rangle \notin \text{int}C.$$

We conclude the equivalence between the solutions of (WGSVVLI) and (WGMVVLI), from Propositions 1 and 2.

Theorem 4 *Let each $\Psi_k,\ k \in \mathcal{K}$ be locally Lipschitz and strongly invex of order s. Let ϑ is skew and affine in first argument, then $\tilde{y}$ is a solution of (WGSVVLI) iff $\tilde{y}$ solves (WGMVVLI).*

Proposition 3 *Let each Ψ_k, $k \in \mathcal{K}$ be strongly invex of order s and $\tilde{y} \in \Omega$ solves (WGSVVLI), then $\tilde{y}$ is strict minimizer of order s w.r.t. ϑ of (NMPP).*

Proof On contrary assume that $\tilde{y} \in \Omega$ solves (WGSVVLI), but not strict minimizer of order s w.r.t. ϑ of (NMPP). Therefore, for some $y \in \Omega$ and for all $\beta \in \text{int}\mathbb{R}^p_+$, we get

$$\Psi(y) - \Psi(\tilde{y}) - \beta||\vartheta(y, \tilde{y})||^s \in \text{int}C. \tag{23}$$

From strong invexity of order s of Ψ_k, $k \in \mathcal{K}$, we can get a constant $\beta \in \text{int}\mathbb{R}^p_+$, such that

$$\langle \zeta_k, \vartheta(y, \tilde{y})\rangle \leq \Psi_k(y) - \Psi_k(\tilde{y}) - \beta_k||\vartheta(y, \tilde{y})||^s, \forall \zeta_k \in \partial_M \Psi_k(\tilde{y}). \tag{24}$$

From (23) and (24), we get

$$\langle \zeta, \vartheta(y, \tilde{y})\rangle \in \text{int}C, \ \forall \zeta \in \partial_M \Psi(\tilde{y}),$$

which is a contradiction.

Proposition 4 *Let each Ψ_k, $k \in \mathcal{K}$ be strongly invex of order s and ϑ is skew. Let $\tilde{y} \in \Omega$ be strict minimizer of order s w.r.t. ϑ of (NMPP), then $\tilde{y}$ solves (WGMVVLI).*

Proof On contrary assume that $\tilde{y} \in \Omega$ is a strict minimizer of order s w.r.t. ϑ of (NMPP), but does not solves (WGMVVLI). Then, we can get $y \in \Omega$, such that

$$\langle \partial_M \Psi(y), \vartheta(y, \tilde{y})\rangle \in \text{int}C. \tag{25}$$

From strong invexity of order s of Ψ_k, $k \in \mathcal{K}$ and using skew property of ϑ, we can get a constant $\beta \in \text{int}\mathbb{R}^p_+$, such that

$$\Psi_k(y) - \Psi_k(\tilde{y}) \leq \langle \xi_k, \vartheta(y, \tilde{y})\rangle - \beta_k||\vartheta(y, \tilde{y})||^s, \ \forall \xi_k \in \partial_M \Psi_k(y), \ k \in \mathcal{K}. \tag{26}$$

Since $\beta_k > 0$, $\forall k \in \mathcal{K}$, from (26), we have

$$\Psi_k(y) - \Psi_k(\tilde{y}) - \beta_k||\vartheta(y, \tilde{y})||^s \leq \langle \xi_k, \vartheta(y, \tilde{y})\rangle - 2\beta_k||\vartheta(y, \tilde{y})||^s \leq \langle \xi_k, \vartheta(y, \tilde{y})\rangle. \tag{27}$$

From (25) and (27), we have

$$\Psi(y) - \Psi(\tilde{y}) - \beta||\vartheta(\tilde{y}, y)||^s \in \text{int}C, \ \forall y \in \Omega,$$

which contradicts our assumption.

For (WGSVVLI), we conclude the following necessary and sufficient condition for a point to be strict minimizer of order s of (NMPP), from Theorem 4 and Propositions 3 and 4.

Theorem 5 *Let each $\Psi_k,\ k \in \mathcal{K}$ be locally Lipschitz and strongly invex of order s. Let ϑ be skew and affine in first argument, then $\tilde{y} \in \Omega$ solves (WGSVVLI) iff $\tilde{y}$ is a strict minimizer of order s w.r.t. ϑ of (NMPP).*

5 Existence Results for (GMVVLI) and (GSVVLI)

In this section, we discuss some conditions for the existence of solutions of (GMVVLI) and (GSVVLI) by employing KKM-Fan theorem.

Definition 10 [21] A map $\Gamma : \Omega \to 2^{\Omega}$ is said to be a KKM map if for each $\{z_1, \ldots, z_l\} \subseteq \Omega$, one has

$$\text{co}\{z_1, \ldots, z_l\} \subseteq \bigcup_{k=1}^{l} \Gamma(z_k),$$

where $\text{co}\{z_1, \ldots, z_l\}$ is the convex hull of $\{z_1, \ldots, z_l\}$.

The following KKM-Fan theorem is from [21].

Lemma 3 *Let Ω be convex set. Let $\Gamma : \Omega \to 2^{\Omega}$ be KKM map and such that $\Gamma(y)$ is closed for each $y \in \Omega$. If for some point $\bar{y} \in \Omega$, $\Gamma(\bar{y})$ is compact, then*

$$\bigcap_{y \in \Omega} \Gamma(y) \neq \emptyset.$$

Theorem 6 *Let $\Psi_k,\ k \in \mathcal{K}$ be lower semicontinuous functions and ϑ is affine in second argument. Let the following conditions hold:*

1. *For every $y_1, y_2 \in \Omega$, $\langle \xi_k, \vartheta(y_2, y_1)\rangle + \langle \zeta_k, \vartheta(y_1, y_2)\rangle \geq 0, \forall \xi_k \in \partial_M \Psi_k(y_2)$ and $\zeta_k \in \partial_M \Psi_k(y_1),\ k \in \mathcal{K}$.*
2. *For all $y \in \Omega$, $\langle \partial_M \Psi(y), \vartheta(y, y)\rangle \nsubseteq -C$.*
3. *The set valued map $\Gamma(y_2) := \{y_1 \in \Omega : \langle \partial_M \Psi(y_2), \vartheta(y_2, y_1)\rangle \nsubseteq C,\ \forall y_2 \in \Omega\}$, is closed for all $y_2 \in \Omega$.*
4. *There exist a compact set $\emptyset \neq G \subset \Omega$ and a compact convex set $\emptyset \neq M \subseteq \Omega$ and for each $y_1 \in \Omega \setminus G$, we can get $y_2 \in M$, such that $y_1 \notin \Gamma(y_2)$.*

Then (GMVVLI) is solvable on Ω.

Proof For all $y_2 \in \Omega$, we define a map

$$\hat{\Gamma}(y_2) := \{y_1 \in \Omega : \langle \partial_M \Psi(y_1), \vartheta(y_1, y_2)\rangle \nsubseteq -C\}.$$

Evidently, $y_2 \in \hat{\Gamma}(y_2)$

Next, we prove that $\hat{\Gamma}(y)$ is KKM map on Ω. On contrary, let $\{u_1, \ldots, u_l\} \subseteq \Omega,\ \mu_k \geq 0,\ k = 1, \ldots, l$, with $\sum_{k=1}^{l} \mu_k = 1$, such that

$$\tilde{y} = \sum_{k=1}^{l} \mu_k u_k \notin \bigcup_{k=1}^{l} \hat{\Gamma}(u_k). \tag{28}$$

Therefore, for each $u_k,\ k = 1, \ldots, l$, we have

$$\langle \partial_M \Psi(\tilde{y}), \vartheta(\tilde{y}, u_k) \rangle \subseteq -C. \tag{29}$$

Since ϑ is affine in second argument, one has

$$0 = \langle \partial_M \Psi(\tilde{y}), \vartheta(\tilde{y}, \tilde{y}) \rangle = \sum_{k=1}^{l} \langle \partial_M \Psi(\tilde{y}), \vartheta(\tilde{y}, \mu_k u_k) \rangle \subseteq -C.$$

Therefore, $\hat{\Gamma}(y)$ is KKM map on Ω. Now, we prove that $\hat{\Gamma}(y) \subseteq \Gamma(y),\ \forall y \in \Omega$. Let $\tilde{y} \notin \Gamma(y)$, for some $y \in \Omega$, then from the definition of $\hat{\Gamma}$, we get

$$\langle \partial_M \Psi(y), \vartheta(y, \tilde{y}) \rangle \subseteq C. \tag{30}$$

Using condition (1), we get

$$\langle \xi_k, \vartheta(y, \tilde{y}) \rangle + \langle \zeta_k, \vartheta(\tilde{y}, y) \rangle \geq 0, \tag{31}$$

for all $\xi_k \in \partial_M \Psi_k(y),\ \zeta_k \in \partial_M \Psi_k(\tilde{y}),\ k \in \mathcal{K}$.

From (30) and (31), we get

$$\langle \partial_M \Psi(\tilde{y}), \vartheta(\tilde{y}, y) \rangle \subseteq -C.$$

Therefore, $\tilde{y} \notin \hat{\Gamma}(y)$. Hence, Γ is a KKM map. From hypotheses, $\Gamma(y)$ is a compact set. Using the KKM-Fan Theorem

$$\bigcap_{y \in \Omega} \Gamma(y) \neq \emptyset.$$

Therefore, we can get a $\tilde{y} \in \Omega$, such that

$$\langle \partial_M \Psi(y), \vartheta(y, \tilde{y}) \rangle \nsubseteq C,\ \forall y \in \Omega.$$

Hence, (GMVVLI) has a solution on Ω.

Example 1 Let $\Psi_1,\ \Psi_2 : [-\frac{1}{2}, \frac{1}{2}] \to \mathbb{R}$ and $\vartheta : [-\frac{1}{2}, \frac{1}{2}] \times [-\frac{1}{2}, \frac{1}{2}] \to \mathbb{R}$ be defined as:

$$\Psi_1(y) = \begin{cases} y+1, & y > 0, \\ 2y, & y \leq 0, \end{cases} \quad \Psi_2(y) = \begin{cases} e^y, & y > 0, \\ y^2 + y, & y \leq 0, \end{cases}$$

$$\text{and } \vartheta(y_2, y_1) = \begin{cases} y_2 - y_1, & y_2 > 0 \text{ and } y_1 > 0, \text{ or } y_2 < 0 \text{ and } y_1 < 0, \\ 0, & y_2 = 0 \text{ and } y_1 = 0, \\ 1 - y_1, & \text{elsewhere.} \end{cases}$$

Now, we can evaluate that

$$\partial_M \Psi(y) = \begin{cases} (1, e^y), & y > 0, \\ \{(t, k) : t \in [2, \infty),\ k \in [1, \infty)\}, & y = 0, \\ (2, 2y+1), & y < 0. \end{cases}$$

For all $\xi_1 \in \partial_M \Psi_1(y_1)$ *and* $\zeta_1 \in \partial_M \Psi_1(y_2)$, we get

$$\langle \xi_1, \vartheta(y_1, y_2) \rangle + \langle \zeta_1, \vartheta(y_2, y_1) \rangle =$$
$$\begin{cases} \langle 1, y_1 - y_2 \rangle + \langle 1, y_2 - y_1 \rangle, & y_1 > 0,\ y_2 > 0; \\ \langle 2, y_1 - y_2 \rangle + \langle 2, y_2 - y_1 \rangle, & y_1 < 0,\ y_2 < 0; \\ \langle 1, 1 - y_2 \rangle + \langle 2, 1 - y_1 \rangle = (1 - y_2) + 2(1 - y_1), & y_1 > 0,\ y_2 < 0; \\ \langle 2, 1 - y_2 \rangle + \langle 1, 1 - y_1 \rangle = 2(1 - y_2) + (1 - y_1), & y_1 < 0,\ y_2 > 0, \end{cases}$$
$$\geq 0.$$

For all $\xi_2 \in \partial_M \Psi_2(y_1)$ *and* $\zeta_2 \in \partial_M \Psi_2(y_2)$, we get

$$\langle \xi_2, \vartheta(y_1, y_2) \rangle + \langle \zeta_2, \vartheta(y_2, y_1) \rangle =$$
$$\begin{cases} \langle e^{y_1}, y_1 - y_2 \rangle + \langle e^{y_2}, y_2 - y_1 \rangle = (e^{y_1} - e^{y_2})(y_1 - y_2), & y_1 > 0,\ y_2 > 0; \\ \langle 2y_1 + 1, y_1 - y_2 \rangle + \langle 2y_2 + 1, y_2 - y_1 \rangle = 2(y_1 - y_2)^2, & y_1 < 0,\ y_2 < 0; \\ \langle e^{y_1}, 1 - y_2 \rangle + \langle 2y_2 + 1, 1 - y_1 \rangle, & y_1 > 0,\ y_2 < 0; \\ \langle 2y_1 + 1, 1 - y_2 \rangle + \langle e^{y_2}, 1 - y_1 \rangle, & y_1 < 0,\ y_1 > 0, \end{cases}$$
$$\geq 0.$$

Furthermore, $\vartheta(y, y) = 0$, for all $y \in [-\frac{1}{2}, \frac{1}{2}]$. Hence,

$$\langle \partial_M \Psi_k(y), \vartheta(y, y) \rangle \nsubseteq -\bar{C}, \text{ where } \bar{C} := \mathbb{R}^2_+ \setminus \{0\}.$$

For all $y \in [-\frac{1}{2}, \frac{1}{2}]$,

$$\Gamma(y) = \begin{cases} [-\frac{1}{2}, y], & y > 0, \\ [-\frac{1}{2}, \frac{1}{2}], & y = 0, \\ [-\frac{1}{2}, y] \cup [0, \frac{1}{2}], & y < 0. \end{cases}$$

Let $G = [0, \frac{1}{2}]$ and $M = (0, \frac{1}{2})$. Clearly, M is convex and for any $y_1 \in [-\frac{1}{2}, \frac{1}{2}] \setminus G$, we can get a $y_2 < y_1$, such that $y_1 \notin \Gamma(y_2)$.

Moreover, we can verify that $\tilde{y} = 0$ solves (GMVVLI).

In the same way as Theorem 6, we conclude the following result for the existence of solution of (GSVVLI).

Theorem 7 *Let $\Psi_k : \Omega \to \mathbb{R},\ k \in \mathcal{K}$ be lower semicontinuous functions and ϑ is affine in second argument. Let the following conditions hold:*

1. *For every $y_1, y_2 \in \Omega,\ \ \langle \xi_k, \vartheta(y_1, y_2)\rangle + \langle \zeta_k, (y_2, y_1)\rangle \geq 0, \forall \xi_k \in \partial_M \Psi_k(y_2)$and $\zeta_k \in \partial_M \Psi_k(y_1),\ k \in \mathcal{K}$.*
2. *For any $y_1 \in \Omega,\ \ \langle \partial_M \Psi(y_1), \vartheta(y_1, y_1)\rangle \nsubseteq -C$.*
3. *The set valued map $\Gamma(y_2) := \{y_1 \in \Omega : \langle \partial_M \Psi(y_1), \vartheta(y_2, y_1)\rangle \nsubseteq C,\ \forall y_2 \in \Omega\}$ is closed for all $y_2 \in \Omega$.*
4. *There exist a compact set $\emptyset \neq G \subset \Omega$ and a compact convex set $\emptyset \neq M \subseteq \Omega$, such that for any $y_1 \in \Omega \setminus G$, we can get a $y_2 \in M$, such that $y_1 \notin \Gamma(y_2)$.*

Then, (GSVVLI) is solvable on Ω.

6 Conclusions

In this article, we have considered generalized vector variational-like inequalities (GMVVLI) and (GSVVLI) with their weaker forms (WGMVVLI) and (WGSVVLI), respectively, in terms of limiting subdifferentials. Under strong invexity of order s hypotheses, we have showed equivalence between the solutions of (GMVVLI), (GSVVLI), and efficient minimizers of order s of (NMPP). We also establish the equivalence among the solutions (WGMVVLI), (WGSVVLI), and strict minimizers of order s of (NMPP). Moreover, we have obtained some existence results for the solutions of (GMVVLI), (GSVVLI), with the help of KKM-Fan theorem. Suitable numerical examples have been given to justify the significance of the obtained results. The results presented in this paper generalize, unify, and sharpen the works of Li and Yu [21], Upadhyay et al. [31] and Upadhyay and Mishra [32, 33].

Acknowledgements The first author is supported by SERB, DST, Government of India, through grant number "ECR/2016/001961".

References

1. Al-Homidan, S., Ansari, Q.H.: Generalized Minty vector variational-like inequalities and vector optimization problems. J. Optim. Theory Appl. **144**, 1–11 (2010)
2. Ansari, Q.H., Yao, J.C.: On nondifferentiable and nonconvex vector optimization problems. J. Optim. Theory Appl. **106**, 475–488 (2000)

3. Auslender, A.: Stability in mathematical programming with nondifferentiable data. SIAM J. Control Optim. **22**, 239–254 (1984)
4. Ben-Israel, A., Mond, B.: What is invexity? J. Austral. Math. Soc. Ser. B **28**, 1–9 (1986)
5. Bhatia, G.: Optimality and mixed saddle point criteria in multiobjective optimization. J. Math. Anal. Appl. **342**, 135–145 (2008)
6. Craven, B.D.: Invex functions and constrained local minima. Bull. Austral. Math. Soc. **24**, 357–366 (1981)
7. Cromme, L.: Strong uniqueness: a far-reaching criterion for the convergence of iterative numerical procedures. Numer. Math. **29**, 179–193 (1978)
8. Deng, S.: On approximate solutions in convex vector optimization. SIAM J. Control Optim. **35**, 2128–2136 (1997)
9. Giannessi, F.: On Minty variational principle. In: Giannessi, F., Komlósi, S., Tapcsák, T. (eds.) New Trends in Mathematical Programming (1998)
10. Giannessi, F.: Theorems of the alternative quadratic programs and complementarity problems. In: Cottle, R.W., Giannessi, F., Lions, J.L. (eds.) Variational Inequalities and Complementarity Problems (1980)
11. Giannessi, F.: Vector Variational Inequalities and Vector Equilibria. Mathematical Theories. Kluwer Academic, Dordrecht (2000)
12. Green, J., Heller, W.P.: Mathematical analysis and convexity with applications to economics. Handb. Math. Econ. **1**, 15–52 (1981)
13. Gupta, D., Mehra, A.: Two types of approximate saddle points. Numer. Funct. Anal. Optim. **29**, 532–550 (2008)
14. Hanson, M.A.: On sufficiency of the Kuhn-Tucker conditions. J. Math. Anal. Appl. **80**, 545–550 (1981)
15. Hartman, P., Stampacchia, G.: On some nonlinear elliptic differential functional equations. Acta Math. **115**, 153–188 (1980)
16. Jabarootian, T., Zafarani, J.: Generalized invariant monotonicity and invexity of non-differentiable functions. J. Global Optim. **36**, 537–564 (2006)
17. Jeyakumar, V., Mond, B.: On generalized convex mathematical programming. J. Austral. Math. Soc. Ser. B **34**, 43–53 (1992)
18. Jiménez, B.: Strict efficiency in vector optimization. J. Math. Anal. Appl. **265**, 264–284 (2002)
19. Karamardian, S., Schaible, S.: Seven kinds of monotone maps. J. Optim. Theory Appl. **66**, 37–46 (1990)
20. Lee, G.M.: On relations between vector variational inequality and vector optimization problem. In: Yang, X.Q., Mees, A.I., Fisher, M.E., Jennings, L.S. (eds.) Progress in Optimization II: Contributions from Australia (2000)
21. Li, R., Yu, G.: A class of generalized invex functions and vector variational-like inequalities. J. Inequal. Appl. **2–14** (2017)
22. Mishra, S.K., Wang, S.Y.: Vector variational-like inequalities and nonsmooth vector optimization problems. Nonlinear Anal. **1939–1945** (2006)
23. Mishra, S.K., Upadhyay, B.B.: Some relations between vector variational inequality problems and nonsmooth vector optimization problems using quasi efficiency. Positivity **17**, 1071–1083 (2013)
24. Mishra, S.K., Upadhyay, B.B.: Pseudolinear Functions and Optimization. Chapman and Hall, CRC Press (2015)
25. Mordukhovich, B.S.: Variational Analysis and Generalized Differentiation I: Basic Theory. Springer, Berlin (2006)
26. Mohan, S.R., Neogy, S.K.: On invex sets and preinvex functions. J. Math. Anal. Appl. **189**, 901–908 (1995)
27. Oveisiha, M., Zafarani, J.: Generalized minty vector variational-like inequalities and vector optimization problems in Asplund spaces. Optim. Lett. **7**, 709–721 (2013)
28. Rahtu, E., Salo, M., Heikkilä, J.: A new convexity measure based on a probabilistic interpretation of images. IEEE Trans. Pattern Anal. Mach. Intell. **28**, 1501–1512 (2006)
29. Reiland, T.W.: Nonsmooth invexity. Bull. Austral. Math. Soc. **42**, 437–446 (1990)

30. Smith, P.: Convexity Methods in Variational Calculus. Research Studies Press, Letchworth (1985)
31. Upadhyay, B.B., Mohapatra, R.N., Mishra, S.K.: On relationships between vector variational inequality and nonsmooth vector optimization problems via strict minimizers. Adv. Nonlinear Var. Inequal. **20**, 1–12 (2017)
32. Upadhyay, B.B., Mishra, P.: On generalized Minty and Stampacchia vector variational-like inequalities and nonsmooth vector optimization problems involving higher order strong invexity. J. Sci. Res. **64**, 182–191 (2020)
33. Upadhyay, B.B., Mishra, P.: On vector variational inequalities and vector optimization problems. In: Soft Computing: Theories and Applications, pp. 257-267. Springer, Singapore (2020)
34. Upadhyay, B.B., Mishra, P., Mohapatra, R.N., Mishra, S.K.: On the applications of nonsmooth vector optimization problems to solve generalized vector variational inequalities using convexificators. Adv. Intell. Syst. Comput. (2019). https://doi.org/10.1007/978-3-030-21803-4_66
35. Ward, D.E.: Characterizations of strict local minima and necessary conditions for weak sharp minima. J. Optim. Theory Appl. **80**, 551–557 (1994)
36. Weir, T., Mond, B.: Pre-invex functions in multiple objective optimization. J. Math. Anal. Appl. **136**, 29–38 (1988)
37. Yang, X.M., Yang, X.Q., Teo, K.L.: Generalized invexity and generalized invariant monotonicity. J. Optim. Theory Appl. **117**, 607–625 (2003)

Weighted Average Approximation in Finite Volume Formulation for One-Dimensional Single Species Transport and the Stability Condition for Various Schemes

S. Prabhakaran and L. Jones Tarcius Doss

Abstract The governing equation for one-dimensional single species transport model in a saturated porous medium with appropriate initial and boundary conditions is discretized by using finite volume formulation. A weighted average approximation is then applied to the integral terms. Twelve different schemes of explicit, semi-implicit, and fully implicit in nature are derived. The stability and convergence of those numerical schemes are also discussed. The numerical experiments are carried out for the single species transport problem with degradation in liquid phase. These numerical results are compared with the analytical solution. It is shown that semi-implicit and fully implicit type schemes are not always unconditionally stable. A novel numerical technique is used to approximate the reaction term of partial differential equation. Taking average for reaction term at different time levels yields a better approximation for upwind scheme. Further, it is proved that the averaging technique gives unconditional stability for implicit nature numerical schemes.

Keywords Finite volume method · Weighted average · Contamination transport · Stability · Consistency · First-order reaction

1 Introduction

In the modernized society, the use of chemicals becomes inevitable in day-to-day life. The chemical producing factories are growing like any other industry. The dumping of chemical waste by these factories spoils the surrounding soil of the earth and groundwater quality. The aquifers beneath the earth surface get contaminated more and more by reactive substances like petroleum hydrocarbons and chlorinated

S. Prabhakaran (✉)
Department of Mathematics with Computer Applications, PSG College of Arts & Science, Civil Aerodrome, Coimbatore 641014, India
e-mail: karan.shivam@gmail.com

L. Jones Tarcius Doss
Department of Mathematics, Anna University, Chennai 600025, India

R. N. Mohapatra et al. (eds.), *Mathematical Analysis and Computing*,
Springer Proceedings in Mathematics & Statistics 344,
https://doi.org/10.1007/978-981-33-4646-8_37

solvents. The chemical particles in the waste react with other minerals and metals beneath the surface. These reactions in the aquifers may pose great danger to nature. Therefore, there is a need to study reactive transport in a porous medium. Further, the study about contamination discharge is very important to protect groundwater, oil, and metals.

Analytical solutions for species transport models have been developed since 1971. Cho [1] described the analytical solution for the transport of ammonium with sequential first-order kinetic reaction. This is considered to be a pioneer paper in species transport. At the end of twentieth century, Sun and Clement [2] and Sun et al. [3] have developed a transport model involving the retardation factor. Latter, Bauer et al. [4] and Clement et al. [5] found an analytical solution for multi species transport with first-order sequential reaction involving distinct retardation factors. The analytical solutions are derived with the main assumption that the data set is continuous. But in many real-life problems, the data varies drastically. Therefore, analytical solutions are not so beneficial and hence computationally simulated solutions with the past information are useful for many physical problems.

The computational aspect purely depends on numerical techniques. There are various numerical techniques like finite difference, finite element, and finite volume used to solve equations arising from physical phenomena. A few well-known numerical techniques are listed below in the arena of species transport models. Clement [6] has applied finite difference method for the transport model and then developed a software RT3D. Many available groundwater simulating software like, MODFLOW, PLASAM, AQUIFEM, and FEFLOW are developed from either finite difference or finite element discretization. There are drawbacks in implementing these methods for property transport problems. The finite difference method is not advisable for complex geometry and flux boundary conditions. The finite element method has the global mass conservation property but not locally. The mass conservation principle is pretty important in any transport problem. The finite volume technique takes care of physical and chemical phenomena of the problem under consideration with its local mass conservation principle. In this article, an attempt has been made to study finite volume formulation for transport equation.

2 Governing Equation

The partial differential equation with the appropriate initial and boundary conditions which describes the single species transport in x-direction in a saturated porous medium is given below (see, [4]).
PDE:

$$R\frac{\partial c}{\partial t} + v\frac{\partial c}{\partial x} - D\frac{\partial^2 c}{\partial x^2} = -kc \qquad 0 < x < \infty, \quad t > 0 \tag{1}$$

Type of boundary conditions:

$$c(0,t) = C_0 \qquad t > 0 \tag{2}$$

$$c(0,t) = f(t) \qquad t > 0 \tag{3}$$

$$\lim_{x\to\infty} c(x,t) = 0 \qquad t > 0 \tag{4}$$

Type of initial conditions:

$$c(x,0) = 0 \qquad 0 < x < \infty \tag{5}$$

$$c(x,0) = (x)\ with\ \lim_{x\to\infty} g(x) = 0 \tag{6}$$

Here, c is the concentration of species [ML^{-3}]; R, the retardation factor; v, the prescribed constant transport velocity in x direction [LT^{-1}]; k, the first-order contaminant destruction rate constant [T^{-1}]; and D, the dispersion coefficient [$L^{-1}T^{-1}$].

The problem is to find the concentration of a species $c(x,t)$ at any distance x measured from the origin in any time t satisfying above PDE, initial, and boundary conditions. The boundary condition (2) represents the source of constant dumping of chemical wastage, whereas (3) represents the varying dumping in time t. The condition (4) indicates the zero concentration of species at the farther end at infinity. Similarly, the initial condition (5) represents that there is no sign of species concentration (i.e., contamination) at the initial time. The alternate condition (6) indicates the initial presence of contamination.

The kinetics of reaction is assumed to be of first order. The radioactive decay is an example for a true first-order process. Also chemical and biological transforms can be approximately treated as first-order reaction. Equation (1) assumes that degradation occurs only in the liquid phase.

The above transport equation (1) is used for solving different types of environmental problems. Bauer et al. [4] utilized to model transport of decay chain in homogeneous porous media. Clement et al. [5] applied the generalized form to model multi-species transport coupled with first-order reaction network with distinct retardation factors. The similar kind of equations are employed to model the fate and transport of chlorinated solvent plumes by Clement et al. [7, 8]. Cho [1] used it to model the fate and transport of nitrate species in soil–water systems; Van Genuchten [9] applied it for modeling radionuclide migration. Elango et al. [10] used for groundwater flow and radionuclide decay-chain transport modeling around a proposed uranium tailing pond in India.

Result 2.1

Analytical solution to species transport equation (1) with conditions (2), (4) and (5) is given by (see, [5]):

$$c(x,t) = \frac{C_0}{2}\exp\left(\frac{vx}{2D}\right)\left[\exp\left(-\frac{mx}{2D}\right)erfc\left(\frac{Rx-mt}{\sqrt{4DRt}}\right)+\exp\left(\frac{mx}{2D}\right)erfc\left(\frac{Rx+mt}{\sqrt{4DRt}}\right)\right], \tag{7}$$

where $m = \sqrt{u^2 + 4kD}$.

3 Derivation of the Numerical Scheme

In this section, a finite volume formulation is presented for the transport equation (2.1) described in previous section. The computational domain is discretized by non overlapping control volumes. The control volume (CV) is given in Fig. 1.

Here, $\Delta V = A\Delta x$. Where A is cross-sectional area and Δx is the spatial discretization length. W, P and E are western, present and eastern nodal points. w and e are western and eastern faces of control volume. Though the control volume is in three-dimensional space, only one-dimensional (i.e., x-direction) transport problem is considered in this paper. Therefore, the other two dimensions are assumed to be negligible.
The vector form of (1) is given by

$$R\frac{\partial c}{\partial t} = \overline{\nabla} \cdot (\overline{\nabla} Dc) - \overline{v} \cdot \overline{\nabla}(c) - kc. \tag{8}$$

Integrating the above PDE over the control volume in the interval $(t, t + \Delta t)$ with time step Δt and then applying Gauss divergence theorem, we obtain

$$R\int_t^{t+\Delta t}\int_{CV}\frac{\partial c}{\partial t}dVdt = \int_t^{t+\Delta t}\int_{CV}\overline{\nabla}\cdot(\overline{\nabla} Dc)dVdt - \int_t^{t+\Delta t}\int_{CV}\overline{v}\cdot\overline{\nabla}(c)dVdt$$
$$- \int_t^{t+\Delta t}\int_{CV} kcdVdt.$$

$$R\int_{CV}(c^{n+1} - c^n)dV = \int_t^{t+\Delta t}\int_S \overrightarrow{n}\cdot\overline{\nabla}(Dc)dSdt - \int_t^{t+\Delta t}\int_S \overrightarrow{n}\cdot(c\overline{v})dSdt$$
$$-C_{avg}k\int_{CV} dV\int_t^{t+\Delta t} dt,$$

where $\overrightarrow{n}$ is the outward normal to the surface S (cross-sectional area A) and C_{avg} is the average concentration c inside CV. One-dimensional formulation of above integral is given by

$$R\int_{CV}(c^{n+1} - c^n)dV = \int_t^{t+\Delta t}\int_S \overrightarrow{n}\cdot\overline{i}\frac{\partial(Dc)}{\partial x}dSdt - \int_t^{t+\Delta t}\int_S \overrightarrow{n}\cdot\overline{i}(cv)dSdt$$
$$-C_{avg}kA\Delta x\Delta t,$$

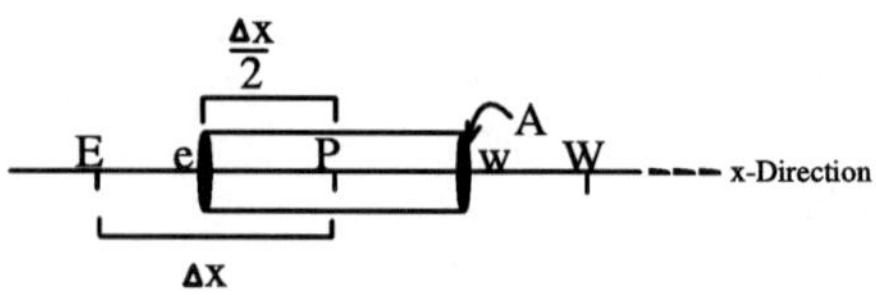

Fig. 1 Control volume (CV)

where C_P^n be the approximation of $c(x, t)$ at the nodal point (x_p, t_n). The parameters D and v are assumed to be constants. Using weighted average for the time integration to obtain

$$R(C_P^{n+1} - C_P^n)\Delta x = \left[(1-\theta)\left[\left(\frac{\partial c}{\partial x}\right)_e^n - \left(\frac{\partial c}{\partial x}\right)_w^n\right] + \theta\left[\left(\frac{\partial c}{\partial x}\right)_e^{n+1} - \left(\frac{\partial c}{\partial x}\right)_w^{n+1}\right]\right] D\Delta t$$
$$-\left[(1-\theta)[c_e^n - c_w^n] - \theta[c_e^{n+1} - c_w^{n+1}]\right] v\Delta t - C_{avg} k \Delta x \Delta t.$$

Using the following central difference approximation for the derivative term

$$\left(\frac{\partial c}{\partial x}\right)_e^n \approx \frac{C_E^n - C_P^n}{\Delta x} \qquad \left(\frac{\partial c}{\partial x}\right)_w^n \approx \frac{C_P^n - C_W^n}{\Delta x},$$

we obtain

$$R(C_P^{n+1} - C_P^n)\Delta x = (1-\theta)\frac{D\Delta t}{\Delta x}\left[C_W^n - 2C_P^n + C_E^n\right] + \theta\frac{D\Delta t}{\Delta x}\left[C_W^{n+1} - 2C_P^n + C_E^{n+1}\right]$$
$$-(1-\theta)v\Delta t\left[c_e^n - c_w^n\right] - \theta v\Delta t\left[c_e^{n+1} - c_w^{n+1}\right] - C_{avg} k \Delta x \Delta t. \quad (9)$$

4 Various Numerical Schemes

The following 12 different numerical schemes are derived for different values of θ and different approximations to C_{avg} and C_{Face}.

θ-values	$C_{Average}$	C_{Face}	Scheme
$\theta = 0$	$C_{avg} = C_P^n$	$C_e^n = \frac{C_E^n + C_P^n}{2}, C_w^n = \frac{C_P^n + C_W^n}{2}$	Scheme 1
		$C_e^n = C_P^n, C_w^n = C_W^n$	Scheme 3
	$C_{avg} = \frac{C_P^n + C_P^{n+1}}{2}$	$C_e^n = \frac{C_E^n + C_P^n}{2}, C_w^n = \frac{C_P^n + C_W^n}{2}$	Scheme 2
		$C_e^n = C_P^n, C_w^n = C_W^n$	Scheme 4
$\theta = \frac{1}{2}$	$C_{avg} = C_P^n$	$C_e^n = \frac{C_E^n + C_P^n}{2}, C_w^n = \frac{C_P^n + C_W^n}{2}$	Scheme 5
		$C_e^n = C_P^n, C_w^n = C_W^n$	Scheme 7
	$C_{avg} = \frac{C_P^n + C_P^{n+1}}{2}$	$C_e^n = \frac{C_E^n + C_P^n}{2}, C_w^n = \frac{C_P^n + C_W^n}{2}$	Scheme 6
		$C_e^n = C_P^n, C_w^n = C_W^n$	Scheme 8
$\theta = 1$	$C_{avg} = C_P^n$	$C_e^n = \frac{C_E^n + C_P^n}{2}, C_w^n = \frac{C_P^n + C_W^n}{2}$	Scheme 9
		$C_e^n = C_P^n, C_w^n = C_W^n$	Scheme 11
	$C_{avg} = \frac{C_P^n + C_P^{n+1}}{2}$	$C_e^n = \frac{C_E^n + C_P^n}{2}, C_w^n = \frac{C_P^n + C_W^n}{2}$	Scheme 10
		$C_e^n = C_P^n, C_w^n = C_W^n$	Scheme 12

The approximations $C_e^n = C_P^n, C_w^n = C_W^n$ and $C_e^n = C_P^n, C_w^n = C_W^n$ are called central difference and upwind, respectively. The average of concentration $C_{avg} = \frac{C_P^n + C_P^{n+1}}{2}$ at n and $n+1$ time level is justified because the control volume is fixed. Following explicit schemes are obtained by substituting $\theta = 0$.

Explicit type schemes ($\theta = 0$)
Scheme 1

$$C_P^{n+1} = \left[\frac{D\Delta t}{R\Delta x^2} + \frac{v\Delta t}{2R\Delta x}\right] C_W^n + \left[1 - \frac{2D\Delta t}{R\Delta x^2} - \frac{k\Delta t}{R}\right] C_P^n + \left[\frac{D\Delta t}{R\Delta x^2} - \frac{v\Delta t}{2R\Delta x}\right] C_E^n. \quad (10)$$

Scheme 2

$$\left[1 + \frac{k\Delta t}{2R}\right] C_P^{n+1} = \left[\frac{D\Delta t}{R\Delta x^2} + \frac{v\Delta t}{2R\Delta x}\right] C_W^n + \left[1 - \frac{2D\Delta t}{R\Delta x^2} - \frac{k\Delta t}{2R}\right] C_P^n + \left[\frac{D\Delta t}{R\Delta x^2} - \frac{v\Delta t}{2R\Delta x}\right] C_E^n. \quad (11)$$

Scheme 3

$$C_P^{n+1} = \left[\frac{D\Delta t}{R\Delta x^2} + \frac{v\Delta t}{R\Delta x}\right] C_W^n + \left[1 - \frac{2D\Delta t}{R\Delta x^2} - \frac{v\Delta t}{R\Delta x} - \frac{k\Delta t}{R}\right] C_P^n + \left[\frac{D\Delta t}{R\Delta x^2}\right] C_E^n. \quad (12)$$

Scheme 4

$$\left[1 + \frac{k\Delta t}{2R}\right] C_P^{n+1} = \left[\frac{D\Delta t}{R\Delta x^2} + \frac{v\Delta t}{R\Delta x}\right] C_W^n + \left[1 - \frac{2D\Delta t}{R\Delta x^2} - \frac{v\Delta t}{R\Delta x} - \frac{k\Delta t}{2R}\right] C_P^n + \left[\frac{D\Delta t}{R\Delta x^2}\right] C_E^n. \quad (13)$$

For $\theta = \frac{1}{2}$, we have the following schemes.

Semi-implicit type schemes ($\theta = \frac{1}{2}$)

Scheme 5

$$-\left[\frac{D\Delta t}{2R\Delta x^2} + \frac{v\Delta t}{4R\Delta x}\right] C_W^{n+1} + \left[1 + \frac{D\Delta t}{R\Delta x^2}\right] C_P^{n+1} - \left[\frac{D\Delta t}{2R\Delta x^2} - \frac{v\Delta t}{4R\Delta x}\right] C_E^{n+1}$$
$$= \left[\frac{D\Delta t}{2R\Delta x^2} + \frac{v\Delta t}{4R\Delta x}\right] C_W^n + \left[1 - \frac{D\Delta t}{R\Delta x^2} - \frac{k\Delta t}{R}\right] C_P^n + \left[\frac{D\Delta t}{2R\Delta x^2} - \frac{v\Delta t}{4R\Delta x}\right] C_E^n. \quad (14)$$

Scheme 6

$$-\left[\frac{D\Delta t}{2R\Delta x^2} + \frac{v\Delta t}{4R\Delta x}\right] C_W^{n+1} + \left[1 + \frac{D\Delta t}{R\Delta x^2} + \frac{k\Delta t}{2R}\right] C_P^{n+1} - \left[\frac{D\Delta t}{2R\Delta x^2} - \frac{v\Delta t}{4R\Delta x}\right] C_E^{n+1}$$
$$= \left[\frac{D\Delta t}{2R\Delta x^2} + \frac{v\Delta t}{4R\Delta x}\right] C_W^n + \left[1 - \frac{D\Delta t}{R\Delta x^2} - \frac{k\Delta t}{2R}\right] C_P^n + \left[\frac{D\Delta t}{2R\Delta x^2} - \frac{v\Delta t}{4R\Delta x}\right] C_E^n. \quad (15)$$

Scheme 7

$$-\left[\frac{D\Delta t}{2R\Delta x^2}+\frac{v\Delta t}{2R\Delta x}\right]C_W^{n+1}+\left[1+\frac{D\Delta t}{R\Delta x^2}+\frac{v\Delta t}{2R\Delta x}\right]C_P^{n+1}-\left[\frac{D\Delta t}{2R\Delta x^2}\right]C_E^{n+1}$$
$$=\left[\frac{D\Delta t}{2R\Delta x^2}+\frac{v\Delta t}{2R\Delta x}\right]C_W^{n}+\left[1-\frac{D\Delta t}{R\Delta x^2}-\frac{v\Delta t}{2R\Delta x}-\frac{k\Delta t}{R}\right]C_P^{n}+\left[\frac{D\Delta t}{2R\Delta x^2}\right]C_E^{n}. \tag{16}$$

Scheme 8

$$-\left[\frac{D\Delta t}{2R\Delta x^2}+\frac{v\Delta t}{2R\Delta x}\right]C_W^{n+1}+\left[1+\frac{D\Delta t}{R\Delta x^2}+\frac{v\Delta t}{2R\Delta x}+\frac{k\Delta t}{2R}\right]C_P^{n+1}-\left[\frac{D\Delta t}{2R\Delta x^2}\right]C_E^{n+1}$$
$$=\left[\frac{D\Delta t}{2R\Delta x^2}+\frac{v\Delta t}{2R\Delta x}\right]C_W^{n}+\left[1-\frac{D\Delta t}{R\Delta x^2}-\frac{v\Delta t}{2R\Delta x}-\frac{k\Delta t}{2R}\right]C_P^{n}+\left[\frac{D\Delta t}{2R\Delta x^2}\right]C_E^{n}. \tag{17}$$

Implicit type schemes ($\theta = 1$)

Scheme 9

$$-\left[\frac{D\Delta t}{R\Delta x^2}+\frac{v\Delta t}{2R\Delta x}\right]C_W^{n+1}+\left[1+\frac{2D\Delta t}{R\Delta x^2}\right]C_P^{n+1}-\left[\frac{D\Delta t}{R\Delta x^2}-\frac{v\Delta t}{2R\Delta x}\right]C_E^{n+1}=\left[1-\frac{k\Delta t}{R}\right]C_P^{n}. \tag{18}$$

Scheme 10

$$-\left[\frac{D\Delta t}{R\Delta x^2}+\frac{v\Delta t}{2R\Delta x}\right]C_W^{n+1}+\left[1+\frac{2D\Delta t}{R\Delta x^2}+\frac{k\Delta t}{2R}\right]C_P^{n+1}-\left[\frac{D\Delta t}{R\Delta x^2}-\frac{v\Delta t}{2R\Delta x}\right]C_E^{n+1}$$
$$=\left[1-\frac{k\Delta t}{2R}\right]C_P^{n}. \tag{19}$$

Scheme 11

$$-\left[\frac{D\Delta t}{R\Delta x^2}+\frac{v\Delta t}{R\Delta x}\right]C_W^{n}+\left[1+\frac{2D\Delta t}{R\Delta x^2}+\frac{v\Delta t}{R\Delta x}\right]C_P^{n}-\left[\frac{D\Delta t}{R\Delta x^2}\right]C_E^{n}=\left[1-\frac{k\Delta t}{R}\right]C_P^{n}. \tag{20}$$

Scheme 12

$$-\left[\frac{D\Delta t}{R\Delta x^2}+\frac{v\Delta t}{R\Delta x}\right]C_W^{n}+\left[1+\frac{2D\Delta t}{R\Delta x^2}+\frac{v\Delta t}{R\Delta x}+\frac{k\Delta t}{2R}\right]C_P^{n}-\left[\frac{D\Delta t}{R\Delta x^2}\right]C_E^{n}=\left[1-\frac{k\Delta t}{2R}\right]C_P^{n}. \tag{21}$$

5 Stability Analysis

In this section, we shall discuss the stability of general form of the explicit, semi-implicit, and implicit finite difference schemes. The general form of explicit, semi-implicit, and implicit schemes are given below:

$$C_m^{n+1} = pC_{m-1}^n + qC_m^n + rC_{m+1}^n \tag{22}$$

$$-pC_{m-1}^{n+1} + qC_m^{n+1} - rC_{m+1}^{n+1} = pC_{m-1}^n + sC_m^n + rC_{m+1}^n \tag{23}$$

$$-pC_{m-1}^{n+1} + qC_m^{n+1} - rC_{m+1}^{n+1} = sC_m^n \tag{24}$$

Definition 5.1 A scheme $U_m^{n+1} = G(U_{m-k}^n, ..., U_m^n, ...U_{m+p}^n)$ is called monotone scheme if G is non-decreasing function of each of its argument.
i.e., $\frac{\partial G}{\partial U_i}(U_{-k}, ..., U_0, ...U_p) \geq 0, i = -k, ..., p.$

Theorem 5.1 *Let $C_m^{n+1} = pC_{m-1}^n + qC_m^n + rC_{m+1}^n$ be the general form of explicit finite difference scheme for the linear time-dependent partial differential equation (1). If $p \geq 0,\ q \geq 0$ and $r \geq 0$ and satisfy $(p+q+r)^2 \leq 1 + 4q(p+r)$, then the scheme is stable and monotone.*

Proof Let $C_m^{n+1} = pC_{m-1}^n + qC_m^n + rC_{m+1}^n$ be the general form of explicit finite difference numerical scheme. Let $C_m^n = B\xi^n e^{im\theta}$. The von Neumann stability analysis for the above difference scheme implies,

$$\xi = pe^{-i\theta} + q + re^{i\theta} = q + (p+r)(\cos\theta) + i(r-p)\sin\theta.$$

The numerical scheme (22) is stable only when $|\xi| \leq 1$ which is equivalently $|\xi|^2 \leq 1$ (Smith, [11]). Therefore,

$$\begin{aligned} & q^2 + (p+r)^2\cos^2\theta + 2q(p+r)\cos\theta + (r-p)^2\sin^2\theta \leq 1 \\ \Leftrightarrow\ & p^2 + q^2 + r^2 + 2pr(\cos^2\theta - \sin^2\theta) + 2q(p+r)\cos\theta \leq 1 \\ \Leftrightarrow\ & (p+q+r)^2 - 2q(p+r)(1-\cos\theta) - 2pr(1-\cos 2\theta) \leq 1 \end{aligned}$$

$$\Leftrightarrow (p+q+r)^2 \leq 1 + 4b(p+r)\sin^2\frac{\theta}{2} + 4pr\sin^2\theta$$

$$\Leftrightarrow (p+q+r)^2 \leq 1 + 4q(p+r)\sin^2\frac{\theta}{2} + 16pr\sin^2\frac{\theta}{2}\cos^2\frac{\theta}{2}$$

$$\Leftrightarrow (p+q+r)^2 + 16pr\sin^4\frac{\theta}{2} \leq 1 + 4q(p+r)\sin^2\frac{\theta}{2} + 16pr\sin^2\frac{\theta}{2}.$$

Maximizing the trigonometric functions in above inequality with respect to their argument, we obtain

$$(p+q+r)^2 \le 1+4q(p+r). \tag{25}$$

Let us assume that p, q and r are greater than or equal to zero $(p, q, r \ge 0)$. Let $G(C^n_{m-1}, C^n_m, C^n_{m+1}) = pC^n_{m-1} + qC^n_m + rC^n_{m+1}$ be the function. From (22), $C^{n+1}_m = G(C^n_{m-1}, C^n_m, C^n_{m+1})$. By Definition (5.1) $\frac{\partial G}{\partial C_i}(C_{-1}, C_0, C_1) \ge 0, i = -1, 0, 1$, which implies that the scheme is monotone. Therefore, the monotone scheme which satisfies (25) is stable.

Theorem 5.2 *Let* $-pC^{n+1}_{m-1} + qC^{n+1}_m - rC^{n+1}_{m+1} = pC^n_{m-1} + sC^n_m + rC^n_{m+1}$ *be the general form of semi-implicit finite difference scheme for the time-dependent linear partial differential equation (1). If* $q + s > 0$, $q \ge s$ *and satisfy* $2(p + r) \le q - s$, *then the scheme is stable.*

Proof Let $-pC^{n+1}_{m-1} + qC^{n+1}_m - rC^{n+1}_{m+1} = pC^n_{m-1} + sC^n_m + rC^n_{m+1}$ be any general semi-implicit scheme. Let $C^n_m = B\xi^n e^{im\theta}$. The von Neumann stability condition $|\xi| \le 1$ implies

$$\begin{aligned}
|\ s + (p+r)(\cos\theta) + i(r-p)\sin\theta\ | &\le |\ q - (p+r)(\cos\theta) + i(p-r)\sin\theta\ | \\
s^2 + 2s(p+r)\cos\theta &\le q^2 - 2q(p+r)\cos\theta \\
2(q+s)(p+r)\cos\theta &\le q^2 - s^2 \\
2(q+s)(p+r)\cos\theta &\le (q+s)(q-s)
\end{aligned}$$

Let us assume that $p + r \ge 0$, $(q + s) \ge 0$ and $q \ge s$. Maximizing with respect to θ, we obtain

$$2(p+r) \le q - s \tag{26}$$

Therefore, any semi-implicit numerical scheme is of the form (23) which satisfies the condition (26) is stable.

Theorem 5.3 *Let* $-pC^{n+1}_{m-1} + qC^{n+1}_m - rC^{n+1}_{m+1} = sC^n_m$ *be the general form of implicit finite difference scheme for the linear time-dependent partial differential equation (1). If* $s^2 \le (p+q+r)^2 - 4q(p+r)$, *then the scheme is stable.*

Proof Let $-pC^{n+1}_{m-1} + qC^{n+1}_m - rC^{n+1}_{m+1} = sC^n_m$ be the general form of implicit finite difference numerical scheme. Let $C^n_m = B\xi^n e^{im\theta}$. The von Neumann stability analysis for the above difference scheme implies

$$\xi(-pe^{-i\theta} + q - re^{i\theta}) = s.$$

$$\begin{aligned}
& s^2 \le (p+r)^2\cos^2\theta - 2q(p+r)\cos\theta + q^2 + (p-r)^2\sin^2\theta \\
\Leftrightarrow\ & s^2 \le p^2 + q^2 + r^2 + 2pr(\cos^2\theta - \sin^2\theta) - 2q(p+r)\cos\theta \\
\Leftrightarrow\ & s^2 \le p^2 + q^2 + r^2 + 2pr(2\cos^2\theta - 1) - 2q(p+r)\cos\theta \\
\Leftrightarrow\ & s^2 \le (p+q+r)^2 + 4pr(\cos^2\theta - 1) - 2q(p+r)(1+\cos\theta)
\end{aligned}$$

$$\Leftrightarrow s^2 + 4pr + 2q(p+r)(1+\cos\theta) \leq (p+q+r)^2 + 4pr\cos^2\theta$$

Maximizing the trigonometric functions in above inequality with respect to their argument, we obtain

$$s^2 \leq (p+q+r)^2 - 4q(p+r). \tag{27}$$

Therefore, any implicit numerical scheme is of the form (24) which satisfies the condition (27) is stable.

Stability for explicit schemes
Comparing *Scheme 1* (10) with general form of explicit schemes (22), we have

$$p = \frac{D\Delta t}{R\Delta x^2} + \frac{v\Delta t}{2R\Delta x} \qquad q = 1 - \frac{2D\Delta t}{R\Delta x^2} - \frac{k\Delta t}{R} \qquad r = \frac{D\Delta t}{R\Delta x^2} - \frac{v\Delta t}{2R\Delta x}$$

$$p+q+r = 1 - \frac{k\Delta t}{R} \qquad 4q(p+r) = \frac{8D\Delta t}{R\Delta x^2} - \frac{16D^2\Delta t^2}{R^2\Delta x^4} - \frac{8Dk\Delta t^2}{R^2\Delta x^2}$$

Stability condition (25) implies

$$\Delta t \leq \frac{R(8D + 2k\Delta x^2)\Delta x^2}{16D^2 + k^2\Delta x^4 + 8Dk\Delta x^2} \tag{28}$$

The above condition is independent of velocity term v. Therefore, the stability behavior of central difference scheme can not judged. It should be noted that the condition (28) coincides with CFL condition for pure diffusion (i.e., $v = 0$ and $k = 0$) process. It is assumed that the coefficients of explicit schemes are greater than or equal to zero. Therefore, the coefficient $r \geq 0$ which is eventually

$$\frac{v\Delta x}{D} \leq 2 \tag{29}$$

The left-side quantity in above is nothing but the Peclet number. Therefore, the central difference scheme is stable for Peclet number less than or equal 2. For a pure reaction process (i.e., $D = 0$ and $v = 0$), the numerical scheme (10) becomes Euler method for first-order differential equation

$$C^{n+1} = \left[1 - \frac{k\Delta t}{R}\right] C^n.$$

And the stability condition (28) coincides with absolutely stable condition for Euler method which is given by

$$\left|1 - \frac{k\Delta t}{R}\right| \leq 1$$

The above condition may lead to produce negative result in the concentration profile in the explicit scheme (10). The positivity of the solution is important. Therefore, it should satisfy

$$0 \leq 1 - \frac{k\Delta t}{R} \leq 1$$

$$\text{i.e.,} \frac{k\Delta t}{R} \leq 1 \tag{30}$$

for (10). The conditions (28) and (30) combined together will produce stable and positive solution for (10). Similarly, the stability condition for (11) is given by

$$p = \frac{\frac{D\Delta t}{R\Delta x^2} + \frac{v\Delta t}{2R\Delta x}}{1 + \frac{k\Delta t}{2R}} \qquad q = \frac{1 - \frac{2D\Delta t}{R\Delta x^2} - \frac{k\Delta t}{2R}}{1 + \frac{k\Delta t}{2R}} \qquad r = \frac{\frac{D\Delta t}{R\Delta x^2} + \frac{v\Delta t}{2R\Delta x}}{1 + \frac{k\Delta t}{2R}}$$

$$\Delta t \leq \frac{R(8D + 4k\Delta x^2)\Delta x^2}{16D^2 + 8Dk\Delta x^2}. \tag{31}$$

$$\frac{v\Delta x}{D} \leq 2$$

and

$$0 \leq 1 - \frac{k\Delta t}{2R} \leq 1$$

i.e.,

$$\frac{k\Delta t}{R} \leq 2.$$

The stability conditions for (12) and (13) can be derived in the similar manner. The conditions for *Scheme 3* is given by

$$\Delta t \leq \frac{R(8D + 4v\Delta x + 2k\Delta x^2)\Delta x^2}{16D^2 + 16Dv\Delta x + 8Dk\Delta x^2 + 4vk\Delta x^3 + 4v^2\Delta x^2 + k^2\Delta x^4}. \tag{32}$$

The above condition satisfies the CFL condition for pure diffusion and advection process. Also it satisfies absolute stability condition for Euler method for pure reaction process. The condition for *Scheme 4* is given by

$$\Delta t \leq \frac{R(8D + 4v\Delta x + 4k\Delta x^2)\Delta x^2}{16D^2 + 16Dv\Delta x + 8Dk\Delta x^2 + 4vk\Delta x^3 + 4v^2\Delta x^2}. \tag{33}$$

In general, the first-order reaction co-efficient (k) is very small. Therefore, the contribution of k in the stability of explicit schemes is negligible. One must ensure that

$\frac{k\Delta t}{R} \leq 1$ for *scheme 1* and *scheme 3* and $\frac{k\Delta t}{R} \leq 2$ for *scheme 2* and *scheme 4* before implementation.

Stability for semi-implicit schemes

The semi-implicit type schemes *scheme 5, 6, 7* and *8* satisfy the condition (26). However the assumptions in Theorem 5.2 are not satisfied by all schemes. Only the two schemes, namely, *scheme 6* and *scheme 8* satisfy all assumptions. Therefore, they are unconditionally stable. *Scheme 5* and *scheme 7* satisfy all assumptions except $q + s \geq 0$. The condition $q + s \geq 0$ implies that

$$\frac{k\Delta t}{R} \leq 2.$$

Therefore, *scheme 5* and *scheme 7* are conditionally stable.

Stability for implicit schemes

Comparing *Scheme 9* (18) with general form of implicit schemes (24), we have

$$p = \frac{D\Delta t}{R\Delta x^2} + \frac{v\Delta t}{2R\Delta x} \qquad q = 1 + \frac{2D\Delta t}{R\Delta x^2} \qquad r = \frac{D\Delta t}{R\Delta x^2} - \frac{v\Delta t}{2R\Delta x} \qquad s = 1 - \frac{k\Delta t}{R}.$$

Stability condition (27) implies that

$$\begin{aligned}
\left(1 - \frac{k\Delta t}{R}\right)^2 &\leq \left(1 + \frac{4D\Delta t}{R\Delta x^2}\right)^2 - 4\left(1 + \frac{2D\Delta t}{R\Delta x^2}\right)\left(\frac{2D\Delta t}{R\Delta x^2}\right) \\
\left(1 - \frac{k\Delta t}{R}\right)^2 &\leq 1 \\
-1 \leq 1 - \frac{k\Delta t}{R} &\leq 1.
\end{aligned}$$

Therefore, *Scheme 9* is stable if $\frac{k\Delta t}{R} \leq 2$. Similarly, *Scheme 11* is also stable if $\frac{k\Delta t}{R} \leq 2$. The stability condition for *Scheme 10* is given by

$$p = \frac{D\Delta t}{R\Delta x^2} + \frac{v\Delta t}{2R\Delta x} \qquad q = 1 + \frac{2D\Delta t}{R\Delta x^2} + \frac{k\Delta t}{2R} \qquad r = \frac{D\Delta t}{R\Delta x^2} - \frac{v\Delta t}{2R\Delta x} \qquad s = 1 - \frac{k\Delta t}{2R}.$$

Stability condition (27) implies that

$$\begin{aligned}
\left(1 - \frac{k\Delta t}{2R}\right)^2 &\leq \left(1 + \frac{4D\Delta t}{R\Delta x^2} + \frac{k\Delta t}{2R}\right)^2 - 4\left(1 + \frac{2D\Delta t}{R\Delta x^2} + \frac{k\Delta t}{2R}\right)\left(\frac{2D\Delta t}{R\Delta x^2}\right) \\
\left(1 - \frac{k\Delta t}{R}\right)^2 &\leq \left(1 + \frac{k\Delta t}{R}\right)^2 \\
0 &\leq \frac{4k\Delta t}{R}.
\end{aligned}$$

Therefore, *Scheme 10* is unconditionally stable. Similarly, *Scheme 12* is also unconditionally stable.

6 Truncation Error and Consistency

The truncation error $T_{i,j}$ at interior nodal point (x_i, t_j) is defined by

$$T_{i,j} = pC_{i-1}^{j+1} + qC_i^{j+1} + rC_{i+1}^{j+1} - aC_{i-1}^j - bC_i^j - dC_{i+1}^j$$

where C_i^j is the solution at (x_i, t_j). Following the usual procedure to obtain the truncation error, we replace numerical solution by exact solution

$$\begin{aligned} T_{i,j} &= pc_{i-1}^{j+1} + qc_i^{j+1} + rc_{i+1}^{j+1} - aC_{i-1}^j - bC_i^j - dC_{i+1}^j \\ &= pc(x_i - \Delta x, t_j + \Delta t) + qc(x_i, t_j + \Delta t) + rc(x_i + \Delta x, t_j + \Delta t) \\ &\quad -ac(x_i - \Delta x, t_j) - bc(x_i, t_j) - dc(x_i + \Delta x, t_j). \end{aligned}$$

Expanding using Taylor series, we have that

$$\begin{aligned} T_{i,j} = &\left\{ (p+q+r-a-b-d)c + (p+q+r)\left[\Delta t \frac{\partial c}{\partial t} + \frac{\Delta t^2}{2}\frac{\partial^2 c}{\partial t^2} + \cdots \right] \right\}_{(x_i,t_j)} \\ &+ \left\{ (r-p-d+a)\left[\Delta x \frac{\partial c}{\partial x} + \frac{\Delta x^3}{6}\frac{\partial^3 c}{\partial x^3} \right] + (r+p-d-a)\frac{\Delta x^2}{2}\frac{\partial^2 c}{\partial x^2} \right\}_{(x_i,t_j)} \\ &+ \left\{ (r-p)\left[\Delta x \Delta t \frac{\partial^2 c}{\partial x \partial t} + \frac{\Delta x \Delta t^2}{2}\frac{\partial^3 c}{\partial x \partial t^2} \right] + (r+p)\frac{\Delta x^2 \Delta t}{2}\frac{\partial^3 c}{\partial x^2 \partial t} + \cdots \right\}_{(x_i,t_j)} \end{aligned} \tag{34}$$

Explicit type schemes

Scheme 1

From (10), $p = r = 0$, $q = 1$, $a + b + d = 1 - \frac{k\Delta t}{R}$, $a - d = \frac{v\Delta t}{R\Delta x}$ and $a + d = \frac{2D\Delta t}{R\Delta x^2}$. Using these in (34), We have that

$$T_{i,j} = \left[\frac{kc\Delta t}{R} + \Delta t\frac{\partial c}{\partial t} + \frac{\Delta t^2}{2}\frac{\partial^2 c}{\partial t^2} + \frac{v\Delta t}{R}\frac{\partial c}{\partial x} - \frac{D\Delta t}{R}\frac{\partial^2 c}{\partial x^2} + \frac{v\Delta t\Delta x^2}{6R}\frac{\partial^3 c}{\partial x^3}\right]_{(x_i,t_j)} + \cdots$$

$$\frac{1}{\Delta t}T_{i,j} = \left[\frac{kc}{R} + \frac{\partial c}{\partial t} + \frac{v}{R}\frac{\partial c}{\partial x} - \frac{D}{R}\frac{\partial^2 c}{\partial x^2}\right]_{(x_i,t_j)} + \left[\frac{\Delta t}{2}\frac{\partial^2 c}{\partial t^2} + \frac{v\Delta x^2}{6R}\frac{\partial^3 c}{\partial x^3}\right]_{(x_i,t_j)} + \cdots \tag{35}$$

The first term of right hand side is the given partial differential equation (1) evaluated at the interior point (x_i, t_j). Therefore, we have that

$$\frac{1}{\Delta t}T_{i,j} = \frac{\Delta t}{2}c_{tt} + \frac{v\Delta x^2}{6R}c_{xxx} + \cdots$$

Hence, the order of truncation error is $O(\Delta t + \Delta x^2)$. If $\Delta t = \Delta x^2$, then the truncation error will be of $O(\Delta x^2)$. There fore, $\|c - C_h\|_\infty = O(h^2)$, where $h = \Delta x$ is order and C_h is numerical solution for the mesh length h. For different mesh lengths h_2 and h_2, we have that

$$\frac{\|c - C_{h_1}\|_\infty}{\|c - C_{h_2}\|_\infty} \approx \left(\frac{h_1}{h_2}\right)^2$$

$$\text{i.e.,}\quad \frac{\log\left(\frac{\|c-C_{h_1}\|_\infty}{\|c-C_{h_2}\|_\infty}\right)}{\log\left(\frac{h_1}{h_2}\right)} \approx 2 \tag{36}$$

Therefore, the order of convergence of *Scheme 1* is two. Let $\Delta t \to 0$ and $\Delta x \to 0$, the truncation error (35) $T_{i,j} \to 0$, the scheme is consistent with the partial differential equation (1).

Scheme 2

From (11), $p = r = 0, q = 1 + \frac{k\Delta t}{2R}, a + b + d = 1 - \frac{k\Delta t}{2R}, a - d = \frac{v\Delta t}{R\Delta x}$ and $a + d = \frac{2D\Delta t}{R\Delta x^2}$ in Eq. (34), We have

$$\frac{1}{\Delta t}T_{i,j} = \left[\frac{kc}{R} + \frac{\partial c}{\partial t} + \frac{v}{R}\frac{\partial c}{\partial x} - \frac{D}{R}\frac{\partial^2 c}{\partial x^2}\right]_{(x_i,t_j)} + \left[\frac{k\Delta t}{2R}\frac{\partial c}{\partial t} + \frac{v\Delta x^2}{6R}\frac{\partial^3 c}{\partial x^3}\right]_{(x_i,t_j)} + \cdots \tag{37}$$

which implies that

$$\frac{1}{\Delta t}T_{i,j} = \frac{k\Delta t}{2R}c_t + \frac{v\Delta x^2}{6R}c_{xxx} + \cdots$$

The local truncation error is $O(\Delta t + \Delta x^2)$ and its order is two. Also the scheme is consistent with the partial differential equation (1).

Scheme 3

Substituting $p = r = 0$, $q = 1$, $a + b + d = 1 - \frac{k\Delta t}{R}$, $a - d = \frac{v\Delta t}{R\Delta x}$ and $a + d = \frac{2D\Delta t}{R\Delta x^2} + \frac{v\Delta t}{R\Delta x}$ in (34), we get truncation error for *Scheme 3*

$$\frac{1}{\Delta t}T_{i,j} = \left[\frac{kc}{R} + \frac{\partial c}{\partial t} + \frac{v}{R}\frac{\partial c}{\partial x} - \frac{D}{R}\frac{\partial^2 c}{\partial x^2}\right]_{(x_i,t_j)} + \left[\frac{\Delta t}{2}\frac{\partial^2 c}{\partial t^2} - \frac{v\Delta x}{2R}\frac{\partial^2 c}{\partial x^2}\right]_{(x_i,t_j)} + \cdots \quad (38)$$

$$\frac{1}{\Delta t}T_{i,j} = \frac{\Delta t}{2}c_{tt} - \frac{v\Delta x}{2R}c_{xx} + \cdots$$

Here, the truncation error is of order $O(\Delta t + \Delta x)$. If $\Delta t = \Delta x$, then the truncation error will be of order $O(\Delta x)$. In a similar manner to central difference scheme, we have

$$\frac{log\left(\frac{\|c - C_{h_1}\|_\infty}{\|c - C_{h_2}\|_\infty}\right)}{log\left(\frac{h_1}{h_2}\right)} \approx 1 \quad (39)$$

It means that *Scheme 3* is of first-order convergence. Also, the truncation error (38) $T_{i,j} \to 0$ as $\Delta t \to 0$ and $\Delta x \to 0$ and the scheme consistent with the partial differential equation (1).

Scheme 4

Substituting $p = r = 0, q = 1 + \frac{k\Delta t}{2R}, a + b + d = 1 - \frac{k\Delta t}{2R}, a - d = \frac{v\Delta t}{R\Delta x}$ and $a + d = \frac{2D\Delta t}{R\Delta x^2} + \frac{v\Delta t}{R\Delta x}$ in (34), we get truncation error

$$\frac{1}{\Delta t}T_{i,j} = \left[\frac{kc}{R} + \frac{\partial c}{\partial t} + \frac{v}{R}\frac{\partial c}{\partial x} - \frac{D}{R}\frac{\partial^2 c}{\partial x^2}\right]_{(x_i,t_j)} + \left[\frac{k\Delta t}{2R}\frac{\partial c}{\partial t} - \frac{v\Delta x}{2R}\frac{\partial^2 c}{\partial x^2}\right]_{(x_i,t_j)} + \cdots \quad (40)$$

$$\frac{1}{\Delta t}T_{i,j} = \frac{k\Delta t}{2R}c_t - \frac{v\Delta x}{2R}c_{xx} + \cdots$$

There fore, *Scheme 4* is of first-order convergence and the scheme is consistent with the partial differential equation (1).

Semi implicit type schemes

Scheme 5

Using $p+r+q=1$, $a+b+d=1-\frac{k\Delta t}{R}$, $a-d=\frac{v\Delta t}{2R\Delta x}$, $r-p=\frac{v\Delta t}{2R\Delta x}$, $r+p=-\frac{D\Delta t}{R\Delta x^2}$ and $a+d=\frac{D\Delta t}{R\Delta x^2}$ in Eq. (34), we have that

$$\frac{1}{\Delta t}T_{i,j}=\left[\frac{kc}{R}+\frac{\partial c}{\partial t}+\frac{v}{R}\frac{\partial c}{\partial x}-\frac{D}{R}\frac{\partial^2 c}{\partial x^2}+\frac{\Delta t}{2}\frac{\partial^2 c}{\partial t^2}+\frac{v\Delta t}{2R}\frac{\partial^2 c}{\partial x\partial t}-\frac{D\Delta t}{2R}\frac{\partial^2 c}{\partial x^2\partial t}+\cdots\right]_{(x_i,t_j)} \tag{41}$$

$$\frac{1}{\Delta t}T_{i,j}=\left(\frac{1}{2}c_{tt}+\frac{v}{2R}c_{xt}-\frac{D}{2R}c_{xxt}\right)\Delta t+\frac{v\Delta x^2}{6R}c_{xxx}+\cdots$$

Hence, the local truncation error is $O(\Delta t+\Delta x^2)$ when $\Delta t\to 0$ and $\Delta x\to 0$. Also, the scheme is consistent with the partial differential equation (1).

Scheme 6

Substituting $p+r+q=1+\frac{k\Delta t}{2R}$, $a+b+d=1-\frac{k\Delta t}{2R}$, $a-d=\frac{v\Delta t}{2R\Delta x}$, $r-p=\frac{v\Delta t}{2R\Delta x}$, $r+p=-\frac{D\Delta t}{R\Delta x^2}$ and $a+d=\frac{D\Delta t}{R\Delta x^2}$ in Eq. (34), we have

$$\frac{1}{\Delta t}T_{i,j}=\left[\frac{kc}{R}+\frac{\partial c}{\partial t}+\frac{v}{R}\frac{\partial c}{\partial x}-\frac{D}{R}\frac{\partial^2 c}{\partial x^2}+\frac{k\Delta t}{2R}\frac{\partial c}{\partial t}+\frac{v\Delta t}{2R}\frac{\partial^2 c}{\partial x\partial t}-\frac{D\Delta t}{2R}\frac{\partial^2 c}{\partial x^2\partial t}+\cdots\right]_{(x_i,t_j)} \tag{42}$$

$$\frac{1}{\Delta t}T_{i,j}=\left(\frac{k}{2R}c_{t}+\frac{v}{2R}c_{xt}-\frac{D}{2R}c_{xxt}\right)\Delta t+\frac{v\Delta x^2}{6R}c_{xxx}+\cdots$$

The scheme is consistent with the partial differential equation (1) and the local truncation error is given by $O(\Delta t+\Delta x^2)$.

Scheme 7

Using $p+q+r=1, a+b+d=1-\frac{k\Delta t}{R}, a-d=\frac{v\Delta t}{2R\Delta x}, r-p=\frac{v\Delta t}{2R\Delta x}, r+p=-\frac{D\Delta t}{R\Delta x^2}-\frac{v\Delta t}{2R\Delta x}$ and $a+d=\frac{D\Delta t}{\Delta x^2}+\frac{v\Delta t}{2R\Delta x}$ in (34), in Eq. (34), we have

$$\frac{1}{\Delta t}T_{i,j}=\left[\frac{kc}{R}+\frac{\partial c}{\partial t}+\frac{v}{R}\frac{\partial c}{\partial x}-\frac{D}{R}\frac{\partial^2 c}{\partial x^2}+\frac{\Delta t}{2}\frac{\partial^2 c}{\partial t^2}+\frac{v\Delta t}{2R}\frac{\partial^2 c}{\partial x\partial t}-\frac{D\Delta t}{2R}\frac{\partial^2 c}{\partial x^2\partial t}+\cdots\right]_{(x_i,t_j)} \tag{43}$$

$$\frac{1}{\Delta t}T_{i,j}=\left(\frac{1}{2}c_{tt}+\frac{v}{2R}c_{xt}-\frac{D}{2R}c_{xxt}\right)\Delta t-\frac{v\Delta x}{2R}c_{xx}+\cdots$$

Therefore, the scheme is consistent with the partial differential equation (1) and the order of truncation error is $O(\Delta t + \Delta x)$.

Scheme 8

Using $p+q+r = 1 + \frac{k\Delta t}{2R}$, $a+b+d = 1 - \frac{k\Delta t}{2R}$, $a-d = \frac{v\Delta t}{2R\Delta x}$, $r-p = \frac{v\Delta t}{2R\Delta x}$, $r+p = -\frac{D\Delta t}{R\Delta x^2} - \frac{v\Delta t}{2R\Delta x}$ and $a+d = \frac{D\Delta t}{\Delta x^2} + \frac{v\Delta t}{2R\Delta x}$ in Eq. (34), We have that

$$\frac{1}{\Delta t}T_{i,j} = \left[\frac{kc}{R} + \frac{\partial c}{\partial t} + \frac{v}{R}\frac{\partial c}{\partial x} - \frac{D}{R}\frac{\partial^2 c}{\partial x^2} + \frac{k\Delta t}{2R}\frac{\partial c}{\partial t} + \frac{v\Delta t}{2R}\frac{\partial^2 c}{\partial x \partial t} - \frac{D\Delta t}{2R}\frac{\partial^2 c}{\partial x^2 \partial t} + \cdots\right]_{(x_i,t_j)}$$
$$\frac{1}{\Delta t}T_{i,j} = \left(\frac{k}{2R}c_t + \frac{v}{2R}c_{xt} - \frac{D}{2R}c_{xxt}\right)\Delta t - \frac{v\Delta x}{2R}c_{xx} + \cdots \tag{44}$$

Hence, it will be of first-order convergence. Also the scheme is consistent with the partial differential equation (1).

Implicit type schemes

Scheme 9

Substituting $p+q+r = 1$, $a = 1 - \frac{k\Delta t}{R}$, $b = d = 0$, $p - r = -\frac{v\Delta t}{R\Delta x}$ and $p + r = -\frac{2D\Delta t}{R\Delta x^2}$ in Eq. (34), We get

$$\frac{1}{\Delta t}T_{i,j} = \left[\frac{kc}{R} + \frac{\partial c}{\partial t} + \frac{v}{R}\frac{\partial c}{\partial x} - \frac{D}{R}\frac{\partial^2 c}{\partial x^2} + \frac{\Delta t}{2}\frac{\partial^2 c}{\partial t^2} + \frac{v\Delta t}{R}\frac{\partial^2 c}{\partial x \partial t} - \frac{D\Delta t}{R}\frac{\partial^2 c}{\partial x^2 \partial t} + \cdots\right]_{(x_i,t_j)}$$
$$\frac{1}{\Delta t}T_{i,j} = \left(\frac{1}{2}c_{tt} + \frac{v}{R}c_{xt} - \frac{D}{R}c_{xxt}\right)\Delta t + \frac{v\Delta x^2}{6R}c_{xxx} + \cdots \tag{45}$$

The scheme is consistent with the partial differential equation (1). If $\Delta t = \Delta x^2$, then the truncation error will be of $O(\Delta x^2)$.

Scheme 10

Using $p+q+r = 1 + \frac{k\Delta t}{2R}$, $a = 1 - \frac{k\Delta t}{2R}$, $b = d = 0$ $p - r = -\frac{v\Delta t}{R\Delta x}$ and $p + r = -\frac{2D\Delta t}{R\Delta x^2}$ in Eq. (34), we have that

$$\frac{1}{\Delta t}T_{i,j} = \left[\frac{kc}{R} + \frac{\partial c}{\partial t} + \frac{v}{R}\frac{\partial c}{\partial x} - \frac{D}{R}\frac{\partial^2 c}{\partial x^2} + \frac{k\Delta t}{2R}\frac{\partial c}{\partial t} + \frac{v\Delta t}{R}\frac{\partial^2 c}{\partial x \partial t} - \frac{D\Delta t}{R}\frac{\partial^2 c}{\partial x^2 \partial t} + \cdots\right]_{(x_i,t_j)}$$
$$\frac{1}{\Delta t}T_{i,j} = \left(\frac{k}{2R}c_t + \frac{v}{R}c_{xt} - \frac{D}{R}c_{xxt}\right)\Delta t + \frac{v\Delta x^2}{6R}c_{xxx} + \cdots \tag{46}$$

The local truncation error is $O(\Delta t + \Delta x^2)$ and the scheme is consistent with the partial differential equation (1).

Scheme 11

Substituting $p+q+r=1, a=1-\frac{k\Delta t}{R}, b=d=0, p-r=-\frac{v\Delta t}{2R\Delta x}$ and $p+r=-\frac{2D\Delta t}{R\Delta x^2}-\frac{v\Delta t}{2R\Delta x}$ in Eq. (34), We have that

$$\frac{1}{\Delta t}T_{i,j}=\left[\frac{kc}{R}+\frac{\partial c}{\partial t}+\frac{v}{R}\frac{\partial c}{\partial x}-\frac{D}{R}\frac{\partial^2 c}{\partial x^2}+\frac{\Delta t}{2}\frac{\partial^2 c}{\partial t^2}+\frac{v\Delta t}{R}\frac{\partial^2 c}{\partial x\partial t}-\frac{D\Delta t}{R}\frac{\partial^2 c}{\partial x^2\partial t}+\cdots\right]_{(x_i,t_j)} \tag{47}$$

$$\frac{1}{\Delta t}T_{i,j}=\left(\frac{1}{2}c_{tt}+\frac{v}{R}c_{xt}-\frac{D}{R}c_{xxt}\right)\Delta t-\frac{v\Delta x}{2R}c_{xx}+\cdots$$

Hence, the order of truncation error is $O(\Delta t+\Delta x)$. Also the scheme is consistent with the partial differential equation (1).

Scheme 12

Using $p+q+r=1+\frac{k\Delta t}{2R}, a=1-\frac{k\Delta t}{2R}, b=d=0, p-r=-\frac{v\Delta t}{2R\Delta x}$ and $p+r=-\frac{2D\Delta t}{R\Delta x^2}-\frac{v\Delta t}{2R\Delta x}$ in Eq. (34), We have that

$$\frac{1}{\Delta t}T_{i,j}=\left[\frac{kc}{R}+\frac{\partial c}{\partial t}+\frac{v}{R}\frac{\partial c}{\partial x}-\frac{D}{R}\frac{\partial^2 c}{\partial x^2}+\frac{k\Delta t}{2R}\frac{\partial c}{\partial t}+\frac{v\Delta t}{R}\frac{\partial^2 c}{\partial x\partial t}-\frac{D\Delta t}{R}\frac{\partial^2 c}{\partial x^2\partial t}+\cdots\right]_{(x_i,t_j)} \tag{48}$$

$$\frac{1}{\Delta t}T_{i,j}=\left(\frac{k}{2R}c_t+\frac{v}{R}c_{xt}-\frac{D}{R}c_{xxt}\right)\Delta t-\frac{v\Delta x}{2R}c_{xx}+\cdots$$

There fore, the scheme is consistent with the partial differential equation (1) and the order of truncation error is $O(\Delta t+\Delta x)$.

7 Results and Discussion

The parameters used in Cho [1], Bauer et al. [4] and Clement et al.[5] are considered for the computation of concentration of species by various numerical schemes presented in Sect. 4. The boundary conditions (2.2), (2.4) and the initial condition (2.5) with $f(t)=C_0$ (a constant dumping) are considered for the test problems. The parameter values used in computation are $C_0=1$ mg/l, $k=0.01\ \text{h}^{-1}$, $R=2$, $v=1\,\text{cm}\ \text{h}^{-1}$, $T=50$ h and $D=0.18$. Analytical solution is obtained from (7).

Table 1 infers the numerical error in maximum norm obtained for various numerical schemes. It is observed that the error in central difference schemes 2, 6, and 10 increases compare to schemes 1, 5, and 9, respectively. The reason for this is that the source term (C_{avg}) in (9) is approximated by C_P^n in odd-numbered schemes 1, 3, 5, 7, 9, and 11, while the same source term is approximated by $\frac{C_P^n+C_P^{n+1}}{2}$ (i.e., average

Table 1 Numerical error

Scheme	L_∞ error $h_1 = 0.3, h_2 = 0.2, h_3 = 0.1$			Order of convergence	
Central	$\|c - C_{h_1}\|_\infty$	$\|c - C_{h_2}\|_\infty$	$\|c - C_{h_3}\|_\infty$	$log\frac{N_1}{N_2}/log\frac{h_1}{h_2}$	$log\frac{N_2}{N_2}/log\frac{h_2}{h_3}$
Scheme 1	0.0151839	0.0065515	0.0016103	2.0730307	2.0244959
Scheme 2	0.0156539	0.0067565	0.0016613	2.0722254	2.0239634
Scheme 5	0.0037391	0.0016503	0.0004113	2.0171596	2.0044653
Scheme 6	0.0043201	0.0019093	0.0004753	2.0138400	2.0061335
Scheme 9	0.0126231	0.0058558	0.0015083	1.8943574	1.9569428
Scheme 10	0.0130631	0.0059716	0.0015196	1.9305645	1.9744258
Upwind	L_∞ error $h_1 = 1, h_2 = 0.5, h_3 = 0.4$			Order of convergence	
Scheme 3	0.0879704	0.0541474	0.0454174	0.7001266	0.7878945
Scheme 4	0.0844378	0.0520764	0.0437294	0.6972593	0.7828649
Scheme 7	0.1459378	0.0958983	0.0823154	0.6057765	0.6844476
Scheme 8	0.1429908	0.0941353	0.0808434	0.6031146	0.6821581
Scheme 11	0.1632868	0.1150218	0.1005767	0.5055008	0.6014114
Scheme 12	0.1606958	0.1135018	0.0992857	0.5016170	0.5996907

where $N_1 = \|c - C_{h_1}\|_\infty$, $N_2 = \|c - C_{h_2}\|_\infty$ and $N_3 = \|c - C_{h_3}\|_\infty$
Central difference: $\Delta t = h^2$ and Upwind: $\Delta t = h$

of C_P taken over n and $n + 1$ time levels) in even-numbered schemes 2, 4, 6, 8, 10, and 12. In the similar manner, the error in upwind schemes 4, 8, and 12 decreases compare to upwind schemes 3, 7, and 11, respectively. Hence, averaging the source term C_{avg} in (9) at n and $n + 1$ time level is a good technique for upwind schemes and a bad choice for central difference schemes.

It is also observed that the implicit type schemes 5, 6, 9, and 10 yield better result in comparing with explicit nature schemes 1 and 2 as far as central difference schemes are concerned, but there is a reverse phenomenon in upwind schemes. That is, the explicit upwind schemes 3 and 4 give better result in comparing with implicit nature upwind schemes 7, 8, 11, and 12.

Theoretically, second- and first-order convergences are obtained for central difference and upwind schemes, respectively. This is validated numerically which can be seen from Table 1. Further, the explicit upwind schemes 3 and 4 converge much faster than implicit nature upwind schemes 7, 8, 11, and 12. Thus, the averaging technique for reaction (i.e., source) term and upwinding for the advection term play a crucial role in numerical schemes for advecton–diffusion–reaction problems.

Table 2 is the summarization of theoretical results from Sects. 5 and 6. In general, explicit and implicit nature schemes are, respectively, conditionally and unconditionally stable for time-dependent problems in the absence of reaction (i.e., source) term. From Table 2, it is clear that the implicit nature schemes 5, 7, 9, and 11 are conditionally stable, while other implicit nature numerical schemes 6, 8, 10, and 12

Table 2 Summary of stability condition and order of convergence from theoretical results obtained from Sects. 5 and 6

Scheme	Stability condition	Order of error
Scheme 1	$\Delta t \leq \frac{R(8D+2k\Delta x^2)\Delta x^2}{16D^2+k^2\Delta x^4+8Dk\Delta x^2}$ $\frac{v\Delta x}{D} \leq 2, \frac{k\Delta t}{R} \leq 1$	Second
Scheme 2	$\Delta t \leq \frac{R(8D+4k\Delta x^2)\Delta x^2}{16D^2+8Dk\Delta x^2}$ $\frac{v\Delta x}{D} \leq 2, \frac{k\Delta t}{R} \leq 2$	Second
Scheme 3	$\Delta t \leq \frac{R(8D+4v\Delta x+2k\Delta x^2)\Delta x^2}{16D^2+16Dv\Delta x+8Dk\Delta x^2+4vk\Delta x^3+4v^2\Delta x^2+k^2\Delta x^4}$ $\frac{k\Delta t}{R} \leq 1$	First
Scheme 4	$\Delta t \leq \frac{R(8D+4v\Delta x+4k\Delta x^2)\Delta x^2}{16D^2+16Dv\Delta x+8Dk\Delta x^2+4vk\Delta x^3+4v^2\Delta x^2}$ $\frac{k\Delta t}{R} \leq 2$	First
Scheme 5	$\frac{k\Delta t}{R} \leq 2$	Second
Scheme 6	Unconditionally stable	Second
Scheme 7	$\frac{k\Delta t}{R} \leq 2$	First
Scheme 8	Unconditionally stable	First
Scheme 9	$\frac{k\Delta t}{R} \leq 2$	Second
Scheme 10	Unconditionally stable	Second
Scheme 11	$\frac{k\Delta t}{R} \leq 2$	First
Scheme 12	Unconditionally stable	First

are unconditionally stable for the advection-diffusion–reaction equation (1). This is due to the fact that, the reaction (i.e., source) term (C_{avg}) in (9) in schemes 6, 8, 10, and 12 is approximated by $\frac{C_P^n+C_P^{n+1}}{2}$, while the same term in schemes 5, 7, 9, and 11 is approximated by C_P^n.

References

1. Cho, C.M.: Convective transport of ammonium with nitrification in soil. Can. J. Soil Sci. **51**, 339–350 (1971)
2. Sun, Y., Clement, T.P.: A generalized decomposition method for solving coupled multi-species reactive transport problems. Transp. Porous Media **37**, 327–346 (1999)
3. Sun, Y., Petersen, J.N., Clement, T.P., Skeen, R.S.: Development of analytical solutions for multispecies transport with serial and parallel reactions. Water Resour. Res. **35**, 185–190 (1999)
4. Bauer, P., Attinger, S., Kinzelbach, W.: Transport of decay chain in homogeneous porous media: analytical solutions. J. Contam. Hydrol. **49**, 217–239 (2001)
5. Clement, T.P., Quezada, C.R., Lee, K.-K.: Generalised solution to multi-dimensional multi species transport equations coupled with a first-order reaction network involving distinct retardation factors. Adv. Water Resour. **27**, 507–520 (2004)

6. Clement, T.P.: A modular computer model for simulating reactive multi species transport in three dimensional groundwater systems. Technical Report, Pacific Northwest National Laboratory, Richland, Washington, PNNL-SA-28967 (1997)
7. Clement, T.P., Johnson, C.D., Sun, Y., Klecka, G.M., Bartlett, C.: Natural attenuation of chlorinated solvent compounds: model development and field-scale application at the Dover site. J. Contam. Hydrol. **42**, 113–140 (2000)
8. Clement, T.P., Sun, Y., Hooker, B.S., Petersen, J.N.: Modeling multispecies reactive transport in groundwater aquifers. Groundw. Monit. Remediat. **18**(2), 79–92 (1998)
9. Van Genutchen, M.T.: Convective-dispersive transport of solutes involved in sequential first-order decay reactions. Comput. Geosci. **11**, 129–147 (1985)
10. Elango, L., Brindha, K., Kalpana, L., Sunny, F., Nair, R.N., Murugan, R.: Groundwater flow and radionuclide decay-chain transport modelling around a proposed uranium tailings pond in India. Hydrogeol. J. (2012)
11. Smith, S.D.: Numerical Solutions of Partial Differential Equations: Finite Difference Methods. Oxford University Press (1985)

Total Variation Diminishing Finite Volume Scheme for Multi Dimensional Multi Species Transport with First Order Reaction Network

S. Prabhakaran and L. Jones Tarcius Doss

Abstract A Total Variation Diminishing (TVD) scheme for multi-species transport with first-order reaction network in multidimensional space is discussed in this article. The partial differential equations which describe this multi-species transport with chain reactions are in the form of a coupled system. This system is then solved by the TVD scheme with various flux limiters. The numerical diffusion controlled by the flux limiters is explained in detail. The stability and consistency conditions of the TVD scheme is also derived. The relation between the flux limiters and mesh parameters is obtained through stability conditions. A necessary condition for a scheme to be TVD is also derived.

Keywords Total variation diminishing · Finite volume method · Contamination transport · Stability · Consistency · Flux limiters

1 Introduction

Groundwater pollution attracted researchers for evaluating the movement of degradable contents in the groundwater system. The transport of these degradable species is governed by advection-diffusion–reaction (ADR) equation. Several researchers have developed an analytical solution for the species transport equation. van Genutchen [1] has given a complete review on analytical solution for one-dimensional advection–diffusion equation under various initial and boundary conditions. Domenico [2] has derived an analytical solution for multidimensional single species transport. Cho [3] has provided an analytical solution for one- dimensional three species transport. Bauer et al. [4], Clement et al. [5–7], and Sun et al. [8, 9] have discussed various

S. Prabhakaran (✉)
Department of Mathematics with Computer Applications, PSG College of Arts & Science, Civil Aerodrome, Coimbatore 641014, India
e-mail: karan.shivam@gmail.com

L. Jones Tarcius Doss
Department of Mathematics, Anna University, Chennai 600025, India

R. N. Mohapatra et al. (eds.), *Mathematical Analysis and Computing*, Springer Proceedings in Mathematics & Statistics 344, https://doi.org/10.1007/978-981-33-4646-8_38

models with analytical solutions for multi-species transport in multi-dimensions. The limitations in deriving an analytical solution and the development of modern sophisticated computing machines lead to serious research on numerical techniques.

Finite difference methods are popular among researchers in earlier days. Sheu et al. [10] and Calvo et al. [11] have used finite difference method for ADR equation. Hundsdorfer et al. [12] have presented a review of ADR equations mostly focusing on finite difference methods. Sibert et al. [13] has modeled fish movement using one-dimensional ADR equation and solved numerically using the finite difference method. The search for providing continuous solution to this problem landed with finite element methods. Houston et al. [14] has used the hp-finite element method for providing a solution to the ADR equation. Ayuso and Donatella Marini [15], Ern et al. [16] and Georgoulis et al. [17] have made contribution to a solution for the same using discontinuous Galerkin method. Idelsohn et al. [18] have applied Petrov-Galerkin method on the ADR equation. Most recently, Mudunuru and Nakshatrala [19] have used a finite element method for solving the ADR equation by enforcing maximum principle with the concern on element-wise species balance.

Finite volume method (FVM) is one of the handy tools for researchers in computational fluid dynamics in recent days due to the property of preservation of exact mass conservation in the local control volume. Eymard et al. [20] have given a comprehensive study on finite volume methods in his monograph. LeVeque [21] has discussed finite volume methods in his book on hyperbolic problems with engineering applications. Ramos [22] has solved reaction–diffusion problem using the finite volume method. Arachchige and Pettet [23] has used FVM for solving ADR equation with linearization in the time domain. ten Thije Boonkkamp and Anthonissen [24] have used a finite volume-complete flux scheme for solving ADR equations. Upwind finite volume schemes have become very popular among researchers due to their advantage over capturing flow direction. The disadvantage of the upwind technique is numerical diffusion. Several searchers tried to control this artificial diffusion and brought it up with Total Variation Diminishing technique (TVD). A concept of flux limiter is introduced in TVD schemes. van Leer [25] is one of the pioneers of this concept. Sweby [26] and Harten [27, 28] have used TVD scheme for solving one-dimensional transport equation. In fact, Sweby has given sufficient conditions for a scheme to be TVD. Jameson and Lax [29] have derived abstract conditions for the construction of total variation diminishing difference schemes. Shu [30] has used Runge–Kutta-type TVD time discretization for transport equation. Thereafter, several researches have been done on TVD schemes.

The governing equation for transport of species with first-order reaction network with liquid phase degradation is given by the following system of advection–diffusion reaction equations [6]:

$$R_k\frac{\partial U_k}{\partial t}-\sum_{i=1}^{3}\frac{\partial}{\partial x_i}\left(D_i\frac{\partial U_k}{\partial x_i}\right)+\sum_{i=1}^{3}\frac{\partial}{\partial x_i}(v_iU_k)=\sum_{d=1}^{k-1}Y_{k/d}K_dU_d-K_kU_k + \sum_{d=k+1}^{r}Y_{k/d}K_dU_d,\quad k=1,2,3...,r \tag{1}$$

where n is the total number of species; U_k the concentration of kth species $[ML^{-3}]$; D_i the dispersion coefficient $[L^2T^{-1}]$; v_i the transport velocity $[LT^{-1}]$; K_k the first-order contaminant destruction rate constant of kth species $[T^{-1}]$; R_k the retardation coefficient, and $Y_{k/d}$ the effective yield factor that describes the mass of a species k produced from another species d $[MM^{-1}]$. The kinetics of reaction is assumed to be of first order. The concentration of kth species U_k $(k = 1, 2, ..., r)$ in the first-order reaction network is to be determined for three-dimensional flow at $X_P = (x_1, x_2, x_3)$ at any given time t.

In case of degradation process occurs in both solid and liquid phase, then the governing equation is given by

$$R_k \frac{\partial U_k}{\partial t} - \sum_{i=1}^{3} \frac{\partial}{\partial x_i}\left(D_i \frac{\partial U_k}{\partial x_i}\right) + \sum_{i=1}^{3} \frac{\partial}{\partial x_i}(v_i U_k) = \sum_{d=1}^{k-1} R_d Y_{k/d} K_d U_d - R_k K_k U_k + \sum_{d=k+1}^{r} R_d Y_{k/d} K_d U_d \quad k = 1, 2, 3..., r \tag{2}$$

The difference between Eqs. (1) and (2) lies in the presence of retardation factor R_k in the reaction term of 2.

2 Numerical Approximation

Let us discuss a control volume before deriving numerical approximation. The letters E, W, N, S, T, and B in the control volume represent east, west, north, south, top, and bottom nodal points located at the middle of their respective sides in the outer cube, respectively. A numerical solution is obtained at the nodal point (P) located at the centroid of the cube. Similarly, the letters e, w, n, s, t_p, and b represent east, west, north, south, top and bottom faces of the inner cube. Spatial step size is the difference between the nodal points. Faces of the control volume are located half-way between nodes. The vector form of a governing equation (2) can be written as

$$R_k \frac{\partial U_k}{\partial t} - \overline{\nabla} \cdot (\overline{\nabla} D U_k) + \overline{\nabla} \cdot (v U_k) = \sum_{d=1}^{k-1} R_d Y_{k/d} K_d U_d - R_k K_k U_k + \sum_{d=k+1}^{r} R_d Y_{k/d} K_d U_d, \tag{3}$$

where $\overline{\nabla} D U_k = (D_1 \frac{\partial U_k}{\partial x_1}, D_2 \frac{\partial U_k}{\partial x_2}, D_3 \frac{\partial U_k}{\partial x_3})$ and $v U_k = (v_1 U_k, v_2 U_k, v_3 U_k)$. Integrating the above equation first with respect to time from t_m to $t_m + \Delta t = t_{m+1}$ and then integrating over a local control volume CV [31], we obtain the following:

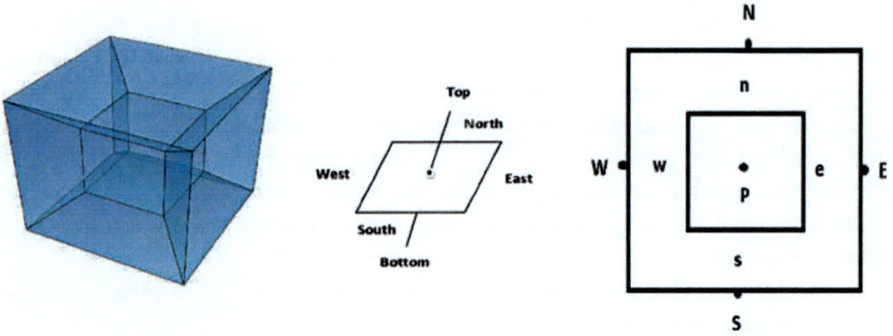

Fig. 1 3D control volume

$$R_k \int_{CV} \int_{t_m}^{t_{m+1}} \frac{\partial U_k}{\partial t} dt dV = \int_{CV} \int_{t_m}^{t_{m+1}} \overline{\nabla} \cdot (\overline{\nabla} D U_k) dt dV - \int_{CV} \int_{t_m}^{t_{m+1}} \overline{\nabla} \cdot (v U_k) dt dV$$
$$+ \int_{CV} \int_{t_m}^{t_{m+1}} \left(\sum_{d=1}^{k-1} R_d Y_{k/d} K_d U_d - R_k K_k U_k + \sum_{d=k+1}^{r} R_d Y_{k/d} K_d U_d \right) dt dV.$$

Using forward Euler for time integration, we obtain

$$R_k \int_{CV} (U_k^{m+1} - U_k^m) dV = \Delta t \int_{CV} \overline{\nabla} \cdot (\overline{\nabla} D U_k^m) dV - \Delta t \int_{CV} \overline{\nabla} \cdot (v U_k^m) dV$$
$$+ \Delta t \int_{CV} \left(\sum_{d=1}^{k-1} R_d Y_{k/d} K_d U_d^m - R_k K_k U_k^m + \sum_{d=k+1}^{r} R_d Y_{k/d} K_d U_d^m \right) dV.$$

The first two volume integrals over CV on the right-hand side can be converted into surface integral by applying the Gauss divergence theorem as follows:

$$R_k \int_{CV} (U_k^{m+1} - U_k^m) dV = \Delta t \int_A \vec{n} . \overline{\nabla} D U_k^m dA - \Delta t \int_A \vec{n} \cdot (v U_k^m) dA$$
$$- \Delta t \int_{CV} R_k K_k U_k^m dV + \Delta t \int_{CV} \sum_{d=1}^{k-1} R_d Y_{k/d} K_d U_d^m dV$$
$$+ \Delta t \int_{CV} \sum_{d=k+1}^{r} R_d Y_{k/d} K_d U_d^m dV,$$

where $\vec{n}$ is the unit outward normal to the surface A of a control volume. The surface integral can be split into six surfaces (S), namely east (e), west (w), north (n), south (s), top (t_p), and bottom (b) as follows:

$$R_k \int_{CV} (U_k^{m+1} - U_k^m) dV = \Delta t \sum \int_S \vec{n} \cdot \nabla D U_k^m dS - \Delta t \sum \int_S \vec{n} \cdot (v U_k^m) dS$$
$$+ \Delta t \int_{CV} \left(\sum_{d=1}^{k-1} R_d Y_{k/d} K_d U_d^m - R_k K_k U_k^m + \sum_{d=k+1}^{r} R_d Y_{k/d} K_d U_d^m \right) dV.$$

$$R_k \int_{CV} (U_k^{m+1} - U_k^m) dV = \Delta t \left[A_e \left(D_1 \frac{\partial U_k^m}{\partial x_1} \right)_e - A_w \left(D_1 \frac{\partial U_k^m}{\partial x_1} \right)_w + A_n \left(D_2 \frac{\partial U_k^m}{\partial x_2} \right)_n \right.$$
$$\left. - A_s \left(D_2 \frac{\partial U_k^m}{\partial x_2} \right)_s + A_{t_p} \left(D_3 \frac{\partial U_k^m}{\partial x_3} \right)_{t_p} - A_b \left(D_3 \frac{\partial U_k^m}{\partial x_3} \right)_b \right] - \Delta t \left[A_e (v_1 U_k^m)_e \right.$$
$$\left. - A_w (v_1 U_k^m)_w + A_n (v_2 U_k^m)_n - A_s (v_2 U_k^m)_s + A_{t_p} (v_3 U_k^m)_{t_p} - A_b (v_3 U_k^m)_b \right]$$
$$+ \Delta t \int_{CV} \left(\sum_{d=1}^{k-1} R_d Y_{k/d} K_d U_d^m - R_k K_k U_k^m + \sum_{d=k+1}^{r} R_d Y_{k/d} K_d U_d^m \right) dV,$$

where $A_e = A_w = \Delta x_2 \Delta x_3$, $A_n = A_s = \Delta x_1 \Delta x_3$ and $A_{t_p} = A_b = \Delta x_1 \Delta x_2$. Let us assume that the control volume is fixed for all time. Taking the average over a control volume at the centroid P for left- and right-hand side integrals, we obtain the following:

$$R_k (U_{kP}^{m+1} - U_{kP}^m) \int_{CV} dV = \Delta t \left[A_e D_1 \left(\frac{\partial U_k^m}{\partial x_1} \right)_e - A_w D_1 \left(\frac{\partial U_k^m}{\partial x_1} \right)_w + A_n D_2 \left(\frac{\partial U_k^m}{\partial x_2} \right)_n \right]$$
$$+ \Delta t \left[-A_s D_2 \left(\frac{\partial U_k^m}{\partial x_2} \right)_s + A_{t_p} D_3 \left(\frac{\partial U_k^m}{\partial x_3} \right)_{t_p} - A_b D_3 \left(\frac{\partial U_k^m}{\partial x_3} \right)_b \right]$$
$$- \Delta t \left[A_e v_1 U_{ke}^m - A_w v_1 U_{kw}^m + A_n v_2 U_{kn}^m - A_s v_2 U_{ks}^m + A_{t_p} v_3 U_{kt_p}^m - A_b v_3 U_{kb}^m \right]$$
$$+ \Delta t \left(\sum_{d=1}^{k-1} R_d Y_{k/d} K_d U_{dP}^m - R_k K_k U_{kP}^m + \sum_{d=k+1}^{r} R_d Y_{k/d} K_d U_{dP}^m \right) \int_{CV} dV,$$

where U_{kP} is average taken over a control volume and $\int_{CV} dV = \Delta V = \Delta x_1 \Delta x_2 \Delta x_3$. Let us apply central difference for the terms $\frac{\partial U_k^m}{\partial x_1}, \frac{\partial U_k^m}{\partial x_2}, \frac{\partial U_k^m}{\partial x_3}$ and bringing $R_k \Delta V$ term to the right- hand side, we obtain

$$U_{kP}^{m+1} - U_{kP}^m = \left[\frac{D_1 \Delta t}{R_k \Delta x_1^2} (U_{kW}^m - 2U_{kP}^m + U_{kE}^m) + \frac{D_2 \Delta t}{R_k \Delta x_2^2} (U_{kS}^m - 2U_{kP}^m + U_{kN}^m) \right.$$
$$\left. + \frac{D_3 \Delta t}{R_k \Delta x_3^2} (U_{kB}^m - 2U_{kP}^m + U_{kT}^m) \right] + \left[-\frac{v_1 \Delta t}{R_k \Delta x_1} U_{ke}^m - \frac{v_2 \Delta t}{R_k \Delta x_2} U_{kn}^m - \frac{v_3 \Delta t}{R_k \Delta x_3} U_{kt_p}^m \right.$$
$$\left. + \frac{v_1 \Delta t}{R_k \Delta x_1} U_{kw}^m + \frac{v_2 \Delta t}{R_k \Delta x_2} U_{ks}^m + \frac{v_3 \Delta t}{R_k \Delta x_3} U_{kb}^m \right]$$
$$+ \left[\left(\sum_{d=1}^{k-1} R_d Y_{k/d} K_d U_{dP}^m - R_k K_k U_{kP}^m + \sum_{d=k+1}^{r} R_d Y_{k/d} K_d U_{dP}^m \right) \frac{\Delta t}{R_k} \right]. \quad (4)$$

The approximation for advection term (that is, second term on right-hand side) is derived by using slope reconstruction technique. Expanding U_{ke}^m in Taylor's series and truncating after second term in x_1 direction, we have

$$U_{ke}^m \approx U_{kP}^m + \frac{\Delta x_1}{2}\left(\frac{\partial U_k^m}{\partial x_1}\right)_P.$$

The flux term $\frac{\partial U_k^m}{\partial x_1}$ in above is called anti-diffusion term. It controls the numerical diffusion. This has been explained in Sect. 4. Introducing flux limiter $\psi(r)$ to control anti-diffusion and using forward difference approximation for flux term, we obtain

$$U_{ke}^m \approx U_{kP}^m + \frac{\psi(r_e)}{2}(U_{kE}^m - U_{kP}^m) \quad \text{where} \quad r_e = \frac{U_{kP}^m - U_{kW}^m}{U_{kE}^m - U_{kP}^m}.$$

Flux limiters for all other directions can be obtained in a similar way,

$$U_{kw}^m \approx U_{kW}^m + \frac{\psi(r_w)}{2}(U_{kP}^m - U_{kW}^m) \quad \text{where} \quad r_w = \frac{U_{kW}^m - U_{kWW}^m}{U_{kP}^m - U_{kW}^m};$$
$$U_{kn}^m \approx U_{kP}^m + \frac{\psi(r_n)}{2}(U_{kN}^m - U_{kP}^m) \quad \text{where} \quad r_n = \frac{U_{kP}^m - U_{kS}^m}{U_{kN}^m - U_{kP}^m};$$
$$U_{ks}^m \approx U_{kS}^m + \frac{\psi(r_s)}{2}(U_{kP}^m - U_{kS}^m) \quad \text{where} \quad r_s = \frac{U_{kS}^m - U_{kSS}^m}{U_{kP}^m - U_{kS}^m};$$
$$U_{kt_p^m} \approx U_{kP}^m + \frac{\psi(r_{t_p})}{2}(U_{kT}^m - U_{kP}^m) \quad \text{where} \quad r_{t_p} = \frac{U_{kP}^m - U_{kB}^m}{U_{kT}^m - U_{kP}^m};$$

and $U_{kb}^m \approx U_{kB}^m + \frac{\psi(r_b)}{2}(U_{kP}^m - U_{kB}^m)$ where $r_b = \frac{U_{kB}^m - U_{kBB}^m}{U_{kP}^m - U_{kB}^m}$.

Substituting the above in (4), we obtain

$$U_{kP}^{m+1} = A_P U_{kP}^m + A_W U_{kW}^m + A_E U_{kE}^m + A_S U_{kS}^m + A_N U_{kN}^m + A_B U_{kB}^m + A_T U_{kT}^m + \left(\sum_{d=1}^{k-1} c_d U_{dP}^m + \sum_{d=k+1}^{n_1} c_d U_{dP}^m\right)\frac{\Delta t}{R_k}, \tag{5}$$

where

$A_P = 1 - \frac{2D_1\Delta t}{R_k\Delta x_1^2} - \frac{2D_2\Delta t}{R_k\Delta x_2^2} - \frac{2D_3\Delta t}{R_k\Delta x_3^2} - \frac{v_1\Delta t}{R_k\Delta x_1} - \frac{v_2\Delta t}{R_k\Delta x_2} - \frac{v_3\Delta t}{R_k\Delta x_3} + \frac{v_1\Delta t}{2R_k\Delta x_1}$
$[\psi(r_e) + \psi(r_w)] + \frac{v_2\Delta t}{2R_k\Delta x_2}[\psi(r_s) + \psi(r_n)] + \frac{v_3\Delta t}{2R_k\Delta x_3}[\psi(r_{t_p}) + \psi(r_b)] - K_k\Delta t$,
$A_W = \frac{D_1\Delta t}{R_k\Delta x_1^2} + \frac{v_1\Delta t}{R_k\Delta x_1} - \frac{v_1\Delta t}{2R_k\Delta x_1}\psi(r_w)$, $A_E = \frac{D_1\Delta t}{R_k\Delta x_1^2} - \frac{v_1\Delta t}{2R_k\Delta x_1}\psi(r_e)$,
$A_N = \frac{D_2\Delta t}{R_k\Delta x_2^2} - \frac{v_2\Delta t}{2R_k\Delta x_2}\psi(r_n)$, $A_S = \frac{D_2\Delta t}{R_k\Delta x_2^2} + \frac{v_2\Delta t}{R_k\Delta x_2} - \frac{v_2\Delta t}{2R_k\Delta x_2}\psi(r_s)$,
$A_T = \frac{D_3\Delta t}{R_k\Delta x_3^2} - \frac{v_3\Delta t}{2R_k\Delta x_3}\psi(r_{t_p})$, $A_B = \frac{D_3\Delta t}{R_k\Delta x_3^2} + \frac{v_3\Delta t}{R_k\Delta x_3} - \frac{v_3\Delta t}{2R_k\Delta x_3}\psi(r_b)$ and
$c_d = R_d Y_{k/d} K_d$.

Table 1 List of flux limiters

Name	Limiter $\psi(r)$	Name	Limiter $\psi(r)$
Upwind	0	Min-Mod	$Max[0, Min(r, 1)]$
Central	1	Superbee	$Max[0, Min(2r, 1), Min(r, 2)]$
van Leer	$(r + \|r\|)/(1 + r)$	Sweby	$Max[0, Min(\beta r, 1), Min(r, \beta)]$
van Albada	$(r + r^2)/(1 + r^2)$	Osher	$Max[0, Min(r, \beta)]\ 1 \le \beta \le 2$
Linear UD	$Min(r, 2)$	Downwind	$Min(2r, 1)$
UMIST	$Max[0, Min(2r, (3 + r)/4, (1 + 3r)/4, 2)]$		

3 Stability Analysis

Let us discuss the stability of the proposed scheme in this section. The general form of explicit scheme for three dimensional is given by $U_{kP}^{m+1} = A_P U_{kP}^m + A_E U_{kE}^m + A_W U_{kW}^m + A_N U_{kN}^m + A_S U_{kS}^m + A_T U_{kT}^m + A_B U_{kB}^m$. The stability condition for this scheme is derived.

Theorem 1
Let $U_{kP}^{m+1} = A_P U_{kP}^m + A_E U_{kE}^m + A_W U_{kW}^m + A_N U_{kN}^m + A_S U_{kS}^m + A_T U_{kT}^m + A_B U_{kB}^m$ be the general form of explicit finite difference scheme for any linear time-dependent partial differential equation in three dimensions with equal mesh length and mesh points in spatial directions. If the coefficients $A_P \ge 0, A_W \ge 0, A_E \ge 0, A_N \ge 0, A_S \ge 0, A_T \ge 0$ and $A_B \ge 0$ and satisfy $(A_P + A_E + A_W + A_N + A_S + A_T + A_B)^2 \le 1 + 4A_P(A_E + A_W + A_N + A_S + A_T + A_B)$, then the scheme is stable.

Proof
Consider the general form of explicit scheme

$$U_{kP}^{m+1} = A_P U_{kP}^m + A_E U_{kE}^m + A_W U_{kW}^m + A_N U_{kN}^m + A_S U_{kS}^m + A_T U_{kT}^m + A_B U_{kB}^m \tag{6}$$

with A_P, A_W, A_E, A_N, A_S, A_T and A_B are greater than or equal to zero. Substituting $U_{kP}^m = B\xi^m e^{i\alpha\theta_1} e^{i\beta\theta_2} e^{i\gamma\theta_3}$ in above explicit scheme, we have

$$\begin{aligned}\xi &= A_P + (A_E + A_W)\cos\theta_1 + (A_N + A_S)\cos\theta_2 + (A_T + A_B)\cos\theta_3 \\ &\quad + i(A_E - A_W)\sin\theta_1 + i(A_N - A_S)\sin\theta_2 + i(A_T - A_B)\sin\theta_3).\end{aligned}$$

von Neumann criteria for stability is given by $|\xi| \le 1$ which implies $|\xi|^2 \le 1$. We therefore have that

$$\begin{aligned}&|A_P + (A_E + A_W)\cos\theta_1 + (A_N + A_S)\cos\theta_2 + (A_T + A_B)\cos\theta_3 \\ &+ i(A_E - A_W)\sin\theta_1 + i(A_N - A_S)\sin\theta_2 + i(A_T - A_B)\sin\theta_3)|^2 \le 1.\end{aligned}$$

Let us assume that $\theta_1 = \theta_2 = \theta_3 = \theta$. That is, the number of nodal points in all directions are equal with the same mesh length. Then, we have

$$
\begin{aligned}
&|A_P + (A_E + A_W + A_N + A_S + A_T + A_B)\cos\theta \\
&\quad + i(A_E - A_W + A_N - A_S + A_T - A_B)\sin\theta)|^2 \le 1; \\
&A_P^2 + (A_E + A_W + A_N + A_S + A_T + A_B)^2\cos^2\theta \\
&\quad + 2A_P(A_E + A_W + A_N + A_S + A_T + A_B)\cos\theta \\
&\quad + (A_E - A_W + A_N - A_S + A_T - A_B)^2\sin^2\theta \le 1; \\
&A_P^2 + A_E^2 + A_W^2 + A_N^2 + A_S^2 \\
&\quad + A_T^2 + A_B^2 + 2(A_E A_W + A_N A_S + A_T A_B)(\cos^2\theta - \sin^2\theta) \\
&\quad + 2A_P(A_E + A_W + A_N + A_S + A_T + A_B)\cos\theta \le 1; \\
&(A_P + A_E + A_W + A_N + A_S + A_T + A_B)^2 \\
&\quad - 2A_P(A_E + A_W + A_N + A_S + A_T + A_B)(1 - \cos\theta) \\
&\quad - 2(A_E A_W + A_N A_S + A_T A_B)(1 - \cos 2\theta) \le 1;
\end{aligned}
$$

$$
\begin{aligned}
(A_P + A_E + A_W + A_N + A_S + A_T + A_B)^2 &\le 1 + 4(A_E A_W + A_N A_S + A_T A_B)\sin^2\theta \\
&\quad + 4A_P(A_E + A_W + A_N + A_S + A_T + A_B)\sin^2\frac{\theta}{2}; \\
(A_P + A_E + A_W + A_N + A_S + A_T + A_B)^2 &\le 1 + 16(A_E A_W + A_N A_S + A_T A_B)\sin^2\frac{\theta}{2}\cos^2\frac{\theta}{2} \\
&\quad + 4A_P(A_E + A_W + A_N + A_S + A_T + A_B)\sin^2\frac{\theta}{2}; \\
(A_P + A_E + A_W + A_N + A_S + A_T + A_B)^2 & \\
16(A_E A_W + A_N A_S + A_T A_B)\sin^4\frac{\theta}{2} &\le 1 + 16(A_E A_W + A_N A_S + A_T A_B)\sin^2\frac{\theta}{2} \\
&\quad + 4A_P(A_E + A_W + A_N + A_S + A_T + A_B)\sin^2\frac{\theta}{2}.
\end{aligned}
$$

Maximizing the trigonometric functions in above inequality with respect to their argument, we obtain

$$
(A_P + A_E + A_W + A_N + A_S + A_T + A_B)^2 \le 1 + 4A_P(A_E + A_W + A_N + A_S + A_T + A_B). \tag{7}
$$

Therefore, any numerical scheme satisfying (7) and positivity on coefficients is stable.

The summation terms $(\sum_{d=1}^{k-1} c_d U_{dP}^m + \sum_{d=k+1}^{n} c_d U_{dP}^m)\frac{\Delta t}{R_k}$ in (5) constitute other species transport except the current species k. Further, c_d, R_k and Δt are all positive. The stability of entire k species depends on the stability of each species transport in the network. Therefore, it is enough to demonstrate the stability of each species transport on the assumption of the other $k-1$ species having stable value. Hence, the stability of a system in (5) depends on the stability of the following for each k

$$
U_{kP}^{m+1} = A_P U_{kP}^m + A_E U_{kE}^m + A_W U_{kW}^m + A_N U_{kN}^m + A_S U_{kS}^m + A_T U_{kT}^m + A_B U_{kB}^m. \tag{8}
$$

The stability condition (7) is derived for equal mesh size in spatial directions. Let us assume that $\Delta x = \Delta x_1 = \Delta x_2 = \Delta x_3$. Further, the value of ψ varies over spatial and temporal directions. Therefore, a general function $\psi(r)$ is considered for stability analysis and truncation error. Substituting the coefficients $A_P, A_E, A_W, A_S, A_N, A_T$, and A_B in (7) with these assumptions, the following stability condition is obtained for (8) and hence for (5)

$$\Delta t \leq \frac{\frac{2D}{R_k \Delta x^2} + \frac{v}{R_k \Delta x} - \frac{v}{R_k \Delta x}\psi(r) + \frac{K_k}{2}}{\frac{4D^2}{R_k^2 \Delta x^4} + \frac{4Dv}{R_k^2 \Delta x^3} + \frac{v^2}{R_k^2 \Delta x^2} + \frac{2DK_k}{R_k \Delta x^2} + \frac{vK_k}{R_k \Delta x} + \frac{K_k^2}{4} - \left[\frac{2v^2}{R_k^2 \Delta x^2} + \frac{4Dv}{R_k^2 \Delta x^3} + \frac{vK_k}{R_k \Delta x}\right]\psi(r) + \frac{v^2}{R_k^2 \Delta x^2}\psi(r)^2} \tag{9}$$

where $D = D_1 + D_2 + D_3$ and $v = v_1 + v_2 + v_3$. The condition on the flux limiter for the total variation diminishing is $0 \leq \psi(r) \leq 2$. Therefore, the stability condition in (9) must satisfy for the least value of $\psi(r)$ (that is, $\psi(r) = 0$). Hence, we get

$$\Delta t_1 \leq \frac{\frac{2D}{R_k \Delta x^2} + \frac{v}{R_k \Delta x} + \frac{K_k}{2}}{\frac{4D^2}{R_k^2 \Delta x^4} + \frac{4Dv}{R_k^2 \Delta x^3} + \frac{v^2}{R_k^2 \Delta x^2} + \frac{2DK_k}{R_k \Delta x^2} + \frac{vK_k}{R_k \Delta x} + \frac{K_k^2}{4}} \tag{10}$$

The stability condition in (9) must also satisfy for the maximum value of $\psi(r)$ (that is, $\psi(r) = 2$).

$$\Delta t_2 \leq \frac{\frac{2D}{R_k \Delta x^2} - \frac{v}{R_k \Delta x} + \frac{K_k}{2}}{\frac{4D^2}{R_k^2 \Delta x^4} - \frac{4Dv}{R_k^2 \Delta x^3} + \frac{v^2}{R_k^2 \Delta x^2} + \frac{2DK_k}{R_k \Delta x^2} - \frac{vK_k}{R_k \Delta x} + \frac{K_k^2}{4}} \tag{11}$$

The above is the relation between spatial step size and temporal step size for stable TVD schemes. Choose always $\Delta t \leq Min(\Delta t_1, \Delta t_2)$ to get a stable scheme.

4 Truncation Error and Consistency

The truncation error for an explicit numerical scheme at interior nodal point (X_P, t_m) is defined by Smith [32]

$$T_{P,m} = \frac{1}{\Delta t}\left[U_k(X_P, t_{m+1}) - U_{kP}^{m+1}\right],$$

where $U_k(X_P, t_{m+1})$ and U_{kP}^{m+1} are the values of exact and numerical solution of kth species U_k at (X_P, t_{m+1}), respectively. From (5)

$$\Delta t T_{P,m} = U_k(X_P, t_{m+1}) - A_P U_{kP}^m - A_E U_{kE}^m - A_W U_{kW}^m - A_N U_{kN}^m - A_S U_{kS}^m - A_T U_{kT}^m - A_B U_{kB}^m - \left(\sum_{d=1}^{k-1} c_d U_{dP}^m + \sum_{d=k+1}^{n} c_d U_{dP}^m\right)\frac{\Delta t}{R_k}.$$

The truncation error for explicit numerical scheme can be obtained by replacing numerical solution with exact solution. Thus, we have

$$\begin{aligned}\Delta t T_{P,m} &= U_k(X_P, t_{m+1}) - A_P U_k(X_P, t_m) - A_E U_k(X_E, t_m) - A_W U_k(X_W, t_m)\\ &\quad - A_N U_k(X_N, t_m) - A_S U_k(X_S, t_m) - A_T U_k(X_T, t_n) - A_B U_k(X_B, t_n)\\ &\quad - \left(\sum_{d=1}^{k-1} c_d U_k(X_P, t_m) + \sum_{d=k+1}^{n} c_d U_k(X_P, t_m)\right)\frac{\Delta t}{R_k}.\end{aligned}$$

Expanding the above using Taylor series, we obtain that

$$\begin{aligned}\Delta t T_{P,m} &= \left[U_{kP}^m + \Delta t\frac{\partial U_{kP}^m}{\partial t} + \frac{\Delta t^2}{2}\frac{\partial^2 U_{kP}^m}{\partial t^2} + \ldots\right] - A_P U_{kP}^m\\ &\quad - A_E\left[U_{kP}^m + \Delta x_1\frac{\partial U_{kP}^m}{\partial x_1} + \frac{\Delta x_1^2}{2}\frac{\partial^2 U_{kP}^m}{\partial x_1^2} + \ldots\right]\\ &\quad - A_W\left[U_{kP}^m - \Delta x_1\frac{\partial U_{kP}^m}{\partial x_1} + \frac{\Delta x_1^2}{2}\frac{\partial^2 U_{kP}^m}{\partial x_1^2} + \ldots\right]\\ &\quad - A_N\left[U_{kP}^m + \Delta x_2\frac{\partial U_{kP}^m}{\partial x_2} + \frac{\Delta x_2^2}{2}\frac{\partial^2 U_{kP}^m}{\partial x_2^2} + \ldots\right]\\ &\quad - A_S\left[U_{kP}^m - \Delta x_3\frac{\partial U_{kP}^m}{\partial x_3} + \frac{\Delta x_3^2}{2}\frac{\partial^2 U_{kP}^m}{\partial x_3^2} + \ldots\right]\\ &\quad - A_T\left[U_{kP}^m + \Delta x_3\frac{\partial U_{kP}^m}{\partial x_3} + \frac{\Delta x_3^2}{2}\frac{\partial^2 U_{kP}^m}{\partial x_3^2} + \ldots\right]\\ &\quad - A_B\left[U_{kP}^m - \Delta x_2\frac{\partial U_{kP}^m}{\partial x_2} + \frac{\Delta x_2^2}{2}\frac{\partial^2 U_{kP}^m}{\partial x_2^2} + \ldots\right] - \left(\sum_{d=1}^{k-1} c_d U_{kP}^m + \sum_{d=k+1}^{n} c_d U_{kP}^m\right)\frac{\Delta t}{R_k}.\end{aligned}$$

We now rearrange terms to obtain

$$\begin{aligned}\Delta t T_{P,m} &= \Delta t\frac{\partial U_{kP}^m}{\partial t} + \frac{\Delta t^2}{2}\frac{\partial^2 U_{kP}^m}{\partial t^2} - (A_P + A_E + A_W + A_N + A_S + A_T + A_B - 1)U_{kP}^m\\ &\quad + (A_W - A_E)\Delta x_1\frac{\partial U_{kP}^m}{\partial x_1} + (A_S - A_N)\Delta x_2\frac{\partial U_{kP}^m}{\partial x_2} + (A_B - A_T)\Delta x_3\frac{\partial U_{kP}^m}{\partial x_3}\\ &\quad - (A_W + A_E)\frac{\Delta x_1^2}{2}\frac{\partial^2 U_{kP}^m}{\partial x_1^2} - (A_S + A_N)\frac{\Delta x_2^2}{2}\frac{\partial^2 U_{kP}^m}{\partial x_2^2} - (A_B + A_T)\frac{\Delta x_3^2}{2}\frac{\partial^2 U_{kP}^m}{\partial x_3^2}\\ &\quad - \left(\sum_{d=1}^{k-1} c_d U_{kP}^m + \sum_{d=k+1}^{n} c_d U_k\right)\frac{\Delta t}{R_k} + \ldots\end{aligned}$$

A general function $\psi(r)$ is considered for truncation error. Hence, we get

$$
\begin{aligned}
T_{P,m} = {} & \frac{\partial U_{kP}^m}{\partial t} + \frac{\Delta t}{2}\frac{\partial^2 U_{kP}^m}{\partial t^2} + \frac{1}{R_k}\Bigg[R_k K_k U_{kP}^m + v_1 \frac{\partial U_{kP}^m}{\partial x_1} + v_2 \frac{\partial U_{kP}^m}{\partial x_2} + v_3 \frac{\partial U_k}{\partial x_3} - D_1 \frac{\partial^2 U_k}{\partial x_1^2} \\
& - D_2 \frac{\partial^2 U_{kP}^m}{\partial x_2^2} - D_3 \frac{\partial^2 U_{kP}^m}{\partial x_3^2} - \sum_{d=1}^{k-1} R_d Y_{k/d} K_d U_{kP}^m - \sum_{d=k+1}^{n} R_d Y_{k/d} K_d U_{kP}^m \Bigg] \\
& + \frac{1}{R_k}\left[[1-\psi(r)]\frac{v_1 \Delta x_1}{2}\frac{\partial^2 U_{kP}}{\partial x_1^2} + [1-\psi(r)]\frac{v_2 \Delta x_2}{2}\frac{\partial^2 U_{kP}}{\partial x_2^2} + [1-\psi(r)]\frac{v_3 \Delta x_3}{2}\frac{\partial^2 U_{kP}}{\partial x_3^2} \right] \\
& + O(\Delta x_1 + \Delta x_2 + \Delta x_3).
\end{aligned} \tag{12}
$$

The flux limiter $\psi(r)$ is associated with the numerical diffusion terms in (12). Numerical diffusion is controlled when $|1-\psi(r)| \leq 1$ (that is, $0 \leq \psi(r) \leq 2$). This is a necessary condition for a scheme to be TVD.

5 Numerical Simulation for Three-Dimensional Test Problems

5.1 Three-Dimensional Multi-Directional Sequential Reactions

Sequential reaction with transport velocity in all three dimensions is considered in this problem. The three-dimensional transport of multi species involved in sequential reaction with solid and liquid phase degradation is governed by the following system of PDE's:

$$
\begin{aligned}
R_1 \frac{\partial U_1}{\partial t} + v_1 \frac{\partial U_1}{\partial x_1} + v_2 \frac{\partial U_1}{\partial x_2} + v_3 \frac{\partial U_1}{\partial x_3} - D_1 \frac{\partial^2 U_1}{\partial x_1^2} - D_2 \frac{\partial^2 U_1}{\partial x_2^2} - D_3 \frac{\partial^2 U_1}{\partial x_3^2} &= -K_1 R_1 U_1 \\
R_2 \frac{\partial U_2}{\partial t} + v_1 \frac{\partial U_1}{\partial x_1} + v_2 \frac{\partial U_1}{\partial x_2} + v_3 \frac{\partial U_1}{\partial x_3} - D_1 \frac{\partial^2 U_2}{\partial x_1^2} - D_2 \frac{\partial^2 U_2}{\partial x_2^2} - D_3 \frac{\partial^2 U_2}{\partial x_3^2} &= K_1 R_1 U_1 - K_2 R_2 U_2 \\
R_3 \frac{\partial U_3}{\partial t} + v_1 \frac{\partial U_1}{\partial x_1} + v_2 \frac{\partial U_1}{\partial x_2} + v_3 \frac{\partial U_1}{\partial x_3} - D_1 \frac{\partial^2 U_3}{\partial x_1^2} - D_2 \frac{\partial^2 U_3}{\partial x_2^2} - D_3 \frac{\partial^2 U_3}{\partial x_3^2} &= K_2 R_2 U_2 - K_3 R_3 U_3 \\
R_4 \frac{\partial U_4}{\partial t} + v_1 \frac{\partial U_1}{\partial x_1} + v_2 \frac{\partial U_1}{\partial x_2} + v_3 \frac{\partial U_1}{\partial x_3} - D_1 \frac{\partial^2 U_4}{\partial x_1^2} - D_2 \frac{\partial^2 U_4}{\partial x_2^2} - D_3 \frac{\partial^2 U_4}{\partial x_3^2} &= K_3 R_3 U_3 - K_4 R_4 U_4
\end{aligned}
$$

with the boundary conditions

$$
\begin{aligned}
U_k(0,0,0,t) &= U_{k0} \\
\lim_{x_1 \to \infty} U_k(x_1, x_2, x_3, t) &= 0 \\
\lim_{x_2 \to \infty} U_k(x_1, x_2, x_3, t) &= 0 \\
\lim_{x_3 \to \infty} U_k(x_1, x_2, x_3, t) &= 0 \; k = 1, 2, 3, 4.
\end{aligned}
$$

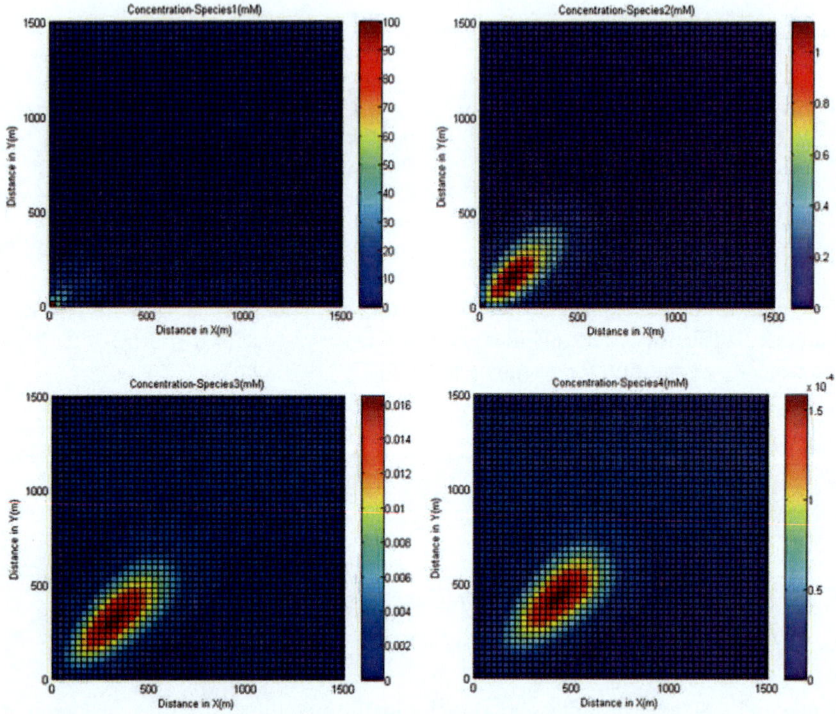

Fig. 2 Simulation for central difference limiter for 1000 days with sequential reaction

A constant amount of 100 mM for species-1 and zero mM for other three species at all times are injected at the origin. The concentration profile of four species transport for 1000 days is simulated numerically. Numerical simulation of this with central difference, van Leer limiter and Sweby limiter is shown in Fig. 2, Fig. 3, and Fig. 4, respectively. The parameters in Table 2 are used for numerical simulation.

5.2 *Three Dimensional Multi-directional Serial–Parallel Reactions*

Three-dimensional transport of multi-species with serial–parallel reversible reaction in both solid and liquid phase degradation is governed by the following system of PDEs:

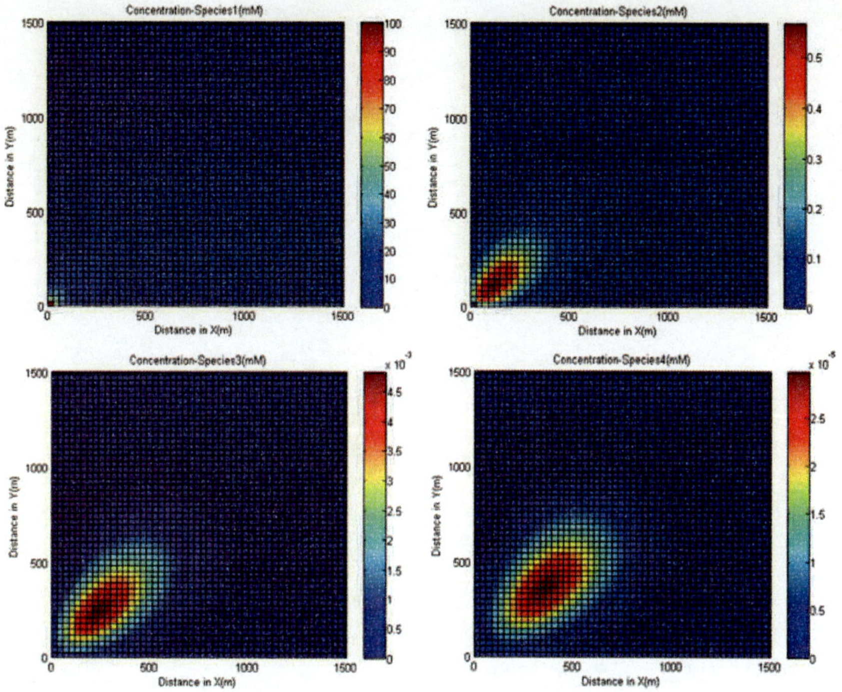

Fig. 3 Simulation for van Leer limiter for 1000 days with sequential reaction

$$
\begin{aligned}
R_1\frac{\partial U_1}{\partial t}+v_1\frac{\partial U_1}{\partial x_1}+v_2\frac{\partial U_1}{\partial x_2}+v_3\frac{\partial U_1}{\partial x_3}-D_1\frac{\partial^2 U_1}{\partial x_1^2}-D_2\frac{\partial^2 U_1}{\partial x_2^2}-D_3\frac{\partial^2 U_1}{\partial x_3^2} &= -K_1R_1U_1\\
R_2\frac{\partial U_2}{\partial t}+v_1\frac{\partial U_1}{\partial x_1}+v_2\frac{\partial U_1}{\partial x_2}+v_3\frac{\partial U_1}{\partial x_3}-D_1\frac{\partial^2 U_2}{\partial x_1^2}-D_2\frac{\partial^2 U_2}{\partial x_2^2}-D_3\frac{\partial^2 U_2}{\partial x_3^2} &= F_{2/1}Y_{2/1}K_1R_1U_1\\
&\quad -K_2R_2U_2\\
R_3\frac{\partial U_3}{\partial t}+v_1\frac{\partial U_1}{\partial x_1}+v_2\frac{\partial U_1}{\partial x_2}+v_3\frac{\partial U_1}{\partial x_3}-D_1\frac{\partial^2 U_3}{\partial x_1^2}-D_2\frac{\partial^2 U_3}{\partial x_2^2}-D_3\frac{\partial^2 U_3}{\partial x_3^2} &= F_{3/1}Y_{3/1}K_1R_1U_1\\
&\quad +F_{3/2}Y_{3/2}K_2R_2U_2\\
&\quad -K_3R_3U_3\\
R_4\frac{\partial U_4}{\partial t}+v_1\frac{\partial U_1}{\partial x_1}+v_2\frac{\partial U_1}{\partial x_2}+v_3\frac{\partial U_1}{\partial x_3}-D_1\frac{\partial^2 U_4}{\partial x_1^2}-D_2\frac{\partial^2 U_4}{\partial x_2^2}-D_3\frac{\partial^2 U_4}{\partial x_3^2} &= F_{4/2}Y_{4/2}K_2R_2U_2\\
&\quad +F_{4/3}Y_{4/3}K_3R_3U_3\\
&\quad -K_4R_4U_4
\end{aligned}
$$

with the boundary conditions

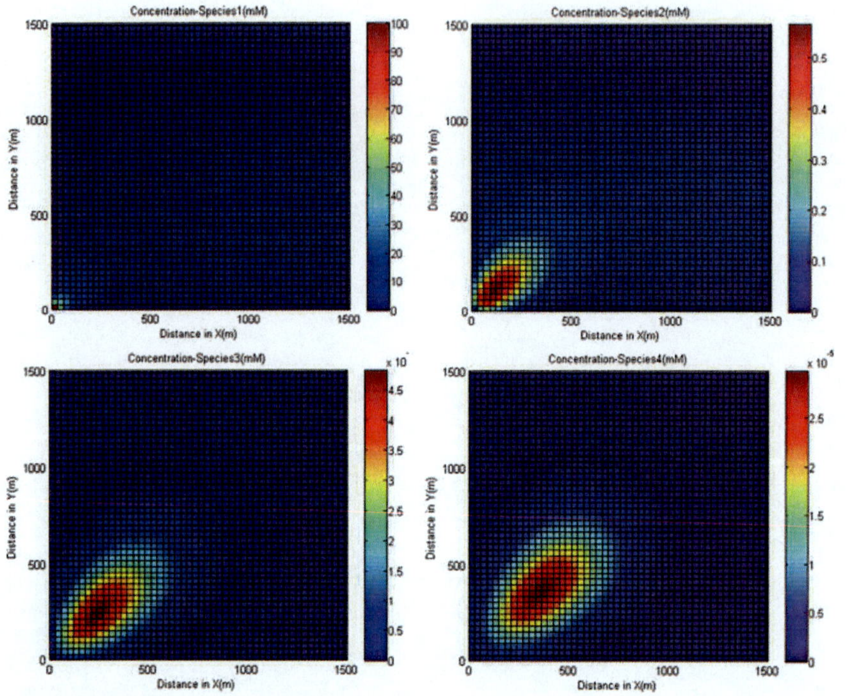

Fig. 4 Simulation for Sweby limiter for 1000 days with sequential reaction

Table 2 Parameters used for numerical simulation

Parameter	Value	Parameter	Value
U_{10}	100 mM (Constant source)	R_4	1.3
U_{10}	100 mM (Instantaneous/Point source)	K_1	$0.0007\,d^{-1}$
U_{20}	0	K_2	$0.0005\,d^{-1}$
U_{30}	0	K_3	$0.00045\,d^{-1}$
U_{40}	0	K_4	$0.00038\,d^{-1}$
D_1	$10\,m^2\,d^{-1}$	$F_{2/1}$	0.75
D_2	$10\,m^2\,d^{-1}$	$F_{3/1}$	0.25
D_3	$0.1\,m^2\,d^{-1}$	$F_{3/2}$	0.5
v_1	$1\,m\,d^{-1}$	$F_{4/2}$	0.5
v_2	$1\,m\,d^{-1}$	$F_{2/3}$	0.9
v_3	$0.1\,m\,d^{-1}$	$F_{4/3}$	0.1
R_1	5.3	$Y_{k/d}$	1 for all of them
R_2	1.9	T	1000 d
R_3	1.2	Δt	10
		Δx	30

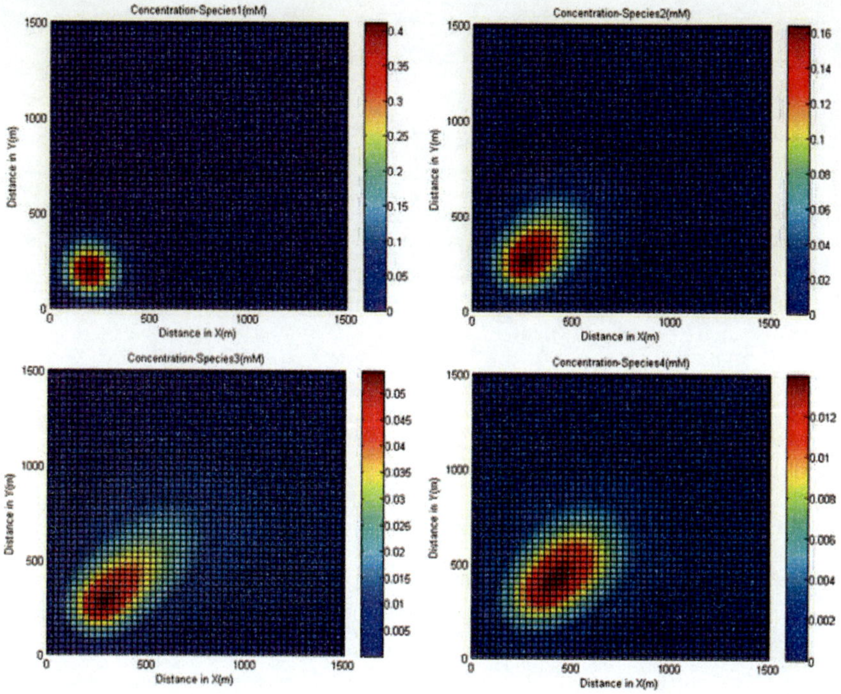

Fig. 5 Instantaneous or point injection simulation for van Albada limiter for 1000 days serial–parallel reversible reaction

$$
\begin{aligned}
U_k(0,0,0,0) &= U_{k0} \\
\lim_{x_1 \to \infty} U_k(x_1, x_2, x_3, t) &= 0 \\
\lim_{x_2 \to \infty} U_k(x_1, x_2, x_3, t) &= 0 \\
\lim_{x_3 \to \infty} U_k(x_1, x_2, x_3, t) &= 0 \; k = 1, 2, 3, 4.
\end{aligned}
$$

Instantaneous injection or point injection at time $t = 0$ is studied in this problem. An amount for 100 mM of species-1 and zero mM for other three species is injected at origin at time $t = 0$ and the movement of species is predicted for 1000 days. Numerical simulation of this with van Albada limiter, and Superbee limiter is shown in Fig. 5 and Fig. 6, respectively. The parameters in Table 2 are used for numerical simulation.

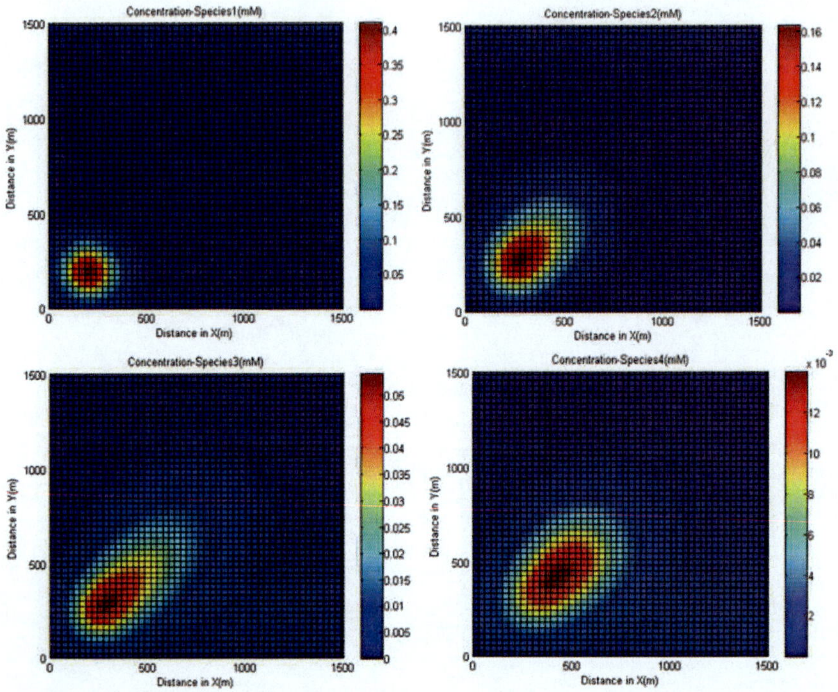

Fig. 6 Instantaneous or point injection simulation for Superbee limiter for 1000 days with serial–parallel reversible reaction

6 Summary and Conclusion

Multi-species transport equation in three dimensions with the first-order reaction network is considered in this article. Total variation diminishing finite volume scheme is applied for this problem. The stability and consistency conditions are derived. The necessary condition for flux limiter for controlling numerical diffusion is also derived. Numerical simulations are carried out for sequential reaction, reversible serial–parallel reaction problems with constant source and instantaneous or point source. Numerical simulations for four species transport are illustrated through graphs.

References

1. van Genutchen, M.T.: Convective-dispersive transport of solutes involved in sequential first-order decay reactions. Comput. Geosci. **11**, 129–147 (1985)
2. Domenico, P.A.: An analytical model for multidimensional transport of a decaying contaminant species. J. Hydrol. **91**, 49–58 (1987)

3. Cho, C.M.: Convective transport of ammonium with nitrification in soil. Can. J. Soil Sci. **51**, 339–350 (1971)
4. Bauer, P., Attinger, S., Kinzelbach, W.: Transport of decay chain in homogeneous porous media: analytical solutions. J. Contam. Hydrol. **49**, 217–239 (2001)
5. Clement, T.P.: A modular computer model for simulating reactive multi species transport in three dimensional groundwater systems. Technical Report, Pacific Northwest National Laboratory, Richland, Washington, PNNL-SA-28967 (1997)
6. Clement, T.P., Quezada, C.R., Lee, K.-K.: Generalised solution to multi-dimensional multi species transport equations coupled with a first-order reaction network involving distinct retardation factors. Adv. Water Resour. **27**, 507–520 (2004)
7. Clement, T.P., Sun, Y., Hooker, B.S., Petersen, J.N.: Modeling multi-species reactive transport in groundwater aquifers. Groundw. Monit. Remediat. J. **18**, 79–92 (1988)
8. Sun, Y., Clement, T.P.: A generalized decomposition method for solving coupled multi-species reactive transport problems. Transp. Porous Media **37**, 32–46 (1999)
9. Sun, Y., Petersen, J.N., Clement, T.P., Skeen, R.S.: Development of analytical solutions for multi-species transport with serial and parallel reactions. Water Resour. Res. **35**, 185–190 (1999)
10. Sheu, T.W.H., Wang, S.K., Lin, R.K.: An implicit scheme for solving the convection–diffusion–reaction equation in two dimensions. J. Comput. Phys. **164**, 123–142 (2000)
11. Calvo, M.P., De Frutos, J., Novo, J.: Linearly implicit Runge-Kutta methods for advection-reaction-diffusion equations. Appl. Numer. Math. **37**, 535–549 (2001)
12. Hundsdorfer, W., Verwer, J.G.: Numerical Solution of Time-dependent Advection-diffusion-reaction Equations, vol. 33. Springer Series in Computational Mathematics (2013)
13. Sibert, J.R., Hampton, J., Fournier, D.A., Bills, P.J.: An advection–diffusion–reaction model for the estimation of fish movement parameters from tagging data, with application to skipjack tuna (Katsuwonus pelamis). Can. J. Fish. Aquat. Sci. **56**, 925–938 (1999)
14. Houston, P., Schwab, C., Suli, E.: Discontinuous hp-finite element methods for advection-diffusion-reaction problems. SIAM J. Appl. Math. **39**, 2133–2163 (2002)
15. Ayuso, B., Donatella Marini, L.: Discontinuous Galerkin element methods for advection-diffusion-reaction problems. SIAM J. Appl. Math. **47**, 1391–1420 (2007)
16. Ern, A., Stephansen, A.F., Zunino, P.: A discontinuous Galerkin method with weighted averages for advection–diffusion equations with locally small and anisotropic diffusivity. IMA J. Numer. Anal. 1–24 (2008)
17. Georgoulis, E.H., Hall, E., Houston, P.: Discontinuous Galerkin methods for advection-diffusion-reaction problems on anisotropically refined meshes. SIAM J. Sci. Comput. **30**, 246–271 (2007)
18. Idelsohn, S., Nigro, N., Storti, M., Buscaglia, G.: A Petrov-Galerkin formulation for advection-reaction-diffusion problems. Comput. Methods Appl. Mech. Eng. **136**, 27–46 (1996)
19. Mudunuru, M.K., Nakshatrala, K.B.: On enforcing maximum principles and achieving element-wise species balance for advection–diffusion–reaction equations under the finite element method. J. Comput. Phys. **305**, 448–493 (2016)
20. Eymard, R., Gallouët, T., Herbin, R.: Handbook of Numerical Analysis (2000)
21. LeVeque, R.: Finite Volume Methods for Hyperbolic Problems. Cambridge University Press, Cambridge (2002)
22. Ramos, J.I.: A finite volume method for one-dimensional reaction-diffusion problems. Appl. Math. Comput. **188**, 739–748 (2007)
23. Arachchige, J.P., Pettet, G.J.: A finite volume method with linearisation in time for the solution of advection–reaction–diffusion systems. Appl. Math. Comput. **231**, 445–462 (2014)
24. ten Thije Boonkkamp, J.H.M., Anthonissen, M.J.H.: The finite volume-complete flux scheme for advection-diffusion-reaction equations. J. Sci. Comput. **46**, 47–70 (2011)
25. van Leer, B.: Towards the ultimate conservative difference scheme. II. Monotonicity and conservation combined in a second-order scheme. J. Comput. Phys. **14**, 361–370 (1974)
26. Sweby, P.K.: High resolution schemes using flux limiters for hyperbolic conservation laws. SIAM J. Numer. Anal. **21**, 995–1011 (1984)

27. Harten, A.: High resolution schemes for hyperbolic conservation laws. J. Comput. Phys. **49**, 357–393 (1983)
28. Harten, A., Lax, P.D.: On a class of high resolution total-variation-stable finite-difference schemes. SIAM J. Numer. Anal. **21**, 1–23 (1984)
29. Jameson, A., Lax, P.D.: Conditions for the construction of multi-point total variation diminishing difference schemes. Appl. Numer. Methods **2**, 335–345 (1986)
30. Shu, C.W.: Total-variation-diminishing time discretizations. SIAM J. Sci. Stat. Comput. **9**, 1073–1084 (1986)
31. Versteg, H.K., Malalasekara, W.: An Introduction to Computational Fluid Dynamics: A Finite Volume Method. Addison-Wesley (1996)
32. Smith, G.D.: Numerical Solutions of Partial Differential Equations: Finite Difference Methods. Oxford University Press (1985)

Existence of Fixed Point Results in C^*-algebra-valued Triple Controlled Metric Type Spaces

Kalpana Gopalan and Sumaiya Tasneem Zubair

Abstract In this study, we first interpret the notion of C^*-algebra-valued triple controlled metric type spaces and derive certain fixed point theorems for Banach and Kannan type contraction mappings of the underlying spaces. Secondly, we furnish an example to show the effectiveness of the proven Banach contraction principle.

Keywords C^*-algebra · C^*-algebra-valued triple controlled metric type spaces · Contractive mapping · Fixed point

Mathematics Subject Classification (2010) 47H10 · 54H25

1 Introduction

Fixed point theory has applications in various branches of mathematics such as nonlinear analysis, differential equation, integral equations, etc. As a generalization of metric spaces, Czerwik [5] initiated the concept of b-metric spaces. In [14], Nabil Mlaiki et al. proved the Banach contraction principle on new type of metric spaces namely controlled metric type spaces which is an extension of b-metric spaces by replacing the constant s by a control function $\delta(x, y)$ where such the variables rely on the equation's right hand side.

In [1], the same authors developed the notion of double controlled metric type spaces by modifying controlled metric type spaces through two control functions $\delta(x, y)$ and $\tau(x, y)$, the parameters of which relies upon the equation's right hand side. The researchers in [18] have recently emphasized the importance of triple

K. Gopalan (✉) · S. T. Zubair
Department of Mathematics, Sri Sivasubramaniya Nadar College of Engineering, Kalavakkam, Chennai 603 110, India
e-mail: kalpanag@ssn.edu.in

S. T. Zubair
e-mail: sumaiyatasneemz@ssn.edu.in

R. N. Mohapatra et al. (eds.), *Mathematical Analysis and Computing*,
Springer Proceedings in Mathematics & Statistics 344,
https://doi.org/10.1007/978-981-33-4646-8_39

controlled metric type spaces and have proven certain fixed point theorems within several contractive conditions.

In [12], Ma proposed the theory of C^*-algebra-valued metric spaces and proved several fixed point theorems with contractive mappings for the self maps and extended the same to C^*-algebra-valued b-metric spaces in [13]. The core idea is to use the set of all the positive elements of a unital C^*-algebra rather than the set of real numbers. However, Kalpana et al. [9], unveiled the idea of C^*-algebra-valued rectangular b-metric spaces and demonstrated certain fixed point theorems in such spaces. For further investigations on the concept of C^*-algebra, the readers can view [3, 4, 6–8, 10, 16, 17, 19].

In this present work, we investigate the hypothesis of C^*-algebra-valued triple controlled metric type spaces and derive certain fixed point theorems for Banach and Kannan type contraction mappings. Lastly, we set an example to demonstrate our main result.

2 Preliminaries

In 2018, Nabil Mlaiki et al. [14] defined a new extension of b-metric spaces, called controlled metric type spaces, and further, combining the definitions of extended b-metric spaces [11] and controlled metric type spaces [14], the researchers in [1] developed the idea of double controlled metric type spaces given below.

Definition 2.1 Let $X \neq \emptyset$ and $\delta, \tau : X \times X \to [1, \infty)$. If $d : X^2 \to [0, \infty)$ satisfies:

(1) $d(\upsilon, \kappa) = 0$ if and only if $\upsilon = \kappa$;
(2) $d(\upsilon, \kappa) = d(\kappa, \upsilon)$;
(3) $d(\upsilon, \kappa) \leq \delta(\upsilon, \omega)d(\upsilon, \omega) + \tau(\omega, \kappa)d(\omega, \kappa)$
for all $\upsilon, \kappa, \omega \in X$. Then d is called a double controlled metric type by α and μ.

In [18], the authors revealed the notion of triple controlled metric type spaces as described in the following:

Definition 2.2 Let $X \neq \emptyset$ and $\alpha, \mu, \delta : X \times X \to [1, \infty)$. A function $d : X \times X \to [0, \infty)$ is called a triple controlled metric type if it satisfies:

(1) $d(\upsilon, \kappa) = 0$ if and only if $\upsilon = \kappa$ for all $\upsilon, \kappa \in X$;
(2) $d(\upsilon, \kappa) = d(\kappa, \upsilon)$ for all $\upsilon, \kappa \in X$;
(3) $d(\upsilon, \kappa) \leq \alpha(\upsilon, \omega)d(\upsilon, \omega) + \mu(\omega, \varsigma)d(\omega, \varsigma) + \delta(\varsigma, \kappa)d(\varsigma, \kappa)$
for all $\upsilon, \kappa \in X$ and for all distinct points $\omega, \varsigma \in X$, each distinct from υ and κ respectively. The pair (X, d) is called a triple controlled metric type space.

We now discuss some essential concepts and results in C^*-algebra.

Let $\mathscr{A}$ denotes an unital C^*-algebra and set $\mathscr{A}_h = \{f \in \mathscr{A} : f = f^*\}$. An element $f \in \mathscr{A}$ is said to be positive, denote it by $\theta_{\mathscr{A}} \preceq f$, if $f \in \mathscr{A}_h$ and $\sigma(f) \subseteq [0, \infty)$, where $\theta_{\mathscr{A}}$ is a zero element in $\mathscr{A}$ and $\sigma(f)$ is the spectrum of f. There is a natural

partial ordering on $\mathscr{A}_h$ given by $f \preceq g$ if and only if $\theta_{\mathscr{A}} \preceq g - f$. We represent the sets $\mathscr{A}_+$ and $\mathscr{A}'$ as $\{f \in \mathscr{A} : \theta_{\mathscr{A}} \preceq f\}$ and $\{f \in \mathscr{A} : fg = gf, \forall g \in \mathscr{A}\}$ and $|w| = (w^*w)^{\frac{1}{2}}$ respectively.
Very recently, Asim et al. [2] initiated the concept of C^*-algebra-valued extended b-metric spaces by extending the notion of extended b-metric spaces, replacing the set of real numbers with $\mathscr{A}_+$.

Definition 2.3 Let $X \neq \emptyset$ and $\delta : X \times X \to \mathscr{A}^I$. The mapping $d : X \times X \to \mathscr{A}$ is called a C^*-algebra-valued extended b-metric on X, if it satisfies the following (for all $\upsilon, \kappa, \omega \in X$):

(1) $\theta_{\mathscr{A}} \preceq d(\upsilon, \kappa)$ for all $\upsilon, \kappa \in X$ and $d(\upsilon, \kappa) = \theta_{\mathscr{A}}$ if and only if $\upsilon = \kappa$;
(2) $d(\upsilon, \kappa) = d(\kappa, \upsilon)$ for all $x, y \in X$;

(3) $d(\upsilon, \kappa) \preceq \delta(\upsilon, \kappa)[d(\upsilon, \omega) + d(\omega, \kappa)]$.
The triplet $(X, \mathscr{A}, d)$ is called a C^*-algebra-valued extended b-metric space.

3 Main Results

In this main section, we first discuss the notion of C^*-algebra valued triple controlled metric type spaces and present an example for the space defined.

For the sake of convenience, we signify the set $\{a \in \mathscr{A} : ab = ba, \forall b \in \mathscr{A}\ and\ a \succeq I\}$ as $\mathscr{A}'_I$, respectively.

Definition 3.1 Let X be a nonempty set and $A_1, A_2, A_3 : X \times X \to \mathscr{A}'_I$. Suppose the mapping $d_c : X \times X \to \mathscr{A}$ satisfies:
(1) $\theta_{\mathscr{A}} \preceq d_c(\upsilon, \kappa)$ for all $\upsilon, \kappa \in X$ and $d_c(\upsilon, \kappa) = \theta_{\mathscr{A}}$ if and only if $\upsilon = \kappa$;
(2) $d_c(\upsilon, \kappa) = d_c(\kappa, \upsilon)$ for all $\upsilon, \kappa \in X$;
(3) $d_c(\upsilon, \kappa) \preceq A_1(\upsilon, \omega)d_c(\upsilon, \omega) + A_2(\omega, \varsigma)d_c(\omega, \varsigma) + A_3(\varsigma, \kappa)d_c(\varsigma, \kappa)$
for all $\upsilon, \kappa, \omega, \varsigma \in X$ and for all distinct points $\omega, \varsigma \in X - \{\upsilon, \kappa\}$. Then d_c is called a C^*-algebra-valued triple controlled metric on X and $(X, \mathscr{A}, d_c)$ is called a C^*-algebra-valued triple controlled metric type space.

Remark 3.2 If we take $\omega = \varsigma$ in the above definition, we get

$$d_c(\upsilon, \kappa) \preceq A_1(\upsilon, \omega)d_c(\upsilon, \omega) + A_3(\omega, \kappa)d_c(\omega, \kappa)\ for\ all\ \upsilon, \omega, \kappa \in X.$$

In this case, d_c is called a C^*-algebra-valued double controlled metric on X and $(X, \mathscr{A}, d_c)$ is called a C^*-algebra-valued double controlled metric type space.

Example 3.3 Let $X = E \cup F$ where $E = \{\frac{1}{n} : n \in N\}$ and F is the set of all positive integers and $\mathscr{A} = M_2(R)$. Define $d_c : X \times X \to \mathscr{A}$ such that $d_c(\upsilon, \kappa) = d_c(\kappa, \upsilon)$ for all $\upsilon, \kappa \in X$ and

$$d_c(\upsilon,\kappa)=\begin{cases}\begin{pmatrix}0&0\\0&0\end{pmatrix}, if\ \upsilon=\kappa\\ \begin{pmatrix}\upsilon+\kappa&0\\0&\upsilon+\kappa\end{pmatrix}, if\ \upsilon,\kappa\in E\\ \begin{pmatrix}6&0\\0&6\end{pmatrix}, if\ \upsilon\in E,\kappa\in\{6,7\}\ or\ \upsilon\in\{6,7\},\kappa\in E\\ \begin{pmatrix}2&0\\0&2\end{pmatrix}, otherwise\end{cases}$$

Let $A_1, A_2, A_3 : X \times X \to \mathscr{A}_I'$ be defined as

$$A_1(\upsilon,\kappa)=\begin{cases}\begin{pmatrix}\upsilon&0\\0&\upsilon\end{pmatrix}, if\ \upsilon,\kappa\in F\\ \begin{pmatrix}1&0\\0&1\end{pmatrix}, otherwise\end{cases}$$

$$A_2(\upsilon,\kappa)=\begin{cases}\begin{pmatrix}\frac{1}{\upsilon}&0\\0&\frac{1}{\upsilon}\end{pmatrix}, if\ \upsilon\in E,\kappa\in F\\ \begin{pmatrix}1&0\\0&1\end{pmatrix}, otherwise\end{cases}$$

and

$$A_3(\upsilon,\kappa)=\begin{cases}\begin{pmatrix}\kappa&0\\0&\kappa\end{pmatrix}, if\ \upsilon\in E,\kappa\in F\\ \begin{pmatrix}1&0\\0&1\end{pmatrix}, otherwise\end{cases}$$

Hence $(X, \mathbb{A}, d_c)$ is a C^*-algebra-valued triple controlled metric type space. However

$$d_c(\frac{1}{2},7)=\begin{pmatrix}6&0\\0&6\end{pmatrix}\succ A_1(\tfrac{1}{2},1)d_c(\tfrac{1}{2},1)+A_2(1,7)d_c(1,7)$$
$$=\begin{pmatrix}4&0\\0&4\end{pmatrix}.$$

Thereby $(X, \mathbb{A}, d_c)$ is not a C^*-algebra-valued double controlled metric type space.

Definition 3.4 A sequence $\{\upsilon_n\}$ in a C^*-algebra-valued triple controlled metric type space $(X, \mathscr{A}, d_c)$ is said to be:
(i) convergent sequence $\Leftrightarrow \exists \upsilon \in X$ such that $d_c(\upsilon_n, \upsilon) \to \theta_{\mathscr{A}} (n \to \infty)$ and we denote it by $\lim_{n\to\infty} \upsilon_n = \upsilon$.
(ii) Cauchy sequence $\Leftrightarrow d_c(\upsilon_n, \upsilon_m) \to \theta_{\mathscr{A}} (n, m \to \infty)$.

Definition 3.5 A C^*-algebra-valued triple controlled metric type space $(X, \mathscr{A}, d_c)$ is said to be complete if every Cauchy sequence is convergent in X with respect to $\mathscr{A}$.

Theorem 3.6 *Let $(X, \mathscr{A}, d_c)$ be a complete C^*-algebra-valued triple controlled metric type space and suppose $T : X \to X$ is a continuous mapping satisfying the following condition:*

$$d_c(T\upsilon, T\kappa) \preceq C^* \, d_c(\upsilon, \kappa) \, C \; for \; all \; \upsilon, \kappa \in X \tag{1}$$

where $C \in \mathscr{A}$ with $\|C\| < 1$. For $\upsilon_0 \in X$, choose $\upsilon_n = T^n \upsilon_0$. Assume that

$$\sup_{m\geq 1} \lim_{i\to\infty} \|A_1(\upsilon_{i+1}, \upsilon_{i+2})A_3(\upsilon_{i+1}, \upsilon_m)\| < \frac{1}{\|C\|^4} \tag{2}$$

and

$$\sup_{m\geq 1} \lim_{i\to\infty} \|A_2(\upsilon_{i+1}, \upsilon_{i+2})A_3(\upsilon_{i+1}, \upsilon_m)\| < \frac{1}{\|C\|^4}. \tag{3}$$

In addition, for each $\upsilon \in X$ and $i = 1,\ 2$ or 3, suppose that

$$\lim_{n\to\infty} \|A_i(\upsilon, \upsilon_n)\| \; and \; \lim_{n\to\infty} \|A_i(\upsilon_n, \upsilon)\| \tag{4}$$

exist and are finite. Then, T has a unique fixed point in X.

Proof Choose an element $\upsilon_0 \in X$ and set $\upsilon_{n+1} = T\upsilon_n = \ldots = T^{n+1}\upsilon_0, n = 1, 2, \ldots$. For simplicity, the elements $d_c(\upsilon_0, \upsilon_1)$ and $d_c(\upsilon_0, \upsilon_2)$ in $\mathscr{A}$ are denoted by C_0 and D_0 respectively. Then

$$d_c(\upsilon_n, \upsilon_{n+1}) \preceq (C^*)^n d_c(\upsilon_0, \upsilon_1)C^n = (C^*)^n C_0 C^n \tag{5}$$

and

$$d_c(\upsilon_n, \upsilon_{n+2}) \preceq (C^*)^n d_c(\upsilon_0, \upsilon_2)C^n = (C^*)^n D_0 C^n. \tag{6}$$

Now, we demonstrate that $\{\upsilon_n\}_{n\in N}$ is a Cauchy sequence i.e., $\lim_{n\to\infty} d_c(\upsilon_n, \upsilon_{n+p}) = \theta_{\mathscr{A}}$, for $p \in N$. Therefore, we separate two cases as follows.
Case 1: Let $p = 2m + 1$, where $m \geq 1$. Then, we obtain

$$
\begin{aligned}
d_c(\upsilon_n, \upsilon_{n+2m+1}) &\preceq A_1(\upsilon_n, \upsilon_{n+1})d_c(\upsilon_n, \upsilon_{n+1}) + A_2(\upsilon_{n+1}, \upsilon_{n+2})d_c(\upsilon_{n+1}, \upsilon_{n+2}) \\
&\quad + \sum_{i=\frac{n}{2}+1}^{\frac{n+2m-2}{2}} A_1(\upsilon_{2i}, \upsilon_{2i+1})d_c(\upsilon_{2i}, \upsilon_{2i+1}) \prod_{j=\frac{n}{2}+1}^{i} A_3(\upsilon_{2j}, \upsilon_{n+2m+1}) \\
&\quad + \sum_{i=\frac{n}{2}+1}^{\frac{n+2m-2}{2}} A_2(\upsilon_{2i+1}, \upsilon_{2i+2})d_c(\upsilon_{2i+1}, \upsilon_{2i+2}) \prod_{j=\frac{n}{2}+1}^{i} A_3(\upsilon_{2j}, \upsilon_{n+2m+1}) \\
&\quad + \prod_{\frac{n}{2}+1}^{\frac{n+2m}{2}} A_3(\upsilon_{2j}, \upsilon_{n+2m+1})d_c(\upsilon_{n+2m}, \upsilon_{n+2m+1}) \\
&\preceq A_1(\upsilon_n, \upsilon_{n+1})d_c(\upsilon_n, \upsilon_{n+1}) + A_2(\upsilon_{n+1}, \upsilon_{n+2})d_c(\upsilon_{n+1}, \upsilon_{n+2}) \\
&\quad + \sum_{i=\frac{n}{2}+1}^{\frac{n+2m}{2}} A_1(\upsilon_{2i}, \upsilon_{2i+1})d_c(\upsilon_{2i}, \upsilon_{2i+1}) \prod_{j=\frac{n}{2}+1}^{i} A_3(\upsilon_{2j}, \upsilon_{n+2m+1}) \\
&\quad + \sum_{i=\frac{n}{2}+1}^{\frac{n+2m-2}{2}} A_2(\upsilon_{2i+1}, \upsilon_{2i+2})d_c(\upsilon_{2i+1}, \upsilon_{2i+2}) \prod_{j=\frac{n}{2}+1}^{i} A_3(\upsilon_{2j}, \upsilon_{n+2m+1}) \\
&\preceq A_1(\upsilon_n, \upsilon_{n+1})(C^*)^n C_0 C^n + A_2(\upsilon_{n+1}, \upsilon_{n+2})(C^*)^{n+1} C_0 C^{n+1} \\
&\quad + \sum_{i=\frac{n}{2}+1}^{\frac{n+2m}{2}} A_1(\upsilon_{2i}, \upsilon_{2i+1})(C^*)^{2i} C_0 C^{2i} \prod_{j=1}^{i} A_3(\upsilon_{2j}, \upsilon_{n+2m+1}) \\
&\quad + \sum_{i=\frac{n}{2}+1}^{\frac{n+2m-2}{2}} A_2(\upsilon_{2i+1}, \upsilon_{2i+2})(C^*)^{2i+1} C_0 C^{2i+1} \prod_{j=1}^{i} A_3(\upsilon_{2j}, \upsilon_{n+2m+1}) \\
&= \left(C_0^{\frac{1}{2}}[A_1(\upsilon_n, \upsilon_{n+1})]^{\frac{1}{2}} C^n\right)^* \left(C_0^{\frac{1}{2}}[A_1(\upsilon_n, \upsilon_{n+1})]^{\frac{1}{2}} C^n\right) \\
&\quad + \left(C_0^{\frac{1}{2}}[A_2(\upsilon_{n+1}, \upsilon_{n+2})]^{\frac{1}{2}} C^{n+1}\right)^* \left(C_0^{\frac{1}{2}}[A_2(\upsilon_{n+1}, \upsilon_{n+2})]^{\frac{1}{2}} C^{n+1}\right) \\
&\quad + \sum_{i=\frac{n}{2}+1}^{\frac{n+2m}{2}} \left(C_0^{\frac{1}{2}}[A_1(\upsilon_{2i}, \upsilon_{2i+1}) \prod_{j=1}^{i} A_3(\upsilon_{2j}, \upsilon_{n+2m+1})]^{\frac{1}{2}} C^{2i}\right)^* \\
&\qquad \left(C_0^{\frac{1}{2}}[A_1(\upsilon_{2i}, \upsilon_{2i+1}) \prod_{j=1}^{i} A_3(\upsilon_{2j}, \upsilon_{n+2m+1})]^{\frac{1}{2}} C^{2i}\right) \\
&\quad + \sum_{i=\frac{n}{2}+1}^{\frac{n+2m-2}{2}} \left(C_0^{\frac{1}{2}}[A_2(\upsilon_{2i+1}, \upsilon_{2i+2}) \prod_{j=1}^{i} A_3(\upsilon_{2j}, \upsilon_{n+2m+1})]^{\frac{1}{2}} C^{2i+1}\right)^*
\end{aligned}
$$

$$
\begin{aligned}
&\Big(C_0^{\frac{1}{2}}[A_2(\upsilon_{2i+1},\upsilon_{2i+2})\prod_{j=1}^{i}A_3(\upsilon_{2j},\upsilon_{n+2m+1})]^{\frac{1}{2}}C^{2i+1}\Big)\\
&= |C_0^{\frac{1}{2}}[A_1(\upsilon_n,\upsilon_{n+1})]^{\frac{1}{2}}C^n|^2 + |C_0^{\frac{1}{2}}[A_2(\upsilon_{n+1},\upsilon_{n+2})]^{\frac{1}{2}}C^{n+1}|^2\\
&+ \sum_{i=\frac{n}{2}+1}^{\frac{n+2m}{2}} |C_0^{\frac{1}{2}}[A_1(\upsilon_{2i},\upsilon_{2i+1})\prod_{j=1}^{i}A_3(\upsilon_{2j},\upsilon_{n+2m+1})]^{\frac{1}{2}}C^{2i}|^2\\
&+ \sum_{i=\frac{n}{2}+1}^{\frac{n+2m-2}{2}} |C_0^{\frac{1}{2}}[A_2(\upsilon_{2i+1},\upsilon_{2i+2})\prod_{j=1}^{i}A_3(\upsilon_{2j},\upsilon_{n+2m+1})]^{\frac{1}{2}}C^{2i+1}|^2\\
&\preceq \|C_0^{\frac{1}{2}}[A_1(\upsilon_n,\upsilon_{n+1})]^{\frac{1}{2}}C^n\|^2 I + \|C_0^{\frac{1}{2}}[A_2(\upsilon_{n+1},\upsilon_{n+2})]^{\frac{1}{2}}C^{n+1}\|^2 I\\
&+ \Big\|\sum_{i=\frac{n}{2}+1}^{\frac{n+2m}{2}} C_0^{\frac{1}{2}}[A_1(\upsilon_{2i},\upsilon_{2i+1})\prod_{j=1}^{i}A_3(\upsilon_{2j},\upsilon_{n+2m+1})]^{\frac{1}{2}}C^{2i}\Big\|^2 I\\
&+ \Big\|\sum_{i=\frac{n}{2}+1}^{\frac{n+2m-2}{2}} C_0^{\frac{1}{2}}[A_2(\upsilon_{2i+1},\upsilon_{2i+2})\prod_{j=1}^{i}A_3(\upsilon_{2j+1},\upsilon_{n+2m+1})]^{\frac{1}{2}}C^{2i+1}\Big\|^2 I\\
&\preceq \|C_0\|\Big[\|A_1(\upsilon_n,\upsilon_{n+1})\|\|C\|^{2n} + \|A_2(\upsilon_{n+1},\upsilon_{n+2})\|\|C\|^{2(n+1)}\\
&+ \sum_{i=\frac{n}{2}+1}^{\frac{n+2m}{2}} \|A_1(\upsilon_{2i},\upsilon_{2i+1})\prod_{j=1}^{i}A_3(\upsilon_{2j},\upsilon_{n+2m+1})\|\|C\|^{4i}\\
&+ \sum_{i=\frac{n}{2}+1}^{\frac{n+2m-2}{2}} \|A_2(\upsilon_{2i+1},\upsilon_{2i+2})\prod_{j=1}^{i}A_3(\upsilon_{2j},\upsilon_{n+2m+1})\|C\|^{4i}\|C\|^2\Big]I.
\end{aligned}
\tag{7}
$$

where I is the unit element in $\mathscr{A}$. Therefore the inequality above indicates

$$
\begin{aligned}
d_c(\upsilon_n,\upsilon_{n+2m+1}) \preceq \|C_0\|\Big[&\|A_1(\upsilon_n,\upsilon_{n+1})\|\|C\|^{2n} + \|A_2(\upsilon_{n+1},\upsilon_{n+2})\|\|C\|^{2(n+1)}\\
&+ (Y_{\frac{n+2m}{2}} - Y_{\frac{n}{2}+1}) + (Z_{\frac{n+2m-2}{2}} - Y_{\frac{n}{2}+1})\|C\|^2\Big]I
\end{aligned}
\tag{8}
$$

where

$$Y_k = \sum_{i=1}^{k} \|C\|^{4i} \|A_1(\upsilon_{2i}, \upsilon_{2i+1}) \prod_{j=1}^{i} A_3(\upsilon_{2j}, \upsilon_{n+2m+1})\|$$

and

$$Z_k = \sum_{i=1}^{k} \|C\|^{4i} \|A_2(\upsilon_{2i+1}, \upsilon_{2i+2}) \prod_{j=1}^{i} A_3(\upsilon_{2j}, \upsilon_{n+2m+1})\|.$$

The ratio test along with (2) and (3) ensures that the limit of the sequences $\{Y_n\}$ and $\{Z_n\}$ exists and so $\{Y_n\}$ and $\{Z_n\}$ are Cauchy. Letting $n \to \infty$ in (8), we get

$$\lim_{n\to\infty} d_c(\upsilon_n, \upsilon_{n+2m+1}) = \theta_{\mathscr{A}}. \tag{9}$$

Case 2: Let $p = 2m$, where $m \geq 2$. Then, we derive

$$\begin{aligned}
d_c(\upsilon_n, \upsilon_{n+2m}) &\preceq A_1(\upsilon_n, \upsilon_{n+2})d_c(\upsilon_n, \upsilon_{n+2}) + A_2(\upsilon_{n+2}, \upsilon_{n+3})d_c(\upsilon_{n+2}, \upsilon_{n+3}) \\
&\quad + \sum_{i=\frac{n+3}{2}}^{\frac{n+2m-1}{2}} A_1(\upsilon_{2i}, \upsilon_{2i+1})d_c(\upsilon_{2i}, \upsilon_{2i+1}) \prod_{j=\frac{n+3}{2}}^{i} A_3(\upsilon_{2j}, \upsilon_{n+2m}) \\
&\quad + \sum_{i=\frac{n+3}{2}}^{\frac{n+2m-3}{2}} A_2(\upsilon_{2i+1}, \upsilon_{2i+2})d_c(\upsilon_{2i+1}, \upsilon_{2i+2}) \prod_{j=\frac{n+3}{2}}^{i} A_3(\upsilon_{2j}, \upsilon_{n+2m}) \\
&\preceq A_1(\upsilon_n, \upsilon_{n+2})\,(C^*)^n D_0 C^n + A_2(\upsilon_{n+2}, \upsilon_{n+3})(C^*)^{n+2} C_0 C^{n+2} \\
&\quad + \sum_{i=\frac{n+3}{2}}^{\frac{n+2m-1}{2}} A_1(\upsilon_{2i}, \upsilon_{2i+1})(C^*)^{2i} C_0 C^{2i} \prod_{j=1}^{i} A_3(\upsilon_{2j}, \upsilon_{n+2m}) \\
&\quad + \sum_{i=\frac{n+3}{2}}^{\frac{n+2m-3}{2}} A_2(\upsilon_{2i+1}, \upsilon_{2i+2})(C^*)^{2i+1} C_0 C^{2i+1} \prod_{j=1}^{i} A_3(\upsilon_{2j}, \upsilon_{n+2m}) \\
&= \left(D_0^{\frac{1}{2}}[A_1(\upsilon_n, \upsilon_{n+2})]^{\frac{1}{2}} C^n\right)^* \left(D_0^{\frac{1}{2}}[A_1(\upsilon_n, \upsilon_{n+2})]^{\frac{1}{2}} C^n\right) \\
&\quad + \left(C_0^{\frac{1}{2}}[A_2(\upsilon_{n+2}, \upsilon_{n+3})]^{\frac{1}{2}} C^{n+2}\right)^* \left(C_0^{\frac{1}{2}}[A_2(\upsilon_{n+2}, \upsilon_{n+3})]^{\frac{1}{2}} C^{n+2}\right)
\end{aligned}$$

$$+\sum_{i=\frac{n+3}{2}}^{\frac{n+2m-1}{2}}\Big(C_0^{\frac{1}{2}}[A_1(\upsilon_{2i},\upsilon_{2i+1})\prod_{j=1}^{i}A_3(\upsilon_{2j},\upsilon_{n+2m})]^{\frac{1}{2}}C^{2i}\Big)^*$$

$$\Big(C_0^{\frac{1}{2}}[A_1(\upsilon_{2i},\upsilon_{2i+1})\prod_{j=1}^{i}A_3(\upsilon_{2j},\upsilon_{n+2m})]^{\frac{1}{2}}C^{2i}\Big)$$

$$+\sum_{i=\frac{n+3}{2}}^{\frac{n+2m-3}{2}}\Big(C_0^{\frac{1}{2}}[A_2(\upsilon_{2i+1},\upsilon_{2i+2})\prod_{j=1}^{i}A_3(\upsilon_{2j},\upsilon_{n+2m})]^{\frac{1}{2}}C^{2i+1}\Big)^*$$

$$\Big(C_0^{\frac{1}{2}}[A_2(\upsilon_{2i+1},\upsilon_{2i+2})\prod_{j=1}^{i}A_3(\upsilon_{2j},\upsilon_{n+2m})]^{\frac{1}{2}}C^{2i+1}\Big)$$

$$=|D_0^{\frac{1}{2}}[A_1(\upsilon_n,\upsilon_{n+2})]^{\frac{1}{2}}C^n|^2+|C_0^{\frac{1}{2}}[A_2(\upsilon_{n+2},\upsilon_{n+3})]^{\frac{1}{2}}C^{n+2}|^2$$

$$+\sum_{i=\frac{n+3}{2}}^{\frac{n+2m-1}{2}}|C_0^{\frac{1}{2}}[A_1(\upsilon_{2i},\upsilon_{2i+1})\prod_{j=1}^{i}A_3(\upsilon_{2j},\upsilon_{n+2m})]^{\frac{1}{2}}C^{2i}|^2$$

$$+\sum_{i=\frac{n+3}{2}}^{\frac{n+2m-3}{2}}|C_0^{\frac{1}{2}}[A_2(\upsilon_{2i+1},\upsilon_{2i+2})\prod_{j=1}^{i}A_3(\upsilon_{2j},\upsilon_{n+2m})]^{\frac{1}{2}}C^{2i+1}|^2$$

$$\preceq\|D_0^{\frac{1}{2}}A_1(\upsilon_n,\upsilon_{n+2})^{\frac{1}{2}}C^n\|^2I+\|C_0^{\frac{1}{2}}A_2(\upsilon_{n+2},\upsilon_{n+3})^{\frac{1}{2}}C^{n+2}\|^2I$$

$$+\Big\|\sum_{i=\frac{n+3}{2}}^{\frac{n+2m-1}{2}}C_0^{\frac{1}{2}}[A_1(\upsilon_{2i},\upsilon_{2i+1})\prod_{j=1}^{i}A_3(\upsilon_{2j},\upsilon_{n+2m})]^{\frac{1}{2}}C^{2i}\Big\|^2I$$

$$+\Big\|\sum_{i=\frac{n+3}{2}}^{\frac{n+2m-3}{2}}C_0^{\frac{1}{2}}[A_2(\upsilon_{2i+1},\upsilon_{2i+2})\prod_{j=1}^{i}A_3(\upsilon_{2j},\upsilon_{n+2m})]^{\frac{1}{2}}C^{2i+1}\Big\|^2I$$

$$\preceq\|D_0\|\|A_1(\upsilon_n,\upsilon_{n+1})\|\|C\|^{2n}I+\|C_0\|\Big[\|A_2(\upsilon_{n+2},\upsilon_{n+3})\|\|C\|^{2(n+2)}$$

$$+\sum_{i=\frac{n+3}{2}}^{\frac{n+2m-1}{2}}\|A_1(\upsilon_{2i},\upsilon_{2i+1})\prod_{j=1}^{i}A_3(\upsilon_{2j},\upsilon_{n+2m})\|\|C\|^{4i}+$$

$$+\sum_{i=\frac{n+3}{2}}^{\frac{n+2m-3}{2}}\|A_2(\upsilon_{2i+1},\upsilon_{2i+2})\prod_{j=1}^{i}A_3(\upsilon_{2j},\upsilon_{n+2m})\|C\|^{4i}\|C\|^2\Big]I.$$

where I is the unit element in $\mathscr{A}$. So the inequality stated above indicates

$$d_c(\upsilon_n, \upsilon_{n+2m}) \preceq D_0\|A_1(\upsilon_n, \upsilon_{n+2})\|\|C\|^{2n}I + \|C_0\|\Big[\|A_2(\upsilon_{n+2}, \upsilon_{n+3})\|\|C\|^{2(n+2)} + (Y_{\frac{n+2m-1}{2}} - Y_{\frac{n+3}{2}}) + (Z_{\frac{n+2m-3}{2}} - Y_{\frac{n+3}{2}})\|C\|^2\Big]I \tag{10}$$

where

$$Y_k = \sum_{i=1}^{k} \|C\|^{4i}\|A_1(\upsilon_{2i}, \upsilon_{2i+1})\prod_{j=1}^{i} A_3(\upsilon_{2j}, \upsilon_{n+2m})\|$$

and

$$Z_k = \sum_{i=1}^{k} \|C\|^{4i}\|A_2(\upsilon_{2i+1}, \upsilon_{2i+2})\prod_{j=1}^{i} A_3(\upsilon_{2j}, \upsilon_{n+2m})\|.$$

The ratio test however with (2) and (3) confirms that the limit of the sequences $\{Y_n\}$ and $\{Z_n\}$ exists and so $\{Y_n\}$ and $\{Z_n\}$ are Cauchy. Letting $n \to \infty$ in (10), we get

$$\lim_{n\to\infty} d_c(\upsilon_n, \upsilon_{n+2m}) = \theta_{\mathscr{A}}. \tag{11}$$

Combining (9) and (11), we obtain the sequence $\{\upsilon_n\}$ to be Cauchy i.e.,

$$\lim_{n\to\infty} d_c(\upsilon_n, \upsilon_{n+p}) = \theta_{\mathscr{A}}. \tag{12}$$

Through completeness, there is an element $\upsilon \in X$ such that $\lim_{n\to\infty} \upsilon_n = \upsilon$. We will now affirm that υ is a fixed point of T . Consider

$$d_c(\upsilon, \upsilon_{n+2}) \preceq A_1(\upsilon, \upsilon_n)d_c(\upsilon, \upsilon_n) + A_2(\upsilon_n, \upsilon_{n+1})d_c(\upsilon_n, \upsilon_{n+1}) + A_3(\upsilon_{n+1}, \upsilon_{n+2})d_c(\upsilon_{n+1}, \upsilon_{n+2}).$$

Hence

$$\begin{aligned}\|d_c(\upsilon, \upsilon_{n+2})\| &\leq \|A_1(\upsilon, \upsilon_n)d_c(\upsilon, \upsilon_n) + A_2(\upsilon_n, \upsilon_{n+1})d_c(\upsilon_n, \upsilon_{n+1}) + A_3(\upsilon_{n+1}, \upsilon_{n+2}) d_c(\upsilon_{n+1}, \upsilon_{n+2})\|\\ &\leq \|A_1(\upsilon, \upsilon_n)\|\|d_c(\upsilon, \upsilon_n)\| + \|A_2(\upsilon_n, \upsilon_{n+1})\|\|d_c(\upsilon_n, \upsilon_{n+1})\| + \|A_3(\upsilon_{n+1}, \upsilon_{n+2})\|\|d_c(\upsilon_{n+1}, \upsilon_{n+2})\|.\end{aligned}$$

We obtain using (4) and (12) that

$$\lim_{n\to\infty} \|d_c(\upsilon, \upsilon_{n+2})\| = 0. \tag{13}$$

Consider

$$\begin{aligned}\|d_c(\upsilon, T\upsilon)\| &\leq \|A_1(\upsilon, \upsilon_{n+2})d_c(\upsilon, \upsilon_{n+2}) + A_2(\upsilon_{n+2}, \upsilon_{n+1})d_c(\upsilon_{n+2}, \upsilon_{n+1}) \\ &\quad + A_3(\upsilon_{n+1}, T\upsilon)d_c(\upsilon_{n+1}, T\upsilon)\| \\ &\leq \|A_1(\upsilon, \upsilon_{n+2})\|\|d_c(\upsilon, \upsilon_{n+2})\| + \|A_2(\upsilon_{n+2}, \upsilon_{n+1})\|\|d_c(\upsilon_{n+2}, \upsilon_{n+1})\| \\ &\quad + \|A_3(\upsilon_{n+1}, T\upsilon)\|\|d_c(T^{n+1}\upsilon, T\upsilon)\|.\end{aligned}$$

Letting $n \to \infty$, we obtain $\|d_c(\upsilon, \upsilon_{n+2})\| \to 0$ by (13). Since $T^n\upsilon \to \upsilon$ and from continuity of T, we get $\|d_c(T^{n+1}\upsilon, T\upsilon)\| \to 0$. Hence $\|d_c(\upsilon, T\upsilon)\| \to 0$ as $n \to \infty$ $\Longleftrightarrow d_c(\upsilon, T\upsilon) \to \theta_{\mathscr{A}}$ as $n \to \infty$. Thereby υ is a fixed point of T.
By employing the inequality (1), it is simple to confirm that υ is a unique fixed of T.

Theorem 3.7 *(Kannan Type) Let $(X, \mathscr{A}, d_c)$ be a complete C^*-algebra-valued triple controlled metric type space and suppose $T : X \to X$ is a mapping satisfying*

$$d_c(T\upsilon, T\kappa) \preceq C[d_c(T\upsilon, \upsilon) + d_c(T\kappa, \kappa)] \text{ for all } \upsilon, \kappa \in X \tag{14}$$

where $C \in \mathscr{A}'_+$ with $\|C\| < \frac{1}{2}$. For $\upsilon_0 \in X$, choose $\upsilon_n = T^n\upsilon_0$. Assume that

$$\sup_{m\geq 1} \lim_{i\to\infty} \|A_1(\upsilon_{i+1}, \upsilon_{i+2})A_3(\upsilon_{i+1}, \upsilon_m)\| < \frac{1}{\|s\|^2} \tag{15}$$

and

$$\sup_{m\geq 1} \lim_{i\to\infty} \|A_2(\upsilon_{i+1}, \upsilon_{i+2})A_3(\upsilon_{i+1}, \upsilon_m)\| < \frac{1}{\|s\|^2} \tag{16}$$

where $s = (I - C)^{-1}C$. In addition, for each $\upsilon \in X$ and $i = 1, 2$ or 3, suppose that

$$\lim_{n\to\infty} \|A_i(\upsilon, \upsilon_n)\| \text{ and } \lim_{n\to\infty} \|A_i(\upsilon_n, \upsilon)\| \tag{17}$$

exist and are finite. Then, T has a unique fixed point in X.

Proof Choose an element $\upsilon_0 \in X$ and set $\upsilon_{n+1} = T\upsilon_n = \ldots = T^{n+1}\upsilon_0, n = 1, 2, \ldots$. For simplicity, the elements $d_c(\upsilon_0, \upsilon_1)$ and $d_c(\upsilon_0, \upsilon_2)$ in $\mathscr{A}$ are denoted by C_0 and D_0 respectively. Then

$$\begin{aligned}d_c(\upsilon_n, \upsilon_{n+1}) &= d_c(T\upsilon_{n-1}, T\upsilon_n) \\ &\preceq C[d_c(T\upsilon_{n-1}, \upsilon_{n-1}) + d_c(T\upsilon_n, \upsilon_n)] \\ &= C[d_c(\upsilon_n, \upsilon_{n-1}) + d_c(\upsilon_{n+1}, \upsilon_n)]\end{aligned}$$

i.e.,

$$\begin{aligned}d_c(\upsilon_n, \upsilon_{n+1}) &\preceq (I - C)^{-1}Cd_c(\upsilon_{n-1}, \upsilon_n) \\ &= sd_c(\upsilon_{n-1}, \upsilon_n)\end{aligned}$$

where $s = (I - C)^{-1}C$. As of $\|C\| < \frac{1}{2}$, $I - C$ is invertible by the lemma in [15]. So, we get

$$d_c(\upsilon_n, \upsilon_{n+1}) \preceq s^n d_c(\upsilon_0, \upsilon_1).$$

Consider

$$\begin{aligned} d_c(\upsilon_n, \upsilon_{n+2}) &= d_c(T\upsilon_{n-1}, T\upsilon_{n+1}) \\ &\preceq C[d_c(T\upsilon_{n-1}, \upsilon_{n-1}) + d_c(T\upsilon_{n+1}, \upsilon_{n+1})] \\ &= C[d_c(\upsilon_n, \upsilon_{n-1}) + d_c(\upsilon_{n+2}, \upsilon_{n+1})] \\ &\preceq C[s^{n-1} d_c(\upsilon_0, \upsilon_1) + s^{n+1} d_c(\upsilon_0, \upsilon_1)] \\ &= C s^{n-1} d_c(\upsilon_0, \upsilon_1)[I + s^2] \\ &= s_0 s^{n-1} d_c(\upsilon_0, \upsilon_1) \end{aligned}$$

where $s_0 = C[I + s^2] \succeq \theta_{\mathscr{A}}$. Now, we prove that $\{\upsilon_n\}_{n \in N}$ is a Cauchy sequence i.e., $\lim\limits_{n \to \infty} d_c(\upsilon_n, \upsilon_{n+p}) = \theta_{\mathscr{A}}$, for $p \in N$. Thus, we distinguish the following two cases.

Case 1: Let $p = 2m + 1$, where $m \geq 1$. Then, we find

$$\begin{aligned} d_c(\upsilon_n, \upsilon_{n+2m+1}) &\preceq A_1(\upsilon_n, \upsilon_{n+1}) d_c(\upsilon_n, \upsilon_{n+1}) + A_2(\upsilon_{n+1}, \upsilon_{n+2}) d_c(\upsilon_{n+1}, \upsilon_{n+2}) \\ &\quad + \sum_{i=\frac{n}{2}+1}^{\frac{n+2m}{2}} A_1(\upsilon_{2i}, \upsilon_{2i+1}) d_c(\upsilon_{2i}, \upsilon_{2i+1}) \prod_{j=\frac{n}{2}+1}^{i} A_3(\upsilon_{2j}, \upsilon_{n+2m+1}) \\ &\quad + \sum_{i=\frac{n}{2}+1}^{\frac{n+2m-2}{2}} A_2(\upsilon_{2i+1}, \upsilon_{2i+2}) d_c(\upsilon_{2i+1}, \upsilon_{2i+2}) \prod_{j=\frac{n}{2}+1}^{i} A_3(\upsilon_{2j}, \upsilon_{n+2m+1}) \\ &\preceq |C_0^{\frac{1}{2}} s^{\frac{n}{2}} [A_1(\upsilon_n, \upsilon_{n+1})]^{\frac{1}{2}}|^2 + |C_0^{\frac{1}{2}} s^{\frac{n+1}{2}} [A_2(\upsilon_{n+1}, \upsilon_{n+2})]^{\frac{1}{2}}|^2 \\ &\quad + \sum_{i=\frac{n}{2}+1}^{\frac{n+2m}{2}} |C_0^{\frac{1}{2}} s^i [A_1(\upsilon_{2i}, \upsilon_{2i+1}) \prod_{j=1}^{i} A_3(\upsilon_{2j}, \upsilon_{n+2m+1})]^{\frac{1}{2}}|^2 \\ &\quad + \sum_{i=\frac{n}{2}+1}^{\frac{n+2m-2}{2}} |C_0^{\frac{1}{2}} s^{\frac{2i+1}{2}} [A_2(\upsilon_{2i+1}, \upsilon_{2i+2}) \prod_{j=1}^{i} A_3(\upsilon_{2j}, \upsilon_{n+2m+1})]^{\frac{1}{2}}|^2 \\ &\preceq \|C_0\| \Big[\|s\|^n \|A_1(\upsilon_n, \upsilon_{n+1})\| + \|s\|^{n+1} \|A_2(\upsilon_{n+1}, \upsilon_{n+2})\| \\ &\quad + \sum_{i=\frac{n}{2}+1}^{\frac{n+2m}{2}} \|s\|^{2i} A_1(\upsilon_{2i}, \upsilon_{2i+1}) \prod_{j=1}^{i} A_3(\upsilon_{2j}, \upsilon_{n+2m+1})\| \\ &\quad + \sum_{i=\frac{n}{2}+1}^{\frac{n+2m-2}{2}} \|s\|^{2i+1} A_2(\upsilon_{2i+1}, \upsilon_{2i+2}) \prod_{j=1}^{i} A_3(\upsilon_{2j}, \upsilon_{n+2m+1})\| \Big] I. \end{aligned}$$

where I is the unit element in $\mathscr{A}$. Further the inequality described above implies

$$d_c(\upsilon_n, \upsilon_{n+2m+1}) \preceq \|C_0\|\Big[\|A_1(\upsilon_n, \upsilon_{n+1})\| \|s\|^n + \|A_2(\upsilon_{n+1}, \upsilon_{n+2})\| \|s\|^{n+1} + (Y_{\frac{n+2m}{2}} - Y_{\frac{n}{2}+1}) + (Z_{\frac{n+2m-2}{2}} - Z_{\frac{n}{2}+1})\|s\|\Big]I \tag{18}$$

where

$$Y_k = \sum_{i=1}^{k} \|s\|^{2i} \|A_1(\upsilon_{2i}, \upsilon_{2i+1}) \prod_{j=1}^{i} A_3(\upsilon_{2j}, \upsilon_{n+2m+1})\|$$

and

$$Z_k = \sum_{i=1}^{k} \|s\|^{2i+1} \|A_2(\upsilon_{2i+1}, \upsilon_{2i+2}) \prod_{j=1}^{i} A_3(\upsilon_{2j}, \upsilon_{n+2m+1})\|.$$

The ratio test along with the inequalities (15) and (16) indicates that the limit of the sequences $\{Y_n\}$ and $\{Z_n\}$ exists and so $\{Y_n\}$ and $\{Z_n\}$ are Cauchy. Letting $n \to \infty$ in (18), we get

$$\lim_{n\to\infty} d_c(\upsilon_n, \upsilon_{n+2m+1}) = \theta_{\mathscr{A}}. \tag{19}$$

Case 2: Let $p = 2m$, where $m \geq 2$. Then, we acquire

$$\begin{aligned}
d_c(\upsilon_n, \upsilon_{n+2m}) &\preceq A_1(\upsilon_n, \upsilon_{n+2})d_c(\upsilon_n, \upsilon_{n+2}) + A_2(\upsilon_{n+2}, \upsilon_{n+3})d_c(\upsilon_{n+2}, \upsilon_{n+3}) \\
&\quad + \sum_{i=\frac{n}{2}+1}^{\frac{n+2m-1}{2}} A_1(\upsilon_{2i}, \upsilon_{2i+1})d_c(\upsilon_{2i}, \upsilon_{2i+1}) \prod_{j=\frac{n}{2}+1}^{i} A_3(\upsilon_{2j}, \upsilon_{n+2m}) \\
&\quad + \sum_{i=\frac{n}{2}+1}^{\frac{n+2m-3}{2}} A_2(\upsilon_{2i+1}, \upsilon_{2i+2})d_c(\upsilon_{2i+1}, \upsilon_{2i+2}) \prod_{j=\frac{n}{2}+1}^{i} A_3(\upsilon_{2j}, \upsilon_{n+2m}) \\
&\preceq s_0|D_0^{\frac{1}{2}} s^{\frac{n}{2}} [A_1(\upsilon_n, \upsilon_{n+2})]^{\frac{1}{2}}|^2 + |C_0^{\frac{1}{2}} s^{\frac{n+2}{2}} [A_2(\upsilon_{n+2}, \upsilon_{n+3})]^{\frac{1}{2}}|^2 \\
&\quad + \sum_{i=\frac{n}{2}+1}^{\frac{n+2m-1}{2}} |C_0^{\frac{1}{2}} s^i [A_1(\upsilon_{2i}, \upsilon_{2i+1}) \prod_{j=1}^{i} A_3(\upsilon_{2j}, \upsilon_m)]^{\frac{1}{2}}|^2 \\
&\quad + \sum_{i=\frac{n}{2}+1}^{\frac{n+2m-2}{2}} |C_0^{\frac{1}{2}} s^{\frac{2i+1}{2}} [A_2(\upsilon_{2i+1}, \upsilon_{2i+2}) \prod_{j=1}^{i} A_3(\upsilon_{2j}, \upsilon_{n+2m})]^{\frac{1}{2}}|^2 \\
&\preceq s_0\|D_0\| \|s\|^n \|A_1(\upsilon_n, \upsilon_{n+2})\| I + \|C_0\|\Big[\|s\|^{n+2} \|A_2(\upsilon_{n+2}, \upsilon_{n+3})\|
\end{aligned}$$

$$+\sum_{i=\frac{n}{2}+1}^{\frac{n+2m-1}{2}} \|s\|^{2i} A_1(\upsilon_{2i}, \upsilon_{2i+1}) \prod_{j=1}^{i} A_3(\upsilon_{2j}, \upsilon_{n+2m})\|$$

$$+\sum_{i=\frac{n}{2}+1}^{\frac{n+2m-3}{2}} \|s\|^{2i+1} A_2(\upsilon_{2i+1}, \upsilon_{2i+2}) \prod_{j=1}^{i} A_3(\upsilon_{2j}, \upsilon_{n+2m})\|\Big]I \tag{20}$$

where I is the unit element in $\mathscr{A}$. Consequently, the latter inequality indicates

$$d_c(\upsilon_n, \upsilon_{n+2m}) \preceq s_0\|D_0\|\|A_1(\upsilon_n, \upsilon_{n+2})\|\|s\|^n + \|C_0\|\Big[\|A_2(\upsilon_{n+2}, \upsilon_{n+3})\|\|s\|^{n+2} + (Y_{\frac{n+2m-1}{2}} - Y_{\frac{n}{2}+1}) + (Z_{\frac{n+2m-3}{2}} - Z_{\frac{n}{2}+1})\|s\|\Big]I \tag{21}$$

where

$$Y_k = \sum_{i=1}^{k} \|s\|^{2i} \|A_1(\upsilon_{2i}, \upsilon_{2i+1}) \prod_{j=1}^{i} A_3(\upsilon_{2j}, \upsilon_{n+2m})\|$$

and

$$Z_k = \sum_{i=1}^{k} \|s\|^{2i} \|A_2(\upsilon_{2i+1}, \upsilon_{2i+2}) \prod_{j=1}^{i} A_3(\upsilon_{2j+1}, \upsilon_{n+2m})\|.$$

The ratio test however with the inequalities (15) and (16) ensures that the limit of the sequences $\{Y_n\}$ and $\{Z_n\}$ exists and so $\{Y_n\}$ and $\{Z_n\}$ are Cauchy. Letting $n \to \infty$ in (21), we get

$$\lim_{n\to\infty} d_c(\upsilon_n, \upsilon_{n+2m}) = \theta_{\mathscr{A}}. \tag{22}$$

Therefore by combining (19) and (22), we acquire that the sequence $\{\upsilon_n\}$ is Cauchy with respect to $\mathscr{A}$, that is

$$\lim_{n\to\infty} d_c(\upsilon_n, \upsilon_{n+p}) = \theta_{\mathscr{A}}. \tag{23}$$

Through completeness, there is an element $\upsilon \in X$ such that $\lim_{n\to\infty} \upsilon_n = \upsilon$. We will now affirm that υ is a fixed point of T. Consider

$$d_c(\upsilon_{n+2}, \upsilon) \preceq A_1(\upsilon_{n+2}, \upsilon_{n+1})d_c(\upsilon_{n+2}, \upsilon_{n+1}) + A_2(\upsilon_{n+1}, \upsilon_n)d_c(\upsilon_{n+1}, \upsilon_n) + A_3(\upsilon_n, \upsilon)d_c(\upsilon_n, \upsilon)$$

Accordingly, we get

$$\begin{aligned}\|d_c(\upsilon_{n+2}, \upsilon)\| &\le \|A_1(\upsilon_{n+2}, \upsilon_{n+1})d_c(\upsilon_{n+2}, \upsilon_{n+1}) + A_2(\upsilon_{n+1}, \upsilon_n)d_c(\upsilon_{n+1}, \upsilon_n) \\ &\quad + A_3(\upsilon_n, \upsilon)d_c(\upsilon_n, \upsilon)\| \\ &\le \|A_1(\upsilon_{n+2}, \upsilon_{n+1})\|\|d_c(\upsilon_{n+2}, \upsilon_{n+1})\| + \|A_2(\upsilon_{n+1}, \upsilon_n)\|\| \\ &\quad d_c(\upsilon_{n+1}, \upsilon_n)\| + \|A_3(\upsilon_n, \upsilon)\|\|d_c(\upsilon_n, \upsilon)\|\end{aligned}$$

We obtain from (17) and (23) that

$$\lim_{n\to\infty} \|d_c(\upsilon_{n+2}, \upsilon)\| = 0. \tag{24}$$

$$\begin{aligned}d_c(T\upsilon, \upsilon) &\preceq A_1(T\upsilon, T\upsilon_n)d_c(T\upsilon, T\upsilon_n) + A_2(T\upsilon_n, T\upsilon_{n+1})d_c(T\upsilon_n, T\upsilon_{n+1}) \\ &\quad + A_3(T\upsilon_{n+1}, \upsilon)d_c(T\upsilon_{n+1}, \upsilon) \\ &\preceq A_1(T\upsilon, T\upsilon_n)[Cd_c(T\upsilon, \upsilon) + Cd_c(T\upsilon_n, \upsilon_n)] + A_2(T\upsilon_n, T\upsilon_{n+1}) \\ &\quad d_c(T\upsilon_n, T\upsilon_{n+1}) + A_3(\upsilon_{n+2}, \upsilon)d_c(\upsilon_{n+2}, \upsilon)\end{aligned}$$

which gives

$$\begin{aligned}d_c(T\upsilon, \upsilon)(I - A_1(T\upsilon, T\upsilon_n)C) &\preceq A_1(T\upsilon, T\upsilon_n)Cd_c(T\upsilon, T\upsilon_{n-1}) \\ &\quad + A_2(T\upsilon_n, T\upsilon_{n+1})d_c(T\upsilon_n, T\upsilon_{n+1}) \\ &\quad + A_3(T\upsilon_{n+1}, \upsilon)d_c(T\upsilon_{n+1}, \upsilon).\end{aligned}$$

Thereby, we get

$$\begin{aligned}d_c(T\upsilon, \upsilon) &\preceq (I - A_1(T\upsilon, T\upsilon_n)C)^{-1}A_1(T\upsilon, T\upsilon_n)Cd_c(T\upsilon, T\upsilon_{n-1}) \\ &\quad + (I - A_1(T\upsilon, T\upsilon_n)C)^{-1}A_2(T\upsilon_n, T\upsilon_{n+1})d_c(T\upsilon_n, T\upsilon_{n+1}) \\ &\quad + (I - A_1(T\upsilon, T\upsilon_n)C)^{-1}A_3(\upsilon_{n+2}, \upsilon)d_c(\upsilon_{n+2}, \upsilon).\end{aligned}$$

This yields

$$\begin{aligned}\|d_c(T\upsilon, \upsilon)\| &\le \|(I - A_1(T\upsilon, T\upsilon_n)C)^{-1}\|\|A_1(T\upsilon, T\upsilon_n)C\|\|d_c(T\upsilon, T\upsilon_{n-1})\| \\ &\quad + \|(I - A_1(T\upsilon, T\upsilon_n)C)^{-1}\|\|A_2(T\upsilon_n, T\upsilon_{n+1})\|\,\|d_c(T\upsilon_n, T\upsilon_{n+1})\| \\ &\quad + \|(I - A_1(T\upsilon, T\upsilon_n)C)^{-1}\|\|A_3(\upsilon_{n+2}, \upsilon)\|\|d_c(\upsilon_{n+2}, \upsilon)\|.\end{aligned}$$

Letting $n \to \infty$ in the above equation, we get $\|d_c(T\upsilon, \upsilon)\| \le 0$. Thus $d_c(T\upsilon, \upsilon) = \theta_{\mathscr{A}}$, this implies $T\upsilon = \upsilon$ i.e., υ is a fixed point of T. We can easily demonstrate, through using (14), that υ is an unique fixed point of T.

Example 3.8 Let $X = \{1, 2, 3, 4\}$ and $\mathscr{A} = M_2(R)$ of all 2×2 matrices. Define $d_c : X \times X \to \mathscr{A}$ as follows:

$d_c(\upsilon, \upsilon) = \begin{pmatrix} 0 & 0 \\ 0 & 0 \end{pmatrix}$ $for\ all\ \upsilon \in X,$

$d_c(1, 2) = d_c(2, 1) = d_c(1, 3) = d_c(3, 1) = \begin{pmatrix} 50 & 0 \\ 0 & 50 \end{pmatrix},$

$d_c(1, 4) = d_c(4, 1) = d_c(2, 3) = d_c(3, 2) = d_c(2, 4) = d_c(4, 2) = \begin{pmatrix} 200 & 0 \\ 0 & 200 \end{pmatrix},$

$d_c(4, 3) = d_c(3, 4) = \begin{pmatrix} 800 & 0 \\ 0 & 800 \end{pmatrix}.$

Let $A_1, A_2, A_3 : X \times X \to \mathscr{A}_I'$ be defined as follows:

$$A_1(\upsilon, \kappa) = \begin{cases} \begin{pmatrix} 1 & 0 \\ 0 & 1 \end{pmatrix}, \forall \upsilon = \kappa \\ \begin{pmatrix} 2 & 0 \\ 0 & 2 \end{pmatrix}, for\ \upsilon = 3,\ \kappa = 4 \\ \begin{pmatrix} \upsilon + \kappa & 0 \\ 0 & \upsilon + \kappa \end{pmatrix}, otherwise \end{cases}$$

$$A_2(\upsilon, \kappa) = \begin{cases} \begin{pmatrix} 1 & 0 \\ 0 & 1 \end{pmatrix}, \forall \upsilon = \kappa \\ \begin{pmatrix} \frac{|\upsilon+\kappa|}{2} & 0 \\ 0 & \frac{|\upsilon+\kappa|}{2} \end{pmatrix}, otherwise \end{cases}$$

and

$$A_3(\upsilon, \kappa) = \begin{cases} \begin{pmatrix} 1 & 0 \\ 0 & 1 \end{pmatrix}, \forall \upsilon = \kappa \\ \begin{pmatrix} 3 & 0 \\ 0 & 3 \end{pmatrix}, for\ \upsilon \in X - \{4\},\ \kappa = 4 \\ \begin{pmatrix} \upsilon + \kappa & 0 \\ 0 & \upsilon + \kappa \end{pmatrix}, otherwise \end{cases}$$

It is clear that $(X, \mathscr{A}, d_c)$ is a complete C^*-algebra-valued triple controlled metric type space. We see that

$$\begin{aligned} d_c(3, 4) = \begin{pmatrix} 800 & 0 \\ 0 & 800 \end{pmatrix} &\succ A_1(3, 1)d_c(3, 1) + A_2(1, 4)d_c(1, 4) \\ &= \begin{pmatrix} 700 & 0 \\ 0 & 700 \end{pmatrix}. \end{aligned}$$

Thereby $(X, \mathscr{A}, d_c)$ is not a C^*-algebra-valued double controlled metric type space. For any $A \in \mathscr{A}$, we define its norm as $\|A\| = \max_{1\leq i\leq 4}\{|a_i|\}$. Let $T : X \to X$ be defined as

$$Tv = \begin{cases} 2, \; if \; v \neq 4 \\ 1, \; if \; v = 4 \end{cases}$$

It is clear that T satisfies (1) with $C = \begin{pmatrix} \frac{1}{2} & 0 \\ 0 & \frac{1}{2} \end{pmatrix}$ and $\|C\| = \frac{1}{2} < 1$. Note that for each $v \in X$, $T^n v = 2$. Thus

$$\sup_{m\geq 1} \lim_{i\to\infty} \|A_1(v_{2i+1}, v_{2i+2})A_3(v_{2i+1}, v_{2i+2})\| = \left\| \begin{pmatrix} 1 & 0 \\ 0 & 1 \end{pmatrix} \right\| = 1 < 2^4 = \frac{1}{\|C\|^4}.$$

Similarly, we get $\sup_{m\geq 1} \lim_{i\to\infty} \|A_2(v_{2i+2}, v_{2i+3})A_3(v_{2i+2}, v_{2i+3})\| = 1 < \frac{1}{\|C\|^4}$. Moreover, for each $v \in X$, we have $\lim_{n\to\infty} \|A_i(v_n, v)\| < \infty, \; for \; i = 1, 2 \; or \; 3$. Therefore, T fulfills all the conditions of Theorem (3.6). Consequently, T has a unique fixed point in X, which is $v = 2$.

4 Conclusion

In this article, we have examined the conception of C^*-algebra-valued triple controlled metric type spaces as a generalization of both C^*-algebra valued rectangular b-metric spaces and double controlled metric type spaces. Further, we proved some fixed point results for Banach and Kannan type contraction mappings on complete C^*-algebra-valued triple controlled metric type spaces. Finally, an illustration is given to demonstrate our main result.

Acknowledgements The authors thank the Principal and management of Sri Sivasubramaniya Nadar College of Engineering, Kalavakkam-603110, for their constant support towards the successful completion of this work.

Conflict of interest The authors declare that they have no conflict of interest.

References

1. Abdeljawad, T., Mlaiki, N., Aydi, H., Souayah, N.: Double controlled metric type spaces and some fixed point results. Mathematics **6**, 320 (2018)
2. Asim, M., Imdad, M.: C^*-algebra-valued extended b-metric spaces and fixed point results with an application, U. P. B. Sci. Bull., Ser. A **82**(1), 207–218 (2020)

3. Bai, C.: Coupled fixed point theorems in C^*-algebra-valued b-metric spaces with application. Fixed Point Theory Appl. **2016**, 70 (2016)
4. Batul, S., Kamran, T.: C^*-valued contractive type mappings. Fixed Point Theory Appl. **2015**, 142 (2015)
5. Czerwik, S.: Nonlinear set-valued contraction mappings in b-metric spaces. Atti Sem. Mat. Fis. Univ. Modena **46**(2), 263–276 (1998)
6. Erden, M., Alaca, C.: C^*-algebra-valued S-metric spaces. Commun. Ser. A1 **67**(2), 165–177 (2018)
7. Kadelburg, Z., Radenović, S.: Fixed point results in $C*$-algebra-valued metric spaces are the direct consequences of their standard metric counterparts. Fixed Point Theory Appl. **2016**, 53 (2016)
8. Kalaivani, C., Kalpana, G.: Fixed point theorems in C^*-algebra-valued S-metric spaces with some applications: U.P.B. Sci. Bull., Ser. A **80**(3) (2018)
9. Kalpana, G., Sumaiya Tasneem, Z.: C^*-algebra-valued rectangular b-metric spaces and some fixed point theorems. Commun. Fac. Sci. Univ. Ank. Ser. A1 Math. Stat. **68**(2), 1–11 (2019)
10. Kalpana, G., Sumaiya Tasneem, Z.: Common fixed point theorems in C^*-algebra-valued hexagonal b-metric spaces. AIP Conf. Proc. 2095, 030012, 030012-1-030012-5 (2019)
11. Kamran, T., Samreen, M., UL Ain, Q.: A generalization of b-metric space and some fixed point theorems. Mathematics **5**, 19 (2017)
12. Ma, Z.H., Jiang, L.N., Sun, H.: C^*-algebra-valued metric spaces and related fixed point theorems. Fixed Point Theory Appl. **2014**, 206 (2014)
13. Ma, Z.H., Jiang, L.N.: C^*-algebra-valued b-metric spaces and related fixed point theorems. Fixed Point Theory Appl. **2015**, 222 (2015)
14. Mlaiki, N., Aydi, H., Souayah, N., Abdeljawad, T.: Controlled metric type spaces and the related contraction principle. Mathematics **6**, 194 (2018)
15. Murphy, G.J.: C^*-Algebras and Operator Theory. Academic Press, London (1990)
16. Radenović, S., Vetro, P., Nastasi, A., Quan, L.T.: Coupled fixed point theorems in C^*-algebra-valued b-metric spaces. Sci. Publ. State Univ. Novi Pazar Ser. A. Appl. Math. Inform. Mech. **9**, 81–90 (2017)
17. Shehwar, D., Batul, S., Kamran, T., Ghiura, A.: Caristis fixed point theorem on C^*-algebra valued metric spaces. J. Nonlinear Sci. Appl. **9**, 584–588 (2016)
18. Sumaiya Tasneem, Z., Kalpana, G., Abdeljawad, T.: A Different approach to fixed point theorems on triple controlled metric type spaces with a numerical experiment. Dyn. Syst. Appl. **30**(1), 111–130 (2021)
19. Xin, Q.L., Jiang, L.N., Ma, Z.H.: Common fixed point theorems in C^*-algebra-valued metric spaces. J. Nonlinear Sci. Appl. **9**, 4617–4627 (2016)

On Fixed Points in the Setting of C^*-Algebra-Valued Controlled F_c-Metric Type Spaces

Kalpana Gopalan and Sumaiya Tasneem Zubair

Abstract In the present article, we first examine the conception of C^*-algebra-valued controlled F_c-metric type spaces as a generalization of F-cone metric spaces over Banach algebra. Further, we prove the fixed point theorem with a specific contractive condition in the framework of C^*-algebra-valued controlled F_c-metric type spaces. Secondly, we furnish an example by means of the acquired result.

Keywords C^*-algebra · C^*-algebra-valued controlled F_c-metric type spaces · Contractive mapping · Fixed point

Mathematics Subject Classification (2010) 47H10 · 54H25

1 Introduction

The conception of b-metric spaces was initiated by Bakhtin [7] as a generalization of metric spaces. In 1994, Matthews [19] proposed the concept of partial metric spaces where the self-distance of any point need not be zero. Tayyab Kamran et al. [14] introduced a new type of metric spaces, namely extended b-metric spaces by replacing the constant s by a function $\theta(x, y)$ depending on the parameters of the left-hand side of the triangle inequality. Nabil Mlaiki et al. [18] proved Banach contraction principle in the setting of controlled metric type spaces which is a generalization of extended b-metric spaces. For more engrossing results in extended b-metric spaces, the readers can view [1–5, 15, 22–24]. In [20], Aiman Mukheimer have recently examined the hypothesis of extended partial S_b-metric spaces.

On the other hand, Fernandez et al. [10] established the notion of F-cone metric spaces over Banach algebras and investigated the existence and uniqueness of the

K. Gopalan (✉) · S. T. Zubair
Sri Sivasubramaniya Nadar College of Engineering, Kalavakkam, Chennai 603 110, India
e-mail: kalpanag@ssn.edu.in

S. T. Zubair
e-mail: sumaiyatasneemz@ssn.edu.in

R. N. Mohapatra et al. (eds.), *Mathematical Analysis and Computing*,
Springer Proceedings in Mathematics & Statistics 344,
https://doi.org/10.1007/978-981-33-4646-8_40

fixed point under the same metric. In [17], Ma initiated the concept of C^*-algebra-valued metric spaces where the set of real numbers is replaced by the set of all positive elements of a unital C^*-algebra. For further probes on C^*-algebra, we refer to [6, 8, 9, 11–13, 16, 21, 25, 26].

As noted above, vigorous research on fixed point results in C^*-algebra-valued metric spaces, extended b-metric spaces, and controlled metric type spaces has been developed in the past few years, we focus our study on the concept of C^*-algebra-valued controlled F_c-metric type spaces in the present paper and prove fixed point theorem with the disparate contractive condition.

2 Preliminaries

To start with, we recollect some necessary definitions which will be utilized in the main theorem.

Throughout this paper, $\mathscr{A}$ signifies the unital C^*-algebra. Set $\mathscr{A}_h = \{z \in \mathscr{A} : z = z^*\}$. We call an element $z \in \mathscr{A}$ a positive element, denote it by $\theta_{\mathscr{A}} \preceq z$, if $z \in \mathscr{A}_h$ and $\sigma(z) \subseteq [0, \infty)$, where $\theta_{\mathscr{A}}$ is a zero element in $\mathscr{A}$ and $\sigma(z)$ is the spectrum of z. There is a natural partial ordering on $\mathscr{A}_h$ given by $z \preceq w$ if and only if $\theta_{\mathscr{A}} \preceq w - z$. We represent $\mathscr{A}_+$ and $\mathscr{A}'$ as the sets $\{z \in \mathscr{A} : \theta_{\mathscr{A}} \preceq z\}$ and $\{z \in \mathscr{A} : zw = wz, \forall w \in \mathscr{A}\}$ and $|z| = (z^*z)^{\frac{1}{2}}$, respectively.

The notion of F-cone metric spaces over Banach algebra was initiated by Fernandez et al. [10] as follows:

Definition 2.1 Let $X \neq \emptyset$. A function $F : X^3 \to A$ is called F-cone metric on X if for any $\alpha, \beta, \gamma, \delta \in X$, the following conditions hold:
1. $\alpha = \beta = \gamma$ if and only if $F(\alpha, \alpha, \alpha) = F(\beta, \beta, \beta) = F(\gamma, \gamma, \gamma) = F(\alpha, \beta, \gamma)$;
2. $\theta \preceq F(\alpha, \alpha, \alpha) \preceq F(\alpha, \alpha, \beta) \preceq F(\alpha, \beta, \gamma)$ for all $\alpha, \beta, \gamma \in X$ with $\alpha \neq \beta \neq \gamma$;
3. $F(\alpha, \beta, \gamma) \preceq s[F(\alpha, \alpha, \delta) + F(\beta, \beta, \delta) + F(\gamma, \gamma, \delta)] - F(\delta, \delta, \delta)$.

Then the pair (X, F) is called an F-cone metric space over Banach algebra A. The number $s \geq 1$ is called the coefficient of (X, F).

In [11], Kalaivani et al. developed the idea of C^*-algebra-valued S_b-metric spaces as a generalization of S_b-metric spaces.

Definition 2.2 Let $X \neq \emptyset$ and $A \in \mathscr{A}'$ such that $A \succeq I_{\mathscr{A}}$. Suppose the mapping $S_b : X \times X \times X \to \mathscr{A}$ satisfies:
1. $\theta_{\mathscr{A}} \preceq S_b(\alpha, \beta, \gamma)$ for all $\alpha, \beta, \gamma \in X$ with $\alpha \neq \beta \neq \gamma \neq \alpha$;
2. $S_b(\alpha, \beta, \gamma) = \theta_{\mathscr{A}}$ if and only if $\alpha = \beta = \gamma$;
3. $S_b(\alpha, \beta, \gamma) \preceq A[S_b(\alpha, \alpha, \delta) + S_b(\beta, \beta, \delta) + S_b(\gamma, \gamma, \delta)]$ for all $\alpha, \beta, \gamma, \delta \in X$.

The triplet $(X, \mathscr{A}, S_b)$ is said to be a C^*-algebra-valued S_b-metric space.

3 Main Results

In this main segment, as a generalization of F-cone metric spaces over Banach algebras, we introduce the notion of C^*-algebra-valued controlled F_c-metric type spaces and furnish an example of the underlying spaces.
Hereinafter, $\mathscr{A}'_I$ will denote the set $\{z \in \mathscr{A} : zw = wz, \forall w \in \mathscr{A} \text{ and } z \succeq I_{\mathscr{A}}\}$, respectively.

Definition 3.1 Let $X \neq \emptyset$ and $C : X \times X \times X \to \mathscr{A}'_I$. Suppose the mapping $F_c : X \times X \times X \to \mathscr{A}$ satisfies:

1. $\varpi = \omega = \tau$ if and only if $F_c(\varpi, \varpi, \varpi) = F_c(\omega, \omega, \omega) = F_c(\tau, \tau, \tau) = F_c(\varpi, \omega, \tau)$;
2. $\theta_{\mathscr{A}} \preceq F_c(\varpi, \varpi, \varpi) \preceq F_c(\varpi, \varpi, \omega) \preceq F_c(\varpi, \omega, \tau)$;
3. $F_c(\varpi, \omega, \tau) \preceq C(\varpi, \varpi, \alpha) F_c(\varpi, \varpi, \alpha) + C(\omega, \omega, \alpha) F_c(\omega, \omega, \alpha) + C(\tau, \tau, \alpha)$
$$F_c(\tau, \tau, \alpha) - F_c(\alpha, \alpha, \alpha)$$

for all $\varpi, \omega, \tau, \alpha \in X$. Then, F_c is called a C^*-algebra-valued controlled F_c-metric type on X and $(X, \mathscr{A}, F_c)$ is a C^*-algebra-valued controlled F_c-metric type space.

Remark 3.2 If $C(\varpi, \varpi, \alpha) = C(\omega, \omega, \alpha) = C(\tau, \tau, \alpha) = C(\varpi, \omega, \tau)$ for all $\varpi, \omega, \tau, \alpha \in X$, then we get

$$F_c(\varpi, \omega, \tau) \preceq C(\varpi, \omega, \tau)[F_c(\varpi, \varpi, \alpha) + F_c(\omega, \omega, \alpha) + F_c(\tau, \tau, \alpha)] - F_c(\alpha, \alpha, \alpha).$$

In this case, F_c is called a C^*-algebra-valued extended F_c-metric on X and $(X, \mathscr{A}, c)$ is called a C^*-algebra-valued extended F_c-metric space.

Example 3.3 Let $X = \{0, 1, 2, \ldots\}$ and $\mathscr{A} = R^2$. If $\varpi, \omega \in \mathscr{A}$ with $\varpi = (\varpi_1, \varpi_2)$, $\omega = (\omega_1, \omega_2)$, then the addition, multiplication and scalar multiplication can be defined as follows:

$$\varpi + \omega = (\varpi_1 + \omega_1, \varpi_2 + \omega_2), k\varpi = (k\varpi_1, k\varpi_2), \varpi\omega = (\varpi_1\omega_1, \varpi_2\omega_2).$$

Now, define the metric $F_c : X \times X \times X \to \mathscr{A}$ and the control function $C : X \times X \times X \to \mathscr{A}'_I$ as

$$F_c(\varpi, \omega, \tau) = \left(\frac{1}{2}(|\varpi + \tau|^2 + |\omega + \tau|^2), \frac{1}{2}(|\varpi + \tau|^2 + |\omega + \tau|^2)\right)$$

and

$$C(\varpi, \omega, \tau) = \left(|\varpi + \omega - \tau + 1|, |\varpi + \omega - \tau + 1|\right).$$

It is easy to verify that F_c is a C^*-algebra-valued controlled F_c-metric type space. Indeed, for $\varpi = 1, \omega = 2, \tau = 3$ *and* $\alpha = 0$, we have

$$F_c(1,2,3) = (20.5, 20.5) \succ (1,1)[(1,1) + (4,4) + (9,9)] - (0,0) = (14,14)$$
$$= C(1,2,3)[F_c(1,1,0) + F_c(2,2,0) + F_c(3,3,0)] - F_c(0,0,0).$$

Therefore, F_c is not a C^*-algebra-valued extended F_c-metric space.

Definition 3.4 A sequence $\{\varpi_n\}$ in a C^*-algebra-valued controlled F_c-metric type space is said to be:
(i) convergent sequence $\Longleftrightarrow \exists \varpi \in X$ such that $F_c(\varpi_n, \varpi_n, \varpi) \to \theta_{\mathscr{A}}$ as $n \to \infty$ and we denote it by $\lim_{n\to\infty} \varpi_n = \varpi$;
(ii) Cauchy sequence $\Longleftrightarrow F_c(\varpi_n, \varpi_n, \varpi_m) \to \theta_{\mathscr{A}}$ as $n, m \to \infty$.

Definition 3.5 A C^*-algebra-valued controlled F_c-metric type space $(X, \mathscr{A}, F_c)$ is said to be complete if every Cauchy sequence is convergent in X with respect to $\mathscr{A}$.

Remark 3.6 In a C^*-algebra-valued controlled F_c-metric type space $(X, \mathscr{A}, F_c)$, if $\varpi, \omega, \tau \in X$ and $F_c(\varpi, \omega, \tau) = \theta$, then $\varpi = \omega = \tau$, but the converse need not be true.

Definition 3.7 A C^*-algebra-valued controlled F_c-metric type space $(X, \mathscr{A}, F_c)$ is said to be symmetric, if it satisfies

$$F_c(\varpi, \varpi, \omega) = F_c(\omega, \omega, \varpi) \; for \; all \; \varpi, \omega \in X.$$

Theorem 3.8 *Let $(X, \mathscr{A}, F_c)$ be a complete symmetric C^*-algebra-valued controlled F_c-metric type space and suppose $T : X \to X$ is a continuous mapping satisfying the following condition, for all $\varpi, \omega \in X$*

$$F_c(T\varpi, T\varpi, T\omega) \preceq P^* F_c(\varpi, \varpi, \omega)P + Q^* F_c(\varpi, \varpi, T\varpi)Q + R^* F_c(\omega, \omega, T\omega)R \tag{1}$$

where $P, Q, R \in \mathscr{A}'$ with $\|P\|, \|Q\|, \|R\| \geq 0$ satisfying $\|P\|^2 + \|Q\|^2 + \|R\|^2 < 1$ and for $\varpi_0 \in X$, choose $\varpi_n = T^n \varpi_0$. Assume that

$$\sup_{m\geq 1} \lim_{i\to\infty} \|C(\varpi_{i+1}, \varpi_{i+1}, \varpi_{i+2}) C(\varpi_{i+1}, \varpi_{i+1}, \varpi_m)\| < \frac{1 - \|R\|^2}{\|P\|^2 + \|Q\|^2}. \tag{2}$$

In addition, for each $\varpi \in X$, suppose that

$$\lim_{n\to\infty} \|C(\varpi, \varpi, \varpi_n)\| \; and \; \lim_{n\to\infty} \|C(\varpi_n, \varpi_n, \varpi)\| \tag{3}$$

exist and are finite. Then, T has a unique fixed point in X.

Proof Let $\varpi_0 \in X$ be arbitrary and define the iterative sequence $\{\varpi_n\}$ by

$$\varpi_{n+1} = T\varpi_n = \ldots = T^{n+1}\varpi_0, n = 1, 2, \ldots. \tag{4}$$

If follows from (1) and (4) that

$$F_c(\varpi_n, \varpi_n, \varpi_{n+1}) = F_c(T\varpi_{n-1}, T\varpi_{n-1}, T\varpi_n) \preceq P^* F_c(\varpi_{n-1}, \varpi_{n-1}, \varpi_n)\, P \\ + Q^* \, F_c(\varpi_{n-1}, \varpi_{n-1}, T\varpi_{n-1})\, Q + R^* \, F_c(\varpi_n, \varpi_n, T\varpi_n)\, R$$

$$\begin{aligned} \|F_c(\varpi_n, \varpi_n, \varpi_{n+1})\| &\leq \|P^* F_c(\varpi_{n-1}, \varpi_{n-1}, \varpi_n)\, P + Q^* \, F_c(\varpi_{n-1}, \varpi_{n-1}, \varpi_n)\, Q + R^* \\ &\quad F_c(\varpi_n, \varpi_n, \varpi_{n+1})\, R\| \\ &\leq \|P^* F_c(\varpi_{n-1}, \varpi_{n-1}, \varpi_n)\, P\| + \|Q^* \, F_c(\varpi_{n-1}, \varpi_{n-1}, \varpi_n)\, Q\| + \\ &\quad \|R^* \, F_c(\varpi_n, \varpi_n, \varpi_{n+1})\, R\| \\ &= (\|P\|^2 + \|Q\|^2)\|F_c(\varpi_{n-1}, \varpi_{n-1}, \varpi_n)\| + \|R\|^2 \|F_c(\varpi_n, \varpi_n, \varpi_{n+1})\|. \end{aligned}$$

This gives

$$\|F_c(\varpi_n, \varpi_n, \varpi_{n+1})\| \leq \frac{\|P\|^2 + \|Q\|^2}{1 - \|R\|^2} \|F_c(\varpi_{n-1}, \varpi_{n-1}, \varpi_n)\|. \tag{5}$$

Let $\|U\| = \left(\frac{1-\|R\|^2}{\|P\|^2+\|Q\|^2}\right)^{\frac{1}{2}}$. Accordingly, we get

$$\|F_c(\varpi_n, \varpi_n, \varpi_{n+1})\| \leq \frac{1}{\|U\|^2} \|F_c(\varpi_{n-1}, \varpi_{n-1}, \varpi_n)\| \tag{6}$$

which gives

$$\begin{aligned} &\|U\|^2 \|F_c(\varpi_n, \varpi_n, \varpi_{n+1})\| \leq \|F_c(\varpi_{n-1}, \varpi_{n-1}, \varpi_n)\| \\ \Rightarrow\ &\|U^*U\| \|F_c(\varpi_n, \varpi_n, \varpi_{n+1})\| = \|U\|^2 \|F_c(\varpi_n, \varpi_n, \varpi_{n+1})\| \leq \|F_c(\varpi_{n-1}, \varpi_{n-1}, \varpi_n)\| \\ \Rightarrow\ &\|U^*U F_c(\varpi_n, \varpi_n, \varpi_{n+1})\| \leq \|U^*U\| \|F_c(\varpi_n, \varpi_n, \varpi_{n+1})\| = \|U\|^2 \|F_c(\varpi_n, \varpi_n, \varpi_{n+1})\| \\ &\leq \|F_c(\varpi_{n-1}, \varpi_{n-1}, \varpi_n)\| \\ \Rightarrow\ &\|U^*U F_c(\varpi_n, \varpi_n, \varpi_{n+1})\| \leq \|F_c(\varpi_{n-1}, \varpi_{n-1}, \varpi_n)\| \end{aligned}$$

that is equivalent to

$$\begin{aligned} &U^*U F_c(\varpi_n, \varpi_n, \varpi_{n+1}) \preceq F_c(\varpi_{n-1}, \varpi_{n-1}, \varpi_n) \\ \Rightarrow\ &F_c(\varpi_n, \varpi_n, \varpi_{n+1}) \preceq \frac{1}{U^*} F_c(\varpi_{n-1}, \varpi_{n-1}, \varpi_n) \frac{1}{U} \\ &= S^* F_c(\varpi_{n-1}, \varpi_{n-1}, \varpi_n) S \end{aligned}$$

where $S = \frac{1}{U}$. Recursively, we find that

$$F_c(\varpi_n, \varpi_n, \varpi_{n+1}) \preceq (S^*)^n F_c(\varpi_0, \varpi_0, \varpi_1) S^n. \tag{7}$$

For any $n \geq 1$ and $q \geq 1$, we find

$$
\begin{aligned}
F_c(\varpi_n, \varpi_n, \varpi_{n+q}) &\preceq C(\varpi_n, \varpi_n, \varpi_{n+1})F_c(\varpi_n, \varpi_n, \varpi_{n+1}) + C(\varpi_n, \varpi_n, \varpi_{n+1})F_c(\varpi_n, \varpi_n, \varpi_{n+1}) \\
&\quad + C(\varpi_{n+q}, \varpi_{n+q}, \varpi_{n+1})F_c(\varpi_{n+q}, \varpi_{n+q}, \varpi_{n+1}) - F_c(\varpi_{n+1}, \varpi_{n+1}, \varpi_{n+1}) \\
&\preceq 2C(\varpi_n, \varpi_n, \varpi_{n+1})F_c(\varpi_n, \varpi_n, \varpi_{n+1}) + C(\varpi_{n+q}, \varpi_{n+q}, \varpi_{n+1}) \\
&\quad F_c(\varpi_{n+1}, \varpi_{n+1}, \varpi_{n+q}) \\
&\preceq 2C(\varpi_n, \varpi_n, \varpi_{n+1})F_c(\varpi_n, \varpi_n, \varpi_{n+1}) + C(\varpi_{n+q}, \varpi_{n+q}, \varpi_{n+1}) \\
&\quad \big[2\,C(\varpi_{n+1}, \varpi_{n+1}, \varpi_{n+2})F_c(\varpi_{n+1}, \varpi_{n+1}, \varpi_{n+2}) + C(\varpi_{n+q}, \varpi_{n+q}, \varpi_{n+2}) \\
&\quad F_c(\varpi_{n+2}, \varpi_{n+2}, \varpi_{n+q})\big] - F_c(\varpi_{n+2}, \varpi_{n+2}, \varpi_{n+2}) \\
&\quad \vdots \\
&= 2C(\varpi_n, \varpi_n, \varpi_{n+1})F_c(\varpi_n, \varpi_n, \varpi_{n+1}) + \\
&\quad 2\sum_{i=n+1}^{n+q-2} C(\varpi_i, \varpi_i, \varpi_{i+1})F_c(\varpi_i, \varpi_i, \varpi_{i+1}) \prod_{j=n+1}^{i} C(\varpi_{n+q}, \varpi_{n+q}, \varpi_j) + \\
&\quad \prod_{i=n+1}^{n+q-1} C(\varpi_{n+q}, \varpi_{n+q}, \varpi_i)F_c(\varpi_{n+q-1}, \varpi_{n+q-1}, \varpi_{n+q}) \\
&\preceq 2C(\varpi_n, \varpi_n, \varpi_{n+1})F_c(\varpi_n, \varpi_n, \varpi_{n+1}) + \\
&\quad 2\sum_{i=n+1}^{n+q-1} C(\varpi_i, \varpi_i, \varpi_{i+1})F_c(\varpi_i, \varpi_i, \varpi_{i+1}) \prod_{j=n+1}^{i} C(\varpi_{n+q}, \varpi_{n+q}, \varpi_j) \\
&\preceq 2C(\varpi_n, \varpi_n, \varpi_{n+1})(S^*)^n S_0 S^n + \\
&\quad 2\sum_{i=n+1}^{n+q-1} C(\varpi_i, \varpi_i, \varpi_{i+1})(S^*)^i S_0 S^i \prod_{j=1}^{i} C(\varpi_{n+q}, \varpi_{n+q}, \varpi_j) \\
&= 2\Big(S_0^{\frac{1}{2}}\, C(\varpi_n, \varpi_n, \varpi_{n+1})^{\frac{1}{2}}\, S^n\Big)^* (S_0^{\frac{1}{2}}\, C(\varpi_n, \varpi_n, \varpi_{n+1})^{\frac{1}{2}}\, S^n\Big) + \\
&\quad 2\sum_{i=n+1}^{n+q-1} \Big(S_0^{\frac{1}{2}}\big[C(\varpi_i, \varpi_i, \varpi_{i+1}) \prod_{j=1}^{i} C(\varpi_{n+q}, \varpi_{n+q}, \varpi_j)\big]^{\frac{1}{2}} S^i\Big)^* \\
&\qquad \Big(S_0^{\frac{1}{2}}\big[C(\varpi_i, \varpi_i, \varpi_{i+1}) \prod_{j=1}^{i} C(\varpi_{n+q}, \varpi_{n+q}, \varpi_j)\big]^{\frac{1}{2}} S^i\Big) \\
&= 2\,|S_0^{\frac{1}{2}}\, C(\varpi_n, \varpi_n, \varpi_{n+1})^{\frac{1}{2}}\, S^n|^2 + \\
&\quad 2\sum_{i=n+1}^{n+q-1} \Big|S_0^{\frac{1}{2}}\big[C(\varpi_i, \varpi_i, \varpi_{i+1}) \prod_{j=1}^{i} C(\varpi_{n+q}, \varpi_{n+q}, \varpi_j)\big]^{\frac{1}{2}} S^i\Big|^2 \\
&\preceq 2\|S_0\|\Big[\|C(\varpi_n, \varpi_n, \varpi_{n+1})\|\ \|S\|^{2n} I_{\mathscr{A}} + \\
&\quad \|C(\varpi_i, \varpi_i, \varpi_{i+1}) \prod_{j=1}^{i} C(\varpi_{n+q}, \varpi_{n+q}, \varpi_j)\|\,\|S\|^{2i} I_{\mathscr{A}}\ \Big]
\end{aligned}
$$

where $F_c(\varpi_0, \varpi_0, \varpi_1) = S_0$, for some $S_0 \in \mathscr{A}$ and $I_{\mathscr{A}}$ is the unit element in $\mathscr{A}$. Consequently, the above inequality implies

$$F_c(\varpi_n, \varpi_n, \varpi_{n+q}) \leq 2\|S_0\|\Big[\|C(\varpi_n, \varpi_n, \varpi_{n+1})\| \, \|S\|^{2n} + (Y_{n+q-1} - Y_n)\Big]I_{\mathscr{A}} \tag{8}$$

where

$$Y_n = \sum_{i=1}^{n} \|S\|^{2i} \, \|C(\varpi_i, \varpi_i, \varpi_{i+1})\prod_{j=1}^{i} C(\varpi_{n+q}, \varpi_{n+q}, \varpi_j)\|.$$

The ratio test jointly with (2) indicates that the limit of the sequence $\{Y_n\}$ exists and so $\{Y_n\}$ is Cauchy. Letting $n \to \infty$ in the inequality above, we get

$$\lim_{n\to\infty} F_c(\varpi_n, \varpi_n, \varpi_{n+q}) = \theta_{\mathscr{A}}. \tag{9}$$

wherefore the sequence $\{\varpi_n\}$ is Cauchy with respect to $\mathscr{A}$. As $(X, \mathscr{A}, F_c)$ is a complete C^*-algebra-valued controlled F_c-metric type space, there exists a point $\varpi \in X$ such that

$$\lim_{n\to\infty} F_c(\varpi_n, \varpi_n, \varpi) = \theta_{\mathscr{A}}. \tag{10}$$

Consider

$$\begin{aligned}
F_c(\varpi, \varpi, \varpi_{n+1}) &\preceq 2C(\varpi, \varpi, \varpi_n)F_c(\varpi, \varpi, \varpi_n) + C(\varpi_{n+1}, \varpi_{n+1}, \varpi_n) \\
&\quad F_c(\varpi_n, \varpi_n, \varpi_{n+1}) - F_c(\varpi_n, \varpi_n, \varpi_n) \\
\iff \|F_c(\varpi, \varpi, \varpi_{n+1})\| &\leq \|2C(\varpi, \varpi, \varpi_n)F_c(\varpi, \varpi, \varpi_n) + C(\varpi_{n+1}, \varpi_{n+1}, \varpi_n) \\
&\quad F_c(\varpi_n, \varpi_n, \varpi_{n+1}) - F_c(\varpi_n, \varpi_n, \varpi_n)\| \\
&\leq 2\|C(\varpi, \varpi, \varpi_n)\|\|F_c(\varpi, \varpi, \varpi_n)\| + \|C(\varpi_{n+1}, \varpi_{n+1}, \varpi_n)\| \\
&\quad \|F_c(\varpi_n, \varpi_n, \varpi_{n+1})\|.
\end{aligned}$$

It yields from (3) and (10) that

$$\lim_{n\to\infty} \|F_c(\varpi, \varpi, \varpi_{n+1})\| = 0. \tag{11}$$

Accordingly,

$$\begin{aligned}
\|F_c(\varpi, \varpi, T\varpi)\| &\leq 2\|C(\varpi, \varpi, \varpi_{n+1})\|\|F_c(\varpi, \varpi, \varpi_{n+1})\| + \|C(T\varpi, T\varpi, \varpi_{n+1})\| \\
&\quad \|F_c(\varpi_{n+1}, \varpi_{n+1}, T\varpi)\| \\
&= 2\|C(\varpi, \varpi, \varpi_{n+1})\|\|F_c(\varpi, \varpi, \varpi_{n+1})\| + \|C(T\varpi, T\varpi, \varpi_{n+1})\| \\
&\quad \|F_c(T^{n+1}\varpi, T^{n+1}\varpi, T\varpi)\|.
\end{aligned}$$

Regarding (11), we get $\|F_c(\varpi, \varpi, \varpi_{n+1})\| \to 0$ as $n \to \infty$. Since $T^n \varpi \to \varpi$ and from the continuity of T, we acquire $T^{n+1}\varpi \to T\varpi$, i.e., $\|F_c(T^{n+1}\varpi, T^{n+1}\varpi, T\varpi)\| \to 0$, as $n \to \infty$. Thus

$$\lim_{n\to\infty} \|F_c(\varpi, \varpi, T\varpi)\| = 0$$

$$\Longleftrightarrow \lim_{n\to\infty} F_c(\varpi, \varpi, T\varpi) = \theta_{\mathscr{A}}.$$

Therefore, $T\varpi = \varpi$, i.e., ϖ is a fixed point of T. To prove an uniqueness, let $\omega \neq \varpi$ be another fixed point of T. Taking the expression (1) into account, we have

$$\begin{aligned} F_c(\varpi, \varpi, \omega) &= F_c(T\varpi, T\varpi, T\omega) \preceq P^*\, F_c(\varpi, \varpi, \omega)\, P + Q^*\, F_c(\varpi, \varpi, T\varpi)\, Q \\ &\quad + R^*\, F_c(\omega, \omega, T\omega)\, R \\ &= P^*\, F_c(\varpi, \varpi, \omega)\, P + Q^*\, F_c(\varpi, \varpi, \varpi)\, Q + R^*\, F_c(\omega, \omega, \omega)\, R \\ &\preceq P^*\, F_c(\varpi, \varpi, \omega)\, P + Q^*\, F_c(\varpi, \varpi, \omega)\, Q + R^*\, F_c(\omega, \omega, \varpi)\, R \\ \Rightarrow \|F_c(\varpi, \varpi, \omega)\| &\le (\|P\|^2 + \|Q\|^2) F_c(\varpi, \varpi, \omega) + \|R\|^2 \|F_c(\omega, \omega, \varpi)\| \\ \Rightarrow \|F_c(\varpi, \varpi, \omega)\| &\le \frac{\|R\|^2}{(1 - \|P\|^2 - \|Q\|^2)} \|F_c(\omega, \omega, \varpi)\| < \|F_c(\omega, \omega, \varpi)\| \\ &= \|F_c(\varpi, \varpi, \omega)\| \end{aligned}$$

which is a contradiction. Thereby the fixed point is unique.

In Theorem (3.8), if we take $Q = R = \theta$, then the above theorem reduces to the following result.

Corollary 3.9 *Let $(X, \mathscr{A}, F_c)$ be a complete C^*-algebra-valued controlled F_c-metric type space and suppose $T : X \to X$ is a continuous mapping satisfying the following condition:*

$$F_c(T\varpi, T\varpi, T\omega) \preceq P^*\, F_c(\varpi, \varpi, \omega)\, P \ \textit{for all}\ \varpi, \omega \in X \tag{12}$$

where $P \in \mathscr{A}'$ with $0 \le \|P\| < 1$ and for $\varpi_0 \in X$, choose $\varpi_n = T^n \varpi_0$. Assume that

$$\sup_{m\ge1} \lim_{i\to\infty} \|C(\varpi_{i+1}, \varpi_{i+1}, \varpi_{i+2}) C(\varpi_{i+1}, \varpi_{i+1}, \varpi_m)\| < \frac{1}{\|P\|^2}. \tag{13}$$

In addition, for each $\varpi \in X$, suppose that

$$\lim_{n\to\infty} \|C(\varpi, \varpi, \varpi_n)\| \ \textit{and}\ \lim_{n\to\infty} \|C(\varpi_n, \varpi_n, \varpi)\| \tag{14}$$

exist and are finite. Then, T has a unique fixed point in X.

Example 3.10 Let $X = [0, 4]$ and $\mathscr{A} = M_2(R)$ be the set of all 2×2 matrices under usual addition, multiplication and scalar multiplication. Define $F_c : X \times X \times X \to \mathscr{A}$ as follows:

$$F_c(\varpi, \omega, \tau) = \begin{pmatrix} max\{\varpi, \tau\} + max\{\omega, \tau\} & 0 \\ 0 & max\{\varpi, \tau\} + max\{\omega, \tau\} \end{pmatrix}.$$

Hence $(X, \mathscr{A}, F_c)$ is a complete C^*-algebra-valued controlled F_c-metric type space with $C(\varpi, \omega, \tau) = 2 + max\{\varpi, \omega, \tau\}$. Now for any $A \in \mathscr{A}$, we define its norm as $\|A\| = \max_{1\le i\le 4}\{|a_i|\}$. Let $T : X \to X$ be defined as $T\varpi = \frac{\varpi}{8}$. Then

$$\begin{aligned} F_c(T\varpi, T\varpi, T\omega) = F_c(\tfrac{\varpi}{8}, \tfrac{\varpi}{8}, \tfrac{\omega}{8}) &= \begin{pmatrix} 2max\{\frac{\varpi}{8}, \frac{\omega}{8}\} & 0 \\ 0 & 2max\{\frac{\varpi}{8}, \frac{\omega}{8}\} \end{pmatrix} \\ &= P^* F_c(\varpi, \varpi, \omega)P \end{aligned}$$

where $P = \begin{pmatrix} \frac{1}{2\sqrt{2}} & 0 \\ 0 & \frac{1}{2\sqrt{2}} \end{pmatrix}$ with $\|P\| = \frac{1}{2\sqrt{2}} < 1$. Now, consider

$$\begin{aligned} C(\varpi_{i+1}, \varpi_{i+1}, \varpi_{i+2}) &= C(T^{i+1}\varpi, T^{i+1}\varpi, T^{i+2}\varpi) = C(\frac{\varpi}{8^{i+1}}, \frac{\varpi}{8^{i+1}}, \frac{\varpi}{8^{i+2}}) \\ &= \begin{pmatrix} 2 + max\{\frac{\varpi}{8^{i+1}}, \frac{\varpi}{8^{i+1}}, \frac{\varpi}{8^{i+2}}\} & 0 \\ 0 & 2 + max\{\frac{\varpi}{8^{i+1}}, \frac{\varpi}{8^{i+1}}, \frac{\varpi}{8^{i+2}}\} \end{pmatrix}. \end{aligned}$$

Similarly,

$$C(\varpi_{i+1}, \varpi_{i+1}, \varpi_m) = \begin{pmatrix} 2 + \max\{\frac{\varpi}{8^{i+1}}, \frac{\varpi}{8^{i+1}}, \frac{\varpi}{8^m}\} & 0 \\ 0 & 2 + \max\{\frac{\varpi}{8^{i+1}}, \frac{\varpi}{8^{i+1}}, \frac{\varpi}{8^m}\} \end{pmatrix}.$$

Thus

$$\begin{aligned} &\lim_{i\to\infty} \|C(\varpi_{i+1}, \varpi_{i+1}, \varpi_{i+2})C(\varpi_{i+1}, \varpi_{i+1}, \varpi_m)\| \\ &= \lim_{i\to\infty} \left\| \begin{pmatrix} (2 + \frac{\varpi}{8^{i+1}})(2 + \max(\frac{\varpi}{8^{i+1}}, \frac{\varpi}{8^m}) & 0 \\ 0 & (2 + \frac{\varpi}{8^{i+1}})(2 + max(\frac{\varpi}{8^{i+1}}, \frac{\varpi}{8^m}) \end{pmatrix} \right\| \\ &= \lim_{i\to\infty} (2 + \frac{\varpi}{8^{i+1}})(2 + \max(\frac{\varpi}{8^{i+1}}, \frac{\varpi}{8^m}) = 4 + \frac{2\varpi}{8^m} \end{aligned}$$

and

$$\sup_{m\ge1} \lim_{i\to\infty} \|C(\varpi_{i+1}, v_{i+1}, \varpi_{i+2})C(\varpi_{i+1}, \varpi_{i+1}, \varpi_m)\| = 4 + \frac{2\varpi}{8} < 8 = \frac{1}{\|P\|^2}.$$

Henceforth, T fulfills all the requirements of Theorem (3.8), and it has a unique fixed point which is $\varpi = 0$.

4 Conclusion

In this manuscript, we have analyzed the structure of C^*-algebra-valued controlled F_c-metric type spaces and acquired a fixed point theorem under different contractive condition of the underlying spaces. Further, an example is conferred to show the effectiveness of the established result.

Acknowledgements The authors thank the Principal and management of Sri Sivasubramaniya Nadar College of Engineering, Kalavakkam-603110, for their constant support toward the successful completion of this work.

Conflict of interest The authors declare that they have no conflict of interest.

References

1. Abdeljawad, T., Karapınar, E., Panda, S.K., Mlaiki, N.: Solutions of boundary value problems on extended-Branciari b-distance. J. Inequal. Appl. **2020**, 103 (2020)
2. Abdeljawad, T., Agrawal, R.P., Karapınar, E., Sumati Kumari, P.: Solutions of the nonlinear integral equation and fractional differential equation using the technique of a fixed point with a numerical experiment in extended b-metric space. Symmetry **11**, 686 (2019)
3. Alqahtani, B., Fulga, A., Karapinar, E.: Common fixed point results on extended b-metric space. J. Inequal. Appl. **2018**, 158 (2018)
4. Alqahtani, B., Karapinar, E., Ozturk, A.: On (α, ψ)-K-contractions in the extended b-metric space. Filomat **32**(15), 5337–5345 (2018)
5. Alqahtani, B., Fulga, A., Karapinar, E.: Non-unique fixed point results in extended b-metric space. Mathematics **6**, 68 (2018)
6. Bai, C.: Coupled fixed point theorems in C^*-algebra-valued b-metric spaces with application. Fixed Point Theory Appl. **2016**, 70 (2016)
7. Bakhtin, A.: The contraction mapping principle in almost metric spaces. Funct. Anal. **30**, 26–37 (1989)
8. Batul, S., Kamran, T.: C^*-valued contractive type mappings. Fixed Point Theory Appl. **2015**, 142 (2015)
9. Erden, M., Alaca, C., C^*-algebra-valued S-metric spaces. Commun. Ser. A1 **67**(2), 165–177 (2018)
10. Fernandez, J., Malviya, N., Radenović, S., Saxena, K.: F-cone metric spaces over banach algebra. Fixed Point Theory Appl. **2017**, 7 (2017)
11. Kalaivani, C., Kalpana, G, Fixed point theorems in C^*-algebra-valued S-metric spaces with some applications, U.P.B. Sci. Bull. Ser. A **80**(3) (2018)
12. Kalpana, G., Sumaiya Tasneem, Z.: C^*-algebra-valued rectangular b-metric spaces and some fixed point theorems, Commun. Fac. Sci. Univ. Ank. Ser. A1 Math. Stat. **68**(2), 2198–2208 (2019)
13. Kalpana, G., Sumaiya Tasneem, Z.: Common fixed point theorems in C^*-algebra-valued Hexagonal b-metric spaces. AIP Conf. Proc. **2095**, 030012-1–030012-5 (2019)
14. Kamran, T., Samreen, M., UL Ain, Q.: A generalization of b-metric space and some fixed point theorems. Mathematics **5**(19) (2017)
15. Karapinar, E., Sumati Kumari, P., Lateef, D.: A new approach to the solution of the Fredholm integral equation via a fixed point on extended b-metric spaces. Symmetry **10**, 512 (2018)
16. Ma, Z.H., Jiang, L.N.: C^*-algebra-valued b-metric spaces and related fixed point theorems. Fixed Point Theory Appl. **2015**, 222 (2015)

17. Ma, Z.H., Jiang, L.N., Sun, H.: C^*-algebra-valued metric spaces and related fixed point theorems. Fixed Point Theory Appl. **2014**, 206 (2014)
18. Mlaiki, N., Aydi, H., Souayah, N., Abdeljawad, T.: Controlled metric type spaces and the related contraction principle. Mathematics **6**, 194 (2018)
19. Matthews, S.G.: Partial metric topology. Ann. N.Y. Acad. Sci. **728**(1), 183–197 (1994)
20. Mukheimer, A.: Extended partial S_b-metric spaces. Axioms **7**, 87 (2018)
21. Murphy, G.J.: C^*-Algebras and Operator Theory. Academic Press, London (1990)
22. Panda, S.K., Tassaddiq, A., Agarwal, R.P.: A new approach to the solution of the non-linear integral equations via various F_{B_e}-contractions. Symmetry **11**, 206 (2019)
23. Samreen, M., Kamran, T., Postolache, M., Extended b-metric space, extended b-comparison function and nonlinear contractions. U. P. B. Sci. Bull. Ser. A **80**(4), 21–28 (2018)
24. Shatanawi, W., Abodayeh, K., Mukheimer, A.: Some fixed point theorems in extended b-metric spaces, U. P. B. Sci. Bull. Ser. A 71–78 (2018)
25. Shehwar, D., Batul, S., Kamran, T., Ghiura, A.: Caristis fixed point theorem on C^*-algebra valued metric spaces. J. Nonlinear Sci. Appl. **9**, 584–588 (2016)
26. Xin, Q.L., Jiang, L.N., Ma, Z.H.: Common fixed point theorems in C^*-algebra-valued metric spaces. J. Nonlinear Sci. Appl. **9**, 4617–4627 (2016)

Detection of Bruises and Flaws in Fruits Using Thermal Imaging

J. Sofia Jennifer, T. Sree Sharmila, H. Sairam, and T. S. Kishorkrishna

Abstract Mechanically damaged fruits and vegetables lead to tremendous economic losses in developing countries. To resolve this issue, it is necessary to identify these damages as increased microbial contamination and accelerated ripening of plant products can result from such damages. In existing scenarios, manual work is done to detect the flaws in fruits. As humans are prone to make errors, the need for the use of thermal imaging to detect bruises and flaws in fruits arises. The proposed work involves the thermal study of fruits like apple, banana, guava, sapota and tomato. Initially, the boundary of the fruits is identified through segmentation and morphological operations. Then, the edges of the bruises are localized using segmentation and edge detection techniques. The percentage area of these bruises spread along the fruit is detected using the temperature variations. Experimental analysis shows that thermal variation of 1–2 °C is identified along the bruised regions.

Keywords Bruise detection · Fractal analysis · Thermal images · Thermography

1 Introduction

Information regarding food quality before consumption is important. Assessing the quality parameters is a big challenge since differentiating impurities or freshness of food is difficult. Bruising is termed as the damage of fruit tissue as a result of

J. Sofia Jennifer (✉) · T. Sree Sharmila · H. Sairam · T. S. Kishorkrishna
Department of Information Technology, Sri Sivasubramaniya Nadar College of Engineering, Kalavakkam, India
e-mail: sofiajenniferj@ssn.edu.in

T. Sree Sharmila
e-mail: sreesharmilat@ssn.edu.in

H. Sairam
e-mail: sai_hari98@hotmail.com

T. S. Kishorkrishna
e-mail: kishorkrishnats@gmail.com

R. N. Mohapatra et al. (eds.), *Mathematical Analysis and Computing*,
Springer Proceedings in Mathematics & Statistics 344,
https://doi.org/10.1007/978-981-33-4646-8_41

external forces that cause physical changes of texture and/or chemical changes of color, smell and taste [1]. Removal of such defective fruits is very much essential to maintain the quality of fruits and to satisfy the consumer's promise. The fruit selling business sector is so large that managing the quality inspection process is tedious, expensive, laborious and unreliable due to its subjective nature. Humans are prone to errors where people inspecting the fruit can miss out on a few defective/rotten fruits. It leads to a lack of quality and loss in business. To resolve this issue, thermal imaging technology is applied where the temperature differences are used to assist in the evaluation for the fruit diagnosis. Thermal imaging holds the advantage of non-invasive, non-contact, radiation-free and non-destructive nature for the assessment of abnormal radiation from objects [2, 3]. In recent years, the research communities have paved the path for developing computer-aided techniques in areas of food quality inspection. Most of the existing research focuses on apples [4, 5]. According to Varith [6], 100% of apple bruises are detected using thermal imaging during the warming of the fruits by discriminating surface temperature between bruised and sound tissues. The thermal image of this bruised tissue portrayed at least 1–2 °C difference from sound tissue. Bennedsen et al. [7] proposed methods to find individual defects and measure the area that ranged 77–91% for the number of defects detected, and 78–92.7% of the total defective area. Baranowski et al. [8] used pulsed-phase thermography where the studied object is heated with an individual thermal pulse and the temperature decay on the surface is analyzed on a pixel-by-pixel basis as a mixture of harmonic waves, thus enabling the computation of phase and amplitude images. Wagh [9] uses image processing and machine vision technology to detect bruises.

2 Bruise Detection System

The bruise detection system consists of the following steps as described in Fig. 1: (i) Image acquisition, (ii) Fruit boundary detection, (iii) Detection of bruised regions and (iv) Spreadness measurement of bruised regions.

2.1 Image Acquisition

The input is captured in a real-time environment through thermal cameras as still images or videos for further processing.

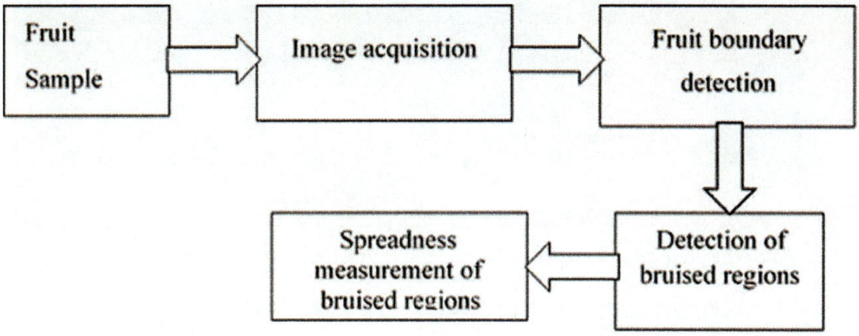

Fig. 1 Bruise detection system

2.2 Fruit Detection

The major concern is to determine the boundary box area of the thermal fruit detection in a less complex and efficient way. The steps are explained in Algorithm 1 as follows:

Algorithm 1: Fruit detection

Input: $I_{thermal}$

Output: Cropped fruit

(i) Apply Otsu's threshold algorithm to eliminate background details as portrayed in Fig. 2b.

(ii) Perform preprocessing steps such as morphological structuring (dilation and closing) to obtain only the object boundary as depicted in Fig. 2c and d.

(iii) Use flood-fill operation to remove the irrelevant artifacts as shown in Fig. 2e.

(iv) Select the largest blob and segment them as the cropped fruit as shown in Fig. 2f.

2.3 Detection of Bruised Regions

On analysis of the thermal properties of both good and bad fruits, it is inferred that good fruits always had a uniform temperature with a minor variation of 0.1°–0.2° whereas, the flawed regions of the fruits (bruises, rotten) shows a lower temperature than the fresh fruit. The lower temperature is due to the effect of thermal emissivity of the region, which changes due to the damage or the exposure of the fruit flesh. Figure 3a shows a sample thermal image of a good guava. The temperature of the fruit is uniform throughout the surface with a maximum temperature of 31.4° and an average temperature of 30.9° It signifies that for every good fruit, the temperature on its surface will be uniform throughout. Figure 3b shows a sample thermal image of a bad guava. This fruit has a temperature of 31.9° around the good region whereas in the damaged region has a relatively lower temperature of 29°. This differentiation

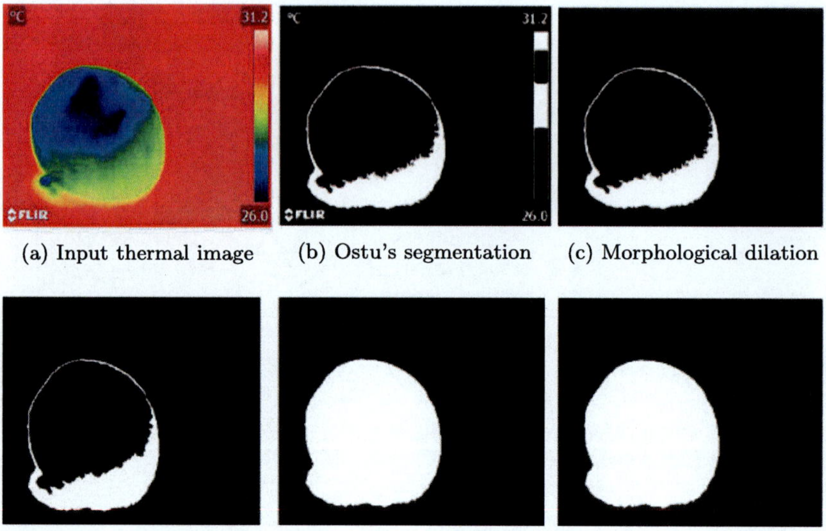

(a) Input thermal image (b) Ostu's segmentation (c) Morphological dilation

(d) Morphological closing (e) Flood Flood-fill operation (f) Fruit segmentation

Fig. 2 Fruit detection

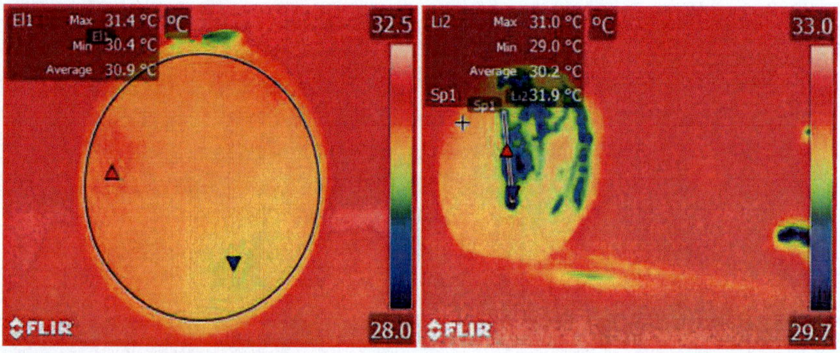

(a) good apple (b) damaged apple

Fig. 3 Thermal image

is noted based on color distribution. It is a well-known fact that the bruised or rotten regions will have a much cooler temperature than the regions which are not damaged. Cooler temperatures will have blue color dominance in the thermal images. Hence, the extraction of the blue channel from the RGB cropped image is done to visually observe the bruised components separately. Then, the image is subjected to a histogram plot to find the extent of the damage.

2.4 Spreadness Measurement of Bruised Regions

After the validation of the detected bruise regions, the spreadness measurement of the extracted region is defined in terms of area and fractal dimension. Areas are calculated as the total number of pixels in the extracted region. The percentage of defect area is calculated using Eq. 1.

$$\text{Percentage of defect area} = \frac{\text{Area of defect}}{\text{Total Area}} * 100 \tag{1}$$

Fractal Dimension is a basic parameter of fractal geometry and it has been applied in texture segmentation. Based on the asymmetric texture differences between bruised and fresh fruits, the fractal dimension is computed using the box-counting technique. It quantifies the space-filling capacity of a fractal pattern. In this method, fractal dimension is the slope of the line when the value of log(N) is plotted on the Y-axis against the value of log(r) on the X-axis. The term N is the number of boxes that cover the pattern, and r is the magnification or the inverse of the box size [10]. The fractal dimension is calculated using Eq. 2.

$$D = \frac{\log N}{\log r} \tag{2}$$

The bruised fruits exhibit irregularity and heterogeneity in both surface color distribution and texture; hence they show higher values of fractal dimension.

3 Experimental Results and Discussions

For image acquisition, thermal samples of different kinds of fruits are captured using the FLIR A35 camera. It is a low-cost solution to monitor the temperatures in fruit monitoring environments that can register differences in temperatures bigger than 50 mK. The thermal videos are recorded with a frame-rate up to 30 images per second, resolution of 320 × 240 pixels, the range of temperature −4 to 662 °F, focal length of 13mm and wavelength of 9–14 μm. For quantitative evaluation purposes, the various fruit samples such as apples, bananas, guavas, sapota and tomatoes are used. All these thermal images are captured in a similar pre-designed environment at room temperature. There are typically two different kinds of fruits in each sample such as damaged and fresh. Both the rotten and fresh samples are considered to explore how the temperature on the fruit surface varies with damage and rot. They are captured and recorded as videos as the particular fruit sample is rotated at 360° to obtain a large amount of dataset. For the testing phase, out of total 200 fruit images, 130 are defective and 70 are sound which are subjected for inspection. The average bruise detection rates for the dataset are tabulated in Table 1. On analysis, the overall result of thermal bruise detection is about 92.458%. The recognition rates for both good

Table 1 Results of average defective rate

S. No.	Fruit sample	Samples under consideration			Average bruise detection rate (%)
		Total count	Bruise fruit	Good fruit	
1	Apple	40	23	17	92.3
2	Bananas	40	17	23	91.45
3	Guavas	40	21	19	93.62
4	Sapota	40	22	18	90.67
5	Tomatoes	40	25	15	94.25

Table 2 Spreadness measurement using area and fractal dimension

S. No.	Fruit sample	Spreadness measurement	
		Area (%)	Fractal dimension
1	Apple	20.213	0.435
2	Bananas	35.412	0.554
3	Guavas	27.32	0.489
4	Sapota	63.8396	0.831
5	Tomatoes	48.034	0.631

(a) Fruit segmentation (b) Blue Channel of the fruit (c) Localize the bruised region

Fig. 4 Damaged banana fruit

and defective fruits show the feasibility and efficiency of using the thermal imaging system.

After bruise detection, the spreadness measurement of the extracted region is computed using area and fractal dimension, and tabulated in Table 2. From the tabulated data it is inferred that the fractal dimension is higher for bruised fruits when compared to fresh fruits as depicted in Figs. 4 and 5.

(a) Fruit segmentation (b) Blue Channel of the fruit (c) Localize the bruised region

Fig. 5 Good tomato fruit

4 Conclusions and Future Work

The proposed method localizes and differentiates the rotten and fresh fruits using thermal imagery. The fruit images are processed into two parts, i.e. bruise detection and the percentage of the bruised regions. The overall thermal bruise detection accuracy is 92.458%. Further research work is intended to study various other fruit samples and extend this idea to a whole bunch of fruit samples. These results will also aid in many real-world applications at the industry level.

References

1. Qiang, L., Mingjie, T.: Detection of hidden bruise on kiwi fruit using hyperspectral imaging and parallelepiped classification. Procedia Environ. Sci. **12**, 1172–1179 (2012)
2. Baranowski, P., Mazurek, W., Walczak, B.W., Slawinski, C.: Detection of early apple bruises using pulsed-phase thermography. Postharv. Biol. Technol. **53**(3), 91–100 (2009)
3. Manisha, S., Samruddhi, D., Vaishnavi, J.: Automatic bruise detection in fruits using thermal images. Int. J. Adv. Res. Comput. Sci. Softw. Eng. **7**, 727–732 (2017)
4. Kenneth, F.: Fractal Geometry: Mathematical Foundations and Applications. Wiley (2004)
5. Chiu, Y.C., Chou, X.L., Grift Tony, Chen, M.T.: Automated detection of mechanically induced bruise areas in golden delicious apples using fluorescence imagery. Trans. ASABE **58**(2), 215–225 (2015)
6. Dey, R., Das, S., Bhowmik, D., Ghosh, M., Anjan, K.: Bruise Detection in Apples Using Infrared Imaging. ICECE, pp. 118–122(2016)
7. Satone, M., Diwakar, S., Joshi, V.: Automatic bruise detection in fruits using thermal images. Int. J. Adv. Res. Comput. Sci. Softw. Eng. **7**(5), 1–5 (2017)
8. Jawale, D.: Real time automatic bruise detection in fruits (apples) using thermal camera. Int. J. Comput. Appl. 26–29 (2018)
9. Varith, J.H., Baritelle, G.M., Fellman, A.L., John Sattabongkot, T.: Non-contact bruise detection in apples by thermal imaging. Innov. Food Sci. Emerg. Technol. **4**, 211–213 (2003)
10. Bennedsen, B.S., Peterson, D.L.: Performance of a system for apple surface defect identification in near-infrared image. Biosyst. Eng. **90**(4), 419–431 (2005)

Heuristics for the Construction of Counterexamples to the Agrawal Conjecture

Amshuman Hegde and P. Devaraj

Abstract The time complexity of the Agrawal-Kayal-Saxena (AKS) primality test (Agrawal et al. in Annals Math. 781–793, 2004) can be reduced by a significantly large amount if the Bhattacharjee-Pandey conjecture (Bhattacharjee and Pandey in Primality testing, 2001) proves to be true. However, Lenstra and Pomerance gave heuristics to construct potential counterexamples for this conjecture (Lenstra and Pomerance in Future directions in algorithmic number theory, 2008). In this paper, we generalize and extend the methods used by Lenstra and Pomerance to provide heuristics for the construction of two additional classes of counterexamples to the Bhattacharjee-Pandey conjecture. We also use some methods of analytic number theory to provide an estimate of the number of such counterexamples within a large interval.

Keywords Number theory · Cryptography · Primality testing

1 Introduction

In a seminal paper [1] by Agrawal et al., a conjecture was put forward which if true would reduce the complexity of their primality testing algorithm by several orders. The conjecture, due to Bhattacharjee and Pandey [2], which we will henceforth refer to as the "Agrawal conjecture", is as follows:

Conjecture: If r is a prime number that does not divide n and if

$$(X-1)^n \equiv X^n - 1(\text{mod} X^r - 1, n),$$

then either n is prime or $n^2 \equiv 1(\text{mod} r)$.

A. Hegde · P. Devaraj (✉)
School of Mathematics, Indian Institute of Science Education and Research Thiruvananthapuram, Maruthamala P.O Vithura, Thiruvananthapuram 695551, India
e-mail: devarajp@iisertvm.ac.in

R. N. Mohapatra et al. (eds.), *Mathematical Analysis and Computing*,
Springer Proceedings in Mathematics & Statistics 344,
https://doi.org/10.1007/978-981-33-4646-8_42

However, a heuristic to construct a class of counterexamples for this conjecture was given by Lenstra and Pomerance [3]. In this paper, we generalize the methods used by Lenstra and Pomerance to provide more classes of such counterexamples. The proposition given by Lenstra and Pomerance is as follows:

Proposition 1 *Let $p_1, p_2, \ldots, p_k$ be k distinct primes and let $n = p_1 \ldots p_k$. Suppose that*

1. $k \equiv 1(mod4)$,
2. $p_i \equiv 3(mod80)$ *for all i,*
3. $(p_i - 1) \mid (n-1)$ *for all i and*
4. $(p_i + 1) \mid (n+1)$ *for all i.*

Then $(X-1)^n \equiv X^n - 1(\, mod\, n, X^5 - 1)$.

We will now proceed to give and prove two propositions that will provide a heuristic for the construction of other counterexamples to the Agrawal conjecture, in the vein of [1].

2 Heuristics for Construction of Counterexamples

Proposition 2 *Let $p_1, p_2, \ldots, p_k$ be k distinct primes, let $n = p_1 \cdots p_k$ and let r be a natural number such that $r \equiv 1(mod4)$. Suppose that*

1. $k \equiv 1(mod\ 4)$,
2. $p_i^2 \equiv -1(mod8r)$ *for all i,*
3. $(p_i - 1) \mid (\, n-1)$ *for all i and*
4. $(p_i + 1) \mid (\, n+1)$ *for all i.*

Then $(X-1)^n \equiv X^n - 1(modn, X^r - 1)$.

Proof By assumptions, it can be easily seen that

$$\begin{aligned} n^2 &= (p_1 p_2 \cdots p_k)^2 \\ &\equiv (\,-1)^k (\text{mod}8r) \\ &\equiv -1(\text{mod}r) \\ &\not\equiv 1(\text{mod}r). \end{aligned}$$

As $(X-1, 1+X+X^2+\cdots+X^{r-1}, n) = (1)$ in the ring $\mathbf{Z}[X]$, in order to prove the identity

$$(X-1)^n \equiv X^n - 1(\text{mod}X^r - 1, n)$$

it suffices to prove the identity

$$(X-1)^n \equiv X^n - 1(\text{mod}1 + X + X^2 + \cdots + X^{r-1}, n).$$

Since r is chosen to be a prime number, we further note that $1 + X + X^2 + \ldots + X^{r-1}$ is simply the rth cyclotomic polynomial which we will henceforth denote as $\Phi(x)$. By the Chinese Remainder Theorem, we have the following isomorphism.

$$\mathbf{Z_n}[X] / \langle \Phi(x) \rangle \cong \prod_{i=1}^{k} \mathbf{Z}_{p_i}[X] / \langle \Phi(x) \rangle. \tag{2.1}$$

We now note that $(p_i, r) = 1$ for all i, therefore $\Phi(x)$ is irreducible in $\mathbf{Z}_{p_i}[X]$ for all i. Then $\mathbf{Z}_{p_i}[X] / \langle \Phi(x) \rangle$ is simply the extension field of $\mathbf{Z}_{p_i}[X]$ that contains the rth primitive root of unity. In light of this, we can now write

$$\mathbf{Z}_{p_i}[X] / \langle \Phi(x) \rangle = \mathbf{Z}_{p_i}(\zeta_r).$$

Substituting this in (2.1), we get

$$\mathbf{Z_n}[X] / \langle \Phi(x) \rangle \cong \prod_{i=1}^{k} \mathbf{Z}_{p_i}(\zeta_r).$$

We must now prove the identity $(X - 1)^n \equiv X^n - 1$ in the field $\mathbf{Z}_{p_i}(\zeta_r)$ for all $i = 1, \ldots, k$. This is equivalent to proving that

$$(\zeta_r - 1)^n \equiv \zeta_r^n - 1 (\text{mod} p_i), \tag{2.2}$$

for all $i = 1, \ldots, k$.

Denoting $p_i := p$, we note that

$$(\zeta_r - 1)^{p^2} = \zeta_r^{p^2} - 1 = \zeta_r^{-1} - 1 = -\zeta_r^{-1}(\zeta_r - 1)$$

and hence

$$(\zeta_r - 1)^{2(p^2-1)} = \zeta_r^{-1}.$$

Therefore

$$(\zeta_r - 1)^{2r(p^2-1)} = 1. \tag{2.3}$$

From Eq. (2.3), it is clear that the order of the element $(\zeta_r - 1)$ divides the number $2r(p^2 - 1)$. We will use this to finish the proof of this proposition. We must prove Eq. (2.2) in order to finish the proof of the proposition, but this will hold if the following is true:

$$n \equiv p_i \ (mod\ 2r(p_i^2 - 1)) \tag{2.4}$$

for all $i = 1, \ldots, k$.

In order for Eq. (2.4) to hold, $n - p_i$ must be divisible by the pairwise co-prime numbers $8r$, $\frac{p_i-1}{2}$, $\frac{p_i+1}{2}$, i.e, we must have

$$8r|(n-p_i), \left(\frac{p_i-1}{2}\right) \Big| (n-p_i) \text{ and } \left(\frac{p_i+1}{2}\right) \Big| (n-p_i).$$

However these hold trivially due to the assumptions made in the proposition. Hence the proposition is proved.

We now proceed to the heuristic for the construction of another class of counterexamples.

Proposition 3 *Let* $p_1^{a_1}, p_2^{a_2} \dots p_k^{a_k}$ *be k distinct prime powers, let* $n = p_1^{a_1} p_2^{a_2} \cdots p_k^{a_k}$. *Suppose that*

1. $k \equiv 1 (mod\ 4)$,
2. $a_i \equiv 1(mod\ 4)$ *for all* $i = 1, \dots, k$,
3. $10 \mid (p_1^{a_1} p_2^{a_2} \dots p_{i-1}^{a_{i-1}} p_{i+1}^{a_{i+1}} \dots p_k^{a_k} - 1)$ *for all* $i = 1, \dots, k$,
4. *There exit* t_i *such that* $(p_i^{t_i} - 1) \mid (p_1^{a_1} p_2^{a_2} \dots p_{i-1}^{a_{i-1}} p_{i+1}^{a_{i+1}} \dots p_k^{a_k}) - 1$ *for all* $i = 1, \dots, k$ *and*
5. $p_i \equiv 3(mod\ 20)$ *for all* $i = 1, \dots, k$.

Then $(X-1)^n \equiv X^n - 1(mod\ n, X^5 - 1)$.

Proof By assumption, we have $n^2 = (p_1^{a_1} p_2^{a_2} \dots p_k^{a_k})^2 \equiv (\ 9)^k (mod\ 20)$. Hence $n^2 \equiv (\ -1)^k (mod\ 5)$, which implies that $n^2 \equiv -1(mod\ 5)$, so $n^2 \not\equiv 1(mod\ 5)$.

We also have

$$(X-1, 1+X+X^2+\cdots+X^4, n) = (1)$$

in the ring $\mathbf{Z}[X]$. Hence in order to prove the identity

$$(X-1)^n \equiv X^n - 1(\mathrm{mod} X^5 - 1, n)$$

it suffices to prove the identity

$$(X-1)^n \equiv X^n - 1(\mathrm{mod} 1 + X + X^2 + \cdots + X^4, n).$$

Since 5 is a prime number, we further note that $1 + X + X^2 + \cdots X^4$ is simply the fifth cyclotomic polynomial which we will henceforth denote as $\Phi_5(x)$.

As a consequence of the Chinese Remainder Theorem, we have the following isomorphism.

$$\mathbf{Z_n}[X] / \langle \Phi_5(x) \rangle \cong \prod_{i=1}^{k} \mathbf{Z}_{p_i^{a_i}}[X] / \langle \Phi_5(x) \rangle. \tag{2.5}$$

We now note that $(p_i, 5) = 1$ for all i, therefore $\Phi_5(x)$ is irreducible in $\mathbf{Z}_{p_i^{a_i}}[X]$ for all i. Then $\mathbf{Z}_{p_i}[X] / \langle \Phi_5(x) \rangle$ is simply the extension field of $\mathbf{Z}_{p_i^{a_i}}[X]$ that contains the fifth primitive root of unity. In light of this, we can now write

$$\mathbf{Z}_{p_i^{a_i}}[X]\,/\langle \Phi_5(x)\rangle = \mathbf{Z}_{p_i^{a_i}}(\zeta_5).$$

Substituting this in Eq. (2.5), we get

$$\mathbf{Z_n}[X]\,/\langle \Phi_5(x)\rangle \cong \prod_{i=1}^{k} \mathbf{Z}_{\mathbf{p_i^{a_i}}}(\zeta_5)$$

We must now prove the identity $(X-1)^n \equiv X^n - 1$ in the field $\mathbf{Z}_{p_i^{a_i}}(\zeta_5)$ for all $i = 1, \ldots, k$.

This is equivalent to proving that

$$(\zeta_5 - 1)^n \; \equiv \zeta_5^n - 1 (mod p_i^{a_i}) \tag{2.6}$$

for all $i = 1, \ldots, k$.

First we note that

$$(\zeta_5 - 1)^{p_i^{t_i}} = \zeta_5^{p_i^{t_i}} - 1 = \zeta_5^{-1} - 1 = -\zeta_5^{-1}(\zeta_5 - 1).$$

In view of the above,

$$(\zeta_5 - 1)^{2(p_i^{t_i}-1)} = \zeta_5^{-1}.$$

Hence

$$(\zeta_5 - 1)^{10(p_i^{t_i}-1)} = 1 \tag{2.7}$$

for all t_i, $i = 1 \ldots k$ as assumed by the proposition.

From Eq. (2.7), it is clear that the order of the element $(\zeta_5 - 1)$ divides the number $10(p_i^{t_i} - 1)$. We will use this to finish the proof of this proposition. We must prove Eq. (2.6) in order to finish the proof of the proposition, but this will hold if the following are true:

$$n \equiv p_i^{a_i} \; (mod\; 10(p_i^{t_i} - 1)) \tag{2.8}$$

$$p_i^{a_i}(p_1^{a_1} p_2^{a_2} \ldots p_{i-1}^{a_{i-1}} p_{i+1}^{a_{i+1}} \ldots p_k^{a_k} - 1) \equiv 0 (mod\; 10(p_i^{t_i} - 1)). \tag{2.9}$$

for all i $= 1, \ldots, k$

In order for Eqs. (2.8) and (2.9) to hold, $p_1^{a_1} p_2^{a_2} \ldots p_{i-1}^{a_{i-1}} p_{i+1}^{a_{i+1}} \ldots p_k^{a_k} - 1$ must be divisible by the pairwise co-prime numbers 20, $\frac{p_i^{t_i}-1}{2}$, i.e., we must have

$$20 \mid \left(p_1^{a_1} p_2^{a_2} \ldots p_{i-1}^{a_{i-1}} p_{i+1}^{a_{i+1}} \ldots p_k^{a_k} - 1\right) \text{ and } \left(\frac{p_i^{t_i} - 1}{2}\right) \Big| \left(p_{i-1}^{a_{i-1}} p_{i+1}^{a_{i+1}} \ldots p_k^{a_k} - 1\right).$$

However, these hold trivially due to the assumptions made in the proposition. Hence the proposition is proved.

By these propositions, we have heuristics that suggest the existence of several counterexamples to the Agrawal conjecture. We now proceed to attempt a rough estimate for the number of counterexamples suggested by [2].

This argument was also used by Lenstra and Pomerance [3], and was earlier outlined by Pomerance to find counterexamples for the Bailie-PSW primality testing algorithm and can be found at

`http://www.pseudoprime.com/dopo.pdf.`

Fix an arbitrarily large integer m and let T be very large. Let $P = P_m(T)$ denote the set of primes p in the interval $[T, T^m]$ such that

1. $p \equiv 3(mod\ 8r)$, where $r \equiv 1(mod\ 4)$ and r is prime;
2. $\frac{p-1}{2}$ is squarefree and divisible only by primes $q \leq T$ with $q \equiv 3(mod\ 4)$;
3. $\frac{p+1}{2}$ is squarefree and divisible only by primes $r \leq T$ with $r \equiv 1(mod\ 4)$.

Clearly, a fraction of all primes (asymptotically) in $[T, T^m]$ satisfies condition 1, we can also prove similarly that a positive fraction of primes in $[T, T^m]$ is such that $\frac{p-1}{2}$ and $\frac{p+1}{2}$ are squarefree. The event that every prime factor of $\frac{(p-1)}{2}$ is congruent to $1(mod\ 4)$ should occur with probability $c(logT)^{-1/2}$, and similarly for every prime factor of $\frac{p+1}{2}$ to be $3(mod\ 4)$ [4], where c is Landau's constant. Thus the cardinality of $P_m(T)$ should asymptotically be, as $T \to \infty$

$$cT^m/log^2T$$

where c is a positive constant that depends on the choice of m.

We now choose k such that $k \equiv 1(mod\ 4)$ and $k < \frac{T^2}{(log(T^m)}$ and we form the squarefree numbers n that run over products of k distinct primes of the set P. The number of choices for n is exactly given by the binomial coefficient $\binom{\#P}{k}$ where $\#P$ is the size of the set P, and we get the lower bound:

$$\begin{aligned}\binom{\#P}{k} &\geq \left(\frac{T^m}{(logT^m)^3(T^2/logT^m)}\right)^{(T^2/logT^m)-4}\\ &> (T^{m-3})^{(T^2/logT^m)-4}\\ &= e^{(1-3/m)T^2-4(m-3)logT}\\ &> e^{1-(4/m)T^2}\end{aligned}$$

for sufficiently large T and a fixed n.

Let Q denote the product of primes $q \leq T$ with $q \equiv 3(mod\ 4)$ and let R denote the product of primes $r \leq T$ with $r \equiv 1(mod\ 4)$. Then Q and R are relatively co-prime so that $QR < e^{2T}$ for a large T. Thus, the number of choices for n that satisfy both $n \equiv 1(mod\ Q)$ and $n \equiv -1(mod\ R)$ should be

$$e^{(1-4/m)T^2}e^{-2T} > e^{T^2(1-\frac{5}{m})}$$

Any such n is a counterexample as proposed in [2]. Thus we see that for ay fixed n and for large T, there should be at least $e^{T^2(1-\frac{5}{m})}$ counterexamples to Agrawal's conjecture below e^{T^2}.

3 Conclusion

Here we have given heuristics to construct a few counterexamples to the Agrawal Conjecture. Further work would be to find the distribution of these counterexamples.

Acknowledgements We would like to thank Dr. H. W. Lenstra and Dr. Carl Pomerance for fruitful discussions that helped to clarify several details mentioned in this paper.

References

1. Agrawal, M., Kayal, N., Saxena, N.: PRIMES is in P. Annals Math. 781–793 (2004)
2. Bhattacharjee, R., Pandey, P.: Primality testing. Technical report, IIT Kanpur (2001). http://www.cse.iitk.ac.in/research/btp2001/primality.html
3. Lenstra, H.W., Pomerance, C.: Future directions in algorithmic number theory, ARCC workshop held at Paolo Alto, California (2008). https://www.aimath.org/WWN/primesinp/articles/html/50a/
4. Landau, E.: ber die Einteilung der positiven ganzen Zahlen in vier Klassen nach der Mindestzahl der zu ihrer additiven Zusammensetzung erforderlichen Quadrate (1909)

Constrained Fractal Trigonometric Approximation: A Sequential Approach

N. Vijender and A. Sathish Kumar

Abstract In this paper, for a given $f \in \mathcal{C}(I)$, we construct a sequence of Bernstein α-fractal functions with variable scaling factors. Bernstein α-fractal functions proposed in this paper are fixed points of the Read-Bajraktarević operators defined on a suitable function space. The convergence of the Bernstein α-fractal function with variable scaling factors towards the original function f follows from the convergence of Bernstein polynomials of f toward f. Owing this reason, Bernstein α-fractal function with variable scaling factors proposed in this paper converge to the original function for any choice of the variable scalings whereas the existing fractal approximants converge to the original function only if the magnitude of the scaling factors goes to zero. Using the Bernstein α-fractal functions, we study the Bernstein α-fractal trigonometric polynomials with variable scaling factors and their constrained approximation properties. For a given sequence $\{f_n(x)\}_{n=1}^{\infty}$ of continuous periodic functions defined on $[-l, l]$ that converges uniformly to a function f, we investigate the existence of a double sequence of Bernstein α-fractal trigonometric polynomials that converges uniformly to f.

Keywords Fractal trigonometric approximation · Variable scaling factors · Convergence · Fractal dimension

Mathematics Subject Classification (2010) 30E10 · 28A80 · 41A30 · 41A17 · 65D017

N. Vijender (✉)
Department of Basic Sciences, Indian Institute of Information Technology Nagpur, Nagpur 440006, India
e-mail: vijendernallapu@gmail.com

A. Sathish Kumar
Department of Mathematics, Visvesvaraya National Institute of Technology, Nagpur 440010, India
e-mail: mathsatish9@gmail.com

R. N. Mohapatra et al. (eds.), *Mathematical Analysis and Computing*,
Springer Proceedings in Mathematics & Statistics 344,
https://doi.org/10.1007/978-981-33-4646-8_43

1 Introduction

Barnsley [1] proposed the concept of a Fractal Interpolation Function (FIF) based on the theory of an Iterated Function System (IFS) [2]. In general, the FIFs are irregular and own a self-referential structure in their graphs. The FIF can be used to construct the fractal analogue of polynomial, rational, and trigonometric interpolants, for details, refer [3–7]. The concept of FIF can be used to associate a family of functions to a given function f defined on a real compact interval I (for instance, see [1, 8]). An element of this family is denoted by f^α, and Navascués [8] called it as α-fractal function associated with f. The α-fractal function f^α interpolates and approximates f simultaneously and the graph of f^α is a union of transformed copies of itself, for details, refer [9–17]. This function f^α contains a set of real parameters, namely scaling factors. The fractality/irregularity associated with f^α can be measured through its fractal dimension [18]. The existing fractal approximants converge to the original only if the magnitude of the scaling factors goes to zero.

In this manuscript, for a given continuous function f defined on a real compact interval, we develop a sequence of Bernstein α-fractal functions with variable scaling factors that converge uniformly to f even if the norm of the variable scaling factors does not go to zero. In the construction of Bernstein α-fractal functions with variable scaling factors, we use Bernstein polynomials of f as base functions. The convergence of Bernstein polynomials of f toward f implies the convergence of Bernstein α-fractal functions toward f. Owing to this reason, Bernstein α-fractal functions proposed in this paper converge to the original function even if scaling factors do not go to zero.

Using the Bernstein α-fractal functions, we define Bernstein α-fractal trigonometric polynomials and show that for each $n \in \mathbb{N}$, $Span\{1, \cos_n^\alpha(\frac{j\pi x}{l}), \sin_n^\alpha(\frac{j\pi x}{l}) : j \in \mathbb{N}\}$ is dense in the space of all $2l$-period continuous function defined on $[-l, l]$, where $\cos_n^\alpha(\frac{j\pi x}{l})$ and $\sin_n^\alpha(\frac{j\pi x}{l})$, respectively, are Bernstein α-fractal functions of $\cos(\frac{j\pi x}{l})$ and $\sin(\frac{j\pi x}{l})$. Approximation of functions from below or above is called constrained approximation. In this paper, we study the constrained approximation by the proposed Bernstein α-fractal trigonometric functions. We need not impose any condition on the scaling vector to obtain constrained Bernstein α-fractal trigonometric approximants. In other words, constrained Bernstein fractal trigonometric approximants proposed in this paper work equally for all the variable scaling factors.

1.1 IFS and Attractor

The following notation and terminologies will be used throughout the article. The set of real numbers will be denoted by $\mathbb{R}$, whilst the set of natural numbers by $\mathbb{N}$. For a fixed $N \in \mathbb{N}$, we shall write $\mathbb{N}_N$ for the set of first N natural numbers. Given real numbers x_1 and x_N with $x_1 < x_N$, let $I = [x_1, x_N]$. Let $(\mathcal{X}, d)$ be a complete metric space, and $\mathcal{H}(\mathcal{X})$ be the set of all nonempty compact subsets of $\mathcal{X}$. Then $\mathcal{H}(\mathcal{X})$ is a complete metric space with respect to the Hausdorff metric h_d, where h_d is defined as $h_d(A, B) = \max\{d(A, B), d(B, A)\}$, and

$$d(A, B) = \max_{x \in A} \min_{y \in B} d(x, y).$$

Let $\vartheta_i : \mathcal{X} \to \mathcal{X}$ be continuous functions for $i \in \mathbb{N}_{N-1}$. The set $\mathcal{I} = \{\mathcal{X}; \vartheta_i, i \in \mathbb{N}_{N-1}\}$ is called an IFS. An IFS $\mathcal{I}$ is called *hyperbolic* if

$$\frac{d(\vartheta_i(x), \vartheta_i(y))}{d(x, y)} \leq c_i < 1 \; \forall x \neq y \in \mathcal{X}.$$

For any $A \in \mathcal{H}(\mathcal{X})$, we define the set valued Hutchinson map V on $\mathcal{H}(\mathcal{X})$ as

$$V(A) = \bigcup_{i \in \mathbb{N}_{N-1}} \vartheta_i(A).$$

If IFS $\mathcal{I}$ is *hyperbolic,* then it is easy to verify that V is a contraction map on $\mathcal{H}(\mathcal{X})$ with the contractive factor $c = \max\{c_i : i \in \mathbb{N}_{N-1}\}$. Then by the Banach Fixed Point Theorem, V has a unique fixed point (say G) and for any starting set A in $\mathcal{H}(\mathcal{X})$ with $V(A) = V^{\circ 1}(A),\; V^{\circ m}(A) = V \circ V^{\circ m-1}(A)$ for $m \geq 2$,

$$\lim_{m \to \infty} V^{\circ m}(A) = G.$$

The set $G \in \mathcal{H}(\mathcal{X})$ is called the attractor or the deterministic fractal of the IFS $\mathcal{I}$.

2 Construction of Bernstein α-Fractal Functions with Variable Scalings

For a given $f \in \mathcal{C}(I)$, consider a partition $\Delta = \{x_1, x_2, \ldots, x_N\}$ of $[x_1, x_N]$ satisfying $x_1 < x_2 < \cdots < x_N$ and the set of data points $\{(x_i, z_i = f(x_i)) : i \in \mathbb{N}_N\}$. Let K be a suitable compact subset of $\mathbb{R}$ such that $z_i \in K, i \in \mathbb{N}_N$. Let $L_i : I \to I_i, i \in \mathbb{N}_{N-1}$ be contractive homeomorphisms defined by $L_i(x) = a_i x + b_i$ such that

$$L_i(x_1) = x_i, \; L_i(x_N) = x_{i+1}. \tag{1}$$

Let $F_{n,i} : I \times K \to K, i \in \mathbb{N}_{N-1}$ be continuous functions defined by

$$F_{n,i}(x, z) = \alpha_i(x) z + f(L_i(x)) - \alpha_i(x) B_n(f)(x), \tag{2}$$

where $\alpha_i : I \to \mathbb{R}$ is a continuous function satisfying $\|\alpha_i\|_\infty < 1$, and $B_n(f)(x)$ is the Bernstein polynomial of f on $I = [x_1, x_N]$, i.e., for all $x \in I, n \in \mathbb{N}$,

$$B_n(f)(x) = \frac{1}{(x_N - x_1)^n} \sum_{k=0}^{n} \binom{n}{k} (x - x_1)^k (x_N - x)^{n-k} f\Big(x_1 + \frac{k(x_N - x_1)}{n}\Big).$$

It is easy to verify that $B_n(f)(x_1) = f(x_1) = z_1$, $B_n(f)(x_N) = f(x_N) = z_N$ for all $n \in \mathbb{N}$. Using this, one can verify from (2) that

$$F_{n,i}(x_1, z_1) = z_i, F_{n,i}(x_N, z_N) = z_{i+1}, i \in \mathbb{N}_{N-1}, n \in \mathbb{N}. \tag{3}$$

Let $\mathcal{G} = \{g \in \mathcal{C}(I) \,|g(x_1) = z_1 \text{ and } g(x_N) = z_N\}$. Then $\mathcal{G}$ is a complete metric space with respect to the uniform metric ρ defined by

$$\rho(g, h) = \max\{|g(x) - h(x)| : x \in I\} \ \forall \ g, h \in \mathcal{G}.$$

Define the Read-Bajraktarević operator T_n on $(\mathcal{G}, \rho)$ as

$$T_n g(x, y) = F_{n,i}\big(L_i^{-1}(x), g(L_i^{-1}(x))\big), \ x \in I_i, \ i \in \mathbb{N}_{N-1}. \tag{4}$$

We prove first in the sequel that T_n maps $\mathcal{G}$ into itself. For $g \in \mathcal{G}$,

$$T_n g(x_1) = F_{n,1}\big(L_1^{-1}(x_1), g(L_1^{-1}(x_1))\big) = F_{n,1}(x_1, g(x_1)) = z_1,$$
$$T_n g(x_N) = F_{n,N-1}\big(L_{N-1}^{-1}(x_N), g(L_{N-1}^{-1}(x_N))\big) = F_{n,N-1}(x_N, g(x_N)) = z_N.$$

Using the properties of L_i, $F_{n,i}$, and (1)–(3), it is easy to verify that $T_n g$ is continuous on the intervals I_i, $i \in \mathbb{N}_{N-1}$, and at each of the points $x_2, \ldots, x_{N-1}$. Also,

$$\rho(T_n g, T_n h) \le \|\alpha\|_\infty \rho(g, h),$$

where $\|\alpha\|_\infty = \max\{\|\alpha_i\|_\infty : i \in \mathbb{N}_{N-1}\} < 1$. Hence, T_n is a contraction map on the complete metric space $(\mathcal{G}, \rho)$. Therefore, by the Banach fixed point theorem, T_n possesses a unique fixed point (say) f_n^α on $\mathcal{G}$, i.e., $(T_n f_n^\alpha)(x) = f_n^\alpha(x)$ for all $x \in I$. According to (4), the function f_n^α satisfies the following functional equation: $x \in I_i, i \in \mathbb{N}_{N-1}$,

$$f_n^\alpha(x) = \alpha_i(L_i^{-1}(x)) f_n^\alpha(L_i^{-1}(x)) + f(x) - \alpha_i(L_i^{-1}(x)) B_n(f)(L_i^{-1}(x)). \tag{5}$$

Also, it is easy to verify that

$$f_n^\alpha(x_i) = z_i \ \forall \ i \in \mathbb{N}_N, n \in \mathbb{N}. \tag{6}$$

Now, define the functions $\omega_{n,i} : I \times K \to I_i \times K, i \in \mathbb{N}_{N-1}$ as

$$\omega_{n,i}(x, z) = \big(L_i(x), F_{n,i}(x, z)\big).$$

Let $\mathcal{X} := I \times K$ and consider the IFS $\mathcal{I}_n = \{\mathcal{X}; \omega_{n,i} : i \in \mathbb{N}_{N-1}\}$. Using [1], it follows that the above IFS $\mathcal{I}_n$ has unique attractor $G_n \in \mathcal{H}(I \times K)$, and G_n is graph of f_n^α. Hence, the above function f_n^α is called the Bernstein α-fractal function with variable scaling factors associated with f. From (5), we have

$$\|f_n^\alpha - f\|_\infty = \|\alpha\|_\infty \|f_n^\alpha - B_n(f)\|_\infty \leq \|\alpha\|_\infty [\|f_n^\alpha - f\|_\infty + \|f - B_n(f)\|_\infty],$$
$$\Longrightarrow \|f_n^\alpha - f\|_\infty \leq \frac{\|\alpha\|_\infty}{1 - \|\alpha\|_\infty} \|f - B_n(f)\|_\infty. \tag{7}$$

From the classical approximation theory [19], we obtain that

$$\|f - B_n(f, .)\|_\infty \to 0. \tag{8}$$

Using (8) in (7), it follows that for every scaling vector $\alpha(x)$, the sequence $\{f_n^\alpha(x)\}_{n=1}^\infty$ of Bernstein α-fractal functions converges uniformly to the function $f \in \mathcal{C}(I)$, and this leads to the following theorem.

Theorem 21 *Let $\mathcal{C}(I)$ be endowed with uniform norm $\|.\|_\infty$ and $f \in \mathcal{C}(I)$. Let $\Delta = \{x_1, x_2, \ldots, x_N\}$ be a partition of $I = [x_1, x_N]$ satisfying $x_1 < x_2 < \cdots < x_N$ and $\alpha(x) = (\alpha_1(x), \alpha_2(x), \ldots, \alpha_{N-1}(x))$. Then, for every scaling vector $\alpha(x)$, the sequence $\{\mathcal{I}_n\}_{n=1}^\infty$ of IFSs determine a sequence $\{f_n^\alpha(x)\}_{n=1}^\infty$ of Bernstein α-fractal functions that converges uniformly to f.*

Remark 21 Under the suitable conditions, the box counting dimension of α-fractal functions with variable scaling factors is studied in [20]. Hence, the fractal dimension of Bernstein α-fractal functions with variable scaling factors can be calculated using the procedure given in [20].

Remark 22 Using the proposed Bernstein α-fractal functions with variable scalings, one can construct fractal approximant with a fixed fractal dimension for a given continuous function. Also, this is not possible using the existing fractal approximants since they converge to the original function if the magnitude of the scaling factors goes to zero.

Remark 23 The convergence of the Bernstein α-fractal functions toward the original function does not depend on the size of the partition Δ of $I = [x_1, x_N]$.

Theorem 22 *Let $\mathcal{C}(I)$ be endowed with uniform norm $\|.\|_\infty$. The α-operator $\mathcal{F}_n^\alpha : \mathcal{C}(I) \to \mathcal{C}(I)$ defined by $\mathcal{F}_n^\alpha(f) = f_n^\alpha$ is linear and bounded.*

Proof Using (5), for each $n \in \mathbb{N}$, we have that for all $x \in I_i$, $i \in \mathbb{N}_{N-1}$,

$$f_n^\alpha(x) = \alpha_i(L_i^{-1}(x)) f_n^\alpha(L_i^{-1}(x)) + f(x) - \alpha_i(L_i^{-1}(x)) B_n(f)(L_i^{-1}(x)),$$

$$g_n^\alpha(x) = \alpha_i(L_i^{-1}(x)) g_n^\alpha(L_i^{-1}(x)) + f(x) - \alpha_i(L_i^{-1}(x)) B_n(g)(L_i^{-1}(x)).$$

Multiplying the first equation by β^* and second equation by γ^*, and the uniqueness of the solution of the fixed point equation defining the FIF gives

$$(\beta^* f + \gamma^* g)_n^\alpha = \beta^* f_n^\alpha + \gamma^* g_n^\alpha \ \ \forall \beta^*, \gamma^* \in \mathbb{R}, n \in \mathbb{N}.$$

Hence, $\mathcal{F}_n^\alpha$ is linear. Using (7), we get

$$\|\mathcal{F}_n^\alpha(f)\|_\infty = \|f_n^\alpha\|_\infty \leq \|f\|_\infty + \frac{\|\alpha\|_\infty}{1 - \|\alpha\|_\infty}\|I_d - B_n(.,.)\|_{\infty^*}\|f\|_\infty, \tag{9}$$

where I_d is the identity operator and $\|.\|_{\infty^*}$ is the operator norm induced by $\|.\|_\infty$. From the classical approximation theory, we can see that $\|I_d - B_n(.,.)\|_{\infty^*} \to 0$ as $n \to \infty$. Therefore, for given $\epsilon = 1$, there exists $N_0 \in \mathbb{N}$ such that

$$\|I_d - B_n(.,.)\|_{\infty^*} < 1 \forall n > N_0.$$

Let $\theta = \max\{\|I_d - B_1(.,.)\|_{\infty^*}, \|I_d - B_2(.,.)\|_{\infty^*}, \ldots \|I_d - B_{N_0}(.,.)\|_{\infty^*}, 1\}$. Then from (9), we get $\|\mathcal{F}_n^\alpha\|_{\infty^*} \leq 1 + \frac{\|\alpha\|_\infty \theta}{1-\|\alpha\|_\infty}$, hence $\mathcal{F}_n^\alpha$ is bounded.

2.1 Examples

In this section, we provide numerical examples to corroborate our findings. For this purpose, let $f(x) = \sin(2\pi x), x \in [0, 1]$. The Bernstein α-fractal functions in Fig. 1a–c are generated with respect to the partition $\Delta = \{0,\ 0.3333,\ 0.6667,\ 1\}$ of $[0, 1]$. The Bernstein α-fractal functions f_3^α, f_{19}^α and f_{89}^α are generated, respectively, in Fig. 1a–c with the choice of the variable scaling factors

$$\alpha_1(x) = \frac{1}{1 + |\sin(-200x^8)|}, \alpha_2(x) = \frac{1}{1 + |\cos(-300x^8)|}, \alpha_3(x) = \frac{1}{1 + e^{-x^2}},$$

for all $x \in [0, 1]$. It is plain to verify that $\|\alpha\|_\infty < 1$. According to Theorem 21, the Bernstein α-fractal function f_{89}^α provides a better approximation for $\sin(2\pi x), x \in [0, 1]$ than that of obtained by f_3^α and f_{19}^α. By observing Fig. 1a–c, one can ask that why are the fractal functions f_3^α, f_{19}^α, and f_{89}^α not having same sort of irregularity even if they are constructed with the same choice of the variable scaling factors? This is due to the following reason: The α-fractal functions f_3^α and f_{19}^α exhibit irregularity on all scales whereas the α-fractal function f_{89}^α exhibits irregularity on small scales. Further, small scales of irregularity of the α-fractal function f_{89}^α can be observed from Fig. 1d which is a part of f_{89}^α under magnification.

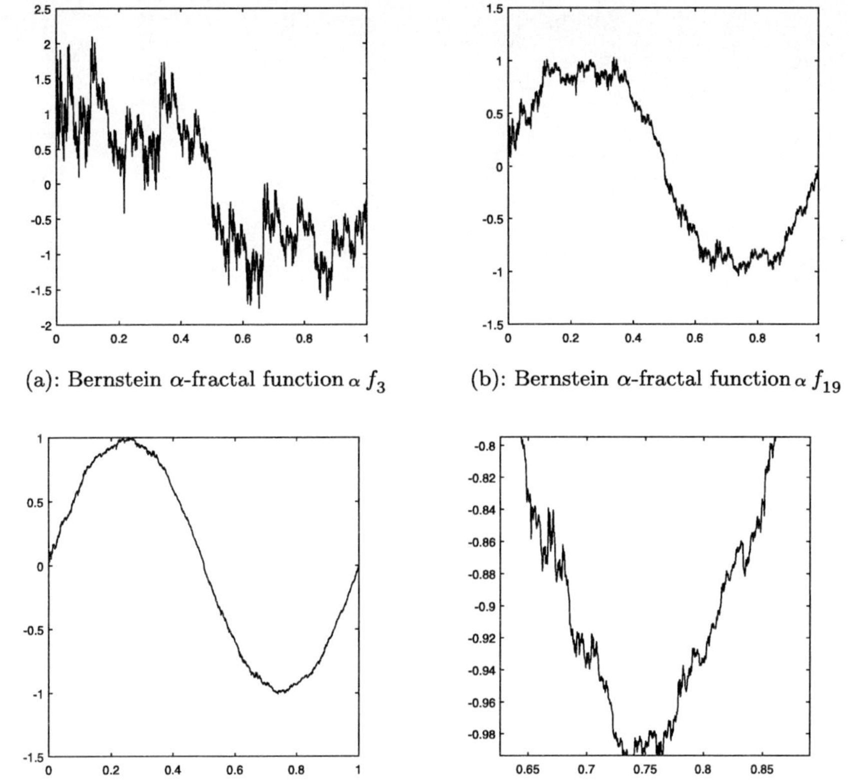

(a): Bernstein α-fractal function αf_3

(b): Bernstein α-fractal function αf_{19}

(c): Bernstein α-fractal function f_{89}^{α}

(d): A part of f_{89}^{α} under magnification

Fig. 1 Bernstein α-fractal functions of $\sin(2\pi x)$, $x \in [0, 1]$

3 Non-smooth Bernstein α-Fractal Functions

In this section, we derive an explicit expression for Bernstein α-fractal function f_n^{α} and use it to study non-differentiability of f_n^{α}.

3.1 Explicit Representation of f_n^{α}

To ease the exposition, throughout the section, we assume $I = [0, 1]$, and $\alpha_i(x) = \alpha_i$, $x \in I$ for all $i \in \mathbb{N}_{N-1}$. Let $\Delta = \{x_1, x_2, \dots, x_N\}$ be a partition of $I = [0, 1]$ satisfying $0 = x_1 < x_2 < \cdots < x_N = 1$ and $x_{i+1} - x_i = h$ for all $i \in \mathbb{N}_{N-1}$. Owing to the above assumptions, the functions L_i, $i \in \mathbb{N}_{N-1}$ adopt the expression

$$L_i(x) = \frac{x}{N-1} + \frac{i-1}{N-1}, x \in [0, 1]. \tag{10}$$

Using [21], it follows that

$$f_n^{\alpha}(\omega) = \sum_{k=0}^{\infty} \Big(\prod_{j=0}^{k} \alpha_{i_j}\Big) r_{n,i_{k+1}} \Big(\sum_{l=1}^{\infty} \frac{i_{k+l+1} - 1}{(N-1)^l}\Big), \tag{11}$$

where

$$r_{n,i_{k+1}}(x) = f(L_{i_{k+1}}(x)) - \alpha_{i_{k+1}} B_n(f)(x), \tag{12}$$

$$\omega = (i_1 i_2 i_3 \ldots) = \sum_{m=1}^{\infty} \frac{i_m - 1}{(N-1)^m}, 1 \le i_m \le N-1 \ \forall m \ge 1. \tag{13}$$

The above expression is the $(N-1)$-adic representation of $\omega \in [0, 1]$. Also, it is assumed in (11) that $\alpha_{i_0} = 1$. Define

$$\sigma^k \omega = (i_{k+1} i_{k+2} i_{k+3} \ldots) = \frac{i_{k+1} - 1}{N-1} + \frac{i_{k+1} - 1}{(N-1)^2} + \cdots$$
$$\sigma^{k+1} \omega = (i_{k+2} i_{k+3} i_{k+4} \ldots) = \frac{i_{k+2} - 1}{N-1} + \frac{i_{k+2} - 1}{(N-1)^2} + \cdots$$

Using the above expression in (11), we get

$$\begin{aligned} f_n^{\alpha}(\omega) &= \sum_{k=0}^{\infty} \Big(\prod_{j=0}^{k} \alpha_{i_j}\Big) r_{n,i_{k+1}} \Big(\sigma^{k+1}\omega\Big) \\ &= \sum_{k=0}^{\infty} \Big(\prod_{j=0}^{k} \alpha_{i_j}\Big) \Big(f(L_{i_{k+1}}(\sigma^{k+1}\omega)) - \alpha_{i_{k+1}} B_n(f)(\sigma^{k+1}\omega)\Big). \end{aligned} \tag{14}$$

From (10), it is easy to see that

$$\begin{aligned} L_{i_{k+1}}(\sigma^{k+1}\omega) &= L_{i_{k+1}}\Big(\frac{i_{k+2} - 1}{N-1} + \frac{i_{k+2} - 1}{(N-1)^2} + \cdots\Big) \\ &= \frac{i_{k+1} - 1}{N-1} + \frac{i_{k+2} - 1}{(N-1)^2} + \cdots = \sigma^k \omega. \end{aligned} \tag{15}$$

Assume that

$$\alpha_{i_j} = a \ \forall \ i \in \mathbb{N}_{N-1}, j \in \mathbb{N}. \tag{16}$$

Using (15) and (16) in (14), we get

$$f_n^\alpha(\omega) = \sum_{k=0}^{\infty} a^k \Big(f(\sigma^k \omega) - a B_n(f)(\sigma^{k+1}\omega) \Big). \tag{17}$$

3.2 Non-differentiability of f_n^α

For $\omega = (i_1 i_2 i_3 \dots i_r \dots)$, we consider $\omega_r \le \omega \le \omega_r^*$, $\omega_r = (i_1 i_2 \dots i_r 1111\dots)$ and $\omega_r^* = (i_1 i_2 \dots i_r (N-1)(N-1)(N-1)\dots)$.

$$\omega_r^* = \sum_{k=1}^{r} \frac{i_k - 1}{(N-1)^k} + \sum_{k=1}^{\infty} \frac{N-2}{(N-1)^k} = \omega_r + \frac{1}{(N-1)^r}. \tag{18}$$

Lemma 31 *If $f \in C^1[0,1]$, $|a| \le \frac{1}{N-1}$, and f_n^α is differentiable at $\omega \in (0,1)$, then*

$$\lim_{k\to\infty} f'(\sigma^k \omega) = z_N - z_1, \ \lim_{k\to\infty} B_n'(f)(\sigma^k\omega) = z_N - z_1. \tag{19}$$

Proof For all $k \ge r$, it is easy to verify that

$$\sigma^k \omega_r = (111\dots) = 0 \tag{20}$$

$$\sigma^k \omega_r^* = ((N-1)(N-1)(N-1)\dots) = 1. \tag{21}$$

From (17), we write

$$\begin{aligned} f_n^\alpha(\omega_r) =& \sum_{k=0}^{r-1} a^k \Big(f(\sigma^k \omega_r) - a B_n(f)(\sigma^{k+1}\omega_r) \Big) \\ &+ \sum_{k=r}^{\infty} a^k \Big(f(\sigma^k \omega_r) - a B_n(f)(\sigma^{k+1}\omega_r) \Big). \end{aligned} \tag{22}$$

Using (20), it follows that

$$\sum_{k=r}^{\infty} a^k \Big(f(\sigma^k \omega_r) - a B_n(f)(\sigma^{k+1}\omega_r) \Big) = \sum_{k=r}^{\infty} a^k z_1 (1-a) = z_1 a^r. \tag{23}$$

Again using (17), we write

$$\begin{aligned} f_n^{\alpha}(\omega_r^*) =&\sum_{k=0}^{r-1} a^k\Big(f(\sigma^k\omega_r^*) - aB_n(f)(\sigma^{k+1}\omega_r^*)\Big) \\ &+\sum_{k=r}^{\infty} a^k\Big(f(\sigma^k\omega_r^*) - aB_n(f)(\sigma^{k+1}\omega_r^*)\Big). \end{aligned} \tag{24}$$

Using (21), it follows that

$$\sum_{k=r}^{\infty} a^k\Big(f(\sigma^k\omega_r^*) - aB_n(f)(\sigma^{k+1}\omega_r^*)\Big) = \sum_{k=r}^{\infty} a^k z_N(1-a) = z_N a^r. \tag{25}$$

Using (22)–(25), we obtain

$$\begin{aligned} f_n^{\alpha}(\omega_r^*) - f_n^{\alpha}(\omega_r) =&(z_N - z_1)a^r + \sum_{k=0}^{r-1} a^k\Big(f(\sigma^k\omega_r^*) - f(\sigma^k\omega_r)\Big) \\ &- a\Bigg[\sum_{k=0}^{r-1} a^k\Big(B_n(f)(\sigma^{k+1}\omega_r^*) - B_n(f)(\sigma^{k+1}\omega_r)\Big)\Bigg]. \end{aligned} \tag{26}$$

Using the mean value theorem, there exist $\xi_k^r, \bar{\xi}_k^r \in (\sigma^k\omega_r, \sigma^k\omega_r^*)$ such that

$$f(\sigma^k\omega_r^*) - f(\sigma^k\omega_r) = f'(\xi_k^r)(\sigma^k\omega_r^* - \sigma^k\omega_r) = f'(\xi_k^r)\frac{1}{(N-1)^{r-k}}. \tag{27}$$

$$\begin{aligned} B_n(f)(\sigma^{k+1}\omega_r^*) - B_n(f)(\sigma^{k+1}\omega_r) &= B_n'(f)(\bar{\xi}_k^r)(\sigma^{k+1}\omega_r^* - \sigma^{k+1}\omega_r) \\ &= B_n'(f)(\bar{\xi}_k^r)\frac{1}{(N-1)^{r-k-1}}. \end{aligned} \tag{28}$$

Substituting (27) and (28) in (26), we get

$$\begin{aligned} f_n^{\alpha}(\omega_r^*) - f_n^{\alpha}(\omega_r) =&(z_N - z_1)a^r + \sum_{k=0}^{r-1} a^k f'(\xi_k^r)\frac{1}{(N-1)^{r-k}} \\ &- a\Bigg[\sum_{k=0}^{r-1} a^k B_n'(f)(\bar{\xi}_k^r)\frac{1}{(N-1)^{r-k-1}}\Bigg]. \end{aligned} \tag{29}$$

Using (18) in (29), we get

$$\begin{aligned} \frac{f_n^{\alpha}(\omega_r^*) - f_n^{\alpha}(\omega_r)}{\omega_r^* - \omega_r} =&(z_N - z_1)\big(a(N-1)\big)^r + \sum_{k=0}^{r-1} f'(\xi_k^r)\big(a(N-1)\big)^k \\ &- \big(a(N-1)\big)\Bigg[\sum_{k=0}^{r-1} B_n'(f)(\bar{\xi}_k^r)\big(a(N-1)\big)^k\Bigg]. \end{aligned} \tag{30}$$

But

$$(z_N - z_1)\big(a(N-1)\big)^r = (z_N - z_1) + \sum_{k=0}^{r-1}(z_N - z_1)\big[a(N-1) - 1\big]\big(a(N-1)\big)^k. \tag{31}$$

Using (31) in (30), we obtain

$$\begin{aligned} \lim_{r\to\infty} \frac{f_n^\alpha(\omega_r^*) - f_n^\alpha(\omega_r)}{\omega_r^* - \omega_r} =& (z_N - z_1) + \lim_{r\to\infty} \sum_{k=0}^{r-1}[f'(\xi_k^r) - (z_N - z_1)]\big(a(N-1)\big)^k \\ & - \lim_{r\to\infty}\Bigg[\sum_{k=0}^{r-1}[B_n'(f)(\bar{\xi}_k^r) - (z_N - z_1)]\big(a(N-1)\big)^{k+1}\Bigg]. \end{aligned} \tag{32}$$

Since f_n^α is differentiable at $\omega \in (0, 1)$, we get

$$\lim_{r\to\infty} \frac{f_n^\alpha(\omega_r^*) - f_n^\alpha(\omega_r)}{\omega_r^* - \omega_r} = (f_n^\alpha)'(\omega).$$

Hence, from (32), we obtain

$$\begin{aligned} (f_n^\alpha)'(\omega) =& (z_N - z_1) + \lim_{r\to\infty} \sum_{k=0}^{r-1}[f'(\xi_k^r) - (z_N - z_1)]\big(a(N-1)\big)^k \\ & - \lim_{r\to\infty}\Bigg[\sum_{k=0}^{r-1}[B_n'(f)(\bar{\xi}_k^r) - (z_N - z_1)]\big(a(N-1)\big)^{k+1}\Bigg]. \end{aligned} \tag{33}$$

Since $|a| < \frac{1}{N-1}$, from (33), it is easy to verify that $(f_n^\alpha)'(\omega)$ is finite if (19) is true. Thus, we complete the proof.

The proof of the following theorem follows from Lemma 31.

Theorem 31 *Let* $f \in C^1[0, 1]$. *If (i)* $\alpha_i(x) = a$ *for all* $x \in [0, 1]$ *and* $i \in \mathbb{N}_{N-1}$, *where* a *is a real number such that* $|a| > \frac{1}{N-1}$, *(ii)* $f'(x)$ *and* $B_n'(f)(x)$ *do not agree with* $z_N - z_1$ *in a open sub-interval of* $[0, 1]$, *then the set of points of non-differentiability of* f_n^α *is dense in* $[0, 1]$.

4 Bernstein α-Fractal Trigonometric Approximation

Lemma 41 ([22]) *The continuous functions* $\psi_1, \psi_2, \ldots, \psi_k$ *defined on* I *is linearly independent if and only if there exist the points* $x_1, x_2, \ldots, x_k$ *in* I *such that* $det(\psi_j(x_i)) \neq 0$.

Let $\mathcal{C}(2l)$ be the space of all continuous functions on $[-l, l]$ of period $2l$, i.e.,

$$\mathcal{C}(2l) = \Big\{ f \in \mathcal{C}[-l, l] : f(x + 2l) = f(x) \; \forall \; x \in [-l, l] \Big\}.$$

Theorem 41 *Let $\Delta = \{x_1, x_2, \ldots, x_N\}$ be a partition of $[-l, l]$ satisfying $-l = x_1 < x_2 < \cdots, < x_N = l$, and $\alpha(x) = (\alpha_1(x), \alpha_2(x), \ldots, \alpha_{N-1}(x))$ be a scaling vector. Let $\Gamma = \{f_j : j \in \mathbb{N}\}$ be a subset of $\mathcal{C}[-l, l]$. If $Span(\Gamma)$ is dense in $\mathcal{C}(2l)$, then $\Gamma_n^\alpha = Span\{f_{j,n}^\alpha : j \in \mathbb{N}\}, n \in \mathbb{N}$ is also dense for each scaling vector function $\alpha(x) = (\alpha_1(x), \alpha_2(x), \ldots, \alpha_{N-1}(x))$.*

Proof We first prove that $\Gamma_n^\alpha = \{f_{j,n}^\alpha : j \in \mathbb{N}\}$ is a linearly independent set. For this it is enough to show that every finite subset of Γ^α is linearly independent. Let $\Gamma_{n,k}^\alpha = \{f_{j,n}^\alpha : j \in \mathbb{N}_k = \{1, 2, \ldots, k\}\} \subset \Gamma_n^\alpha$. Since f_j, $j \in \mathbb{N}_k$ are linearly independent, by previous lemma, there exist $\bar{x}_1, \bar{x}_2, \ldots, \bar{x}_k$, in $[-l, l]$ such that $det(f_j(x_i)) \neq 0$. Let us consider the partition $\Delta = \{x_1, \bar{x}_1, \bar{x}_2, \ldots, \bar{x}_k, x_N\}$ of I. Bearing in mind that the fractal function f_n^α coincides with the original function f at partition points, we have $det(f_{n,k}^\alpha(x_i)) = det(f_j(x_i)) \neq 0$. Thus, it follows that $\Gamma_{n,k}^\alpha$ is linearly independent for every $k \in \mathbb{N}$, and hence Γ_n^α is linearly independent. Let $\epsilon > 0$. Since $Span(\Gamma)$ is dense in $\mathcal{C}(2l)$, for every $f \in \mathcal{C}(2l)$, there exists a function $P \in Span(\Gamma)$ such that

$$\|f - P\|_\infty < \frac{\epsilon}{2}. \tag{34}$$

Without loss of generality, let us assume that $P = \sum_{j=1}^{s} \beta_j f_j$, $\beta_j \in \mathbb{R}$. Using linearity of $\mathcal{F}_n^\alpha$, we get

$$\mathcal{F}_n^\alpha(P) = P_n^\alpha = \sum_{j=1}^{s} \beta_j f_{j,n}^\alpha.$$

It is plain to verify that

$$\|P - P_n^\alpha\|_\infty \leq \sum_{j=1}^{s} |\beta_j| \, \|f_j - f_{j,n}^\alpha\|_\infty. \tag{35}$$

Using Theorem 21, we obtain that for $j = 1, 2, \ldots, s$, the sequence $\{f_{j,n}^\alpha\}_{n=1}^\infty$ converges to f_j. Hence for a given $\epsilon > 0$, there exists $n_j \in \mathbb{N}$ such that

$$\|f_j - f_{j,n}^\alpha\|_\infty < \frac{\epsilon}{s|\beta_j|2} \forall n \geq n_j. \tag{36}$$

Using (36) in (35), we get

$$\|P - P_n^\alpha\|_\infty < \frac{\epsilon}{2} \forall n \geq \max\{n_j : j = 1, 2, \ldots, s\}. \tag{37}$$

Next, using (34) and (37) with the triangular inequality

$$\|f - P_n^\alpha\|_\infty \leq \|f - P\|_\infty + \|P - P_n^\alpha\|_\infty$$

we get $\|f - P_n^\alpha\|_\infty < \epsilon$ for all $n \geq \max\{n_j : j = 1, 2, \ldots, s\}$, and it gives the desired result.

We know that a trigonometric polynomial of degree (or order) k would be in the form

$$\psi = \sum_{j=1}^{k} \beta_j \cos(\frac{j\pi x}{l}) + \gamma_j \sin(\frac{j\pi x}{l}), \beta_j, \gamma_j \in \mathbb{R}.$$

Then,

$$\mathcal{F}_n^\alpha(\psi) = \psi_n^\alpha = \sum_{j=1}^{k} \beta_j \cos_n^\alpha(\frac{j\pi x}{l}) + \gamma_j \sin_n^\alpha(\frac{j\pi x}{l}).$$

is called the Bernstein α-fractal trigonometric polynomial of degree k. The next theorem reveals that the set of all Bernstein α-fractal trigonometric polynomials is dense in $\mathcal{C}(2l)$.

Theorem 42 *Let $\Delta = \{x_1, x_2, \ldots, x_N\}$ be a partition of $[-l, l]$ satisfying $x_1 = -l < x_2 < \cdots < x_N = l$. For every $f \in \mathcal{C}(2l)$ and every scaling vector α, there exists a sequence of Bernstein α-fractal trigonometric polynomials converging to f.*

Proof From [23], we get that $\mathbb{T} = \{1, \cos(\frac{j\pi x}{l}), \sin(\frac{j\pi x}{l}) : j \in \mathbb{N}\}$ is dense in $\mathcal{C}(2l)$. By using the previous theorem, we obtain that $\mathbb{T}_n^\alpha = \{1, \cos_n^\alpha(\frac{j\pi x}{l}), \sin_n^\alpha(\frac{j\pi x}{l}) : j \in \mathbb{N}\}$ is dense in $\mathcal{C}(2l)$ for all $n \in \mathbb{N}$ and hence there exists a sequence of Bernstein α-fractal trigonometric polynomials that converge to $f \in \mathcal{C}(2l)$.

Theorem 43 *Let $f \in \mathcal{C}(2l)$. Let $\Delta = \{x_1, x_2, \ldots, x_N\}$ be a partition of $[-l, l]$ satisfying $x_1 = -l < x_2 < \cdots < x_N = l$. Then, for every scaling vector $\alpha(x) = (\alpha_1(x), \alpha_2(x), \ldots (x), \alpha_{N-1}(x))$, there exists a sequence $\{G_n^\alpha\}_{n=1}^\infty$ of Bernstein fractal trigonometric polynomials that converge uniformly to f such that $G_n^\alpha(x) \geq f(x)$ for all $x \in [-l, l]$ and $n \geq N_0 \in \mathbb{N}$.*

Proof For given $\epsilon > 0$, $f \in \mathcal{C}(I)$, and scaling vector $\alpha(x)$, using Theorem 42, there exist a sequence $\{g_n^\alpha\}_{n=1}^\infty$ of Bernstein α-fractal trigonometric polynomials such that

$$\|g_n^\alpha - f\|_\infty < \frac{\epsilon}{2} \; \forall n \geq N_0 \in \mathbb{N}.$$

Define

$$G_n^\alpha(x) = g_n^\alpha(x) + \frac{\epsilon}{2} \; \forall x \in I, \forall n \in \mathbb{N}.$$

Next,

$$G_n^\alpha(x) = g_n^\alpha(x) + \frac{\epsilon}{2} = f(x) + g_n^\alpha(x) + \frac{\epsilon}{2} - f(x) \geq f(x) + \frac{\epsilon}{2} - \|g_n^\alpha - f\|_\infty \geq f(x)\ \forall x \in I,$$

and

$$\|f - G_n^\alpha\|_\infty \leq \|f - g_n^\alpha\|_\infty + \|g_n^\alpha - G_n^\alpha\|_\infty < \epsilon.$$

It establishes the desired result.

The next theorem is an immediate consequence of the previous theorem.

Theorem 44 (Positivity preserving Bernstein α-fractal approximation) *Let $f \in \mathcal{C}(2l)$ be such that $f(x) \geq 0$ for all $x \in I$. Let $\Delta = \{x_1, x_2, \ldots, x_N\}$ be a partition of $[-l, l]$ satisfying $x_1 = -l < x_2 < \cdots < x_N = l$. Then, for every scaling vector $\alpha(x) = (\alpha_1(x), \alpha_2(x), \ldots (x), \alpha_{N-1}(x))$, there exists a sequence $\{G_n^\alpha\}_{n=1}^\infty$ of non-negative Bernstein fractal trigonometric polynomials that converge uniformly to f.*

Theorem 45 *Let $f \in \mathcal{C}(2l)$. Let $\Delta = \{x_1, x_2, \ldots, x_N\}$ be a partition of $[-l, l]$ satisfying $x_1 = -l < x_2 < \cdots < x_N = l$. Then, for every scaling vector $\alpha(x) = (\alpha_1(x), \alpha_2(x), \alpha_{N-1}(x))$, there exists a sequence $\{H_n^\alpha\}_{n=1}^\infty$ of Bernstein fractal trigonometric polynomials that converge uniformly to f such that $H_n^\alpha(x) \leq f(x)$ for all $x \in [-l, l]$ and $n \geq N_0 \in \mathbb{N}$.*

Proof For given $\epsilon > 0$, $f \in \mathcal{C}(I)$, and scaling vector $\alpha(x)$, Theorem 42 ensures the existence of a sequence $\{h_n^\alpha\}_{n=1}^\infty$ of Bernstein α-fractal functions of f such that

$$\|h_n^\alpha - f\|_\infty < \frac{\epsilon}{2}\ \forall n \geq N_0 \in \mathbb{N}.$$

Define

$$H_n^\alpha(x) = h_n^\alpha(x) - \frac{\epsilon}{2}\ \forall x \in I, \forall n \in \mathbb{N}.$$

Next,

$$H_n^\alpha(x) = h_n^\alpha(x) - \frac{\epsilon}{2} = f(x) + h_n^\alpha(x) - \frac{\epsilon}{2} - f(x) \leq f(x) - \frac{\epsilon}{2} + \|h_n^\alpha - f\|_\infty \leq f(x)\ \forall x \in I,$$

and

$$\|f - H_n^\alpha\|_\infty \leq \|f - h_n^\alpha\|_\infty + \|h_n^\alpha - H_n^\alpha\|_\infty < \epsilon.$$

This completes the proof.

Theorem 46 *Let $\{f_m(x)\}_{m=1}^\infty$ be a sequence of functions in $\mathcal{C}(2l)$ that converge uniformly to a function f. Let $\Delta = \{x_1, x_2, \ldots, x_N\}$ be a partition of $[-l, l]$ satisfying $-l = x_1 < x_2 < \cdots < x_N = l$. Then, there exist a double sequence $\left\{\{W_{m,n}^\alpha(x)\}_{n=1}^\infty\right\}_{m=1}^\infty$ of Bernstein fractal trigonometric polynomials that converge uniformly to f.*

Proof Since $\{f_m(x)\}_{m=1}^{\infty}$ be a sequence of trigonometric functions in $C(2l)$ that converge uniformly to f, for a given ϵ there exists $N_1 \in \mathbb{N}$ such that

$$\|f_m - f\|_\infty < \frac{\epsilon}{2} \; \forall m \geq N_1. \tag{38}$$

From Theorem 42, for each $m \in \mathbb{N}$, there exists a sequence $\{W^{\alpha}_{m,n}\}_{n=1}^{\infty}$ of Bernstein α-fractal trigonometric polynomials that converge uniformly to f_m. Therefore, there exists $N_2 \in \mathbb{N}$ such that

$$\|W^{\alpha}_{m,n} - f_m\|_\infty < \frac{\epsilon}{2} \; \forall n \geq N_2. \tag{39}$$

Using (38) and (39) with the triangular inequality

$$\|W^{\alpha}_{m,n} - f\|_\infty \leq \|W^{\alpha}_{m,n} - f_m\|_\infty + \|f_m - f\|_\infty,$$

for given $\epsilon > 0$, we get

$$\|W^{\alpha}_{m,n} - f\|_\infty < \epsilon \; \forall m, n \geq N = \max\{N_1, N_2\}. \tag{40}$$

The above inequality provides the desired double sequence. Thus, we complete the proof.

The following theorem can be proved using Theorem 46 and the arguments that are similar to those used in Theorem 43. Hence, it is omitted.

Theorem 47 *Let $\{f_m(x)\}_{m=1}^{\infty}$ be a sequence of functions in $C(2l)$ that converges uniformly to a function f. Let $\Delta = \{x_1, x_2, \ldots, x_N\}$ be a partition of $[-l, l]$ satisfying $-l = x_1 < x_2 < \cdots < x_N = l$. Then, for every scaling vector $\alpha(x) = (\alpha_1(x), \alpha_2(x), \ldots, \alpha_{N-1}(x))$, there exists a double sequence $\{\{U^{\alpha}_{m,n}\}_{m=1}^{\infty}\}_{n=1}^{\infty}$ of Bernstein fractal trigonometric polynomials that converge uniformly to f such that $U^{\alpha}_{m,n}(x) \geq f(x)$ for all $x \in [-l, l]$ and $n \geq \max\{N_1, N_2\}$.*

The following theorem can be proved using Theorem 46 and the arguments that are similar to those used in Theorem 45. Hence, it is omitted.

Theorem 48 *Let $\{f_m(x)\}_{m=1}^{\infty}$ be a sequence of functions in $C(2l)$ that converge uniformly to a function f. Let $\Delta = \{x_1, x_2, \ldots, x_N\}$ be a partition of $[-l, l]$ satisfying $-l = x_1 < x_2 < \cdots < x_N = l$. Then, for every scaling vector function $\alpha(x) = (\alpha_1(x), \alpha_2(x), \ldots (x), \alpha_{N-1}(x))$, there exists a double sequence $\{\{V^{\alpha}_{m,n}\}_{m=1}^{\infty}\}_{n=1}^{\infty}$ of Bernstein fractal trigonometric polynomials that converge uniformly to f such that $V^{\alpha}_{m,n}(x) \leq f(x)$ for all $x \in [-l, l]$ and $n \geq \max\{N_1, N_2\}$.*

References

1. Barnsley, M.F.: Fractal functions and interpolation. Constr. Approx. **2**, 303–329 (1986)
2. Hutchinson, J.E.: Fractals and self similarity. Indiana Univ. J. Math. **30**, 713–747 (1981)
3. Chand, A.K.B., Kapoor, G.P.: Generalized cubic spline fractal interpolation functions. SIAM J. Num. Anal. **44**(2), 655–676 (2006)
4. Chand, A.K.B., Viswanathan, P.: A constructive approach to cubic Hermite fractal interpolation function and its constrained aspects. BIT Numer. Math. **53**(4), 841–865 (2013)
5. Chand, A.K.B., Vijender, N., Navascués, M.A.: Shape preservation of scientific data through rational fractal splines. Calcolo **51**, 329–362 (2013)
6. Chand, A.K.B., Navascués, M.A., Viswanathan, P., Katiyar, S.K.: Fractal trigonometric polynomials for restricted range approximation. Fractals **24**, 1650022 (2016)
7. Chand, A.K.B., Tyda, K.R.: Partially blended constrained rational cubic trigonometric fractal interpolation surfaces. Fractals **24**, 1650027 (2016)
8. Navascués, M.A.: Fractal polynomial interpolation. Z. Anal. Anwend. **24**(2), 401–418 (2005)
9. Navascués, M.A.: Non-smoothpolynomials. Int. J. Math. Anal. **1**, 159–174 (2007)
10. Navascués, M.A.: Fractal approximation. Complex Anal. Oper. Theory **4**, 953–974 (2010)
11. Navascués, M.A.: Fractal bases of L_p spaces. Fractals **20**, 141–148 (2012)
12. Navascués, M.A., Chand, A.K.B.: Fundamental sets of fractal functions. Acta Appl. Math. **100**, 247–261 (2008)
13. Navascués, M.A., Sebastián, M.V.: Smooth fractal interpolation. J. Inequal. Appl. Article ID 78734 (2006)
14. Vijender, N.: Bernstein fractal trigonometric approximation. Acta Appl. Math. **159**(1) (2018)
15. Vijender, N.: Bernstein fractal rational approximants with no condition on scaling vectors. Fractals **26**(4) (2018)
16. Vijender, N.: Fractal perturbation of shaped functions: convergence independent of scaling. Mediterr. J. Math. **15**(211) (2018)
17. Vijender, N.: Bernstein fractal approximation and fractal full Müntz theorems. Electron. Trans. Numer. Anal. (ETNA) **51**, 1–24 (2019)
18. MD. Nasim Akhtar, M. Guru Prem Prasad, and M. A. Navascués, Box dimensions of α-fractal functions, Fractals 24(3), (2016)
19. Gal, S.G.: Shape Preserving Approximation by Real and Complex Polynomials. Birkhäuser, Boston (2008)
20. Nasim Akhtar, M.D., Guru Prem Prasad, M., Navascués, M.A.: Box dimension of α-fractal function with variable scaling factors in subintervals. Chaos Solitons Fractals **103**, 440–449 (2016)
21. Chen, S.: The non-differentiability of a class of fractal interpolation functions. Acta Math. Sci. **19**(4), 425–430 (1999)
22. Cheney, E.W.: Introduction to Approximation Theory. AMS Chelsea Publishing, Providence, RI (1982)
23. Isaacson, E., Keller, H.B.: Analysis of Numerical Methods. Wiley, New York (1996)

Max-Product Type Exponential Neural Network Operators

Shivam Bajpeyi and A. Sathish Kumar

Abstract In the present article, we extend the theory of exponential sampling type neural network operators to the max-product setting. The approximation properties of these operators activated by the sigmoidal functions have been studied by using the moment type approach. We establish the point-wise and uniform approximation theorem for these operators along with the quantitative estimate of the order of convergence using the modulus of continuity. Consequently, we discuss the convergence of exponential sampling type quasi-interpolation operators of the max-product kind. At the end, we provide a few examples of the sigmoidal functions satisfying the assumptions of the presented theory.

Keywords Exponential sampling operators · Max-product operators · Neural networks · Quasi interpolation operators

Mathematics Subject Classification (2010) 41A35 · 47A58 · 94A20 · 41A25

1 Introduction

The theory of neural network (NN) operators has been introduced to provide a constructive approach for approximating functions by a neural process. The mathematical formulation of these NN operators can be defined as

$$N_p(x) = \sum_{j=0}^{p} c_j \sigma(\langle \alpha_j, x \rangle + \beta_j), \quad x \in \mathbb{R}^n \text{ and } n \in \mathbb{N}.$$

S. Bajpeyi · A. S. Kumar (✉)
Department of Mathematics, Visvesvaraya National Institute of Technology,
Nagpur 440010, India
e-mail: mathsatish9@gmail.com

S. Bajpeyi
e-mail: shivambajpai1010@gmail.com

R. N. Mohapatra et al. (eds.), *Mathematical Analysis and Computing*,
Springer Proceedings in Mathematics & Statistics 344,
https://doi.org/10.1007/978-981-33-4646-8_44

Here, $\beta_j \in \mathbb{R}$ are threshold values and $\alpha_j \in \mathbb{R}^n$ are the weight functions for $0 \leq j \leq p$. Moreover, $\langle \alpha_j.x \rangle$ denotes the usual inner product of α_j and x, whereas σ be the activation function of the network. Anastassiou [1] initiated the study of approximating continuous functions on compact subset of $\mathbb{R}$ by these NN operators. This theory was motivated by the results obtained by Cardalignet and Euvrard in [20]. Costarelli and Spigler extended the results proved by Anastassiou in [27, 29]. Many researchers studied the NN-operators in various aspects, see, e.g., [5, 28, 30–32, 34, 36, 37].

Recently, a constructive approach has been discussed to approximate the class of functions having sample values that are exponentially spaced by exploiting the theory of NN operators in [7]. The exponential sampling is a useful technique to handle the problems arising in the wide areas of mathematics as well as engineering, see, e.g., [15, 21, 38, 40]. Butzer and Jansche [19] pioneered the mathematical study of the theory of exponential sampling using the tools of Mellin analysis. Several authors have contributed to the advancement of the Mellin theory, see e.g., [11, 12, 18] and the references therein. We also refer some of the recent developments related to the theory of exponential sampling [6, 8, 9, 13, 14].

The max-product NN operators represent neuro-processing models in which the global behavior of the network is mainly determined by one of the artificial neurons of the network. The main advantage of studying the max-product operators is that it provides a better order of convergence when non-negative continuous functions are approximated as compared to the corresponding linear ones. Coroianu and Gal pioneered the work related to the max-product type operators in [22–25]. But the first instance of extension of the theory of NN operators to the max-product setting can be found in [2]. Costarelli and Vinti [33] generalized the results obtained in [2] by avoiding the assumption of compact support on the density functions and also improved the rate of convergence. Gal [39] established the convergence of the max-product Bernstein operators by using the possibility theory, which is a mathematical theory dealing with certain types of uncertainties and is considered as an alternative to probability theory. Bede et.al [16] studied the max-product version of the Bernstein polynomials and to other linear Bernstein-type operators, like those of Favard-Szász-Mirakjan operators, Baskakov operators, etc. Recently, Costarelli et al. [26] studied the approximation properties of max-product analogues of the generalized sampling series introduced by Butzer [17]. This paper investigates the approximation behavior of the family of exponential sampling type neural network operators of the Max-product kind.

2 Preliminaries

Let $I = [a, b]$ be any compact subset of $\mathbb{R}^+$, where $\mathbb{R}^+$ denotes the set of all positive real numbers. We denote $C(I)$ the space of all uniformly continuous functions defined on the set I, equipped with the usual sup-norm $\|f\|_\infty = \sup_{x \in I} |f(x)|$. Further, $C_+(I)$ denotes the subspace of $C(I)$ which contains non-negative functions defined on $\mathbb{R}^+$. A function $f : \mathbb{R}^+ \to \mathbb{R}$ is called log-uniformly continuous if for any given $\epsilon > 0$, there exists $\delta > 0$ such that $|f(x) - f(y)| < \epsilon$ whenever $|\log x - \log y| \leq \delta$, for any $x, y \in \mathbb{R}^+$. We denote the space of all log-uniformly continuous functions defined on I by $\mathcal{C}(I)$ and analogously $\mathcal{C}_+(I)$ represents the space of all non-negative log-uniformly continuous functions. In general, a log-uniformly continuous function may not be uniformly continuous and vice-versa. But, there is an equivalence between these two notions on any compact subset of $\mathbb{R}^+$. For any index set $\mathcal{I} \subseteq \mathbb{Z}$, we have

$$\bigvee_{k \in \mathbb{Z}} J_k := \sup\{J_k : k \in \mathcal{I}\}.$$

Note that if the set $\mathcal{I}$ is finite then the notion of supremum can be replaced by the maximum.

In what follows, a function $\sigma : \mathbb{R} \to \mathbb{R}$ is sigmoidal function if and only if $\lim_{u \to -\infty} \sigma(u) = 0$ and $\lim_{u \to \infty} \sigma(u) = 1$. In addition, we assume that the sigmoidal function satisfies the following conditions:

(a) $\sigma \in C^{(2)}(\mathbb{R})$ and concave on $\mathbb{R}$.
(b) $\sigma(u) = \mathcal{O}(|u|^{-1-\nu})$, as $u \to -\infty$, for some $\nu > 0$.
(c) The function $(\sigma(u) - \frac{1}{2})$ is an odd function.

Now we define the density function activated by the sigmoidal function σ using the linear combination of the sigmoidal functions as follows:

$$\chi_\sigma(u) := \frac{1}{2}\left[\sigma(\log u + 1) - \sigma(\log u - 1)\right], \quad u \in \mathbb{R}^+.$$

Some important properties and results related to the above density function χ_σ can be found in [7]. We consider the density function χ_σ as the kernel function which satisfies the following assumptions:

(i) $\hat{M}_\nu(\chi_\sigma) < +\infty$ for some $\nu > 0$ and $\forall\, u \in \mathbb{R}^+$.
(ii) For every $u \in I$, $\bigvee_{k=\lceil na \rceil}^{\lfloor nb \rfloor} \chi_\sigma(e^{-k}u) \geq \beta(u)$, *where* β *is a constant depending on* u.

Next, we introduce the notion of *generalized absolute moment* of the density function χ_σ. The *generalized absolute moment* of order ν can be defined as

$$\hat{M}_\nu(\chi_\sigma) := \sup_{u \in \mathbb{R}^+} \bigvee_{k \in \mathbb{Z}} |\chi_\sigma(e^{-k}u)|\, |k - \log u|^\nu.$$

Remark 1 It can be observed that $M_\nu(\chi_\sigma) \geq \hat{M}_\nu(\chi_\sigma)$, where

$$M_\nu(\chi_\sigma) := \sup_{u \in \mathbb{R}^+} \sum_{k \in \mathbb{Z}} |\chi_\sigma(e^{-k}u)|\, |k - \log u|^\nu.$$

This concludes that the generalized absolute moments provide the sharper estimates as compared to the absolute moments in general.

Now we have the following Lemma.

Lemma 1 *Let $\chi_\sigma : \mathbb{R}^+ \to \mathbb{R}$ be bounded such that $\hat{M}_\nu(\chi_\sigma) < +\infty$, for some $\nu > 0$ then $\hat{M}_\eta(\chi_\sigma) < +\infty$, for every $0 \le \eta \le \nu$.*

Proof For any fixed $\eta > 0$ such that $\eta \le \nu$ and for every $u \in \mathbb{R}^+$, we can write

$$\begin{aligned}\bigvee_{k\in\mathbb{Z}} |\chi_\sigma(e^{-k}u)||k-\log u|^\eta \le & \bigvee_{|k-\log u|<\delta} |\chi_\sigma(e^{-k}u)||k-\log u|^\eta \\ & + \bigvee_{|k-\log u|\ge\delta} |\chi_\sigma(e^{-k}u)||k-\log u|^\eta \\ & \le \delta\|\chi_\sigma\|_\infty + \bigvee_{|k-\log u|\ge\delta^\eta} |\chi_\sigma(e^{-k}u)|\frac{|k-\log u|^\nu}{|k-\log u|^{\nu-\eta}} \\ & \le \|\chi_\sigma\|_\infty + \hat{M}_\nu(\chi_\sigma) \ < +\infty.\end{aligned}$$

Lemma 2 *For every $\delta > 0$ and $u \in \mathbb{R}^+$, the following condition holds*

$$\lim_{n\to\infty} \bigvee_{|k-\log u|>n\delta} \chi_\sigma(e^{-k}u) = 0.$$

Proof The proof can be established by using the similar arguments as in the proof of Lemma 3 of [7].

Now we present the approximation results related to the family of operators $(P_n^{\chi_\sigma}(f,.))_{n>0}$ for $f \in C_+(I)$.

3 Main Results

Let $f : I \to \mathbb{R}^+$ be any bounded function and $n \in \mathbb{N}$ such that $\lceil na \rceil \le \lfloor nb \rfloor$. Then, the max-product type exponential NN operators are defined as

$$P_n^{\chi_\sigma}(f,x) := \frac{\bigvee_{k=\lceil na\rceil}^{\lfloor nb\rfloor} \chi_\sigma(e^{-k}x^n)\, f\left(e^{\frac{k}{n}}\right)}{\bigvee_{k=\lceil na\rceil}^{\lfloor nb\rfloor} \chi_\sigma(e^{-k}x^n)}, \qquad x \in \mathbb{R}^+. \tag{1}$$

Here the symbols $\lceil . \rceil$ and $\lfloor . \rfloor$ represent the *ceiling function* and the *floor function* respectively. The above operators are well defined in the case of bounded functions on I.

The following Lemma is useful in order to derive the convergence result for the family of operators $(P_n^{\chi_\sigma}(f,.))_{n>0}$.

Lemma 3 *Assume that $f_1, f_2 : I \to \mathbb{R}^+$ be the bounded functions. Then we have the following properties:*

1. *If $f_1(x) \leq f_2(x)$, $\forall x \in \mathbb{R}^+$, we have $(P_n^{\chi_\sigma} f_1)(x) \leq (P_n^{\chi_\sigma} f_2)(x)$, $\forall x \in \mathbb{R}^+$.*
2. *For every $x \in \mathbb{R}^+$, we have $|P_n^{\chi_\sigma}(f_1, x) - P_n^{\chi_\sigma}(f_2, x)| \leq P_n^{\chi_\sigma}(|f_1 - f_2|, x)$.*
3. *The operators $(P_n^{\chi_\sigma} f_i)$ are sub-linear operators for $i = 1, 2$.*

Proof Since f_1, f_2 are non-negative, the properties 1 and 3 can easily be observed from the definition of the operator $(P_n^{\chi_\sigma}(f,.))_{n>0}$. For 2, we write $|f_1| < |f_1 - f_2| + |f_2|$ and similarly $|f_2| < |f_2 - f_1| + |f_1|$, for $x \in I$. Using these inequalities, we obtain

$$P_n^{\chi_\sigma}(f_1, x) \leq P_n^{\chi_\sigma}(|f_1 - f_2|, x) + P_n^{\chi_\sigma}(f_2, x) \quad \text{and}$$

$$P_n^{\chi_\sigma}(f_2, x) \leq P_n^{\chi_\sigma}(|f_2 - f_1|, x) + P_n^{\chi_\sigma}(f_1, x), \quad \text{for } x \in I.$$

On exploiting the above inequalities, we get the desired result.

Now we present the point-wise and uniform convergence theorem for these operators.

Theorem 1 *Let $f : I \to \mathbb{R}^+$ be a bounded function. Then*

$$\lim_{n\to\infty} P_n^{\chi_\sigma}(f, x) = f(x)$$

at the point of log-continuity $x \in \mathbb{R}^+$. Moreover, if $f \in C_+(I)$, we have

$$\lim_{n\to\infty} \|P_n^{\chi_\sigma} f - f\|_\infty = 0.$$

Proof We see that

$$\begin{aligned}|P_n^{\chi_\sigma}(f, x) - f(x)| &\leq |P_n^{\chi_\sigma}(f, x) - f(x)P_n^{\chi_\sigma}(1, x)| + |f(x)P_n^{\chi_\sigma}(1, x) - f(x)| \\ &\leq |P_n^{\chi_\sigma}(f, x) - f(x)P_n^{\chi_\sigma}(1, x)| + |f(x)\,(P_n^{\chi_\sigma}(1, x) - 1)|.\end{aligned}$$

For the sake of convenience, we put $f(x) := f_x$. Now, using the property 2 of Lemma 3, we obtain

$$|P_n^{\chi_\sigma}(f, x) - f(x)| \leq P_n^{\chi_\sigma}(|f - f_x|, x).$$

This gives

$$P_n^{\chi_\sigma}(|f - f_x|, x)$$

$$\leq \frac{\bigvee_{k=\lceil na \rceil}^{\lfloor nb \rfloor} |\chi_\sigma(e^{-k}x^n)| \; |f(e^{\frac{k}{n}}) - f(x)|}{\bigvee_{k=\lceil na \rceil}^{\lfloor nb \rfloor} |\chi_\sigma(e^{-k}x^n)|}$$

$$\leq \frac{1}{\bigvee_{k=\lceil na \rceil}^{\lfloor nb \rfloor} |\chi_\sigma(e^{-k}x^n)|} \max\Bigg\{ \bigvee_{|\frac{k}{n}-\log(x)|\leq\delta} |\chi_\sigma(e^{-k}x^n)| \, |f(e^{\frac{k}{n}}) - f(x)|,$$

$$\bigvee_{|\frac{k}{n}-\log(x)|>\delta} |\chi_\sigma(e^{-k}x^n)| \, |f(e^{\frac{k}{n}}) - f(x)| \Bigg\}$$

$$:= \frac{1}{\beta(x)} \max\{J_1, J_2\}, \text{ where } \beta(x) \leq \bigvee_{k=\lceil na \rceil}^{\lfloor nb \rfloor} \chi_\sigma(e^{-k}x^n).$$

Since $f \in C_+(I)$, we have $|f(e^{\frac{k}{n}}) - f(x)| < \epsilon$ whenever $|\frac{k}{n} - \log(x)| \leq \delta$. So, we have $|J_1| < \epsilon \|\chi_\sigma\|_\infty$. Now from Lemma 2, we have

$$|I_2| \leq 2 \, \|f\|_\infty \epsilon.$$

From the estimates of J_1 and J_2, we conclude that

$$|P_n^{\chi_\sigma}(f, x) - f(x)| \leq \frac{\epsilon}{\beta(x)} (\|\chi_\sigma\|_\infty + 2\|f\|_\infty).$$

As $\epsilon > 0$ is arbitrary, we obtain the desired result.

3.1 *Quantitative Estimate*

We estimate the order of convergence for the operators $(P_n^{\chi_\sigma}(f, .))_{n>0}$ in terms of *logarithmic modulus of continuity*. For $f \in C(\mathbb{R}^+)$ and $x, y \in \mathbb{R}^+$, the notion of logarithmic modulus of continuity is defined as

$$\omega(f, \delta) := \sup\{|f(x) - f(y)|; \text{ whenever } |\log(x) - \log(y)| \leq \delta, \;\; \delta \in \mathbb{R}^+\}.$$

The properties of logarithmic modulus of continuity are similar to the properties of usual modulus of continuity (see [6, 10]).

Theorem 2 *Let χ_σ be the kernel and $f \in C_+(I)$. Then we have the following estimate:*

$$|P_n^{\chi_\sigma}(f,x) - f(x)| \leq \frac{1}{\beta(x)}\, \omega\left(f, \frac{1}{n}\right) (\hat{M}_0(\chi_\sigma) + \hat{M}_1(\chi_\sigma)).$$

Proof In view of Theorem 1, we have

$$\begin{aligned}|P_n^{\chi_\sigma}(f,x) - f(x)| &\leq \frac{1}{\beta(x)} \bigvee_{k=\lceil na \rceil}^{\lfloor nb \rfloor} |\chi_\sigma(e^{-k}x^n)|\ |f(e^{\frac{k}{n}}) - f(x)| \\ &\leq \frac{1}{\beta(x)} \bigvee_{k=\lceil na \rceil}^{\lfloor nb \rfloor} |\chi_\sigma(e^{-k}x^n)|\, \omega\left(f, |\frac{k}{n} - \log x|\right).\end{aligned}$$

Using the property $\omega(f, \lambda\delta) \leq (\lambda + 1)\, \omega(f, \delta)$, we obtain

$$|P_n^{\chi_\sigma}(f,x) - f(x)| \leq \frac{\omega(f, \frac{1}{n})}{\beta(x)} \bigvee_{k=\lceil na \rceil}^{\lfloor nb \rfloor} |\chi_\sigma(e^{-k}x^n)|\ (|k - n\log x| + 1).$$

This gives the required estimate.

4 Max-Product Type Quasi-interpolation Operators

The main aim of this section is to extend the aforementioned theory for the functions defined on whole $\mathbb{R}^+$. Let $f : \mathbb{R}^+ \to \mathbb{R}^+$ be any bounded function and χ_σ be the kernel function. Then, we define the class of quasi-interpolation operators of the family $(P_n^{\chi_\sigma}(f,.))_{n>0}$ by

$$T_n^{\chi_\sigma}(f,x) := \frac{\bigvee_{k=-\infty}^{\infty} f(e^{\frac{k}{n}})\chi_\sigma(e^{-k}x^n)}{\bigvee_{k=-\infty}^{\infty} \chi_\sigma(e^{-k}x^n)}, \qquad k \in \mathbb{Z}.$$

Remark 2 The properties derived in Lemma 3 for the operator $P_n^{\chi_\sigma}(f,.)$ can be extended for the family of operators $(T_n^{\chi_\sigma} f)_{w>0}$ using the similar arguments as in the proof of Lemma 3.

Now we state the convergence results for the aforementioned family of operators.

Theorem 3 *Let $f : \mathbb{R}^+ \to \mathbb{R}^+$ be a bounded function. Then*

$$\lim_{n\to\infty} T_n^{\chi_\sigma}(f, x) = f(x)$$

at the point of log-continuity $x \in \mathbb{R}^+$. Moreover, for $f \in \mathcal{C}_+(\mathbb{R}^+)$ we have

$$\lim_{n\to\infty} \|T_n^{\chi_\sigma} f - f\|_\infty = 0.$$

Theorem 4 *Let χ_σ be the kernel and $f \in \mathcal{C}_+(\mathbb{R}^+)$. Then we have*

$$|T_n^{\chi_\sigma}(f, x) - f(x)| \leq \frac{1}{\beta(x)}\, \omega\left(f, \frac{1}{n}\right) (\hat{M}_0(\chi_\sigma) + \hat{M}_1(\chi_\sigma)).$$

Remark 3 The proofs of Theorems 3 and 4 can be established by following along the lines of proof of Theorems 1 and 2.

5 Examples of Sigmoidal Function

In this section, we discuss a few examples of the sigmoidal functions $\sigma(x)$ satisfying the assumptions of the presented theory. We start with a well-known example of the sigmoidal function, namely *logistic function* defined as

$$\sigma_l(x) = \frac{1}{1 + e^{-x}}, \quad x \in \mathbb{R}.$$

The corresponding density function χ_{σ_l} can be written as follows:

$$\begin{aligned}\chi_{\sigma_l}(x) &= \frac{1}{2}\big[\sigma_l(\log x + 1) - \sigma_l(\log x + 1))\big] \\ &= \frac{(e^2 - 1)x}{2(ex + 1)(x + e)}.\end{aligned}$$

Indeed, the logistic function satisfies all the required conditions (a)–(c) (see [27]).

The next example of an smooth sigmoidal function is *hyperbolic tangent* sigmoidal function [3], defined as

$$\sigma_h(x) := \frac{1}{2}(\tanh\, x + 1), \quad x \in \mathbb{R}.$$

We can obtain the corresponding density function for the hyperbolic tangent function using the definition of $\chi_\sigma(x)$.

Further, the example of sigmoidal function can also be constructed using the well-known *B-spline functions* [27] of order n by defining as follows:

$$\sigma_{M_n}(x) := \int_{-\infty}^{x} M_n(t)\, dt, \quad x \in \mathbb{R}.$$

Indeed, for every $x \in \mathbb{R}^+$, the sigmoidal function $\sigma_{M_n}(x)$ is non-decreasing on $\mathbb{R}^+$ and $0 \leq \sigma_{M_n}(x) \leq 1$, $\forall n \in \mathbb{N}$. One can see that

$$\sigma_{M_1}(x) = \begin{cases} 0, & \text{if, } x < -\frac{1}{2} \\ x + \frac{1}{2} & \text{if, } -\frac{1}{2} \leq x \leq \frac{1}{2} \\ 1, & \text{if, } x > \frac{1}{2} \end{cases}$$

The above function $\sigma_{M_1}(x)$ is an example of a non-smooth sigmoidal function. This function is also known as *ramp-function*, see [3, 4, 35].

6 Final Remarks and Conclusion

We analyzed the max-product version of the family of operators $(E_n^{\chi_\sigma}(f, .))_{n>0}$ introduced in [7] to approximate the functions having exponentially spaced data by exploiting NN operators. The main approximation results related to the proposed family of operators have been derived for $f \in C_+([a, b])$ in Sect. 3. It can be observed that the the order of approximation obtained by both the family of operators $(P_n^{\chi_\sigma}(f, .))_{n>0}$ and $(E_n^{\chi_\sigma}(f, .))_{n>0}$ is the same, but the max-product version produces the sharper estimates as compared to the classical one. In Sect. 4, we introduced associated quasi-interpolation operators to furnish an useful tool to approximate non-negative functions defined on $\mathbb{R}^+$, unlike the operators $(P_n^{\chi_\sigma} f)_{n>0}$ which are restricted to the compact subsets of $\mathbb{R}^+$. Moreover, it is important to note that the denominator in the definition of operators $T_n^{\chi_\sigma}(f, .)_{n>0}$ cannot be removed unlike in the definition of exponential NN quasi-interpolation operators in [7] since $\bigvee_{k=-\infty}^{\infty} \chi_\sigma(e^{-k}x^n) \neq 1$ in general, or in other words, the sup-operator $\bigvee$ is not a *generalized approximation identity*. Finally, in Sect. 5, few examples of the sigmoidal functions have been discussed which satisfy the assumptions of our theory.

Acknowledgements The authors are thankful to the reviewers for making valuable suggestions leading to a better presentation of the article. The first author is thankful to the "Ministry of Human Resource and Development", India for financial support to carry out his research work. The second author is supported by DST, SERB, India, Project file no: EEQ/2017/000201.

References

1. Anastassiou, G.A.: Rate of convergence of some neural network operators to the unit-univariate case. J. Math. Anal. Appl. **212**(1), 237–262 (1997)
2. Anastassiou, G.A., Coroianu, L., Gal, S.G.: Approximation by a nonlinear Cardaliaguet-Euvrard neural network operator of max-product kind. J. Comput. Anal. Appl. **12**, 396–406 (2010)
3. Anastassiou, G.A.: Univariate hyperbolic tangent neural network approximation. Math. Comput. Model. **53**, 1111–1132 (2011)
4. Anastassiou, G.A.: Multivariate hyperbolic tangent neural network approximation. Comput. Math. Appl. **61**(4), 809–821 (2011)
5. Anastassiou, G.A.: Multivariate sigmoidal neural network approximation. Neural Netw. **24**, 378–386 (2011)
6. Angamuthu, S.K., Bajpeyi, S.: Direct and inverse results for Kantorovich type exponential sampling series. Results Math **75**, 119 (2020)
7. Bajpeyi, S., Kumar, A.S.: Approximation by exponential sampling type neural network operators. arXiv preprint arXiv:1911.05587 (2019)
8. Balsamo, S., Mantellini, I.: On linear combinations of general exponential sampling series. Res. Math. **74**, 4 (2019)
9. Bardaro, C., Butzer, P.L., Mantellini, I.: The exponential sampling theorem of signal analysis and the reproducing kernel formula in the Mellin transform setting. Sampl. Theory Signal Image Process. **13**(1), 35–66 (2014)
10. Bardaro, C., Mantellini, I.: On Mellin convolution operators: A direct approach to the asymptotic formulae. Integral Transforms Spec. Funct. **25**(3), 182–195 (2014)
11. Bardaro, C., Butzer, P.L., Mantellini, I.: The Mellin-Parseval formula and its interconnections with the exponential sampling theorem of optical physics. Integral Transforms Spec. Funct. **27**(1), 17–29 (2016)
12. Bardaro, C., Butzer, P.L., Mantellini, I., Schmeisser, G.: On the Paley-Wiener theorem in the Mellin transform setting. J. Approx. Theory. **207**, 60–75 (2016)
13. Bardaro, C., Faina, L., Mantellini, I.: A generalization of the exponential sampling series and its approximation properties. Math. Slovaca. **67**(6), 1481–1496 (2017)
14. Bardaro, C., Mantellini, I., Schmeisser, G.: Exponential sampling series: convergence in Mellin-Lebesgue spaces. Results Math **74**(3), (2019)
15. Bertero, M., Pike, E.R.: Exponential-sampling method for Laplace and other dilationally invariant transforms. II. Examples in photon correlation spectroscopy and Fraunhofer diffraction. Inverse Prob. **7**(1), 21–41 (1991)
16. Bede,B., Coroianu, L., Gal, S.G.: Approximation by Max-Product Type Operators. Springer, Cham (2016). xv+458 pp. ISBN: 978-3-319-34188-0; 978-3-319-34189-7
17. Butzer P.L., Stens, R.L.: Linear prediction by samples from the past. In: Marks II, R.J. (ed.) Advanced Topics in Shannon Sampling and Interpolation Theory. Springer Texts Electrical Eng., Springer, New York (1993)
18. Butzer, P.L., Jansche, S.: A direct approach to the Mellin transform. J. Fourier Anal. Appl. **3**(4), 325–376 (1997)
19. Butzer, P.L., Jansche, S.: The exponential sampling theorem of signal analysis, Dedicated to Prof. C. Vinti (Italian) (Perugia, 1996), Atti. Sem. Mat. Fis. Univ. Modena, Suppl. **46**, 99–122(1998)
20. Cardaliaguet, P., Euvrard, G.: Approximation of a function and its derivative with a neural network. Neural Netw. **5**(2), 207–220 (1992)
21. Casasent, D.: Optical Data Processing, pp. 241–282. Springer, Berlin, Heidelberg (1978)
22. Coroianu, L., Gal, S.G.: Approximation by nonlinear generalized sampling operators of max-product kind. Sampl. Theory Signal Image Process. **9**, 59–75 (2010)
23. Coroianu, L., Gal, S.G.: Approximation by max-product sampling operators based on sinc-type kernels. Sampl. Theory Signal Image Process. **10**, 211–230 (2011)

24. Coroianu, L., Gal, S.G.: Saturation results for the truncated max-product sampling operators based on sinc and Fejér-type kernels. Sampl. Theory Signal Image Process. **11**, 113–132 (2012)
25. Coroianu, L., Gal, S.G.: Saturation and inverse results for the Bernstein max-product operator. Period. Math. Hungar. **69**, 126–133 (2014)
26. Coroianu, L., Costarelli, D., Gal, S.G., Vinti, G.: The max-product generalized sampling operators: convergence and quantitative estimates. Appl. Math. Comput. **355**, 173–183 (2019)
27. Costarelli, D., Spigler, R.: Approximation results for neural network operators activated by sigmoidal functions. Neural Netw. **44**, 101–106 (2013)
28. Costarelli, D., Spigler, R.: Constructive approximation by superposition of sigmoidal functions. Anal. Theory Appl. **29**(2), 169–196 (2013)
29. Costarelli, D., Spigler, R.: Multivariate neural network operators with sigmoidal activation functions. Neural Netw. **48**, 72–77 (2013)
30. Costarelli, D., Spigler, R.: Convergence of a family of neural network operators of the Kantorovich type. J. Approx. Theory. **185**, 80–90 (2014)
31. Costarelli, D.: Neural network operators: constructive interpolation of multivariate functions. Neural Netw. **67**, 28–36 (2015)
32. Costarelli, D., Spigler, R., Vinti, G.: A survey on approximation by means of neural network operators. J. Neurotechnol. **1**(1), 29–52 (2016)
33. Costarelli, D., Vinti, G.: Max-product neural network and quasi-interpolation operators activated by sigmoidal functions. J. Approx. Theory. **209**, 1–22 (2016)
34. Costarelli, D., Vinti, G.: Quantitative estimates involving K-functionals for neural network-type operators. Appl. Anal. **98**(15), 2639–2647 (2019)
35. Cheang, Gerald H.L.: Approximation with neural networks activated by ramp sigmoids. J. Approx. Theory. **162**(8), 1450–1465 (2010)
36. Cybenko, G.: Approximation by superpositions of sigmoidal function. Math. Control Signals Syst. **2**, 303–314 (1989)
37. Funahashi, K.I.: On the approximate realization of continuous mappings by neural networks. Neural Netw. **2**, 183–192 (1989)
38. Gori, F.: Sampling in Optics. Advanced Topics in Shannon Sampling and Interpolation Theory, pp. 37–83. Springer Texts Electrical Engrg., Springer, New York (1993)
39. Gal, S.G.: A possibilistic approach of the max-product Bernstein operators. Results Math **65**, 453–462 (2014)
40. Ostrowsky, N., Sornette, D., Parke, P., Pike, E.R.: Exponential sampling method for light scattering polydispersity analysis. Opt. Acta. **28**, 1059–1070 (1994)

Local Weighted Average Sampling Over Multiply Generated Spline Spaces of Polynomial Growth

S. Yugesh and P. Devaraj

Abstract We analyse the following average sampling problem over multiply generated spline spaces of polynomial growth: Let h be a nonnegative integrable function supported in $\left[-\frac{1}{2}, \frac{1}{2}\right]$. Given a sequence of samples $\{y_n\}_{n\in\mathbb{Z}}$, of polynomial growth, find a spline f having polynomial growth such that $\int_{-\frac{1}{2}}^{\frac{1}{2}} f(n-y)h(y)dy = y_n, n \in \mathbb{Z}$. It is shown that this problem has a unique solution for certain subspaces of the multiply generated spline spaces of polynomial growth.

Keywords Multiply generated spline · Multiply generated spline space · Average sampling

1 Introduction and Preliminaries

In signal and image processing, continuous signals need to be represented by their discrete samples. A significant problem in signal processing is how to represent a continuous signal in terms of its discrete samples. One of the important themes of sampling theory is, to recover a continuous function from its discrete sample values. The sampling theorems provide the reconstruction formulas and hence such theorems become the most useful tool in the field of signal and image processing. The famous Shannon sampling theorem states that finite energy bandlimited signals are completely characterized by their sample values [2–5, 10]. Moreover, Shannon gave the following reconstruction formula

S. Yugesh (✉)
Department of Mathematics, Sri Sivasubramaniya Nadar College of Engineering, Kalavakkam, Chennai 603 110, Tamil Nadu, India
e-mail: mathsyugesh@gmail.com; yugeshs@ssn.edu.in

P. Devaraj
School of Mathematics, Indian Institute of Science Education and Research Thiruvananthapuram, Vithura, Thiruvananthapuram 695551, India
e-mail: devarajp@iisertvm.ac.in

R. N. Mohapatra et al. (eds.), *Mathematical Analysis and Computing*, Springer Proceedings in Mathematics & Statistics 344,
https://doi.org/10.1007/978-981-33-4646-8_45

$$f(x) = \sum_{k\in\mathbb{Z}} f\left(\frac{k\pi}{\Omega}\right) \frac{\sin(\Omega x - k\pi)}{\Omega x - k\pi}.$$

In the Shannon reconstruction formula, $sinc$ function is used and is in fact a scaling function of a multiresolution analysis used in wavelet theory, since the bandlimited condition forces the signal to be of infinite duration which is not always realistic. Further, the $sinc$ function itself is not very suitable for rapid communications. Some of these constraints necessitate the need for investigating other function classes for which the sampling theorem holds. Many generalizations of the classical Shannon sampling theorem have been proposed. Moreover, in [2–5, 10, 14], the sampling procedure in shift-invariant spaces as well as spline spaces has been analysed. The requirement is that the signal to be bandlimited can be avoided by considering signals in spaces like the wavelet subspaces, shift-invariant spaces and spline subspaces.

Recently, the sampling and reconstruction technique was investigated for multiply generated shift-invariant spaces and spline subspaces in [1, 6, 9, 11–13]. In these literatures, they consider finite energy signals. In this paper, we consider the space of functions having polynomial growth. In [11, 12], the multiply generated spline space is defined as

$$\mathcal{S} = \left\{ f : f = \sum_{i=1}^{r} \sum_{n\in\mathbb{Z}} a_i(n)\beta_{d_i}(t-n) \right\}$$

with suitable coefficients $a_i(n)$, where β_{d_i} is the cardinal central B-spline of degree d_i and is defined by

$$\beta_{d_i} = \chi_{[-\frac{1}{2},\frac{1}{2}]} \star \chi_{[-\frac{1}{2},\frac{1}{2}]} \star \cdots \star \chi_{[-\frac{1}{2},\frac{1}{2}]} \quad (d_i + 1 \text{ terms}).$$

We consider the following subspace of the multiply generated spline space:

$$\mathcal{S}_N := \left\{ f : f = \sum_{n\in\mathbb{Z}} a_n \sum_{i=1}^{r} \beta_{d_i}(t-n) \right\}$$

If $M = max\{d_1, d_2, \ldots, d_r\}$ and $m = min\{d_1, d_2, \ldots, d_r\}$, then $f \in \mathcal{S}_N$ provided that $f(x) \in \mathbb{C}^{m-1}$ and that the restriction of $f(x)$ to any interval between consecutive knots is identical with a polynomial of degree not exceeding M. If d_i's are distinct, then $\sum_{i=1}^{r} \beta_{d_i}(. - n)$, $n \in \mathbb{Z}$, are globally linearly independent (Lemma 1).

Let

$$\mathcal{S}_{N,\gamma} = \{f(t) \in \mathcal{S}_N : f(t) = O(|t|^{\gamma}) \ as \ t \to \pm\infty\}$$

and

$$\mathcal{D}_{\gamma} = \{\{y_n\} : y_n = O(|n|^{\gamma}) \ as \ n \to \pm\infty\}.$$

For $\gamma \geq 0$, a given sequence of numbers $\{y_n\}_{n\in\mathbb{Z}} \in \mathcal{D}_\gamma$ the following problem: Find a spline $f \in \mathcal{S}_{\mathcal{N},\gamma}$, satisfying $f(n) = y_n$, $n \in \mathbb{Z}$, has a unique solution. However, in many real applications, sampling points are not always measured exactly. Sometimes, the sampling steps need to be fluctuated according to the signals so as to reduce the number of samples and computational complexity. Therefore average sampling have been analysed in the literature [1, 2, 5, 7, 10–14]. In most of these studies, the sequence of samples $\{y_n\}$ is assumed to be in ℓ^2 or ℓ^p. In [8], we considered the average samples $\{f \star h(n)\}$ of polynomial growth and analysed the following problem.

Problem:

Given a sequence of numbers $\{y_n\}_{n\in\mathbb{Z}} \in \mathcal{D}_\gamma$, find a multiply generated spline $f \in \mathcal{S}_{\mathcal{N},\gamma}$, such that

$$f \star h(n) = y_n, \quad n \in \mathbb{Z},$$

where h satisfies

$$supp(h) \subseteq \left[-\frac{1}{2}, \frac{1}{2}\right], \qquad h(t) \geq 0, t \in \mathbb{R} \tag{1}$$

$$0 < \int_{-\frac{1}{2}}^{0} h(t)dt < \infty \quad \text{and} \quad 0 < \int_{0}^{\frac{1}{2}} h(t)dt < \infty. \tag{2}$$

In [8], it is shown that this problem has a unique solution for $d_1 = 1, d_2 = 2, d_3 = 3$ and $d_4 = 4$. In this paper, we analyse all the possible cases for $d_i \leq 4$ and show that the solution to the local average sampling problem is unique.

Lemma 1 *Let $d_i \in \mathbb{N}$ be distinct. Then the set $\left\{\sum_{i=1}^{r} \beta_{d_i}(. - k) : k \in \mathbb{Z}\right\}$ of functions is globally linearly independent on $\mathbb{R}$.*

Proof For $N \in \mathbb{N}$, consider $S_N = \left\{f|_{[-N,N]} : f \in S\right\}$. That is the restriction to $[-N, N]$ of the functions in S. We shall show that the set

$$\left\{\sum_{i=1}^{r} \beta_{d_i}(. - k) : k = -(N - M), -(N - M) + 1, \ldots, N - M\right\}$$

is linearly independent on $[-N, N]$, where $M = Max(d_1, d_2, \ldots, d_r)$. Without loss of generality, we may assume that $d_1 < d_2 < \ldots < d_r$. Let

$$\sum_{k=-(N-M)}^{N-M} c_k \sum_{i=1}^{r} \beta_{d_i}(x - k) = 0 \tag{3}$$

for $x \in [-N, N]$. For $x = N - \frac{1}{2}$ and $k = N - M$, $-d_r < x - k < d_r$ and

$$\sum_{i=1}^{r} \beta_{d_i}(x-k) = 0$$

for $-(N-M) \leq k \leq N-M-1$ and hence by substituting this x in Eq. (3) we get $c_{N-M} = 0$. Similarly, by choosing suitable x and substituting in (3), we get $c_k = 0$ for $-(N-M) \leq k \leq N-M$. As $S = \bigcup_{N \in \mathbb{N}} S_N$, we get that $\{\sum_{i=1}^{r} \beta_{d_i}(.-k) : k \in \mathbb{Z}\}$ is linearly independent on $\mathbb{R}$. □

2 Average Sampling Theorem for Multiply Generated Spline Space

Theorem 1 (Main Theorem) *Let $d_i \leq 4$ and $h(t)$ be an integrable function satisfying conditions 1, 2. Then for a given sequence of numbers $\{y_n\}_{n\in\mathbb{Z}} \in \mathcal{D}_\gamma$, there exists a unique $f \in \mathcal{S}_{N,\gamma}$, such that*

$$f \star h(n) = y_n, n \in \mathbb{Z}. \tag{4}$$

The Generalized Euler-Frobenius Laurent polynomial is defined as

$$G_h(z) = \sum_{i=1}^{r} G_{h,d_i}(z) = \sum_{i=1}^{r} \sum_{n\in\mathbb{Z}} h \star \beta_{d_i}(n) z^n.$$

It is easy to see that

$$G_{h,d_i}(z) = \int_{-\frac{1}{2}}^{\frac{1}{2}} h(t) \Upsilon_{z,d_i}(t) dt, \tag{5}$$

where $\Upsilon_{z,d_i}(t)$ is the exponential Euler spline and is defined as

$$\Upsilon_{z,d_i}(t) := \sum_{n\in\mathbb{Z}} z^n \beta_{d_i}(n-t).$$

Therefore

$$G_h(z) = \int_{-\frac{1}{2}}^{\frac{1}{2}} h(t) \Upsilon_z(t) dt,$$

where $\Upsilon_z(t) = \sum_{i=1}^{r} \Upsilon_{z,d_i}(t)$.

We need some properties of exponential Euler splines.

Lemma 2 *[8] For $d_i \in \mathbb{N}, n \in \mathbb{Z}$ and $z \in \mathbb{C} \setminus \{0\}$, we have*

(i) $\Upsilon_{z^{-1}}(-t) = \Upsilon_z(t)$,
(ii) $\Upsilon_z(t+n) = (z)^n \Upsilon_z(t)$,
(iii) $\frac{d}{dt}(\Upsilon_{z,d_i+1}(t)) = \left(1 - \frac{1}{z}\right) \Upsilon_{z,d_i}\left(t + \frac{1}{2}\right)$,
(iv) $\Upsilon_{-1,d_i}\left(\frac{1}{2}\right) = 0$ *and* $\Upsilon_{-1,d_i}(t) > 0$ *for* $t \in \left(-\frac{1}{2}, \frac{1}{2}\right)$.

In [8], it is shown that if the roots of $G_h(z)$ are simple and there is no root on the unit circle, then the local average sampling problem has a unique solution.

Theorem 2 *[8] Let $d_i \in \mathbb{N}$ and $h(t)$ be an integrable function satisfying conditions 1 and 2. If the roots of $G_h(z)$ are simple and there are no roots on the unit circle $|z| = 1$, then for a given sequence of numbers $\{y_n\}_{n\in\mathbb{Z}} \in \mathcal{D}_\gamma$, there exists a unique $f \in \mathcal{S}_{N,\gamma}$, such that*

$$f \star h(n) = y_n, n \in \mathbb{Z}. \tag{6}$$

Moreover, the solution can be written as

$$f(t) = \sum_{n\in\mathbb{Z}} y_n L(t-n),$$

where the reconstruction function L is given by $L(t) := \sum_{i=1}^r L_i(t) := \sum_{i=1}^r \sum_{n\in\mathbb{Z}} c_n \beta_{d_i}(t-n)$ and c_n are the coefficients of the Laurent series expansion of $G_h(z)^{-1}$. Further the reconstruction function L is of exponential decay.

3 Behaviour of $G_h(z)$

Proof (Main Theorem)

As a consequence of Theorem 2, it is sufficient to show that for $d_i \leq 4$ all the roots of $G_h(z)$ are simple and none of them are on the unit circle $|z| = 1$.

The Generalized Euler-Frobenius Laurent polynomial $G_h(z) = \sum_{i=1}^r G_{h,d_i}(z)$ can be written as

$$G_h(z) = \sum_{i=1}^r z^{\frac{-l_i}{2}} P_i(z)$$

where $l_i := \begin{cases} d_i + 1 & \text{if } d_i \text{ is odd} \\ d_i & \text{if } d_i \text{ is even} \end{cases}$ and $P_i(z)$ is a polynomial of degree l_i. Hence

$$G_h(z) = z^{\frac{-m}{2}} \sum_{i=1}^r z^{\frac{m-l_i}{2}} P_i(z) = z^{\frac{-m}{2}} P(z),$$

where $P(z)$ is a polynomial of degree $m = max(l_1, l_2, \ldots, l_r)$.

For $d_i \leq 4$, we get $m = 4$ and we can write

$$\begin{aligned}P(z) &= z^2 G_h(z)\\ &= z^4\left\{h\star\beta_{d_4}(2)+h\star\beta_{d_3}(2)\right\}+z^3\left\{h\star\beta_{d_4}(1)+h\star\beta_{d_3}(1)+h\star\beta_{d_2}(1)+h\star\beta_{d_1}(1)\right\}\\ &\quad+z^2\left\{h\star\beta_{d_4}(0)+h\star\beta_{d_3}(0)+h\star\beta_{d_2}(0)+h\star\beta_{d_1}(0)\right\}\\ &\quad+z\left\{h\star\beta_{d_4}(-1)+h\star\beta_{d_3}(-1)+h\star\beta_{d_2}(-1)+h\star\beta_{d_1}(-1)\right\}\\ &\quad+\left\{h\star\beta_{d_4}(-2)+h\star\beta_{d_3}(-2)\right\}.\end{aligned}$$

We obtain $P(0) > 0$ and $P(1) > 0$.
We can write $P(z)$ as

$$P(z) = z^2 \sum_{i=1}^{4} \int_{-\frac{1}{2}}^{\frac{1}{2}} h(t) \Upsilon_{z,d_i}(t) dt. \tag{7}$$

By lemma 2 and Eq. (7), we obtain

$$P(-1) = \sum_{i=1}^{4} \int_{-\frac{1}{2}}^{\frac{1}{2}} h(t) \Upsilon_{-1,d_i}(t) dt > 0.$$

Since $\lim_{z \longrightarrow \infty} P(z) = \infty$, it suffices to find $z_0 \in (-1, 0)$ such that

$$\sum_{i=1}^{4} \Upsilon_{z_0,d_i}(t) < 0, \text{ for all} t \in \left(-\frac{1}{2}, \frac{1}{2}\right), \tag{8}$$

since for such a z_0, we have

$$P(z_0) = z_0^2 \sum_{i=1}^{4} \int_{-\frac{1}{2}}^{\frac{1}{2}} h(t) \Upsilon_{z_0,d_i}(t) dt < 0, z_0 \in (-1, 0)$$

and

$$P\left(\frac{1}{z_0}\right) = \frac{1}{z_0^2} \sum_{i=1}^{4} \int_{-\frac{1}{2}}^{\frac{1}{2}} h(t) \Upsilon_{z_0^{-1},d_i}(t) dt = \frac{1}{z_0^2} \sum_{i=1}^{4} \int_{-\frac{1}{2}}^{\frac{1}{2}} h(t) \Upsilon_{z_0,d_i}(-t) dt < 0, z_0^{-1} \in (-\infty, -1).$$

By solving $\sum_{i=1}^{4} \Upsilon_{z_0,d_i}\left(\frac{1}{2}\right) = 0$, we get a unique $z_0 \in (-1, 0)$.

In the next table, we obtain all possible cases of $d_i \leq 4$.

S. No.	d_1	d_2	d_3	d_4	$\sum_{i=1}^{4} \Upsilon_{z_0,d_i}\left(\frac{1}{2}\right) = 0$	Possible solutions
1	1	2	3	4	$\frac{3}{48}z_0^3 + \frac{93}{48}z_0^2 + \frac{93}{48}z_0 + \frac{3}{48}$	$z_0 = -1, -15 - 4\sqrt{14}, -15 + 4\sqrt{14}.$
2	1	–	3	4	$\frac{1}{16}z_0^3 + \frac{69}{48}z_0^2 + \frac{69}{48}z_0 + \frac{1}{16}$	$z_0 = -1, -11 + 2\sqrt{30}, -11 - 2\sqrt{30}.$
3	1	2	–	4	$\frac{1}{24}z_0^3 + \frac{35}{24}z_0^2 + \frac{35}{24}z_0 + \frac{1}{24}$	$z_0 = -1, -17 + 12\sqrt{2}, -17 - 12\sqrt{2}.$
4	–	2	3	4	$\frac{1}{16}z_0^3 + \frac{69}{48}z_0^2 + \frac{69}{48}z_0 + \frac{1}{16}$	$z_0 = -1, -11 + 2\sqrt{30}, -11 - 2\sqrt{30}.$
5	1	2	3	–	$\frac{1}{48}z_0^3 + \frac{71}{48}z_0^2 + \frac{71}{48}z_0 + \frac{1}{48}$	$z_0 = -1, -35 + 6\sqrt{34}, -35 - 6\sqrt{34}.$
6	–	2	–	4	$\frac{1}{24}z_0^3 + \frac{23}{24}z_0^2 + \frac{23}{24}z_0 + \frac{1}{24}$	$z_0 = -1, -11 + 2\sqrt{30}, -11 - 2\sqrt{30}.$
7	–	–	3	4	$\frac{1}{16}z_0^3 + \frac{45}{48}z_0^2 + \frac{45}{48}z_0 + \frac{1}{16}$	$z_0 = -1, -7 + 4\sqrt{3}, -7 - 4\sqrt{3}.$
8	–	2	3	–	$\frac{1}{48}z_0^3 + \frac{47}{48}z_0^2 + \frac{47}{48}z_0 + \frac{1}{48}$	$z_0 = -1, -23 + 4\sqrt{33}, -23 - 4\sqrt{33}.$
9	1	–	–	4	$\frac{1}{24}z_0^3 + \frac{23}{24}z_0^2 + \frac{23}{24}z_0 + \frac{1}{24}$	$z_0 = -1, -11 + 2\sqrt{30}, -11 - 2\sqrt{30}.$
10	1	–	3	–	$\frac{1}{48}z_0^3 + \frac{47}{48}z_0^2 + \frac{47}{48}z_0 + \frac{1}{48}$	$z_0 = -1, -23 + 4\sqrt{33}, -23 - 4\sqrt{33}.$
11	1	2	–	–	$1 + z_0$	$z_0 = -1.$

By this table, we get a unique solution $z_0 \in (-1, 0)$. For such a z_0 value, we obtain

$$\sum_{i=1}^{4} \Upsilon_{z_0,d_i}(t) < 0, \text{ for all} t \in \left(-\frac{1}{2}, \frac{1}{2}\right).$$

Thus all the roots of $G_h(z)$ are simple and there are no roots on the unit circle for $d_i \leq 4$. □

Acknowledgements The author would like to thank the Management of Sri Sivasubramaniya Nadar College of Engineering, Kalavakkam-603 110, Tamil Nadu, India.

References

1. Aldroubi, A., Sun, Q. and Tang, W.S.: Nonuniform average sampling and reconstruction in multiply generated shift-invariant spaces. Constr. Approx. **20**, 173–189 (2004)
2. Aldroubi, A.: Non-uniform weighted average sampling and exact reconstruction in shift invariant and wavelet spaces. Appl. Comput. Harmon. Anal. **13**, 151–161 (2002)
3. Aldroubi, A., Feichtinger, H.: Exact iterative reconstruction algorithm for multivariate irregularly sampled functions in spline-like spaces: the LP theory. Proc. Am. Math. Soc. **126**(9), 2677–2686 (1998)
4. Aldroubi, A., Gröchenig, K.: Beurling-Landau-type theorems for nonuniform sampling in shift-invariant spaces. J. Fourier Anal. Applications **6**(1), 91–101 (2000)
5. Aldroubi A., Gröchenig, K.: Nonuniform sampling and reconstruction in shift-invariant spaces. SIAM Rev. **43**(4), 562–585 (2001)
6. de Boor, C., Devore, R., Ron, A.: The structure of finitely generated shift invariant spaces in $L^2(R^d)$. J. Funct. Anal. **119**, 37–78 (1994)
7. Devaraj, P., Yugesh, S.: Existence and uniqueness of spline reconstruction from local weighted average samples. Rendiconti del Circolo Matematico di Palermo **63**, 97–108 (2014)

8. Devaraj, P., Yugesh, S.: Reconstruction of multiply generated splines from local average samples. Math. Anal. Its Appl., Springer Proc. Math. Stat. **143**, 63–72 (2015)
9. Garcia, A.G., Hernandez-Medina, M.A., Perez-Villalon, G.: Generalized sampling in shift-invariant spaces with multiple stable generators. J. Math. Anal. Appl. **337**(1), 69–84 (2008)
10. Gröchenig, K.: Reconstruction algorithms in irregular samplings. Math. Comput. **59**, 181–194 (1992)
11. Xian, J., Zhu, J.H., Wang, J.: Local sampling and reconstruction in spline signal spaces and multiply generated spline signal spaces. In: 2009 International Joint Conference on Computational Sciences and Optimization
12. Xian, J., Sun, W.: Local sampling and reconstruction in shift-invariant spaces and their applications in spline subspaces. Numer. Funct. Anal. Optim. **31**(3), 366–386 (2010)
13. Xian, J., Qiang, X.: Non-uniform sampling and reconstruction in weighted finitely generated shift-invariant spaces. Far East. J. Math. Sci. **8**(3), 281–293 (2003)
14. Perez-Villalon, G., Portal, A.: Reconstruction of splines from local average samples. Appl. Math. Lett. **25**, 1315–1319 (2012)
15. Schoenberg, I.J.: Cardinal spline interpolation. In: SIAM Regional Conference Seriesin Applied Mathematics (1973)

An Insight into the Frames in Hilbert C^*-modules

Nabin K. Sahu and Ekta Rajput

Abstract Hilbert C^*-modules generalize Hilbert spaces which allow the inner product to take values in a C^*-algebra rather than in the field of real or complex numbers. A frame is a more flexible substitute for a basis, and it allows us to represent a vector as linear combinations of frame elements in multiple ways. Frank and Larson, in 2002 introduced the notion of frames in Hilbert C^*-modules. Here, we present a thorough discussion on frames in Hilbert C^*-modules. We discuss different types of frames, such as K-frames, controlled frames, fusion frames and weaving frames with examples. We also define weaving K-frames and explain this new concept with examples.

Keywords Frames · Hilbert C^*-module · Frames in Hilbert C^*-modules · Weaving frames

MSC 2010 42C15 · 06D22 · 46H35

1 Introduction

In the study of vector spaces, the basis is one of the most important concepts which allows each element in the space to be written as a linear combination of the elements in the basis. However, the linear independence condition between the elements is very restrictive, and thus one might look for a more flexible substitute. Frames are such tools as linear independence between the frame elements is not required. Frames in Hilbert spaces were first proposed by Duffin and Schaeffer [13] in 1952 while studying nonharmonic Fourier series, which can be viewed as more flexible sub-

N. K. Sahu (✉) · E. Rajput
Dhirubhai Ambani Institute of Information and Communication Technology,
Gandhinagar, India
e-mail: nabindaiict@gmail.com

E. Rajput
e-mail: ekta.rajput6@gmail.com

R. N. Mohapatra et al. (eds.), *Mathematical Analysis and Computing*,
Springer Proceedings in Mathematics & Statistics 344,
https://doi.org/10.1007/978-981-33-4646-8_46

stitutes of bases in Hilbert spaces. In 1985, as the wavelet era began, Daubechies et al. [11] reintroduced and developed the theory of frames in 1986. Due to their remarkable structure, the subject drew the attention of many mathematicians, physicists, and engineers because of its broad application in various well-known fields like signal processing [16], coding and communications [28], image processing [6], sampling [14, 15], numerical analysis, filter theory [5]. Nowadays, it has emerged as an essential tool in compressive sensing, data analysis, and other areas. L. Găvruta [18] proposed the notion of K-frames in Hilbert space to study the atomic systems with respect to a bounded linear operator K. Balazs [3] introduced controlled frames in Hilbert spaces to improve the numerical efficiency of iterative algorithms for inverting the frame operator. In recent years, many mathematicians have obtained significant results by generalizing the frame theory in Hilbert spaces to frame theory in Hilbert C^*-modules that enrich the theory of frames. Kaplansky [21] offered and investigated initially Hilbert C^*-modules as generalizations of Hilbert spaces by allowing the inner product to take values in a C^*-algebra rather than in the field of real or complex numbers. Frank and Larson [17] defined the concept of standard frames in finitely or countably generated Hilbert C^*-modules over a unital C^*-algebra. In [26], Najati et al. introduced the atomic system concepts for operators and K-frames in Hilbert C^*-modules. Rashidi and Rahimi [25] presented controlled frames in Hilbert C^*-modules and showed that they share many useful properties with their corresponding notions in a Hilbert space. Recently, Controlled K-frames in Hilbert C^*-modules were introduced in [27]. The authors established an equivalent condition for controlled K-frame and investigated some operator theoretic characterizations of controlled K-frames and controlled Bessel sequences. They also proved a perturbation result for the controlled K-frame in Hilbert C^*-modules.

The notion of weaving frames in Hilbert space was introduced in [4] and investigated in [9, 10]. The concept of weaving frames is partially motivated by pre-processing of Gabor frames and has probable implementations in wireless sensor networks that require distributed processing under different frames, as well as pre-processing of signals using Gabor frames. In 2018, Deepshikha and Lalit K. Vashisht [12] studied the weaving properties of K-frames in Hilbert space. They presented necessary and sufficient conditions for weaving K-frames in Hilbert spaces and sufficient conditions for Paley–Wiener type perturbation. Also, it is shown that woven K-frames and weakly woven K-frames are equivalent. Woven frames for finitely or countably generated Hilbert C^*-modules were introduced and studied [19]. Authors have investigated some properties of woven frames and obtained some conditions on a perturbed family of sequences. In [23], Khosravi and Khosravi introduced fusion frames and g-frames in Hilbert C^*-modules that share many useful properties with their corresponding notions in Hilbert space. They also generalize a perturbation result in frame theory to g-frames in Hilbert spaces.

2 Frames in Hilbert C^*-modules

In this section, we give some elementary definitions related to Hilbert C^*-modules and frames in Hilbert C^*-modules. Hilbert C^*-modules are the generalization of Hilbert spaces by allowing the inner product to take values in C^*-algebra rather than $\mathbb{R}$ or $\mathbb{C}$.

Definition 2.1 Let $\mathscr{A}$ be a C^*-algebra. An inner product $\mathscr{A}$-module is a complex vector space $\mathscr{H}$ such that

(i) $\mathscr{H}$ is a right $\mathscr{A}$-module, i.e., there is a bilinear map

$$\mathscr{H} \times \mathscr{A} \to \mathscr{A} : (x, a) \to x \cdot a$$

satisfying $(x \cdot a) \cdot b = x \cdot (ab)$ and $(\lambda x) \cdot a = x \cdot (\lambda a)$, and $x \cdot 1 = x$ where $\mathscr{A}$ has a unit 1.

(ii) There is a map $\mathscr{H} \times \mathscr{H} \to \mathscr{A} : (x, y) \to \langle x, y \rangle$ satisfying

1. $\langle x, x \rangle \geq 0$
2. $\langle x, y \rangle^* = \langle y, x \rangle$
3. $\langle ax, y \rangle = a\langle x, y \rangle$
4. $\langle x + y, z \rangle = \langle x, z \rangle + \langle y, z \rangle$
5. $\langle x, x \rangle = 0$ if and only if $x = 0$ (for every $x, y, z \in \mathscr{H}, a \in \mathscr{A}$).

Definition 2.2 A Hilbert C^*-module over $\mathscr{A}$ is an inner product $\mathscr{A}$-module with the property that $(\mathscr{H}, \| \cdot \|_{\mathscr{H}})$ is complete with respect to the norm $\|x\| = \|\langle x, x \rangle\|_{\mathscr{A}}^{\frac{1}{2}}$, where $\|.\|_{\mathscr{A}}$ denotes the norm on $\mathscr{A}$.

Let $\mathscr{A}$ be a C^*-algebra, $\mathbb{J} \subseteq \mathrm{N}$ be a finite or countable index set and consider

$$l^2(\mathscr{A}) = \{\{a_j\}_{j \in \mathbb{J}} \subseteq \mathscr{A} : \sum_{j \in \mathbb{J}} a_j a_j^* \text{ converges in norm}\}$$

where the sum converges in norm in $\mathscr{A}$. It is easy to see that $l^2(\mathscr{A})$ is a Hilbert C^*-module with pointwise operations and the inner product defined as

$$\langle \{a_j\}, \{b_j\} \rangle = \sum_{j \in \mathbb{J}} a_j b_j^*, \quad \{a_j\}, \{b_j\} \in l^2(\mathscr{A})$$

and

$$\|\{a_j\}\| = \sqrt{\|\sum_{j \in \mathbb{J}} a_j a_j^*\|}.$$

Definition 2.3 [20] Let $\mathscr{A}$ be a unital C^*-algebra and $j \in \mathbb{J}$ be a finite or countable index set. A sequence $\{x_j\}_{j\in\mathbb{J}}$ of elements in a Hilbert $\mathscr{A}$-module $\mathscr{H}$ is said to be a frame if there exist two constants $C, D > 0$ such that

$$C\langle x, x\rangle \leq \sum_{j\in\mathbb{J}} \langle x, x_j\rangle\langle x_j, x\rangle \leq D\langle x, x\rangle, \forall x \in \mathscr{H}. \tag{1}$$

The frame $\{x_j\}_{j\in\mathbb{J}}$ is said to be a tight frame if $C = D$, and is said to be Parseval or a normalized tight frame if $C = D = 1$.
Suppose that $\{x_j\}_{j\in\mathbb{J}}$ is a frame of a finitely or countably generated Hilbert C^*-module $\mathscr{H}$ over a unital C^*-algebra $\mathscr{A}$. The operator $T: \mathscr{H} \to l^2(\mathscr{A})$ defined by

$$Tx = \{\langle x, x_j\rangle\}_{j\in\mathbb{J}}$$

is called the analysis operator.
The adjoint operator $T^*: l^2(\mathscr{A}) \to \mathscr{H}$ is given by

$$T^*\{c_j\}_{j\in\mathbb{J}} = \sum_{j\in\mathbb{J}} c_j x_j$$

T^* is called a pre-frame operator or the synthesis operator.
By composing T and T^*, we obtain the frame operator $S: \mathscr{H} \to \mathscr{H}$

$$Sx = T^*Tx = \sum_{j\in\mathbb{J}} \langle x, x_j\rangle x_j \tag{2}$$

We now give an example of a frame in the Hilbert C^*-module.

Example 2.1 Let l^∞ be the unitary C^*-algebra of all bounded complex-valued sequences with the following operations.

$$uv = \{u_i v_i\}_{i\in\mathbb{N}},\ u^* = \{\overline{u_i}\}_{i\in\mathbb{N}},\ \forall\, u = \{u_i\}_{i\in\mathbb{N}}, v = \{v_i\}_{i\in\mathbb{N}} \in l^\infty$$

Let $\mathscr{H} = C_0$ be the set of all sequences converging to zero. Then, C_0 is a Hilbert l^∞-module with l^∞-valued inner product

$$\langle u, v\rangle = uv^* = \{u_i v_i^*\}_{i\in\mathbb{N}} = \{u_i \overline{v_i}\}_{i\in\mathbb{N}}\ \forall\ u, v \in C_0$$

Let $\mathbb{J} = \mathbb{N}$ and we define $\{x_j\}_{j\in\mathbb{J}} \in C_0$ as follows:

$$\{x_j\}_{j\in\mathbb{J}} = \{e_1, e_2, e_3, e_4, e_5, \ldots\}$$

where $\{e_j\}_{j\in\mathbb{J}}$ be the standard orthonormal basis for $\mathscr{H}$.
Let $x = \{\alpha_1, \alpha_2, \alpha_3, \alpha_4, \alpha_5, \ldots\} \in \mathscr{H}$. Then, $\langle x, x\rangle = \{\alpha_1\alpha_1^*, \alpha_2\alpha_2^*, \alpha_3\alpha_3^*, \alpha_4\alpha_4^*, \ldots\}$. Here, partial ordering $' \leq '$ means pointwise comparison.

Now, for the upper bound, we have

$$\begin{aligned}
&\sum_{j\in\mathbb{J}}\langle x, x_j\rangle\langle x_j, x\rangle \\
&= \langle x, e_1\rangle\langle e_1, x\rangle + \langle x, e_2\rangle\langle e_2, x\rangle + \langle x, e_3\rangle\langle e_3, x\rangle + \langle x, e_4\rangle\langle e_4, x\rangle + \cdots \\
&= \{\alpha_1\alpha_1^*, 0, 0, ...\} + \{0, \alpha_2\alpha_2^*, 0, 0, \ldots\} + \{0, 0, \alpha_3\alpha_3^*, 0, 0, \ldots\} \\
&+\{0, 0, 0, \alpha_4\alpha_4^*, 0, 0, \ldots\} + \cdots = \{\alpha_1\alpha_1^*, \alpha_2\alpha_2^*, \alpha_3\alpha_3^*, \alpha_4\alpha_4^*, ...\} \\
&= \sum_{j\in\mathbb{J}}\langle x, e_j\rangle\langle e_j, x\rangle \\
&= \langle x, x\rangle
\end{aligned}$$

On the other hand, we can write x as $x = \sum_{j\in\mathbb{J}} \alpha_j e_j$. Thus, we have

$$\begin{aligned}
\langle x, x\rangle &= \left\langle \sum_{j\in\mathbb{J}} \alpha_j e_j, \sum_{j\in\mathbb{J}} \alpha_j e_j \right\rangle \\
&= \sum_{j\in\mathbb{J}}\langle x, x_j\rangle\langle x_j, x\rangle
\end{aligned}$$

Hence, $\{x_j\}_{j\in\mathbb{J}}$ is a Parseval or a normalized tight frame with frame bound 1.

In the following Proposition, we present an equivalent condition for a sequence $\{x_j\}_{j\in\mathbb{J}} \subseteq \mathscr{H}$ to be a frame. The main advantage of the equivalent condition of frame in the Hibert C^*-module is that it is much easier to compare the norms of two elements rather than to compare two elements in C^*-algebras.

Proposition 2.1 *[20] Let $\mathscr{H}$ be a finitely or countably generated Hilbert $\mathscr{A}$-module $\mathscr{H}$ over a unital C^*-algebra $\mathscr{A}$ and $\{x_j\}_{j\in\mathbb{J}} \subseteq \mathscr{H}$ a sequence. Then, $\{x_j\}_{j\in\mathbb{J}}$ is a frame of $\mathscr{H}$ with bounds C and D if and only if*

$$C\|x\|^2 \leq \|\sum_{j\in\mathbb{J}}\langle x, x_j\rangle\langle x_j, x\rangle\| \leq D\|x\|^2$$

for all $x \in \mathscr{H}$.

3 *K*-frames in Hilbert C^*-modules

Najati et al. [26] defined atomic system and a K-frame in Hilbert C^*-module .

Definition 3.1 [26] A sequence $\{x_j\}_{j\in\mathbb{J}}$ of $\mathscr{H}$ is called an atomic system for $K \in L(\mathscr{H})$ if the following statement holds:

(1) The series $\sum_{j\in\mathbb{J}} c_j x_j$ converges for all $c = \{c_j\}_{j\in\mathbb{J}} \in l^2(\mathscr{A})$;

(2) There exists $C > 0$ such that for every $x \in \mathscr{H}$ there exists $\{a_{j,x}\}_{j\in\mathbb{J}} \in l^2(\mathscr{A})$ such that $\sum_{j\in\mathbb{J}} a_{j,x}a_{j,x}^* \leq C\langle x, x\rangle$ and $Kx = \sum_{j\in\mathbb{J}} a_{j,x}x_j$.

In the above definition, condition (1) says that $\{x_j\}_{j\in\mathbb{J}}$ is a Bessel sequence.

Theorem 3.1 *[26] If $K \in L(\mathscr{H})$, then there exists an atomic system for K.*

Definition 3.2 [26] A sequence $\{x_j\}_{j\in\mathbb{J}}$ of elements in a Hilbert $\mathscr{A}$-module $\mathscr{H}$ is said to be a K-frame ($K \in L(\mathscr{H})$) if there exist constants $C, D > 0$ such that

$$C\langle K^*x, K^*x\rangle \leq \sum_{j\in\mathbb{J}} \langle x, x_j\rangle\langle x_j, x\rangle \leq D\langle x, x\rangle, \forall x \in \mathscr{H}. \tag{3}$$

We now give an example of K-frame in the Hilbert C^*-module.

Example 3.1 Let l^∞ be the unitary C^*-algebra of all bounded complex-valued sequences with the following operations

$$uv = \{u_i v_i\}_{i\in\mathbb{N}},\ u^* = \{\overline{u_i}\}_{i\in\mathbb{N}},\ \forall\ u = \{u_i\}_{i\in\mathbb{N}},\ v = \{v_i\}_{i\in\mathbb{N}} \in l^\infty$$

Let $\mathscr{H} = C_0$ be the set of all sequences converging to zero. Then, C_0 is a Hilbert l^∞-module with l^∞-valued inner product

$$\langle u, v\rangle = uv^* = \{u_i v_i^*\}_{i\in\mathbb{N}} = \{u_i\overline{v_i}\}_{i\in\mathbb{N}}\ \ \forall\ \ u, v \in C_0$$

Let $\mathbb{J} = \mathbb{N}$ and we define $\{x_j\}_{j\in\mathbb{J}} \in C_0$ as follows:

$$\{x_j\}_{j\in\mathbb{J}} = \{0, e_1, 0, e_2, 0, e_3, 0, e_4, 0, e_5, \ldots\}$$

where $\{e_j\}_{j\in\mathbb{J}}$ be the standard orthonormal basis for $\mathscr{H}$.
Let K be the orthogonal projection from $\mathscr{H}$ onto $\overline{span}\{e_j\}_{j=3}^{\infty}$ and $x = \{\alpha_1, \alpha_2, \alpha_3, \alpha_4, \alpha_5, ...\} \in \mathscr{H}$. Then $\langle x, x\rangle = \{\alpha_1\alpha_1^*, \alpha_2\alpha_2^*, \alpha_3\alpha_3^*, \alpha_4\alpha_4^*, \ldots\}$. Here, partial ordering $'\leq'$ means pointwise comparison.
Now, we have

$$\begin{aligned}
&\sum_{j\in\mathbb{J}}\langle x, x_j\rangle\langle x_j, x\rangle \\
&= \langle x, e_1\rangle\langle e_1, x\rangle + \langle x, e_2\rangle\langle e_2, x\rangle + \langle x, e_3\rangle\langle e_3, x\rangle + \langle x, e_4\rangle\langle e_4, x\rangle + \cdots \\
&= \{\alpha_1\alpha_1^*, 0, 0, \ldots\} + \{0, \alpha_2\alpha_2^*, 0, 0, \ldots\} + \{0, 0, \alpha_3\alpha_3^*, 0, 0, \ldots\} \\
&+\{0, 0, 0, \alpha_4\alpha_4^*, 0, 0, \ldots\} + \cdots \\
&= \{\alpha_1\alpha_1^*, \alpha_2\alpha_2^*, \alpha_3\alpha_3^*, \alpha_4\alpha_4^*, \ldots\} \\
&= \sum_{j\in\mathbb{J}}\langle x, e_j\rangle\langle e_j, x\rangle \\
&= \langle x, x\rangle
\end{aligned}$$

On the other hand, we can write x as $x = \sum_{j=1}^{\infty}\alpha_j e_j$. Thus, we have

$$\begin{aligned}
\langle K^*x, K^*x\rangle &= \left\langle K^*\left(\sum_{j=1}^{\infty}\alpha_j e_j\right), K^*\left(\sum_{j=1}^{\infty}\alpha_j e_j\right)\right\rangle \\
&= \left\langle \sum_{j=3}^{\infty}\alpha_j e_j, \sum_{j=3}^{\infty}\alpha_j e_j\right\rangle \\
&= \sum_{j=3}^{\infty}\langle x, e_j\rangle\langle e_j, x\rangle \\
&\leq \sum_{j\in\mathbb{J}}\langle x, x_j\rangle\langle x_j, x\rangle
\end{aligned}$$

Hence, $\{x_j\}_{j\in\mathbb{J}}$ is a K-frame with lower and upper frame bound 1.

Najati et al. [26] also established an equivalent condition of K-frame which is given in the following result.

Theorem 3.2 *[26] Let $\{x_j\}_{j\in\mathbb{J}}$ be a Bessel sequence for $\mathscr{H}$ and $K \in L(\mathscr{H})$. Suppose that $T \in L(\mathscr{H}, l^2(\mathscr{A}))$ is given by $T(x) = \{\langle x, x_j\rangle\}_{j\in\mathbb{J}}$ and $\overline{R(T)}$ is orthogonally complemented then the following statements are equivalent:*

(i) *$\{x_j\}_{j\in\mathbb{J}}$ is an atomic system for K;*

(ii) *There exist $A, B > 0$ such that*

$$A\|K^*x\|^2 \leq \|\sum_{j\in\mathbb{J}}\langle x, x_j\rangle\langle x_j, x\rangle\| \leq B\|x\|^2, \ \forall x \in \mathscr{H};$$

(iii) *There exist $D \in L(\mathscr{H}, l^2(\mathscr{A}))$ such that $K = T^*D$.*

Next, we study the action of an operator on a K-frame.

Lemma 3.1 *[1] Let $\mathscr{H}$ be a Hilbert $\mathscr{A}$-module and $\{x_j\}_{j\in\mathbb{J}}$ be a Bessel sequence, then $\{Mx_j\}_{j\in\mathbb{J}}$ is a Bessel sequence for every $M \in L(\mathscr{H})$.*

The above lemma shows that the action of an adjointable operator on a Bessel sequence is again a Bessel sequence.

Lemma 3.2 *[1] Let $\mathscr{H}$ be a Hilbert $\mathscr{A}$-module, $K \in L(\mathscr{H})$ and $\{x_j\}_{j\in\mathbb{J}}$ be a K-frame for $\mathscr{H}$. Let $M \in L(\mathscr{H})$ with $R(M) \subset R(K)$ and $\overline{R(K^*)}$ orthogonally complemented. Then, $\{x_j\}_{j\in\mathbb{J}}$ is an M-frame for $\mathscr{H}$.*

Theorem 3.3 *[1] Let $K \in L(\mathscr{H})$ and $\{x_j\}_{j\in\mathbb{J}}$ be a K-frame for $\mathscr{H}$. If $T \in L(\mathscr{H})$ with closed range such that $\overline{R(TK)}$ is orthogonal complemented and $KT = TK$. Then, $\{Tx_j\}_{j\in\mathbb{J}}$ is a K-frame for $R(T)$.*

Theorem 3.4 *[1] Let $\mathscr{H}$ be a Hilbert $\mathscr{A}$-module, $K \in L(\mathscr{H})$ and $\{x_j\}_{j\in\mathbb{J}}$ be a K-frame for $\mathscr{H}$, and $T \in L(\mathscr{H})$ is a co-isometry such that $R(T^*K^*) \subset R(K^*T^*)$ with $\overline{R(TK)}$orthogonal complemented. Then, $\{Tx_j\}_{j\in\mathbb{J}}$ is a K-frame for $\mathscr{H}$.*

4 Controlled Frames in Hilbert C^*-modules

In [25], Rashidi and Rahimi introduced controlled frames in Hilbert C^*-modules and revealed that they share multiple useful properties with their corresponding notions in a Hilbert space.

Definition 4.1 [25] Let $\mathscr{H}$ be a Hilbert C^*-module and $C \in GL(\mathscr{H})$. A frame controlled by the operator C or C -controlled frame in Hilbert C^*-module $\mathscr{H}$ is a family of vectors $\{x_j\}_{j\in\mathbb{J}}$, such that there exist two constants $A, B > 0$ satisfying

$$A\langle x, x\rangle \leq \sum_{j\in\mathbb{J}} \langle x, x_j\rangle\langle Cx_j, x\rangle \leq B\langle x, x\rangle, \forall x \in \mathscr{H} \tag{4}$$

Likewise, $\{x_j\}_{j\in\mathbb{J}}$ is called a C-controlled Bessel sequence with bound B, if there exists $B > 0$ such that

$$\sum_{j\in\mathbb{J}} \langle x, x_j\rangle\langle Cx_j, x\rangle \leq B\langle x, x\rangle, \forall x \in \mathscr{H}, \tag{5}$$

where the sum in the above inequalities converges in the norm.
If $A = B$, we call $\{x_j\}_{j\in\mathbb{J}}$ a C-controlled tight frame, and if $A = B = 1$, it is called a C-controlled Parseval frame.
If $C = I$, then controlled frames are nothing but standard frames in Hilbert C^*-module $\mathscr{H}$. Hence, controlled frames are generalizations of frames.
We now give an example of a C-controlled frame in the Hilbert C^*-module.

Example 4.1 Let l^∞ be the unitary C^*-algebra of all bounded complex-valued sequences with the following operations

$$uv = \{u_i v_i\}_{i\in\mathbb{N}},\ u^* = \{\overline{u_i}\}_{i\in\mathbb{N}},\ \forall\, u = \{u_i\}_{i\in\mathbb{N}}, v = \{v_i\}_{i\in\mathbb{N}} \in l^\infty$$

Let $\mathscr{H} = C_0$ be the set of all sequences converging to zero. Then, C_0 is a Hilbert l^∞-module with l^∞-valued inner product

$$\langle u, v\rangle = uv^* = \{u_i v_i^*\}_{i\in\mathbb{N}} = \{u_i \overline{v_i}\}_{i\in\mathbb{N}}\ \ \forall\ u, v \in C_0$$

Let $\mathbb{J} = \mathbb{N}$ and we define $\{x_j\}_{j\in\mathbb{J}} \in C_0$ as follows:

$$\{x_j\}_{j\in\mathbb{J}} = \{0, 0, e_3, e_4, e_5, \ldots\}$$

where $\{e_j\}_{j\in\mathbb{J}}$ be the standard orthonormal basis for $\mathscr{H}$.
Let $C \in GL(\mathscr{H})$ be such that

$$C(e_i) = \begin{cases} e_1 + e_2, & i = 1 \\ e_i, & \text{otherwise} \end{cases}$$

Let $x = \{\alpha_1, \alpha_2, \alpha_3, \alpha_4, \alpha_5, \ldots\} \in \mathscr{H}$. Then $\langle x, x\rangle = \{\alpha_1\alpha_1^*, \alpha_2\alpha_2^*, \alpha_3\alpha_3^*, \alpha_4\alpha_4^*, \ldots\}$.
Here, partial ordering $'\leq'$ means pointwise comparison.
Now, for the upper bound, we have

$$\begin{aligned}
&\sum_{j\in\mathbb{J}} \langle x, x_j\rangle\langle Cx_j, x\rangle \\
&= \langle x, x_1\rangle\langle C(x_1), x\rangle + \langle x, x_2\rangle\langle C(x_2), x\rangle \\
&+ \langle x, x_3\rangle\langle C(x_3), x\rangle + \langle x, x_4\rangle\langle C(x_4), x\rangle + \cdots \\
&= 0 + 0 + \langle x, e_3\rangle\langle C(e_3), x\rangle + \langle x, e_4\rangle\langle C(e_4), x\rangle + \langle x, e_5\rangle\langle C(e_5), x\rangle + \cdots \\
&= \langle x, e_3\rangle\langle e_3, x\rangle + \langle x, e_4\rangle\langle e_4, x\rangle + \langle x, e_5\rangle\langle e_5, x\rangle + \cdots \\
&= \{0, 0, \alpha_3\alpha_3^*, 0, 0, \ldots\} + \{0, 0, 0, \alpha_4\alpha_4^*, 0, 0, \ldots\} \\
&+ \{0, 0, 0, 0, \alpha_5\alpha_5^*, 0, 0, \ldots\} + \cdots \\
&= \{0, 0, \alpha_3\alpha_3^*, \alpha_4\alpha_4^*, \alpha_5\alpha_5^*, \ldots\} \\
&\leq \sum_{j\in\mathbb{J}} \langle x, e_j\rangle\langle e_j, x\rangle \\
&= \langle x, x\rangle
\end{aligned}$$

On the other hand, we can write x as $x = \sum_{j\in\mathbb{J}} \alpha_j e_j$. Thus, we have

$$\begin{aligned}\langle x, x\rangle &= \left\langle \sum_{j\in\mathbb{J}} \alpha_j e_j, \sum_{j\in\mathbb{J}} \alpha_j e_j \right\rangle \\ &\leq \sum_{j\in\mathbb{J}} \langle x, x_j\rangle \langle Cx_j, x\rangle\end{aligned}$$

Hence, $\{x_j\}_{j\in\mathbb{J}}$ is a Parseval or a normalized tight C-controlled frame with frame bound 1.

Lemma 4.1 *[25] Let $\mathcal{H}$ be a Hilbert C^*-module and $C \in GL(\mathcal{H})$. A sequence $\{x_j\}_{j\in\mathbb{J}}$ is C-controlled Bessel sequence in Hilbert C^*-module $\mathcal{H}$ if and only if the operator*

$$S_C x = \sum_{j\in\mathbb{J}} \langle x, x_j\rangle C x_j$$

is well defined and there exists constant $M < \infty$ such that

$$\sum_{j\in\mathbb{J}} \langle x, x_j\rangle \langle Cx_j, x\rangle \leq M\langle x, x\rangle$$

for every $x \in \mathcal{H}$.

Definition 4.2 [25] Let $\mathcal{H}$ be a Hilbert C^*-module and $C \in GL(\mathcal{H})$. Assume that the sequence $\{x_j\}_{j\in\mathbb{J}}$ is the C-controlled frame in Hilbert C^*-module $\mathcal{H}$. The operator

$$S_C x = \sum_{j\in\mathbb{J}} \langle x, x_j\rangle C x_j$$

is called the C-controlled frame operator.

The main properties of C-controlled frame operators in Hilbert C^*-modules and controlled frame operators in Hilbert spaces are similar. The C-controlled frame operator S_C is also invertible, positive, adjointable, and self-adjoint.
Now, we present below the equivalent condition for a sequence $\{x_j\}_{j\in\mathbb{J}}$ in $\mathcal{H}$ to be a C-controlled frame.

Theorem 4.1 *[25] Let $\mathcal{H}$ be a Hilbert C^*-module and $C \in GL(\mathcal{H})$. A sequence $\{x_j\}_{j\in\mathbb{J}}$ is C-controlled frame in Hilbert C^*-module $\mathcal{H}$ if and only if there exist constants $A > 0$ and $B < \infty$ such that*

$$A\|x\|^2 \leq \|\sum_{j\in\mathbb{J}} \langle x, x_j\rangle \langle Cx_j, x\rangle\| \leq B\|x\|^2$$

for all $x \in \mathcal{H}$.

5 Controlled K-frames in Hilbert C^*-modules

Recently, [27] proposed the notions of Controlled K-frames in Hilbert C^*-module. First, we give the definition of Controlled K-frames in Hilbert C^*-module and then we study them from operator theoretic approach.

Definition 5.1 [27] Let $\mathscr{H}$ be a Hilbert $\mathscr{A}$-module over a unital C^*-algebra, $C \in GL^+(\mathscr{H})$ and $K \in L(\mathscr{H})$. A sequence $\{x_j\}_{j\in\mathbb{J}}$ in $\mathscr{H}$ is said to be a C-controlled K-frame if there exist two constants $0 < A \leq B < \infty$ such that

$$A\langle C^{\frac{1}{2}}K^*x, C^{\frac{1}{2}}K^*x\rangle \leq \sum_{j\in\mathbb{J}}\langle x, x_j\rangle\langle Cx_j, x\rangle \leq B\langle x, x\rangle, \forall x \in \mathscr{H}. \tag{6}$$

If $C = I$, the C-controlled K-frame $\{x_j\}_{j\in\mathbb{J}}$ is simply K-frame in $\mathscr{H}$. The sequence $\{x_j\}_{j\in\mathbb{J}}$ is called a C-controlled Bessel sequence with bound B, if there exists $B > 0$ such that

$$\sum_{j\in\mathbb{J}}\langle x, x_j\rangle\langle Cx_j, x\rangle \leq B\langle x, x\rangle, \forall x \in \mathscr{H}, \tag{7}$$

where the sum in the above inequalities converges in the norm.
We now give an example of a C-controlled K-frame in the Hilbert C^*-module.

Example 5.1 Let l^∞ be the unitary C^*-algebra of all bounded complex-valued sequences with the following operations

$$uv = \{u_iv_i\}_{i\in\mathbb{N}},\ u^* = \{\overline{u_i}\}_{i\in\mathbb{N}},\ \forall\ u = \{u_i\}_{i\in\mathbb{N}}, v = \{v_i\}_{i\in\mathbb{N}} \in l^\infty$$

Let $\mathscr{H} = C_0$ be the set of all sequences converging to zero. Then, C_0 is a Hilbert l^∞-module with l^∞-valued inner product

$$\langle u, v\rangle = uv^* = \{u_iv_i^*\}_{i\in\mathbb{N}} = \{u_i\overline{v_i}\}_{i\in\mathbb{N}}\ \ \forall\ \ u, v \in C_0$$

Let $\mathbb{J} = \mathbb{N}$ and we define $\{x_j\}_{j\in\mathbb{J}} \in C_0$ as follows:

$$\{x_j\}_{j\in\mathbb{J}} = \{0, 0, e_3, e_4, e_5, \ldots\}$$

Let K be the orthogonal projection from $\mathscr{H}$ onto $\overline{span}\{e_j\}_{j=3}^{\infty}$ and $C \in GL^+(\mathscr{H})$ be such that

$$C(e_i) = \begin{cases} e_1 + e_2, & i = 1 \\ e_i, & \text{otherwise} \end{cases}$$

Let $x = \{\alpha_1, \alpha_2, \alpha_3, \alpha_4, \alpha_5, \ldots\} \in \mathscr{H}$. Then, $\langle x, x\rangle = \{\alpha_1\alpha_1^*, \alpha_2\alpha_2^*, \alpha_3\alpha_3^*, \alpha_4\alpha_4^*, \ldots\}$. Here, partial ordering $' \leq '$ means pointwise comparison.
Now, for the upper bound, we have

$$
\begin{aligned}
&\sum_{j\in\mathbb{J}}\langle x, x_j\rangle\langle Cx_j, x\rangle \\
&= \langle x, x_1\rangle\langle C(x_1), x\rangle + \langle x, x_2\rangle\langle C(x_2), x\rangle \\
&+\langle x, x_3\rangle\langle C(x_3), x\rangle + \langle x, x_4\rangle\langle C(x_4), x\rangle + \cdots \\
&= 0 + 0 + \langle x, e_3\rangle\langle C(e_3), x\rangle + \langle x, e_4\rangle\langle C(e_4), x\rangle + \langle x, e_5\rangle\langle C(e_5), x\rangle + \cdots \\
&= \langle x, e_3\rangle\langle e_3, x\rangle + \langle x, e_4\rangle\langle e_4, x\rangle + \langle x, e_5\rangle\langle e_5, x\rangle + \cdots \\
&= \{0, 0, \alpha_3\alpha_3^*, 0, 0, \ldots\} + \{0, 0, 0, \alpha_4\alpha_4^*, 0, 0, \ldots\} \\
&+\{0, 0, 0, 0, \alpha_5\alpha_5^*, 0, 0, \ldots\} + \cdots \\
&= \{0, 0, \alpha_3\alpha_3^*, \alpha_4\alpha_4^*, \alpha_5\alpha_5^*, \ldots\} \\
&\le \sum_{j\in\mathbb{J}}\langle x, e_j\rangle\langle e_j, x\rangle \\
&= \langle x, x\rangle
\end{aligned}
$$

On the other hand, we can write x as $x = \sum_{j=1}^{\infty}\alpha_j e_j$. Thus, we have

$$
\begin{aligned}
\langle C^{\frac{1}{2}}K^*x, C^{\frac{1}{2}}K^*x\rangle &= \langle CK^*x, K^*x\rangle \\
&= \left\langle CK^*\left(\sum_{j=1}^{\infty}\alpha_j e_j\right), K^*\left(\sum_{j=1}^{\infty}\alpha_j e_j\right)\right\rangle \\
&= \left\langle C\left(\sum_{j=3}^{\infty}\alpha_j e_j\right), \sum_{j=3}^{\infty}\alpha_j e_j\right\rangle \\
&= \left\langle \sum_{j=3}^{\infty}\alpha_j e_j, \sum_{j=3}^{\infty}\alpha_j e_j\right\rangle \\
&= \sum_{j=3}^{\infty}\langle x, e_j\rangle\langle e_j, x\rangle \\
&\le \sum_{j\in\mathbb{J}}\langle x, x_j\rangle\langle Cx_j, x\rangle
\end{aligned}
$$

Hence, $\{x_j\}_{j\in\mathbb{J}}$ is a C-controlled K-frame with lower and upper frame bound 1.

Let $\{x_j\}_{j\in\mathbb{J}}$ be a C-controlled Bessel sequence for Hilbert module $\mathscr{H}$ over $\mathscr{A}$. The operator $T: \mathscr{H} \to l^2(\mathscr{A})$ defined by

$$Tx = \{\langle x, x_j\rangle\}_{j\in\mathbb{J}}, x \in \mathscr{H} \tag{8}$$

is called the analysis operator. The adjoint operator $T^*: l^2(\mathscr{A}) \to \mathscr{H}$ is given by

$$T^*(\{c_j\})_{j\in\mathbb{J}} = \sum_{j\in\mathbb{J}} c_j C x_j \tag{9}$$

is called a pre-frame operator or the synthesis operator. By composing T and T^*, we obtain the C-controlled frame operator $S_C : \mathscr{H} \to \mathscr{H}$ as

$$S_C x = T^* T x = \sum_{j\in\mathbb{J}} \langle x, x_j\rangle C x_j. \tag{10}$$

The following theorem presents an equivalence condition for the C-controlled K-frame in a Hilbert C^*-module $\mathscr{H}$.

Theorem 5.1 *[27] Let $\mathscr{H}$ be a finitely or countably generated Hilbert $\mathscr{A}$-module over a unital C^*-algebra $\mathscr{A}$, $\{x_j\}_{j\in\mathbb{J}} \subset \mathscr{H}$ be a sequence, $C \in GL^+(\mathscr{H})$, $K \in L(\mathscr{H})$, $KC = CK$ and $R(C^{\frac{1}{2}}) \subseteq R(K^* C^{\frac{1}{2}})$ with $\overline{R((C^{\frac{1}{2}})^*)}$ be orthogonally complemented. Then, $\{x_j\}_{j\in\mathbb{J}}$ is a C-controlled K-frame in Hilbert C^*-module if and only if there exist constants $0 < A \leq B < \infty$ such that*

$$A\|C^{\frac{1}{2}} K^* x\|^2 \leq \|\sum_{j\in\mathbb{J}} \langle x, x_j\rangle\langle C x_j, x\rangle\| \leq B\|x\|^2, \ \forall x \in \mathscr{H}.$$

The following theorem gives a characterization of the C-controlled Bessel sequence.

Theorem 5.2 *[27] Let $\{x_j\}_{j\in\mathbb{J}}$ be a sequence of a finitely or countably generated Hilbert $\mathscr{A}$-module $\mathscr{H}$ over a unital C^*-algebra $\mathscr{A}$. Suppose that C commutes with the controlled frame operator S_C and $R(S_C^{\frac{1}{2}}) \subseteq R((C S_C)^{\frac{1}{2}})$ with $\overline{R((S_C^{\frac{1}{2}})^*)}$ is orthogonally complemented. Then $\{x_j\}_{j\in\mathbb{J}}$ is a C-controlled Bessel sequence with bound B if and only if the operator $U : l^2(\mathscr{A}) \to \mathscr{H}$ defined by*

$$U\{a_j\}_{j\in\mathbb{J}} = \sum_{j\in\mathbb{J}} a_j C x_j$$

is a well-defined bounded operator from $l^2(\mathscr{A})$ into $\mathscr{H}$ with $\|U\| \leq \sqrt{B}\|C^{\frac{1}{2}}\|$.

Now, we give a perturbation result for C-controlled K-frame in Hilbert C^*-module.

Theorem 5.3 *[27] Let $F = \{f_j\}_{j\in\mathbb{J}}$ be a C-controlled K-frame for $\mathscr{H}$, with controlled frame operator S_C. Suppose $K \in L(\mathscr{H})$, $KC = CK$, $R(C^{\frac{1}{2}}) \subseteq R(K^* C^{\frac{1}{2}})$ with $\overline{R((C^{\frac{1}{2}})^*)}$ is orthogonally complemented. If $G = \{g_j\}_{j\in\mathbb{J}}$ is a non-zero sequence in $\mathscr{H}$, and $E = T_F - T_G$ be a compact operator, where $T_G(\{c_j\}_{j\in\mathbb{J}}) = \sum_{j\in\mathbb{J}} c_j g_j$ for $\{c_j\}_{j\in\mathbb{J}} \in l^2(\mathscr{A})$, then $G = \{g_j\}_{j\in\mathbb{J}}$ is a C-controlled K-frame for $\mathscr{H}$.*

6 Weaving Frames in Hilbert C^*-modules

The concept of weaving frames in Hilbert space is motivated by a problem in distributed signal processing. Bemrose et al. introduced weaving frames in Hilbert space. In [4], the authors developed fundamental properties of woven frames in Hilbert space. Woven frames for finitely or countably generated Hilbert C^*-modules were introduced and studied in [19]. The authors investigated some properties of woven frames in Hilbert C^*-module.

Definition 6.1 [19] A family $\{\{\phi_{ij}\}_{i\in I}\}_{j\in\mathbb{J}}$ of frames for $\mathscr{H}$ is called woven if there exist universal constants $0 < A < B < \infty$ such that for every partition $\{\sigma_j\}_{j\in\mathbb{J}}$ of I, the family $\{\{\phi_{ij}\}_{i\in I}\}_{j\in\mathbb{J}}$ is a frame for $\mathscr{H}$ with lower and upper frame bounds A and B, respectively. Each family $\{\{\phi_{ij}\}_{i\in\sigma_i}\}_{j\in\mathbb{J}}$ is called weaving.

We now give an example of woven frames in Hilbert C^*-module.

Example 6.1 Let l^∞ be the unitary C^*-algebra of all bounded complex-valued sequences with the following operations

$$uv = \{u_i v_i\}_{i\in\mathbb{N}},\ u^* = \{\overline{u_i}\}_{i\in\mathbb{N}},\ \forall\, u = \{u_i\}_{i\in\mathbb{N}},\, v = \{v_i\}_{i\in\mathbb{N}} \in l^\infty$$

Let $\mathscr{H} = C_0$ be the set of all sequences converging to zero. Then, C_0 is a Hilbert l^∞-module with l^∞-valued inner product

$$\langle u, v\rangle = uv^* = \{u_i v_i^*\}_{i\in\mathbb{N}} = \{u_i \overline{v_i}\}_{i\in\mathbb{N}}\ \ \forall\ u, v \in C_0$$

Let $\mathbb{J} = \mathbb{N}$ and we define $\phi = \{\phi_{1j}\}_{j=1}^\infty \in \mathscr{H}$ and $\psi = \{\phi_{2j}\}_{j=1}^\infty \in \mathscr{H}$ as follows:

$$\begin{aligned}\{\phi_{1j}\}_{j=1}^\infty &= \{e_1, e_2, 0, e_3, 0, e_4, 0, e_5, \ldots\}\\ \{\phi_{2j}\}_{j=1}^\infty &= \{0, e_2, e_2, e_3, e_3, e_4, e_4, e_5, e_5, \ldots\}\end{aligned}$$

where $\{e_j\}_{j=1}^\infty$ be the standard orthonormal basis for $\mathscr{H}$.
Let $f = \{\alpha_1, \alpha_2, \alpha_3, \alpha_4, \alpha_5, \ldots\} \in \mathscr{H}$. Then, $\langle f, f\rangle = \{\alpha_1\alpha_1^*, \alpha_2\alpha_2^*, \alpha_3\alpha_3^*, \alpha_4\alpha_4^*, \ldots\}$. Here, partial ordering $'\leq'$ means pointwise comparison.
For any subset σ of $\mathbb{N}$, we have

$$\sum_{j\in\sigma}\langle f, \phi_{1j}\rangle\langle\phi_{1j}, f\rangle + \sum_{j\in\sigma^c}\langle f, \phi_{2j}\rangle\langle\phi_{2j}, f\rangle \leq 2\sum_{j=1}^\infty\langle f, e_j\rangle\langle e_j, f\rangle = 2\langle f, f\rangle$$

On the other hand, let $f \in \mathscr{H}$. Then, we have

$$\begin{aligned}\langle f, f\rangle &= \sum_{j=1}^\infty\langle f, e_j\rangle\langle e_j, f\rangle\\ &\leq \sum_{j\in\sigma}\langle f, \phi_{1j}\rangle\langle\phi_{1j}, f\rangle + \sum_{j\in\sigma^c}\langle f, \phi_{2j}\rangle\langle\phi_{2j}, f\rangle\end{aligned}$$

Hence, ϕ and ψ are woven frames with universal lower and upper frame bounds 1 and 2, respectively.

The following proposition shows that every weaving automatically has a universal upper frame bound. To verify that a family of frames is woven, it is enough to check that there exists a universal lower frame bound.

Proposition 6.1 *[19] If $\{\{\phi_{ij}\}_{i\in I}\}_{j\in\mathbb{J}}$ is a family of Bessel sequences with bounds B_j for $j \in \mathbb{J}$, then every weaving is a Bessel sequence with Bessel bound $\sum_{j\in\mathbb{J}} B_j$.*

Proposition 6.2 *[19] Let $\{\{\phi_{ij}\}_{i\in I}\}_{j\in\mathbb{J}}$ be a woven family of frames for $\mathscr{H}$ and $P\colon \mathscr{H} \to \mathscr{H}$ be an adjointable operator. Then, $\{\{P\phi_{ij}\}_{i\in I}\}_{j\in\mathbb{J}}$ are woven frames if and only if P is surjective.*

7 Weaving K-frames in Hilbert C^*-modules

We have introduced the notion of weaving K-frames in the Hilbert C^*-module. We studied weaving K-frames from the operator theoretic point of view and defined an atomic system for weaving K-frames in Hilbert C^*-module. We have also presented an equivalent definition of weaving K-frames and characterization theorems of weaving K-frames in terms of operator theory in the Hilbert C^*-module.

Definition 7.1 Let $\mathscr{H}$ be a Hilbert $\mathscr{A}$-module over a unital C^*-algebra. A family of K-frames $\{\{f_{ij}\}_{j=1}^{\infty} : i \in I\}$ for $\mathscr{H}$ is said to be K-woven if there exist universal positive constants A and B such that for any partition $\{\sigma_i\}_{i\in I}$ of $\mathbb{N}$, the family $\bigcup_{i\in I}\{f_{ij}\}_{j\in\sigma_i}$ is a K-frame for $\mathscr{H}$ with lower and upper K-frame bounds A and B, respectively. Each family $\bigcup_{i\in I}\{f_{ij}\}_{j\in\sigma_i}$is called weaving.

The woven frame is called a tight woven frame if $A = B$, and it is called a normalized woven tight frame if $A = B = 1$.
For any partition $\{\sigma_i\}_{i\in I}$ of $\mathbb{N}$, space is defined as

$$\bigoplus_{i\in I} l^2(\sigma_i) = \Big\{\{c_{ij}\}_{j\in\sigma_i, i\in I} | c_{ij} \in \mathscr{A}, \sum_{i\in I}\sum_{j\in\sigma_i} c_{ij}c_{ij}^* \text{ converges in } \|\cdot\|_{\mathscr{A}}\Big\}$$

with the inner product

$$\langle \{c_{ij}\}_{j\in\sigma_i, i\in I}, \{d_{ij}\}_{j\in\sigma_i, i\in I}\rangle = \sum_{i\in I}\sum_{j\in\sigma_i} c_{ij}d_{ij}^*$$

Let the family of K-frames $\{F_i = \{f_{ij}\}_{j\in J} : i \in I\}$ be woven for $\mathscr{H}$, for any partition $\{\sigma_i\}_{i\in I}$ of J and $W = \{f_{ij}\}_{j\in\sigma_i, i\in I}$ be a K-frame for $\mathscr{H}$, then the corresponding

synthesis operator, analysis operator and frame operator defined as follows:
The operator $T_W \colon \bigoplus_{i\in I} l^2(\sigma_i) \to \mathscr{H}$ defined by

$$T_W(\{c_{ij}\}) = \sum_{i\in I} T_{F_i} D_{\sigma_i}(\{c_{ij}\}) = \sum_{i\in I}\sum_{j\in\sigma_i} c_{ij} f_{ij} \tag{11}$$

is called the synthesis or pre-frame operator, where T_{F_i} is the synthesis operator of F_i, and D_{σ_i} is a $|J| \times |J|$ diagonal matrix with $d_{jj} = 1$ for $j \in \sigma_i$ and otherwise 0. The adjoint operator $T_W^* \colon \mathscr{H} \to \bigoplus_{i\in I} l^2(\sigma_i)$ is given by

$$T_W^*(f) = \sum_{i\in I} D_{\sigma_i} T_{F_i}^{\sigma_i *}(f) = \{\langle f, f_{ij}\rangle\}_{j\in\sigma_i, i\in I} \tag{12}$$

and is called the analysis operator.
By composing T_W and T_W^*, the frame operator is defined as $S_W \colon \mathscr{H} \to \mathscr{H}$

$$S_W(f) = T_W T_W^*(f) = \left(\sum_{i\in I} T_{F_i} D_{\sigma_i}\right)\left(\sum_{i\in I} T_{F_i} D_{\sigma_i}\right)^* = \sum_{i\in I}\sum_{j\in\sigma_i} \langle f, f_{ij}\rangle f_{ij} \tag{13}$$

Now, in the following results, we state some of the essential properties of the synthesis operator, analysis operator, and frame operator of weaving K-frames in the Hilbert C^*-module.

Lemma 7.1 *Let $\{\{f_{ij}\}_{j=1}^{\infty} : i \in I\}$ be a woven Bessel sequence, then the synthesis operator T_W is linear and bounded.*

Lemma 7.2 *Let $\{\{f_{ij}\}_{j=1}^{\infty} : i \in I\}$ be a woven frame for $\mathscr{H}$ with universal bounds A and B. Then, the frame operator S_W is self-adjoint, positive, bounded, and invertible on $\mathscr{H}$.*

We now give an example of woven K-frames in the Hilbert C^*-module.

Example 7.1 Let l^∞ be the unitary C^*-algebra of all bounded complex-valued sequences with the following operations

$$uv = \{u_i v_i\}_{i\in\mathbb{N}},\ u^* = \{\overline{u_i}\}_{i\in\mathbb{N}},\ \forall\, u = \{u_i\}_{i\in\mathbb{N}}, v = \{v_i\}_{i\in\mathbb{N}} \in l^\infty$$

Let $\mathscr{H} = C_0$ be the set of all sequences converging to zero. Then C_0 is a Hilbert l^∞-module with l^∞-valued inner product

$$\langle u, v\rangle = uv^* = \{u_i v_i^*\}_{i\in\mathbb{N}} = \{u_i \overline{v_i}\}_{i\in\mathbb{N}} \ \forall\ u, v \in C_0$$

Let $\mathbb{J} = \mathbb{N}$ and we define $\phi = \{\phi_{1j}\}_{j=1}^\infty \in \mathscr{H}$ and $\psi = \{\phi_{2j}\}_{j=1}^\infty \in \mathscr{H}$ as follows:

$$\{\phi_{1j}\}_{j=1}^\infty = \{0, e_3, 0, e_4, 0, e_5, 0, e_6, \ldots\}$$
$$\{\phi_{2j}\}_{j=1}^\infty = \{0, e_3, e_3, e_4, e_4, e_5, e_5, e_6, e_6, \ldots\}$$

where $\{e_j\}_{j=1}^\infty$ be the standard orthonormal basis for $\mathscr{H}$.
Let $f = \{\alpha_1, \alpha_2, \alpha_3, \alpha_4, \alpha_5, \ldots\} \in \mathscr{H}$. Then, $\langle f, f\rangle = \{\alpha_1\alpha_1^*, \alpha_2\alpha_2^*, \alpha_3\alpha_3^*, \alpha_4\alpha_4^*, \ldots\}$. Here, partial ordering $'\leq'$ means pointwise comparison.
For any subset σ of $\mathbb{N}$, we have

$$\sum_{j\in\sigma}\langle f, \phi_{1j}\rangle\langle\phi_{1j}, f\rangle + \sum_{j\in\sigma^c}\langle f, \phi_{2j}\rangle\langle\phi_{2j}, f\rangle \leq 2\sum_{j=1}^{\infty}\langle f, e_j\rangle\langle e_j, f\rangle = 2\langle f, f\rangle$$

On the other hand, let $f \in \mathscr{H}$. Then, $f = \sum_{j=1}^{\infty}\alpha_j e_j$. Thus, we have

$$\begin{aligned}
\langle K^*f, K^*f\rangle &= \left\langle K^*(\sum_{j=1}^{\infty}\alpha_j e_j), K^*(\sum_{j=1}^{\infty}\alpha_j e_j)\right\rangle \\
&= \left\langle \sum_{j=3}^{\infty}\alpha_j e_j, \sum_{j=3}^{\infty}\alpha_j e_j\right\rangle \\
&= \sum_{j=3}^{\infty}\langle f, e_j\rangle\langle e_j, f\rangle \\
&\leq \sum_{j\in\sigma}\langle f, \phi_{1j}\rangle\langle\phi_{1j}, f\rangle + \sum_{j\in\sigma^c}\langle f, \phi_{2j}\rangle\langle\phi_{2j}, f\rangle
\end{aligned}$$

Hence, ϕ and ψ are K-woven frame with universal lower and upper frame bounds 1 and 2, respectively.

Definition 7.2 The sequence $\{\{f_{ij}\}_{j=1}^\infty : i \in I\}$ of $\mathscr{H}$ is said to be a woven atomic system for $K \in L(\mathscr{H})$, if for any partition $\{\sigma_i\}_{i\in I}$ of $\mathbb{N}$, the family $\bigcup_{i\in I}\{f_{ij}\}_{j\in\sigma_i}$ is a woven atomic system for K, i.e., the following statements hold:

(i) The series $\sum_{i\in I}\sum_{j\in\sigma_i} c_{ij} f_{ij}$ converges for all $\{c_{ij}\}_{j\in\sigma_i, i\in I} \in l^2(\mathscr{A})$.

(ii) There exist $C > 0$ such that for every $f \in \mathscr{H}$, there exists $\{a_{ij,f}\}_{j\in\sigma_i, i\in I} \in l^2(\mathscr{A})$ such that $\sum_{i\in I}\sum_{j\in\sigma_i} a_{ij,f}a^*_{ij,f} \le C\langle f, f\rangle$ and $Kf = \sum_{i\in I}\sum_{j\in\sigma_i} a_{ij,f} f_{ij}$.

Theorem 7.1 *If $K \in L(\mathscr{H})$, then there exists a woven atomic system for K.*

The following result gives an equivalent condition of weaving K-frames in the Hilbert C^*-module as it is much easier to be applied.

Theorem 7.2 *For any partition $\{\sigma_i\}_{i\in I}$ of $\mathbb{N}$, let the family $\bigcup_{i\in I}\{f_{ij}\}_{j\in\sigma_i}$ be a woven Bessel sequence for $\mathscr{H}$ and $K \in L(\mathscr{H})$. Suppose that $T^* \in L(\mathscr{H}, l^2(\mathscr{A}))$ given by $T^*(f) = \{\langle f, f_{ij}\rangle\}_{j\in\sigma_i, i\in I}$ and $\overline{R(T^*)}$ is orthogonally complemented then the following statements are equivalent:*

(i) The sequence $\{\{f_{ij}\}_{j=1}^{\infty} : i \in I\}$ of $\mathscr{H}$ is a woven atomic system for K.

(ii) There exist $A, B > 0$ such that

$$A\|K^*f\|^2 \le \|\sum_{i\in I}\sum_{j\in\sigma_i}\langle f, f_{ij}\rangle\langle f_{ij}, f\rangle\| \le B\|f\|^2, \forall f \in \mathscr{H}.$$

(iii) There exist $D \in L(\mathscr{H}, l^2(\mathscr{A}))$ such that $K = TD$.

The following theorem gives a characterization of weaving K-frames in terms of a bounded linear operator in the Hibert C^*-module.

Theorem 7.3 *For each $i \in I$, suppose $\{\{f_{ij}\}_{j=1}^{\infty} : i \in I\}$ is a K-frame for $\mathscr{H}$ with bounds A_i and B_i. Then, the following conditions are equivalent:*

(i) The family $\{\{f_{ij}\}_{j=1}^{\infty} : i \in I\}$ is K-woven.

(ii) There exists $A > 0$ such that for any partition $\{\sigma_i\}_{i\in I}$ of $\mathbb{N}$, there exists a bounded linear operator $M_\sigma : l^2(\mathscr{A}) \to \mathscr{H}$ such that

$$M_\sigma(e_j) = \begin{cases} f_{1j}, & j \in \sigma_1 \\ f_{2j}, & j \in \sigma_2 \\ \cdot & \\ \cdot & \\ f_{mj}, & j \in \sigma_m \end{cases}$$

and $AKK^* \le M_\sigma M_\sigma^*$, where $\{e_j\}_{j=1}^{\infty}$ is the canonical orthonormal basis.

8 Fusion Frames in Hilbert C^*-modules

The fusion frame is a generalization of frames introduced by Cassaza and Kutyniok [7] in 2003 and investigated in [2, 8, 22, 24]. The purpose of introducing the fusion frame or frame of the subspace is first to construct local components and then build a global frame from these, i.e., construction of global frames from local frames in Hilbert space. Fusion frames have vast applications in various fields such as distributed sensing, parallel processing, and packet encoding. In [23], fusion frames in Hilbert C^*modules are introduced, and authors showed that they share many useful properties with their corresponding notions in Hilbert space.

A closed submodule M of a Hilbert C^*-module X is complemented if for some closed submodule N of X we have $X = M \bigoplus N$ and $\pi_M : X \to M$ is the orthogonal projection of M. We say M is orthogonally complemented if $X = M \bigoplus M^{\perp}$ and in this case $\pi_M \in \text{End}^*_A(X, M)$.

Definition 8.1 [23] Let $\mathscr{A}$ be a unital C^*-algebra, X be a Hilbert $\mathscr{A}$-module, I be finite or countable index set and let $\{v_i : i \in I\}$ be a family of weights in $\mathscr{A}$, i.e., each v_i is a positive invertible element from the center of the C^*-algebra $\mathscr{A}$. A sequence of closed submodules $\{M_i : i \in I\}$ is a frame of submodules if every M_i is orthogonally complemented and there exist real constants $0 < C \leq D < \infty$ such that

$$C\langle x, x\rangle \leq \sum_{i\in I} v_i^2 \langle \pi_{M_i}(x), \pi_{M_i}(x)\rangle \leq D\langle x, x\rangle \quad \text{for} \quad x \in X \tag{14}$$

C and D are the lower and upper bounds of the frame of submodules and $\{(M_i, v_i): i \in I\}$ is called a fusion frame. If $C = D = \lambda$, the family $\{M_i : i \in I\}$ is called a λ-tight frame of submodules with respect to $\{v_i : i \in I\}$ and if $C = D = 1$, it is called a Parseval or a normalized tight frame of submodules.

Example 8.1 Let $\{M_i : i \in I\}$ be a sequence of Hilbert $\mathscr{A}$-modules and

$$X = \bigoplus_{i\in I} M_i = \{x = (x_i) : x_i \in M_i \text{ and } \sum_{i\in I} \langle x_i, x_i\rangle \text{ is norm convergent in } \mathscr{A}\}.$$

Then, X is a Hilbert $\mathscr{A}$-module with $\mathscr{A}$-valued inner product $\langle x, y\rangle = \sum_{i\in I} \langle x_i, y_i\rangle$ where $x = (x_i)_{i\in I}$ and $y = (y_i)_{i\in I}$, pointwise operations and the norm defined by $\|a\| = \|\langle a, a\rangle\|_{\mathscr{A}}^{\frac{1}{2}}$. Then, $\{M_i : i \in I\}$ is a standard Parseval frame of submodules of X with respect to $\{v_i : i \in I\}$, where $v_i = 1$ for each $i \in I$.

Theorem 8.1 *[23] Let $\{v_i : i \in I\}$ be a family of weights in $\mathscr{A}$. Let for each $i \in I$, M_i be an orthogonally complemented submodule of X, and let $\{f_{ij} : j \in I_i\}$ be a frame for M_i with bounds C_i and D_i. Suppose $0 < C = \inf C_i \leq D = \sup_i D_i < \infty$. Then, the following conditions are equivalent* *(a)* $\{v_i f_{ij} : i \in I, j \in I_i\}$*is a frame for* X,

(b) $\{(M_i, v_i) : i \in I\}$ *is a fusion frame for* X.

References

1. Abbaspour, T.G., Arefijamaal, A.A., Mahmoudieh, M.: K-frames in Hilbert C^*-modules. arXiv: 1802.01947 [math.OA] (2018)
2. Asgari, M.S., Khosravi, A.: Frames and bases of subspaces in Hilbert spaces. J. Math. Anal. Appl. **308**, 541–553 (2005)
3. Balazs, P., Antoine, J.-P., Grybos, A.: Weighted and controlled frames. Int. J. Walvelets Multi. Inf. Process. **8(1)**, 109–132 (2010)
4. Travis, B., Casazza, P.G., Gröchenig, K., Lammers, M.C., Lynch, R.G.: Weaving frames. Oper. Matrices **10(4)**, 1093–1116 (2016)
5. Bolcskei, H., Hlawatsch, F., Feichtinger, H.G.: Frame-theoretic analysis of oversampled filter banks. IEEE Trans. Signal Process. **46**(12), 3256–3268 (1998)
6. Candes, E.J., Donoho, D.L.: New tight frames of curvelets and optimal representations of objects with piecewise C^2 singularities. Commun. Pure Appl. Math. **57**(2), 219–266 (2004)
7. Casazza, P.G., Kutyniok, G.: Frames of Subspaces. Contemp. Math. Am. Math. Soc. **345**, 87–113 (2004)
8. Casazza, P.G., Kutyniok, G., Li, Sh: Fusion frames and distributed processing. Appl. Comput. Harmon. Anal. **25**, 114–132 (2008)
9. Casazza, P.G., Lynch, R.G.: Weaving properties of hilbert space frames. Sampling Theory and Applications (SampTA). IEEE, pp. 110–114 (2015)
10. Casazza, P.G., Freeman, D., Lynch, R.G.: Weaving schauder frames. J. Approx. Theory **211**, 42–60 (2016)
11. Daubechies, I., Grossmann, A., Meyer, Y.: Painless non orthogonal expansions. J. Math. Phys. **27**, 1271–1283 (1986)
12. Deepshikha, V.L.K.: Weaving K-frames in Hilbert spaces. Res. Math. **73**, 81 (2018)
13. Duffin, R.J., Schaeffer, A.C.: A class of nonharmonic Fourier series. Trans. Math. Soc. **72**, 341–366 (1952)
14. Eldar, Y.C.: Sampling with arbitrary sampling and reconstruction spaces and oblique dual frame vectors. J. Fourier. Anal. Appl. **9**(1), 77–96 (2003)
15. Eldar, Y.C., Werther, T.: General framework for consistent sampling in Hilbert spaces. Int. J. Walvelets Multi. Inf. Process. **3**(3), 347–359 (2005)
16. Ferreira P.J.S.G.: Mathematics for multimedia signal processing II: Discrete finite frames and signal reconstruction. In: Byrnes, J.S. (Ed.) Signals Processing for Multimedia. pp. 35–54 (1999)
17. Frank, M., Larson, D.R.: Frames in Hilbert C^*-modules and C^*-algebras. J. Oper. Theory **48**, 273–314 (2002)
18. Găvruta, L.: Frames for operators. Appl. Comput. Harmon. Anal. **32**, 139–144 (2012)
19. Ghobadzadeh, F., Najati, A., Anastassiou, G.A., Park, C.: Woven frames in Hilbert C^*-module spaces. J. Comput. Anal. Appl. **25**, 1220–1232 (2018)
20. Jing W.: Frames in Hilbert C^*-modules. Ph.D. Thesis, University of Central Frorida (2006)
21. Kaplansky, I.: Algebra of type I. Ann. Math. **56**, 460–472 (1952)
22. Khosravi, A., Asgari, M.S.: Frames of subspaces and approximation of the inverse frame operator. Houston. J. Math. **33**, 907–920 (2007)
23. Khosravi, A., Khosravi, B.: Fusion frames and g-frames in Hilbert C^*-modules. Int. J. Wavelets Multiresolut. Inf. Process. **6**, 433–466 (2008)

24. Khosravi, A., Musazadeh, K.: Fusion frames and g-frames. J. Math. Anal. Appl. **342**, 1068–1083 (2008)
25. Kouchi, M.R., Rahimi, A.: On controlled frames in Hilbert C^*-modules. Int. J. Walvelets Multi. Inf. Process. **15**(4), 1750038 (2017)
26. Najati, A., Saem, M.M., Gavruta, P.: Frames and operators in Hilbert C^*-modules. Oper. Matrices **10**(1), 73–81 (2016)
27. Rajput, E., Sahu, N.K.: Controlled K-frames in Hilbert C^*-modules. arXiv:1903.09928 (2019)
28. Strohmer, T., Heath Jr., R.: Grassmanian frames with applications to coding and communications. Appl. Comput. Harmon. Anal. **14**, 257–275 (2003)

A Class of Variational-Like Inequalities in Banach Spaces

Variational-Like Inequalities

R. N. Mohapatra, Bijaya Kumar Sahu, and Sabyasachi Pani

Abstract In this paper, we find some properties of generalized relaxed $\eta - \alpha$ quasimonotone mappings and generalized relaxed $\eta - \alpha$ properly quasimonotone mappings. With these generalized monotone mappings, we have proved some existence results for the variational-like inequalities. Some relations between Minty variational-like inequality problem and Stampacchia variational-like inequality problem were then established. We have also established some existence of solutions for the variational-like inequalities with densely relaxed $\eta - \alpha$ pseudomonotone operators.

Keywords Variational-like inequalities · Generalized relaxed $\eta - \alpha$ quasimonotonicity · Generalized relaxed $\eta - \alpha$ properly quasimonotonicity · Densely relaxed $\eta - \alpha$ pseudomonotone operators · KKM mapping

Mathematics Subject Classification (2010) 47H05 · 47J20 · 47J30 · 49J40

R. N. Mohapatra
Department of Mathematics, University of Central Florida,
Orlando, FL 32816, USA
e-mail: ramm1627@gmail.com

B. K. Sahu (✉) · S. Pani
School of Basic Sciences, Indian Institute of Technology Bhubaneswar,
Bhubaneswar 752050, India
e-mail: bks10@iitbbs.ac.in

S. Pani
e-mail: spani@iitbbs.ac.in

R. N. Mohapatra et al. (eds.), *Mathematical Analysis and Computing*,
Springer Proceedings in Mathematics & Statistics 344,
https://doi.org/10.1007/978-981-33-4646-8_47

1 Introduction

The variational inequality problem, that was given by Italian mathematician Stampacchia [15] in the year 1966, in connection with the work on regularity of solutions for elliptic equations now has many refinements along with applications to plasticity, fluid dynamics, filtration, etc., see for instance [1, 4, 8, 14]. The variational-like inequalities were introduced by Fang et al. [12] in 2003, where they have used the relaxed $\eta - \alpha$ monotone mappings and proved that the variational-like inequalities have solutions in reflexive Banach spaces. The problem was then solved in more general settings by Bai et al. [6] in 2006. They have defined relaxed $\eta - \alpha$ pseudomonotone mappings and used it to find the solutions of the variational-like inequalities in reflexive Banach spaces. Again in 2007, Bai et al. [7] defined densely relaxed μ pseudomonotonicity and proved that the variational inequalities have solutions. This problem was then studied by various authors, see for instance [2, 17, 19, 23]. The operators associated with the variational inequalities play key roles in its generalizations. For instance, Verma [24] defined $p-$monotone type maps and proved that the nonlinear variational inequality (NVI) problems have solutions, Sahu et al. [22] defined (A, η)-maximal monotonicity, Bai et al. [6] defined relaxed $\eta - \alpha$ pseudomonotonicity, Pany et al. [20] defined generalized weakly relaxed $\eta - \alpha$ monotonicity and proved that the variational inequality problems have solutions. The researchers like Chadli et al. [9] and Sahu and Pani [21] extended and applied the variational inequalities into equilibrium problems.

In 1999, Daniilidis and Hadjisavvas [11] defined properly quasimonotonicity of an operator T and proved that the variational inequality problem (2) has solutions for T to be a multivalued mapping. In 2004, Aussel and Hadjisavvas [3] further generalized the results of Hadjisavvas and Schaible [16] by considering multivalued mapping T to be upper sign-continuous. In 2013, Chen and Luo [10] established the existence results for the variational-like inequality problem (1) for relaxed $\eta - \alpha$ quasimonotone mappings.

Very recently Sahu et al. [23] defined the following variational-like inequality problem: Find a vector $x \in K$ such that

$$\langle Tx, \eta(y, x)\rangle \geq 0, \forall y \in K, \tag{1}$$

where K is a nonempty subset of real reflexive Banach space X with its dual X^* and $\eta : K \times K \to X, T : K \to X^*$ are the mappings. If we take $\eta(y, x) = y - x$, then the problem (1) reduces to classical variational inequality problem: Find a vector $x \in K$ such that

$$\langle Tx, y - x\rangle \geq 0, \forall y \in K. \tag{2}$$

Inspired and motivated by these researches, we have established certain relations between generalized relaxed $\eta - \alpha$ quasimonotone mappings and generalized relaxed $\eta - \alpha$ properly quasimonotone mappings. Some existence results for the variational-like inequality problem (1) with generalized relaxed $\eta - \alpha$ quasimonotone mappings are then studied. Lastly, we have found some solutions for the variational-like inequalities with generalized relaxed $\eta - \alpha$ properly quasimonotone mappings. The results that we have proved in this paper extend and generalize many results of Luc [18], Sahu et al. [23], Bai and Hadjisavvas [5] and Daniilidis and Hadjisavvas [11].

The paper is organized as follows. In Sect. 2, we provide some definitions and preliminaries that are required in the sequel along with generalized relaxed $\eta - \alpha$ quasimonotone mappings and generalized relaxed $\eta - \alpha$ properly quasimonotone mappings. Section 3 is devoted to study the existence of solutions for the variational-like inequalities with generalized relaxed $\eta - \alpha$ quasimonotone mappings. We proceed to find some relations between Minty variational-like inequality problem and Stampacchia variational-like inequality problem. In Sect. 4, we focus our study on variational-like inequalities in reflexive Banach spaces with generalized relaxed $\eta - \alpha$ properly quasimonotone mappings and generalized densely relaxed $\eta - \alpha$ pseudomonotone mappings.

2 Preliminaries

In this section, we recall some preliminary concepts and results that are needed in sequel. Assume X be a real normed space and X^* be its dual. Let K be a nonempty subset of X and $\langle \cdot, \cdot \rangle$ denotes the pairing between X^* and X. For $y_1, y_2, ..., y_n$ of n elements in K, the convex hull of $y_1, y_2, ..., y_n$ is denoted by $co\{y_1, y_2, ..., y_n\}$. A real-valued function f defined on a convex subset K of X is said to be positively homogeneous if $f(\lambda x) = \lambda f(x)$ for $\lambda > 0$.

Definition 1 Let K be a convex subset of X and $f : K \to \mathbb{R}$ be a real-valued function, then f is said to be

(i) hemicontinuous if for any $x, y \in K$ fixed,

$$\lim_{t \to 0^+} f(x + t(y - x)) = f(x);$$

(ii) lower semicontinuous at $x \in X$ if for any sequence $\{x_n\}_{n \in \mathbb{N}} \subset X$ converging to x, we have $f(x) \leq \liminf_{n \to \infty} f(x_n)$;
(iii) lower hemicontinuous if for any $x, y \in K$, the functional $t \mapsto f(x + t(y - x))$ is lower semicontinuous at 0^+.

Definition 2 [6] Let $T : K \to X^*$ and $\eta : K \times K \to X$ be two mappings. T is said to be η-hemicontinuous if, for any fixed $x, y \in K$, the mapping $f : [0, 1] \to \mathbb{R}$ defined by $f(t) = \langle T(x + t(y - x)), \eta(y, x) \rangle$ is continuous at 0^+.

The relaxed $\eta - \alpha$ pseudomonotone operator defined by Arunchai et al. [2] is the following.

Definition 3 [2] An operator $T : K \to X^*$ is said to be relaxed $\eta - \alpha$ pseudomonotone if there exists a function $\eta : K \times K \to X$ and a function $\alpha : X \to \mathbb{R}$ with $\lim_{t \to 0^+} \frac{\alpha(t\eta(x, y))}{t} = 0, \forall (x, y) \in K \times K$ such that for any $x, y \in K$, we have

$$\langle Tx, \eta(y, x)\rangle \geq 0 \implies \langle Ty, \eta(y, x)\rangle \geq \alpha(\eta(y, x)). \tag{3}$$

From now onwards in this paper, we shall use the term generalized relaxed $\eta - \alpha$ pseudomonotone operator for the above relaxed $\eta - \alpha$ pseudomonotone operator to distinguish from the one defined by Bai et al. [6]. They defined the relaxed $\eta - \alpha$ pseudomonotone operator in the following way.

Definition 4 [6] An operator $T : K \to X^*$ is said to be relaxed $\eta - \alpha$ pseudomonotone if there exists a function $\eta : K \times K \to X$ and a function $\alpha : X \to \mathbb{R}$ with $\alpha(tz) = t^p \alpha(z)$ for all $t > 0$ and $z \in X$ such that for any $x, y \in K$, we have

$$\langle Tx, \eta(y, x)\rangle \geq 0 \implies \langle Ty, \eta(y, x)\rangle \geq \alpha(y - x). \tag{4}$$

The following lemma given by Arunchai et al. [2] will be needed in the sequel.

Lemma 1 ([2], Theorem 3.1) *Let K be a nonempty closed convex subset of a real reflexive Banach space X and $T : K \to X^*$ and $\eta : K \times K \to X$ be mappings. Assume that*

(i) T is η-hemicontinuous and generalized relaxed $\eta - \alpha$ pseudomonotone;
(ii) $\eta(x, x) = 0, \forall x \in K$;
(iii) $\eta(tx + (1 - t)z, y) = t\eta(x, y) + (1 - t)\eta(z, y), \ \forall x, y, z \in K$ and $t \in [0, 1]$. Then, $x \in K$ is a solution of variational-like inequality problem (1) if and only if

$$\langle Ty, \eta(y, x)\rangle \geq \alpha(\eta(y, x)), \ \forall y \in K. \tag{5}$$

Let K be a convex set in X and K_0 be a subset of K. Luc [18] said that the set K_0 is a *segment-dense* in K if for each $x \in K$, there exits $x_0 \in K_0$ such that x is a cluster point of the set $[x, x_0] \cap K_0$.

The operator $T : K \to X^*$ is said to be generalized densely relaxed $\eta - \alpha$ pseudomonotone mappings on K if there exists a segment-dense subset $K_0 \subset K$ such that T is generalized relaxed $\eta - \alpha$ pseudomonotone at every point of K_0.

The following lemma given by Sahu et al. [23] for the variational-like inequality problem (1) with generalized densely relaxed $\eta - \alpha$ pseudomonotone will be needed in the sequel.

Lemma 2 ([23], Theorem 3.1) *Let K be a nonempty, convex and compact subset of a normed space X and $T : K \to X^*$ be an $\eta-$hemicontinuous and generalized densely relaxed $\eta - \alpha$ pseudomonotone on K. Suppose that*

(i) $\eta(x, y) + \eta(y, x) = 0$, for all $x, y \in K$;

(ii) $\eta(tx + (1-t)z, y) = t\eta(x, y) + (1-t)\eta(z, y)$, *for all* $x, y, z \in K$ *and* $t \in [0, 1]$;

(iii) *For each fixed* $w, z \in K$, *the mapping* $y \in K \mapsto \alpha(\eta(y, z))$ *is lower hemicontinuous and the mapping* $x \in K \mapsto \alpha(\eta(w, x))$ *is lower semicontinuous;*

(iv) *For each fixed* $x, z \in K$, *the mapping* $y \in K \mapsto \langle Tx, \eta(y, z)\rangle$ *is lower semicontinuous.*

Then, the variational-like inequality problem (1) has a solution.

Definition 5 An operator $T : K \to X^*$ is said to be generalized relaxed $\eta - \alpha$ quasimonotone, if there exists a function $\eta : K \times K \to X$ and a function $\alpha : X \to \mathbb{R}$ with $\lim\limits_{t\to 0^+} \dfrac{\alpha(t\eta(x, y))}{t} = 0, \forall (x, y) \in K \times K$ such that for any $x, y \in K$, we have

$$\langle Tx, \eta(y, x)\rangle > 0 \implies \langle Ty, \eta(y, x)\rangle \geq \alpha(\eta(y, x)). \tag{6}$$

If $\alpha(\eta(y, x)) = \alpha(y - x)$, then (6) implies that T is relaxed $\eta - \alpha$ quasimonotone mapping established by Chen and Luo [10] in 2013:

$$\langle Tx, \eta(y, x)\rangle > 0 \implies \langle Ty, \eta(y, x)\rangle \geq \alpha(y - x).$$

Again if $\eta(y, x) = y - x$ and $\alpha(u) = -\mu\|u\|^2$, then (6) implies that T is relaxed μ quasimonotone mapping given by Bai et al. [7] in 2007:

$$\langle Tx, y - x\rangle > 0 \implies \langle Ty, y - x\rangle \geq -\mu\|y - x\|^2.$$

Taking $\eta(y, x) = y - x$ and $\alpha(u) = 0$ in (6), we see that our generalized relaxed $\eta - \alpha$ quasimonotone mapping reduces to quasimonotone mapping given by Luc [18] in 2001:

$$\langle Tx, y - x\rangle > 0 \implies \langle Ty, y - x\rangle \geq 0.$$

Definition 6 An operator $T : K \to X^*$ is said to be generalized relaxed $\eta - \alpha$ properly quasimonotone, if there exists a function $\eta : K \times K \to X$ and a function $\alpha : X \to \mathbb{R}$ with $\lim\limits_{t\to 0^+} \dfrac{\alpha(t\eta(x, y))}{t} = 0, \forall (x, y) \in K \times K$ such that for any $y_1, y_2, ..., y_n \in K$ and $x \in co\{y_1, y_2, ..., y_n\}$, there exists $i \in \{1, 2, ..., n\}$ such that

$$\langle Ty_i, \eta(y_i, x)\rangle \geq \alpha(\eta(y_i, x)). \tag{7}$$

Remark 1 a. If we take $\alpha(\eta(y, x)) = \alpha(y - x)$, then the generalized relaxed $\eta - \alpha$ properly quasimonotonicity reduces to the relaxed $\eta - \alpha$ properly quasimonotonicity given by Chen and Luo in [10].

b. If we take $\eta(y, x) = y - x$ and $\alpha(u) = 0$, then we will get properly quasimonotonicity given by Daniilidis and Hadjisavvas [11] in case of T single valued.

Let K be a convex set in X and K_0 a subset of K. According to Luc [18], the set K_0 is said to be *segment-dense* in K if for each $x \in K$, there exits $x_0 \in K_0$ such that x is a cluster point of the set $[x, x_0] \cap K_0$.

Definition 7 [23] Let $T : K \to X^*$ and $\eta : K \times K \to X$ be mappings, and let $\alpha : X \to \mathbb{R}$ be a function such that $\lim\limits_{t \to 0^+} \dfrac{\alpha(t\eta(x, y))}{t} = 0$, for all $(x, y) \in K \times K$. The operator $T : K \to X^*$ is said to be generalized densely relaxed $\eta - \alpha$ pseudomonotone mappings on K if there exists a segment-dense subset $K_0 \subset K$ such that T is generalized relaxed $\eta - \alpha$ pseudomonotone at every point of K_0.

An element $u \in X^*$ is $\eta-$perpendicular to K if for all x, y in K, $\langle u, \eta(x, y)\rangle = 0$. The set of all $\eta-$perpendiculars to K is denoted by $K^{\perp}$. If $T : K \to X^*$ is an operator, then we shall denote $T(K) = \{T(x) : x \in K\}$.

Definition 8 [19] A point $x_0 \in K$ is said to be an $\eta-$positive point of the mapping $T : K \to X^*$ on K if for all $x \in K$, either $T(x) \in K^{\perp}$ or there exists a vector $y \in K$ such that $\langle Tx, \eta(y, x_0)\rangle > 0$. The set of all $\eta-$positive points of T on K is denoted by K_T.

Definition 9 [13] The set-valued mapping $f : K \to 2^X$ is said to be a KKM mapping if for any finite subset $\{y_1, y_2, ..., y_n\}$ of K, we have $co\{y_1, y_2, ..., y_n\} \subset \bigcup_{i=1}^{n} f(y_i)$.

The following lemma given by Fan [13] will be needed in the sequel.

Lemma 3 [13] *Let K be a nonempty subset of a Hausdorff topological vector space X and let $f : K \to 2^X$ be a KKM mapping. If $f(y)$ is closed in X for all $y \in K$ and compact for some $y \in K$, then*

$$\bigcap_{y \in K} f(y) \neq \phi.$$

3 Variational-Like Inequalities with Generalized Relaxed $\eta - \alpha$ Quasimonotone Mappings

In this section, we establish some existence results for the variational-like inequalities (1) with the generalized relaxed $\eta - \alpha$ quasimonotone mappings. We then establish some relations between Minty variational-like inequality problem and Stampacchia variational-like inequality problem.

Lemma 4 *Let K be a nonempty convex subset of a normed space X with its dual X^*. Assume that the mapping $T : K \to X^*$ is η-hemicontinuous and generalized relaxed $\eta - \alpha$ quasimonotone. Suppose that*

(i) $\eta(tx + (1-t)z, y) = t\eta(x, y) + (1-t)\eta(z, y)$, *for all* $x, y, z \in K$ *and* $t \in [0, 1]$;

(ii) *For each fixed* $w, z \in K$, *the mapping* $y \in K \mapsto \alpha(\eta(y, z))$ *is lower hemicontinuous.*

If for each $y \in K$, *there exists an* $x \in K$ *such that* $\langle T(x), \eta(y, x)\rangle \geq 0$, *then either* $\langle T(y), \eta(y, x)\rangle \geq \alpha(\eta(y, x))$ *or* $\langle T(x), \eta(z, x)\rangle \leq 0$, *for all* $z \in K$.

Proof Suppose $\langle T(x), \eta(z, x)\rangle > 0$, for some $z \in K$. Let $y_t = tz + (1-t)y$, for $t \in (0, 1]$, then $y_t \in K$. Consider

$$\langle T(x), \eta(y_t, x)\rangle = t\langle T(x), \eta(z, x)\rangle + (1-t)\langle T(x), \eta(y, x)\rangle > 0.$$

Since T is generalized relaxed $\eta - \alpha$ quasimonotone, we have

$$\langle T(y_t), \eta(y_t, x)\rangle \geq \alpha(\eta(y_t, x)).$$

$$\Longrightarrow t\langle T(y_t), \eta(z, x)\rangle + (1-t)\langle T(y_t), \eta(y, x)\rangle > \alpha(\eta(tz + (1-t)y, x)).$$

Since T is η-hemicontinuous, letting $t \to 0$ and using condition (ii), we deduce that

$$\langle T(y), \eta(y, x)\rangle \geq \alpha(\eta(y, x)), \tag{8}$$

which establishes the first inequality. To prove the second inequality, let us assume that

$$\langle T(y), \eta(y, x)\rangle < \alpha(\eta(y, x)).$$

Suppose on contrary $\langle T(x), \eta(z, x)\rangle > 0$, for some $z \in K$. Then, we get relation (8), which contradicts our assumption. Hence,

$$\langle T(x), \eta(z, x)\rangle \leq 0, \text{ for all } z \in K,$$

and completes the proof. □

Remark 2 Taking $\eta(y, x) = y - x$ and $\alpha(u) = -\mu\|u\|^2$, we see that Lemma 4 reduces to Lemma 2.1 of Bai et al. [7]. Therefore, Lemma 4 generalizes Lemma 2.1 of Bai et al. [7]. Lemma 4 also generalizes Lemma 3.1 of Hadjisavvas and Schaible [16] from quasimonotone operators to generalized relaxed $\eta - \alpha$ quasimonotone operator.

Lemma 5 *Let* X *be a normed space with its dual* X^* *and* K *be a nonempty convex subset of* X. *Assume that the mapping* $T : K \to X^*$ *is* η*-hemicontinuous and generalized relaxed* $\eta - \alpha$ *quasimonotone on* K. *Suppose that*

(i) $T(K) \cap K^{\perp} = \phi$;
(i) $\eta(tx + (1-t)z, y) = t\eta(x, y) + (1-t)\eta(z, y)$, *for all* $x, y, z \in K$ *and* $t \in [0, 1]$;
(ii) *For each fixed* $w, z \in K$, *the mapping* $y \in K \mapsto \alpha(\eta(y, z))$ *is lower hemicontinuous.*

Then, T is generalized relaxed $\eta - \alpha$ pseudomonotone on K_T.

Proof Suppose $x \in K_T$ and $y \in K$ be any point. Let $\langle Tx, \eta(y, x)\rangle \geq 0$. Since all the conditions of Lemma 4 are satisfied, from Lemma 4 we have

$$\text{either } \langle T(y), \eta(y, x)\rangle \geq \alpha(\eta(y, x)) \text{ or } \langle T(x), \eta(z, x)\rangle \leq 0, \text{ for all } z \in K. \quad (9)$$

Suppose

$$\langle T(x), \eta(z, x)\rangle \leq 0, \text{ for all } z \in K. \quad (10)$$

Since x is η−positive and $T(K) \cap K^{\perp} = \phi$, there exists a point $u \in K$ such that

$$\langle T(x), \eta(u, x)\rangle > 0,$$

which contradicts relation (10). Thus, from (9), we get

$$\langle T(y), \eta(y, x)\rangle \geq \alpha(\eta(y, x)).$$

Therefore, T is generalized relaxed $\eta - \alpha$ pseudomonotone on K_T. □

Remark 3 Since the generalized relaxed $\eta - \alpha$ quasimonotone mapping is the proper generalization relaxed μ quasimonotone mapping of Bai et al. [7] and quasimonotone mapping of Luc [18], we deduce that Lemma 5 generalizes Proposition 2.1 of Bai et al. [7] and Proposition 3.2 of Luc [18].

Theorem 1 *Let X be a normed space and X^* be its dual. Suppose K be a nonempty compact and convex subset of X and $T : K \to X^*$ is η-hemicontinuous and generalized relaxed $\eta - \alpha$ quasimonotone on K. Furthermore, assume that*

(i) *K_T is segment-dense in K;*
(ii) *$\eta(x, y) + \eta(y, x) = 0$, for all $x, y \in K$;*
(iii) *$\eta(tx + (1 - t)z, y) = t\eta(x, y) + (1 - t)\eta(z, y)$, for all $x, y, z \in K$ and $t \in [0, 1]$;*
(iv) *For each fixed $w, z \in K$, the mapping $y \in K \mapsto \alpha(\eta(y, z))$ is lower hemicontinuous and the mapping $x \in K \mapsto \alpha(\eta(w, x))$ is lower semicontinuous;*
(v) *For each fixed $x, z \in K$, the mapping $y \in K \mapsto \langle Tx, \eta(y, z)\rangle$ is lower semicontinuous.*

Then, the variational-like inequality problem (1) has a solution.

Proof If for some $x_0 \in K$, $T(x_0) \in K^{\perp}$, then x_0 is a solution of the variational-like inequality problem (1). Thus, we may assume that $T(K) \cap K^{\perp} = \phi$. By Lemma 5, T is generalized relaxed $\eta - \alpha$ pseudomonotone at every point of K_T. Therefore, by Lemma 2, the variational-like inequality problem (1) has a solution. □

Remark 4 Theorem 1 is a proper generalization of the Theorem 3.2 of Bai et al. [7] from relaxed μ quasimonotone operators to generalized relaxed $\eta - \alpha$ quasimonotone operators and also Corollary 4.4 of Luc [18] from quasimonotone operators to generalized relaxed $\eta - \alpha$ quasimonotone operators.

Lemma 6 *Let K be a nonempty closed and convex subset of a normed space X and K_0 be segment-dense subset of K. If the mapping $T : K \to X^*$ is generalized relaxed $\eta - \alpha$ quasimonotone on K_0 and η-hemicontinuous on K. Furthermore, if the following conditions hold*

(i) $\eta(x, y) + \eta(y, x) = 0$, for all $x, y \in K$;
(ii) For each fixed $z \in K$, the mapping $y \in K \mapsto \alpha(\eta(y, z))$ is lower hemicontinuous.

Then, the mapping T is generalized relaxed $\eta - \alpha$ quasimonotone on K.

Proof For $x, y \in K$, let

$$\langle Tx, \eta(y, x)\rangle > 0.$$

Since K_0 is segment-dense in K, we can find $y_0 \in K_0$ and a sequence $\{y_n\}_{n\in\mathbb{N}}$ in $[y, y_0] \cap K_0$ such that $\lim_{t\to\infty} y_n = y$. Thus, $y_n = y + t_n(y_0 - y) \in K_0$ with $t_n \in [0, 1]$ and $t_n \to 0$. Assume that,

$$\langle Tx, \eta(y_n, x)\rangle > 0, \ \forall n \in \mathbb{N}.$$

Since T is generalized relaxed $\eta - \alpha$ quasimonotone at y_n, we have

$$\langle Ty_n, \eta(y_n, x)\rangle \geq \alpha(\eta(y_n, x)).$$

By condition (i),

$$(1 - t_n)\langle Ty_n, \eta(y, x)\rangle + t_n\langle Ty_n, \eta(y_0, x)\rangle \geq \alpha(\eta(y_n, x)).$$

Since T is η-hemicontinuous on K, by condition (ii)

$$\langle Ty, \eta(y, x)\rangle \geq \alpha(\eta(y_n, x)).$$

Therefore, T is generalized relaxed $\eta - \alpha$ quasimonotone at x, which completes the proof. □

Remark 5 If K_T is segment-dense in K by Lemma 6, T is generalized relaxed $\eta - \alpha$ quasimonotone on K. Therefore, Theorem 1 is still true if the assumption generalized relaxed $\eta - \alpha$ quasimonotonicity of $T : K \to X^*$ on K is replaced by the following weaker assumption:
The mapping $T : K \to X^*$ is generalized relaxed $\eta - \alpha$ quasimonotone on K_T.

Remark 6 Lemma 6 generalizes Proposition 3.5 of Luc [18] from the case of quasimonotone operator to the case of more general settings of generalized relaxed $\eta - \alpha$ quasimonotone operators.

In Lemma 6, we have proved that if a mapping $T : K \to X^*$ is generalized relaxed $\eta - \alpha$ quasimonotone on a segment-dense subset K_0 of K, then it is generalized relaxed $\eta - \alpha$ quasimonotone on K. Since every generalized relaxed $\eta - \alpha$ pseudomonotone operator is generalized relaxed $\eta - \alpha$ quasimonotone, the natural question arises whether the above property holds for generalized relaxed $\eta - \alpha$ pseudomonotone operator. Unfortunately, we have no such result at this moment in our hand, but in the following corollary we are going to prove that if a mapping T is generalized relaxed $\eta - \alpha$ pseudomonotone on a segment-dense subset of K, then it is generalized relaxed $\eta - \alpha$ pseudomonotone on the set of all η−positive points of T.

Corollary 1 *Let K be a nonempty, compact and convex subset of a normed space X and $T : K \to X^*$ be an η−hemicontinuous and generalized densely relaxed $\eta - \alpha$ pseudomonotone on K. Furthermore, if the following conditions hold*

(i) $T(K) \cap K^{\perp} = \phi$;
(ii) $\eta(x, y) + \eta(y, x) = 0$, *for all* $x, y \in K$;
(iii) $\eta(tx + (1 - t)z, y) = t\eta(x, y) + (1 - t)\eta(z, y)$, *for all* $x, y, z \in K$ *and* $t \in [0, 1]$;
(iv) *For each fixed* $w, z \in K$, *the mapping* $y \in K \mapsto \alpha(\eta(y, z))$ *is lower hemicontinuous;*
(v) *For each fixed* $x, z \in K$, *the mapping* $y \in K \mapsto \langle Tx, \eta(y, z)\rangle$ *is lower semicontinuous.*

Then, T is generalized relaxed $\eta - \alpha$ pseudomonotone on K_T.

Proof Since T is generalized densely relaxed $\eta - \alpha$ pseudomonotone on K, it is generalized relaxed $\eta - \alpha$ pseudomonotone on segment-dense subset of K_0 of K. Thus, T is generalized relaxed $\eta - \alpha$ quasimonotone on K_0. By Lemma 6, T is generalized relaxed $\eta - \alpha$ quasimonotone on K. Therefore, the result follows from Lemma 5. □

Now, we define different concepts of solutions for the variational-like inequalities (1). The dual problem of the variational-like inequality problem (1) is called as *Minty variational-like inequality problem*: Find a vector $x \in K$ such that

$$\langle Ty, \eta(y, x)\rangle \geq 0, \ \forall y \in K. \tag{11}$$

We shall denote by $\mathbb{S}_d(T, K)$ the solution set of the dual problem (11). Corresponding to the above problem, we consider the following relaxed dual problem, called also *relaxed Minty variational-like inequality problem*: Find a vector $x \in K$ such that

$$\langle Ty, \eta(y, x)\rangle \geq \alpha(\eta(y, x)), \ \forall y \in K. \tag{12}$$

We shall denote by $\mathbb{S}_{r,d}(T, K)$ the solution set of the relaxed dual problem (12). An element $x \in K$ is called *local solution* of the dual problem (11), if there exists a neighborhood U of x such that $x \in \mathbb{S}_d(T, K \cap U)$. The set of all local solutions of the dual problem (11) will be denoted by $\mathbb{S}_{d,loc}(T, K)$.

Lemma 7 *Let K be a nonempty closed and convex subset of the Banach space X, $T : K \to X^*$ be a η-hemicontinuous mapping on K and $\eta : K \times K \to X$ be a mapping such that $\eta(x, x) = 0$ and $\eta(tx + (1-t)y, z) = t\eta(x, z) + (1-t)\eta(y, z)$ for all $x, y, z \in K$ and $t \in [0, 1]$. Then, $\mathbb{S}_{d,loc}(T, K) \subset \mathbb{S}(T, K)$.*

Proof Let $x \in \mathbb{S}_{d,loc}(T, K)$. Then, there exists a neighborhood U of x such that $x \in \mathbb{S}_d(T, K \cap U)$. Therefore, $\langle Ty, \eta(y, x)\rangle \geq 0$ for all $y \in K \cap U$. Let $y \in K$, then there exists $\tilde{y} \in]x, y[$ such that $[x, \tilde{y}] \subset K \cap U$. Hence, for $y_t = t\tilde{y} + (1-t)x$ with $t \in [0, 1]$, we have

$$\langle Ty_t, \eta(y_t, x)\rangle \geq 0. \tag{13}$$

But

$$\begin{aligned}\langle Ty_t, \eta(y_t, x)\rangle &= \langle Ty_t, \eta(t\tilde{y} + (1-t)x, x)\rangle \\ &= t\langle Ty_t, \eta(\tilde{y}, x)\rangle + (1-t)\langle Ty_t, \eta(x, x)\rangle.\end{aligned}$$

Thus, from (13), we get

$$\langle Ty_t, \eta(\tilde{y}, x)\rangle \geq 0.$$

Since, T is η-hemicontinuous, we have

$$\langle Tx, \eta(\tilde{y}, x)\rangle \geq 0. \tag{14}$$

Since $\tilde{y} \in]x, y[$, we have that $\tilde{y} = \lambda x + (1-\lambda)y$ for some $\lambda \in]0, 1[$. Using the above argument, from (14), we have

$$\langle Tx, \eta(y, x)\rangle \geq 0.$$

Therefore, for each $y \in K$, there exists $x \in K$ such that $\langle Tx, \eta(y, x)\rangle \geq 0$ and hence $x \in \mathbb{S}(T, K)$. □

Lemma 8 *Let K be a nonempty closed and convex subset of a Banach space X and $\eta : K \times K \to X$ be a mapping. Suppose $T : K \to X^*$ is a generalized relaxed $\eta - \alpha$ properly quasimonotone operator and the following conditions holds*

(i) $\eta(x, y) + \eta(y, x) = 0$, for all $x, y \in K$;
(ii) For each fixed $w, z \in K$, the mapping $x \in K \mapsto \alpha(\eta(w, x))$ is lower semicontinuous;
(iii) For each fixed $x, z \in K$, the mapping $y \in K \mapsto \langle Tx, \eta(y, z)\rangle$ is lower semicontinuous;
(iv) Either K is weakly compact or there exists a weakly compact subset C of K and $y_0 \in C$ such that

$$\forall x \in K \setminus C, \quad \langle Ty_0, \eta(y_0, x)\rangle < \alpha(\eta(y_0, x)). \tag{15}$$

Then, the relaxed dual variational-like inequality problem (12) *has a solution, i.e.,* $\mathbb{S}_{r,d}(T, K) \neq \phi$.

Proof Define the set-valued mapping $G : K \to 2^X$ by

$$G(y) = \{x \in K : \langle Ty, \eta(y, x)\rangle \geq \alpha(\eta(y, x))\}.$$

For any $y_1, y_2, ..., y_n \in K$, let $y \in co\{y_1, y_2, ..., y_n\}$. Since T is generalized relaxed $\eta - \alpha$ properly quasimonotone, we have $y \in \bigcup_{i=1}^{n} f(y_i)$. Thus, G is a KKM mapping. Consider a sequence $\{x_n\}_{n\in\mathbb{N}}$ in $G(y)$ converging to x. From the definition of G, we have

$$\langle Ty, \eta(y, x_n)\rangle \geq \alpha(\eta(y, x_n)).$$

Hence, from (i), we get

$$\alpha(\eta(y, x_n)) + \langle Ty, \eta(x_n, y)\rangle \leq 0.$$

By using (ii) and (iii), we get

$$\begin{aligned}\alpha(\eta(y, x)) + \langle Ty, \eta(x, y)\rangle &\leq \liminf \alpha(\eta(y, x_n)) + \liminf \langle Ty, \eta(x_n, y)\rangle\\ &\leq \liminf\, [\alpha(\eta(y, x_n)) + \langle Ty, \eta(x_n, y)\rangle]\\ &\leq 0.\end{aligned}$$

Thus, we get

$$\langle Ty, \eta(y, x)\rangle \geq \alpha(\eta(y, x)).$$

Thus, $x \in G(y)$ and hence $G(y)$ is closed for each $y \in K$. Therefore, if K is weakly compact, $G(y)$ is weakly compact in K for each $y \in K$ and hence from Lemma 3, we have $\bigcap_{y\in K} G(y) \neq \phi$. Otherwise from (15), we have $G(y_0) \subset C$. Since C is compact and $G(y_0)$ is closed subset of C, $G(y_0)$ is weakly compact. Therefore, again from Lemma 3, we have $\bigcap_{y\in K} G(y) \neq \phi$. Thus, $\mathbb{S}_{r,d}(T, K) \neq \phi$. □

Theorem 2 *Let K be a nonempty closed and convex subset of a Banach space X with topological dual space X^*. Let $\eta : K \times K \to X$, $\alpha : X \to \mathbb{R}$ be mappings and $T : K \to X^*$ be an η-hemicontinuous and generalized relaxed $\eta - \alpha$ quasimonotone operator. Suppose that the following properties hold:*

(i) $\eta(x, y) + \eta(y, x) = 0$, *for all* $x, y \in K$;
(ii) $\eta(tx + (1 - t)z, y) = t\eta(x, y) + (1 - t)\eta(z, y)$, *for all* $x, y, z \in K$ *and* $t \in [0, 1]$;
(iii) For each fixed $w, z \in K$, *the mapping* $x \in K \mapsto \alpha(\eta(w, x))$ *is lower semicontinuous;*
(iv) For each $x \in K$ *and* $w \in X$, *the mappings* $\eta(x, \cdot)$ *and* $\alpha(\eta(x, \cdot))$ *are continuous;*
(v) For each fixed $x, z \in K$, *the mapping* $y \in K \mapsto \langle Tx, \eta(y, z)\rangle$ *is lower semicontinuous;*

(vi) (Coercivity condition) There exists $r_0 > 0$ such that for each $x \in K \setminus \bar{\mathbb{B}}(0, r_0)$, there exists $y \in K$ with $\|y\| < \|x\|$ satisfying: $\langle Tx, \eta(x, y)\rangle \geq 0$.
Moreover, suppose that there exists $r_1 > r_0$ such that $K \cap \bar{\mathbb{B}}(0, r_1)$ is nonempty and weakly compact.

Then, the variational-like inequality problem (1) *has a solution, i.e., $\mathbb{S}(T, K) \neq \emptyset$.*

Proof Let us set $K_{r_1} := K \cap \bar{\mathbb{B}}(0, r_1)$ which is a nonempty convex and weakly compact set. From Lemma 10 ,we have either T is properly $\eta - \alpha$ quasimonotone or $\mathbb{S}_{d,loc}(T, K) \neq \emptyset$.
If $\mathbb{S}_{d,loc}(T, K) \neq \emptyset$, from Lemma 7 we conclude that $\mathbb{S}(T, K) \neq \emptyset$.
If T is properly $\eta - \alpha$ quasimonotone, then applying Lemma 8 we get $\mathbb{S}_{r,d}(T, K_{r_1}) \neq \emptyset$. By using Lemma 1, we derive that $\mathbb{S}(T, K_{r_1}) \neq \emptyset$. Let $x_0 \in \mathbb{S}(T, K_{r_1})$, then

$$\langle Tx_0, \eta(y, x_0)\rangle \geq 0, \ \forall y \in K_{r_1}. \tag{16}$$

Now, we have two cases:

Case 1 If $\|x_0\| < r_1$, then for any $y \in K$ we can find $t \in]0, 1[$ such that $y_t := ty + (1 - t)x_0 \in K_{r_1}$.
From (16),

$$\langle Tx_0, \eta(y_t, x_0)\rangle \geq 0. \tag{17}$$

By using (i) and (ii), we get

$$\begin{aligned}\langle Tx_0, \eta(y_t, x_0)\rangle &= \langle Tx_0, \eta(ty + (1 - t)x_0, x_0)\rangle \\ &= t\langle Tx_0, \eta(y, x_0)\rangle + (1 - t)\langle Tx_0, \eta(x_0, x_0)\rangle \\ &= t\langle Tx_0, \eta(y, x_0)\rangle.\end{aligned}$$

Thus, from (17), we have

$$\langle Tx_0, \eta(y, x_0)\rangle \geq 0.$$

Hence, $x_0 \in \mathbb{S}(T, K)$.

Case 2 If $\|x_0\| = r_1$, then from condition (vi) we deduce that there exists $y_0 \in K$ with $\|y_0\| < \|x_0\|$ such that

$$\langle Tx_0, \eta(x_0, y_0)\rangle \geq 0. \tag{18}$$

Since $y_0 \in K_{r_1}$, from (16) we get $\langle Tx_0, \eta(y_0, x_0)\rangle \geq 0$. From (i), it follows

$$\langle Tx_0, \eta(x_0, y_0)\rangle \leq 0. \tag{19}$$

Hence, from (18) and (19), we deduce that $\langle Tx_0, \eta(x_0, y_0)\rangle = \langle Tx_0, \eta(y_0, x_0)\rangle = 0$. Thus, y_0 is a minimum of the function $g(y) = \langle Tx_0, \eta(y_0, x_0)\rangle$ on K_{r_1}. This implies that y_0 is a global minimum of g on K. Therefore, $y_0 \in \mathbb{S}(T, K)$ and hence $\mathbb{S}(T, K) \neq \emptyset$. □

Remark 7 Theorem 2 generalizes Proposition 2.3 and Theorem 2.1 of Bai and Hadjisavvas [5] from the case of μ quasimonotone operator to the case of more general settings of generalized relaxed $\eta - \alpha$ quasimonotone operators.

4 Variational-Like Inequalities with Generalized Relaxed $\eta - \alpha$ Properly Quasimonotone Mappings

In this section, we first establish a relation between generalized relaxed $\eta - \alpha$ quasimonotone operator and generalized relaxed $\eta - \alpha$ properly quasimonotone operator. We then prove some existence results for solutions of variational-like inequalities with generalized relaxed $\eta - \alpha$ properly quasimonotone operator.

The following lemma shows how the generalized relaxed $\eta - \alpha$ properly quasimonotonicity implies the generalized relaxed $\eta - \alpha$ quasimonotonicity.

Lemma 9 *Let K be a nonempty convex subset of a normed space X and X^* be the dual of X. Suppose $T : K \to X^*$ is a generalized relaxed $\eta - \alpha$ properly quasimonotone mapping and following properties holds*

(i) $\alpha(tz) = t^p \alpha(z)$, *for all* $t > 0$ *with some constant* $p > 1$;
(ii) $\eta(x, y) + \eta(y, x) = 0$, *for all* $x, y \in K$;
(iii) $\eta(tx + (1-t)z, y) = t\eta(x, y) + (1-t)\eta(z, y)$, *for all* $x, y, z \in K$ *and* $t \in [0, 1]$.

Then, T is generalized relaxed $\eta - \alpha$ quasimonotone.

Proof Suppose $x, y \in K$ and

$$\langle Tx, \eta(y, x) \rangle > 0. \tag{20}$$

Let $x_t = x + t(y - x)$, $t \in (0, 1)$. Since T is generalized relaxed $\eta - \alpha$ properly quasimonotone, we have either

$$\langle Ty, \eta(y, x_t) \rangle \geq \alpha(\eta(y, x_t)). \tag{21}$$

or

$$\langle Tx, \eta(x, x_t) \rangle \geq \alpha(\eta(x, x_t)). \tag{22}$$

From (22), we get

$$t\langle Tx, \eta(x, y) \rangle \geq \alpha(t\eta(x, y)).$$

$$\langle Tx, \eta(x, y) \rangle \geq t^{p-1} \alpha(\eta(x, y)).$$

Taking $t \to 0^+$, we get

$$\langle Tx, \eta(x, y) \rangle \geq 0.$$

$$\Longrightarrow \langle Tx, \eta(y, x)\rangle \le 0,$$

which contradicts (20) and therefore relation (21) is true. Hence, from (21), we have

$$\langle Ty, \eta(y, x)\rangle \ge (1-t)^{p-1}\alpha(\eta(y, x)).$$

Taking $t \to 0^+$, we get

$$\langle Ty, \eta(y, x)\rangle \ge \alpha(\eta(y, x)).$$

Therefore, T is generalized relaxed $\eta - \alpha$ quasimonotone. □

Remark 8 Lemma 9 generalizes the Theorem 2.1 of Chen et al. [10] and Proposition 2.1 of Bai et al. [5].

In the next lemma, we will prove that if T is generalized relaxed $\eta - \alpha$ quasimonotone then it is generalized relaxed $\eta - \alpha$ properly quasimonotone.

Lemma 10 *Let K be a nonempty convex subset of a normed space X and X^* be the dual of X. Suppose $T : K \to X^*$ be a generalized relaxed $\eta - \alpha$ quasimonotone mapping. Furthermore, assume that*

(i) $\eta(x, y) + \eta(y, x) = 0$ *for all* $x, y \in K$*;*
(ii) $\eta(tx + (1-t)y, z) = t\eta(x, z) + (1-t)\eta(y, z)$ *for all* $t \in [0, 1]$ *and* $x, y, z \in K$*;*
(iii) For each $x \in K$, the mappings $\eta(x, \cdot)$ and $\alpha(\eta(x, \cdot))$ are continuous.

Then, either T is generalized relaxed $\eta - \alpha$ properly quasimonotone or there exists a solution of the problem: Find $x \in K$ such that

$$\langle Ty, \eta(y, x)\rangle \ge 0, \ \forall y \in K \cap U, \ \textit{for some neighborhood } U \textit{ of } x. \qquad (23)$$

Proof Suppose T is not generalized relaxed $\eta - \alpha$ properly quasimonotone, then there exists $x_1, x_2, ..., x_n \in K$ and $y \in co\{x_1, x_2, ..., x_n\}$ such that

$$\langle Tx_i, \eta(x_i, y)\rangle < \alpha(\eta(x_i, y)), \ \forall i \in \{1, 2, ..., n\}.$$

By conditions (iii), there exists a neighborhood U of y such that for any $z \in U \cap K$, we have

$$\langle Tx_i, \eta(x_i, z)\rangle < \alpha(\eta(x_i, z)), \ \forall i \in \{1, 2, ..., n\}.$$

As T is generalized relaxed $\eta - \alpha$ quasimonotone, we have

$$\langle Tz, \eta(x_i, z)\rangle \le 0, \ \forall i \in \{1, 2, ..., n\}.$$

Since $y \in co\{x_1, x_2, ..., x_n\}$, we get

$$\langle Tz, \eta(y, z)\rangle \le 0.$$

$$\Longrightarrow \langle Tz, \eta(z, y)\rangle \geq 0, \ \forall z \in K \cap U.$$

Hence, $y \in K$ is a solution of the problem (23).

Suppose the problem (23) has no solution, that is, there is no $x \in K$ such that

$$\langle Ty, \eta(y, x)\rangle \geq 0, \ \forall y \in K \cap U, \ \text{ for some neighborhood } U \text{ of } x.$$

Suppose on contrary T is not a generalized relaxed $\eta - \alpha$ properly quasimonotone mapping. Then by the above argument, the problem (23) has a solution, which contradicts assumption. Hence, T is generalized relaxed $\eta - \alpha$ properly quasimonotone. □

Remark 9 Lemma 10 generalizes the Theorem 3.2 of Chen et al. [10] and Proposition 2.2 of Bai et al. [5].

Theorem 3 *Let K be a nonempty closed, convex and bounded subset of a real reflexive Banach space X and X^* be the dual of X. Assume that $T : K \to X^*$ is an $\eta-$hemicontinuous and generalized relaxed $\eta - \alpha$ properly quasimonotone operator. Furthermore, assume the following conditions*

(i) $\eta(x, y) + \eta(y, x) = 0$, *for all* $x, y \in K$;
(ii) $\eta(tx + (1 - t)z, y) = t\eta(x, y) + (1 - t)\eta(z, y)$, *for all* $x, y, z \in K$ *and* $t \in [0, 1]$;
(iii) *For each fixed* $w, z \in K$, *the mapping* $x \in K \mapsto \alpha(\eta(w, x))$ *is lower semicontinuous;*
(iv) *For each fixed* $x, z \in K$, *the mapping* $y \in K \mapsto \langle Tx, \eta(y, z)\rangle$ *is lower semicontinuous.*

Then, variational-like inequality problem (1) has a solution.

Proof Define the set-valued mapping $G : K \to 2^X$ by

$$G(y) = \{x \in K : \langle Ty, \eta(y, x)\rangle \geq \alpha(\eta(y, x))\}.$$

For any $y_1, y_2, ..., y_n \in K$, let $y \in co\{y_1, y_2, ..., y_n\}$. Since T is generalized relaxed $\eta - \alpha$ properly quasimonotone, we have $y \in \bigcup_{i=1}^{n} f(y_i)$. Thus, G is a KKM mapping. Consider a sequence $\{x_n\}_{n\in\mathbb{N}}$ in $G(y)$ converging to x. From the definition of G, we have

$$\langle Ty, \eta(y, x_n)\rangle \geq \alpha(\eta(y, x_n)).$$

Hence, from (i), we get

$$\alpha(\eta(y, x_n)) + \langle Ty, \eta(x_n, y)\rangle \leq 0.$$

By using (iii) and (iv), we get

$$\begin{aligned}\alpha(\eta(y,x)) + \langle Ty, \eta(x,y)\rangle &\le \liminf \alpha(\eta(y,x_n)) + \liminf \langle Ty, \eta(x_n,y)\rangle \\ &\le \liminf\,[\alpha(\eta(y,x_n)) + \langle Ty, \eta(x_n,y)\rangle] \\ &\le 0.\end{aligned}$$

Thus, we get

$$\langle Ty, \eta(y,x)\rangle \ge \alpha(\eta(y,x)).$$

Therefore, $x \in G(y)$ and hence $G(y)$ is closed for each $y \in K$. Since K is closed, convex and bounded in the real reflexive Banach space X, K is weakly compact. Thus, $G(y)$ is weakly compact in K for each $y \in K$. Therefore, from Lemma 3, we have

$$\bigcap_{y\in K} G(y) \ne \phi.$$

Thus, there exists $x \in K$ such that

$$\langle Ty, \eta(y,x)\rangle \ge \alpha(\eta(y,x)),\ \forall y \in K.$$

Hence, by Lemma 1, x is a solution of the problem (1). □

Remark 10 Theorem 3 generalizes the Theorem 5.1 of Daniilidis and Hadjisavvas [11].

Lemma 11 *Let K be a nonempty convex subset of a normed space X and $T : K \to X^*$ be an $\eta-$hemicontinuous and generalized densely relaxed $\eta - \alpha$ pseudomonotone on K. Furthermore, assume that*

(i) $\eta(x,y) + \eta(y,x) = 0$, for all $x, y \in K$;
(ii) For each fixed $w, z \in K$, the mapping $y \in K \mapsto \alpha(\eta(y,z))$ is lower hemicontinuous and the mapping $x \in K \mapsto \alpha(\eta(w,x))$ is lower semicontinuous;
(iii) For each fixed $x, z \in K$, the mapping $y \in K \mapsto \langle Tx, \eta(y,z)\rangle$ is lower semicontinuous and $y \in K \mapsto \langle Tx, \eta(z,y)\rangle$ is lower hemicontinuous.

Then, T is generalized relaxed $\eta - \alpha$ properly quasimonotone.

Proof Suppose T is not generalized relaxed $\eta - \alpha$ properly quasimonotone, then there exists $x_1, x_2, ..., x_n \in K$, $y = \sum_{i=1}^n \lambda_i x_i$, with $\lambda_i > 0$ and $\sum_{i=1}^n \lambda_i = 1$, such that

$$\langle Tx_i, \eta(x_i,y)\rangle < \alpha(\eta(x_i,y)),\ \forall i \in \{1,2,...,n\}. \tag{24}$$

Since T is generalized densely relaxed $\eta - \alpha$ pseudomonotone on K, we can find a vector $u \in X$ and a sequence $\{t^k\}_{k\in\mathbb{N}}$ converging to zero such that T is generalized relaxed $\eta - \alpha$ pseudomonotone at $y^k = y + t^k u$, $k \in \mathbb{N}$. Setting $x_i^k = x_i + t^k u$, we see that $\sum_{i=1}^n \lambda_i x_i^k = y^k$ and when $k \to \infty$, $x_i^k \to x_i$ and $y^k \to y$. Using $\eta-$hemicontinuity of T and condition (iii), we have

$$\begin{aligned}\langle Tx_i, \eta(y, x_i)\rangle &= \lim \langle Tx_i^k, \eta(y, x_i)\rangle \\ &\leq \liminf \langle Tx_i^k, \eta(y, x_i^k)\rangle \\ &\leq \liminf \langle Tx_i^k, \eta(y^k, x_i^k)\rangle, \ \forall i \in \{1, 2, ..., n\}.\end{aligned}$$

$$\langle Tx_i, \eta(x_i, y)\rangle \geq \limsup \langle Tx_i^k, \eta(x_i^k, y^k)\rangle, \ \forall i \in \{1, 2, ..., n\}. \tag{25}$$

Again using condition (ii), we get

$$\begin{aligned}\alpha(\eta(x_i, y)) &\leq \liminf \alpha(\eta(x_i^k, y)) \\ &\leq \liminf \alpha(\eta(x_i^k, y^k)), \ \forall i \in \{1, 2, ..., n\}.\end{aligned} \tag{26}$$

From relations (24), (25), and (26), we have

$$\limsup \langle Tx_i^k, \eta(x_i^k, y^k)\rangle < \liminf \alpha(\eta(x_i^k, y^k)), \ \forall i \in \{1, 2, ..., n\}.$$

Thus, for sufficiently large k, we have

$$\langle Tx_i^k, \eta(x_i^k, y^k)\rangle < \alpha(\eta(x_i^k, y^k)), \ \forall i \in \{1, 2, ..., n\}.$$

Since T is generalized relaxed $\eta - \alpha$ pseudomonotone at y^k, we have

$$\langle Ty^k, \eta(x_i^k, y^k)\rangle < 0, \ \forall i \in \{1, 2, ..., n\}, \tag{27}$$

which implies that

$$\sum_{i=1}^{n} \lambda_i \langle Ty^k, \eta(x_i^k, y^k)\rangle < 0.$$

$$\Longrightarrow \ \langle Ty^k, \eta(y^k, y^k)\rangle < 0,$$

which is a contradiction and therefore T is generalized relaxed $\eta - \alpha$ properly quasimonotone. □

Remark 11 The Lemma 11 generalizes Proposition 2.4 of Bai et al. [5] from densely relaxed μ pseudomonotone operator to generalized densely relaxed $\eta - \alpha$ pseudomonotone operator.

Corollary 2 *Let K be a nonempty bounded closed and convex subset of a real reflexive Banach space X. If $T : K \to X^*$ be an η−hemicontinuous and generalized densely relaxed $\eta - \alpha$ pseudomonotone on K. Assume that*

(i) $\eta(x, y) + \eta(y, x) = 0$, *for all* $x, y \in K$;
(ii) $\eta(tx + (1 - t)z, y) = t\eta(x, y) + (1 - t)\eta(z, y)$, *for all* $x, y, z \in K$ *and* $t \in [0, 1]$;
(iii) For each fixed $w, z \in K$, *the mapping* $y \in K \mapsto \alpha(\eta(y, z))$ *is lower hemicontinuous and the mapping* $x \in K \mapsto \alpha(\eta(w, x))$ *is lower semicontinuous;*

(iv) For each fixed $x, z \in K$, the mapping $y \in K \mapsto \langle Tx, \eta(y, z)\rangle$ is lower semicontinuous and $y \in K \mapsto \langle Tx, \eta(z, y)\rangle$ is lower hemicontinuous.

Then, variational-like inequalities (1) *has a solution.*

Proof Since T is generalized densely relaxed $\eta - \alpha$ pseudomonotone on K, by Lemma 11, T is generalized relaxed $\eta - \alpha$ properly quasimonotone. Thus, all the conditions of Theorem 3 are satisfied and therefore by Theorem 3, the variational-like inequalities (1) has a solution. □

Remark 12 In the proof of Corollary 2, we have used an approach which is different from the one that was used in the proof of Theorem 3.1 of Sahu et al. [23]. Therefore, Corollary 2 provides an alternative proof of Theorem 3.1 of Sahu et al. [23]. Since generalized densely relaxed $\eta - \alpha$ pseudomonotonicity is a proper generalization of densely pseudomonotonicity and densely relaxed μ pseudomonotonicity, therefore Corollary 2 also generalizes Theorem 4.3 of Luc [18] and Theorem 3.1 of Bai et al. [7].

5 Conclusion

In this paper, we have established some existence of solutions for variational-like inequalities with generalized relaxed $\eta - \alpha$ quasimonotone mappings, generalized relaxed $\eta - \alpha$ properly quasimonotone mappings, and densely relaxed $\eta - \alpha$ pseudomonotone mappings. In the future, we plan to find out the existence of solutions for the variational-like inequalities in more general settings.

Acknowledgements The second author wishes to thank the UGC (University Grants Commission), New Delhi, India for the grant of research fellowship (UGC-F.2-4/2013(SA-I)) and IIT Bhubaneswar for providing research facilities.

Conflict of interest The authors have no conflict of interest.

References

1. Antman, S.S.: The influence of elasticity on analysis: modern developments. Bull. Am. Math. Soc. **9**, 267–291 (1983)
2. Arunchai, A., Plubtieng, S., Wen, C.F.: Existence theorems for relaxed η-α pseudomonotone and strictly η-quasimonotone generalized variational-like inequalities. J. Inequal. Appl. **2014**, 442 (2014)
3. Aussel, D., Hadjisavvas, N.: On quasimonotone variational inequalities. J. Optim. Theory Appl. **121**, 445–450 (2004)
4. Baiocchi, C., Capelo, A.: Variational and Quasivariational Inequalities: Applications to Free Boundary Problems. Wiley, New York (1984)
5. Bai, M.R., Hadjisavvas, N.: Relaxed quasimonotone operators and relaxed quasiconvex functions. J. Optim. Theory Appl. **138**, 329–339 (2008)

6. Bai, M.R., Zhou, S.Z., Ni, G.Y.: Variational-like inequalities with relaxed η- α pseudomonotone mappings in Banach spaces. Appl. Math. Lett. **19**, 547–554 (2006)
7. Bai, M.R., Zhou, S.Z., Ni, G.Y.: On the generalized monotonicity of variational inequalities. Comput. Math. Appl. **53**, 910–917 (2007)
8. Brézis, H.: Equations et inéquations nonlinéaires dans les espaces vectoriels en dualité. Ann. Inst. Fourier **18**(1), 115–175 (1968)
9. Chadli, O., Ansari, Q.H., Suliman, A.H.: Existence of solutions and algorithms for Bilevel vector equilibrium problems: an auxiliary principle technique. J. Optim. Theory Appl. **172**, 726–758 (2017)
10. Chen, Q., Luo, J.: Relaxed η-α quasimonotone and application to the generalized variational-like inequality problem. J. Inequal. Appl. **2013**, 488 (2013)
11. Daniilidis, A., Hadjisavvas, N.: Characterization of nonsmooth semistrictly quasiconvex and strictly quasiconvex functions. J. Optim. Theory Appl. **102**, 525–536 (1999)
12. Fang, Y.P., Huang, N.J.: Variational-like inequalities with generalized monotone mappings in Banach spaces. J. Optim. Theory Appl. **118**, 327–338 (2003)
13. Fan, K.: A generalization of Tychonoff's fixed point theorem. Math. Ann. **142**, 305–310 (1961)
14. Giannessi, F., Maugeri, A., Pardalos, P. (eds.): Equilibrium Problems Nonsmooth Optimization and Variational Inequality Models. Kluwer Academic Publishers, Dordrecht-Boston-London (2004)
15. Hartman, P., Stampacchia, G.: On some non-linear elliptic differential-functional equations. Acta Mathematica **115**, 271–310 (1966)
16. Hadjisavvas, N., Schaible, S.: Quasimonotone variational inequalities in Banach spaces. J. Optim. Theory Appl. **90**(1), 95–111 (1996)
17. Karamardian, S., Schaible, S.: Seven kinds of monotone maps. J. Optim. Theory Appl. **66**, 37–46 (1990)
18. Luc, D.T.: Existence results for densely pseudomonotone variational inequalities. J. Math. Anal. Appl. **254**, 291–308 (2001)
19. Mahato, N.K., Nahak, C.: Variational-like inequalities and equilibrium problems with generalized monotonicity in Banach spaces. Adv. Oper. Res. **2012**, 15 (2012)
20. Pany, G., Mohapatra, R.N., Pani, S.: A class of mixed variational-like inequalities and equilibrium problems in Banach spaces. J. Inequal. Spec. Funct. **7**, 241–252 (2016)
21. Sahu, B.K., Pani, S.: Equilibrium problems with generalized relaxed $\eta - \alpha$-pseudomonotone and strictly η-quasimonotone mappings in Banach spaces. Commun. Appl. Nonlinear Anal. **25**, 80–91 (2018)
22. Sahu, N.K., Mahato, N.K., Mohapatra, R.N.: System of nonlinear variational inclusion problems with (A, η)-maximal monotonicity in Banach spaces. Stat. Optim. Inf. Comput. **5**, 244–261 (2017)
23. Sahu, B.K., Nguyen, G., Pany, G., Chadli, O.: Densely relaxed pseudomonotone and quasimonotone generalized variational-like inequalities. Optimization 1–23 (2020)
24. Verma, R.U.: Nonlinear variational inequalities on convex subsets of Banach spaces. Appl. Math. Lett. **10**, 25–27 (1997)

An Alternative Approach to Convolutions of Harmonic Mappings

Chinu Singla, Sushma Gupta, and Sukhjit Singh

Abstract Convolutions or Hadamard products of analytic functions are a well-explored area of research, and many nice results are available in literature. On the other hand, very little is known in general about the convolutions of univalent harmonic mappings. So, researchers started exploring properties of convolutions of some specific univalent harmonic mappings and while doing so, they have used well known '*Cohn's rule*' or/and '*Schur-Cohn's algorithm*', which involves computations that are very cumbersome. The main objective of this article is to present an alternative approach, which requires very less computational efforts and allows us to prove more general results. Most of the earlier known results follow as particular cases of the results proved herein.

Keywords Harmonic mapping · Convolution · Right half plane mapping · Convexity in one direction

Mathematics Subject Classification (2010) Primary 30C45 · Secondary 30C80

1 Introduction

Let us consider a class S_H of complex valued univalent harmonic functions f in the open unit disc $\mathbb{D} = \{z \in \mathbb{C} : |z| < 1\}$, having canonical representation $f = h + \overline{g}$ and normalized by the conditions $h(0) = g(0) = 0 = h'(0) - 1$. The Jacobian $J_f(z)$ of f is given by

$$J_f(z) = |f_z(z)|^2 - |f_{\overline{z}}(z)|^2 = |h'(z)|^2 - |g'(z)|^2.$$

This work was completed with the financial support of UGC—CSIR, New Delhi in the form of Senior Research Fellowship vide Award Letter Number—2061540842.

C. Singla (✉) · S. Gupta · S. Singh
Department of Mathematics, Sant Longowal Institute of Engineering and Technology, Longowal, Punjab, India
e-mail: chinusingla204@gmail.com

R. N. Mohapatra et al. (eds.), *Mathematical Analysis and Computing*,
Springer Proceedings in Mathematics & Statistics 344,
https://doi.org/10.1007/978-981-33-4646-8_48

Lewy [9] proved that a harmonic function $f = h + \overline{g}$ is locally univalent and sense preserving in $\mathbb{D}$ if and only if $J_f(z) > 0$ in $\mathbb{D}$. This is equivalent to the existence of an analytic function $w(z) = \frac{g'(z)}{h'(z)}$, satisfying $|w(z)| < 1$ in $\mathbb{D}$. Here, w is called the dilatation of f (see [5]).

Denote by S_H^0 the subclass of S_H having functions f with additional normalization condition $f_{\overline{z}}(0) = 0$. The subclass of S_H (or S_H^0) containing convex functions is denoted by K_H (or K_H^0). Let $S \subset S_H$ be the class of analytic and univalent functions and let K, S^*, and C be the usual subclasses of S containing convex, starlike, and close-to-convex functions, respectively. A domain E in $\mathbb{C}$ is said to be convex in the direction $\psi, 0 \leq \psi < \pi$, if every line parallel to the line through 0 and $e^{i\psi}$ has an empty or connected intersection with E. A function f is said to be convex in the direction ψ if it maps $\mathbb{D}$ onto the domain convex in the direction $\psi, 0 \leq \psi < \pi$. If $\psi = 0$, then f is said to be convex in the direction of the real axis and if $\psi = \pi/2$, then f is said to be convex in the direction of the imaginary axis.

The convolution or Hadamard product of two analytic functions $f(z) = z + \sum_{n=2}^{\infty} a_n z^n$ and $F(z) = z + \sum_{n=2}^{\infty} A_n z^n$ is denoted by $f * F$ and is defined as $(f * F)(z) = z + \sum_{n=2}^{\infty} a_n A_n z^n$. Let $f = h + \overline{g}$ and $F = H + \overline{G}$ be two harmonic mappings, then their convolution is denoted by $f \tilde{*} F$ and is defined as $f \tilde{*} F = h * H + \overline{g * G}$. Many pleasant results are available in literature on the convolution of analytic functions. For example, Ruscheweyh and Sheil-Small [12] proved that

1. $f * g \in K$ for all $f, g \in K$;
2. $f * g \in S^*$ for all $f \in K$ and $g \in S^*$;
3. $f * g \in C$ for all $f \in K$ and $g \in C$.

On the other hand, not much is known about the properties of convolutions of harmonic functions and there are no general results of the kind stated above in this case. For example, for $F, G \in K_H$, $F \tilde{*} G$ may not be univalent in $\mathbb{D}$ even (see [11]). However, some researchers started exploring the nature of convolutions of some specific harmonic mappings. We mentioned some of them below.

It is well known that if $f = h + \overline{g} \in S_H^0$ maps $\mathbb{D}$ onto the right half plane $R = \{w \in \mathbb{C} : Re w > -1/2\}$, then it must satisfy

$$h(z) + g(z) = \frac{z}{1 - z},$$

and such mappings are called right half plane mappings. In 2001, Dorff [3] started study of convolutions of right half plane harmonic mappings and presented the following result.

Theorem 1 *Let $f_i = h_i + \overline{g_i}$ be the harmonic right half plane mappings with $h_i(z) + g_i(z) = \frac{z}{1-z}$ for $i = 1, 2$. Then, $f_1 \tilde{*} f_2 \in S_H^0$ and is convex in the direction of the real axis provided, $f_1 \tilde{*} f_2$ is locally univalent and sense preserving in $\mathbb{D}$.*

In the same paper, he defined a family of harmonic mappings $V_\beta = u_\beta + \overline{v_\beta} \in S_H^0$, obtained from the analytic strip mappings

$$u_\beta(z) + v_\beta(z) = \frac{1}{2i \sin \beta} \log\left(\frac{1 + ze^{i\beta}}{1 + ze^{-i\beta}}\right), 0 < \beta < \pi, \tag{1}$$

by using shearing technique of Clunie and Sheil-Small [2] and established the following result.

Theorem 2 *Let $f = h + \overline{g}$ be the harmonic right half plane mappings with $h(z) + g(z) = \frac{z}{1-z}$ and $V_\beta = u_\beta + \overline{v_\beta} \in S_H^0$ as defined in* (1) *with $\pi/2 \le \beta < \pi$. Then, $f \tilde{*} V_\beta \in S_H^0$ and is convex in the direction of the real axis provided, $f \tilde{*} V_\beta$ is locally univalent and sense preserving in $\mathbb{D}$.*

Let $F_0 = H_0 + \overline{G_0}$ be the harmonic right half plane mapping given by

$$H_0 + G_0 = z/(1 - z), \qquad G_0'/H_0' = -z. \tag{2}$$

Then, F_0 is called the standard right half plane mapping. Dorff et al. [4] were able to drop the requirement of local univalence and sense preserving from the above results as under.

Theorem 3 *Let $F = H + \overline{G}$ be the harmonic right half plane mapping with $H(z) + G(z) = \frac{z}{1-z}$ and $G'(z)/H'(z) = e^{i\theta} z^n (\theta \in \mathbb{R}, n \in \mathbb{N})$. Then, for $n = 1, 2$, $F_0 \tilde{*} F \in S_H^0$ and is convex in the direction of the real axis. Here, F_0 is the standard right half plane mapping as defined in* (2).

Theorem 4 *Let $V_\beta = u_\beta + \overline{v_\beta} \in S_H^0$ be the harmonic mapping as defined in* (1) *with $v_\beta'(z)/u_\beta'(z) = e^{i\theta} z^n (\theta \in \mathbb{R}, n \in \mathbb{N})$. Then, for $n = 1, 2$, $F_0 \tilde{*} V_\beta \in S_H^0$ and is convex in the direction of the real axis.*

For a real number $\gamma, 0 \le \gamma < 2\pi$, the mapping $f_\gamma = h_\gamma + \overline{g_\gamma} \in S_H$, given by

$$h_\gamma(z) + e^{-2i\gamma} g_\gamma(z) = \frac{z}{1 - e^{i\gamma} z} \tag{3}$$

is called slanted right half plane mapping and it maps the unit disc $\mathbb{D}$ onto slanted right half plane given by $H(\gamma) = \{w \in \mathbb{C} : Re(e^{i\gamma} w) > -1/2\}$ (see [4]). We denote this subclass of S_H by $S_{H(\gamma)}$. In 2013, Li and Ponnusamy [10] proved the following.

Theorem 5 *Let $f_\gamma \in S_{H(\gamma)}$ be as defined above with dilatation $g_\gamma'(z)/h_\gamma'(z) = e^{i\theta} z^n, \theta \in \mathbb{R}, n = 1, 2$. Then, $F_0 \tilde{*} f_\gamma \in S_H^0$ and is convex in the direction $-\gamma$.*

In 2015, Kumar et al. [7] generalized Theorem 3 by taking F_a instead of F_0, where $F_a = H_a + \overline{G_a} \in K_H$ is the right half plane mapping given by

$$H_a(z) + G_a(z) = \frac{z}{1 - z}, \qquad \frac{G_a'(z)}{H_a'(z)} = \frac{a - z}{1 - az}, a \in (-1, 1) \tag{4}$$

and proved the following.

Theorem 6 *Let $F = H + \overline{G}$ be the harmonic right half plane mapping with $H(z) + G(z) = \frac{z}{1-z}$ and $G'(z)/H'(z) = e^{i\theta}z^n (\theta \in \mathbb{R}, n \in \mathbb{N})$. Then, $F_a \tilde{*} F \in S_H$ and is convex in the direction of the real axis for all $a \in [(n-2)/(n+2), 1)$.*

Similarly, in an attempt to generalize Theorem 4, Kumar et al. [8] proposed the following conjecture, which they themselves proved for $0 < \beta < \pi$ and $n = 1, 2, 3, 4$; and also for the case when $\beta = \pi/2$ and n is a natural number, see [6].

Theorem 7 *Let $V_\beta = u_\beta + \overline{v_\beta} \in S_H$ be the harmonic mapping as defined in (1) with $v'_\beta(z)/u'_\beta(z) = e^{i\theta}z^n (\theta \in \mathbb{R}, n \in \mathbb{N})$. Then, $F_a \tilde{*} V_\beta \in S_H$ and is convex in the direction of the real axis for all $a \in [(n-2)/(n+2), 1)$, where, F_a is the harmonic right half plane mapping as defined in (4).*

Recently, Beig et al. [1] considered more general slanted right half plane mappings $F_{(a,\alpha)} = H_{(a,\alpha)} + \overline{G_{(a,\alpha)}} \in S_{H(\alpha)}$, essentially given by

$$H_{(a,\alpha)}(z) + e^{-2i\alpha}G_{(a,\alpha)}(z) = \frac{z}{1 - ze^{i\alpha}}, \quad \frac{G'_{(a,\alpha)}(z)}{H'_{(a,\alpha)}(z)} = e^{2i\alpha}\left(\frac{a - ze^{i\alpha}}{1 - aze^{i\alpha}}\right), \tag{5}$$

where $a \in (-1, 1)$, $\alpha \in [0, 2\pi)$ and studied its convolution with another slanted right half plane mapping $f_\gamma = h_\gamma + \overline{g_\gamma}$ given by (3) with dilatation

$$g'_\gamma(z)/h'_\gamma(z) = e^{i\theta}z^n, \theta \in \mathbb{R}, n \in \mathbb{N}. \tag{6}$$

They obtained the following result.

Theorem 8 *Let $F_{(a,\alpha)} = H_{(a,\alpha)} + \overline{G_{(a,\alpha)}} \in S_{H(\alpha)}$ be given by (5) and $f_\gamma = h_\gamma + \overline{g_\gamma}$ be given by (3) with dilatation as given in (6). Then, $F_{(a,\alpha)} \tilde{*} f_\gamma \in S_H$ and is convex in the direction $-(\alpha + \gamma)$ for all $a \in [(n-2)/(n+2), 1)$.*

In 2016, Wang et al. [14] considered a new family of harmonic mappings $f_0 = h_0 + \overline{g_0} \in S_H^0$, convex in horizontal direction given by

$$h_0(z) - g_0(z) = \frac{z}{1-z} \quad \frac{g'_0(z)}{h'_0(z)} = z \tag{7}$$

and presented the following result.

Theorem 9 *For $n \in \mathbb{N}$, let $f_n = h_n + \overline{g_n} \in S_H^0$, be harmonic mappings with*

$$h_n(z) - g_n(z) = \frac{1}{2i \sin \psi} \log\left(\frac{1 + ze^{i\psi}}{1 + ze^{-i\psi}}\right), \pi/2 \le \psi < \pi$$

and $\frac{g'_n(z)}{h'_n(z)} = e^{i\theta}z^n (\theta \in \mathbb{R})$. If $n = 1, 2$, then $f_0 \tilde{*} f_n \in S_H^0$ and is convex in the direction of the real axis, where f_0 is given by (7).

It has been observed that most of the results listed out above have been proved by using '*Cohn's rule*' or/and '*Schur-Cohn's algorithm*' and computations involved are extremely cumbersome. The main objective of this article is to present a technique, which is simple to apply and involves very less computations. Our technique enables us to prove more general results and all the results stated above deduce as particular cases of the results obtained herein.

2 Convolution of Harmonic Univalent Functions

We begin this section by stating the following lemma ([13], Lemma 4.4), which will be required to establish main theorems in this paper.

Lemma 1 *Let k and k' be real numbers with $k' - k > 0$. Then, for $w \in \mathbb{C}$,*

$$\left|\frac{k+w}{k'+w}\right| < 1$$

if and only if

$$Re(w) > -\left(\frac{k+k'}{2}\right).$$

We shall also need the following result whose proof runs on the same lines as that of Theorem 2 in [4] and hence is omitted.

Lemma 2 *Let $F_1 = H_1 + \overline{G_1} \in S_{H(\alpha)}$, $F_2 = H_2 + \overline{G_2} \in S_H$ be two harmonic functions with $H_1(z) + e^{-2i\alpha}G_1(z) = \frac{z}{1-ze^{i\alpha}}$, $\alpha \in [0, 2\pi)$ and $H_2(z) + e^{-2i\gamma}G_2(z) = f(z)$, $\gamma \in [0, 2\pi)$, where*

$$zf'(z) = \frac{z}{(1+ze^{i(\eta+\gamma)})(1+ze^{-i(\eta-\gamma)})}$$

for some $\eta \in \mathbb{R}$. Then, $F_1 \tilde{*} F_2 \in S_H$ and is convex in the direction $-(\alpha+\gamma)$, provided $F_1 \tilde{*} F_2$ is locally univalent and sense preserving in $\mathbb{D}$.

Now, consider a family of harmonic mappings $T_{(\eta,\gamma)} = R_{(\eta,\gamma)} + \overline{S_{(\eta,\gamma)}} \in S_H$, given by

$$R_{(\eta,\gamma)}(z) + e^{-2i\gamma}S_{(\eta,\gamma)}(z) = f(z), \qquad \frac{S'_{(\eta,\gamma)}(z)}{R'_{(\eta,\gamma)}(z)} = e^{i\theta}z^n, \tag{8}$$

where f is the analytic mapping in $\mathbb{D}$ given by

$$zf'(z) = \frac{z}{(1+ze^{i(\eta+\gamma)})(1+ze^{-i(\eta-\gamma)})}$$

for some $\eta \in \mathbb{R}$ and $\gamma \in [0, 2\pi)$. We now prove the following.

Theorem 10 *Let $F_{(a,\alpha)} = H_{(a,\alpha)} + \overline{G_{(a,\alpha)}} \in S_{H(\alpha)}$ be given by (5) and $T_{(\eta,\gamma)} \in S_H$ be given by (8). Then, $F_{(a,\alpha)} \tilde{*} T_{(\eta,\gamma)} \in S_H$ and is convex in the direction $-(\alpha+\gamma)$ for $a \in \left[\frac{n-2}{n+2}, 1\right)$.*

Proof From (5), we get

$$H_{(a,\alpha)}(z) = \frac{1}{2}\left[\frac{z}{(1-ze^{i\alpha})} + \left(\frac{1-a}{1+a}\right)\frac{z}{(1-ze^{i\alpha})^2}\right]$$

and

$$G_{(a,\alpha)}(z) = \frac{1}{2}\left[\frac{ze^{2i\alpha}}{(1-ze^{i\alpha})} - \left(\frac{1-a}{1+a}\right)\frac{ze^{2i\alpha}}{(1-ze^{i\alpha})^2}\right].$$

In view of Lemma 2, it is enough to prove that dilatation $W(z) = \frac{(G_{(a,\alpha)} * S_{(\eta,\gamma)})'(z)}{(H_{(a,\alpha)} * R_{(\eta,\gamma)})'(z)}$ of $F_{(a,\alpha)} \tilde{*} T_{(\eta,\gamma)}$ satisfies $|W(z)| < 1$ in $\mathbb{D}$. As

$$W(z) = e^{2i\alpha}\frac{(G_{(a,0)} * S_{(\eta,\gamma)})'(ze^{i\alpha})}{(H_{(a,0)} * R_{(\eta,\gamma)})'(ze^{i\alpha})} = e^{2i\alpha}\widehat{w}(ze^{i\alpha})(say),$$

it is therefore enough to prove that $|\widehat{w}(z)| < 1$ in $\mathbb{D}$. Further, note that

$$\widehat{w}(z) = \frac{2aS'_{(\eta,\gamma)}(z) - (1-a)zS''_{(\eta,\gamma)}(z)}{2R'_{(\eta,\gamma)}(z) + (1-a)zR''_{(\eta,\gamma)}(z)}. \tag{9}$$

From (8), we get

$$S'_{(\eta,\gamma)}(z) = e^{i\theta}z^n R'_{(\eta,\gamma)}(z)$$

and so

$$S''_{(\eta,\gamma)}(z) = e^{i\theta}z^n R''_{(\eta,\gamma)}(z) + ne^{i\theta}z^{n-1}R'_{(\eta,\gamma)}(z).$$

Putting these values of $S'_{(\eta,\gamma)}$ and $S''_{(\eta,\gamma)}$ in (9), we have

$$\widehat{w}(z) = -e^{i\theta}z^n\left[\frac{\frac{n-(n+2)a}{1-a} + \frac{zR''_{(\eta,\gamma)}(z)}{R'_{(\eta,\gamma)}(z)}}{\frac{2}{1-a} + \frac{zR''_{(\eta,\gamma)}(z)}{R'_{(\eta,\gamma)}(z)}}\right]. \tag{10}$$

Now, $|\widehat{w}(z)| < 1$ if

$$\left|\frac{\frac{n-(n+2)a}{1-a} + \frac{zR''_{(\eta,\gamma)}(z)}{R'_{(\eta,\gamma)}(z)}}{\frac{2}{1-a} + \frac{zR''_{(\eta,\gamma)}(z)}{R'_{(\eta,\gamma)}(z)}}\right| \le 1. \tag{11}$$

For $a = \frac{n-2}{n+2}$, we note that left-hand side of (11) is equal to 1 and for $a \in \left(\frac{n-2}{n+2}, 1\right)$,

$$\left(\frac{2}{1-a}\right)-\left(\frac{n-(n+2)a}{1-a}\right)>0.$$

Therefore, in view of Lemma 1, it is sufficient to prove that

$$Re\left\{\frac{zR''_{(\eta,\gamma)}(z)}{R'_{(\eta,\gamma)}(z)}\right\}>-\frac{n+2}{2}. \tag{12}$$

From (8), we have

$$R'_{(\eta,\gamma)}(z)=\frac{f'(z)}{1+e^{i(\theta-2\gamma)}z^n},$$

which gives

$$\frac{zR''_{(\eta,\gamma)}(z)}{R'_{(\eta,\gamma)}(z)}=\frac{zf''(z)}{f'(z)}-\frac{ne^{i(\theta-2\gamma)}z^n}{1+e^{i(\theta-2\gamma)}z^n},$$

or equivalently

$$n+2+\frac{2zR''_{(\eta,\gamma)}(z)}{R'_{(\eta,\gamma)}(z)}=2\left(1+\frac{zf''(z)}{f'(z)}\right)+n\left(\frac{1-e^{i(\theta-2\gamma)}z^n}{1+e^{i(\theta-2\gamma)}z^n}\right).$$

Now, $Re\left(1+\frac{zf''(z)}{f'(z)}\right)>0$ in $\mathbb{D}$ because f is convex in $\mathbb{D}$ (as zf' is starlike in $\mathbb{D}$) and thus for $|z|<1$, we have $Re\left(\frac{1-e^{i(\theta-2\gamma)}z^n}{1+e^{i(\theta-2\gamma)}z^n}\right)>0$. This gives

$$Re\left\{n+2+\frac{2zR''_{(\eta,\gamma)}(z)}{R'_{(\eta,\gamma)}(z)}\right\}>0$$

in $\mathbb{D}$, which in turn, shows that (12) is true.

Remark 1 By assigning suitable values to the parameters α, γ and η in Theorem 10, we easily deduce most of the results listed in Sect. 1 as under.

1. If we take $\eta=\pi$ in (8), we get

$$R_{(\pi,\gamma)}(z)+e^{-2i\gamma}S_{(\pi,\gamma)}(z)=\frac{z}{1-ze^{i\gamma}}$$

   and thus Theorem 10 reduces to Theorem 8.
2. If we take $\alpha=\gamma=0$ and $0<\eta<\pi$ in Theorem 10, we get Theorem 7. In addition, if we take $a=0$ also, then we get Theorem 4.
3. If we take $\alpha=a=0$ and $\eta=\pi$ in Theorem 10, we get Theorem 5.
4. If we take $\alpha=\gamma=0$ and $\eta=\pi$ in Theorem 10, we get Theorem 6. In addition, if we take $a=0$ also, we get Theorem 3.

Next, we study convolutions of harmonic functions convex in one direction. To proceed further, we need following result whose proof is omitted as it follows similarly as the proof of Theorem 2 in [4].

Lemma 3 *Let $f_1 = h_1 + \overline{g_1} \in S_H$ and $f_2 = h_2 + \overline{g_2} \in S_H$ be two harmonic functions with $h_1(z) - e^{-2i\alpha} g_1(z) = \frac{z}{1-ze^{i\alpha}}$, $\alpha \in [0, 2\pi)$ and $h_2(z) - e^{-2i\gamma} g_2(z) = f(z)$, $\gamma \in [0, 2\pi)$, where*

$$zf'(z) = \frac{z}{(1 + ze^{i(\eta+\gamma)})(1 + ze^{-i(\eta-\gamma)})}$$

for some $\eta \in \mathbb{R}$. Then, $f_1 \tilde{*} f_2 \in S_H$ and is convex in the direction $-(\alpha + \gamma)$, provided $f_1 \tilde{*} f_2$ is locally univalent and sense preserving in $\mathbb{D}$.

Now, consider harmonic mapping $f_{b,\alpha} = h_{b,\alpha} + \overline{g_{b,\alpha}} \in S_H$, with

$$h_{b,\alpha}(z) - e^{-2i\alpha} g_{b,\alpha}(z) = \frac{z}{1 - ze^{i\alpha}}, \qquad \frac{g'_{b,\alpha}(z)}{h'_{b,\alpha}(z)} = e^{2i\alpha} \left(\frac{b + ze^{i\alpha}}{1 + bze^{i\alpha}} \right), \tag{13}$$

where $b \in (-1, 1)$, $\alpha \in [0, 2\pi)$ and another harmonic mapping $t_{\eta,\gamma} = r_{\eta,\gamma} + \overline{s_{\eta,\gamma}} \in S_H$, given by

$$r_{\eta,\gamma}(z) - e^{-2i\gamma} s_{\eta,\gamma}(z) = f(z), \qquad \frac{s'_{\eta,\gamma}(z)}{r'_{\eta,\gamma}(z)} = e^{i\theta} z^n, \tag{14}$$

where $\gamma \in [0, 2\pi)$, $\theta \in \mathbb{R}$, $n \in \mathbb{N}$ and

$$zf'(z) = \frac{z}{(1 + ze^{i(\eta+\gamma)})(1 + ze^{-i(\eta-\gamma)})}$$

for some $\eta \in \mathbb{R}$. We present the following result.

Theorem 11 *Let $f_{b,\alpha} = h_{b,\alpha} + \overline{g_{b,\alpha}} \in S_H$ be given by (13) and $t_{\eta,\gamma} = r_{\eta,\gamma} + \overline{s_{\eta,\gamma}} \in S_H$ be given by (14). Then, $f_{b,\alpha} \tilde{*} t_{\eta,\gamma} \in S_H$ and is convex in the direction $-(\alpha + \gamma)$ for all $b \in \left(-1, \frac{-(n-2)}{n+2}\right]$.*

Proof From (13), we get

$$h_{b,\alpha}(z) = \frac{1}{2} \left[\left(\frac{1+b}{1-b} \right) \frac{z}{(1 - ze^{i\alpha})^2} + \frac{z}{(1 - ze^{i\alpha})} \right]$$

and

$$g_{b,\alpha}(z) = \frac{1}{2} \left[\left(\frac{1+b}{1-b} \right) \frac{ze^{2i\alpha}}{(1 - ze^{i\alpha})^2} - \frac{ze^{2i\alpha}}{(1 - ze^{i\alpha})} \right].$$

If W denotes the dilatation of the function $f_{b,\alpha} \tilde{*} t_{\eta,\gamma} = h_{b,\alpha} * r_{\eta,\gamma} + \overline{g_{b,\alpha} * s_{\eta,\gamma}}$, we get

$$W(z) = \frac{(g_{b,\alpha} * s_{\eta,\gamma})'(z)}{(h_{b,\alpha} * r_{\eta,\gamma})'(z)}.$$

In view of Lemma 3, we only need to show that $|W(z)| < 1$ in $\mathbb{D}$. As

$$W(z) = e^{2i\alpha}\frac{(g_{b,0} * s_{\eta,\gamma})'(ze^{i\alpha})}{(h_{b,0} * r_{\eta,\gamma})'(ze^{i\alpha})} = e^{2i\alpha}\widehat{w}(ze^{i\alpha})(say),$$

it is therefore enough to prove that $|\widehat{w}(z)| < 1$ in $\mathbb{D}$. Further, note that

$$\widehat{w}(z) = \frac{2bs'_{\eta,\gamma}(z) + (1+b)zs''_{\eta,\gamma}(z)}{2r'_{\eta,\gamma}(z) + (1+b)zr''_{\eta,\gamma}(z)}. \tag{15}$$

From (14), we have

$$s'_{\eta,\gamma}(z) = e^{i\theta}z^n r'_{\eta,\gamma}(z)$$

and therefore

$$s''_{\eta,\gamma}(z) = e^{i\theta}z^n r''_{\eta,\gamma}(z) + ne^{i\theta}z^{n-1}r'_{\eta,\gamma}(z).$$

Putting these values of $s'_{\eta,\gamma}$ and $s''_{\eta,\gamma}$ in (15), we have

$$\widehat{w}(z) = e^{i\theta}z^n\left[\frac{\frac{n+(n+2)b}{1+b} + \frac{zr''_{\eta,\gamma}(z)}{r'_{\eta,\gamma}(z)}}{\frac{2}{1+b} + \frac{zr''_{\eta,\gamma}z)}{r'_{\eta,\gamma}(z)}}\right]. \tag{16}$$

So, it is enough to prove that

$$\left|\frac{\frac{n+(n+2)b}{1+b} + \frac{zr''_{\eta,\gamma}(z)}{r'_{\eta,\gamma}(z)}}{\frac{2}{1+b} + \frac{zr''_{\eta,\gamma}(z)}{r'_{\eta,\gamma}(z)}}\right| \le 1. \tag{17}$$

We get equality in (17) for $b = -\frac{n-2}{n+2}$ and for $b \in \left(-1, -\frac{n-2}{n+2}\right)$,

$$\left(\frac{2}{1+b}\right) - \left(\frac{n+(n+2)b}{1+b}\right) > 0.$$

Therefore, in view of Lemma 1, it is sufficient to prove that

$$Re\left\{\frac{zr''_{\eta,\gamma}(z)}{r'_{\eta,\gamma}(z)}\right\} > -\frac{n+2}{2}$$

in $\mathbb{D}$, or equivalently

$$Re\left\{n+2+\frac{2zr''_{\eta,\gamma}(z)}{r'_{\eta,\gamma}(z)}\right\}>0 \tag{18}$$

in $\mathbb{D}$. From (14), we have

$$r'_{\eta,\gamma}(z)=\frac{f'(z)}{1-e^{i(\theta-2\gamma)}z^n},$$

which gives

$$\frac{zr''_{\eta,\gamma}(z)}{r'_{\eta,\gamma}(z)}=\frac{zf''(z)}{f'(z)}+\frac{ne^{i(\theta-2\gamma)}z^n}{1-e^{i(\theta-2\gamma)}z^n}$$

or

$$n+2+\frac{2zr''_{\eta,\gamma}(z)}{r'_{\eta,\gamma}(z)}=2\left(1+\frac{zf''(z)}{f'(z)}\right)+n\left(\frac{1+e^{i(\theta-2\gamma)}z^n}{1-e^{i(\theta-2\gamma)}z^n}\right).$$

Now, $Re\left(1+\frac{zf''(z)}{f'(z)}\right)>0$ in $\mathbb{D}$ because f is convex in $\mathbb{D}$ (as zf' is starlike in $\mathbb{D}$) and for $|z|<1$, we have $Re\left(\frac{1+e^{i(\theta-2\gamma)}z^n}{1-e^{i(\theta-2\gamma)}z^n}\right)>0$. Thus (18) is true.

Remark 2 Theorem 9 can be obtained from Theorem 11 by setting $b=\alpha=\gamma=0$ and choosing η in $[\pi/2,\pi)$.

References

1. Beig, S., Ravichandran, V.: Convolution and convex combination of harmonic mappings. Bull. Iran. Math. Soc. **45**(3), 1467–1486 (2019)
2. Clunie, J., Sheil-Small, T.: Harmonic univalent functions. Ann. Acad. Sci. Fenn. Ser. A I Math. **9**, 3–25 (1984)
3. Dorff, M.: Convolutions of planar harmonic convex mappings. Complex Var. Elliptic Equ. **45**(3), 263–271 (2001)
4. Dorff, M., Nowak, M., Woloszkiewicz, M.: Convolutions of harmonic convex mappings. Complex Var. Elliptic Equ. **57**(5), 489–503 (2012)
5. Duren, P.: Harmonic Mappings in the Plane. Cambridge University Press (2004)
6. Kumar, R., Gupta, S., Singh, S., Dorff, M.: On harmonic convolutions involving a vertical strip mapping. Bull. Korean Math. Soc. **52**(1), 105–123 (2015)
7. Kumar, R., Dorff, M., Gupta, S., Singh, S.: Convolution properties of some harmonic mappings in the right half-plane. Bull. Malays. Math. Sci. Soc. **39**(1), 439–455 (2016)
8. Kumar, R., Gupta, S., Singh, S., Dorff, M.: An application of Cohn's rule to convolutions of univalent harmonic mappings. Rocky Mt. J. Math. **46**(2), 559–570 (2016)
9. Lewy, H.: On the non vanishing of the Jacobian in certain one to one mappings. Bull. Am. Math. Soc. **42**, 689–692 (1936)
10. Li, L., Ponnusamy, S.: Convolutions of slanted half-plane harmonic mappings. Analysis (Munich) **33**(2), 159–176 (2013)
11. Nagpal, S., Ravichandran, V.: Convolution properties of the harmonic Koebe function and its connection with 2-starlike mappings. Complex Var. Elliptic Equ. **60**(2), 191–210 (2015)

12. Ruscheweyh, S., Sheil-Small, T.: Hadamard products of Schlicht functions and the *Pólya*-Schoenberg conjecture. Comment. Math. Helv. **48**(1), 119–135 (1973)
13. Singla, C., Gupta, S., Singh, S.: Some new harmonic mappings convex in one direction and their convolution. Filomat **33**(19), 6131–6139 (2019)
14. Wang, Z., Liu, Z., Li, Y.: On convolutions of harmonic univalent mappings convex in the direction of the real axis. J. Appl. Anal. Comput. **6**(1), 145–155 (2016)

A Survey on the Theory of Integral and Related Operators in Geometric Function Theory

Om P. Ahuja and Asena Çetinkaya

Abstract A brief tour of more than one hundred years of the historical development of some of the popular integral and related operators in Geometric Function Theory (GFT) is given in this article. The strengths and discovery of the methods used in these operators lie in their ability to unify a large number of diverse operators and results. We also address some of the $q-$analogues of the integral operators in GFT. Since there are several surveys and books in GFT, we present here only a selection of the results related to our precise objectives.

Keywords Univalent functions · Analytic functions · Convolution · Integral operators · Differential operators · Hypergeometric functions · Starlike functions · Convex functions · Close-to-convex functions · Quantum calculus · q-integral operator · q-derivative operator · q-difference operator

Mathematics Subject Classification (2010) 30C45 · 30C55 · 81Q99

1 Introduction

Let $\mathcal{A}$ denote the class of functions of the form

$$f(z) = z + \sum_{n=2}^{\infty} a_n z^n \tag{1.1}$$

O. P. Ahuja (✉)
Department of Mathematical Sciences, Kent State University,
Kent, OH 44021, USA
e-mail: oahuja@kent.edu

A. Çetinkaya
Department of Mathematics and Computer Science, İstanbul Kültür University,
Istanbul, Turkey
e-mail: asnfigen@hotmail.com

R. N. Mohapatra et al. (eds.), *Mathematical Analysis and Computing*,
Springer Proceedings in Mathematics & Statistics 344,
https://doi.org/10.1007/978-981-33-4646-8_49

that are analytic in the open unit disk $\mathbb{D} := \{z : |z| < 1\}$ with the normalization $f(0) = f'(0) - 1 = 0$ and let $\mathcal{S}$ denote the subset of $\mathcal{A}$ consisting of all univalent functions in $\mathbb{D}$. In 1907, Koebe [39] discovered an important function k given by

$$k(z) = \frac{z}{(1-z)^2} = z + 2z^2 + 3z^3 + \ldots + nz^n + \ldots = \sum_{n=1}^{\infty} nz^n.$$

He discovered that this function k is the largest function in $\mathcal{S}$, because it is impossible to add to its image domain $k(\mathbb{D})$ any open set of points without destroying univalence.

The following subclasses of $\mathcal{A}$ or $\mathcal{S}$ are well-known in Geometric Function Theory (GFT):

$$\begin{aligned}
\mathcal{K} : &= \{f \in \mathcal{S} : f(\mathbb{D})\ is\ convex\} \\
&= \left\{f \in \mathcal{S} : Re\left(1 + \frac{zf''(z)}{f'(z)}\right) > 0,\ z \in \mathbb{D}\right\},\ [69]. \\
\mathcal{S}^* : &= \{f \in \mathcal{S} : f(\mathbb{D})\ is\ starlike\ with\ respect\ to\ the\ origin\} \\
&= \left\{f \in \mathcal{S} : Re\left(\frac{zf'(z)}{f(z)}\right) > 0,\ z \in \mathbb{D}\right\},\ [7]. \\
\mathcal{C} : &= \{f \in \mathcal{A} : f(\mathbb{D})\ is\ close-to-convex\} \\
&= \left\{f \in \mathcal{A} : Re\left(\frac{f'(z)}{g'(z)}\right) > 0,\ g \in \mathcal{K},\ z \in \mathbb{D}\right\},\ [36, 55]. \\
\mathcal{SP}^{\lambda} : &= \{f \in \mathcal{A} : f(\mathbb{D})\ is\ \lambda - spiral - like\ for\ |\lambda| < \pi/2\} \\
&= \left\{f \in \mathcal{A} : Re\left(e^{i\lambda}\frac{zf'(z)}{f(z)}\right) > 0,\ |\lambda| < \pi/2,\ z \in \mathbb{D}\right\},\ [50, 67].
\end{aligned}$$

It is well-known in the literature that $f \in \mathcal{K} \Leftrightarrow zf' \in \mathcal{S}^*$ and

$$\mathcal{K} \subset \mathcal{S}^* \subset \mathcal{SP}^{\lambda} \subset \mathcal{S}, \quad \mathcal{K} \subset \mathcal{S}^* \subset \mathcal{C} \subset \mathcal{S}.$$

For further details and references, one may refer to [3, 5, 23, 27].

If $f \in \mathcal{A}$, then $f(z)/z \neq 0$ in $\mathbb{D}$. Suppose γ is any real or complex number. There are two types of integral operators. An operator $J_\gamma : \mathcal{A} \to \mathcal{A}$ is called *Type-1 integral operator* if

$$J_\gamma(f)(z) = \int_0^z \left(\frac{f(t)}{t}\right)^\gamma dt. \tag{1.2}$$

On the other hand, an operator $I_\gamma : \mathcal{A} \to \mathcal{A}$ defined by

$$I_\gamma(f)(z) = \int_0^z (f'(t))^\gamma dt \tag{1.3}$$

is called *Type-2 integral operator*.

Though there are several motivations for study of integral operators in GFT, some of the motivations for researchers have been based on the following main questions.

Question 1 Do integral operators J_γ and I_γ belong to the classes $\mathcal{S}^*, \mathcal{K}, \mathcal{SP}^\lambda, \mathcal{C}, \mathcal{S}$ or many of their subclasses where $f \in \mathcal{A}$ or $\mathcal{S}$ or their subclasses?

Question 2 Investigate values of γ for which Type-1 and Type-2 integral operators belong to the classes $\mathcal{S}^*, \mathcal{K}, \mathcal{SP}^\lambda, \mathcal{C}, \mathcal{S}$ or several of their subclasses where $f \in \mathcal{A}$ or $\mathcal{S}$ or their subclasses?

Nevanlinna [51] in 1920 found the largest disk of radius R such that $f(\mathbb{D}_R)$ is convex for all $f \in \mathcal{S}$. He discovered that $R_\mathcal{K}(\mathcal{S}) = 2 - \sqrt{3}$, where $R_\mathcal{K}(\mathcal{S})$ denotes the radius of convexity in class $\mathcal{S}$ (or $\mathcal{K}$-radius in $\mathcal{S}$). Motivated by Nevanlinna, several researchers in the last hundred years investigated radius problems for numerous subclasses of $\mathcal{A}$ and $\mathcal{S}$. In particular, they studied the following type of question related to integral operators.

Question 3 Let $\mathcal{M}$ be a family of functions generated by the integral operators of Type-1 or Type-2. Investigate radius problems in $\mathcal{M}$; for example, find $R_\mathcal{S}(\mathcal{M})$, $R_{\mathcal{S}^*}(\mathcal{M})$, and $R_{\mathcal{SP}^\lambda}(\mathcal{M})$. Also, investigate radius problems of type $R_\mathcal{M}(\mathcal{S})$, $R_\mathcal{M}(\mathcal{S}^*)$, and $R_\mathcal{M}(\mathcal{SP}^\lambda)$.

Note that if we are given 100 different subclasses of $\mathcal{A}$ or $\mathcal{S}$ and 100 different properties, then the determination of the radius of each property P or class P in each class $\mathcal{M}$ gives 10,000 radius problems. However, we note that (1) some of the radius problems may be meaningless or trivial; (2) some of the radius problems may be different and are already known; and (3) some of the radius problems may still be open.

Before going to Question 4, we need to introduce two important concepts of GFT, called the convolution (or Hadamard product) and convolution operator.

If $f, g \in \mathcal{A}$ with

$$f(z) = \sum_{n=1}^{\infty} a_n z^n \text{ and } g(z) = \sum_{n=1}^{\infty} b_n z^n,$$

then the *convolution* or *Hadamard product* of f and g, denoted by $f * g$, is the function h defined by

$$h(z) = (f * g)(z) = z + \sum_{n=2}^{\infty} a_n b_n z^n.$$

The term *convolution* arises from the integral formula

$$h(r^2 e^{i\theta}) = \frac{1}{2\pi} \int_0^{2\pi} f(re^{i(\theta - t)}) g(re^{it}) dt, \ (r < 1).$$

Given $f \in \mathcal{A}$, we can define the *convolution operator* $\Delta : \mathcal{A} \to \mathcal{A}$ by $\Delta(g) = f * g$. If $g(z) = z/(1-z)$, $z \in \mathbb{D}$, then obviously $f * g = f$ for all $f \in \mathcal{A}$. Again, if $g(z) = z/(1-z)^2$, then it is straightforward to notice that $f * g = zf'$ for all $f \in \mathcal{A}$.

Using the convolution techniques of Ruscheweyh [60], Ruscheweyh and Sheil-Small [61] and motivated by Barnard and Kellog [13], several researchers studied the following type of question related to integral operators.

Question 4 Study integral operators of Type-1, Type-2 or a mixed type of integral operators that can be written as convolution operators. For example, for an integral operator J_γ (or I_γ), γ real or complex, find the function g such that $\Delta(g) \equiv J_\gamma = f * g$ (or $I_\gamma = f * g$) and study its properties.

In this paper, we mainly focus on a brief survey of more than one hundred years of historical development of some of the integral and related operators in GFT. The strengths and discovery of the methods used in these operators lie in their ability to unify a large number of diverse integral operators and their properties. We also briefly survey $q-$analogues of the integral operators in GFT. Since there are several surveys and books in GFT and also on integral operators (see [3, 5, 10, 23, 27, 45, 64]), we present here only a selection of the results related to our precise objectives.

2 Alexander-Type Integral Operators

In 1915, Alexander in [7] defined an integral operator $J_1 : \mathcal{A} \to \mathcal{A}$ given by

$$J_1(f)(z) = \int_0^z \frac{f(t)}{t} dt. \tag{2.1}$$

This operator is called the Alexander integral operator. In [7], it was proved that J_1 preserves starlikeness, convexity, and close-to-convexity; that is, $J_1(\mathcal{S}^*) \subset \mathcal{S}^*$, $J_1(\mathcal{K}) \subset \mathcal{K}$, and $J_1(\mathcal{C}) \subset \mathcal{C}$. Also, $J_1(\mathcal{S}^*) \subset \mathcal{K}$. But J_1 does not preserve univalence. In 1960, Biernacki [16] claimed that $f \in \mathcal{S}$ implies $J_1(f) \in \mathcal{S}$, but this turned out to be wrong. In 1963, a counterexample was given by Krzyz and Lewandowski [41] who showed that $J_1(\mathcal{S}) \not\subset \mathcal{S}$ because the function

$$f(z) = \frac{z}{(1-iz)^{1-i}} \in \mathcal{SP}^\lambda \subset \mathcal{S}$$

for $\lambda = \pi/4$, but $J_1(f) \notin \mathcal{S}$. This negative result resulted into several radius problems explained in Question 3 in Introduction.

In view of (2.1), Alexander [7] observed the following elementary and beautiful relationship between convex and starlike functions.

Theorem 1 ([7]) *If $f \in \mathcal{A}$, then $f \in \mathcal{K} \Leftrightarrow zf'(z) \in \mathcal{S}^*$.*

According to this theorem, if $f \in \mathcal{A}$, then

(i) $f \in \mathcal{K} \Rightarrow zf'(z) \in \mathcal{S}^*$.
(ii) $f \in \mathcal{S}^* \Rightarrow \int_0^z \frac{f(t)}{t} dt \in \mathcal{K}$.

The Alexander integral operator J_1, defined in (2.1), motivated several researchers in more than one hundred years to discover many new subclasses of $\mathcal{S}$. For example, if a function F is in any class, we can look for a normalized function f such that $F(z) = zf'(z)$ or integral operator J_1 defined by (2.1). Motivated by this idea, Robertson [58] in 1969 defined a new class $\mathcal{KSP}^\lambda$ of convex λ−spiral functions as follows:

$$F(z) \in \mathcal{SP}^\lambda \Leftrightarrow f(z) = \int_0^z \frac{F(t)}{t} dt \in \mathcal{KSP}^\lambda.$$

Using analytic charaterization of class $\mathcal{SP}^\lambda$ defined in Sect. 1, Robertson [58] found that

$$\mathcal{KSP}^\lambda = \left\{ f \in \mathcal{A} : Re\left(e^{i\lambda} \left(1 + \frac{zf''(z)}{f'(z)} \right) \right) > 0, \ |\lambda| < \pi/2, \ z \in \mathbb{D} \right\}.$$

In fact, he observed that $\mathcal{KSP} = \bigcup_\lambda \mathcal{KSP}^\lambda$.

Motivated by Theorem 1 and the definition of spiral-like functions of order α, $(0 \leq \alpha < 1)$ and type β, $(0 < \beta \leq 1)$ defined in [49], the first author [1] in 1983 defined and studied a family of λ−Robertson functions of order α and type β.

Since 1915, many papers appeared concerning non-linear integral operators of Type-1 given by (2.1); that is,

$$J_\gamma(f)(z) = \int_0^z \left(\frac{f(t)}{t} \right)^\gamma dt,$$

where γ is a real or complex number. The operator J_γ is called the Cauchy integral operator. For $\gamma = 1$, this gives the Alexander integral operator J_1. A number of papers (including paper [19] by Causey in 1967) have appeared that determine choices of γ and subclasses $\mathcal{M}$ of $\mathcal{S}$ such that $J_\gamma(\mathcal{M}) \subset \mathcal{S}$. For example, we have the following known results.

Theorem 2 ([19]) *$J_\gamma(\mathcal{S}) \subset \mathcal{S}$ for $0 \leq \gamma \leq (\sqrt{5} - 2)/4$.*

Theorem 3 ([54])

(i) $J_\gamma(\mathcal{S}^*) \subset \mathcal{S}$ for $\gamma \in [0, 3/2]$,
(ii) $J_\gamma(\mathcal{K}) \subset \mathcal{S}$ for $\gamma \in [0, 3]$.

Theorem 4 ([44])

(i) $J_\gamma(\mathcal{S}^*) \subset \mathcal{C}$ for $\gamma \in [-1/2, 3/2]$; and if $\gamma \notin [-1/2, 3/2]$, then $J_\gamma(\mathcal{S}^*) \not\subset \mathcal{S}$.
(ii) $J_\gamma(\mathcal{K}) \subset \mathcal{C}$ for $\gamma \in [-1, 3]$; and if $\gamma \notin [-1, 3]$, then $J_\gamma(\mathcal{K}) \not\subset \mathcal{S}$.
(iii) $J_\gamma(\mathcal{C}) \subset \mathcal{C}$ for $\gamma \in [-1/2, 1]$; and if $\gamma \notin [-1/2, 1]$, then $J_\gamma(\mathcal{C}) \not\subset \mathcal{C}$.

Theorem 5 ([37])

(i) $J_\gamma(\mathcal{S}) \subset \mathcal{S}$ *for* $|\gamma| \leq 1/4$, *however, the bound 1/4 is not best possible.*
(ii) $J_\gamma(\mathcal{SP}^\lambda) \not\subset \mathcal{S}$ *for* $|\gamma| > 1/2$.

These papers motivated many researchers to investigate other problems related to the Cauchy integral operator J_γ; see, for example, [20, 21]. Also, one may refer to a survey article by Merkes [45].

3 Bernardi-Type Integral Operators

In 1969, Bernardi [15] defined an integral operator $V_\gamma : \mathcal{A} \to \mathcal{A}$ given by

$$V_\gamma(f)(z) = \frac{1+\gamma}{z^\gamma} \int_0^z t^{\gamma-1} f(t)dt, \tag{3.1}$$

where $\gamma = 1, 2, 3, \ldots$. In fact, Bernardi was motivated by the integral operator

$$V_1(f)(z) = \frac{2}{z} \int_0^z f(t)dt. \tag{3.2}$$

Note that (3.2) is a special case of (3.1) because for $\gamma = 1$, (3.1) reduces to (3.2). The operator (3.2) was defined and studied by Libera [42] in 1965 and is, therefore, called the Libera integral operator. However, the operator given by (3.1) is called the Bernardi integral operator. We note that the Libera operator V_1 is the solution of the first-order linear differential equation

$$zg'(z) + g(z) = 2f(z).$$

Moreover, Libera showed that $V_1(\mathcal{S}^*) \subset \mathcal{S}^*$, $V_1(\mathcal{K}) \subset \mathcal{K}$, and $V_1(\mathcal{C}) \subset \mathcal{C}$. However, Libera demonstrated that $V_1(\mathcal{SP}^\lambda) \not\subset \mathcal{SP}^\lambda$; that is, there exists a function $f \in \mathcal{SP}^\lambda$ such that $V_1(f)$ may not be a $\lambda-$spiral-like function in $\mathbb{D}$.

Bernardi [15] extended the corresponding results discovered by Libera [42] for the operator V_γ.

Theorem 6 ([15]) *If* $\gamma = 1, 2, 3, \ldots$, *then*

(i) $V_\gamma(\mathcal{S}^*) \subset \mathcal{S}^*$,
(ii) $V_\gamma(\mathcal{K}) \subset \mathcal{K}$,
(iii) $V_\gamma(\mathcal{C}) \subset \mathcal{C}$.

Using the well-known Ruscheweyh derivative operator $D^n f : \mathcal{A} \to \mathcal{A}$ defined by (see [60])

$$D^n f(z) = \frac{z}{(1-z)^{n+1}} * f(z), \;\; (n \in \mathbb{R}, n > -1), \tag{3.3}$$

the first author [2] in 1985 investigated integral operator $V_\gamma(f)$ given by (3.1), where γ is a complex number and $f \in R_n(\alpha)$, where

$$R_n(\alpha) = \left\{ f \in \mathcal{A} : Re\left(\frac{z(D^n f(z))'}{D^n f(z)} \right) > \alpha,\ 0 \leq \alpha < 1,\ z \in \mathbb{D} \right\}. \tag{3.4}$$

Theorem 7 ([2]) *Under certain restrictions on complex number γ and α, we have*

(i) $V_\gamma(R_n(\alpha)) \subset R_n(\alpha)$,
(ii) $V_\gamma(R_n) \subset R_n(\alpha)$,

where $R_n \equiv R_n(0)$.

Corollary 8 ([2]) $V_1(\mathcal{S}^*) \subset \mathcal{S}^*((\sqrt{17}-3)/4)$, *where* $V_1(f) = \frac{2}{z}\int_0^z f(t)dt$ *is the Libera integral operator defined in (3.2).*

Theorem 6 was further generalized by several researchers for real or complex γ, and when f is in different subclasses of $\mathcal{A}$ and $\mathcal{S}$; for example, see [8] and [56].

4 Miller-Mocanu-Reade-Type Integral Operators

Miller et al. [47] in 1974 extended the Alexander operator in another direction. They studied the integral operator

$$MJ_\beta(f)(z) = \left[\beta \int_0^z \frac{f^\beta(t)}{t} dt \right]^{1/\beta}, \tag{4.1}$$

where $\beta > 0$. Note that $MJ_1 \equiv J_1$ is the Alexander operator. They proved several results for the operator $MJ_\beta(f)$; for example,

Theorem 9 ([47]) $MJ_\beta(\mathcal{S}^*) \subset \mathcal{S}^*$.

In 1973, almost at the same time, Singh [66] defined the integral operator $J_{\gamma,\beta} : \mathcal{A} \to \mathcal{A}$ given by

$$J_{\gamma,\beta}(f)(z) = \left[\frac{\beta+\gamma}{z^\gamma} \int_0^z f^\beta(t) t^{\gamma-1} dt \right]^{1/\beta}, \tag{4.2}$$

where $\gamma, \beta = 1, 2, 3, \ldots$.

Theorem 10 ([66]) $J_{\gamma,\beta}(\mathcal{S}^*) \subset \mathcal{S}^*$.

Remark 11 If $\gamma > -1$ and $\beta > 0$, we get the following special cases of the operator $J_{\gamma,\beta}$:

(i) $J_{0,1} \equiv J_1$, (Alexander operator)

(ii) $J_{1,1} \equiv V_1$, (Libera operator)
(iii) $J_{0,\beta} \equiv MJ_\beta$, (Miller-Mocanu-Reade operator).

In 2007, Breaz and Breaz [18] investigated the operator $MJ_\beta(f)$ for a complex number β and when f satisfies certain condition.

In 1978, Miller et al. [46] defined a more general operator $J : \mathcal{A} \to \mathcal{A}$ given by

$$J(f)(z) = \left[\frac{\beta+\gamma}{z^\gamma \Phi(z)} \int_0^z f^\alpha(t)\varphi(t)t^{\delta-1}dt\right]^{1/\beta}, \tag{4.3}$$

where $\beta > 0$, with suitable analytic functions $\Phi(z)$ and $\varphi(z)$, and where $\alpha, \beta, \gamma, \delta$ are restricted to be real. The operator J is called the Miller-Mocanu-Reade integral operator.

Remark 12 Most of the integral operators studied before 1978 became special cases of the operator J. In particular, the operators J_1, J_γ, MJ_β, V_1 and V_γ are special cases of the operator J.

Theorem 13 ([46]) *Under suitable restrictions on analytic functions Φ and φ, and for suitable choices of the real $\alpha, \beta, \gamma, \delta$, we have*

$$J(\mathcal{S}^*) \subset \mathcal{S}^*.$$

For several other results and proofs, a reader may refer to the paper in [46].

In 1979, Bajpai [12] defined a generalized Bernardi integral operator $F : \mathcal{A} \to \mathcal{A}$ given by

$$F(f)(z) = \left[\frac{\gamma + e^{i\lambda}}{z^\gamma} \int_0^z t^{\gamma-1}(f(t))^{e^{i\lambda}}dt\right]^{e^{-i\lambda}},$$

where $Re\gamma > -\alpha\cos\lambda$, $|\lambda| < \pi/2$, $0 \leq \alpha < 1$, and $z \in \mathbb{D}$. Note that for $\lambda = 0$, operator $F \equiv V_\gamma$ gives the Bernardi integral operator. He studied the operator for the class $\mathcal{SP}^\lambda(\alpha)$ of $\lambda-$spiral-like function of complex order α, $(0 \leq \alpha < 1)$ defined by

$$\mathcal{SP}^\lambda(\alpha) = \left\{f \in \mathcal{A} : Re\left(e^{i\lambda}\frac{zf'(z)}{f(z)}\right) > \alpha\cos\lambda,\ |\lambda| < \pi/2,\ z \in \mathbb{D}\right\}.$$

Theorem 14 ([12]) $F(\mathcal{SP}^\lambda(\alpha)) \subset \mathcal{SP}^\lambda(\alpha)$ *whenever* $Re\gamma > -\alpha\cos\lambda$.

Remark 15 Exact analogues of this theorem are false for Libera and Bernardi operators:

$$V_1(\mathcal{SP}^\lambda) \not\subset \mathcal{SP}^\lambda \text{ and } V_\gamma(\mathcal{SP}^\lambda) \not\subset \mathcal{SP}^\lambda.$$

Remark 16 The operator F and Theorem 14 extend the corresponding results given by Libera [42] and Bernardi [15].

In 1991, Miller and Mocanu [48] investigated the operator $J(f)$ defined in (4.3), where $f \in \mathcal{A}$ and where Φ and φ are suitable analytic functions and $\alpha, \beta, \gamma, \delta$ are complex parameters. They proved the following result.

Theorem 17 ([48]) *For certain condition on parameters $\alpha, \beta, \gamma, \delta$, and functions Φ and φ, we have*

(i) $J(f)(\mathcal{S}^*) \subset \mathcal{S}^*$,
(ii) $J(f)(\mathcal{K}) \subset \mathcal{S}^*$.

Corollary 18 *If $\Phi \equiv 1$ then $J(f)(\mathcal{S}^*) \subset \mathcal{S}^*[(\beta - \gamma)/2\beta]$.*

In 2002, Bulboaca [17] investigated integral operators of the type (4.2) using superordination.

5 Type-2 Integral Operators

Type-2 integral operators are defined in (1.3); that is,

$$I_\gamma(f)(z) = \int_0^z (f'(t))^\gamma dt,$$

where γ is complex and the power is defined via the branch of $\log f'(w)$ for which $\log f'(0) = 0$. Many researchers worked on Type-2 integral operators I_γ.

In [27], Goodman looked at univalent functions f for which $I_\gamma(f)$ is not univalent in $\mathbb{D}$. In fact, he proved the following:

Lemma 19 *([27]) If $|\gamma| > 1/3$ and $\gamma \neq 1$, then there is a function $f \in \mathcal{S}$ for which $I_\gamma(f)$ is not univalent in $\mathcal{S}$.*

Here are some important inclusion properties.

Theorem 20 ([24]) *$I_\gamma(\mathcal{S}) \subset \mathcal{S}$ for $0 \leq \gamma \leq (\sqrt{5} - 2)/3$.*

Theorem 21 ([59])

(i) *$I_\gamma(\mathcal{S}) \not\subset \mathcal{S}$ for all complex numbers for each complex $\gamma \neq -1$ in the range $|\gamma| > 1/3$.*
(ii) *$I_\gamma(\mathcal{C}) \subset \mathcal{C}$ for all $0 \leq \gamma \leq 1$.*

Theorem 22 ([14])

(i) *$I_\gamma(\mathcal{S}^*) \subset \mathcal{S}$ for all complex numbers γ such that $|\gamma| \leq 1/4$.*
(ii) *$I_\gamma(\mathcal{S}) \subset \mathcal{S}$ for all complex numbers γ such that $|\gamma| \leq 1/6$.*

In 1975, Pfaltzgraff showed the following univalence criteria:

Theorem 23 ([57]) *$I_\gamma(\mathcal{S}) \subset \mathcal{S}$ for all complex γ such that $|\gamma| \leq 1/4$.*

For more details and problems, one may also refer to [28].

Badghaish et al. [11] in 2011 further extended Type-1 integral operators and studied closure properties on the Ma-Minda type of the starlike and convex functions.

In 2012, Kim and Sugawa [38] investigated an operator K_γ defined by

$$K_\gamma(f)(z) = z\left(\frac{f(z)}{z}\right)^\gamma$$

for a complex number γ. Such an operator K_γ is called the power deformation of f with exponent γ. These authors studied several properties related to K_γ, Type-1, and Type-2 integral operators.

6 Convolution Operators as Integral Operators

Barnard and Kellog [13] in 1980 observed that many of the integral operators can be written as convolution operators. We notice that convolution operators help us to understand the geometric properties of differential and integral operators.

Barnard and Kellog [13] observed that Alexander, Libera, and Bernardi integral operators can be written as convolution operators.

Example 24 The Alexander integral operator J_1 defined in (2.1) can be written as a convolution operator

$$J_1(f)(z) = \int_0^z \frac{f(t)}{t}dt = (h_1 * f)(z),$$

where

$$h_1(z) = z + \sum_{n=2}^{\infty} \frac{1}{n} z^n = -\log(1-z).$$

Example 25 The Libera integral operator V_1 defined in (3.2) can be written as a convolution operator

$$V_1(f)(z) = \frac{2}{z}\int_0^z f(t)dt = (h_2 * f)(z),$$

where

$$h_2(z) = z + \sum_{n=2}^{\infty} \frac{2}{n+1} z^n = -\frac{2[z + \log(1-z)]}{z}.$$

Example 26 The Bernardi integral operator V_γ defined in (3.1) can be written as a convolution operator

$$V_\gamma(f)(z) = \frac{1+\gamma}{z^\gamma}\int_0^z t^{\gamma-1} f(t)dt = (h_\gamma * f)(z),$$

where $\gamma > -1$ and

$$h_\gamma(z) = z_2F_1(1, 1+\gamma; 2+\gamma; z)$$

is a special case of the Gaussian hypergeometric function ${}_2F_1(a, b; c; z)$ defined by

$$ {}_2F_1(z) = {}_2F_1(a, b; c; z) = \sum_{n=0}^{\infty} \frac{(a)_n (b)_n}{(c)_n} \frac{z^n}{n!}, \tag{6.1}$$

and where $(\alpha)_n$ is the Pochhammer symbol given by

$$(\alpha)_n = \frac{\Gamma(\alpha+n)}{\Gamma(\alpha)} = \alpha(\alpha+1)\ldots(\alpha+n-1),\ (n > 1) \text{ and } (\alpha)_0 = 1.$$

Here, Γ denotes the Gamma function and it is defined as an extension of the factorial to a complex or real number argument. It is related to the factorial given by $\Gamma(n) = (n-1)!$.

Carlson and Shaffer [22] in 1984 defined a linear operator $L(a, c) : \mathcal{A} \to \mathcal{A}$ given by

$$L(a, c)f(z) = \varphi(a; c; z) * f(z), \tag{6.2}$$

where incomplete beta function $\varphi(a; c; z)$ is defined by

$$\varphi(a; c; z) = \sum_{n=0}^{\infty} \frac{(a)_n}{(c)_n} z^{n+1} = z + \sum_{n=2}^{\infty} \frac{(a)_{n-1}}{(c)_{n-1}} z^n$$
$$= z + \sum_{n=2}^{\infty} \frac{\Gamma(c)\Gamma(a+n-1)}{\Gamma(a)\Gamma(c+n-1)} z^n, \quad (c \neq 0, -1, \ldots).$$

By using the Gaussian hypergeometric functions given by (6.1), we obtain

$$\varphi(a; c; z) = z_2F_1(a, 1; c; z)$$

and therefore

$$L(a, c)f(z) = z_2F_1(a, 1; c; z) * f(z).$$

There are the following special cases of the Carlson-Shaffer operator $L(a, c)$.

(i) $L(a, a)f(z) = f(z)$ is the identity operator.
(ii) $L(2, 1)f(z) = zf'(z)$ is the Alexander differential operator given in [7].
(iii) $L(3, 1)f(z) = \frac{f(z)+zf'(z)}{2}$ is the Livingston differential operator found by Livingston [43] in 1966.

(iv) $L(n+1,1)f(z) = D^n f(z) := \frac{z}{(1-z)^{n+1}} * f(z)$, $(n \in \mathbb{R}, n > -1)$, where $D^n f$ is the Ruscheweyh convolution operator [60] given by (3.3).

Shanmugan et al. [63] in 2008 found some sandwich theorems for certain subclasses of analytic functions associated with the Carlson-Shaffer operator $L(a, c)$.

Motivated by the Carlson-Shaffer operator, Hohlov [29] in 1984 introduced the generalized convolution operator $H_{a,b,c} : \mathcal{A} \to \mathcal{A}$ defined by

$$H_{a,b,c}f(z) = z_2F_1(a, b; c; z) * f(z), \tag{6.3}$$

where ${}_2F_1(a, b; c; z)$ is the Gaussian hypergeometric function given by (6.1). Also, Hohlov in the same paper [29] showed that the linear operator $H_{a,b,c}$ maps the class $\mathcal{K}$ onto $\mathcal{S}$. We note that for different values of parameters a, b, and c, the operator $H_{a,b,c}$ reduces to several known linear or differential operators; for example, for $b = 1$ the operator $H_{a,b,c}$ reduces to the Carlson-Shaffer linear operator $L(a, c)$ defined in (6.2).

In 1989, the first author and Silverman [4] investigated convolution integral operators expressed as a hypergeometric function.

Theorem 27 ([4]) *Let*

$$H(z) =_{m+1} F_m(n+1, n+1, \ldots, n+1, 1; n+2, n+2, \ldots, n+2; z)$$

be a hypergeometric function. Then

$$f \in R_n(\alpha) \Leftrightarrow f * zH(z) \in R_{n+m}(\alpha)$$

for any $m = 1, 2, \ldots$, *where* $R_n(\alpha)$ *is defined in (3.4).*

For additional references on convolution operators, one may refer to a survey article by Shareef et al. [64] and references given therein.

7 Fournier-Ruscheweyh Integral Operator

In 1994, Fournier and Ruscheweyh [25] introduced the linear integral operator $F_\lambda :$ $\mathcal{A} \to \mathcal{A}$ given by

$$F_\lambda(f)(z) = \int_0^1 \lambda(t)\frac{f(tz)}{t}dt, \tag{7.1}$$

where $\lambda : [0, 1] \to \mathbb{R}$ is non-negative with $\int_0^1 \lambda(t)dt = 1$. The operator F_λ reduces to various well-known integral operators for specific choices of λ.

Example 28 (i) Letting $\lambda(t) = (1+\gamma)t^\gamma$ and $\gamma > -1$, we get the Bernardi integral operator

$$F_\lambda(f)(z) \equiv V_\gamma(f)(z) = \frac{1+\gamma}{z^\gamma}\int_0^z t^{\gamma-1}f(t)dt.$$

(ii) Letting $\lambda(t) = \frac{(\gamma+1)^\delta}{\Gamma(\delta)} t^\gamma (\log(1/t))^{\delta-1}$, $\gamma > -1$ and $\delta \geq 0$, we get the Komatu operator [40] given by

$$F_\lambda(f)(z) = \frac{(1+\gamma)^\delta}{\Gamma(\delta)} \int_0^1 \left(\log\frac{1}{t}\right)^{\delta-1} t^{\gamma-1} f(tz)dt.$$

In 1994, Fournier and Ruscheweyh [25] proved the following result.

Theorem 29 ([25]) *If $\beta = \beta(\lambda)$, then $F_\lambda(P_\beta) \subset \mathcal{S}$ where $\beta(\lambda) < 1$ is given by*

$$\frac{\beta(\lambda)}{1-\beta(\lambda)} = -\int_0^1 \lambda(t)\frac{1-t}{1+t}dt,$$

and the class P_β defined by

$$P_\beta = \{f \in \mathcal{A} : \exists\phi \in \mathbb{R}\ s.t.\ Re(e^{i\phi}(f'(z) - \beta)) > 0,\ z \in \mathbb{D}\}.$$

Ali et al. [9] in 2012 investigated the starlikeness of $F_\lambda(f)$ for the function in the class $W_\beta(\alpha, \gamma)$, where

$$W_\beta(\alpha,\gamma) = \Bigg\{ f \in \mathcal{A} : \exists\ \phi \in \mathbb{R}\ and\ z \in \mathbb{D}\ such\ that$$

$$Re\ e^{i\phi}\left((1-\alpha+2\gamma)\frac{f(z)}{z} + (\alpha-2\gamma)f'(z) + \gamma z f''(z) - \beta\right) > 0\Bigg\},$$

and where $\alpha \geq 0,\ \gamma \geq 0,\ \beta < 1$. We note that $W_\beta(1,0) \equiv P_\beta$, which was obtained in [25].

8 Integral Operators in GFT Associated with Quantum Calculus

Quantum calculus is that version of calculus where we do not take a limit. In fact, quantum calculus is a theory of calculus where smoothness is not required. In quantum calculus, the derivative of a function f is just a difference quotient and the integral of f is the sum. In 1909, Jackson in [31] defined $q-$derivative of a function f defined in $\mathbb{D}$, denoted by $(D_q f)(z)$, by

$$(D_q f)(z) = \frac{d_q f(z)}{d_q z} = \frac{f(qz) - f(z)}{(q-1)z},$$

where $q \in (0, 1)$. Note that $\lim_{q\to 1^-}(D_q f)(z) = f'(z)$ if f is differentiable at z.

At the same time, Jackson [32] also defined $q-$antiderivative. A function F is called the $q-$antiderivative of f if $D_qF(z) = f(z)$. This integral is called the Jackson integral of f and is denoted by $\int f(x)d_qx$. In [32], Jackson found that it can be defined as

$$\int f(x)d_qx = x(1-q)\sum_{n=0}^{\infty} q^n f(xq^n),$$

provided that the series on right-hand side converges absolutely.

For definitions, notations, and properties of $q-$derivative and $q-$integral operators, one may refer to [6, 26, 30–35].

In 1989, Srivastava [68] initiated research in the use of $q-$calculus in GFT. He defined and studied the $q-$Libera integral operator.

Definition 30 ([68]) An operator $V_1^q : \mathcal{A} \to \mathcal{A}$ is called a $q-$Libera integral operator if

$$V_1^q(f)(z) = \frac{[2]_q}{z}\int_0^z f(t)d_qt = \sum_{n=1}^{\infty}\left(\frac{[2]_q(1-q)}{1-q^{n+1}}\right)a_nz^n, \tag{8.1}$$

where $q \in (0, 1)$ and $[n]_q = \frac{1-q^n}{1-q}$ is $q-$bracket or $q-$number of n.

Motivated by the $q-$Libera integral operator, Noor et al. [52] in 2017 introduced the $q-$Bernardi integral operator $V_\gamma^q : \mathcal{A} \to \mathcal{A}$ given by

$$\begin{aligned} V_\gamma^q(f)(z) &= \frac{[1+\gamma]_q}{z^\gamma}\int_0^z t^{\gamma-1}f(t)d_qt \\ &= \sum_{n=1}^{\infty}\left(\frac{[1+\gamma]_q}{[n+\gamma]_q}\right)a_nz^n, \ (\gamma = 1, 2, \ldots). \end{aligned} \tag{8.2}$$

We note that for $\gamma = 1$, the operator V_γ^q reduces to the operator V_1^q. The researchers in [52] studied the operator V_γ^q when f is in $\mathcal{S}_q^*(A, B)$ or $\mathcal{C}_q(A, B)$.

Definition 31 ([52]) A function f is in $\mathcal{S}_q^*(A, B)$ if and only if

$$\mathcal{S}_q^*(A, B) = \left\{f \in \mathcal{A} : \frac{zD_qf(z)}{f(z)} = \frac{1+Aw(z)}{1+Bw(z)},\ -1 \le B < A \le 1, z \in \mathbb{D}\right\},$$

where w is analytic with $w(0) = 0$ and $|w(z)| < 1$.

Definition 32 ([52]) A function f belongs to $\mathcal{C}_q(A, B)$ if there exists a function $g \in \bigcap_{0<q<1}\mathcal{S}_q^*(A, B)$ such that

$$\mathcal{C}_q(A, B) = \left\{f \in \mathcal{A} : \frac{zD_qf(z)}{g(z)} = \frac{1+Aw(z)}{1+Bw(z)},\ -1 \le B < A \le 1, z \in \mathbb{D}\right\},$$

where w is analytic with $w(0) = 0$ and $|w(z)| < 1$.

Theorem 33 ([52]) *If $f \in S_q^*(A, B)$, then the function V_γ^q defined by (8.2) belongs to the class $\cap_{0<q<1} S_q^*(A, B)$, where $-1 \le B < A \le 1, z \in \mathbb{D}$.*

Theorem 34 ([52]) *Let $f \in \mathcal{A}$ and let $f \in \mathcal{C}_q(A, B)$ be q-close-to-convex functions with respect to g, $g \in \cap_{0<q<1} S_q^*(A, B)$. If*

$$F(z) = \frac{[1+\gamma]_q}{z^\gamma} \int_0^z t^{\gamma-1} f(t) d_q t, \quad G(z) = \frac{[1+\gamma]_q}{z^\gamma} \int_0^z t^{\gamma-1} g(t) d_q t,$$

then F belongs to the class of $q-$close-to-convex functions with respect to G in $\mathbb{D}$.

The study of $q-$analogues of integral operators in GFT is a new area of research. For some of the recent articles in this area, one may refer to [53, 62, 65].

9 Concluding Remarks

In this article, we have made an attempt to present a survey of integral operators in Geometric Function Theory. In an effort to cut down the size of this survey article, we have omitted integral operators in several research areas in GFT, such as multivalent functions, meromorphic functions, and harmonic univalent functions. However, the list of potential areas of further research related to integral operators in GFT includes problems for known and new classes concerning (i) radius problems, (ii) convolution properties, (iii) inclusion properties, (iv) differential subordination, (v) neighborhood problems, (vi) negative coefficients, and others. We also omitted integral operators in several research areas in Quantum calculus, for example, $q-$integral operators in multivalent functions, meromorphic functions, $q-$fractional calculus, and (p, q)-calculus.

We hope this survey article may be a useful guide for new and old researchers working in Geometric Function Theory and related research areas.

Acknowledgements The authors are grateful to the referees for their helpful suggestions and insights that helped us to improve the quality and clarity of this manuscript.

References

1. Ahuja, O.P.: Certain generalizations of the Robertson functions. Yokohama Math. J. **31**, 5–11 (1983)
2. Ahuja, O.P.: Integral operators of certain univalent functions. Int. J. Math. Math. Sci. **8**, 653–662 (1985)
3. Ahuja, O.P.: The Bieberbach conjecture and its impact on the developments in geometric function theory. Math. Chronicle **15**, 1–28 (1986)
4. Ahuja, O.P., Silverman, H.: Function classes related to Ruscheweyh derivatives. J. Austral. Math. Soc. **47**, 438–444 (1989)

5. Ahuja, O.P., Silverman, H.: A survey on spiral-like and related function classes. Math. Chronicle **20**, 39–66 (1991)
6. Ahuja, O.P., Çetinkaya, A.: Use of quantum calculus approach in mathematical sciences and its role in geometric function theory. In: National Conference on Mathematical Analysis and Computing 2018, AIP Conference Proceedings, vol. 2095, p. 020001 (2019)
7. Alexander, J.W.: Functions which map the interior of the unit circle upon simple regions. Ann. Math. **17**, 12–22 (1915–1916)
8. Ali, R.M., Ravichandran, V., Seenivasagan, N.: On Bernardi's integral operator and the Briot-Bouquet differential subordination. J. Math. Anal. Appl. **324**, 663–668 (2006)
9. Ali, R.M., Badghaish, A.O., Ravichandran, V., Swaminathan, A.: Starlikeness of integral transforms and duality. J. Math. Anal. Appl. **385**, 808–822 (2012)
10. Arif, M., Noor, K.I., Ghani, F.: Some properties of an integral operator defined by convolution. J. Inequal. Appl. 13 (2012)
11. Badghaish, A.O., Ali, R.M., Ravichandran, V.: Closure properties of operators on the Ma-Minda type starlike and convex functions. Appl. Math. Comput. **218**, 667–672 (2011)
12. Bajpai, S.K.: An analogue of a R. J. Libera's result. Rendiconti di Mathematica **12**, 285–289 (1979)
13. Barnard, R.W., Kellog, C.: Applications of convolution operators to problems in univalent function theory. Michigan Math. J. **27**, 81–94 (1980)
14. Becker, J.: Löwnersche Differentialgleichung und quasi-konform fortsetzbare schlichte funktionen. J. Reine. Angew. Math. **255**, 23–43 (1972)
15. Bernardi, S.D.: Convex and starlike univalent functions. Trans. Amer. Math. Soc. **135**, 429–446 (1969)
16. Biernacki, M.: Sur l'intégrale des fonctions univalentes. Bull. Acad. Polon. Sci. Ser. Sci. Math. Astr. Phys. **8**, 29–34 (1960)
17. Bulboaca, T.: A class of superordination-preserving integral operators. Indag. Math. **13**(3), 301–311 (2002)
18. Breaz, D., Breaz, N.: An integral univalent operator. Acta Math. Univ. Comenian **7**, 137–142 (2007)
19. Causey, W.M.: The close-to-convexity and univalence of an integral. Math. Zeitschr. **99**, 207–212 (1967)
20. Causey, W.M.: The univalence of an integral. Proc. Amer. Math. Soc. **27**, 500–503 (1971)
21. Causey, W.M., Reade, M.O.: On the univalence of functions defined by certain integral operators. J. Math. Anal. Appl. **89**, 28–39 (1982)
22. Carlson, B.C., Shaffer, D.B.: Starlike and prestarlike hypergeometric functions. SIAM J. Math. Anal. **15**, 737–745 (1984)
23. Duren, P.L.: Univalent Functions. Grundlehren der Mathematischen Wissenschaften, vol. 259. Springer, Newyork (1983)
24. Duren, P.L., Shapiro, H.S., Shields, A.L.: Singular measures and domains not of Smirnov type. Duke Math. J. **33**, 247–254 (1966)
25. Fournier, R., Ruscheweyh, S.: On two extremal problems related to univalent functions. Rocky Mountain J. Math. **24**(2), 529–538 (1994)
26. Gasper, G., Rahman, M.: Basic Hypergeometric Series. Cambridge University Press (2004)
27. Goodman, A.W.: Univalent Functions, vols. I and II. Polygonal Pub. House (1983)
28. Hayman, W.K.: Research Problems in Function Theory. Athlone Press, London (1967)
29. Hohlov, Y.E.: Hadamard convolution, hypergeometric functions and linear operators in the class of univalent functions. Dokl. Akad. Nauk Ukr. SSR, Ser. A **7**, 25–27 (1984)
30. Ismail, M.E.H., Merkes, E., Steyr, D.: A generalization of starlike functions. Complex Variables Theory Appl. **14**(1), 77–84 (1990)
31. Jackson, F.H.: On $q-$functions and a certain difference operator. Trans. Royal Soc. Edinburgh **46**(2), 253–281 (1908)
32. Jackson, F.H.: On $q-$definite integrals. Quart. J. Pure Appl. Math. **41**, 193–203 (1910)
33. Jackson, F.H.: Basic integration. Quart. J. Math. (Oxford) **2**(2), 1–16 (1951)
34. Jackson, F.H.: $q-$difference equations. Amer. J. Math. **32**(4), 305–314 (1910)

35. Kac, V., Cheung, P.: Quantum Calculus. Springer, New-York (2002)
36. Kaplan, W.: Close-to-convex schlicht functions. Michigan Math. J. **1**, 169–185 (1952)
37. Kim, Y.J., Merkes, E.P.: On an integral of powers of a spiral-like function. Kyungpook Math. J. **12**, 249–252 (1972)
38. Kim, Y.C., Sugawa, T.: On power deformations of univalent functions. Monatsh. Math. **167**, 231–240 (2012)
39. Koebe, P.: Über die Uniformisierung beliebiger analytischer kurven. Nachr. Acad. Wiss. Göttingen, Math. Phys. KI. 191–210 (1907)
40. Komatu, Y.: On analytic prolongation of family of operators. Mathematica (Cluj) **39**(55), 141–145 (1990)
41. Krzyz, J., Lewandowski, Z.: On the integral of univalent functions. Bull. Acad. Polon. Sci. Ser. Sci. Math. Astr. Phys. **11**, 447–448 (1963)
42. Libera, R.J.: Some classes of regular univalent functions. Proc. Amer. Math. Soc. **16**, 755–758 (1965)
43. Livingston, A.E.: On the radius of univalence of certain analytic functions. Proc. Amer. Math. Soc. **17**, 352–357 (1966)
44. Merkes, E.P., Wright, D.J.: On the univalence of a certain integral. Proc. Amer. Math. Soc. **27**, 97–100 (1971)
45. Merkes, E.P.: Univalence of an integral transform. Topics in complex analysis (Fairfield, Conn., 1983), 113–119. Contemp. Math. **38**, Amer. Math. Soc., Providence, RI (1985)
46. Miller, S.S., Mocanu, P.T., Reade, M.O.: Starlike integral operators. Pacific J. Math. **79**(1), 157–168 (1978)
47. Miller, S.S., Mocanu, P.T., Reade, M.O.: Bazilevic functions and generalized convexity. Rev. Roumaine Math. Pures Appl. **19**, 213–224 (1974)
48. Miller, S.S., Mocanu, P.T.: Classes of univalent integral operators. J. Math. Anal. Appl. **157**, 147–165 (1991)
49. Mogra, M.L., Ahuja, O.P.: On spiral-like functions of order α and type β. Yokohama Math. J. **29**, 145–156 (1981)
50. Montel, P.: Leçons sur les fonctions univalentes ou multivalentes. Gauthier-Villars, Paris (1933)
51. Nevanlinna, R.: Über die schlichten Abbildungen des Einheitskreises. Översikt Finska Vetenskaps-Soc. Förh. **62A**, 1–14 (1920)
52. Noor, K.I., Riaz, S., Noor, M.A.: On $q-$Bernardi integral operator. TWMS J. Pure Appl. Math. **8**(1), 3–11 (2017)
53. Noor, K.I.: Some classes of analytic functions associated with $q-$Ruscheweyh differential operators. Facta Universitatis, Ser. Math. Inform. **33**(4), 531–538 (2018)
54. Nunokawa, M.: On the univalence of a certain integral. Trans. Amer. Math. Soc. **146**, 439–446 (1969)
55. Ozaki, S.: On the theory of multivalent functions. Sci. Rep. Tokyo Bunrika Daigaku, A **40**, 167–188 (1935)
56. Parvatham, R.: On Bernardi's integral operators of certain classes of functions. Kyungpook Math. J. **42**(2), 437–441 (2002)
57. Pfaltzgraff, J.A.: Univalence of the integral of $f'(z)^{\alpha}$. Bull. London Math. Soc. **7**, 254–256 (1975)
58. Robertson, M.S.: Univalent functions $f(z)$ for which $zf'(z)$ is spiral-like. Michigan Math. J. **16**, 97–101 (1969)
59. Royster, W.C.: On the univalence of a certain integral. Michigan Math. J. **12**, 385–387 (1965)
60. Ruscheweyh, S.: New criteria for univalent functions. Proc. Amer. Math. Soc. **49**, 109–115 (1975)
61. Ruscheweyh, S., Sheil-Small, T.: Hadamard products of schlicht functions and the Pólya-Schoenberg conjecture. Comment. Math. Helv. **48**, 119–135 (1973)
62. Shah, S.A., Noor, K.I.: Study on the $q-$analogue of a certain family of linear operators. Turkish J. Math. **43**, 2707–2714 (2019)
63. Shanmugan, T.N., Srikandan, S., Frasin, B.A., Kavitha, S.: On Sandwich Theorems for certain subclasses of analytic functions involving Carlson-Shaffer operator. J. Korean Math. Soc. **45**, 611–620 (2008)

64. Shareef, Z., Hussain, S., Darus, M.: Convolution operators in the geometric function theory. J. Inequal. Appl. 213 (2012)
65. Shi, L., Raza, M., Javad, K., Hussain, S., Arif, M.: Class of analytic functions defined by $q-$integral operator in a symmetric region. Symmetry **11**(8), 1042 (2019)
66. Singh, R.: On Bazilevic functions. Proc. Amer. Math. Soc. **18**, 261–271 (1973)
67. Špaček, L.: Contribution à la théorie des fonctions univalentes. Časopis Pěst. Mat. **62**, 12–19 (1932)
68. Srivastava, H.M.: Univalent functions, fractional calculus, and associated generalized hypergeometric functions. In: Srivastava, H.M., Owa, S. (eds.) Univalent Functions, Fractional Calculus, and Their Applications. Halsted Press (Ellis Horwood Limited, Chichester), pp. 329–354. Wiley, New York, Chichester, Brisbane and Toronto (1989)
69. Study, E.: Vorlesungen über ausgewählte Gegenstände der Geometrie. 2. Heft, Leipzig und Berlin (1913)

Integral Operators on a Class of Analytic Functions

S. Sunil Varma, Thomas Rosy, Atulya K. Nagar, and K. G. Subramanian

Abstract A class $SD(\alpha)$, $\alpha \geq 0$, of analytic functions is considered and functions in this class are shown to be univalent and starlike of order $(1 - \frac{1}{\alpha})$, for $\alpha \geq 1$. For functions $f(z)$ to belong to the class $SD(\alpha)$, a sufficient condition is obtained. For functions $f(z)$ satisfying this condition, the functions $F(z)$ defined by several integral operators on $f(z)$ are shown to be in the class $SD(\alpha)$. For a hypergeometric function to belong to the class $SD(\alpha)$, a sufficiency condition is also obtained.

Keywords Analytic functions · Starlike function of order α · Integral operators

1 Introduction

Let A be the class of analytic functions

$$f(z) = z + \sum_{n=2}^{\infty} a_n z^n \tag{1}$$

defined on the unit disk $\Delta = \{z \in \mathcal{C} : |z| < 1\}$. Let $S \subset A$ be the class of analytic univalent functions. Ruscheweyh [22] considered a subclass $D \subset S$ consisting of

S. S. Varma · T. Rosy
Department of Mathematics, Madras Christian College, Tambaram,
Chennai 600059, India
e-mail: sunu_79@yahoo.com

T. Rosy
e-mail: thomas.rosy@gmail.com

A. K. Nagar · K. G. Subramanian (✉)
Department of Mathematics, Computer Science and Engineering,
Liverpool Hope University, Liverpool L16 9JD, USA
e-mail: kgsmani1948@gmail.com

A. K. Nagar
e-mail: atulya.nagar@hope.ac.uk

R. N. Mohapatra et al. (eds.), *Mathematical Analysis and Computing*,
Springer Proceedings in Mathematics & Statistics 344,
https://doi.org/10.1007/978-981-33-4646-8_50

convex functions f for which $Re\{f'(z)\} \geq |zf''(z)|$. Motivated by the class D, a family $UCD(\alpha), \alpha \geq 0$, was introduced in [21] connecting various subclasses of convex functions, especially, the subclass (UCV) of uniformly convex functions (see, for example, [2] for an excellent survey on UCV). A function f given by (1) is in $UCD(\alpha)$ if $Re\{f'(z)\} \geq \alpha|zf''(z)|,\ z \in \Delta, \alpha \geq 0$. A family $SD(\alpha)$ related to $UCD(\alpha)$ was introduced in [20]. Recently, this class has been studied in [9, 25]. Also several authors have considered different integral operators on functions in S and its subclasses (see, for example, [1, 4, 5, 7, 11–14, 16, 17, 26]). In this paper, functions in the class $SD(\alpha)$ are shown to be univalent and starlike of order $(1 - \frac{1}{\alpha})$, for $\alpha \geq 1$. For a function f to belong to the class $SD(\alpha)$ [20], a sufficient condition is obtained. For functions f satisfying this condition, it is shown that the functions $F(z)$ defined by various integral operators on $f(z)$ belong to the class $SD(\alpha)$. Also, for a hypergeometric function to belong to the class $SD(\alpha)$, a condition of sufficiency is obtained.

2 The Class $SD(\alpha)$

In [20, 21], a class $UCD(\alpha),\ \alpha \geq 0$, consisting of functions satisfying the condition $Ref'(z) \geq \alpha|zf''(z)|,\ z \in \Delta$ was introduced and various properties of this class were obtained. Subsequently, this class has been considered by several authors [4, 5, 10, 24] in the context of different studies. A related class $SD(\alpha)$ motivated by the class $UCD(\alpha)$ was considered in [21], which is recalled here.

Definition 1 [21] A function f of the form (1) is said to be in the class $SD(\alpha)$ if

$$Re\left\{\frac{f(z)}{z}\right\} \geq \alpha\left|f'(z) - \frac{f(z)}{z}\right| \tag{2}$$

for $\alpha \geq 0$.

We note that $f \in UCD(\alpha)$ if and only if $zf'(z) \in SD(\alpha)$.

Remark 1 Chichra [6, pp. 41 and 42] has considered a class $\mathcal{G}(\alpha)$ of analytic functions f of the form (1) satisfying the condition

$$Re\left\{(1-\alpha)\frac{f(z)}{z} + \alpha f'(z)\right\} \geq 0 \tag{3}$$

for $\alpha \geq 0$ and has shown that functions in $\mathcal{G}(\alpha)$ are univalent, if $\alpha \geq 1$. Hence the functions f in $SD(\alpha)$ are univalent for $\alpha \geq 1$, since $Re\left\{\frac{f(z)}{z}\right\} \geq -\alpha Re\left(f'(z) - \frac{f(z)}{z}\right)$ if $f \in SD(\alpha)$ so that $Re\left\{(1-\alpha)\frac{f(z)}{z} + \alpha f'(z)\right\} \geq 0$.

The class $S^*(\alpha)$ of starlike functions of order α [19] is well-known and consists of functions f satisfying the analytic condition $Re\{\frac{zf'(z)}{f(z)}\} > \alpha$, for

$0 \le \alpha < 1$, $z \in \Delta$. The class $SD(\alpha)$ is now related with the class $S^*(\alpha)$ in the following theorem.

Theorem 1 $SD(\alpha) \subseteq S^*(1 - \frac{1}{\alpha}), \alpha \ge 1.$

Proof Let $f \in SD(\alpha)$, $\alpha \ge 1$. Then f is univalent and

$$\left|\frac{f(z)}{z}\right| \ge Re\left\{\frac{f(z)}{z}\right\} \ge \alpha\left|f'(z) - \frac{f(z)}{z}\right| = \alpha\left|\frac{f(z)}{z}\right|\left|\frac{zf'(z)}{f(z)} - 1\right|$$

so that

$$\left|\frac{zf'(z)}{f(z)} - 1\right| \le \frac{1}{\alpha}.$$

Now

$$Re\left\{\frac{zf'(z)}{f(z)}\right\} = Re\left\{\left(\frac{zf'(z)}{f(z)} - 1\right) + 1\right\} \ge 1 - \left|\frac{zf'(z)}{f(z)} - 1\right| \ge 1 - \frac{1}{\alpha}.$$

Hence $f \in S^*\left(1 - \frac{1}{\alpha}\right)$.

The following theorem gives a sufficient condition for f of the form (1) to be in the class $SD(\alpha)$.

Theorem 2 *A function f of the form (1) is in the class $SD(\alpha)$ if*

$$\sum_{n=2}^{\infty}[1 + \alpha(n-1)]|a_n| \le 1 \tag{4}$$

Proof For $|z| < 1$,

$$Re\left\{\frac{f(z)}{z}\right\} - \alpha\left|f'(z) - \frac{f(z)}{z}\right| \ge 1 - \left|\frac{f(z)}{z} - 1\right| - \alpha\left|f'(z) - \frac{f(z)}{z}\right|$$

$$\ge 1 - \sum_{n=2}^{\infty}|a_n| - \alpha\sum_{n=2}^{\infty}(n-1)|a_n| = 1 - \sum_{n=2}^{\infty}[1 + \alpha(n-1)]|a_n| \ge 0$$

by (4). Hence

$$Re\left\{\frac{f(z)}{z}\right\} - \alpha\left|f'(z) - \frac{f(z)}{z}\right| \ge 0$$

which implies that $f \in SD(\alpha)$.

Theorem 3 *Let $f \in A$ be given by (1) and satisfy the condition (4). Then the function*

$$F(z) = \frac{1+\gamma}{z^\gamma} \int_0^z f(t) t^{\gamma-1} dt, \ \gamma \geq -1$$

defined by the Bernardi operator belongs to $SD(\alpha)$ for all $\alpha \geq 0$.

Proof Since $f \in A$, $F(z) = z + b_2 z^2 + \cdots$ where $b_n = \frac{\gamma+1}{\gamma+n} a_n, n \geq 2$. Now

$$\sum_{n=2}^{\infty} [1 + \alpha(n-1)]|b_n| = \sum_{n=2}^{\infty} [1 + \alpha(n-1)]\left(\frac{\gamma+1}{\gamma+n}\right)|a_n|$$

$$< \sum_{n=2}^{\infty} [1 + \alpha(n-1)]|a_n| < 1, \text{ since } \gamma + 1 < \gamma + n.$$

Thus, by Theorem 2, $F \in SD(\alpha)$, for all $\alpha \geq 0$.

On substituting $\gamma = 0$ and $\gamma = 1$ in the Bernardi operator, we obtain the Alexander transformation and Libera operator, respectively, and so we have the following Corollary of Theorem 3.

Corollary 1 *Let $f \in A$ be given by (1) and satisfy the condition (4). Then*

1. *the function*

$$F(z) = \int_0^z \frac{f(t)}{t} dt$$

   defined by the Alexander transformation belongs to $SD(\alpha)$, for all $\alpha \geq 0$.
2. *the function*

$$F(z) = \frac{2}{z} \int_0^z f(t) dt$$

   defined by the Libera operator belongs to $SD(\alpha)$, for all $\alpha \geq 0$.

Theorem 4 *Let $f \in SD(\alpha)$ and*

$$F(z) = \frac{1+\gamma}{z^\gamma} \int_0^z f(t) t^{\gamma-1} dt. \tag{5}$$

Then

$$|\gamma F(z) + zF'(z)| \geq \alpha |z^2 F''(z) + \gamma z F'(z) - \gamma F(z)| \tag{6}$$

Proof By (5),

$$f(z) = \frac{\gamma}{\gamma+1} F(z) + \frac{zF'(z)}{1+\gamma} \tag{7}$$

and

$$f'(z) = F'(z) + \frac{1}{1+\gamma} zF''(z) \tag{8}$$

Since $f \in SD(\alpha)$,

$$Re\left\{\frac{f(z)}{z}\right\} \geq \alpha\left|f'(z) - \frac{f(z)}{z}\right|$$

which implies

$$\left|\frac{f(z)}{z}\right| \geq \alpha\left|f'(z) - \frac{f(z)}{z}\right| \tag{9}$$

Using (7) and (8) in (9), we obtain

$$\left|\frac{\gamma}{1+\gamma}\frac{F(z)}{z} + \frac{F'(z)}{1+\gamma}\right| \geq \alpha\left|\frac{\gamma F'(z)}{1+\gamma} + \frac{zF''(z)}{1+\gamma} - \frac{\gamma F(z)}{1+\gamma}\right|$$

which is equivalent to

$$|\gamma F(z) + zF'(z)| \geq \alpha|z^2F''(z) + \gamma zF'(z) - \gamma F(z)|$$

Corollary 2 *If $f \in SD(\alpha)$, then*

$$\left|\left(log\frac{f(z)}{z}\right)'\right| \leq \frac{1}{\alpha|z|}\ for\ all\ z \in \Delta.$$

Proof By Theorem 4, (6) can be written as

$$\left|\frac{z^2F''(z) + \gamma zF'(z) - \gamma F(z)}{\gamma F(z) + zF'(z)}\right| \leq \frac{1}{\alpha}$$

In terms of $f(z)$, the above inequality becomes

$$\left|\frac{f'(z)}{f(z)} - \frac{1}{z}\right| \leq \frac{1}{\alpha|z|}$$

which implies

$$\left|\left(log\frac{f(z)}{z}\right)'\right| \leq \frac{1}{\alpha|z|}\ for\ all\ z \in \Delta.$$

Theorem 5 *Let $f \in A$ be given by (1) and satisfy the condition (4). Then the function*

$$F(z) = \frac{a^\lambda}{\Gamma(\lambda)}\int_0^1 t^{a-2}\left(log\frac{1}{t}\right)^{\lambda-1} f(tz)dt, a > 0, \lambda \geq 0$$

defined by the Komatu operator belongs to $SD(\alpha)$, for all $\alpha \geq 0$.

Proof Here $F(z) = z + b_2 z^2 + \cdots$ where $b_n = \left(\frac{a}{a+n-1}\right)^{\lambda} a_n, a > 0, \lambda \geq 0$.

Now

$$\sum_{n=2}^{\infty}[1+\alpha(n-1)]|b_n| = \sum_{n=2}^{\infty}[1+\alpha(n-1)]\left|\frac{a}{a+n-1}\right|^{\lambda}|a_n|$$

$$< \sum_{n=2}^{\infty}[1+\alpha(n-1)]|a_n| < 1, \forall\alpha \geq 0,$$

since $a < a + n - 1 \; for \; n \geq 2$. Thus by Theorem 2, $F \in SD(\alpha)$ for all $\alpha \geq 0$.

Theorem 6 *Let $f \in A$ be given by (1) and satisfy the condition (4). Then the function*

$$F(z) = \frac{2^{\lambda}}{z\Gamma(\lambda)}\int_0^z \left(log\frac{z}{t}\right)^{\lambda-1} f(t)dt, \alpha > 0$$

defined by the Jung-Kim-Srivastava operator I belongs to $SD(\alpha)$ for all $\alpha \geq 0$.

Proof Here $F(z) = z + b_2 z^2 + \cdots$ where $b_n = \left(\frac{2}{n+1}\right)^{\lambda} a_n, \alpha > 0$.

Now

$$\sum_{n=2}^{\infty}[1+\alpha(n-1)]|b_n| = \sum_{n=2}^{\infty}[1+\alpha(n-1)]\left|\frac{2}{n+1}\right|^{\lambda}|a_n|$$

$$< \sum_{n=2}^{\infty}[1+\alpha(n-1)]|a_n| < 1, \forall\alpha \geq 0,$$

since $n \geq 2$. Thus by Theorem 2, $F \in SD(\alpha)$ for all $\alpha \geq 0$.

Theorem 7 *Let $f \in A$ be given by (1) and satisfy the condition (4). Then the function*

$$F(z) = \frac{\Gamma(\alpha+\beta+1)}{\Gamma(\alpha+1)\Gamma(\beta+1)}\frac{\alpha}{z^{\beta}}\int_0^z \left(1-\frac{t}{z}\right)^{\alpha-1} t^{\beta-1} f(t)dt, \beta > 0,$$

defined by the Jung-Kim-Srivastava operator II belongs to $SD(\alpha)$ for all $\alpha \geq 0$.

Proof Here $F(z) = z + b_2 z^2 + \cdots$ where $b_n = \frac{\Gamma(\beta+n)\Gamma(\alpha+\beta+1)}{\Gamma(\alpha+\beta+n)\Gamma(\beta+1)} a_n$.
Now

$$\sum_{n=2}^{\infty}[1+\alpha(n-1)]|b_n| = \sum_{n=2}^{\infty}[1+\alpha(n-1)]\frac{\Gamma(\beta+n)\Gamma(\alpha+\beta+1)}{\Gamma(\alpha+\beta+n)\Gamma(\beta+1)}|a_n|$$

$$= \sum_{n=2}^{\infty}[1+\alpha(n-1)]\frac{\Gamma(\beta+n)}{\Gamma(\beta)}\frac{\Gamma(\alpha+\beta)}{\Gamma(\alpha+\beta+n)}\frac{\alpha+\beta}{\beta}$$

$$< \sum_{n=2}^{\infty}[1+\alpha(n-1)]|a_n|\frac{(\beta+n-1)\cdots(\beta+1)\beta}{(\alpha+\beta+n-1)\cdots(\alpha+\beta)}\frac{\alpha+\beta}{\beta}$$

$$= \sum_{n=2}^{\infty}[1+\alpha(n-1)]|a_n|\frac{(\beta+n-1)(\beta+n-2)\cdots(\beta+1)}{(\alpha+\beta+n-1)(\alpha+\beta+n-2)\cdots(\alpha+\beta+1)}$$

$< 1,\ since\ \beta+i < \alpha+\beta+i,\ for\ i = 1, 2, 3, \cdots, n-1.$

Thus, by Theorem 2, $F \in SD(\alpha)$ for all $\alpha \geq 0$.

Several studies on the problem of deriving conditions for different forms of hypergeometric functions to belong to various subclasses of analytic functions in the unit disk have been done (see, for example, [3, 8, 15, 23, 24]). Here we consider the Gaussian hypergeometric function $F(\xi, \eta, \zeta; z)$ given by

$$F(\xi, \eta, \zeta; z) = \sum_{n=0}^{\infty}\frac{(\xi)_n(\eta)_n}{(\zeta)_n(1)_n}z^n,\ z \in \Delta \tag{10}$$

where ξ, η, ζ are complex numbers such that $\zeta \neq -n,\ n \in \{0, 1, 2, \cdots\}$, $(\xi)_0 = 1$, for $\xi \neq 0$ and for each positive integer n, $(\xi)_n = \xi(\xi+1)(\xi+2)\cdots(\xi+n-1)$ is the Pochhammer symbol. We derive a sufficient condition in terms of a hypergeometric inequality for $zF(\xi, \eta, \zeta; z)$ to belong to the class $SD(\alpha)$. We make use of the Gauss summation formula [18] given by

$$F(\xi, \eta, \zeta; 1) = \sum_{n=0}^{\infty}\frac{(\xi)_n(\eta)_n}{(\zeta)_n(1)_n} = \frac{\Gamma(\zeta-\xi-\eta)\Gamma(\zeta)}{\Gamma(\zeta-\xi)\Gamma(\zeta-\eta)}$$

if $Re(\zeta-\xi-\eta) > 0$.

Theorem 8 *Let ξ, η be two non-zero complex numbers and ζ be a real number such that $\zeta > |\xi| + |\eta| + 1$. Let $f \in A$ be of the form given by (1). Then $zF(\xi, \eta, \zeta; z) \in SD(\alpha)$ if the following hypergeometric inequality holds:*

$$\frac{\Gamma(\zeta-|\xi|-|\eta|-1)\Gamma(\zeta)}{\Gamma(\zeta-|\xi|)\Gamma(\zeta-|\eta|)}[(\zeta-|\xi|-|\eta|-1)+\alpha|\xi\eta|] < 2. \tag{11}$$

Proof In view of Theorem 2 and the series representation of $zF(\xi, \eta, \zeta; z)$ given by

$$zF(\xi, \eta, \zeta; z) = z + \sum_{n=2}^{\infty} \frac{(\xi)_{n-1}(\eta)_{n-1}}{(\zeta)_{n-1}(1)_{n-1}} z^n, \ z \in \Delta, \tag{12}$$

it is enough to prove that

$$S = \sum_{n=2}^{\infty} (1 + \alpha(n-1)) \left| \frac{(\xi)_{n-1}(\eta)_{n-1}}{(\zeta)_{n-1}(1)_{n-1}} \right| < 1. \tag{13}$$

Using the fact that $|(\xi)_n| \leq (|\xi|)_n$ and noticing that ζ is a positive real number, we have

$$\begin{aligned} S &\leq \sum_{n=2}^{\infty} (1 + \alpha(n-1)) \frac{(|\xi|)_{n-1}(|\eta|)_{n-1}}{(\zeta)_{n-1}(1)_{n-1}} \\ &= \sum_{n=2}^{\infty} \frac{(|\xi|)_{n-1}(|\eta|)_{n-1}}{(\zeta)_{n-1}(1)_{n-1}} + \alpha \sum_{n=2}^{\infty} (n-1) \frac{(|\xi|)_{n-1}(|\eta|)_{n-1}}{(\zeta)_{n-1}(1)_{n-1}} \\ &= \sum_{n=2}^{\infty} \frac{(|\xi|)_{n-1}(|\eta|)_{n-1}}{(\zeta)_{n-1}(1)_{n-1}} + \alpha \sum_{n=2}^{\infty} \frac{(|\xi|)_{n-1}(|\eta|)_{n-1}}{(\zeta)_{n-1}(1)_{n-2}} \end{aligned}$$

Thus using the property $(\xi)_n = \xi(1+\xi)_{n-1}$, we have

$$\begin{aligned} S &\leq \sum_{n=2}^{\infty} \frac{(|\xi|)_{n-1}(|\eta|)_{n-1}}{(\zeta)_{n-1}(1)_{n-1}} + \alpha \frac{|\xi||\eta|}{\zeta} \sum_{n=2}^{\infty} \frac{(1+|\xi|)_{n-2}(1+|\eta|)_{n-2}}{(1+\zeta)_{n-2}(1)_{n-2}} \\ &= F(|\xi|, |\eta|, \zeta; 1) - 1 + \alpha \frac{|\xi||\eta|}{\zeta} F(1+|\xi|, 1+|\eta|, 1+\zeta; 1) \end{aligned}$$

An application of the Gauss summation formula in (9) yields the result.

Acknowledgements The authors thank the reviewers for their very pertinent comments and for pointing out corrections, especially in Theorems 1 and 4, which helped us to revise and improve the paper with a correct formulation of the results. The authors are grateful to Prof. V. Ravichandran for bringing to our notice the paper [6] in the list of references and for pointing out the univalence of the functions in the class $SD(\alpha)$ for $\alpha \geq 1$ which helped us to formulate Theorem 1 in a correct form.

References

1. Ahuja, O.P.: Integral operators of certain univalent functions. Int. J. Math. Math. Sci. **8**, 653–662 (1985)
2. Ali, R.M., Ravichandran, V.: Uniformly convex and uniformly starlike functions. Ramanujan Math. Newslett. **21**(1), 16–30 (2011)
3. Aouf, M.K., Darwish, H.E.: Subclasses of analytic functions associated with the generalized hypergeometric function. Comput. Math. Appl. **57**, 356–366 (2009)
4. Breaz, D.: Integral operators on the UCD(α)-class. In: Proceedings of International Conference on Theory and Applications of Mathematics and Informatics (ICTAMI2003), Alba Lulia, pp. 61–65 (2003)
5. Breaz, D.: Integral operators on the TUCD(α)-class. In: Proceedings of International Conference on Theory and Applications of Mathematics and Informatics (ICTAMI2004), Thessaloniki, Greece, pp. 61–66 (2004)
6. Chichra, P.N.: New subclasses of the class of close-to-convex functions. Proc. American Math. Soc. **62**(1), 37–43 (1977)
7. Cho, N.E., Owa, S.: Integral operators preserving certain analytic functions. J. Math. Anal. Appl. **286**, 168–176 (2003)
8. Dziok, J., Srivastava, H.M.: Certain subclasses of analytic functions associated with the generalized hypergeometric function. Inte. Trans. Spec. Fun. **14**, 7–18 (2003)
9. Ezhilarasi, R., Sudharsan, T.V., Mohd, M.H., Subramanian, K.G.: Connections between certain subclasses of analytic univalent functions based on operators. J. Complex Anal. 5. Article ID 6104210 (2017). https://doi.org/10.1155/2017/6104210
10. Fadipe-Joseph, O.A., Oladipo, A.T., Ejieji, C.N.: Sufficient condition for a class of analytic and univalent functions of complex order. Acta Universitatis Apulensis **36**, 101–108 (2013)
11. Gordji, M.E., Ebadian, A., Aghalary, R., Darus, M.: Starlikeness and univalence criterion for non-linear operators. Math. Rep. **13**(63). No. 2, 141–149 (2011)
12. Li, J.L.: Some properties of two integral operators. Soochow J. Math. **25**(1), 91–96 (1999)
13. Liu, J.L.: Notes on Jung-Kim-Srivastava integral operator. J. Math. Anal. Appl. **294**, 96–103 (2004)
14. Miller, S.S., Mocanu, P.T.: Integral operators on certain classes of analytic functions. In: Univalent Functions, Fractional Calculus and their Applications, vol. 153–166. Halstead Press, Wiley (1989)
15. Murugusundaramoorthy, G., Magesh, N.: On certain subclasses of analytic functions associated with hypergeometric functions. Appl. Math. Lett. **24**, 494–500 (2011)
16. Noor, K.I., Noor, M.A.: On integral operators. J. Math. Anal. Appl. **238**, 341–352 (1999)
17. Ong, K.W., Tan, S.L., Tu, Y.E.: Integral operators and univalent functions. Tamkang J. Math. **43**, 215–221 (2012)
18. Rainville, E.D.: Special Functions. Chelsea Publishing Company, New York (1960)
19. Robertson, M.S.: On the theory of univalent functions. Annals Math. **37**, 374–408 (1936)
20. Rosy, T.: Studies on subclasses of starlike and convex functions. Ph.D. thesis, University of Madras (2001)
21. Rosy, T., Stephen, B.A., Subramanian, K.G., Silverman, H.: Classes of convex functions. Int. J. Math. Math. Sci. **23**(12), 819–825 (2000)
22. Ruscheweyh, S.: Extension of Szegö's theorem on the sections of univalent functions. SIAM J. Math. Anal. **19**(6), 1442–1449 (1988)
23. Silverman, H.: Starlike and convexity properties for hypergeometric functions. J. Math. Anal. Appl. **172**, 574–581 (1993)
24. Sivasubramanian, S., Sokol, J.: Hypergeometric transforms in certain classes of analytic functions. Math. Comput. Model. **54**, 3076–3082 (2011)
25. Varma, S.S., Rosy, T.: Certain properties of a subclass of univalent functions with finitely many fixed coefficients. Khayyam J. Math. **3**(1), 26–33 (2017)
26. Tehranchi, A.: Study on integral operators by using komato operator on a new class of univalent functions. J. Math. Res. **3** (2011)